Few-
Body
Systems

Suppl. 6

INFN　　

　　**ICTP**

Few-Body Systems

Editor: H. Mitter, Graz   Managing Editor: W. Plessas, Graz

Supplementum 6

Few-Body Problems in Physics

Proceedings of the XIIIth European Conference on Few-Body Physics, Marciana Marina, Isola d'Elba, Italy, September 9—14, 1991

Edited by
C. Ciofi degli Atti, E. Pace, G. Salmè, and *S. Simula*

Springer-Verlag Wien New York

Prof. Dr. Claudio Ciofi degli Atti
Department of Physics
University of Perugia
Perugia, Italy

Prof. Dr. Emanuele Pace
Department of Physics
University of Rome "Tor Vergata"
Rome, Italy

Dr. Giovanni Salmè
Dr. Silvano Simula
INFN, Sezione Sanità
Rome, Italy

With 260 Figures and 1 Frontispiece

ISSN 0177-8811
ISBN-13:978-3-7091-7583-5   e-ISBN-13:978-3-7091-7581-1
DOI: 10.1007/ 978-3-7091-7581-1

Organizing Committee

C. Ciofi degli Atti, University of Perugia and INFN, Sezione Sanità
Conference Chairman

E. Pace, University of Roma and INFN, Tor Vergata, Roma
G. Salmè, INFN, Sezione Sanità, Roma
S. Simula, INFN, Sezione Sanità, Roma

International Advisory Committee

Y. Akaishi, Sapporo
R. Arnold, SLAC
J. Arvieux, Saturne
V. B. Belyaev, Dubna
P. K. A. de Witt Huberts, NIKHEF
M. Fabre de la Ripelle, Paris
L. D. Faddeev, Leningrad
A. Faessler, Tübingen
S. Fantoni, Trieste
H. W. Fearing, TRIUMF
A. Fonseca, Lisboa
J. Friar, Los Alamos
B. Frois, Saclay
F. Gross, William and Mary, CEBAF
E. M. Henley, Washington
R. Holt, Argonne
F. Iachello, Yale
N. Isgur, CEBAF
I. Lovas, Budapest
L. Lovitch, Ferrara
I. E. McCarthy, Flinders
J. S. McCarthy, Charlottesville

S. Merkuriev, Leningrad
E. Moniz, MIT
P. Picozza, Roma
W. Pisent, Padova
W. Plessas, Graz
L. Ponomarev, Moscow
D. Prosperi, Roma
A. Rinat, Weizmann
K. Rith, Heidelberg
S. Rosati, Pisa
W. Sandhas, Bonn
T. Sasakawa, Tsukuba
E. W. Schmid, Tübingen
B. Schoch, Bonn
I. Sick, Basel
Yu. Simonov, Moscow
I. Šlaus, Zagreb
A. Sgamellotti, Perugia
A. Thomas, Adelaide
J. Tjon, Utrecht
Th. Walcher, Mainz
J. Žofka, Řež

Conference Secretariat

Simona Ceccarelli, INFN, Sezione Sanità
Antonella Sapere, EIPC

Sponsored
by

European Physical Society (EPS)
Istituto Nazionale di Fisica Nucleare (INFN), Italy
International Centre for Theoretical Physics (ICTP), Trieste, Italy
University of Perugia, Italy

Site

Elba International Physics Center (EIPC)
Marciana Marina (Isola d'Elba), Livorno, Italy

Foreword

The Thirteenth European Conference on Few-Body Problems in Physics (European Few-Body Problems XIII) was held at the Elba International Physics Centre (EIPC) in Marciana Marina, Isola d'Elba, Italy, during September 9—14, 1991. The previous Conferences of the series, promoted by the European Few-Body Physics Research Committee, took place in Budapest (1972), Graz (1973), Tübingen (1975), Vlieland (1976), Uppsala (1977), Dubna (1979), Sesimbra (1980), Ferrara (1981), Tbilisi (1984), Balatonfüred (1985), Fontevraud (1987), and Uzhgorod (1990).

The European Few-Body Conferences represent a relevant opportunity for European scientists interested in few-body problems, of summarizing and updating, together with colleagues from countries all over the world, the status of art in this field of research, which ranges from the study of atomic and molecular structure, to nuclear and particle physics. The success of this series of Conferences, which also represent a bridge between the triennial IUPAP International Conferences on Few-Body Problems in Physics, testifies the relevance reached by few-body physics in various fields and the important theoretical and experimental contributions provided by the European few-body community.

During 1991 and January 1992, a wealth of activity in few-body physics occurred all over the world: worth being mentioned, in this regard, are the Workshop on Electromagnetic Interactions and Very Light Nuclei (Institute for Nuclear Theory, Seattle, USA, February—June, 1991), the Symposium on Clusters in Hadrons and Nuclei (Tübingen, Germany, June 15—17, 1991), the Symposium on Mesons and Light Nuclei (Prague, Czechoslovakia, September 1—6, 1991) and, finally, the most relevant event, viz. the XIIIth International Conference on Few-Body Problems in Physics (Adelaide, Australia, January 5—11, 1992). The format of the Elba Conference was conceived having also in mind these related initiatives; it was decided, in particular, to have Plenary Sessions only, based on invited talks and Discussion Sessions organized by a Convener, and to place special emphasis on the discussion of a limited number of topics, such as the structure of hadrons and hadronic interactions, the interplay between the quark and meson-nucleon descriptions of few-hadronic systems, the electro- and photo-disintegration of the two-body systems, the novel approaches in few-body problems in atomic, molecular, nuclear and particle physics, the response of few-body systems to electromagnetic and hadronic probes, including also polarization degrees of freedom, the physics programs of the running

VIII

and proposed experimental facilities and their impact on few-body problems.

This Volume contains all of the invited talks and all of the contributions to the Discussion Sessions.

A total number of 140 participants from 22 countries in Europe, USA, Africa and Australia, attended the Conference, with the largest delegations from Italy (31), Germany (21), CSI (18), USA (13) and France (10).

Many people and Institutions have contributed to the organization of the Conference. First of all we would like to thank Professor Nicola Cabibbo, President of the Istituto Nazionale di Fisica Nucleare (INFN) (the Italian Agency for Nuclear Physics), whose generous financial contribution made the organization of the Conference possible, Professor Stefano Fantoni, Director of the Elba International Physics Centre, where the Conference was held, Professor Giancarlo Dozza, Rector of the Perugia University, whose printing facilities were largely used during the preparation of the Conference. The sponsorship of the European Physical Society (EPS) and the International Centre for Theoretical Physics (ICTP) is also very much appreciated. Our thanks are extended to the members of the International Advisory Committee, who responded with many suggestions on the programme and the speakers; to the Conveners, who enthusiastically organized their Discussion Sessions; to the Chairmen of the Sessions, who smartly kept the Sessions within the planned time; to Professor Ian McCarthy, who kindly tackled the difficult task of delivering the Closing Remarks, even if asked about that the very last day of the Conference.

We are deeply indebted to all those people who before, during and after the Conference provided the necessary support to smooth out all the difficulties one usually encounters during the organization of a Conference of this size. The Organizing Committee is particularly indebted to the members of the Conference Secretariat, Mrs. Simona Ceccarelli (INFN, Sezione Sanità, Rome), who smartly prepared all circulars prior the Conference and the Booklet of Abstracts, and Mrs. Antonella Sapere (Elba International Physics Centre), who helped us very much in solving many of the logistic problems. The invaluable help of Mrs. Paola Di Ciaccio (Physics Laboratory, Istituto Superiore di Sanità, Rome) and Mr. Giacomo Monteleone (Physics Laboratory, Istituto Superiore di Sanità, Rome), whose continuous assistance made the task of the Organizing Committee much easier in many respects, is warmly acknowledged. Special thanks are also extended to Mrs. Bruna Ceccarelli (INFN, Sezione Sanità), for her precious administrative help.

Finally, we would like to express our appreciation to all speakers for providing us with the manuscript of their papers in the very due time.

In planning the social activities of the Conference an important event was foreseen: the celebration of the 60th birthday of two eminent scientists, whom the Few-Body community owe very much: Professors Erich Schmid (Tübingen, Germany) and Ivo Šlaus (Zagreb, Croatia); unfortunately, both of them, at the last moment could not reach us. Particularly sad was the reason of Ivo's absence: the dramatic events which occurred in his country. To both of them this volume is dedicated.

Perugia, January 1992 Claudio Ciofi degli Atti
for the Organizing Committee

Contents

SESSION 2
The Quark and Meson-Nucleon Descriptions
of Few-Body Systems

Chairman: *E. O. Alt*

Chairman: *T. S. H. Lee*

Chairman: *G. Karl*

Discussion Session: Electro- and Photodisintegration of the Deuteron

Convener: *H. Arenhövel*

Discussion Session: Few-Body Problems in Atomic and Molecular
Physics

Dynamics and Reactions

Convener: *A. Sgamellotti*

Bound States and Molecular Structure

Listed in Current Contents

Few-Body Systems, Suppl. 6, 1—8 (1992)

Few-
Body
Systems
© by Springer-Verlag 1992

THE NAÏVE QUARK MODEL AND BEYOND

G. Karl

Department of Physics, University of Guelph,

Guelph, Ontario N1G 2W1, Canada

(E-mail: PHYGKARL@VM.UOGUELPH.CA)

Abstract: I comment on the (lack of) connection between the Naïve Quark Model and Quantum Chromodynamics. The relation between magnetic moments of baryons and the problem of the spin of the proton is discussed in more detail; in particular it is emphasized that data on the magnetic moments and axial couplings of baryons are now much more accurate than in the past, and are less encouraging for the Naïve Quark Model than in the past.

The world of baryons and mesons has been studied in ever increasing detail for nearly fifty years. Although there is now wide consensus that these systems are composite, there is much discussion about details, especially the relevant degrees of freedom and so the situation is still not satisfactory. The most optimistic viewpoint is that we know the relevant Lagrangian (at short distances), called QCD, and only mathematical details are missing. The most extreme version of this view is that only the backwardness of present day computers prevents a complete numerical solution of all questions experimenters might be able to measure in the future. Since the present author believes that the future tends to resemble the past, he does not subscribe to such an extreme view. The past way of development in physics consisted in finding (usually painstakingly slowly) the right approximation to use, (even when the Lagrangian was known) by using some help from experiment. It is likely that hadron physics will follow the same pattern, even if we know the Lagrangian of QCD.

Part of the specific difficulty with hadron physics, is that the observed hadrons themselves are not the degrees of freedom at short distances. So we have to rely on approximate descriptions of bound states. I will try to list some of the approximations used, with a short description to them.

I should first list the Naïve Quark Model, (NQM) since it is the main topic of my talk. It is indeed one of the models widely used to describe hadrons. It is hard to say, in a mathematical sense, that the NQM is a well defined approximation. The best derivation we have of the NQM comes from the limit of QCD as the number of colors N_c becomes large. Witten [1] showed that in this limit baryons are well described by NQM wavefunctions. To be more precise the limit of QCD as N_c is large becomes identical to the limit of the NQM as N_c is large. In the real world N_c equals three and the NQM is an approximation; the usefulness of this approximation is non-uniform. In some applications such as spectroscopy, the NQM works very well, while in others, such as the "spin of the proton" it fails. Aside from the large N_c limit we have no justification for the Naïve Quark Model, and so its use tends to get mixed reviews [2]. Generally, the Bag Model is considered as a different "ad hoc" model, but I take it to be just another version of NQM, which only differs by using relativistic quarks. There are many other relativistic potential models, harmonic oscillators, etc, but they all can be thought of as examples of NQM. Their common feature is the fact that they have $q\bar{q}$ and qqq as allowed states.

These models should be distinguished sharply from QCD on the lattice, which is a well defined and justified theory of hadrons which however suffers from the problem of being still in search of reasonable approximations. The practitioners of lattice QCD are crying for more computing power, while probably good insight into the physics of QCD would be equally welcome. The gap between data and lattices will be bridged not entirely by brute force, one suspects.

An entirely different and rigorous approach to hadron physics at low energies is based on chiral symmetry breaking, and it consists of deriving properties of hadrons which follow from the smallness of light quark masses, relative to the QCD scale. These properties are supposed to follow rigorously from the QCD Lagrangian, although there is some ambiguity in relating the parameters of the fundamental Lagrangian to the parameters of the chiral Lagrangian [3]. However chiral models are quite limited to the very low energy sector of the hadronic spectrum, pion physics and perhaps ground state baryons. In these models the nucleon is mimicked by so called chiral solitons — the skyrmion. The application of skyrmions to the whole baryon octet is much more problematic than that to the nucleon itself. In particular the claim that skyrmions quantized as SU_3 flavor octets correspond in some way to the large N_c limit of QCD [4], is simply wrong, since in the large N_c limit the baryons are not members of SU_3 octets [5].

Another approach in hadron physics is based on the importance of instantons in the ground state of QCD. This approach concentrates on the computation of correlation functions in hadronic matter [10].

On the theoretical side there is a great deal of support for the ideas of Quantum Field Theory. In hadron physics the relevant field theory is Q.C.D. which has been studied in a variety of ways, but primarily on the lattice. This work is certainly ongoing, and it is certainly making progress, although there are still many difficulties, both conceptual and computational. This work is important because the nature of the vacuum of Q.C.D. is still not fully understood and different from the vacuum of QED, so that the discussion of bound states in QCD is still unclear.

The numerical work is already at the stage of providing fits of hadron properties which are almost competitive with phenomenological models, especially for nonstrange hadrons [9]. Of course the major strength of lattice QCD computations is that they contain no arbitrary parameters, so that even if the results are still imperfect they nevertheless can be said to reflect the properties of QCD. Aside from this, much conceptual progress has been provided by lattice calculations which should form the basis of newer and better phenomenological models. In particular chiral symmetry breaking can be studied in numerical models in addition to so called chiral Lagrangians. It also seems probable, at the present time, that the chiral phase transition is independent of the existence confinement, and it accounts for the constituent quark model, along lines suggested by Georgi and Manohar [6]. To be more explicit, although quarks have small masses at short distances, due to CSB they acquire a "constituent" quark mass, while remaining confined. There is also recent work on using QCD on the light cone to obtain hadron wave functions [7].

In summary then, we have many approximations in hadronic physics, but the connection between them and their connection to the underlying theory (QCD_{3+1}) is still not clear. In particular the Naïve Quark MOdel for 3 colors is not an approximation to QCD. Thus hadron physics still lacks a secure foundation. Therefore, this is a good subject for further study, with lots of room for progress.

In the remainder of this review, I shall be much more specific and deal with the topic of baryon magnetic moments and their connection to the distribution of the proton spin. This is an important topic because the data contradicts our expectations based on NQM. Therefore we are given information which we should use to construct models which improve NQM. This has not yet been done — but it is important to know the weaknesses of a model if we wish to improve it. I now discuss problems connected with the spin of the proton. In particular, I wish to discuss how, starting from magnetic moments, one can show that the contribution of the helicity of the quarks and antiquarks to the helicity of the proton is but a small fraction. This conclusion contradicts the NQM where the entire proton helicity is carried by the quarks. This conclusion is based on magnetic moments of baryons in the octet, which are by now known experimentally reasonably well.

The argument is based on a set of equations which I call the Generalized Sehgal Equations (GSE) which connect the magnetic moments to the contributions of the quark spins:

$$M_z(P) = \mu_u \Delta u + \mu_d \Delta d + \mu_s \Delta s \tag{1}$$

where μ_u, μ_d, μ_s are effective quark magnetic moments, and Δs for example is the contribution of the u-quark helicity to the proton helicity:

$$\Delta u \equiv \langle P_\uparrow | \bar{u} \gamma_z \gamma_5 u | P_\uparrow \rangle = n(u_\uparrow) - n(u_\downarrow) + n(\bar{u}_\uparrow) - n(\bar{u}_\downarrow) \tag{2}$$

Equation (1) holds [8] in a class of models in which all quarks and antiquarks of a given flavor are in the same shell of j = 1/2 of some cavity or potential, and the proof involves a simple transformation between the linear combination natural to magnetic moments $[n(u_\uparrow)-n(u_\downarrow)-n(\bar{u}_\uparrow)+n(\bar{u}_\downarrow)]$ and the quantity Δu. In other words, equation (1) is not generally valid, but only in a specific class of models which generalize the NQM, where $\Delta u = 4/3$, $\Delta d = -1/3$ and $\Delta s = 0$.

If we assume SU_3 flavor symmetry, we can obtain the magnetic moments of all baryons in the octet from equation (1). For example, for the neutron $N_\uparrow$

$$\langle N_\uparrow | \bar{u} \gamma_z \gamma_5 u | N_\uparrow \rangle = \langle P_\uparrow | \bar{d} \gamma_z \gamma_5 d | P_\uparrow \rangle \equiv \Delta d$$
$$\langle N_\uparrow | \bar{d} \gamma_z \gamma_5 d | N_\uparrow \rangle = \langle P_\uparrow | \bar{u} \gamma_z \gamma_5 u | P_\uparrow \rangle \equiv \Delta u \tag{3}$$

etc.

where we used only isospin, so that the analogous equation to (1) is

$$M_z(N) \equiv \mu_d \Delta u + \mu_u \Delta d + \mu_s \Delta s \tag{4}$$

Similar application of SU_3 symmetry, for example to the Σ^+ implies:

$$\langle \Sigma_{\uparrow}^{+}|\bar{u}\gamma_{z}\gamma_{5}u|\Sigma_{\uparrow}^{+}\rangle = \Delta u$$

$$\langle \Sigma_{\uparrow}^{+}|\bar{s}\gamma_{z}\gamma_{5}s|\Sigma_{\uparrow}^{+}\rangle = \Delta d \qquad (5)$$

$$\langle \Sigma_{\uparrow}^{+}|\bar{d}\gamma_{z}\gamma_{5}d|\Sigma_{\uparrow}^{+}\rangle = \Delta s$$

so that

$$M_z(\Sigma^+) = \mu_u\Delta u + \mu_d\Delta d + \mu_s\Delta s \qquad (6)$$

and similar equations for the other baryons in the octet.

Therefore we have a set of equations for the magnetic moments of P, N, Σ^+, Σ^-, Λ, Ξ^0, Ξ^-, and the transition moment $\Sigma^0 \to \Lambda$ parametrized in terms of Δu, Δd, Δs and μ_u, μ_d, μ_s. In fact the GSE are quite symmetric, and they do not depend on six parameters but only five, as the linear combinations $(\Delta u + \Delta d + \Delta s)$ and $(\mu_u + \mu_d + \mu_s)$ only appear as a product with each other. Moreover, we have to prevent the relative rescaling of Δu, Δd, Δs w.r.t. μ_u, μ_d, and μ_s, which we can do by choosing as a normalization, for example:

$$\Delta u - \Delta d = 1.26 \qquad (7)$$

though this choice is arbitrary — from the point of view of the G.S.E.

Therefore there are four parameters left: $(\mu_u + \mu_d + \mu_s)$ $(\Delta u + \Delta d + \Delta s)$, $\Delta u + \Delta d - 2\Delta s$, $\mu_u - \mu_d$, $\mu_u + \mu_d - 2\mu_s$. If we make the substitution:

$$\mu_u = 2\mu_d = 0 \qquad (8)$$

there is no loss of generality, as the number of parameters remains four, but now we can choose as parameters μ_d, μ_s, $(\Delta u + \Delta d + \Delta s)$ and $(\Delta u + \Delta d - 2\Delta s)$, which are more interesting. So we can parametrize eight magnetic moments in terms of four parameters. A special set of our parameters corresponds to the NQM which has three parameters μ_u, μ_d, μ_s. We can now go to the data and see what it tells us. We find that the NQM parametrization fits at the level of 15% whereas the full GSE parametrization fits at the level of about 7%. Moreover the best GSE fit corresponds to $(\Delta u + \Delta d + \Delta s) \simeq 0.2 \pm 0.2$, therefore a small fraction of the proton helicity is carried by quark helicities.

I should explain a little better the way one obtains these fits. Since these equations (GSE) do not fit the magnetic moments perfectly, we want to fit them without favoring any particular baryon in the octet. This can be accomplished most simply by pretending that each

Best Fits to Magnetic Moments (in Nucl. Magnetons)

Particle	Magn. Moment	Fit 1 GSE	Fit 2 NQM
P	2.79 ± 0.10	2.69	2.68
N	-1.91 ± 0.10	-1.85	-1.92
Σ^+	2.48 ± 0.11	2.59	2.55
Σ^-	-1.16 ± 0.10	-1.22	-1.13
Ξ°	-1.25 ± 0.10	-1.33	-1.40
Ξ^-	-0.68 ± 0.10	-0.61	-0.48
Λ	-0.61 ± 0.10	-0.59	-0.59
$\Lambda\Sigma$	-1.6 ± 0.13	-1.53	-1.60
χ^2	-----	4.42/4df	7.53/5df
$a^{(1)} = \Delta u + \Delta d + \Delta s$:	0.12 ± 0.17	0.28	1.00 (input)
$a^{(8)} = \Delta u + \Delta d - 2\Delta s$:	0.60 ± 0.05	0.86	1.00 (input)
$g_A = \Delta u - \Delta d$:	1.26 ± 0.01	1.26 (input)	1.67 (input)
μ_u		-2.42	1.76
μ_d:	-----	-1.21	-1.00
μ_s:	-----	-0.71	-0.61

magnetic moment has a "theoretical" error of ±0.1 n.m. and doing a χ^2 fit to all of them, minimizing the total χ^2. Such a fit is presented in Table I. The main conclusion is that the data on magnetic moments prefers quantitatively a set of parameters Δu, Δd, Δs which is nowhere near the NQM solution. The NQM fit to magnetic moments is worse, and we should try to rethink our views about the NQM on the basis of these fits. We should try to construct quark models which correspond to these fits. Such models would have many $q\bar{q}$ pairs in the proton including strange pairs, and these pairs would be polarized. Many simple questions are raised by such models; in particular how come that these pairs do not contribute to the spectrum of excitations of the proton at low energies. In other words, how come the proton seems to be a 3-body system when kicked, but has many more degrees of freedom when analyzed with a magnet?

Acknowledgements: This report was written at the Aspen Center for Physics and the author is very grateful for the opportunity to visit the Center. I also thank for conversations and lectures on these topics E. Shuryak, J. Negele, S. Brodsky, K. Wilson, and other participants at Aspen workshops in the summer of 1991.

References:

[1]   E. Witten, Nucl. Phys. B160, 57 (1979).

[2]   for recent work on the NQM see e.g.

       S. Capstick and N. Isgur, Phys. Rev. D34, 2809 (1986).

       S. Godfrey and N. Isgur, Phys. Rev. D32, 189 (1985).

[3]   see e.g. J. Gasser and H. Leutwyler, Phys. Reports 87, 77 (1982).

       D.B. Kaplan and A.V. Manohar, Phys. Rev. Let. 56, 2004 (1986).

       K. Choi et al, Phys. Rev. Letts. 61, 794 (1988).

       K. Maltman, T. Goldman and G.L. Stephenson Jr., Phys. Letters B234, 158 (1990).

[4]   see e.g. S.J. Brodsky, J. Ellis and M. Karliner, Phys. Letters B206, 309 (1988).

       J. Bijnens, H. Sonoda and M. Wise, Phys. Letters B140, 421 (1984).

[5]   J. Bijnens, H. Sonoda and M. Wise, Can. J. Phys. 64, 1 (1986).

       G. Karl, G. Patera and S. Perantonis, Phys. Letters B172, 49 (1986).

       G. Karl and H.J. Lipkin (to be published).

[6]   H. Georgi and A. Manohar, Nucl. Phys. B234, 189 (1984).

[7]   S. Pinsky, K.G. Wilson et al: work described at the Aspen workshop on light cone physics.

[8]   G. Karl, Guelph preprint GWP2-PP-91-02. Earlier references includes:

      – L.M. Sehgal, Phys. Rev. D10, 1663 (1974).

      – J. Bartelski and R. Rodenberg, Phys. Rev. D41, 2800 (1990).

      – R. Decker, M. Nowakowski and J. Stahov, Nucl. Phys. A512, 626 (1990).

      – S.M. Gerasimov, Int. Conf. at Alushta (1987) Dubna preprint E2-88-122.

      – G. Karl and M.D. Scadron, Proceedings of MRST Conf. (1990).

[9]   See for example M.C. Chu, M. Lissia and J.W. Negele, to appear in Nucl. Phys. B (1991), who discuss extensive computations of correlation functions in the π, ρ and nucleon.

[10]   See for example E.V. Shuryak, Nucl. Phys. B328, 85 (1989) and references therein.

Few-Body Systems, Suppl. 6, 9—25 (1992)

Few-
Body
Systems
© by Springer-Verlag 1992

ALGEBRAIC APPROACH TO HADRONIC STRUCTURE

F. Iachello

Center for Theoretical Physics, Sloane Laboratory,

Yale University, New Haven, Connecticut 06511

Abstract

The algebraic approach to hadronic structure based on the algebra of U(4) to describe string excitations is briefly reviewed. Results for ($q\bar{q}$) configurations are presented. Work in progress to describe (qqq) and ($qq\bar{q}\bar{q}$) configurations is also discussed.

1. Introduction

In the last few years, quantum chromodynamics (QCD) has emerged as the theory of strong interactions. In this theory, the observable quantities of hadronic structure, masses, decay widths and form factors, should, in principle, all be obtained from a knowledge of the quark masses, M_i, and the coupling constant, g_s. In practice, only partial analytic results have been obtained in 1 + 1 dimensions, and only partial numerical results within the framework of the lattice formulation of QCD. As a result, hadronic properties have been mostly studied by resorting to models, some of "collective" nature, such as the bag and the string model, some of "single-particle" nature, such as the constituent quark model.

Once a model has been chosen, there remains still the problem of how to find the corresponding spectral properties. In the traditional approach a differential equation is written whose eigenvalues, E_i, and eigenvectors, ψ_i, give the spectrum and wave functions. Matrix elements of operators $\langle\psi_j|\hat{T}|\psi_i\rangle$ are then evaluated to give decay widths and form factors. There are some difficulties with this approach. First, the situation is highly relativistic (at least for light quarks) and thus the use of a Schrödinger equation is not appropriate. One may circumvent this problem by introducing semi-relativistic equations, but part of the difficulty still remains. Second, in view of the relativistic nature of the problem, the concept of a local two-body quark-quark interaction may not be appropriate. The actual "potential" may actually be non-local and contain large three-body terms. Finally, the differential approach becomes very cumbersome after the three-body problem.

An alternative approach was suggested long ago by Dothan, Gell-Mann and Ne'eman [1] and Barut and Böhm [2]. In this approach the energy operator and other physical operators are expanded into elements, G_α, of an algebras, $\mathcal{G}$. Hence, the name algebraic approach given to the method. Usually this algebra is a Lie algebra, although other, more complicated, algebras have also been considered. Generically, this can be written as

$$0 - f(G_\alpha) \quad , \quad G_\alpha \in \mathcal{G} \quad ; \tag{1.1}$$

where 0 represents any operator. A particular case of (1.1) is that in which the elements G_α contained in the energy operators are of a special type, i.e. they are the invariant (Casimir) operators, G_i, of $\mathcal{G}$ and its subalgebras $\mathcal{G} \supset \mathcal{G}' \supset \mathcal{G}'' \supset \ldots$,

$$0 - f(G_i) \quad . \tag{1.2}$$

In this case, called dynamic symmetry, all results can be obtained in closed, analytic, form and thus are particularly useful for an analysis of experimental data.

The use of algebraic methods bypasses the difficulties of the traditional approach since it does not commit itself to any non-relativistic or semi-relativistic equation. In this article, applications of the algebraic method to hadronic structure will be briefly reviewed.

2. Hadronic configurations

In this paper, I will consider applications of the algebraic method to a string-like model. In this model, the Fock space representation of the wave function of a hadron, for example a meson $|M\rangle$, is written in terms of configurations involving a certain number of quarks, q, antiquarks, $\bar{q}$, and gluons, g,

$$|M\rangle = c_0|q\bar{q}\rangle + c_1|q\bar{q}q\bar{q}\rangle + \ldots \quad +$$

$$+ c_0'|q\bar{q}g\rangle + \ldots \quad +$$

$$+ c_0''|gg\rangle + \ldots \quad . \tag{2.1}$$

The string picture for some of these configurations is shown in Fig. 1. A complete spectroscopy would involve the study of all configurations and their mixings. In practice, only few configurations, those whose amplitude is large, are included. It is the purpose of the algebraic method to study, in a systematic way, as many configurations as possible, including configurations with only quarks, with only gluons and hybrid quark-gluon configurations.

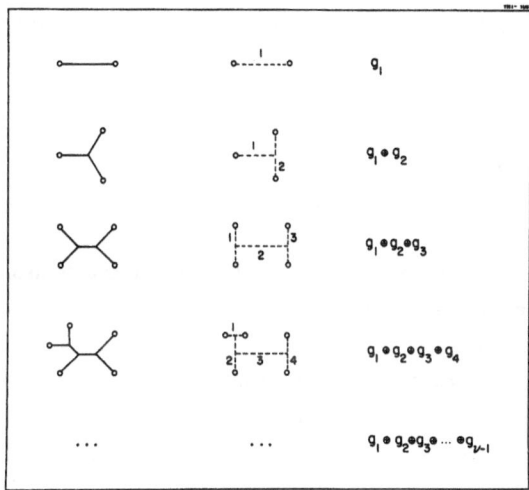

Fig.1. Hadronic configurations composed of ν quarks and antiquarks and their description in terms of spectrum generating algebras, g_i (i=1,2,...,ν-1).

3. Spectrum generating algebra of hadronic structure

The key ingredient of the algebraic method is the choice of the algebra $\mathcal{G}$ in which to expand all operators (spectrum generating algebra). The problem is complicated by the fact that one needs to describe both the internal (spin-flavor-color) part and the space (geometric) part. Thus $\mathcal{G}$ must be of the form

$$\mathcal{G} = \mathcal{R} \otimes \mathcal{G}_s \otimes \mathcal{G}_f \otimes \mathcal{G}_c \quad , \qquad (3.1)$$

where the indices s,f, and c denote spin-flavor and color and $\mathcal{R}$ denotes the space part. There is no doubt what to use for the internal part. The spin algebra is obviously $SU_s(2)$. The flavor algebra is $SU_f(n)$, for n flavors, and the color algebra is $SU_c(3)$.

Several suggestions have been made for the space part. A condition on $\mathcal{R}$ is that it must contain the angular momentum algebra $SO_L(3)$. It has been suggested [3] that for string-like models of hadronic structure one takes

$$\mathcal{R} = \prod_{i=1}^{\nu-1} \otimes U_i(3,1) \quad , \qquad (3.2)$$

where ν is the number of constituent particles in the hadrons. With two constituent particles, as in $(q\bar{q})$ configurations, $\mathcal{R} = U(3,1)$. The algebra in (3.2) is non-compact and its use is rather cumbersome. It is more convenient to make use of the correspondence between the discrete infinite dimensional representations of $U(3,1)$ and those of $U(4)$ in the limit in which the quantum number(s) characterizing the representations, N, goes to infinity, $N \to \infty$, and to use thus

$$\mathcal{R} = \prod_{i=1}^{\nu-1} \otimes U_i(4) \quad . \qquad (3.3)$$

The algebra $\mathcal{G}$ is therefore

$$\mathcal{R} = \prod_{i=1}^{\nu-1} \otimes U_i(4) \otimes SU_s(2) \otimes SU_f(n) \otimes SU_c(3) \quad , \tag{3.4}$$

where the explicit form of the spin, flavor and color part has been written down.

In addition to the choice of $\mathcal{G}$ one must specify also the representation of $\mathcal{G}$ that one is considering. Again for the internal part this is obvious. For the space part, the condition is that all the states of a given configuration be contained in a single representation of $\mathcal{G}$. As it will be seen below, this implies that we choose the totally symmetric representations of U(4), characterized by the Young tableau [N,0,0,0].

It is worthwhile at this stage discussing briefly the two dynamic symmetries of U(4) and the associated classification scheme. U(4) admits two chains:

$$U(4) \begin{array}{l} \diagup \quad U(3) \supset SO(3) \supset SO(2) \quad , \quad \text{(I)} \\ \\ \diagdown \quad SO(4) \supset SO(3) \supset SO(2) \quad . \quad \text{(II)} \end{array} \tag{3.5}$$

The classification scheme for chain (I) is

$$\left| \begin{array}{cccc} U(4) \supset & U(3) \supset & SO(3) \supset & SO(2) \\ \downarrow & \downarrow & \downarrow & \downarrow \\ N & n & L & M_L \end{array} \right\rangle \quad , \quad \text{(I)} \quad . \tag{3.6}$$

The values of the quantum numbers n, L and M_L contained in a given representation, $[N,0,0,0] = N$, are

$$n = N, N-1, \ldots, 1, 0 \quad ; \quad L = n, n-2, n-4, \ldots, 1 \text{ or } 0 \text{ (}n = \text{odd or even)} \quad ;$$

$$-L \le M_L \le +L \quad . \tag{3.7}$$

The classification scheme for chain (II) is

$$\left| \begin{array}{cccc} U(4) \supset SO(4) \supset SO(3) \supset SO(2) \\ \downarrow \quad\quad \downarrow \quad\quad\; \downarrow \quad\quad\; \downarrow \\ N \quad\quad \omega \quad\quad\; L \quad\quad\; M_L \end{array} \right\rangle \quad , \quad (II) \; . \qquad (3.8)$$

The values of the quantum numbers ω, L, M_L contained in N are

$$\omega = N, N-2, \ldots, 1 \text{ or } 0 \text{ (N = odd or even)} \quad ; \quad L = \omega, \omega-1, \ldots, 1, 0 \quad ;$$

$$- L \leq M_L \leq + L \quad . \tag{3.9}$$

Here L represents the angular momentum and M_L its z-component. For reasons which will become apparent below, it is convenient to introduce the quantum number, v, related to ω by

$$v = \frac{N-\omega}{2} \quad . \tag{3.10}$$

The quantum number v represents the vibrational number of the quantized string.

4. Meson (q$\bar{q}$) configurations

The study of each hadronic configuration is done, within the algebraic method, using the following logical scheme. First expand the energy operator into elements of $\mathcal{G}$. In non-relativistic physics, this expansion is of the type

$$H = E_0 + \sum_{\alpha} \epsilon_{\alpha} G_{\alpha} + \sum_{\alpha\beta} \frac{1}{2} u_{\alpha\beta} G_{\alpha} G_{\beta} + \ldots \quad ; \quad G_{\alpha} \in \mathcal{G} \; . \tag{4.1}$$

For relativistic systems, it is more convenient to expand the mass-squared operator, M^2 rather than H. Furthermore, as a result of relativistic dynamics, one is forced to use more complicated expansions, involving square root operators, which can be written generically as

$$M^2 = f(G_\alpha) \quad ; \quad G_\alpha \in \mathcal{G} \quad . \tag{4.2}$$

Diagonalization of M^2 within the space of $\mathcal{G}$ gives the mass spectrum and eigenfunctions.

A dynamic symmetry corresponds to the special situation in which M^2 is only a function of the invariants G_i,

$$M^2 = f(G_i) \quad . \tag{4.3}$$

It turns out, as it will be seen below, that the meson spectrum has an almost exact dynamic symmetry. It will therefore be conveniently analyzed by using a M^2 operator of the type

$$M^2 = f(G_i) + \text{small symmetry breaking interactions.} \tag{4.4}$$

Concretely, the algebra $\mathcal{G}$ of $(q\bar{q})$ mesons is

$$\mathcal{G} = U(4) \otimes SU_s(2) \otimes SU_f(n) \otimes SU_c(3) \quad . \tag{4.5}$$

For those configurations, color does not play any non-trivial role, and will be henceforth deleted. Consider then the class of dynamic symmetry corresponding to chain II, Eq. (3.8), of $U(4)$, of the type

$$M^2 = \sum_i A_i (G_i)^{\alpha_i} \quad . \tag{4.6}$$

In particular, consider the form

$$M^2 = M_0^2 + A_1 G(SO(4)) + B\left[\sqrt{G(SO(3)) + \frac{1}{4}} - \frac{1}{2}\right] \qquad . \qquad (4.7)$$

This form is diagonal in the basis (3.8) with eigenvalues

$$M^2(N,v,L,M_L) = M_0^2 - 4A_1(N+1)\left[v - \frac{1}{N+1} v^2\right] + BL \qquad . \qquad (4.8)$$

When $N \to \infty$ with $-4A_1(N+1) = \text{const} = A$, this leads to the simple mass formula

$$M^2 (v,L,M_L) = M_0^2 + Av + BL \qquad . \qquad (4.9)$$

Including spin, S, and total angular momentum J, one obtains [4]

$$M^2 (v,L; S; J,M_J) = M_0^2 + Av + BL + CS + DJ \qquad . \qquad (4.10)$$

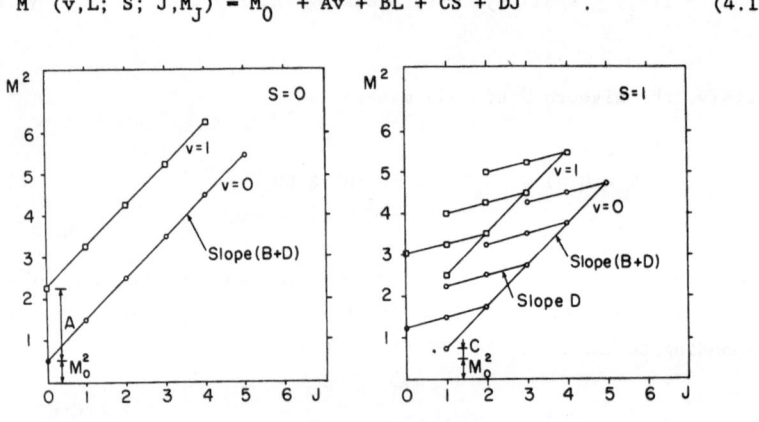

Fig. 2. Spectrum of the mass squared operator of $q\bar{q}$ mesons according to
Eq.(4.10) (left panel, S=0; right panel, S=1).

The mass spectrum corresponding to this mass formula is shown schematically in Fig. 2. For each "vibrational" number, v, there is a sequence of "rotational" states (Regge trajectories). Each Regge trajectory forms a representation of SO(4) and has M^2 increasing linearly with angular momentum. The extent to which the experimental data are described by Eq. (4.10) is evidence for the occurrence of this dynamic symmetry. Examples are shown in Fig. 3.

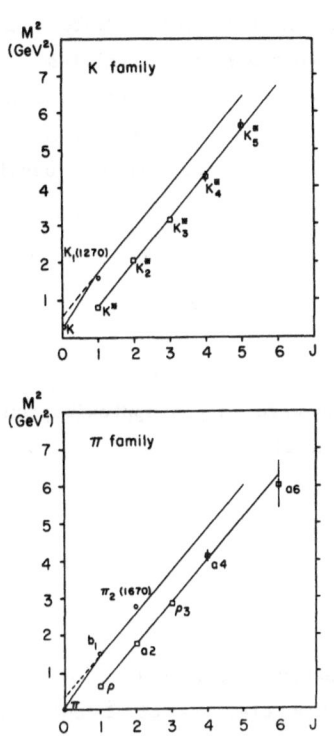

Fig.3. "Rotational" Regge trajectories for π and K families.

One can also study the flavor dependence of M^2, using similar methods. For configurations composed of a quark of flavor i and an antiquark of flavor j one finds [4]

$$M^2(q_i\bar{q}_j;v,L;S;J,M_J) = (M_0^2)_{ij} + A_{ij}v + B_{ij}L + C_{ij}S + D_{ij}J \quad ;$$

$$A_{ij} = a + a'M_{ij} \quad , \qquad B_{ij} = b + b'M_{ij} \quad ,$$

$$C_{ij} = c + c'M_{ij} \quad , \qquad D_{ij} = d + d'M_{ij} \quad ,$$

$$(M_0^2)_{ij} = eM_{ij} + M_{ij}^2 \quad ;$$

$$M_{ij} = M_i + M_j \quad . \tag{4.11}$$

In this formula, the flavor dependence arises entirely from the quark masses, M_i. Symmetry breaking interactions can be studied by adding them as in Eq. (4.4) and diagonalizing them in the basis provided by the SO(4) symmetry. A full analysis of the presently available experimental situation has been done and is reported in Ref. [4]. Table I shows a portion of this analysis, that referring to the π and K families.

The next step is the computation of decay widths and form factors. In order to do this computation, one needs to expand the transition operators into elements of $\mathcal{G}$. In non-relativistic physics, this expansion is usually linear

$$\hat{T} = \sum_\alpha t_\alpha G_\alpha + \ldots \qquad ; \qquad G_\alpha \in \mathcal{G} \ . \qquad (4.12)$$

In the present situation, since the long-wave length approximation is no longer valid, one must consider expansions of the type

$$\hat{T} = \sum_\alpha t_\alpha G_\alpha \ e^{i\sum_\beta k_\beta G_\beta} \ . \qquad (4.13)$$

The long wave-length approximation corresponds to $k_\beta \to 0$. In the generic situation, the calculation of the matrix elements of $\hat{T}$ must be done numerically. However, whenever a dynamic symmetry exists, Eq. (4.3), the calculation can be done in closed, analytic form, since then the states are irreducible representations of $\mathcal{G}$ and one knows how the G_α's act on them. A new technical problem arises with the evaluation of exponential operators, but this problem has been fully solved. A typical matrix element to evaluate is

$$\langle N,\omega,L,M_L | e^{i\alpha\hat{D}_0} | N,\omega',L',M'_L \rangle \ , \qquad (4.14)$$

where $\hat{D}_0$ is an element of SO(4). This is a group element of U(4). In the limit of large N, the form of the matrix elements is particularly simple.

TABLE I. A selection of meson masses in the algebraic approach [4].

π Family

$M^2(\text{GeV}^2)$

Meson	Expt.	SO(4) Symmetry	v	L	S	J^{PC}
π	0.019±0.000	0.022	0	0	0	0^{-+}
$\rho(770)$	0.590±0.001	0.586	0	0	1	1^{--}
$b_1(1235)$	1.520±0.025	1.437	0	1	0	1^{+-}
$a_1(1260)$	1.588±0.076	1.636	0	1	1	1^{++}
$a_2(1320)$	1.738±0.002	1.680	0	1	1	2^{++}
$\rho(1450)$	2.103±0.023	1.884	1	0	1	1^{--}
$\pi_2(1670)$	2.772±0.067	2.531	0	2	0	2^{-+}
$\rho_3(1690)$	2.859±0.017	2.775	0	2	1	3^{--}
$\rho(1700)$	2.890±0.068	2.686	0	2	1	1^{--}

K Family

$M^2(\text{GeV}^2)$

Meson	Expt.	SO(4) Symmetry	v	L	S	J^{P}
K	0.246±0.000	0.231	0	0	0	0^{-}
$K^*(892)$	0.795±0.000	0.815	0	0	1	1^{-}
$K_1(1270)$	1.613±0.025	1.737	0	1	0	1^{+}
$K^*(1370)$	1.869±0.148	2.234	1	0	1	1^{-}
$K_1(1400)$	1.966±0.020	1.943	0	1	1	1^{+}
$K^*_0(1430)$	2.042±0.017	1.887	0	1	1	0^{+}
$K^*_2(1430)$	2.032±0.004	1.999	0	1	1	2^{+}
$K^*(1680)$	2.816±0.215	3.072	0	2	1	1^{-}
$K_2(1770)$	3.126±0.050	3.128	0	2	1	2^{-}
$K^*_3(1780)$	3.147±0.028	3.184	0	2	1	3^{-}
$K^*_4(2045)$	4.182+0.037	4.368	0	3	1	4^{+}

For example,

$$\langle N, \omega{-}N, L, 0 | e^{i\alpha\hat{D}_0} | N, \omega{-}N, 0, 0 \rangle =$$

$$= i^L \sqrt{2L{+}1} \; j_L(N\alpha) \qquad . \qquad (4.15)$$

Once the matrix elements of the transition operators $\hat{T}$ have been evaluated, the decay widths are obtained in the usual way. For example, the radiative decay $M \to M'{+}\gamma$ is calculated as

$$\Gamma(M \to M'{+}\gamma) = (\text{Phase space}) \times \sum_{m_i, m_f} |\langle M | \hat{T}_\gamma | M' \rangle|^2 \qquad . \qquad (4.16)$$

TABLE II. Comparison of light meson experimental decay widths with those computed in the U(4) ⊃ SO(4) model [5].

Decay	Γ (KeV)	
$\Delta S{=}1$, $\Delta L{=}0$	Experiments	SO(4) Symmetry
$\rho^\pm(770) \to \gamma\pi^\pm$	68 ± 7	66
$\rho^0(770) \to \gamma\pi(549)$	62 ± 17	90
$\omega(783) \to \gamma\pi^0$	717 ± 51	609
$\omega(783) \to \gamma\eta(549)$	4.0 ± 1.9	9.9
$\eta'(958) \to \gamma\rho^0(770)$	62 ± 9	72
$\eta'(958) \to \gamma\omega(783)$	6.3 ± 1.2	9.4
$\phi(1020) \to \gamma\pi^0$	5.78 ± 0.67	5.9
$\phi(1020) \to \gamma\eta(549)$	56.4 ± 3.5	32
$\phi(1020) \to \gamma\eta'(958)$	< 1.8	0.5
$K^{*0}(902) \to \gamma K^0(498)$	117 ± 10	114
$K^{*\pm}(892) \to \gamma K^\pm(494)$	50 ± 5	55

Some results [5] of the calculation of electromagnetic decays of mesons are shown in Table II. It is for the calculation of these decay widths that the algebraic method offers an advantage over the traditional method, especially when a dynamic symmetry exists, since in that case all results can be obtained in explicit form, and thus become very transparent.

5. Baryon (qqq) configurations

For these configurations one uses the same logical scheme of the previous section. The algebra R of (qqq) baryons is

$$R = U_1(4) \otimes U_2(4) \qquad . \qquad (5.1)$$

Each U(4) describes the quantized excitations of the relative coordinates. It is customary to introduce as independent coordinates the Jacobi coordinates $\vec{\rho}$ and $\vec{\lambda}$, related to the three quark coordinates $\vec{r}_1$, $\vec{r}_2$ and $\vec{r}_3$ by

$$\vec{\rho} = \frac{1}{\sqrt{2}} \, (\vec{r}_1 - \vec{r}_2) \qquad ,$$

$$\vec{\lambda} = \frac{1}{\sqrt{6}} \, (\vec{r}_1 + \vec{r}_2 - 2\vec{r}_3) \qquad\qquad (5.2)$$

The algebra $U_\rho(4) \otimes U_\lambda(4)$ admits more dynamic symmetries the U(4). They can be obtained by considering all its possible branchings. Of particular importance is the symmetry corresponding to string-like configurations with classification scheme

$$\left| \begin{array}{ccccccc}
U_\rho(4) \otimes U_\lambda(4) & \supset & SO_\rho(4) \otimes SO_\lambda(4) & \supset & SO(4) \supset & SO(3) \supset & SO(2) \\
\downarrow & \downarrow & \downarrow & \downarrow & \downarrow & \downarrow & \downarrow \\
N_\rho & N_\lambda & \omega_\rho & \omega_\lambda & (\tau_1, \tau_2) & L & M_L
\end{array} \right\rangle \qquad . \quad (5.3)$$

The representations to be used here are the symmetric representations of $U_\rho(4)$ and $U_\lambda(4)$ characterized by $N_\rho = [N_\rho, 0, 0, 0]$ and $N_\lambda = [N_\lambda, 0, 0, 0]$ in the limit in which $N_\rho = N_\lambda \to \infty$. One can verify that the state (5.3) contains all the quantum numbers of the three-body problem. In the string-like picture, Fig. 4, there are three vibrational quantum numbers v_1, v_2 and v_3, plus the projection of the angular momentum on the body-fixed axis, K, the angular momentum, L, and its component on a laboratory fixed axis, M_L. These "physical" quantum numbers can be obtained from the "algebraic" quantum numbers by substituting

$$v_1 = \frac{N_\rho - \omega_\rho}{2} \qquad ,$$

$$v_2 = \frac{N_\lambda - \omega_\lambda}{2} \qquad ,$$

$$v_3 = \frac{\omega_\rho + \omega_\lambda - \tau_1 - \tau_2}{2} \qquad ,$$

$$K = \tau_2 \qquad . \qquad\qquad (5.4)$$

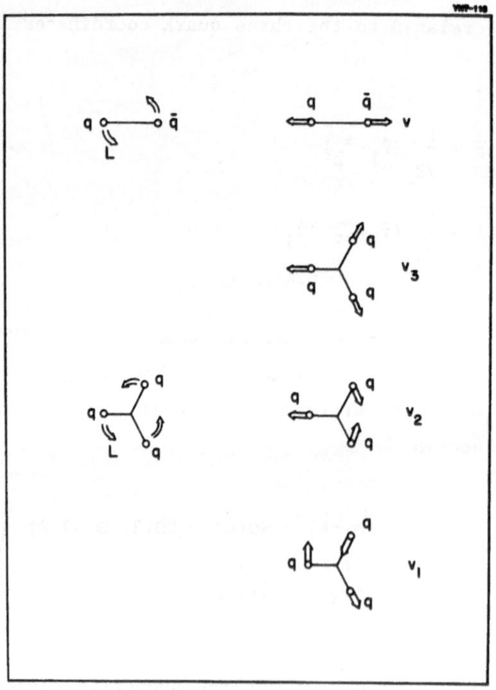

Fig.4. Schematic representation of the rotation-vibration degrees of freedom of ($q\bar{q}$) and (qqq) configurations.

A new problem arises here when combining $\mathcal{R}$ with the internal algebra to form

$$\mathcal{G} = \mathcal{R} \otimes SU_s(2) \otimes SU_f(n) \otimes SU_c(3) \qquad . \qquad (5.5)$$

The wave functions are of the type

$$\psi = \psi_{\mathcal{R}} \times \psi_s \times \psi_f \times \psi_c \qquad . \qquad (5.6)$$

Since we are dealing with three identical fermions, ψ must be antisymmetric. But the color piece is already antisymmetric (color singlet). Thus the product $\psi_{\mathcal{R}} \times \psi_s \times \psi_f$ must be __symmetric__. This implies that appropriate representations of $\mathcal{R} = U_\rho(4) \otimes U_\lambda(4)$ must be combined with appropriate representations of $SU_s(2) \otimes SU_f(n)$. This problem has been solved so far only for harmonic oscillator wave functions $U_\rho(3) \otimes U_\lambda(3)$ in the notation used here [6]. We plan to solve this problem in general for any wave function, in particular for the $SO_\rho(4) \otimes SO_\lambda(4)$ symmetry which presumably is the relevant symmetry of light quarks in baryons. The work on (qqq) configuration is in progress, including masses, decay widths and form factors, and is expected to be completed in course of next year.

5. Meson $(q\bar{q}q\bar{q})$ configurations

These configurations appear as additional states in the meson spectrum. The algebra $\mathcal{R}$ is here

$$\mathcal{R} = U_1(4) \otimes U_2(4) \otimes U_3(4) \qquad . \qquad (6.1)$$

A convenient choice of coordinates is that indicated in Fig. 1. There are no new technical problems associated with these configurations and a calculation similar to that already done for $(q\bar{q})$ should be feasible, although tedious.

7. Gluonic configurations

Configurations involving gluons are much more difficult to deal with. In the string-like model, gluonic configurations are represented by closed strings (or toroidal flux tubes). These closed strings do not have end points and thus they must be treated as continuous structures. A possibility here is the introduction of infinite dimensional (Kac-Moody) algebras. Another possibility, which we plan to follow since it corresponds more closely to what is done in the constituent quark and gluon model, is that of discretizing the string into a certain number of pieces, k, and then letting k→∞. In other words we take the infinite product

$$\mathcal{R}(\text{closed string}) = \prod_{i=1}^{\infty} \otimes\, U_i(4) \quad . \tag{7.1}$$

8. Conclusions

I have reviewed here the main lines of the algebraic approach to hadronic structure. The application of this approach to $(q\bar{q})$ configurations has been just completed (two-body problem). Application to (qqq) configurations is in progress (three-body problem). We plan to explore also in the future $q\bar{q}q\bar{q}$ configurations (four-body problem) and some gluonic states.

The main result obtained so far is that the meson spectrum displays a relatively good SO(4) dynamic symmetry, which implies that all of its properties can be obtained in a simple form (mass formula, formulas for widths, ...), without the need for a numerical solution of a Schrödinger-like equation. Whether this result will persist for the baryon spectrum remains to be seen. If so, it will allow to make simple predictions for baryonic properties that could eventually be tested at new facilities, such as the CEBAF facility presently under construction. The "goodness" of the SO(4) space symmetry in the low-lying mesons (π,K,η families) [4] appears to be comparable, if not better, than that of the internal symmetry (Gürsey-Radicati SU(6) ⊃ $SU_s(2) \otimes SU_f(3)$) [7].

Aknowledgements

This work was supported in part by D.O.E. Grant No. DE-FG02-91ER40608.

References

[1] Y. Dothan, M. Gell-Mann and Y. Ne'eman, Phys. Lett. <u>17</u>, 148 (1965).

[2] A.O. Barut and A. Böhm, Phys. Rev. <u>B139</u>, 1107 (1965).

[3] F. Iachello, Nucl. Phys. <u>A497</u>, 23c (1989); Nucl. Phys. <u>A518</u>, 173 (1990).

[4] F. Iachello, N.C. Mukhopadhyay and L. Zhang, Phys. Lett. <u>B256</u>, 295 (1991); Phys. Rev. <u>D44</u>, 898 (1991).

[5] F. Iachello and D. Kusnezov, Phys. Lett. <u>B255</u>, 493 (1991) and to be published.

[6] For a review see, A.J.G. Hey and R.L. Kelly, Phys. Rep. <u>96</u>, 71 (1983).

[7] F. Gürsey and L.A. Radicati, Phys. Rev. Lett. <u>13</u>, 173 (1964).

Few-Body Systems, Suppl. 6, 26—49 (1992)

ANTIPROTON REACTIONS AND CHARM
(WITH AND WITHOUT NUCLEI)

Kamal K. Seth

Northwestern University, Evanston, IL 60208, USA

The availability of pure, high intensity, ultra-high energy resolution antiproton beams in the 0.1 to 10 GeV/c range has made it possible to make precision studies of few-quark systems and the interactions between them. Proton-antiproton interaction can now be studied with precision heretofore achieved only in the study of the proton-proton interaction. It has become possible to test scaling predictions of perturbative QCD, e.g., for the proton form-factor in the time-like region. Precision measurements of the spectra of $q\bar{q}$ mesons with charm and beauty quarks are leading to deeper understanding of the quark-quark interactions. Rather unusual and exotic effects are predicted when charmonium is embedded in nuclei. Some of these new developments are described and the feasibility of several nuclear experiments is examined.

1. INTRODUCTION

Some people may have private doubts about it, but most of us will agree that in hadronic physics the name of the game is QCD. It is doubtful that classical nuclear physics, which is for the most part concerned with nuclear structure, needs any help from QCD. However, it is obvious that the more fundamental aspects of nuclear physics, the basic interactions, the few-body problems, and the high momentum transfer phenomena have to be reconsidered in terms of QCD degrees of freedom and QCD symmetries.

As is well known, the three characteristics properties of QCD are:

- Asymptotic freedom, i.e., the weakness of the quark-quark, quark-gluon interaction at very short distances, or for large momentum transfers. In this regime perturbation theory applies.

- Confinement, i.e., the feature of the interaction which keeps quarks and gluons forever confined in very small volumes. The strength of the interaction increases with increasing separation between individual quarks and gluons and prevents their appearance as free particles. In this regime momentum transfers are small, and perturbation theory can not be used. This is unluckily the regime in which most of the nuclear physics resides.

*Supported in part by the U.S. Department of Energy.

- <u>Gluon self-interaction</u>, This is a unique consequence of the non-Abelian nature of the theory which makes the quanta of the field, the gluons, interact strongly with each other, leading to the possibility of pure-glue structures, the 'glueballs'.

The validity of the concept of asymptotic freedom has been confirmed by many experimental observations. Such is, however, not the case with phenomenon in the non-perturbative domain. Here few predictions can be made with confidence, and fewer still are borne out by experiments. The belief in QCD in this domain is largely based on faith and aesthetic appeal. Asfar as Glueballs are concerned, despite the best efforts, none have been found so far. It follows that much remains to be done to establish QCD as the correct theory of strong interactions.

There are two main ways of studying QCD structures: e^+e^- annihilations, and $p\bar{p}$ annihilations. In the present talk I will confine myself to antiproton annihilations as the means to study QCD structures and QCD predictions.

There are good reasons for talking about antiproton experiments at this time. There is a resurgence of interest in antiproton physics at intermediate energies, as evidenced by the proceedings of numerous international conferences at LEAR, FERMILAB, and KAON, etc. This interest has been catalyzed by two presently active programs. The first is the excellent and diverse program of antiproton physics at low energies currently in progress at LEAR at CERN. The second is the program of ultra-high resolution studies of Charmonium at Fermilab (experiment E760). The exciting physics coming out of these programs has given rise to at least two new, ambitious projects-the plans for SuperLear at CERN and for KAON at TRIUMF. In Table 1 we summarize the relevant parameters of the antiproton beams available and planned at these facilities.

Table 1. Characteristics of Antiproton beams at various facilities,in operation and planned.

Facility	Momenta GeV/c	$\Delta p/p$	Intensity $\bar{p}$/s $\mathcal{L}$ in cm^{-2}s^{-1}	Purity %
LEAR	0.05-2.0	10^{-3}	10^5-10^6	100
FNAL Accum	2.0-9.0	10^{-4}	$\mathcal{L}=10^{31}$	100
FNAL Upgrade*	2.0-9.0	10^{-4}	$\mathcal{L}=10^{32}$	100
SUPERLEAR*	1.5-12.0	10^{-3}	$\mathcal{L}=10^{32}$	100
KAON*	0.4-20	10^{-2}-10^{-3}	10^7-10^5	≥50
*Planned				

In this talk I want to introduce antiproton reactions via the Fermilab experiment E-760 which is devoted to high resolution Charmonium spectroscopy.[1] This is however not the main subject of this talk. I use it primarily as a case study--to illustrate what is being done today and what may be done tomorrow. Most of my talk will be devoted to experiments beyond Charmonium spectroscopy, including antiproton experiments with nuclear targets. I will examine their physics interest and their feasibility.

2. CHARMONIUM SPECTROSCOPY

All two-body interactions in QCD, gg, qq, $q\bar{q}$, qq are flavor-independent. This means that in principle they can be equally well studied with quarks of any flavor. In practice, two considerations determine the choice. Relativistic effects and gluon self-interaction effects are worse for the lightest quarks. On the other hand, cross sections are the worst (smallest) for the heaviest quarks. For example quarkonium ($q\bar{q}$) formation cross sections in $p\bar{p}$ annihilations are proportional to m_q^{-8}. It follows that for very pragmatic reasons, the extremes are to be avoided. Charmed quark structures are therefore the best ones to study.

We know a lot about Charmonium.[2] The energy level spectrum of Charmonium is shown in Fig. 1. The first thing to notice is the incredible richness of this two-body spectrum. [Just

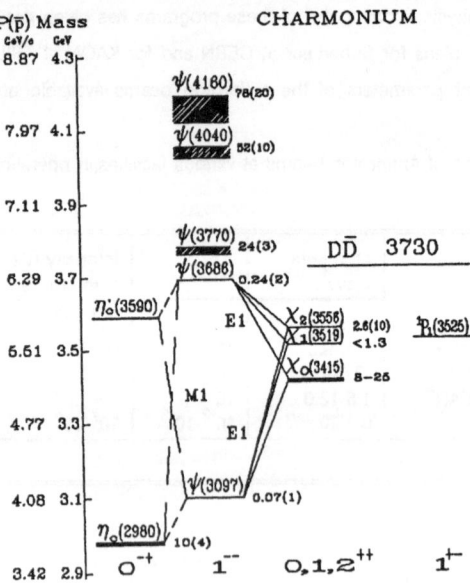

Fig. 1 Mass spectrum of Charmonium states.

imagine how much we could learn about the NN interaction if the deuteron had such a structure, with eight bound states instead of one.] All that is shown in Fig. 1 is essentially from e^+e^- experiments. Since only the 1^{--} ((ψ, ψ', ψ'') states are directly populated in these experiments, information about other states, the 0^{-+} (η_c, η'_c) $^1S_0$ states, the 0^{++}, 1^{++}, 2^{++} (χ_J) ^3P_J states, the 1^{+-} $^1P_1$ state, is rather poor. Thus, for example, the widths of most of these states were very poorly determined in e^+e^- experiments: $\Gamma(\chi_0) \approx$ 8-21 MeV, $\Gamma(\chi_1) <$ 3.8 MeV., $\Gamma(\chi_2)$ = 0.84-4.9 MeV, $\Gamma(\eta_c)$ = 11 ± 4 MeV. η'_c was not convincingly identified, and no evidence at all was found for the $^1P_1$. The reason for poor width determinations lay primarily in the inherently poor energy resolution, 20-30 MeV, of the γ-rays detected in the famous crystal-ball.[3]

As shown in Fig. 2, unlike e^+e^- annihilations which proceed via a single photon, and can therefore directly populate only 1^{--} states ($\psi,\psi',...$), $p\bar{p}$ annihilations proceed via two and three

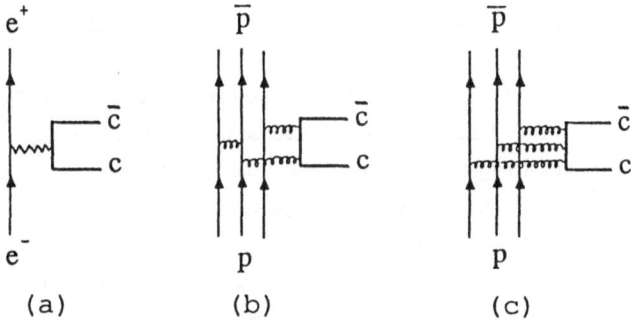

Fig. 2. (a) Production of vector (1^{--}) states of $c\bar{c}$ in e^+e^- annihilation via a virtual photon; (b), (c) production of charmonium states in $p\bar{p}$ annihilation via to and three gluons.

gluons, and can directly populate states with all J^{PC}. Thus the excellent energy-resolution of accelerator beams can be brought to bear on all states. This makes $p\bar{p}$ annihilations a very versatile tool for the study of Charmonium.

What do we expect to learn from high resolution studies of Charmonium states? First, contrary to the popular belief the $q\bar{q}$ potential is not well established. It has been shown that many completely different central potentials, which track each other in the 0.3 to 0.8 fm range, provide equally good fits to the energy spectra of the known Charmonium and Bottonium states.[4,5] Lattice QCD calculations are nowhere near providing a choice. The range of possible potentials can only be narrowed by precision experiments. Second, our knowledge of the spin-dependent part of the potential is even poorer. We do now know whether the confinement potential is scalar or vector. We do not know the effect of relativistic corrections and channel coupling. Only precision experiments identifying the η'_c, $^1P_1$, $^3D_2$, $^1D_2$ states can shed light on some of these important problems. It appears that p$\bar{p}$ annihilations, with their ability to directly populate all these states, and measure their widths, might very well provide the ideal tool for these purposes. These considerations led to the first p$\bar{p}$ experiment, R-704 at the ISR at CERN[6] and its success, in turn, led to the E760 experiment at Fermilab.[1]

In order to do precision Charmonium spectroscopy at Fermilab, a collaboration of physicists from Fermilab, Ferrara, Genoa, Irvine, Northwestern, Penn State and Torino proposed E760, to be done with a gas jet target in the Fermilab antiproton accumulator. The experiment was approved in December 1985 and its first data taking runs were made in June-August, 1990. The first results are just going out for publication.

A number of modifications have been made in the Fermilab accumulator, which was originally designed as the antiproton source for the Tevatron[7]. Instead of operating at a fixed momentum of 8.9 GeV/c, the accumulator can now decelerate (or accelerate) the stored antiprotons by small energy steps, cool them stochastically to $\Delta p/p \leq 10^{-4}$, and provide circulating beams of $\sim$ 2.5×10^{11} antiprotons (25 mA equivalent) at any momentum between about 2 to 9 GeV/c. A gas-jet internal target has been installed in a straight section of the accumulator ring. For hydrogen it currently provides a target density of $\sim 0.5\times10^{4}$ atoms/cm^2, leading to a p$\bar{p}$ luminosity of 0.8×10^{31} cm^{-2}s^{-1}. A detector system with large acceptance has been designed (see Fig. 3) primarily to detect electromagnetic final states, p$\bar{p}$ --> X --> e^+e^-, $e^+e^-\gamma$, $\gamma\gamma$, ... with good efficiency, excellent particle discrimination, and extremely low levels of background (<10 picobarns/MeV)[8]. Unprecedented mass resolutions, Γ = 0.4 - 1.0 MeV/c^2 are realized for beam momenta from 2 - 9 GeV/c.

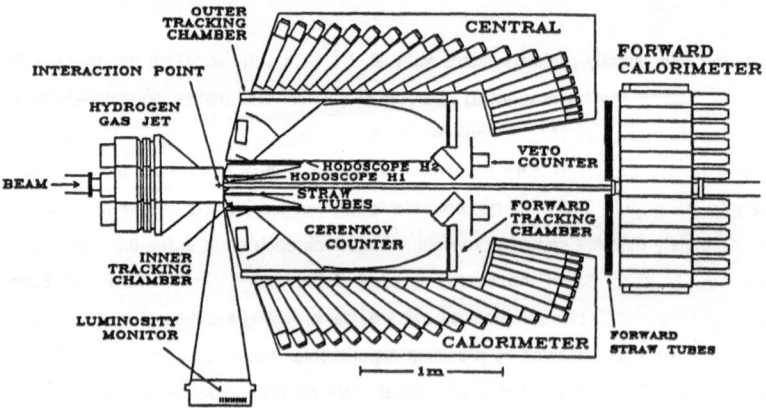

Fig. 3. Schematic representations of the Fermilab E760 detector system.

Figures 4 and 5 illustrate typical results obtained in the first experiments with $p\bar{p}\rightarrow(c\bar{c})_R\rightarrow e^+e^-$, or $e^+e^-\gamma$. The energy resolution achieved at the J/ψ resonance ($\sqrt{s}$ = 3097 MeV) was FWHM = 400 KeV. With a factor ~50 better energy resolution than that achieved with γ-ray detection in e^+e^- experiments the widths of χ_1 and χ_2 states were directly measured. Background away from the resonances was measured to be ≤ 10 picobarns.

During the 1991 running, which began in July, we are concentrating on finding the missing $^1P_1$ state, on verifying the existence of the doubtful η'_c, on improving our knowledge about η_c, and on multipole analysis of the radiative decays of χ_1 and χ_2.

3. BEYOND CHARMONIUM

While the primary focus of E-760 is Charmonium spectroscopy, a number of very interesting parallel experiments are being done on the side. Let me describe some of these.

3.1 THE NUCLEON-ANTINUCLEON INTERACTION

The understanding of the nucleon-nucleon interaction has rightly been the most important and the most basic of the pursuits of nuclear physics. Great progress has been made in terms of meson-exchange models in explaining NN scattering below 400 MeV, i.e., below the single-pion production threshold. Above this energy severe impediments are encountered. There are hints that above these energies one must face the dynamics of the underlying quark-gluon degrees of freedom.

The nucleon-antinucleon interaction has been studied much less. Experiments are scarce and theoretical understanding is even more scarce. At low energies (< 1 GeV/c) new data from LEAR exist. However, little beyond the G-parity approach has been tried so far. In this approach one takes the best one-boson exchange potential one can find, flips the sign of odd G-parity exchange terms (π,σ, ρ..), and adds an ad-hoc imaginary potential to account for annihilation. The approach works only at the lowest energies. Of course, little or no data exist above 1 GeV/c (1 < p < 10 GeV/c).

If one subscribes to the belief that "no understanding of the NN interaction can be considered fundamental and complete unless it leads to a simultaneous understanding of N$\bar{N}$ interaction", it follows that the data-base for N$\bar{N}$ scattering has to be increased many-fold, especially in the 1-10 GeV/c region.

One of the most important parameters which describe elastic scattering is the so called ρ = Ref(0^0)/Imf(0^0). This parameter can be analyzed in terms of dispersion relations. Kroll has done several such analyses.[10] For $p\bar{p}$ scattering the existing experimental data display considerable structure in ρ (see Fig. 6) and suggestions have been made that these indicate the

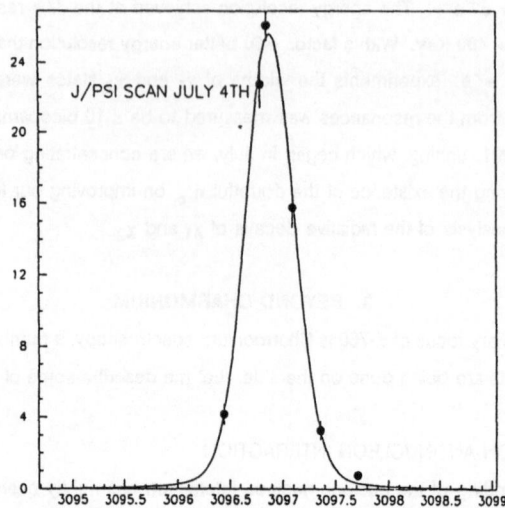

Fig. 4 An excitation function scan for the J/ψ (3097) resonance from E760. Notce that the c.m. energy resolution is FWHM ~ 0.4 MeV.

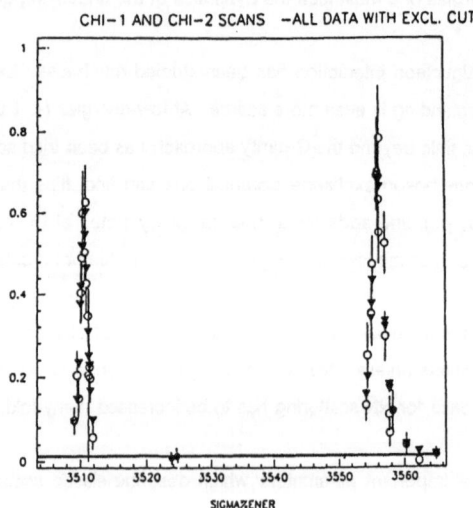

Fig. 5 A composite of E760 scans of χ_1(3510), and χ_2(3556), and the background level in between.

existence of resonant states in the p$\bar{\text{p}}$ system. It is obviously very important to submit these conjectures to careful scrutiny. In the 3-10 GeV/c region only one set of prior measurements exists.[11] These measurements suggest a rather narrow oscillation of ρ, which, if real, would be highly significant (see Fig. 6). In the E-670 experiment we have developed a unique capability to make precision measurements of ρ in a rather symbiotic relationship with the main experiment.

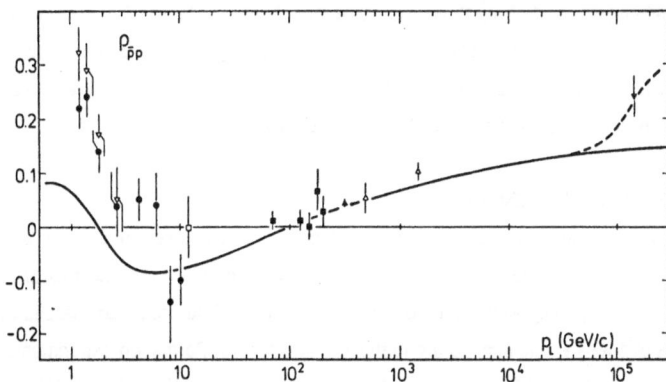

Fig. 6 The ρ-parameter for p$\bar{\text{p}}$ elastic scattering. The curve is from a dispersion relation cacluation due to Kroll.[10]

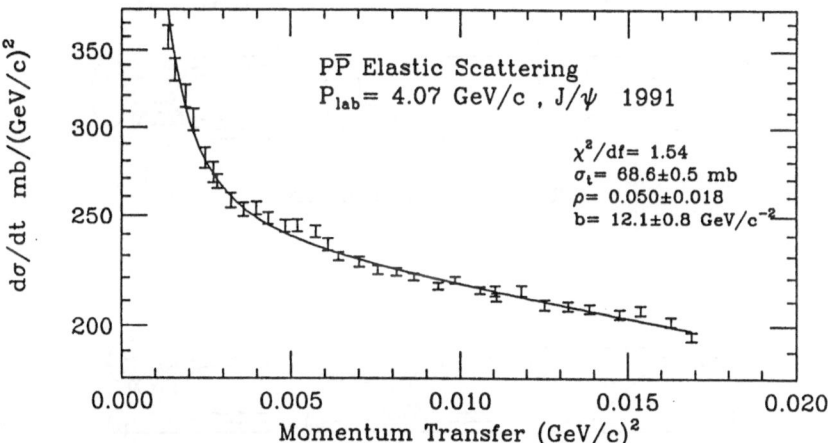

Fig. 7 A typical measurement of small t elastic p$\bar{\text{p}}$ scattering. Preliminary results of anlaysis are shown in the insert.

For the Charmonium experiment it was necessary to design a luminosity monitor. We designed this to be a monitor of protons from p̄p elastic scattering recoiling near 90°. Because the recoil energies are only a few MeV, we can detect these protons at θ = 82°-90° in several solid state detectors mounted on a carriage. The recoil protons correspond to antiproton scattering at very small t = 0.0015-0.0270 $(GeV/c)^2$. This includes the region of Coulomb-nuclear interference and therefore allows very precise normalization of the cross sections. An example of the preliminary results from these measurements is shown in Fig. 7. It may be noted that the precision with which we are able to determine ρ (generally ±0.014) is nearly a factor five better than the best of the old measurements. We expect to have equally precise results at the energies of all Charmonium states-from 3.68 GeV/c to 8.9 GeV/c.

3.2 PURSUIT OF THE EXOTICA

Fig. 8 illustrates that because p̄p annihilations are copious source of glue, they may be the ideal means of creating Glueballs (g^3, g^2) and hybrids (q̄qg). Also di-mesons or cryptoexotics ($q^2\bar{q}^2$) can be hopefully easily formed in p̄p annihilations. The predicted masses of the lowest members all these exotic families are in the 1-3 GeV range. Formation experiments in this range are best suited for LEAR. However, they can also be investigated at FNAL when they are produced in association with one or more mesons. Indeed, at E760 we may have already observed $A_x(1550)$ which is a candidate for being a $q^2\bar{q}^2$ exotic.[12]

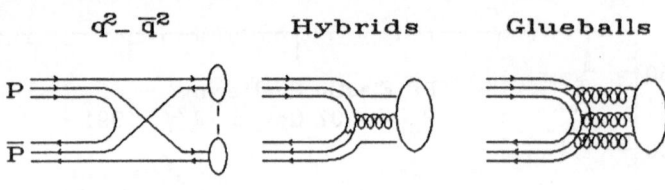

Resonant production

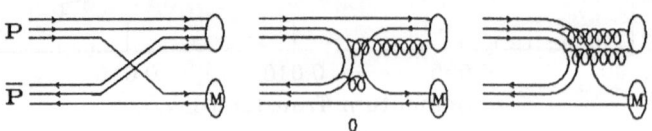

Production with a Meson

Fig. 8 Production of exotica in p̄p collisions: (left) p̄p → $q^2\bar{q}^2$, (center) p̄p → q̄qg, (right) p̄p → glueball.

Time does not permit me to go into this topic in any detail. However, let me point out that one of the reasons why the usual exotics are difficult to identify is that in the 1-3 GeV mass range there is already an over-abundance of 'normal' particles. Perhaps, prospects will be better at higher masses with charmed quark hybrids and charm containing di-mesons. Theoretical help in terms of ideas about distinctive decay modes of these objects would be most welcome.

3.3 QCD SCALING RULES

Brodsky and collaborators[13] have made a series of predictions about exclusive processes based on their factorization hypothesis. The contention is that the reaction rates at large momentum transfers can be factorized into a distribution amplitude, which contains all the non-perturbative dynamics, and an amplitude which can be calculated perturbatively. This leads to scaling laws which depend on the number of 'elementary constituents' (quarks, leptons, photons). Some of these for $p\bar{p}$ annihilations are:

$$p\bar{p} \to \gamma\gamma, \qquad \sigma(\theta) \propto p_T^{-10} \, f(\ell n p_T, \theta) \tag{1}$$

$$p\bar{p} \to \gamma M, \qquad \sigma(\theta) \propto p_T^{-12} \, f(\ell n p_T, \theta) \tag{2}$$

$$p\bar{p} \to M\bar{M}, \qquad \sigma(\theta) \propto p_T^{-14} \, f(\ell n p_T, \theta) \tag{3}$$

$$p\bar{p} \to B\bar{B}, \qquad \sigma(\theta) \propto p_T^{-18} \, f(\ell n p_T, \theta) \tag{4}$$

While some scaling laws have been tested for pp collisions, to the best of my knowledge no such tests have been made for the above predictions for $p\bar{p}$ annihilations. It is difficult to test these predictions over an extended range of p_T. However, at E760 we hope to test at least some of them. The $p\bar{p} \to \gamma\gamma$ cross sections have been found to be very small ($\leq$ a few pb). However, we are able to test $p\bar{p} \to \pi^\circ \pi^\circ$. It may also be possible to test $p\bar{p} \to p\bar{p}$. We expect to have the first results soon.

A similar scaling law prediction concerns the reaction

$$p\vec{p} \to e^+ e^- \tag{5}$$

which allows one to measure the proton form-factor in the time-like region. Here the prediction for the two form-factors $F_1(s)$ and $F_2(s)$ is that

$$F_1(s) \propto \frac{f(\ell n s)}{s^2}, \qquad F_2(s) \propto \frac{f(\ell n s) \cdot M^2}{s^3} \tag{6}$$

In terms of the more familiar form-factors, G_E and G_M

$$\frac{d\sigma}{d\Omega} = \frac{\alpha^2}{16Ep}\left[|G_M|^2(1+\cos^2\theta) + \frac{4M^2}{Q^2}|G_E|^2\sin^2\theta\right] . \qquad (7)$$

Because of its more rapid fall with increasing Q, we may neglect the contribution due to the second term, and therefore determine $|G_M|$. The scaling prediction is that $Q^4|G_M(p)|$ should reach a constant value asymptotically. In the space-like region this value is ~ 1 (GeV/c)^4. One would expect that the asymptotic value in the time-like region is the same.

Few measurements of the proton form-factor in the time-like region are available.[14] Recently, precision measurements of $p\bar{p} \rightarrow e^+e^-$ have been made at LEAR at $Q^2 < 4.2$ (GeV/c)^2[15]. They indicate an unexpected structure, a minimum in $Q^4|G_M|$ near $Q^2 \sim 3.7$ (GeV/c)^2. Clearly, it is extremely important to extend these measurements to as large Q^2 as possible. At E760 we have the capability to go up to $Q^2 \sim 19$ (GeV/c), and we intend to analyze our data for $|G_M|$ at all energies where the contribution of ψ, ψ', ψ'' is not too large. In Fig. 9 we show the preliminary result at $Q^2 = 12.5$ (GeV/c)^2, along with the other results.

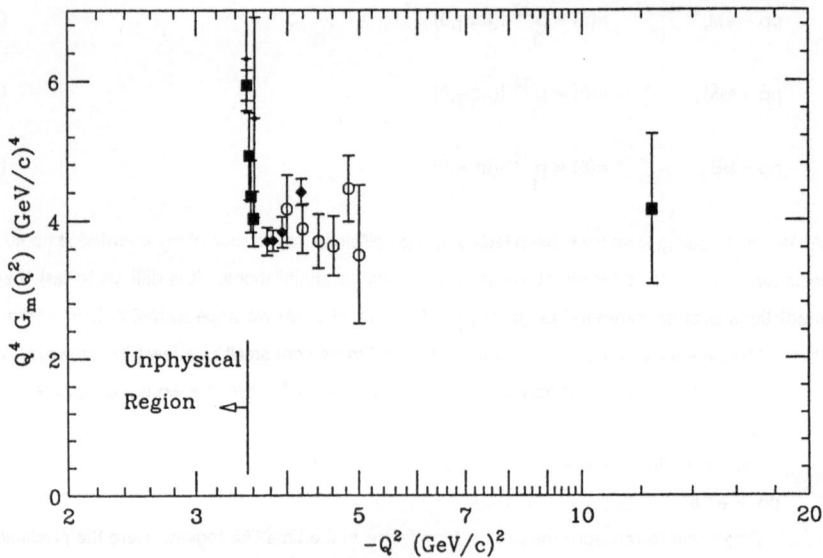

Fig. 9   Proton form-factor in the time-like region plotted as $Q^4|G_M|(p)|$. Open symbols are from $e^+e^- \rightarrow p\bar{p}$. Solid symbols are from $p\bar{p} \rightarrow e^+e^-$. A typical E760 datum point is shown at 12.5 (GeV/c)^2.

I now want to change my direction completely and talk about Charmonium in nuclei. This is a completely unchartered territory. As we know from the example of hyperons in nuclei, when a "non-nucleon" is immersed in the nuclear environment completely new phenomena may occur. Indeed, some very interesting effects are predicted. These relate to Charmonium-nucleon cross sections and color transparency, nuclear-bound Charmonium, resonance-narrowing in nuclei and charmed particle production in nuclei. I will now discuss some of these topics.

4. CHARMONIUM NUCLEON INTERACTION

Despite its obvious and fundamental importance, no reliable measures of Charmonium-nucleon cross sections exist. There is a number which everybody uses. It is σ = 1 - 2 mb. It comes from a Glauber-model analysis of the high-energy experiments on nuclei (photo-production[16], E(J/ψ) ~20 GeV; hadroproduction[17] E(J/ψ) $\leq$ 200 GeV). However, as Brodsky and Muller[18] were the first to point out, these estimates of total cross section "have little to do with J/ψ scattering on a nucleon". The reason lies in the phenomenon of "color transparency".

Color transparency owes its origin to the fact that in a large momentum transfer (p_t) process the only part of a hadron's wave function which plays a role is that in which all the valence quarks (the color constituents) are close together, within a relative impact parameter of the order of 1/p_t. These configurations correspond to a small size and have a small color dipole moment because of large color cancellations. Such an object interacts weakly with other nucleons in the nucleus. If this compact hadron has enough energy, it can escape the nuclear medium without much interaction or attenuation. In other words, it finds the nucleus transparent. The process is schematically illustrated in Fig. 10. If the data for the production of J/ψ under such circumstances

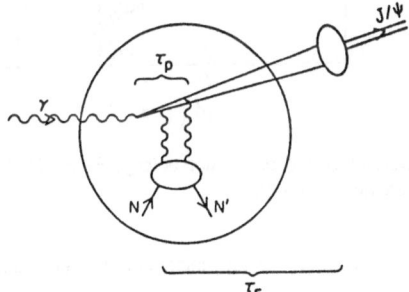

Fig. 10 A schematic illustration of J/ψ photoproduction at 100 GeV. The production time, τ_p p_γ/M_ψ^2 = 2 fm is short, compared to the formation time, τ_f = r_ψ/v_t = 1 fm x p_γ (GeV) = 100 fm. (From Ref.18).

were analyzed in the Glauber model one would incorrectly infer a very small interaction cross section. This, in essence, is the basis of Brodsky and Muller's critique of the results of the high-energy photoproduction experiments.

The fundamental importance of determining the 'true' Charmonium-nucleon cross section is beyond question. This is best illustrated by an example.

One of the hottest questions in all of nuclear and particle physics is whether there exists a quark-gluon plasma phase of matter. It is conjectured that such a phase can be realized in the collision of heavy ions at relativistic energies. However, it is far from clear what would constitute a convincing signal for the formation of the quark-gluon plasma. Currently, the most popular signal is the observation of a characteristic suppression of J/ψ production in such collisions[19]. Such suppression has been actually observed in the CERN experiment NA-38[20]. The data are shown in Fig. 11. These data have been widely cited as evidence for the formation of quark-gluon plasma.[21] However, in the words of the authors of the experiment themselves[20], "alternative explanations can not be ruled out". Let us look at the situation a bit closer.

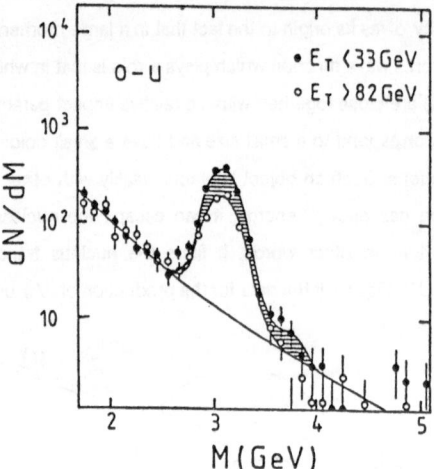

Fig. 11 The observed suppression of J/ψ production at large E_T in CERN experiment NA38, with GeV/nucleon oxygen incident on uranium.

It is conjectured that at high density strongly interacting matter undergoes a transition from a many-body system of hadrons to a quark-gluon plasma. The confining part of the quark-quark potential transforms from one proportional to r to a Debye-screened Coulomb type potential:

$$V(r) \propto r \ \longrightarrow \ V(r) \propto (1/r) \exp(-r/r_D) \tag{8}$$

At high temperatures r_D, which is proportional to $1/T$, can become smaller than the radius of bound $c\bar{c}$, for example, the J/ψ. Then $c\bar{c}$ can no longer bind to form J/ψ. This is essentially the argument for the expected suppression of J/ψ production in presence of a quark-gluon plasma.

There are, of course, more conventional reasons for the suppression of J/ψ[21]. They invariably invoke attenuation of the J/ψ in the normal nuclear medium, and therefore depend on the J/ψ-nucleon cross section. The current claim is that if this cross section is as small as 1-2 mb, as the high-energy experiments suggest, nuclear attenuation alone "fails to explain by nearly a factor two" the suppression observed. However, as discussed earlier, the existing measurements of J/ψ-N cross sections are at best lower limits. It is necessary that the 'true' J/ψ nucleon cross section be measured in an experiment in which the effects of color transparency can be carefully controlled. Only then can one judge if the alternatives to the QG plasma explanation really fail.

Brodsky and Muller[18] have suggested that "J/ψ production by antiprotons on a nuclear target .....could provide direct information on the J/ψ-nucleon cross section, σ(J/ψ,N)".[11] The idea is simple. Unlike photoproduction or hadroproduction, in the resonant production of J/ψ by $\bar{p}$ incident on a nucleus,

$$\bar{p} + A \rightarrow J/\psi + (A\text{-}1) \tag{9}$$

the reaction takes place on a quasi-free proton at the laboratory momentum of 4.07 GeV/c. The outgoing J/ψ (mass = 3.1 GeV) has a laboratory momentum of 4.07 GeV/c, i.e., it is very nonrelativistic. [Both momenta are smeared by $\approx \pm$ 0.15 GeV/c because of the Fermi motion of the bound proton in the nucleus A.] For J/ψ with such small velocities one does not expect any color transparency effects. A Glauber-model analysis of its attenuation should therefore lead directly to σ(J/ψ, N). Recently Ferrar et al.[22] have made a detailed analysis of J/ψ production of $\bar{p}$ incident on a nucleus. They model the incident antiproton's attenuation after the proton attenuation observed in the proton quasielastic scattering experiment of Carroll et al.[23]. Since in absence of any attenuation the J/ψ production cross section on nuclei should scale as Z, the measured cross section can be expressed as

$$\sigma(J/\psi,A) = Z_{eff}\ \sigma(J/\psi,/N). \tag{10}$$

The quantity Z_{eff}/Z provides a measure of the net attenuation. Ferrar et al.[9] express their results in terms of this quantity. They find that none of the color transparency models used for ($c\bar{c}$) expansion cause significant differences from the pure Glauber-model result. On the other hand, they find that Z_{eff}/Z is very sensitively dependent on σ(J/ψ,N). This means that a measurement of Z_{eff}/Z can be used to determine σ(J/ψ,N) with good accuracy.

Let us now address the question: Is such a measurement possible with the E760 set-up?

This question has several parts.

1. What luminosity can we get with a nuclear gas target?

   We have studied this question. Without going into details, the answer is that for light gases (Ne, O, N, CH_3, ^4He, ^3He) if we reduce the gas-jet atomic density by a factor $= Z$, (so that the proton density is the same as with the hydrogen target), the lifetimes of the $\bar{p}$ beams are expected to be approximately the same as with the present hydrogen jet, but with a momentum spread which is $\sim$ 3 times wider, i.e., $\Gamma_{beam}^{lab} \sim 3$ MeV.

2. Can we use the present e^+e^- decay trigger with advantage in these measurements?

   The answer is yes, because the kinematics is absolutely identical to the $p\bar{p} \to J/\psi \to e^+e^-$ kinematics except for the fact that the target proton in the nucleus now has a Fermi momentum spread which has a half width of $\sim$25 MeV.

3. What does the Fermi motion do to our yield?

   The peak cross section for $p\bar{p} \to J/\psi \to e^+e^-$ is $\sigma_0 = 365$ nb. In the present $p\bar{p}$ experiment, done with $\Gamma(beam)/\Gamma(J/\psi) \sim 7$, the measured peak cross section for the J/ψ becomes $\sigma_{peak} = \sigma_0/7 = 52$ nb. For the Fermi smeared J/ψ produced with a nuclear target the peak cross section is reduced by a factor $\sim \Gamma(Fermi)/\Gamma(J/\psi) = 25/0.068 = 368$, i.e., $\sigma_{peak} \sim 1$ nb (the effect of the beam width is now negligible). Thus the $\bar{p}$ experiment would go nearly a factor 50 slower than the present $p\bar{p}$ experiment. For the maximum luminosity (25 mA stacked beam) this means J/ψ (detected and analyzed) = 8 events/hr. One beam fill with a half-life of 48 hours would yield 270 events and therefore lead to a (J/ψ, A) cross section determination with a statistical error of < 5%. Thus one could measure σ (J/ψ,A) for four or five nuclear targets in less than a month of running including plenty of overhead time ($\sim$ 50%) for changing target gases.

4. With peak cross section reduced by a factor 100, what about the backgrounds?

   In the present experiment the background levels are < 20 pb. There is no reason to believe that for the nuclear targets the background level should be any higher. With the nuclear, Fermi smeared J/ψ peak cross section of 1 nb, the signal to noise ratio is still 50:1. There is no problem here.

5. How about the ψ'-nucleon cross section?

   This is a very interesting and logical extension of the ψ-nucleon measurements. From the QCD-inspired potential models of Charmonium one finds that the radius of ψ' ($r \sim 0.85f$) is about twice as large as that of J/ψ ($r \sim 0.45f$). Thus one expects that if the geometrical arguments are correct $\sigma(\psi',N) \sim 4\sigma(J/\psi,N)$. It would be extremely interesting to find out if this prediction is indeed true. Because the basic ψ' production cross section is a factor

30 smaller, this measurement will yield only 40 events in one beam-fill. On the other hand, to establish the truth of the geometrical argument one does not have to measure five nuclear targets. One or two would be quite sufficient. Two $\sigma(\psi')$ measurements with ±10% errors can be made for two nuclear targets in less than three weeks, since no additional overhead beyond that required for $\sigma(\psi)$ measurements would be involved.

6. How about mapping the Fermi momentum distribution of the nucleons in the nucleus which participate in J/ψ production?

Brodsky and Muller[18] suggested that this is an interesting thing to investigate. This would be a prohibitively time-consuming measurement at Fermilab. On the other hand, this investigation would be done much more efficiently with an external antiproton beam with an order of magnitude poorer energy resolution, for example, at KAON.

5. NUCLEAR BOUND CHARMONIUM

It is rather odd that mesons exist, and baryons exist, but bound states of mesons with baryons do not seem to exist. No rules of QCD forbid this, but no such states have so far been found. One must ask: Is this simply because the numbers do not quite add up, or does this indicate a serious lack in our present understanding of QCD? On the other hand, if meson-baryon systems do exist, their discovery would be like the discovery of a new species - fantastic!

There are two necessary ingredients for binding a meson to a nucleon, an attractive interaction of sufficient strength, and a large enough mass of the meson. In terms of elementary quantum mechanics, in order to 'just bind' a particle of reduced mass m in an attractive square well of depth V and radius R, the condition which must be satisfied is:

$$\cot KR = 0, \qquad \text{or } KR = \pi/2, \qquad \text{where } K = (2mV/h^2)^{1/2} \qquad (11)$$

Let us see how this works out in the cases which have already been investigated.

5.1 Pion-Nucleus Bound States

Many attempts have been made to look for bound states of pions in nuclei. A priori, one would not expect pions to bind because their mass (~140 MeV) is too small. It takes a potential well ~50 MeV deep to bind a proton and a neutron rather loosely in a deuteron. It follows that in order to bind a pion, with 1/7 the mass of a nucleon, to another nucleon in a deuteron-size structure, the potential depth required would be ~ 7 x 50 = 350 MeV. Such strong s-wave attractive interaction just does not exist in the π-N system. However, from time to time theoreticians have come up with scenarios involving the p-wave πN attractive interaction which predict bound states of pions and nuclei. Recently, a great amount of experimental activity was spurred by one such prediction for a T = 2 bound state of the πNN system. In a high sensitivity, high energy resolution pion double charge exchange experiment, $\pi^- + d \to \pi^+ (\pi^-NN)$, done at the Los Alamos meson factory, we searched for such a state and did not find it. Upper limits for the formation cross sections of 0-5 nb/sr were established[24].

5.2 Eta-Nucleus Bound States

Recently, Liu and collaborators[25], and others[26] have claimed that the s-wave η-nucleon interaction is attractive. The mass of η is large (550 MeV), and η should bind to nuclei. Liu et al.[25] calculated binding energies, widths, and formation cross sections for η-nucleus states in several nuclei. Since η-production near the threshold is dominated by the pion-nucleon resonance N^* (1535), they suggested the (π^+,p) reaction to produce the η-nucleus bound states.

Chrien et al.[27] at Brookhaven have carried out a search for η-nucleus bound states by means of (π^+,p) reactions on Li, C, water, Al, and polyethylene. The incident π^+ momentum was 800 MeV/c and the outgoing proton energy resolution was 2.5-4.0 MeV. No structures were found in the outgoing proton spectra for any of the targets. For ^{16}O they were able to compare their results with a specific prediction of Liu et al.[25] They established a 95% confidence upper limit of 8.7 μb/sr MeV for the peak production cross section, whereas the prediction of Liu et al.[25] was ~25 μb/sr MeV. It would, however, appear that a far more sensitive search, at 1 nb/sr MeV level, needs to be made in order to settle the case of η-nucleus bound states independently of the specific predictions of Liu et al.

5.3 Charmonium-Nucleus Bound States

Despite the failures to find π-nucleus or η-nucleus bound states, there are good reasons to believe that Charmonium-nucleus bound states might very well exist. The first reason is the large increase in mass in going from the η to (cc̄). The second reason is provided by the arguments of Brodsky, Schmidt, and de Teramond[10] for a strong attractive (cc̄)-nucleus interaction.

Brodsky et al.[28] point out that since there are no common quarks between the (cc̄) states and a nucleus (A), the interaction is entirely mediated by a multi-gluon exchange. This is the QCD version of the van der Waals potential (see Fig. 12). There is therefore no short-range repulsive interaction due to quark exchange (or equivalently, effective meson exchange), or due to Pauli blocking. However, as mentioned earlier, the (cc̄)-nucleon interaction is itself unknown. BST are therefore forced to choose the (cc̄)-nucleus interaction on the basis of phenomenological arguments. They parametrize the interaction as a vector interaction of the Yukawa form:

$$V(c\bar{c},A) = (\alpha/r)\exp(-\mu r) \tag{12}$$

In order to estimate the coupling strength parameter α, and the range parameter μ^{-1}, they identify

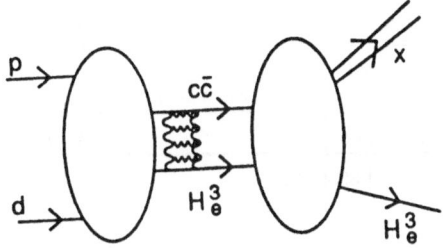

Fig. 12. Formation of Charmonium-nucleus bound state by multi-gluon van der Waals type interaction, and the subsequent decay of the system $(c\bar{c}) A \rightarrow A + \gamma + \gamma$, or $(c\bar{c}) A \rightarrow A + e^+ + e^-$.

their gluon-exchange potential model with the Pomeron exchange model of Donnachie and Landshoff[29] for high-energy forward angle meson-nucleus scattering. They note that since $(c\bar{c})$ is much smaller than the nuclear size, $<R_A^2>^{1/2}$, the range parameter μ^{-1}, which is determined by the slope (at t = 0) of the forward angle cross section, is given by

$$\mu^{-2} = <R_A^2>/6, \tag{13}$$

and the coupling strength, which is related to the magnitude of the cross section (at t = 0), is given by

$$\alpha = 3A\beta^2\mu^2/2\pi \tag{14}$$

where $\beta = 1.85$ GeV^1 is the flavor-independent Pomeron-quark coupling constant. BST distinguish between three different strengths of the interaction: unscreened - with α as above, fully screened - with $\alpha/2$, and partially screened - with $\alpha/\sqrt{2}$. They use a variational wave function

$$\psi(r) = (\gamma^3/\pi)^{1/2} \exp(-\gamma r) \tag{15}$$

to calculate the binding energies in their Yukawa potential, for which the condition for binding is that $(\alpha m_{reduced}/\mu) > 1$.

Brodsky et al.[28] find that neither η_c nor J/ψ bind to the proton or the deuteron. Finite binding energy is predicted for ^{3}He. Since α increases linearly with A, greater and greater binding is predicted as nuclei become heavier. In Table 2 we present the results for the binding energies as calculated by us using the formalism of BST. From this table several things are obvious:

(1) η_c and J/ψ binding energies are nearly the same.

(2) Screening effects on binding energies are very large. In absence of any real knowledge of the interaction, this introduces large uncertainty in where to look for the bound states.

(3) The interaction model obviously breaks down for heavy nuclei, with 2 GeV binding predicted for argon 40.

Table 2. Binding energies (in MeV) for Charmonium-nucleus states for unscreened and partially screened interactions.

Binding Nucleus, A	$\alpha m/\mu$ (unscreened)		B.E. for unscreened interaction		B.E. for partially screened interaction	
	η_c	J/ψ	η_c	J/ψ	η_c	J/ψ
He-3	1.83	1.85	18	20	3	4
He-4	3.23	3.32	143	149	49	52
C-12	9.25	9.53	581	608	258	269
O-16	11.72	12.09	747	780	344	358
Ar-40	25.25	26.16	2043	2168	984	1024

It follows that in order to stay within the range of the validity of the present interaction model, and in order to have a manageable energy range for searching for the bound states, one should stay with the light nuclei.

The next question is the all important one. Charmonium-nucleus bound states might very well exist, but are they reachable by practical means. The question boils down to estimates of formation cross sections in typical reactions:

$$p + A \rightarrow (c\bar{c}) \bullet (A + 1)$$
$$\bar{p} + A \rightarrow (c\bar{c}) \bullet (A - 1),$$

which we abbreviate as

$$h + A \rightarrow (c\bar{c}) \bullet (A'). \tag{16}$$

We write the corresponding elementary reactions on proton target as

$$h + p \rightarrow (c\bar{c}) \bullet X. \tag{17}$$

In order to obtain an estimate of the cross-section for the nuclear reaction (eq. 16) we must relate it to the elementary nucleonic reaction (eq. 17). It is not quite clear how to do this. Brodsky et al.[28] use the following reasoning. The probability that a composite object, like the nucleus A continues to hang together even after a four-momentum transfer Q, is given by the form factor at that moment transfer, $F_A^2 (Q^2)$. In a reaction like (eq. 16) the collision partners h and A have to slow down in their center of mass from their initial

$$(E_{cm}, P_{cm})_{A,h} = [(M^2 + P_{cm}^2)^{1/2}, P_{cm}]_{A,h} \tag{18}$$

to rest, i.e., to

$$(M,0)_{A,h} . \tag{19}$$

The momentum transfer involved is

$$Q_{A,h}^2 = [(M^2 + P_{cm}^2)^{1/2} - M]_{A,h}^2 - [P_{cm} - 0]_{A,h}^2 \tag{20}$$

Thus the joint probability of both A and h maintaining their identity despite momentum transfers

Q_A^2, and Q_h^2, and coming to rest together in their common center of mass is

$$F_A^2(Q_A^2) \bullet F_h^2 (Q_h^2) . \tag{21}$$

The corresponding probability for the nucleonic reaction (eq. 17) is

$$F_h^2(Q_h^2) \bullet F_p^2 (Q_p^2) = F_N^4 (Q_N^2). \tag{22}$$

At this point it is assumed that once the collision partners are at rest with respect to each other, the cross section for the $(c\bar{c})$ resonance production is the same in the nuclear case as it is in the nucleonic case. Therefore,

$$\sigma[h + A \rightarrow (c\bar{c}) \bullet A^I] = \sigma [h + \rightarrow (c\bar{c}) \bullet X] S_A. \tag{23}$$

where the 'sticking factor' S_A is just the ratio of the form-factors in eqs. 21 and 22,

$$S_A = F_A^2 (Q_A^2) \bullet F_h^2 (Q_h^2)/F_N^4 (Q_N^2) = S \bullet F_A^2 (Q_A^2) \tag{24}$$

where $S = F_h^2 (Q_h^2)/F_N^4 (Q_N^2)$.

We must now face the problem of how to obtain $F_A^2 (Q_A^2)$. The usual assumption is that this is the charge form-factor measured in electron elastic scattering. This assumption is questionable, because at the large momentum transfers $(Q^2 = 1\text{-}5 \ (GeV/c)^2)$ at which we need $F_A^2(Q^2)$, electron elastic scattering is known to have many complications, including meson-exchange current contributions. The interpretation of the measured elastic scattering cross section $\sigma(Q^2)/\sigma(Mott)$ as the $F_A^2 (Q^2)$ which we need, may therefore be not justifiable.

Even if we choose to use $F_A^2 (Q^2)$ from electron scattering, we have to face the problem that at present <u>no data exist</u> on electron elastic scattering from nuclei with $A > 4$ and $Q^2 > 0.5$ $(GeV/c)^2$. Can extrapolations from existing data be made to large Q^2 with any confidence? The answer is no. An example will illustrate the point better than a lot of hand-waving arguments. Until June 1976 elastic scattering data for ^4He existed only for $Q^2 < 0.8 \ (GeV/c)^2$. A popular extrapolation, based on nucleon dipole form factor, gave $F^2[Q^2 = 2.0 \ (GeV/c)^2] \approx 4 \times 10^{-6}$. A more sophisticated extrapolation was made by Brodsky and Chertok[30] on the basis of dimensional quark-counting rules. This gave $F^2[Q^2 = 2.0(GeV/c)^2] \approx 4 \times 10^{-8}$. Three months later, a 'tour de force' measurement on ^4He was made by Arnold et al.[31] at SLAC. It gave $F^2[Q^2 = 2.0 \ (GeV/c)^2] \approx 6 \times 10^{-9}$. In other words, the nucleonic extrapolation turned out to be wrong by a factor 10^{-3}, and the quark-model extrapolation turned out to be wrong by a factor 15. To me this is a convincing illustration of the pitfalls of any extrapolations for nuclei with $A > 4$. Actually, as we shall see later, things are already so tough for ^3He and ^4He that one does not have to bother about any heavier nuclei.

In Table 3 we summarize the rsults for some calculations with hydrogen and helium. The final answers for production cross sections are rather depressing. Brodsky et al.[28] assumed a $1\mu b$ cross section for resonant production of η_c in pp collisions. This seems to be a very large overestimate when we consider that η_c resonant production in $\bar{p}p$ collisions has a production

Table 3. Production cross section estimates for bound states of charmonium in nuclei.

Reaction $h + A \to (c\bar{c}) \bullet A'$	p(lab) GeV/c	Q_h^2, Q_A^2 $(GeV/c)^2$	S_N	$F_A^2 (Q^2)$	S_A	σ (prod) pico barns
$p+p \to \eta_c$ (pp)	11.59	2.80, 2.80	$*(F_N)^{-2}$	F_N^2	1	$\sim 10^6$(?)
$p+d \to \eta_c\,^3He$	7.64	3.22, 4.62	19.7	$\sim 1 \times 10^{-8}$	$\sim 2 \times 10^{-7}$	~ 0.2
$\bar{p} + p \to \eta_c$	3.67	1.03, 1.03	$(F_N)^{-2}$	F_N^2	1	$\sim 0.9 \times 10^6$
$\bar{p} + \,^4He \to \eta_c\,^3He$	2.30	1.54, 2.13	3.64	2.5×10^{-9}	$\sim 9 \times 10^{-9}$	$\sim 0.8 \times 10^{-2}$
$\bar{p} + p \to \psi$	4.07	1.15, 1.15	$(F_N)^{-2}$	F_N^2	1	$\sim 5.2 \times 10^6$
$\bar{p} + \,^4He \to \psi\,^3He$	2.53	1.71, 2.44	4.00	$\sim 1.6 \times 10^{-9}$	$\sim 6 \times 10^{-9}$	$\sim 3 \times 10^{-2}$

cross section of only 0.9 ± 0.2 μb. Even with this large elementary cross section the $p + d \to \eta_c\,^3He$ reaction has sub-picobarn cross section.[32] For $\bar{p} + \,^4He$ even the favored production of J/ψ has the terribly low cross section of 3×10^{-2} picobarn (or 3×10^{-38} cm^2). At this level it does not look possible to make the measurement at an existing (or likely to exist) facility. One would have to invoke an enhancement factor of at least 1000 in order to bring the cross sections at a measurable level of tens of picobarns. At E760 at the Fermilab we have indeed measured cross sections at this level. Also, backgrounds are expected to be at the level of $\sim$ 1 picobarn, so that signal/background should not be a problem. But, do we have any credible arguments for expecting such enhancements. I do not believe so. However, some people do[33].

Having drawn a rather disappointing conclusion for E760, let us examine if the proposed experiment is feasible with an extracted beam. Assuming that a detector with the efficiency of the E760 detector is available, the exercise is simply one of comparing luminosities with beams of momentum resolution comparable to that of E760. At the K2.5 channel of the proposed KAON a 2.5 GeV $\bar{p}$ beam of intensity 1.1×10^8 $\bar{p}$/s is expected in a 4% momentum bite. Assuming that one can cut into the beam phase space linearly, we may obtain a beam of 0.1% momentum bite ($\Delta p = 2.5$ MeV/c) with an intensity of $\sim 3 \times 10^6$ $\bar{p}$/s[26]. With this intensity we need a 2 meter long liquid helium target to achieve the luminosity of 10^{31} cm^{-2} s^{-1}, comparable to that available at E760. How practical is this? You be the judge! In any case, the preceding discussion of cross section still applies, and the experiment remains impossible unless we can come up with some plausible reasons for very large enhancements of the production cross sections.

6. NARROWING OF ψ" (3770)

In a recent paper D.Kharzeev[34] suggested that in the presence of dynamical quarks and gluons, as are present in nuclear matter, there is a partial suppression of quark and gluon condensates of the QCD vacuum. Kharzeev argues that this should lead to a decrease in quarkonium masses, especially for states with large n or ℓ quantum numbers. In particular, he argues that the mass of ψ" (3770 MeV) should decrease in the nuclear environment.

Normal ψ" (3770 MeV) is unbound by 40 MeV with respect to the $D\bar{D}$ breakup threshold at 3730 MeV. This permits it to have OZI allowed decay to $D\bar{D}$ and gives it a large width of ~25 MeV, in contrast to the 68 keV width of J/ψ, and the 240 keV width of ψ'. ψ" is known to be mostly $^3D_1$. It is therefore most likely to have its mass driven down if it is submerged in the nuclear medium. Kharzeev makes the conjecture that ψ" might very well have its mass drop down below the $D\bar{D}$ threshold. In that case, of course, it can not decay to $D\bar{D}$, and its width should become of the same order as that of J/ψ or ψ', i.e., of the order of 100's of keV. This would be fantastic indeed!

Kharzeev is well aware of the fact that in the presence of the Fermi motion of the protons in the target nucleus, the observed resonance will be broadened. He points out, however, that since the angle between the e^+ and e^- from ψ" decay depends on the relative momenta of the incident antiproton and the annihilating proton in the nucleus, by selecting the e^+e^- opening angle one can effectively become free of Fermi broadening.

Unfortunately, there is a very serious problem in Kharzeev's arguments. He assumes that while the ψ" mass is depressed, nothing much happens to the mass of the D and therefore to the threshold energy. He has two arguments for this. One is that the mass of D is mostly the mass of the c-quark, and as long as quark masses do not change, the D mass should change little. His second argument is that "in the QCD sum-rules approach gluon condensate does not contribute to the heavy-light meson mass in the heavy quark limit." This interpretation of Shuryak's work[35] is questionable. As a matter of fact, Brown[36] has argued that the effect may be exactly opposite, because the condensate effects are largest on light quarks and ψ" contains none. So it might be that it is the DD' threshold which is lowered relative to ψ", and ψ'. If it goes below ψ' we may actually expect a broadening of ψ' and not a narrowing of ψ". This may be more difficult to identify in presence of Fermi broadening.

REFERENCES

1. Fermilab experiment E-760, "A Proposal to Investigate the Formation of Charmonium States Using the pbar Accumulator Ruig", Collaboration: Fermilab, Universities of Errara, Genoa, and Torino, University of California (Irvine), Northwestern University and Pennsylvania State University.

2. "Review of Particle Properties" Physics Letters B239 (1990) 1.

3. E.D. Bloom and C.W. Peck, Ann. Rev. Nucl. Part. Sci. 33 (1983) 143.

4. J.Lee-Franzini, Nucle Phys. B (Proc. Suppl.) 3 (1988) 139.

5. K.K. Seth, Proc. Int. Conf. on Medium and High Energy Nuclear Physics, edited by W.-Y Pauchy Hwang, K.-F. Liu and Y. Tzeng, World Scientific (Singapore 1989) pp. 773-788.

6. C. Baglin et al., Phys. Lett B171 (1986) 135; B172 (1986) 455, B187 (1987) 191, B195 (1987) 85; B225(1989) 296; B231 (1989) 557; also Nucl. Phys. B286 (1987) 592.

7. Design Report: Tevatron I Project, Fermi National Accelerator Laboratory, Batavia, IL (unpublished), Sept. 1984;
    J. Peoples, Proc. Workshop on the Design of a Low Energy Antiproton Facility, edited by D. Cline, World Scientific (Singapore 1986) p.

8. L. Bartoszek et al., Nucl. Instr. Meth. A301 (1991) 47. M.A. Hasan et al., Nucl. Instr. Meth. A295 (1990) 73. M. Sarmiento et al., Nucl. Instr. Meth. (to be published).

9. T.A. Armstrong et al., Phys. Rev. Lett., submitted for publication.

10. P. Kroll and W. Schweiger, Nucl. Phys. A503 (1989) 865.

11. P. Jenni et al., Nucl Phys. B129 (1977) 232.

12. Y.L. Zhang et al., Proc. Hadron '91, University of Maryland, Aug. 1991.

13. S.J. Brodsky and G.R. Farrar, Phys. Rev. Lett. 31 (1973) 1153; Phys. Rev. D11 (1975) 1309.

14. G. Bassompierre et al., Phys. Lett. 68B (1977) 477, also Nuovo Cimento 73A (1983) 347; B. Delcourt et al., Phys. Lett. 86B (1979) 395; D. Bisello et al., Nucl. Phys. B224 (1983) 379.

15. G. Bardin et al., Phys. Lett. B255 (1991) 149.

16. R. Anderson et al., Phys. Rev. Lett. 38 (1977) 263.

17. U. Camerini et al., Phys. Rev. Lett. 35 (1975) 1040; B. Knapp et al., Phys. Rev. Lett. 34 (1975) 1040; J. Branson et al., Phys. Rev. Lett. 38 (1977) 1334, M.D. Sokoloff et al., Phys. Rev. Lett. 57 (1986) 3003.

18. S.J. Brodsky and A.H. Mueller, Phys. Lett. B206 (1988) 685.

19. T. Matsui and H. Satz, Phys. Lett. B179 (1986) 416.

20. C. Baglin et al., Phys. Rev. Lett. B220 (1989) 471; B251 (1990) 465; B255 (1991) 459.

21. H. Satz, Nucl. Phys. A. 488 (1988) 511c; V. Cerny et al., Z. Phys. C46 (1990) 481, and references therein.

22. G.R. Farrar, L.L. Frankfurt, M.I. Strikman and H. Liu, Nucl. Phys. B345 (1990) 125.

23. A.S. Caroll et al., Phys. Rev. Lett. 61 (1988) 1698.

24. B. Parker et al., Phys. Rev. Lett. 63 (1989) 1570.

25. R.S. Bhalerao and L.C. Liu, Phys. Rev. Lett. 54 (1985) 865; L.C. Liu and Q. Haider, Phys. Rev. C345 (1986) 1845; Q. Haider and L.C. Liu, Phys. Lett. B195 (1986) 515.

26. G.L. Li, W.K. Cheng, and T.T.S. Kuo, Phys. Lett. B195 (1987) 515.

27. R.E. Chrien et al., Phys. Rev. Lett 60 (1988) 2595.

28. S.J. Brodsky, I. Schmidt, and G.F. de Teramond, Phys. Rev. Lett. 64 (1990) 1011.

29. A Donnachie and P.V. Landshoff; Nucl. Phys. B244 (1984) 322.

30. S.J. Brodsky and B. Chertok, Phys. Rev. D14 (1976) 3003.

31. R. Arnold et al., Phys. Rev. Lett. 40 (1978) 1429.

32. Our estimate of $S \approx 2 \times 10^{-7}$ is substantially different from that quoted by Brodsky et al.[28]. We have used experimentally measured F_d^2 (4.62 GeV/c^2) $\approx 1 \times 10^{-8}$. due to Arnold et al. (Phys. Rev. Lett. 35 (1975) 776).

33. K. Maruyama, Proc. Workshop on Science at the KAON factory, TRIUMF (1990) vol. 2. Maruyama makes the mistake of assuming a beam intensity of 1.1×10^8 $\bar{p}$/s with a 0.1% energy resolution. Actually, this intensity is projected for $\Delta p/p = 4\%$. Maruyama also assumes formation cross sections of 10^{-32} cm^2, which are six orders of magnitude larger than our calculated value of 10^{-38} cm^2. The only way such a large discrepancy can arise is if Mazuyama has mistakenly used F_A (Q^2) instead of $F_A^2(Q^2)$.

34. D. Kharzeev, INFN (Pavia), preprint FNT/T-90/22, Dec. 1990.

35. E.V. Shuryak, Nucl. Phys. B198 (1982) 83.

36. G.E. Brown, priv. comm.

Few-Body Systems, Suppl. 6, 50—65 (1992)

© by Springer-Verlag 1992

THE STRUCTURE OF SCALAR AND VECTOR MESONS

J. Speth*, K. Holinde, B.C. Pearce, and R. Tegen[1]

Institut für Kernphysik, Forschungszentrum Jülich GmbH., 5170 Jülich, Germany

[1] permanent address: Department of Physics, University of the Witwatersrand,
PO Wits, 2050 Johannesburg, RSA

ABSTRACT

We review the dynamical model for meson—meson scattering developed by our group. This model is based on the meson—exchange picture and reproduces quantitatively the $\pi\pi$ and $K\pi$ phase shifts. These results can be used for the calculation of the pion electromagnetic form factor as well as the scalar form factor of the pion. We also discuss our recent calculation of the nucleon sigma term within the same model.

1. Introduction

While quantum chromodynamics (QCD) is generally thought to be the underlying theory of strong interactions, the mathematical intractibility of the theory in the low energy (i.e., non—perturbative) domain makes a phenomenological description of low energy processes a practical necessity. Indeed, there are theoretical indications that low energy phenomena can be well described in terms of the quantities observed directly — mesons and nucleons — and their interactions by meson exchange.

*Invited talk presented at 13th European Conference on Few—Body Problems in Physics, Elba, Italy, September 9–14, 1991.

Clearly, the meson exchange model of hadronic interactions can only be applicable in a limited energy range, which must be related to the size of the confinement region of the quarks. This size is not necessarily the same as the empirical size of a hadron, since the size of a hadron includes its meson cloud. The range of applicability of meson exchange models may then be somewhat larger than simple estimates based on RMS radii. The interesting question is: how large is this range? At what point *must* one consider quark effects?

In many models based on quarks and gluons the effects of the meson cloud have been ignored. For example, the non–relativistic quark model [1] describes states only in terms of valence quarks. In that model the structure of mesons is entirely $q\bar{q}$ yet presumably, in nature it must have a component that is $qq\bar{q}\bar{q}$. If we consider e.g. the electromagnetic form factor of the pion, which is known experimentally in the space– and time–like region, one can easily reproduce the space–like form factor within a pure $q\bar{q}$ picture, where the bag–radius is relatively large. Such a model, however, does not give the form factor in the time–like region which is known to be dominated by the ρ–meson (vector dominance). This fact clearly indicates that the pion must have a large component of $2q2\bar{q}$ which we shall describe in the following as a correlated two pion configuration. We will also demonstrate that the electric rms–radius and the time–like form factor in such a picture can be quantitatively reproduced with a very short–ranged strong interaction form factor ($\rho\pi\pi$ form factor). If we qualitatively interpret this strong interaction form factor with the confining region of the $q\bar{q}$ valence quarks than one obtains an extension which is less than 1/3 of the electric rms radius.

The scalar–isoscalar form factor of the pion and the nucleon is even more dominated by the correlated two pions and two kaons. Indeed we shall show that the scalar meson $f_0(975)$ is not a genuine $q\bar{q}$ meson but rather a $K\bar{K}$ molecule. The genuine scalar meson might be between 1200–1600 MeV. It is not possible to make a more precise statement without including all relevant thresholds that influence this region.

The paper is organized as follows. We first introduce our model which includes the interaction between two pseudoscalar mesons in the framework of meson–exchange. This model is than applied to the $\pi\pi$– and $K\pi$–scattering processes. The electric and scalar form factors of the pion are discussed in the fifth chapter. With the help of these results we are able to calculate the scalar form factor of the nucleon and the nuclear σ–term. In the last chapter we draw some conclusions.

2. The Model

The meson exchange model is obtained by a) writing down the Lagrangian describing the basic meson–meson–meson and meson–meson–baryon vertices, b) defining phenomenological form factors for each vertex (parameterised in terms of a cutoff mass) to account for the omitted underlying quark–gluon structure, c) constructing a potential V by computing as many of the resulting two–particle–irreducible meson exchange diagrams as possible, d) summing all of the two–particle diagrams by solving a Bethe–Salpeter – Lippmann–Schwinger type integral equation with V as the driving term.

The exact solution to the field theory would be obtained by solving the Bethe–Salpeter equation with a complete kernel (i.e., including *all* two–particle irreducible diagrams in step c) above). This is of course impossible in practice so we make two approximations. First, the infinite set of two–particle irreducible diagrams constituting the potential is truncated to include only t– and s–channel meson exchanges. Secondly, we utilise either the Time Ordered Perturbation Theory (TOPT) or Blankenbeclar–Sugar (BbS) approaches to reduce the dimensionality of the integral equation from four to three, which makes it more amenable to solution [2].

The starting point of all the calculations is the Bethe–Salpeter equation for the t–matrix T

$$T(p,p',P) = V(p,p',P) + \int d^4k V(p,k,P) \, G(k,P) \, T(k,p',P) \qquad (1)$$

where the variables denote 4–momenta and G is the two–body propagator. In cases where several exit channels are possible and the corresponding coupling is strong we have to solve a coupled channel problem and in that case eq. (1) denotes a matrix equation. In the scalar–isoscalar channel of the $\pi\pi$–system also the $K\bar{K}$–channel has to be included. The details of this calculation already appeared in Ref. 3. Here we shall mainly discuss the results and the physics involved.

To completely specify the calculation we require an effective meson Lagrangian to constrain the vertices. For this purpose we use the Lagrangian of Bando et al. [4]. In this model, the ρ meson emerges as the gauge boson of a hidden local gauge symmetry in the non–linear sigma–model. After gauge fixing, this model is identical to the Weinberg Lagrangian [5]. For us, the important ingredients of the model (after extension to SU(3)) are that it incorporates the well known symmetries of QCD and nature (e.g. chiral symmetry) and provides us with pseudoscalar–pseudoscalar–vector vertices in which the coupling constants are constrained by the symmetry. It also allows inclusion of couplings to the photon in a natural way.

The relevant t–channel meson exchange diagrams that will contribute to V are (a) ρ–exchange between $\pi\pi$ and $\pi\pi$ states, (b) K*(890) exchange between $\pi\pi$ and K$\overline{\text{K}}$ states, and (c) ρ, ω and ϕ exchange between K$\overline{\text{K}}$ and K$\overline{\text{K}}$ states. These are illustrated in Figs. 1(a)–1(c). To these we must also add s–channel pole diagrams as illustrated in

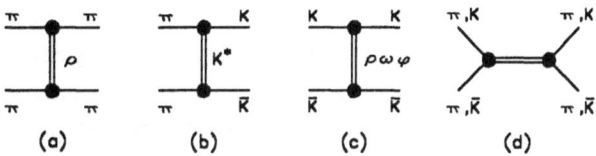

Figure 1:
Meson exchange contributions to the coupled
$\pi\pi$–K$\overline{\text{K}}$ system. (a)–(c) t–channel. (d) s–channel.

Fig. 1(d). These correspond to the genuine $q\overline{q}$ states while the t–channel exchanges provide the (possibly strong) background. In solving the integral equation, many diagrams are generated that renormalise both the mass and coupling constants of these s–channel poles (this can be seen by iterating Eq. (1) with the potential of Fig. 1). One example of such a diagram is given in Fig. 2. Hence, the physically observed state is a combination of bare

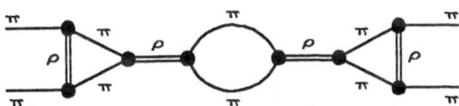

Figure 2
An example of a diagram generated by solving Eq. (1)
that renormalises Fig.1(d).

$q\overline{q}$ state and two–meson dressing. Rather than explicitly calculating the mass shift and change in the coupling constant due to renormalisation, we introduce the bare mass and bare coupling constant as free parameters which are adjusted to fit the resonant phases.

3. $\pi\pi$ Scattering

The scalar–isoscalar channel with quantum numbers $I^{G}(I^{PC})=0^{+}(0^{++})$ of the $\pi\pi$ scattering amplitude is of special interest because it provides the intermediate attraction in the baryon–baryon interaction. There is a long standing discussion whether or not the lowlying "σ"–meson introduced into these kind of calculations is a real meson. Indeed we will show that the genuine scalar–isoscalar meson is between 1.2–1.6 GeV and that the phase shifts in the scalar–isoscalar channel are dominated up to 800 MeV by the t–channel interaction between the two pions. The narrow resonance, the $f_0(975)$ (which was once identified as a member of the scalar $q\bar{q}$ nonet but it is now generally agreed that its mass and decay properties are inconsistent with that assignment) will turn out to be of more complicated structure.

The phase shifts and inelasticities resulting from the meson–exchange coupled–channel calculation for this partial wave are shown in the solid curves of Fig. 3. If the coupling

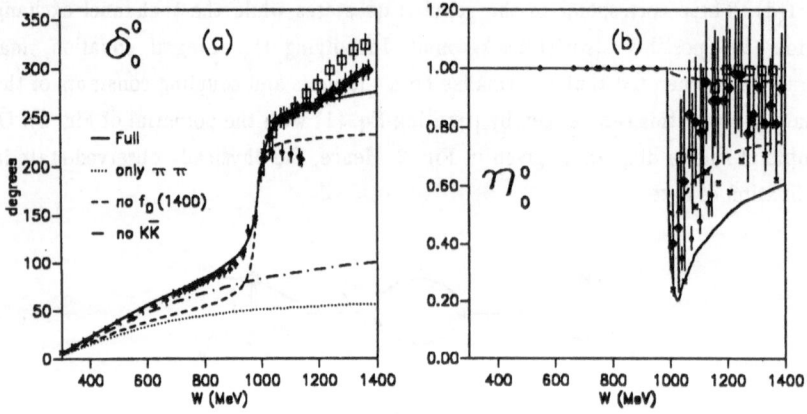

Figure 3

Results for the phase shifts (a) and inelasticity (b)
in the $I^{G}(J^{PC}) = 0^{+}(0^{++})$ partial wave. For references
to the data, see Ref. 3.

to the $K\bar{K}$ channel is omitted then we have a single–channel problem with the potential given by ρ exchange (Fig. 1(a)). This results in the dotted curve. If the channel coupling is turned on by including the diagrams of Figs. 1(b) and (c) then resonant structure at the $f_0(975)$ energy is immediately obvious in the dashed curve. In fact, if we calculate $K\bar{K}$ scattering with the potential given only by Fig. 1(c) (i.e., no coupling to the $\pi\pi$ channel) then we find typical bound state phase shifts. This confirms that the structure observed in Fig. 3 is due to a bound $K\overline{K}$ state. It comes about because the ρ, ω and ϕ exchange contributions of Fig. 1(c) are each attractive, providing enough attraction to bind. To obtain complete agreement with the empirical $\pi\pi$ phase shifts, it is necessary to include an s–channel pole corresponding to the broad

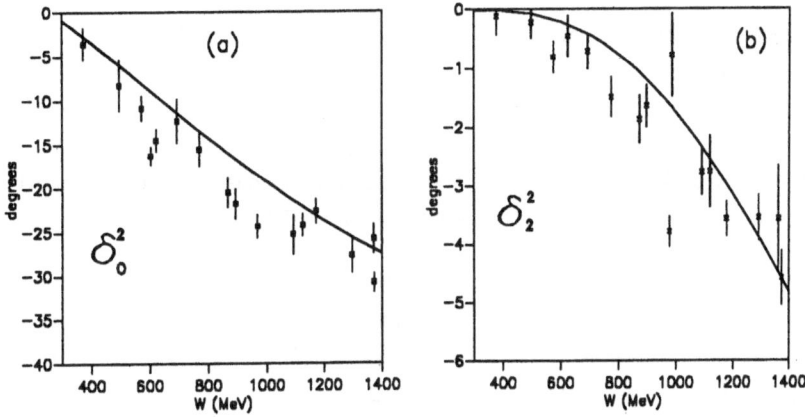

Figure 4

Results for the non–resonant, T = 2 partial waves. (a) $I^G(J^{PC}) =$ $2^+(0^{++})$ and (b) $2^+(2^{++})$. For references to the data, see Ref. 3.

$f_0(1400)$. The bare mass and coupling constant are taken as free parameters which are adjusted to fit the data, resulting in the solid curve. The dash–dotted curve is the result of turning off the diagram of Fig. 1(c), which, since Fig. 1(c) is the mechanism for creating the bound state, illustrates the strong background to the $f_0(975)$ provided by t–channel exchanges and the $f_0(1400)$.

The scalar nonet requires two non–strange, isoscalar states (a mixture of the singlet and octet representations) although we have only included one. We should, in principle, also include the $f_0(1590)$ but we have omitted it at this stage since it is at a mass where effects of coupling to the $\rho\rho$ and $N\overline{N}$ channels may become important.

The partial waves with isospin 2 and spin 0 and 2 ($2^+(0^{++})$ and $2^+(2^{++})$) provide a useful check on the model. There are no experimentally observed resonances in these channels, so the phase shifts are completely determined within the model by the t–channel processes, whose parameters have already been constrained by the $0^+(0^{++})$ channel. Also, isospin 2 means there is no coupling to the $K\overline{K}$ channel. The results in Fig. 4 show excellent agreement with the data.

Results for the partial waves $1^+(1^{--})$ and $0^+(2^{++})$ are shown in Fig. 5. These waves contain the ρ and the $f_2(1270)$ respectively, which both decay predominantly to $\pi\pi$. In both cases it was necessary to include an s–channel pole (corresponding to a genuine $q\bar{q}$ state) in order to reproduce the data. The results of just the t–channel processes are indicated in the figures by the dashed lines.

From these calculations we may conclude that the resonances in the various channels are in general not $q\bar{q}$ states but strong mixtures of a genuine resonance ($q\bar{q}$) and correlated two pion configurations ($2q2\bar{q}$). The specific $f_0(975)$ resonance is mainly a $K\bar{K}$ molecule whereas the genuine scalar–isoscalar meson is beyond 1.2 GeV. This conclusion is strongly supported by the scalar–isoscalar mesons with strangeness which can be seen in $K\pi$ scattering and which we discuss in the following chapter.

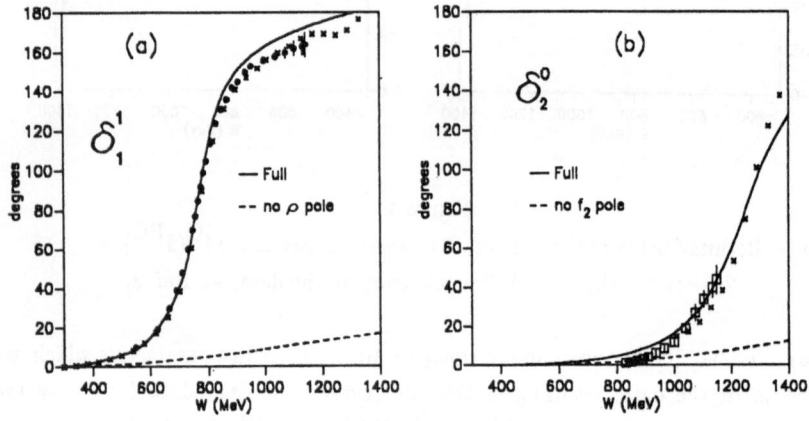

Figure 5
Results for the resonant (a) $I^G(J^{PC}) = 1^+(1^{--})$ and (b) $0^+(2^{++})$
partial waves. For references to the data, see Ref. 3.

4. $K\pi$ Scattering

It is straightforward to extend the model of the $\pi\pi$ system to the $K\pi$ interaction. Since all of the t–channel exchange processes are completely determined by the $\pi\pi$ system, it provides a useful consistency check of the model. Only the bare masses and coupling constants of the s–channel poles remain to be adjusted in the cases where they are necessary. This system, indeed, is much simpler because there is no other channel which strongly couples. The possible $K\eta$ channel couples only very weakly to the $I = \frac{1}{2}$ $K\pi$–system. Therefore we expect here a very similar situation as in the scalar–isoscalar channel, without however, the complications due to the coupled $K\bar{K}$–channel.

In Fig. 6 I show the results for the resonant partial waves, $I(J^P) = \frac{1}{2}(0^+)$ and $\frac{1}{2}(1^-)$. These correspond respectively to the $K_0^*(1430)$ and $K^*(892)$ states listed by the Particle Data Group. The dashed curves are the results using only the t–channel exchange driving terms while the solid curves include s–channel states (with bare masses and couplings adjusted to reproduce the data as usual). As indicated above, the $K_0^*(1430)$ is of special interest since it is a member of the lowest lying scalar nonet which we have just demonstrated should not contain the $f_0(975)$. In fact, as can be seen, the model requires a genuine $q\bar{q}$ state in order to agree with experiment. However, the t–channel exchanges provide a strong, non–negligible background. Similarly, the $\frac{1}{2}(1^-)$ requires an s–channel pole term in order to reproduce the vector $K^*(892)$, although the background is much less in that case.

Once again, the non–resonant partial waves provide a useful check of the self–consistency of the model. In this case, all the parameters are determined in the $\pi\pi$ sector. As can be seen in Fig. 7, the results are very reassuring.

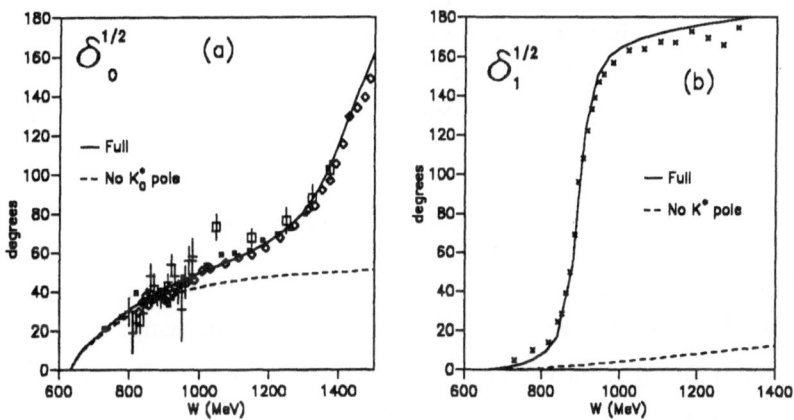

Figure 6

Results for the resonant (a) $I(J^P) = \frac{1}{2}(0^+)$ and
(b) $\frac{1}{2}(1^-)$ partial waves of the $K\pi$ system. For
references to the data, see Ref. 3.

It is clear that the model described is very successful at describing the low–lying states that couple strongly to $\pi\pi$ and $K\pi$. The $f_0(975)$, about which there is some controversy, emerges as a $K\bar{K}$ bound state (or molecule). This interpretation agrees in spirit (although not in the details) with the non–relativistic quark model calculation of Weinstein and Isgur [6] but is at variance with the analysis of Morgan and Pennington [7]. All other resonance states examined ($\rho(770)$, $f_2(1270)$, $K_0^*(1430)$ and $K^*(892)$) required the inclusion of s–channel poles. It was also necessary to include a state corresponding to the $f_0(1400)$. To extend the calculation to higher energies it will be necessary to include coupling to other channels (for example, $\rho\rho$ and $N\bar{N}$).

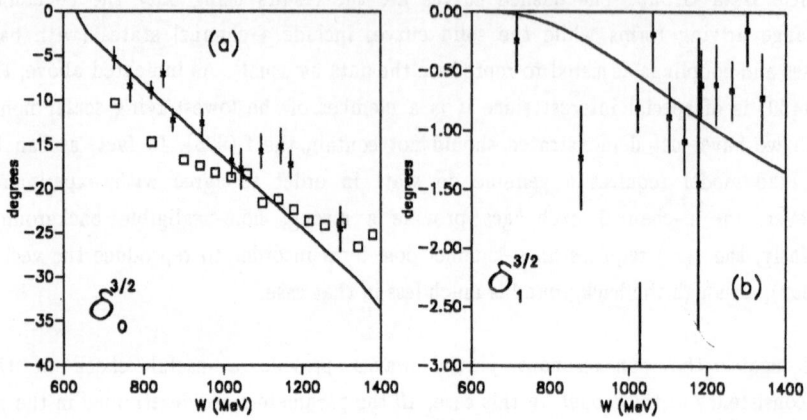

Figure 7

Results for the non−resonant (a) $I(J^P) = 3/2(0^+)$ and
(b) $3/2(1^-)$ partial waves of the $K\pi$ system. For references
to the data, see Ref. 3.

5. Form Factors of the Pion and Nucleon

The form factors of the lightest hadron, the pion, have attracted much attention in the past years. The electromagnetic form factor $F_\pi(q^2)$ provides information on the pionic root mean squared (rms) radius, $<r^2>_\pi^{\frac{1}{2}}$ (which together with the pion decay constant f_π leads to the Van Royen−Weisskopf paradox [8], still representing a major challenge for quark models and other models of the hadrons). The scalar form factor $\Gamma_\pi(q^2)$, on the other hand is an important ingredient in the determination of the σ−term of the nucleon from πN data. In order to explain the magnitude of the empirical $\Sigma_{\pi N}$−term one has to assume either a large strangeness content of the nucleon or a strong q^2−dependence of the nucleonic scalar form factor $\sigma(q^2)$ between $q^2 = 0$ and $q^2 = 2m_\pi^2$, i.e. a very large scalar radius of the nucleon. We will show in the following that this strong q^2−dependence has its natural origin in the strongly correlated $\pi\pi$−contribution in the scalar channel of the $\pi\pi$−scattering phase shifts. Here we shall report on calculations of both form factors $F_\pi(q^2)$ and $\Gamma_\pi(q^2)$ within the same model which reproduces quantitatively the phase shifts as discussed in the previous chapters. In addition, we also discussed the q^2−dependence of the scalar form factor of the nucleon.

5.a. Pion Electromagnetic Form Factor

Effective meson Lagrangians can easily be extended to include couplings to photons [4] enabling us to compute, for example, the electromagnetic form factor of the pion. The important new ingredient for us is a diagram in which the photon transforms to a ρ–meson. In the usual vector dominance picture, the pion electromagnetic form factor would then be obtained from the photon transforming to a ρ which then decays to two pions. By using a Breit–Wigner propagator for the ρ to account for its width and adjusting the overall strength, one can obtain agreement with the data in the ρ resonance region but it is not possible to simultaneously fit the ρ resonance and the normalisation requirement $F_\pi(q^2{=}0){=}1$. However, in the model of the $\pi\pi$ system just described, the $\pi\pi$ self energy corrections to the ρ means that the physical ρ propagator is more complicated than a simple Breit–Wigner.

If we write $\Gamma_{\gamma\rho}$ for the amplitude arising from the Lagrangian for the process $\gamma{\rightarrow}\rho$, d for the bare ρ propagator and f_B for the bare $\rho{\rightarrow}\pi\pi$ vertex, then, up to appropriate normalisations, the π electromagnetic form factor is written (in operator form) as

$$F_\pi = \Gamma_{\gamma\rho}df_B + \Gamma_{\gamma\rho}df_B GT, \qquad (2)$$

where T is the coupled channel t–matrix discussed earlier and G is the two–body propagator. This equation is illustrated diagramatically in Fig. 8. The first term is just

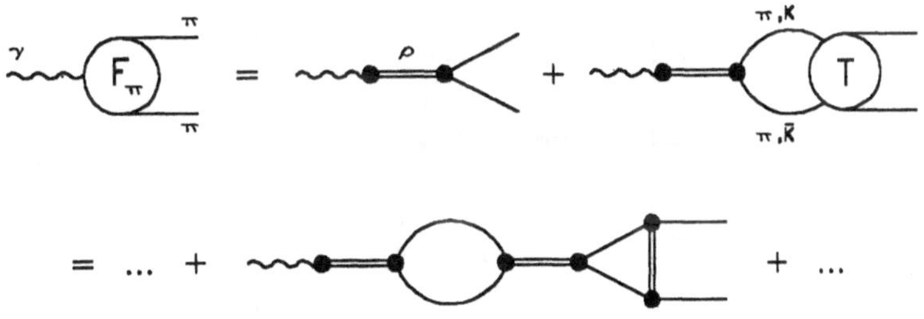

Figure 8

The equation for the π electromagnetic form factor.

the diagram usually written down in the vector dominance picture, except that now the ρ propagator is simply the Feynman propagator for the bare ρ appearing in the Lagrangian. The second term is the final state interaction that dresses this bare ρ giving it its physical mass and width. The second line in Fig. 8 gives an example of a diagram that is included by applying the final state interaction.

Since we take the γ-ρ interaction to be pointlike (i.e., it has no form factor associated with it) the only new parameter introduced is the γ-ρ coupling constant. This is fixed by requiring $F_\pi(0)=1$. However, it turns out that the $\pi\pi$ results are somewhat insensitive to the bare $\pi\pi\rho$ cutoff mass. This is illustrated in Fig. 9(a), where we show the $\pi\pi$ phase shifts in the ρ channel for four values of the cutoff mass. In each case, the bare ρ mass is readjusted to fit the data but all other parameters are kept fixed. In Fig. 9(b) the corresponding results for the electromagnetic form factor are shown, where we see a strong dependence on the $\pi\pi\rho$ cutoff mass. Hence, the electromagnetic form factor

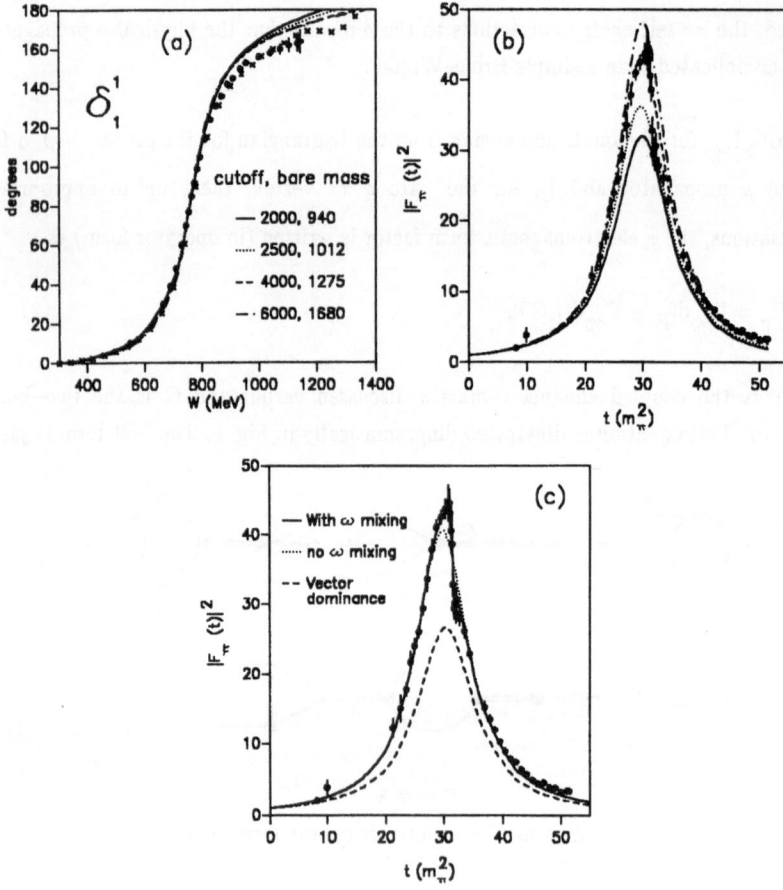

Figure 9

(a) The dependence of the $\pi\pi$ phase shifts on the $\pi\pi\rho$ cutoff mass.

(b) Corresponding dependence of the electromagnetic form factor.

(c) π electromagnetic form factor including ρ-ω mixing.

essentially provides a measurement of the off–shell $\pi\pi\rho$ vertex. In Fig. 9(c) we show the results using the optimal value of the cutoff mass (3300 MeV). The solid curve includes the addition of a ρ–ω mixing contribution obtained by adding to the calculation the usual Breit–Wigner vector (ω) dominance contribution. The coupling constants are fixed by known decays ($\omega\to\pi\pi$ and $\omega\to e^{+}e^{-}$), with only the phase adjusted to fit the data. If this ω contribution is neglected, we obtain the dashed curve. The dotted curve is a Breit–Wigner with the ρ mass and width and the strength adjusted to ensure $F_{\pi}(0)=1$ (the naive vector dominance model). The full calculation results in an r.m.s. radius of 0.647 fm. This is much closer to the experimental value [9] of 0.66 ± 0.01 fm than the value 0.62 fm obtained using the vector dominance model.

It is an assumption of the meson exchange approach that the form factors that are introduced are a means of parametrising the underlying quark–gluon structure of the hadrons. Since the electromagnetic form factor seems to be so sensitive to the form factor at the vertex where two pions couple to a bare ρ, it is tempting to interpret this as a measure of the size of the confining region. Yielding to such temptation, the required cutoff mass of 3300 MeV would imply an r.m.s. radius of only 0.15 fm. However, in the analysis so far we have kept the parameter a defined in the Lagrangian of Bando et al. [4] fixed at the "vector dominance" value of 2. This means the photon does not couple directly to the π but only via the bare ρ. Relaxing this constraint introduces some ambiguity, making such a prediction less reliable.

5.b Scalar Form Factors and the Nucleon σ Term

The nucleon σ term, $\sigma(t)$, can be defined in terms of the matrix element [10]

$$\sigma(t)\bar{u}(p')u(p) = \frac{m_u+m_d}{4m_N}<p'|(\bar{u}u+\bar{d}d)|p>\tag{3}$$

A low energy theorem relates $\sigma(t)$ defined in this way, to the isospin even πN amplitude $\check{D}(t)$ at the Cheng–Dashen point [11]

$$\sigma(2m_{\pi}^2) \approx f_{\pi}^2\check{D}(2m_{\pi}^2) \ (\equiv \Sigma) \ .\tag{4}$$

The interest in the σ term was generated when analysis of the πN data revealed $\Sigma = 64 \pm 8$ MeV [12] yet naive estimates based on baryon mass differences gave $\sigma(0) \approx 30$ MeV. It was recently pointed out [10] that the strong s–wave $\pi\pi$ interaction produces a significant t dependence of $\sigma(t)$. That is, the usual assumption that $\sigma(2m_{\pi}^2) \approx \sigma(0)$ may not be valid.

To investigate this effect using the $\pi\pi$ model discussed above, we consider $\sigma(t)$ to be given by

$$\sigma = \sigma_B + \Gamma_B Gv + \Gamma_B GTGv. \tag{5}$$

Here, σ_B represents the "bare nucleon" σ term. We do not attempt to calculate this contribution, concentrating instead on the q^2 dependence of the remaining pieces. The next two terms involve the matrix element of $\bar{u}u+\bar{d}d$ between pion (kaon) states. The bare part (i.e., without $\pi\pi(K\bar{K})$ rescattering which we put in explicitly) of this process is denoted by Γ_B. The pions (kaons) then connect to the nucleon directly or via multiple scattering. The elementary $\pi\pi(K\bar{K}) \to N\bar{N}$ amplitude is denoted by v. For the elementary pion (kaon) matrix elements (Γ_B) we simply couple the two pseudoscalar fields to a scalar with a pointlike vertex. The coupling constants of these vertices account for the strength of the matrix elements they represent and are taken as parameters, which we will discuss in the following.

Equation (5) is represented graphically in Fig. 10. The dotted line does not represent a particle line but represents the operator $\bar{u}u + \bar{d}d$ appearing in Eq. (3). We have drawn it as a line to draw the obvious parallel between this and the electromagnetic form factor.

Embedded in Eq. (5) is the σ term of the pion (kaon), $\Gamma(t)$, which is given in this model by

$$\Gamma = \Gamma_B + \Gamma_B GT. \tag{6}$$

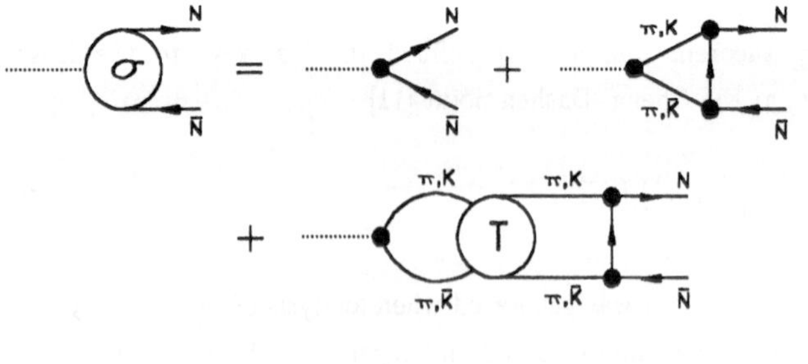

Figure 10
Pionic corrections to the nucleon σ term.

This illustrated in Fig. 11. For $q^2 = 0$ this can be expressed as derivatives of the pion (kaon) masses with respect to the quark masses [13]. By requiring that the solution of

Eq. (6) agrees with these predictions at $q^2 = 0$ we are able to fix the two coupling constants mentioned above. This means we have no free parameters in the calculation.

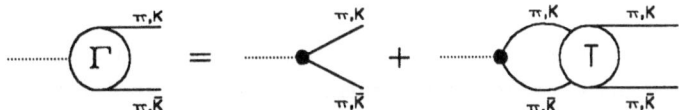

Figure 11

The pion (and kaon) scalar form factor.

The result for the pion form factor $\Gamma_\pi(t)$ is shown in the solid curve of Fig. 12(a). The effect of the $f_0(975)$ is clearly evident. The r.m.s. radius is 0.75 fm. For comparison, we also show the results of a one–loop chiral perturbation theory calculation [10, 13] (dotted curve) and, in the dashed curve, a recent calculation by Donoghue, Gasser and Leutwyler [10, 13].

The latter calculation uses a dispersion relation approach using the empirical $\pi\pi$ phase shifts (solving the Muskhelisvili–Omnes equations) to impose unitarity on $\Gamma(q^2)$. In a sense, our calculation is a close parallel of theirs. It differs in that we use the solution of the scattering integral equation to impose unitarity rather than dispersion theory, and we use a dynamical model of the $\pi\pi$–$K\overline{K}$ system that includes off–shell effects rather than the empirical phase shifts. We use the same method to fix the two coupling constants described above as they use to fix the two subtraction constants in their

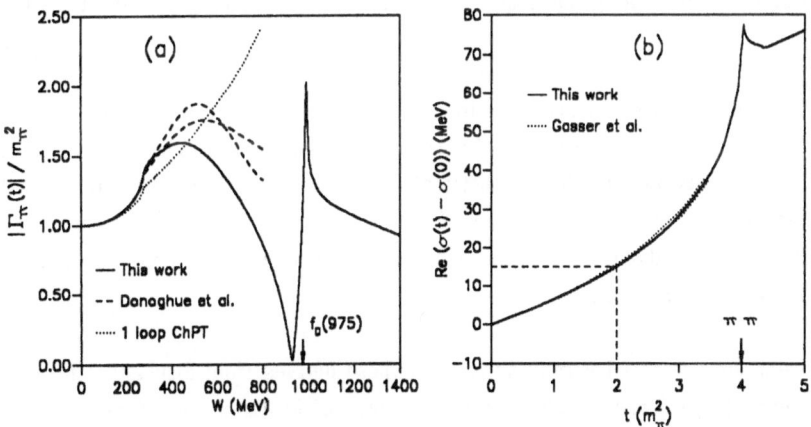

Figure 12

(a) The pion scalar form factor. The dashed and dotted curves are from Ref. 13.
(b) q^2 dependence of the nucleon σ term. The dotted curve is from Ref. 10.

analysis and the same unitarity relation they use is automatically embedded in Eq. (6). It is interesting that our calculation agrees with theirs below threshold but differs beyond that.

Having fixed the coupling constants and confirmed that our calculation of $\Gamma(q^2)$ is sensible, we can proceeed to the nucleon σ term. In Fig. 12(b) we show the difference $\sigma(t) - \sigma(0)$ in the vicinity of the Cheng–Dashen point (assuming the contribution of the "bare nucleon" σ term, σ_B is approximately a constant function of (q^2). Our results are in almost exact agreement with the disperion relation calculation of Gasser, Leutwyler and Sainio [10] (dotted curve) throughout the range of their calculation. (Again, embedded in our calculation is the same unitarity relation that drives their dispersion integrals.) In particular, we confirm their result that there is a strong q^2 dependence of $\sigma(q)^2$ near the Cheng–Dashen point, giving rise to a 15 MeV contribution to $\sigma(2m_\pi^2) - \sigma(0)$ and a scalar r.m.s. radius of 1.25 fm.

6. Conclusion

We have reviewed a dynamical model for meson–meson scattering which is driven by s– and t–channel meson exchanges with the vertices obtained from an SU(3) symmetric Lagrangian. Only the lowest mass vector mesons (ρ, ω, ϕ, K*) are used in the t–channel driving terms. We obtain quantitative agreement with the experimental phase shifts. In the scalar–isoscalar channel the $f_0(975)$ emerges in our model as a $K\bar{K}$ bound state whereas the genuine $q\bar{q}$ scalar isoscalar meson has a mass of at least 1.2 GeV.

Within the same model we were able to calculate the pion electromagnetic and scalar form factors and the q^2–dependence of nucleon scalar form factor. The electromagnetic form factor results provide tighter constraints on the bare $\rho\pi\pi$ form factor than do the $\pi\pi$ phase shifts. The value ob tained for this cutoff parameter corresponds to a confinement size of about 0.15 fm, although this may change if the photon is allowed to couple directly to the pions.

The scalar form factors basically agree with those derived from dispersion theory by Gasser et al. [10, 13] with excellent agreement obtained for the nucleon form factor. We obtain a 15 MeV contribution to the nucleon σ term arising from the difference $\sigma(2m_\pi^2) - \sigma(0)$. In our approach however, we do not use the experimental phase shifts as an input but we presented a theoretical framework which allows to calculate all these quantities simultaneously.

References

1. S. Godfrey and N. Isgur, Phys. Rev. D32 (1985) 189.
2. Reference 3 utilised the TOPT approach. However, in order to obtain the correct analytic behaviour of the scalar form factor in the unphysical region, it was necessary to redo the calculation using BbS. Both methods gave similar results in the physical region after some readjustment of cutoff parameters. In particular, the conclusions regarding ther nature of the $f_0(975)$ were unchanged.
3. D. Lohse, J.W. Durso, K.Holinde and J. Speth, Nucl. Phys. A516 (1990) 513.
4. M. Bando, T. Kugo, S. Uehara, K. Yamawaki and T. Yanagida, Phys. Rev. Lett. 54 (1985) 1215.
5. S. Weinberg, Phys. Rev. 166 (1968) 1568.
6. J. Weinstein and N. Isgur, Phys.Rev. D41 (1990) 2236.
7. D. Morgan and M.R. Pennington, Phys. Lett. B258 (1991) 444.
8. R. Van Royen and V. Weisskopf, Nuov. Cim. 50A (1967) 617.
9. S.R. Amendolia et al., Phys. Lett. 138B (1984) 545; Nucl. Phys. B277 (1986) 186.
10. J. Gasser, H. Leutwyler and M.E. Sainio, Phys. Lett. B253 (1991) 260.
11. J. Gasser, H. Leutwyler and M.E. Sainio, Phys. Lett. B253 (1991) 252 and references therein.
12. R. Koch, Z. Phys. C15 (1982) 161.
13. J.F. Donoghue, J. Gasser and H. Leutwyler, Nucl. Phys. B343 (1990) 341.

Few-Body Systems, Suppl. 6, 66—76 (1992)

Few-
Body
Systems
© by Springer-Verlag 1992

SOME MEASUREMENTS FOR DETERMINING STRANGENESS
MATRIX ELEMENTS IN THE NUCLEON

Ernest M. Henley, T. Frederico[a], S.J. Pollock
S. Ying, G. Krein[b] and A.G. Williams[c]
Department of Physics, FM-15 and Institute for Nuclear Theory, HN-12
University of Washington, Seattle, Washington 98195

Abstract

Some experiments to measure strangeness matrix elements of the proton are proposed. Two of these suggestions are described in some detail, namely electro-production of phi mesons and the difference between neutrino and antineutrino scattering for isospin zero targets such as deuterium.

1. Introduction

The constituent quark model of the nucleon has been very successful in predicting some of its static properties. The model assumes that the nucleon is composed solely of three valence u and d quarks (uud or udd). However, recent measurements, particularly of the spin structure of the nucleon by the EMC group [1] reveal that we do not fully understand the structure of the proton. The measurements suggest that the quarks may only be responsible for a small fraction of the proton's spin and that the strange quarks may contribute as much as 20% to the spin. The latter conclusion has been challenged by many theorists [2]. By contrast, there are many fewer proposals to experimentally determine the strange quark matrix elements in the nucleon [3-5].

The production of pure strange hadrons from a nucleon is inhibited by the OZI rule in the absence of strangeness in the structure. Ellis, Gabathuler, and Karliner [4] noted that the OZI rule appeared to be violated in phi production from nuclear targets. Parity-violating elastic electron scattering from hydrogen has been proposed to measure the vector matrix element $\langle N | \bar{s} \gamma^\mu s | N \rangle$ [6]. We have suggested [3] electron- and neutrino-production of ϕ mesons from a hydrogen target to measure the same matrix element and the difference between neutrino and antineutrino cross sections on an isospin zero target to obtain the isoscalar axial current matrix element $\langle N|\bar{s}\gamma^{\mu}\gamma^{5}s|N\rangle$. In this talk I will describe both of these proposals in some detail.

2. Electro-production of ϕ Mesons

If the OZI rule is exact, and the ϕ meson is a pure $s\bar{s}$ state, then the ϕ cannot be produced "directly" from a hydrogen target. We use "directly" because the ϕ can be obtained "indirectly" through a vector dominance process in electro- and neutrino-production. Despite this shortcoming, we believe that the reactions $e^-p \rightarrow e^-p\phi$ and $\nu p \rightarrow \nu p\phi$ offer opportunities for seeking vector strangeness matrix elements in the proton. First of all, the interactions of the probes with the target are well known and understood. Second, we believe that a separation of the desired "knock-out" process from the vector meson background, although difficult, is feasible. Since electromagnetic probes provide a much higher count rate than parity-violating processes, we believe that the electro-production of ϕ mesons may open another avenue for seeking vector matrix elements, $\langle p|\bar{s}\gamma_{\mu}s|p\rangle$, due to strangeness in the nucleon. We have therefore analyzed this reaction in some detail.

A constituent quark model is used to estimate the knockout production cross section of ϕ mesons near threshold and for squared momentum transfers $Q^2 = -q^2$ that are not too large, i.e. $Q^2 \lesssim 1(\text{GeV/c})^2$. Fig. 1a illustrates the kinematics of the reaction and Fig. 1b the knock-out process. We assume, for simplicity, a two-component Fock-space proton wavefunction

$$|p\rangle = A|(uud)_{1/2}\rangle + B|\left[a_0(uud)_{1/2}(s\bar{s})_0 + a_1(uud)_{1/2}(s\bar{s})_1\right]>_{1/2}, \qquad (1)$$

with $|A|^2 + |B|^2 = a_0^2 + a_1^2 = 1$. In Eq. (1) B determines the admixture of strangeness in the proton, and $a_0(a_1)$ is the fraction of spin 0 (spin 1) $s\bar{s}$. Other components, and more complex wavefunctions are possible, but Eq. (1) is adequate for our purposes. In order for the parity of the proton to be +, we require the $s\bar{s}$ to be in a relative p-wave ($\ell = 1$) about the uud valence quarks. The knock-out cross section for ϕ mesons is proportional to A^2B^2. We use harmonic oscillator spatial wavefunctions in order to treat the c.m. effects correctly. However, this choice has the effect of underestimating the high momentum components of the wavefunction. We base oscillator parameters on known electromagnetic charge radii, $\beta^2 = (3/2)\langle r^2\rangle^{-1}$; we choose the square roots of mean square mass-distribution radii $\sqrt{\langle r^2\rangle}$ to be 1.2 (set 1) and 1.5 (set 2) times smaller than these charge radii. If we write

$$\psi \sim e^{-\beta^2 r^2/2}$$

then the values of β used are given in Table I.

Table 1

		Harmonic Oscillator Parameters	
Wavefunction	β	Set 1 (fm^{-1})	Set 2 (fm^{-1})
Relative motion of s and $\bar{s}$ in proton	β_s	1.63	2.04
Motion of $s\bar{s}$ c.m. in proton	β_{3s}	1.94	2.43
ϕ	β_ϕ	1.63	2.04

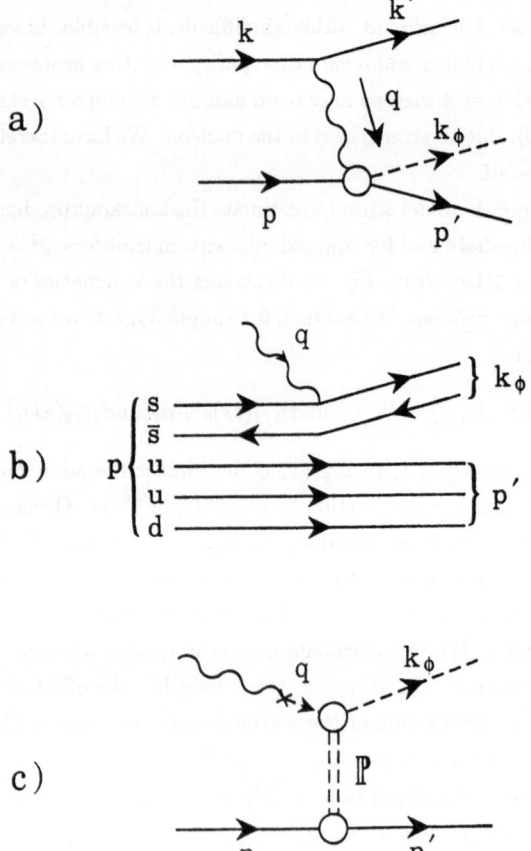

Fig. 1. Leptoproduction of ϕ mesons: a) kinematics, b) knock-out contribution, and c) diffractive production via pomeron exchange.

In the laboratory frame, we obtain for the cross section

$$\frac{d^3\sigma}{dW\,dQ^2\,dt} = \frac{W}{M}\,\frac{(E-\nu)}{2E}\,\frac{E'_p}{2M}\,\frac{1}{Q^4}\,(16\pi^2\alpha^2)Q_s^2\,A^2B^2(2a_0)^2|F|^2$$

$$\times \frac{|f|}{|\vec{q}|}\theta(1-|h|)\left(\frac{|\vec{q}|}{2m_s}\right)^2\frac{1}{4\pi}[1-\hat{q}\cdot\hat{k}'\,\hat{q}\cdot\hat{k}]\,, \tag{2a}$$

$$f \equiv \nu + M - E'_p\,, \tag{2b}$$

$$h \equiv \frac{m_\phi^2 + \vec{q}^{\,2} + \vec{p}^{\,\prime 2} - f^2}{2|\vec{q}||\vec{p}^{\,\prime}|}\,, \tag{2c}$$

$$|F|^2 \equiv \frac{8}{3\sqrt{\pi}}\,\frac{1}{(\beta_\phi\beta_s\beta_{3s})^3}\left(\frac{2\beta_\phi^2}{1+(\beta_\phi/\beta_s)^2}\right)^3\exp\left\{-\frac{|\vec{q}|^2}{4(\beta_\phi^2+\beta_s^2)}\right\}$$

$$\times \left(\frac{|\vec{p}^{\,\prime}|^2}{\beta_{3s}^2}\right)\exp\left\{-\frac{|\vec{p}^{\,\prime}|^2}{\beta_{3s}^2}\right\}\,, \tag{2d}$$

with $m_\phi = 1020$ MeV and $\hat{q} = \vec{q}/|\vec{q}|$, $\nu = E - E'$; $Q_s = -\frac{1}{3}$, $m_s \approx 500$ MeV/c^2 are the strange quark charge and mass; $W^2 = (p+q)^2 = (p'+k_\phi)^2$ is the invariant squared mass of the photon-hadron system; and $t \equiv (p'-p)^2 \approx -\vec{p}^{\,\prime 2}$ for small t. It should be noted that a_1 does not contribute to the triple differential cross section. The cross section is plotted in Figs. 2 and 3 for 20% strangeness, $B^2 = 0.2$, and with $a_0^2 = a_1^2 = 0.5$.

Electroproduction studies of ϕ mesons have been carried out [7], but not with sufficient accuracy and detail to establish the knock-out component. The experiments have been analyzed on the basis of a vector dominance model, illustrated in Fig. 1c. Here the virtual photon is transformed into a vector ϕ meson, which scatters diffractively (Pomeron exchange) from the hydrogen target. The observed Q^2 behavior agrees with the model, within errors of $\sim$ 15%, but the W-dependence and width of the forward ($t \approx t_{min}$) peak are in less satisfactory agreement. The cross section can be written as

$$\frac{d^3\sigma_{diff}}{dW\,dQ^2\,dt} = \frac{\alpha}{4\pi}\,\frac{W}{ME^2}\,\frac{W^2-M^2}{MQ^2}\,\frac{b_\phi}{1-\varepsilon}\,\frac{\sigma(0,W)}{[1+(Q^2/m_\phi^2)]^2}\left[\frac{p_\gamma^*(0)}{p_\gamma(Q^2)}\right](1+\varepsilon R_\phi)$$

$$\times \exp\left\{-b_\phi(|t|-|t_-(Q^2)|)\right\}\,, \tag{3}$$

where ε is the virtual photon polarization parameter, $\sigma(0,W)$ is the observed photoproduction cross section, $p_\gamma^*(Q^2)$ is the virtual photon flux in the photon-hadron c.m., and $|t_\pm|$ are the kinematic limits of t. A term $\propto e^{-b_\phi|t_+|}$ has been dropped. The values used to fit the data are $b_\phi = 3.46 \pm 0.22$ GeV^{-2}, $\sigma(0,W) = 0.45 \pm 0.02\mu b$ for $W \gtrsim 2.4$ GeV and $\sigma(0,W) = 0.22 \pm 0.03$ at $W \approx 2.1$ GeV due to the reduced phase space near threshold; ε and $p_\gamma^*(0)/p_\gamma(Q^2)$ are given by expressions in Ref. 8. Plots of the diffractive cross section are also shown in Figs. 2 and 3.

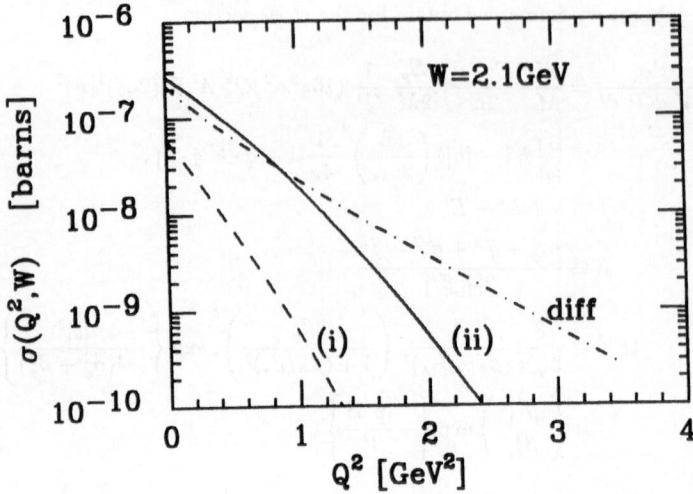

Fig. 2.   Virtual photoproduction cross section $\sigma(Q^2, W)$ for $W = 2.1$ GeV as a function of Q^2. The dash-dot curve is the diffractive cross section. The dashed and solid curves are the predicted knock-out cross section for $B^2 = 0.2$ and cases 1 and 2 (see text), respectively.

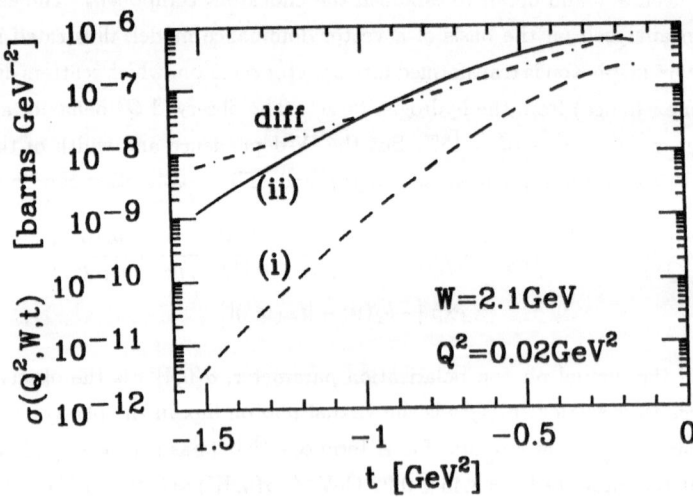

Fig. 3.   The t-dependence of the electroproduction cross section for $W = 2.1$ GeV and $Q^2 = 0.02$ (GeV/c)^2. The curves have the same correspondence as in Fig. 1.

The major difference between the two cross sections, which are of the same order of magnitude for a 20%-strange quark presence ($B^2 = 0.2$) occurs for small $t \equiv (p' - p)^2 \approx -\vec{p}\,'^2$. For the knock-out cross section there is an extra factor of t, whereas the diffractive one is purely exponential. Because the latter cross section is due to a natural parity (0^+) exchange and is s-channel helicity conserving, a polarization experiment would help to isolate the knock-out contribution more clearly.

In conclusion, accurate measurements of electro-phi production cross section, especially with polarized targets, are a means to obtain information on the vector strangeness matrix elements in the proton.

The analysis for neutrino production of phi mesons is quite similar, but is also sensitive to the axial vector strangeness matrix elements of the proton.

3. Neutrino and Antineutrino Elastic Scattering from ^{2}H

The coupling, $g_A^{(0)}$, of the axial vector current to the nucleon is poorly known. We define it by $\frac{1}{2}(<p|\gamma^\mu\gamma^5|p> + <n|\gamma^\mu\gamma^5|n>)$, and not $\frac{1}{2} <p|\bar{u}\gamma^\mu\gamma^5 u + \bar{d}\gamma^\mu\gamma^5 d|p>$ [9]. In terms of the u, d, s quark structure and the standard model $g_A^{(p)} = \frac{1}{2}(<p|\bar{u}\gamma^\mu\gamma^5 u - \bar{d}\gamma^\mu\gamma^5 d - \bar{s}\gamma^\mu\gamma^5 s|p>)$; with the assumption of isospin symmetry $g_A^{(0)}$ is then given by $g_A^{(0)} = - <p|\bar{s}\gamma^\mu\gamma^5 s|p>$. The EMC measurements [1] suggest $g_A^{(0)} \approx -0.22 \pm .07$, whereas the axial anomaly [10] predicts $|g_A^{(0)}| \approx 0.05\text{-}0.15$ [11]. It is clearly of interest to measure $g_A^{(0)}$ directly. This can be carried out for a proton target [12], but the term is isolated most directly for an isospin zero target since the cross section vanishes unless $g_A^{(0)} \neq 0$. Because the leading non-relativistic term for the axial vector current is the spin, we need a target with spin > 0. The deuteron is the lightest such target. The difference between neutrino and antineutrino elastic scattering cross sections on deuterium is directly proportional to $g_A^{(0)}$ for small $g_A^{(0)}$. The neutrino and antineutrino fluxes could be normalized by measuring their scattering cross sections on ^{12}C.

We have calculated the ratio

$$R = \frac{\frac{d\sigma}{dQ^2}(\nu d \rightarrow \nu d) - \frac{d\sigma}{dQ^2}(\bar{\nu} d \rightarrow \bar{\nu} d)}{\frac{d\sigma}{dQ^2}(\nu d \rightarrow \nu d) + \frac{d\sigma}{dQ^2}(\bar{\nu} d \rightarrow \bar{\nu} d)}, \tag{4}$$

where $Q^2 = -q^2$ is the negative of the 4-momentum transfer to the deuteron. Our calculation uses a multipole decomposition and an extended Siegert theorem [13] at low Q^2, a covariant formalism [14] together with a nonrelativistic deuteron wavefunction for intermediate $Q^2 [.01(\text{GeV/c})^2 \lesssim Q^2 \lesssim 1(\text{GeV/c})^2]$ and a light-cone impulse approximation at the highest Q^2 [15]. These methods overlap and agree with each other in their applicable regions. I shall illustrate our development for the covariant method. The weak current matrix elements can be written as

$$\langle D'|J_\mu|D\rangle = \frac{-1}{\sqrt{4D_0 D_0'}} \Bigg[G_1(Q^2)(\xi'^* \cdot \xi) P_\mu$$

$$+ G_2(Q^2)[\xi_\mu(\xi'^* \cdot q) - \xi_\mu'^*(\xi \cdot q)]$$

$$- G_3(Q^2)(\xi \cdot q)(\xi'^* \cdot q) P_\mu / 2M^2 \Bigg] ,$$

$$\langle D'|J_\mu^{(5)}|D\rangle = \frac{-1}{\sqrt{4D_0 D_0'}} \Bigg[iG_4(Q^2)\varepsilon_{\mu\alpha\beta\gamma}\xi'^{*\alpha}\xi^\beta P^\gamma$$

$$+ \frac{iG_5(Q^2)}{M^2}\varepsilon_{\mu\alpha\beta\gamma}q^\alpha P^\beta[\xi_\mu(\xi'^* \cdot q) - \xi_\mu'^*(\xi \cdot q)] \Bigg] , \qquad (5)$$

where $P_\mu = D_\mu' + D_\mu$, with $D(D')$ the initial (final) deuteron momentum, M is the mass of the deuteron, $D_0 = \sqrt{M^2 + Q^2/4}$ is the deuteron energy in the Breit frame, and $\xi_\mu(\xi_\mu')$ is the polarization 4-vector for the initial (final) deuteron. The neutrino-deuteron elastic scattering cross section can be written as

$$\frac{d\sigma}{dQ2}\begin{pmatrix} \nu d \\ \bar\nu d \end{pmatrix} = \frac{G_F^2}{2\pi} r \left[2W_1 \sin^2\frac{\theta}{2} + W_2 \cos^2\frac{\theta}{2} \mp \frac{2}{M}(E_\nu + E_\nu')W_8 \sin^2\frac{\theta}{2} \right] \qquad (6)$$

where θ is the scattering angle, $(1 - \cos\theta) = Q^2/(2E_\nu E_\nu')$, and r is the nuclear recoil factor, $r = [1 + Q^2/2ME_\nu']$. The response functions are

$$W_1(Q^2) = \frac{Q^2}{6M^2}(1 + Q^2/4M^2)G_M^2 + \tfrac{2}{3}(1 + Q^2/4M^2)^2 G_A^2 ,$$

$$W_2(Q^2) = \left[G_C^2 + \frac{Q^2}{6M^2}G_M^2 + \frac{Q^4}{18M^4}G_Q^2 + \tfrac{2}{3}(1 + Q^2/4M^2)G_A^2 \right] ,$$

$$W_8(Q^2) = \tfrac{2}{3}(1 + Q^2/4M^2)[G_M G_A] , \qquad (7)$$

where G_C, G_Q, G_M, and G_A are the charge, quadrupole magnetic, and axial deuteron form factors. They are related to $G_1 - G_5$ by

$$G_C(Q^2) \equiv G_1(Q^2) - (Q^2/6M^2)G_Q(q^2) ,$$

$$G_Q(Q^2) \equiv G_1(Q^2) - G_2(Q^2) + (1 - Q^2/4M^2)G_3(Q^2) ,$$

$$G_M(Q^2) \equiv G_2(Q^2) ,$$

$$G_A(Q^2) \equiv G_4(Q^2) + (Q^2/M^2)G_5(Q^2) . \qquad (8)$$

These form factors are evaluated in the Breit frame with a non-relativistic deuteron wavefunction and non-relativistic reduction of nucleon-current operators. In most of our work the deuteron wavefunction has been taken to be that obtained from the Paris

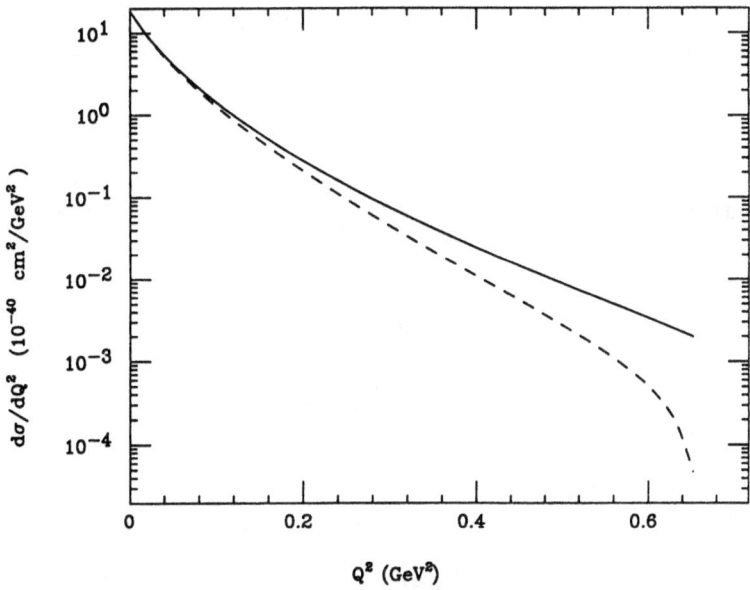

Fig. 4.   Neutrino (solid) and antineutrino (dashed) elastic differential cross sections as a function of Q^2 for $E_\nu = E_{\bar{\nu}} = 500$ MeV, $g_A^{(0)} = 0.2$ (solid) and 0.1 (dashed).

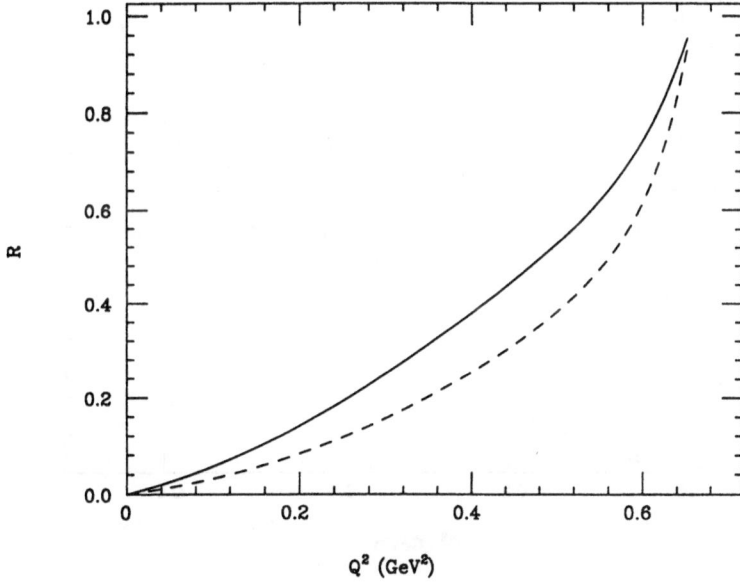

Fig. 5.   The ratio R as a function of Q^2 at $E_\nu = E_{\bar{\nu}} = 500$ MeV for $g_A^{(0)} = 0.2$ (solid) and 0.1 (dashed).

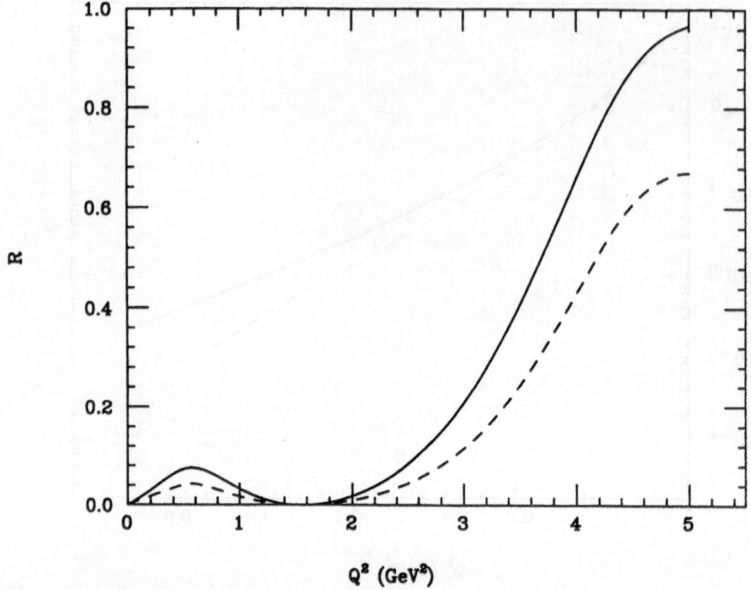

Fig. 6. Same as Fig. 5, but at $E_\nu = E_{\bar\nu} = 2$ GeV.

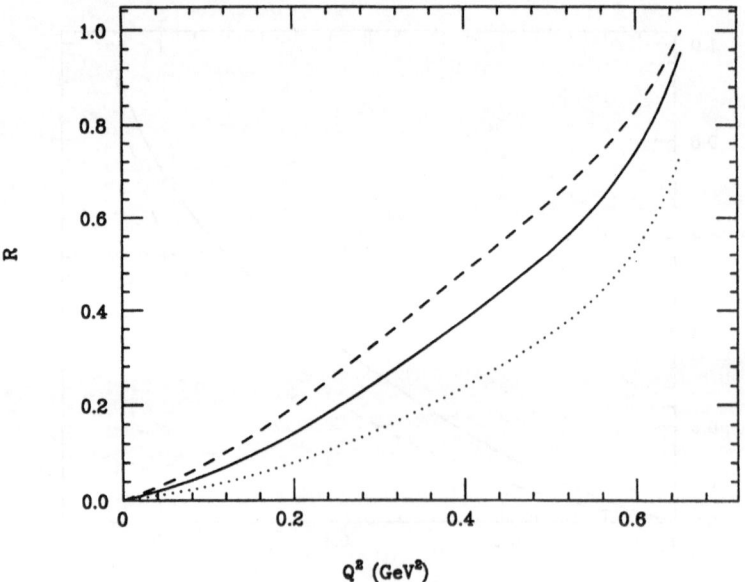

Fig. 7. The ratio R at $E_\nu = E_{\bar\nu} = 500$ MeV for $g_A^{(0)} = 0.2$ and $g_M^{(s)} = 0.2$ (dashed), 0 (solid), and -0.2 (dots).

potential [17]. The isoscalar axial vector form factor is assumed to be similar to that of its isospin 1 partner, i.e., of the dipole form $(1 + Q^2/M_A^2)^2$, with $M_A = 1.03$ GeV/c^2.

Results are shown in Figs. 4-7. Fig. 4 shows the differential cross sections for ν and $\bar{\nu}$ elastic scattering as a function of Q^2. Fig. 5 shows the features we wish to highlight. Here we plot R as a function of Q^2 for $E_\nu = E_{\bar{\nu}} = 500$ MeV for $g_A^{(0)} = 0.1$ and 0.2. Of course, $R = 0$ if $g_A^{(0)} = 0$; for $g_A^{(0)} = 0.1$ and 0.2, R grows smoothly and monotonically from 0 to 1 and is larger than 1/4 for $Q^2 > 0.4$ (GeV/c)^2. For lower energies the dependence of R on Q^2 is comparable to that shown, but for $E_\nu \gtrsim 1$ GeV, the growth of R with Q^2 is stunted by the zeroes of the deuteron form factor. Fig. 6 shows R vs Q^2 at $E_\nu = E_{\bar{\nu}} = 2$ GeV. Thus, average energies of several hundreds of MeV seem preferable. It is also better to stay away from the maximum possible Q^2, because the sensitivity to the numerical value of $g_A^{(0)}$ is reduced at 180°. In Fig. 7, we show that the results also depend on the s-quark contribution to the nucleon's magnetic form factor, which is presently unconstrained by experiment [6]. The contribution enters by modifying the deuteron's weak magnetic form factor with which the axial one interferes. The dependence of R on $g_M^{(s)}$ is similar to that of a variation of $g_A^{(0)}$. The two effects may be separated at small Q^2, where the contribution from $g_M^{(s)}$ is suppressed by a factor of Q^2/M^2.

Despite the shortcoming of the necessity to isolate $g_M^{(s)}$, or to measure it separately from $g_A^{(0)}$, we believe that the neutrino experiments are feasible at LAMPF, TRIUMF or similar accelerators. We hope they will be undertaken.

In summary, I have described in some detail two difficult, but feasible experiments to measure the presence of strangeness in nucleons.

This work was supported in part by the U.S. Department of Energy and in part by the National Science Foundation.

References

Permanent address:

^{a} Instituto de Estudos Avançados, Centro Técnico Aeroespacial, 12.225 São José dos Campos, São Paulo, Brazil

^{b} Instituto de Física Téorica, Universidade Estadual Paulista, Rua Pamplona, 145, São Paulo, Brazil

^{c} Department of Physics and the Supercomputer Computations Research Institute, Florida State University, Tallahassee, FL 32306-3016

1. Ashman, J. *et al.*: Phys. Lett. B206, 364 (1988) and Nucl. Phys. B328, 1 (1989).

2. Nowak, M.A., Verbaarshot, J.J.M., Zahed, I.: Phys. Lett. B217, 157 (1989); Fritzch. H.: Phys. Lett. B229, 122 (1989) and Mod. Phys. Lett. A5, 1815 (1990); Anselmino, R., Scadron, M.D.: Phys. Lett. B229, 117 (1989); Stern, J., Cle'ment, G.: Phys. Lett. B231, 471 (1989); Bernard, V., Meissner, U.: Phys. Lett. 216, 392 (1989) and 223, 439 (1989); Jaffe, R.L., Manohar, A.: Nucl. Phys. B337, 509 (1990).

3. Henley, E.M., Pollock, S.J., Krein, G., Williams, A.G.: Phys. Lett. B (to be published).

4. Ellis, J., Gabathuler, E., Karliner, M.: Phys. Lett. B217, 173 (1989).

5. Beck, D.: Phys. Rev. D39, 3248 (1989); Decker, R., Leize, Th.: Phys. Lett. B246, 233 (1990); Zuddin, Fayya, Zuddin, Ria, Phys. Rev. D42, 794 (1990); McKeown, R.D.: Phys. Lett. B219, 140 (1989); E.J. Beise, E.J., McKeown, R.D.: Comm. Nucl. Part. Phys. 20, 105 (1991); Napolitano, J.: Phys. Rev. C43, 1473 (1991).

6. Beck, D.H.: Phys. Rev. D39, 3248 (1989; McKeown, R.D.: Phys. Lett. 219B, 140 (1989); Beise, E.J., McKeown, R.D.: Comm. Nucl. Part. Phys. 20, 104 (1991).

7. Dakin, J.T. et al.: Phys. Rev. D8, 687 (1973); Dakin, J.T. et al.: Phys. Rev. Lett. 30, 142 (1973); Dixon, R. et al.: Phys. Rev. D19, 3185 (1979); Dixon, R. et al.: Phys. Rev. Lett. 39, 516 (1977); Cassel, D.G. et al.: Phys. Rev. D24, 2787 (1981); Atkinson, M. et al.: Z. Phys. C30, 521 (1986); Behrend, H.-J. et al.: Nucl. Phys. B144, 22 (1978); ABBHHM Collaboration, Phys. Rev. 175, 1669 (1968); Bauer, T.H., Spital, R.D., Yennie, D.R., Pipkin, F.M.: Rev. Mod. Phys. 50, 261 (1978).

8. Fraas, H., Schildknecht, D.: Nucl. Phys. B14, 543 (1969).

9. Belyaev, V.M., Ioffe, B.L., Kogan, Ya.I.: Phys. Lett. 151B, 290 (1985); Gupta, S., Murthy, M.V.M., Pasupathy, J.: Phys. Rev. D39, 2547 (1989); Henley, E.M., Hwang, W-Y.P., Kisslinger, L.S.: (private communication).

10. Adler, S.L., Phys. Rev. 177, 2426 (1969); Bell, J.S., Jackiw, R.: Nuovo Cim. 60A, 47 (1969).

11. Collins, J., Wilczek, F., Zee, A.: Phys. Rev. D18, 242 (1978); Wolfenstein, L.: Phys. Rev. D19, 3450 (1979); Kunz, J., Mulders, P.J., Pollock, S.: Phys. Lett. B222, 481 (1989).

12. Ahrens, L. et al.: Phys. Rev. D35, 785 (1987); Garvey, G.: (private communication).

13. Ying, S., Henley, E.M., Miller, G.A.: Phys. Rev. C38, 1584 (1988); Ying, S., Haxton, W.C., Henley, E.M.: Phys. Rev. C40, 3211 (1989).

14. Pollock, S.J., Phys. Rev. D42, 3010 (1990).

15. Frederico, T., Henley, E.M., Miller, G.A.: Nucl. Phys. (to be published).

16. Lacombe, M. et al., Phys. Lett. 101B, 139 (1981).

Few-Body Systems, Suppl. 6, 77—82 (1992)

A COVARIANT MODEL OF MESONS THAT DECOUPLES CONFINEMENT AND CHIRAL SYMMETRY BREAKING

Joseph Milana

College of William & Mary

Department of Physics

Williamsburg, VA. 23185

Abstract

New solutions to a recently proposed model of mesons as quark-antiquark bound states are presented. The model is covariant, confining, and chirally symmetric. That the two main features of low-energy QCD, confinement and chiral symmety breaking, enter as completely distict components of the model is emphasized. In particular, the confining potential can be taken purely scalar.

I am here to discuss the work I've been doing with Franz Gross developing a new model of mesons that is covariant, chirally symmetric and confining. Within this limited time frame, I will first highlight the main ingredients and results of the model, then move onto more of the details, and finally present new numerical solutions of the model. Much of the formalism has already been published[1] and we have a second manuscript[2] with a fuller development of the arguements to be now presented and with also furthur details concerning the new numerical solutions.

Let me now give an overview of the approach:

First, the model is covariant, i.e. we include both relativistic kinematics and dynamics into our model building. This is an important component for at least two reasons: 1) in the case of the light mesons, one expects important relativistic effects, and 2) it will be essential in the application to exclusive or semiexclusive events at low to moderate energy transfers as for example will occur at CEBAF. Let me remind you this is an ingredient lacking in most of the familiar models (e.g. the various bag models, nonrelativisic potential models, and soliton models) of the hadronic spectrum.

Second, we include chiral symmetry and confinement-arguably the two most salient features of low-energy QCD. The interesting and we think unique feature of our model is that these two elements completely decouple in our approach. That is, we can use two separate scales and interactions. In particular, the confining potential can be chosen to be purely scalar as suggested by lattice studies and heavy quark phenomenology, while still obtaining a zero-mass pion in the pure chiral limit. This decoupling is furthur suggested by the fact that there appears to be two distinct low-energy scales: $\Lambda_{QCD} \sim 200$ Mev in the case of confinement and; $4\pi f_\pi \sim 1$ Gev from chiral perturbation theory[3]. Indeed, in the case of a quark-gluon plasma, it has become common place to discuss the possibility of two seperate phase transitions, one associated with deconfinement and one with chiral restoration. Independent of the ultimate correctness of this decoupling, a model in which this seperation is explicitly realized shows that at least in principle we could be discussing two seperate, independent manifestations of low-energy QCD.

The model is a covariant generalization of nonrelativistic linear potential models[4] that includes chiral symmetry breaking by dynamically generating a constituent, quark mass[5]. The light mesons are viewed as bound states of these dynamically generated, massive quark-antiquark pairs. A self-consistency condition then ensures that in the chiral limit when the current quark mass is zero, the pion appears as a zero-mass, pseudoscalar, Goldstone boson. Schematically, we are solving the self-consistent equations for the quark self-energy and vertex function, figures (1a) and (1b) below.

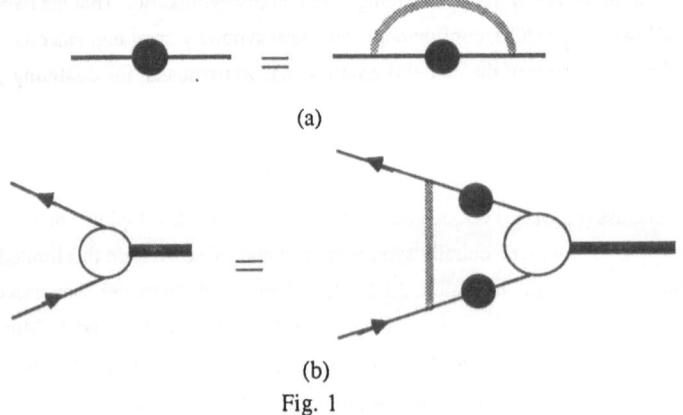

(a)

(b)

Fig. 1

The heavy dashed lines represent the quark potential, modeled as an exchange interaction (as would occur in a simple boson-exchange picture), involving two three-point vertices with the exchanged momentum determined by energy-momentum conservation. In both equations, the kernel is further defined by restricting some of the quarks to their mass shell. In the vertex equation, two channels are created, one with the *quark* restricted to its positive-energy mass shell, and one with the *antiquark* restricted to its negative-energy mass shell. The resulting two channel bound-state equation is shown in Fig. 2 in which a line with a cross signifies that the quark is onshell. These restrictions mean that even though the equations are exactly covariant, they depend, like non-relativistic equations, on the relative three-momentum only, and have a smooth non-relativistic limit. The second (antiquark) channel is necessary for a consistent description of deeply bound states but is negligible for loosley bound, heavy quark systems.Finally, restricting both the internal and external quarks to their mass shell reduces the self energy equation (Fig. 1a) to an algebraic self consistency condition between the bare (current) and dynamical (constituent) quark masses, and the parameters of the same kernel appearing in the bound equation.

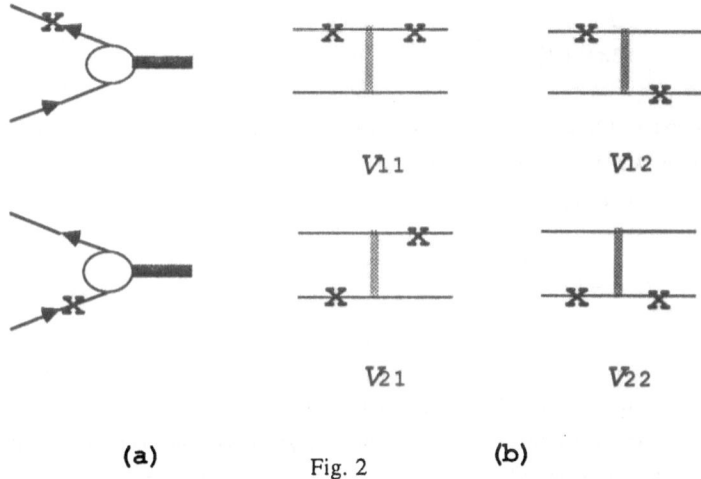

V_{11} V_{12}

V_{21} V_{22}

(a)   Fig. 2   (b)

Our effective quark interaction $V_{eff}(k)$, contains two components: one piece is a covariant generalization of a linear potential and provides confinement, and a second piece is a covariant generalization of a nonrelativistic constant potential:

$$V_{eff}(k) = V_L(k) \sum_i O_1^i \, O_2^i + V_C(k) \sum_i \tilde{O}_1^i \, \tilde{O}_2^i \ . \tag{1}$$

The Dirac matrices O and $\tilde{O}$ operate on the Dirac indices of quarks 1 and 2, and describe the spin dependent structure of each of the two pieces of the effective interaction. They are, in general, distinct. The covariant scalar functions $V_L(k)$ and $V_C(k)$ contain the momentum dependence of the two pieces of the effective interaction.

An essential feature which makes our equations tractable is the infrared regularized,

Fourier transformed, linear potential $V(r) = \sigma r$. In momentum space, the linear potential behaves as $1/q^4$ plus an infrared subtraction that regulates the potential at $q^2 = 0$ and ensures that $V(r = 0) = 0$. The covariant generalization of this condition satisfied by the confining potential $V_L(k)$ is:

$$\int \frac{d^3k}{(2\pi)^3} \left(\frac{m}{E_k} \right) V_L(k) = 0, \tag{2}$$

where $E_k = \sqrt{m^2 + k^2}$, and m is the quark constituent mass. (Note that in the nonrelativistic limit, the term in brackets is absent and eq.(2) merely reduces to the fourier transform of the linear potential at $r = 0$.) We likewise define a covariant generalization of a nonrelativistic constant potential, $V_C(k)$, which satisfies:

$$\int \frac{d^3k}{(2\pi)^3} \left(\frac{1}{E_k} \right) V_C(k) = C. \tag{3}$$

By placing both quark legs on-shell in the self-energy diagram fig.(1a), we obtain the following relation between the constituent quark mass m, the bare quark mass m_o, and the strength of the constant potential, C:

$$C = -\left(1 - \frac{m_0}{m} \right). \tag{4}$$

Notice that the strength of the linearly confining potential does not enter this relation. This is because the contributions of our two potentials in the mass shift equation effectively reduces to the defining relations, equations (2) and (3). The linear confining potential thereby makes no contribution to the quark's mass generation; it arises solely from our constant potential. It is this decoupling that allows us to chose the confining potential to be purely scalar. In nonrelativistic models the constant piece provides an overall mass shift to the potential, suggesting a relationship to the vacuum. In our covariant generalization, we see that there is a deep relationship, as it is this constant potential that is crudely modeling the vacuum and providing dynamical chiral symmetry breaking.

Our bound state equations for finite pion mass are in general, complicated two channel equations. The equations in general must be solved numerically. However, in the zero mass pion limit with zero bare quark mass, we obtain for our two wavefunctions the analytic solution $\Psi_1(p) = \Psi_2(p) = N / E_p$ and $C = -1$, as required by chiral symmetry, Eq. (4). The linear potential again completely decouples in the chiral limit.

Sample solutions for finite pion mass are shown in figs.(2) and (3). For the initial numerical studies, we chose to work with a particularly convenient form for the Dirac matrices, O and $\tilde{O}$:

$$\sum_i O_i O_i = \sum_i \tilde{O}_i \tilde{O}_i = \frac{1}{2}(1 - \gamma_1^5 \gamma_2^5 - \gamma_1^\mu \gamma_{2\mu}).$$

(5)

This form was chosen because it allows a bound state solution which is a pure pseudoscalar for the vertex function: $\Gamma(p,\mu) = \Gamma_o(p)\tau\gamma^5$, where μ is the bound state rest mass. While the choice (5) is convenient, it is certainly not best from a phenomeno-logical point of view, and the optimal form for O and $\tilde{O}$ will be deferred to a later work when we use this model to fit the physical spectrum.

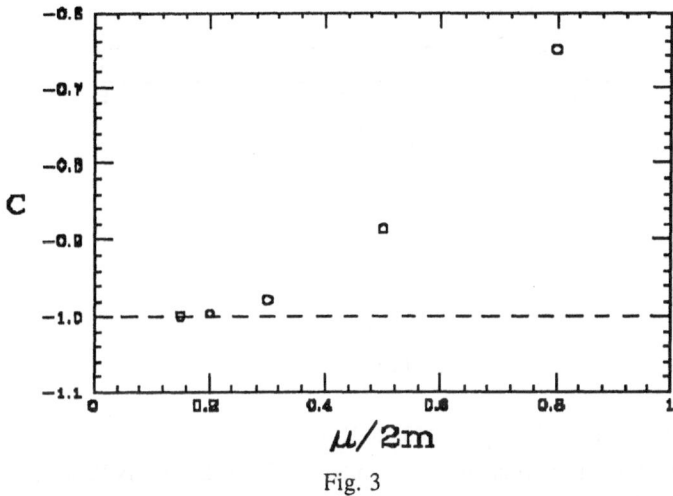

Fig. 3

Fig. 3 is a plot of our results for the constant C as a function of the bound-state mass. Notice how close the value of C is to the chiral limit at the physical pion mass ($\mu / 2m \sim 0.2$). Fig. 4 contains our solutions for the wave functions $\Psi_1(p)$ (solid curves) and $\Psi_2(p)$ (dashed curves) for a family of values of the bound-state mass ($\mu / 2m = 0.8, 0.5, 0.3,$ and 0.15). Notice how the wave functions grow in momentum space as the bound-state mass decreases, and how the second channel smoothly approaches the first.

82

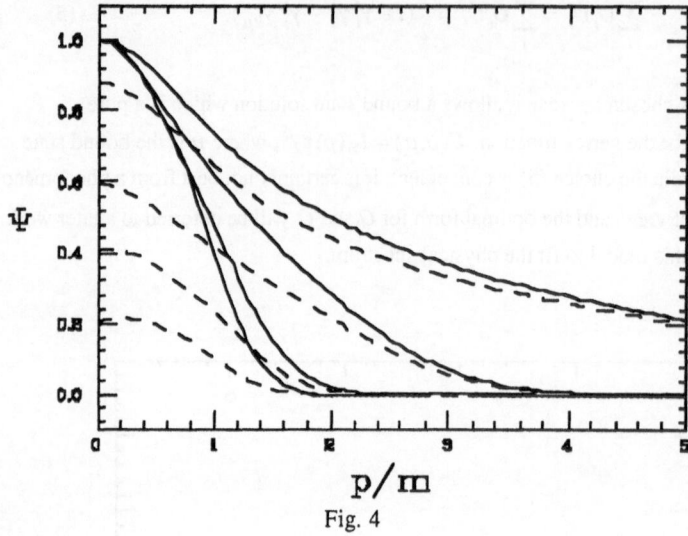

$$p/m$$

Fig. 4

References:

1) F. Gross and J. Milana, Phys. Rev. **D43**, 2401, (1991).
2) F. Gross and J. Milana, "Decoupling Confinement and Chiral Symmetry Breaking: an Explicit Model", William & Mary preprint W&M-91-110 and CEBAF preprint CEBAF-TH-91-15, (1991).
3) J. Gasser and H. Leutwyler, Ann. Phys. **158**, 142 (1984); A. Manohar and H. Georgi, Nucl. Phys. **B234**, 189 (1984).
4) S. Godfrey and N. Isgur, Phys. Rev. **D32**, 189 (1985).
5) *ala* Y. Nambu and G. Jona-Lasinio, Phys. Rev. **122**, 345 (1964).

Few-Body Systems, Suppl. 6, 83—88 (1992)

COMPACT COLORLESS QUARK-GLUON CLUSTERS IN IAE BAG MODEL AND INTERMEDIATE STRUCTURE OF SU(3) HADRONS. PION AND NUCLEON ELECTROMAGNETIC FORM FACTORS

Yu. E. Pokrovsky

I.V.Kurchatov Institute of Atomic Energy

123182, Moscow, USSR

Within a framework of modified version of a bag model (IAE bag) which leads to a new physical picture of effective hadron structure, possibly appearing at 4-momentum transfer $|t|$=1-10 GeV², electromagnetic form factors of pion and nucleons are considered.

1. INTRODUCTION

One of the challenging problems of nuclear physics today is to establish the link between the hadron-nuclear phenomena and the strong interaction of quarks and gluons in QCD. QCD is believed to be the proper theory of the strong interaction. But only very limited phenomena of strongly interacting systems are quantitatively understood within this theory: phenomena which can be calculated by the perturbative QCD. Nuclear interactions do not belong to this category of processes and are intimately related to the physics of confinement. Confinement is the outstanding open problem of the theory of strong interaction now.

Using boundary condition of MIT bag model [1] as a model of confinement with assumption about complete break of QCD-vacuum condensates in the bag I have shown [2-5] that at "intermediate values" of 4-momentum transfer $|t|$=1-10 GeV² and at arbitrary values of invariant masses (s) the role of fundamental constituents in light-hadron structure can play the compact baryon-meson-like colorless quark-gluon clusters but not quarks or gluons.

In framework of this version of the model (IAE bag [2-5]), mean square radii of these clusters are as little as $R_{clust} \approx 0.2$ fm and ones are much less than the charge radii of the light hadrons ($R^p_{ch} \approx 0.9$ fm, $R^{\pi}_{ch} \approx 0.7$ fm).

Note that $R_{clust} \simeq 0.7 R_{bag}$. Such strong compression of the bags takes place in the model due to two reasons: 1) Strong surface tension that was calculated in [1] and which is proportional to a sum of vacuum condensates of the light-quark fields. 2) The center-of-mass corrections [3] which reduce the quark-gluon pressure inside the bag.

The mentioned above inequality $R_{clust} \ll R_{ch}^{h}$ means that the light physical hadrons must be composed average from a few such elementary clusters. Therefore qqq or $q\bar{q}$-state contributions to amplitudes of any processes with light baryons and mesons must be negligible up to $|t| \simeq (\pi/R_{clust})^2 \simeq 10 \ GeV^2$.

One of the main predictions of proposing picture is that there are three qualitatively different levels of physical phenomena in hadron physics which appear in three different regions of values of $|t|$: $0 < |t| < (\pi/R_{ch})^2 \simeq 1$ GeV^2, $(\pi/R_{ch})^2 < |t| < (\pi/R_{clust})^2 \simeq 10 \ GeV^2$ and $|t| > (\pi/R_{clust})^2$. These regions of $|t|$ correspond to appearance of multicluster hadrons, one-cluster hadrons, and pure quark-gluon states in any hadronic processes. Each such structure level is characterized by different properties of constituents – such properties as masses, spins, charges, form factors etc.

To verify this picture all more informative hadron data have been qualitatively discussed [4]: hadron's spectroscopy, $e^{+}e^{-}$-annihilation to hadrons, elastic and deep inelastic (unpolarized and polarized) lepton-hadron and hadron-hadron interactions, and ultrarelativistic heavy ion collisions. More detail calculations and discussion of the electromagnetic form factors of pion and nucleon is the subject of this paper.

2. CLUSTER DECOMPOSITION OF MULTICLUSTER STATES

All necessary ingredients for calculating masses and radii of the colorless clusters in the model are quark kinetic energy with center-of-mass corrections, Casimir energy, one-gluon-exchange energy, surface and volume contributions [2,3,4]. There are no free parameters in this model. Using one-loop expression for α_s with $\Lambda_{QCD} = 0.1 \ GeV$ one can obtain following values for masses and bag radii of colorless clusters with quantum numbers of pion, ρ, ω-meson, nucleon and Δ-isobar:

$$M_{\pi'} = 2.877 \ GeV, \ R_{\pi'} = 0.301 \ fm, \quad M_{\rho'} = 3.025 \ GeV, \ R_{\rho'} = 0.305 \ fm,$$
$$M_{N'} = 4.091 \ GeV, \ R_{N'} = 0.355 \ fm, \quad M_{\Delta'} = 4.291 \ GeV, \ R_{\Delta'} = 0.360 \ fm.$$

To build the light hadrons from such massive constituents one needs to suppose that there is very strong attractive cluster-cluster interaction. Therefore one-cluster components would be negligible in the light hadron structure and hadrons should be multicluster states with the following relations for decomposition coefficients $|C_{N'}|^2 \ll |C_{N'\rho'}|^2 \simeq |C_{N'\pi'\pi'}|^2 \ldots \ll 1,$

$|C_{\pi'}|^2 \ll |C_{\pi'\rho'}|^2 \simeq |C_{\pi'\pi'\pi'}|^2 \ldots \ll 1$. These coefficients must satisfy to the ordinary conditions of normalization: $|C_{N'}|^2 + |C_{N'\rho'}|^2 + |C_{N'\pi'\pi'}|^2 + \ldots = 1$, $|C_{\pi'}|^2 + |C_{\pi'\rho'}|^2 + |C_{\pi'\pi'\pi'}|^2 + \ldots = 1$.

The dominant contributions to electromagnetic form factors of hadrons at intermediate $|t|$ should be given then by following simplest states: $|N\rangle = C_{N'}|N'\rangle + C_{N'\rho'}|N'\rho'\rangle + C_{N'\pi'\pi'}|N'\pi'\pi'\rangle + C_{N'\omega'}|N'\omega'\rangle + C_{N'\pi'\pi'\pi'}|N'\pi'\pi'\pi'\rangle$ for the nucleon and $|\pi\rangle = C_{\pi'}|\pi'\rangle + C_{\pi'\rho'}|\pi'\rho'\rangle + C_{\pi'\pi'\pi'}|\pi'\pi'\pi'\rangle$ for the pion. Here N', π', ρ', ω' - are one-cluster qqq, $q\bar{q}$ states with quantum numbers of nucleons, pions or ρ, ω-mesons. The contributions of more complicated states should be suppressed by masses of constituents in energy denominators of propagators of meson exchanges.

3. ELECTROMAGNETIC FORM FACTORS OF PION AND NUCLEON

There are one electromagnetic form factor $F^{\pi}(t)$ for charge pions and two form factors for nucleons - electric form factor $G_E^N(t)$ and magnetic one $G_M^N(t)$. Here $N=n,p$ for neutron and proton correspondingly. But $G_E^N(t)$ has not been measured at high $|t| > 5$ GeV^2 and don't consider in this paper.

Contributions to F^{π} and G_M^N are different in the three regions of $|t|$: $\pi, \rho, \omega, \varphi$ states give dominant contributions in the first region of $|t|$. In the second region of $|t|$ $\pi', \rho', \omega', \varphi'$ states give dominant contributions with relatively hard form factors of strong cluster-cluster couplings $F'(t) = (\Lambda_{V'}^2 - m_{V'}^2)/(\Lambda_{V'}^2 - t)$, where $m_{V'} \simeq 2$ GeV is a mass of one-cluster vector-meson state, $\Lambda_{V'}^2 \simeq (\pi/R_{clust})^2 \simeq 10$ GeV^2 at $R_{clust} \simeq 0.2$ fm. The ρ, ω, φ- contributions are strongly suppressed in this region of $|t|$ because the form factors of strong couplings of the multicluster ρ, ω, φ-mesons should be relatively soft: $F(t) = (\Lambda_V^2 - m_V^2)/(\Lambda_V^2 - t)$, where $m_V \simeq 1$ GeV is a mass of vector meson, $\Lambda_V^2 \simeq (\pi/R_V)^2 \simeq 1$ GeV^2 at $R_V \simeq 0.7$ fm (mean radius of multicluster hadron). But direct quark contributions are strongly suppressed by small value of a ratio of the cluster radii to the virtual photon wave length. In the third $|t|$-region pure quark-gluon contributions must be dominant over all others.

To obtain quantitative predictions for $F^{\pi}(t)$ and $G_M^N(t)$ within proposing picture let us use the cluster decomposition mentioned above and take into account few first terms which are dominant at $|t| \simeq 1$-10 GeV^2. Next more complicated terms are depressed at this values of t by higher powers t in denominators which arise from propagators of meson-like cluster exchanges.

Electromagnetic form factors at $|t| \simeq 1$-10 GeV^2 can be calculated in Breit system with relativistic kinematics corrections [6] taken into account.

In this case best fit of calculated π, p, n form factors to recent data *corresponds to the following admixtures of first five multicluster states:*

1. $\quad |C_{\pi'}|^2 \quad =0.02 \qquad |C_{p'}|^2 \qquad =0.003 \qquad |C_{n'}|^2 \qquad =0.003$

2. $\quad |C_{\pi'\rho'}|^2 =0.71 \qquad |C_{p'v'}|^2 \qquad =0.201 \qquad |C_{n'v'}|^2 \qquad =0.160$

3. $\quad |C_{\pi'\pi'\pi'}|^2 =0.00 \qquad |C_{p'\pi'\pi'}|^2 \quad =0.393 \qquad |C_{n'\pi'\pi'}|^2 \quad =0.554 \qquad (1)$

4. $\quad |C_{\pi'\rho'\rho}|^2 =0.27 \qquad |C_{p'\pi'\rho}|^2 \quad =0.094 \qquad |C_{n'\pi'\rho}|^2 \quad =0.034$

5. $\qquad\qquad\qquad\qquad\quad |C_{p'\pi'\pi'\pi'\rho'}|^2 =0.309 \qquad |C_{n'\pi'\pi'\pi'\rho'}|^2 =0.249$

So, $q\bar{q}$ and qqq state probabilities in pion and proton are ~1% only. This values of decomposition coefficients are in a good agreement with assumption of collective multicluster structure of pions and nucleons: $|C_{\pi'}|^2 \ll |C_{\pi'\rho'}|^2 \ll 1$ and $|C_{N'}|^2 \ll |C_{N'v'}|^2 \ll 1$. In this case average numbers of constituents in nucleon ($<n_N>=\sum_i C_i n_i =2.7\approx3$) and in pion ($<n_\pi>=\sum_i C_i n_i =2.1\approx2$) are closed to main assumption of simple quark model.

These results for $G_M^p(t)$, $G_M^n(t)$ and $F^\pi(t)$ are plotted in Fig.1-4. They scaled by t^2/μ_p, t^2/μ_n and $|t|$ correspondingly, where μ_p and μ_n are the proton and neutron magnetic moments (+2.7928474 and -1.91304275 n.m.). The numbers near the curves correspond to contributions of the states (1).

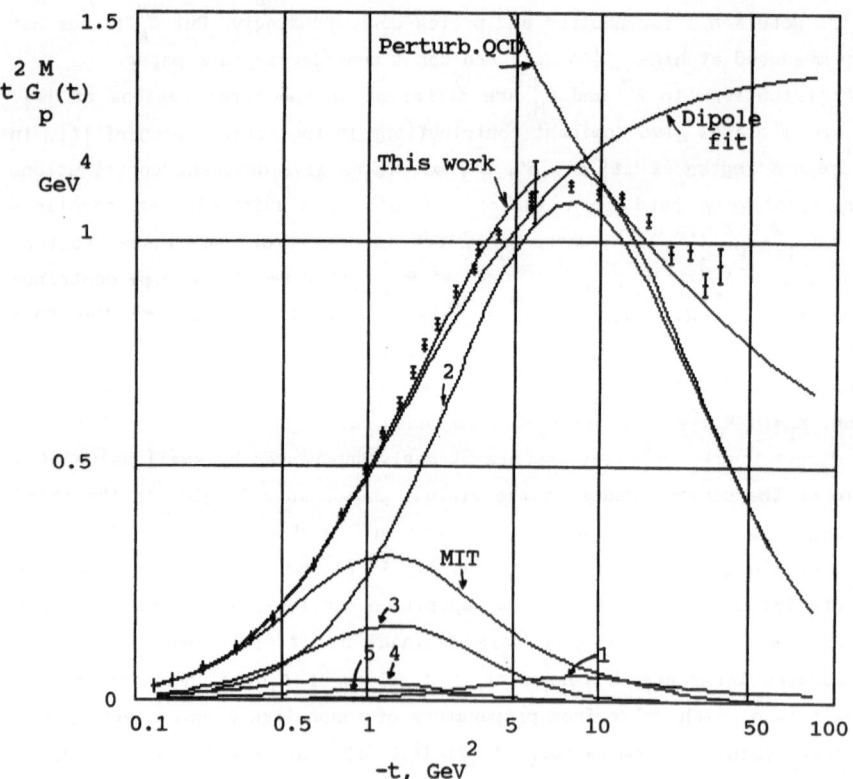

Fig.1. The data [7] for $t^2 G_M^p(t)/\mu_p$ in comparison with prediction of this work, dipole fit $G_M^p(t)=\mu_p/(1-t/0.71)^2$ and perturbative QCD [8].

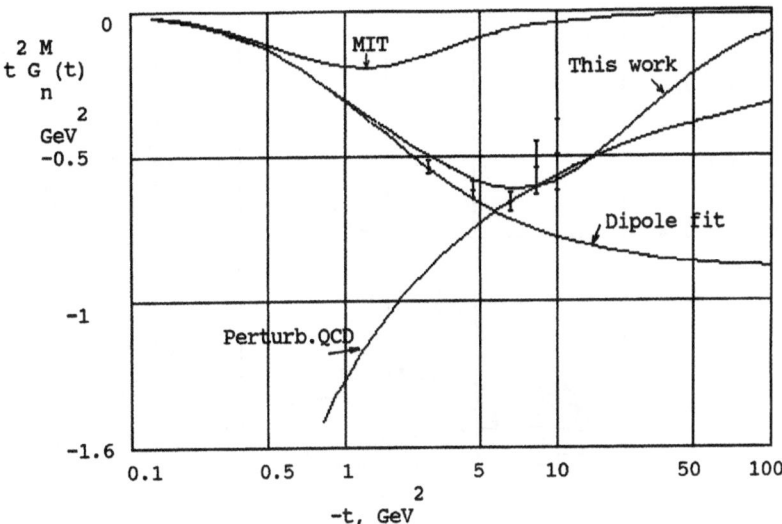

Fig.2. The data [9] for $t^2 G_M^n(t)/\mu_n$ in comparison with prediction of this work, dipole fit $G_M^n(t)=\mu_n/(1-t/0.71)^2$ and perturbative QCD [8].

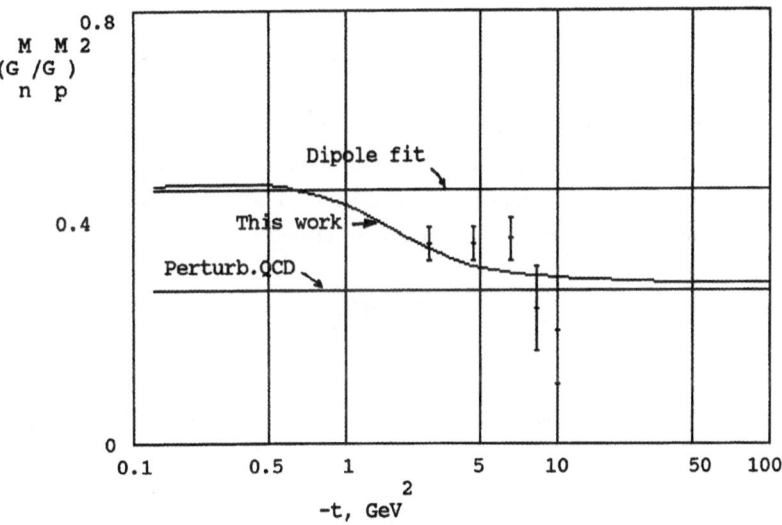

Fig.3. The data [9] for $(G_M^n(t)/G_M^p(t))^2$ in comparison with prediction of this work, dipole fit $\approx(\mu_n/\mu_p)^2=0.47$ and perturbative QCD [8].

So, the proposing picture leads to clear predictions for e-π experiments - the electromagnetic form factor of pion, scaled by -t, must have a broad peak near $-t \approx 2$ GeV^2. Then this form factor must significantly decrease with increasing -t (similar to proton magnetic form factor, scaled by t^2, which have a broad peak at $t \approx 8$ GeV^2 [7]) to reach QCD asymptotic at $|t| \approx 20$ GeV^2.

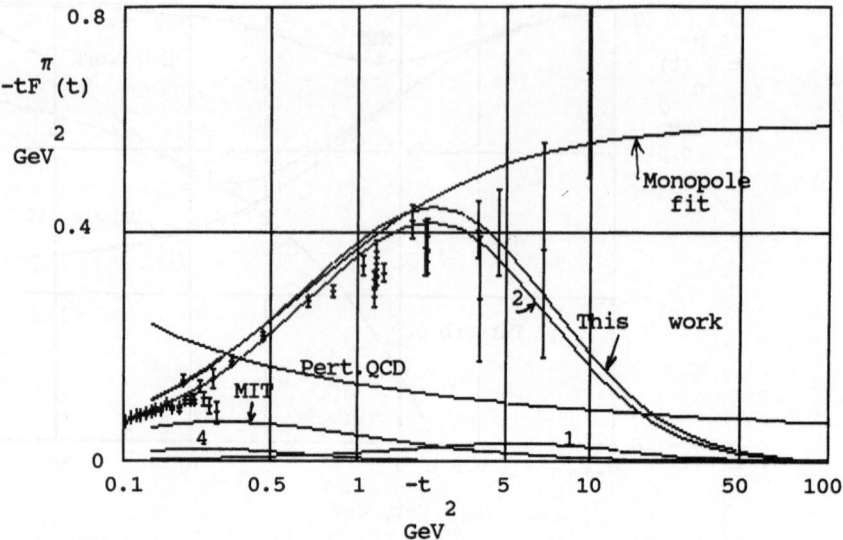

Fig.4. The data [10] for $-tF^{\pi}(t)$ in comparison with predictions of this work, VDM model [11] and perturbative QCD [8].

4. CONCLUSION

So, within proposing picture the value $R_{clust} \approx 0.2$ fm is in quantitatively agreement with the recent data for elastic e-p and e-π scattering. And therefore, one should expect a dominant excitation of the baryon-meson-like degrees of freedom only in all nuclear phenomena up to $|t| \approx 10$ GeV^2.

REFERENCES

1) A. Chodos, R.L. Jaffe, K. Johnson et al. Phys. Rev. **D9**, 3471 (1974)

2) Yu.E.Pokrovsky, Sov. J. Nucl. Phys. **50**, 907 (1989)

3) Yu.E.Pokrovsky, Sov. J. Nucl. Phys. **40**, 670 (1984)

4) Yu.E. Pokrovsky, Proc. XXIX Int. Winter. Meeting on Nuclear Physics. Edited by I.Iori. Bormio (Italy) 14-19 January 1991. UNIVERSITA DEGLI STUDI DI MILANO, **83**, 338 (1991)

5) Yu.E.Pokrovsky, Nucl. Phys. **A525**, 673 (1991)

6) D.P. Stanley and D. Robson, Phys. Rev. **D26**, 223 (1982)

7) SLAC Collab.: R.G. Arnold et al., Phys. Rev. Lett. **57**, 174 (1986)

8) V.L. Chernyak, A.R. Zhitnitsky, Pisma v ZhETF. **25**, 544 (1977)
    G.P. Leparge, S.J. Brodsky, Phys. Rev. **D22**, 2157 (1980)

9) S. Rock, R.G. Arnold et al., Phys. Rev. Lett., **49**, 1139 (1982)

10) C.J. Bebek et al., Phys. Rev. **D17**, 1693 (1978)

11) M. Gell-Mann and F. Zachariazen, Phys. Rev. **124**, 953 (1961)

Few-Body Systems, Suppl. 6, 89—94 (1992)

HYPERON STRUCTURE IN THE SOLITON APPROACH

N.N. Scoccola

Niels Bohr Institute,

Blegdamsvej 17, DK-2100 Copenhagen, Denmark

ABSTRACT

The structure of hyperons is studied within the context of the bound state soliton model. In this approach strange hyperons are described as bound states of an $SU(2)$ topological soliton and kaons. Masses and electromagnetic properties of the strange baryons are calculated in a version of the model where chiral symmetry breaking effects due to the different empirical values of the pion and kaon decay constants are considered. The possibility of extending this scheme to the description of other massive-quark baryons like charmed hyperons is also discussed.

INTRODUCTION

The main difficulty in describing the structure of the hyperons within the soliton approach is the existence of explicit chiral symmetry breaking terms associated with the massive flavor degrees of freedom. In the bound state approach [1] this problem is treated by considering isospin and flavored fluctuations around the soliton in a different way. Hyperons are then composed by a flavored pseudoscalar doublet (i.e. kaons) bound to an $SU(2)$ soliton. Although the model was originally applied to the study of strange hyperons [2] it has recently been shown [3] that the properties of hyperons with heavier flavors (i.e. charm) are equally well described by it. In this contribution we discuss some of the hyperon properties as described by the bound state approach.

THE BOUND-STATE MODEL OF HYPERONS

We start with the effective action for the simple Skyrme model with an appropriate symmetry breaking, expressed in terms of the $SU(3)$-valued chiral field $U(x)$ as

$$\Gamma = \int d^4x \left\{ \frac{F_\pi^2}{16} \text{Tr} \left[\partial_\mu U \partial^\mu U^\dagger \right] + \frac{1}{32e^2} \text{Tr} \left[[U^\dagger \partial_\mu U, U^\dagger \partial_\nu U]^2 \right] \right\} + \Gamma_{WZ} + \Gamma_{SB}, \qquad (1)$$

where F_π is the pion decay constant ($= 186 \ MeV$ empirically) and e is the so–called Skyrme parameter. In Eq.(1), Γ_{WZ} is the Wess-Zumino action and Γ_{SB} is responsible for the explicit symmetry breaking of chiral symmetry. We use the following form for Γ_{SB}

$$\Gamma_{SB} = \int d^4x \left\{ \frac{F_\pi^2 m_\pi^2 + 2F_K^2 m_K^2}{48} \mathrm{Tr}\left[U + U^\dagger - 2\right] + \frac{F_\pi^2 m_\pi^2 - F_K^2 m_K^2}{24} \mathrm{Tr}\left[\sqrt{3}\lambda^8 \left(U + U^\dagger\right)\right] \right.$$
$$\left. - \frac{F_K^2 - F_\pi^2}{48} \mathrm{Tr}\left[\left(1 - \sqrt{3}\lambda^8\right)\left(U\partial_\mu U^\dagger \partial^\mu U + U^\dagger \partial_\mu U \partial^\mu U^\dagger\right)\right] \right\} \tag{2}$$

where λ^8 is the eighth Gell-Mann matrix and m_π and m_K represent the pion and kaon masses respectively. Eq.(2) accounts not only for the finite mass of the pseudoscalar mesons but also for the empirical difference between their decay constants. In previous calculations [2] the kaon was found to be overbound to the soliton. It was recently shown [4] that this defect can be mostly eliminated if the difference in the decay constants is properly taken into account as done in Eq.(2).

We continue by introducing the Callan-Klebanov (CK) ansatz for the chiral field [1]

$$U_{CK} = \begin{pmatrix} \exp\left[\frac{i}{2}\vec{\tau}\cdot\hat{r}\theta(r)\right] & 0 \\ 0 & 1 \end{pmatrix} \exp\left[\frac{i2\sqrt{2}}{F_\pi}\begin{pmatrix} 0 & K \\ K^\dagger & 0 \end{pmatrix}\right] \begin{pmatrix} \exp\left[\frac{i}{2}\vec{\tau}\cdot\hat{r}\theta(r)\right] & 0 \\ 0 & 1 \end{pmatrix} \tag{3}$$

Inserting Eq.(3) into Eq.(1) and expanding to second order in $K = \frac{1}{\chi_1}\begin{pmatrix} K^+ \\ K^0 \end{pmatrix}$, we can obtain the Lagrangian density of the soliton-kaon system. The meson decay constant ratio $\chi_1 = F_K/F_\pi$ is introduced in order to recover the canonical form of the free kaon Lagrangian when the interaction with the soliton is turned off. We determine the soliton profile $\theta(r)$ by minimizing the soliton energy. Then we proceed to solve the eigenvalue equation of kaons moving in the background potential provided by the soliton. This determines the kaon energy ω which is of $O(N_c^0)$ in N_c counting and its wavefunction $k(r)$. Finally, to obtain the (hyperfine) splitting between states with same strange quantum number but different spin-isospin quantum numbers, the soliton has to be rotated in the $SU(2)$ isospace. This provides the $O(1/N_c)$ contribution to the mass. Details of this procedure can be found in Ref.[2].

As proposed first in Ref.[3], charmed baryons can be described in the present model by introducing the D-meson doublet $D = \frac{1}{\chi_2}\begin{pmatrix} \bar{D}^0 \\ D^- \end{pmatrix}$ in analogy with the kaon doublet. Here $\chi_2 = F_D/F_\pi$. Since interactions between the K and D can be ignored within the quadratic approximation, the equations of motion for the K and D are formally identical, the only differences being in the meson masses and the constants χ's. We take the experimental meson masses, $m_K = 495 \ MeV$ and $m_D = 1867 \ MeV$. The values for the meson decay constant ratios χ_i will be given later. The mass formula for the baryons is

$$M(I, J, n_1, n_2, J_1, J_2, J_m) = M_{\mathrm{sol}} + n_1\omega_1 + n_2\omega_2 + M_{\mathrm{rot}}, \tag{4}$$

$$M_{\mathrm{rot}} = \frac{1}{2\mathcal{I}}\left\{ I(I+1) + (c_1 - c_2)[c_1 J_1(J_1 + 1) - c_2 J_2(J_2 + 1)] + c_1 c_2 J_m(J_m + 1) \right.$$
$$\left. + \left[J(J+1) - J_m(J_m+1) - I(I+1)\right]\left[\frac{c_1 + c_2}{2} + \frac{c_1 - c_2}{2}\frac{J_1(J_1+1) - J_2(J_2+1)}{J_m(J_m+1)}\right]\right\}, \tag{5}$$

where M_{sol} is the soliton mass and $\mathcal{I}$ the $SU(2)$ moment of inertia. Here n_1 is the absolute value of strangeness, n_2 the charm quantum number and ω_1 and ω_2 are, respectively, the bound-state energies of the K and D mesons. In addition, c_1 is the hyperfine splitting constant corresponding to K and c_2 the one corresponding to D. The angular momenta J_1 and J_2 are defined as $J_i = n_i j_i$ with j_i being the angular momentum of the bound state orbital and J_m is given by $J_m = J_1 + J_2, \cdots, \mid J_1 - J_2 \mid$. The total angular momentum $\vec{J}$ is the sum of the the rotor spin $\vec{R}$ and the total meson spin $\vec{J}_m$. The quantum numbers of the physical hyperons can be obtained by using the quantization rules described in Ref.[5].

THE HYPERON SPECTRUM

In Table 1 we show the masses of the low-lying positive parity hyperons predicted by our model in comparison with the available experimental data [6] and with the quark model predictions of Refs.[7, 8]. In the soliton model these states are obtained by populating the $K-$ and/or $D-$meson ground states (which have quantum numbers $l = 1$, $j = \frac{1}{2}$).

Table 1
Masses (in MeV) of the low-lying positive parity strange and charmed hyperons.

Particle	Exp.	SET I	SET II	Ref.[7]	Ref.[8]
Λ	1116	1106	1086		
Σ	1193	1203	1205		
Σ^*	1385	1350	1320		
Ξ	1318	1332	1311		
Ξ^*	1530	1480	1425		
Ω	1672	1621	1549		
Λ_c	2285	2172	2209	2200	2260
Σ_c	2453	2327	2379	2360	2440
Σ_c^*	?	2387	2417	2420	2510
Ξ_{cc}	?	3513	3601	3550	
Ξ_{cc}^*	?	3574	3639	3610	
Ω_{ccc}	?	4791	4898	4810	
Ξ_c	2470	2381	2426	2420	2480
Ξ_c'	?	2509	2514	2523	2575
Ξ_c^*	?	2524	2539	2531	2645
Ω_c	2740	2643	2647	2680	2730
Ω_c^*	?	2674	2662	2720	2790
Ω_{cc}	?	3700	3764	3730	
Ω_{cc}^*	?	3730	3778	3770	

In our model we have considered two sets of parameters. SET I corresponds to

$$m_\pi = 0, \quad F_\pi = 129\ MeV, \quad e = 5.45, \quad \chi_1 = 1.22, \quad \chi_2 = 1.8,$$

while SET II corresponds to

$$m_\pi = 138\ MeV, \quad F_\pi = 108\ MeV, \quad e = 4.84, \quad \chi_1 = 1.22, \quad \chi_2 = 2.0.$$

In SET I we consider the chiral limit in the $SU(2)$ sector while in SET II we consider the case of massive pions. In both cases we fit F_π and e to reproduce the empirical nucleon and $\Delta_{3,3}$ masses and set χ_1 to its empirical value. On the other hand the empirical value of χ_2 is not very well established. In SET I we choose $\chi_2 = 1.8$ that falls well within the range given in Ref.[9], while in SET II we use a slightly larger value in order to get a better D-meson binding energy.

We observe that our model, with or without pion mass, works very well for both strange and charmed baryons. Moreover the similarity between our predictions and those of the quark model is striking. Both models predict the same number of low-lying states as well as the same energy ordering. As mentioned in Ref.[3], our model even predicts a certain number of mass formulae that agree with those of quark model.

Hyperon resonances can also be described in our model. In the strange sector there is only one excited bound state which has $l = 0$, $j = \frac{1}{2}$. It provides a natural description of the odd parity resonance $\Lambda(1405)$. Our predictions for the mass of this state are around 100 MeV too low (1325 MeV for SET I and 1297 MeV for SET II). This situation contrasts with typical quark model calculations where this state appears too high in energy. On the other hand our model predicts a far richer spectrum for the charmed hyperons. The D-meson has bound states in all partial waves with $l \leq 3$ and additional excited states in the P and S-waves. As discussed in Ref.[3] the spectra single D-meson (Λ_c, Σ_c) states are in good qualitative agreement with quark-model predictions [8]. Like in the strange sector however negative-parity states are too low as compared with the quark model calculations.

THE HYPERON MAGNETIC MOMENTS

In this section we will concentrate on the magnetic moment of the hyperons. This provides a further check of the validity of our approach. Given the electromagnetic current derived from our effective Lagrangian by means of Noether's theorem, one can readily obtain the magnetic moment operator [10]. To calculate the matrix elements of this operator we use the hyperon wave functions written in terms of the rotor and bound meson wave functions. In this way we obtain the magnetic moment of the different hyperons.

Before presenting our results it is worthwhile noting that soliton models with Skyrme term stabilization are known to predict a somewhat small value for the proton magnetic moment [11] ($\mu_p = 1.88$ using SET I and $\mu_p = 1.97$ using SET II to be compared with the empirical value $\mu_p^{emp} = 2.79$). On the other hand, the ratio μ_n/μ_p is predicted quite accurately. As we shall see, a similar situation occurs for the hyperon magnetic moments.

In Table 2 we show the values of the magnetic moments of the spin 1/2 strange and charmed baryons as obtained in the bound state approach with the two sets of parameters given in the previous section. Also listed in Table 2 are the available empirical results [6] and the quark model and bag model predictions of Ref.[12] and Ref.[13] respectively. All the values are given relative to the proton magnetic moment. Similar results are obtained for the spin 3/2 hyperons magnetic moment [14]. As in the case of the hyperon masses the agreement of our predicted magnetic moments with experiments and quark model predictions is quite remarkable. The discrepancies appearing in the magnetic moments of the non-zero isospin hyperons are due to the rather small value of the meson contribution to the isovector meson magnetic moment. The explicit incorporation of vector mesons in the effective action is expected to increase such a contribution.

Table 2
Magnetic moments of the spin 1/2 low-lying strange and charmed hyperons.

Particle	Exp.	SET I	SET II	Quark Model	Bag Model
Λ^0	−0.22	−0.21	−0.27	−0.21	−0.25
Σ^+	0.87	1.07	1.10	0.96	0.97
Σ^0	−	0.27	0.34	0.29	0.31
Σ^-	−0.41	−0.54	−0.42	−0.38	−0.36
Ξ^0	−0.45	−0.58	−0.66	−0.50	−0.56
Ξ^-	−0.24	−0.07	−0.19	−0.16	−0.23
Λ_c^+	−	0.11	0.10	0.13	0.18
Σ_c^{++}	−	0.98	0.99	0.85	0.70
Σ_c^+	−	0.16	0.21	0.18	0.13
Σ_c^0	−	−0.65	−0.56	−0.49	−0.44
Ξ_{cc}^{++}	−	−0.17	−0.17	−0.04	0.06
Ξ_{cc}^+	−	0.35	0.32	0.29	0.31
Ξ_c^+	−	0.11	0.10	0.13	0.18
Ξ_c^0	−	0.11	0.10	0.13	0.18
$\Xi_c'^+$	−	0.44	0.39	0.26	0.17
$\Xi_c'^0$	−	−0.59	−0.57	−0.41	−0.39
Ω_c^0	−	−0.31	−0.40	−0.32	−0.35
Ω_{cc}^+	−	0.21	0.23	0.25	0.30

CONCLUSIONS

We have shown that the masses and magnetic moments of strange and charmed hyperons are well described in the bound meson-soliton model. It is particularly noteworthy that with only two parameters needed for light-quark (up and down) systems and masses and decay constants taken from empirical sources, the model is able to fit not only the masses but also the magnetic moments of strange and charmed baryons. The agreement with experiments and quark-model predictions is quite remarkable and suggests strongly that the model is close to nature in its physics content. At first sight, this is surprising since the skyrmion model looks so different from the quark description. The crucial feature of the model is that a massive scalar doublet carrying the flavor quantum number of the massive quark gets bound to and wrapped by the $SU(2)$ soliton, the quantum numbers arising through topologically induced transmutation. As discussed in Ref.[15], the essential dynamics can be understood in terms of a hierarchy of induced gauge (Berry) connections generated in integrating out layers of length scales. Finally, it should be noted that in order to obtain realistic predictions for the hyperon energies it is crucial to use different values for the decay constants of the different pseudoscalar mesons.

References

[1] Callan, C.G., Klebanov, I.: Nucl. Phys. **B262**, 365 (1985)

[2] Scoccola, N.N., Nadeau, H., Nowak, M., Rho, M.: Phys. Lett. **B201**, 425 (1988); Callan, C.G., Hornbostel, K., Klebanov, I.: Phys. Lett. **B202**, 269 (1988); Blom, U., Dannbom, K., Riska, D.O.: Nucl. Phys. **A493**, 384 (1989); Scoccola, N.N., Min, D.P., Nadeau, H., Rho M.: Nucl. Phys. **A505**, 497 (1989)

[3] Rho, M., Riska, D.O., Scoccola, N.N.: Phys. Lett. **B251**, 597 (1990); Rho, M., Riska, D.O., Scoccola, N.N.: Z.Phys. **A**, in press

[4] Riska, D.O., Scoccola, N.N.: Phys. Lett. **B265**, 188 (1991)

[5] Scoccola, N.N., Wirzba, A.: Phys.Lett. **B258**, 451 (1991)

[6] Particle Data Group: Phys. Lett. **B239**, 1 (1990)

[7] De Rújula, A., Georgi, H., Glashow, S.L.: Phys. Rev. **D12**, 147 (1975)

[8] Copley, L.A., Isgur, N., Karl, G.: Phys. Rev. **D20**, 768 (1979); Maltman, K., Isgur, N.: Phys. Rev. **D22**, 1701 (1980); Capstick, S., Isgur, N.: Phys. Rev. **D34**, 2809 (1986)

[9] Dominguez, C.A., Paver, N.: Phys. Lett. **B197**, 423 (1987)

[10] Nyman, E., Riska, D.O.: Nucl. Phys. **B325**, 593 (1989); Kunz, J., Mulders, P.J.: Phys. Lett. **B231**, 335 (1989); Min, D.P., Koh, Y.S., Oh, Y., Lee, H.K.: Nucl. Phys. **A530**, 698 (1991)

[11] Adkins, G.S., Nappi, C.R., Witten, E.: Nucl. Phys. **B228**, 552 (1983); Adkins, G.S., Nappi, C.R.: Nucl. Phys. **B233**, 109 (1984)

[12] Lichtenberg, D.B.: Phys. Rev. **D15**, 345 (1977)

[13] DeGrand, T., Jaffe, R.L., Johnson, K., Kiskis, J.: Phys. Rev. **D12**, 2060 (1975); Bose, S.K., Singh, L.P.: Phys. Rev. **D22**, 773 (1980)

[14] Oh, Y., Min, D.P., Rho, M., Scoccola, N.N.: Nucl. Phys. **A**, in press

[15] Rho, M.: Modern Phys. Lett. **A**, in press

Few-Body Systems, Suppl. 6, 95—104 (1992)

ELECTROPRODUCTION OF STRANGENESS IN THE NUCLEON

S. Frullani
Physics Laboratory - Istituto Superiore di Sanità and
Sezione Sanità - I.N.F.N.- 00161 Rome. Italy

M. Sotona
Nuclear Physics Institute, Czechoslovak Academy of Sciences
25068 Rez. Czechoslovakia

Abstract

In the light of the interest raised by the strangeness content of the proton, some questions posed by the theoretical description of kaon photo- and electro-production and by the available data are addressed.

1. Introduction

Recently combined experimental and theoretical investigations seem to indicate that even when the nucleon is studied with small momentum transfer probe (large distances in the nucleon are averaged) its structure cannot be understood without requiring a substantial presence of strange sea quark pairs.

Main indications come from:

- large σ-term of the πN scattering [1] gives a sizeable value [2] of the matrix element of the scalar operator $\bar{s}s$ between proton states $<p\,|\bar{s}s\,|p>$

- analyses of EMC measurements [3] of spin dependent structure functions as well as elastic neutrino proton scattering [4] suggest non vanishing strange matrix element of the axial current $<p\,|\bar{s}\gamma_\mu\gamma_5 s\,|p>$ [5]

- analyses of the electromagnetic form factor of the proton [6] indicate that the pole fit requires a contribution at a mass compatible with the ϕ meson, suggesting a large ϕNN vector coupling due to a coupling of the $\bar{s}s$ quark state to NN; theoretical evaluation of baryons magnetic moments gives better agreement with experimental values when a strange quark content different from that predicted by a naive quark model [7]; these arguments imply non negligible value for the strange vector current matrix element $<p \,|\bar{s}\gamma_\mu s\,|p>$ [8].

The usual description of the origin of the strange sea is the perturbative gluon splitting or flux tube breaking but the relatively large component of strange quarks that seems to be required to explain even static properties or low energy, low transfer momentum data could imply that the strange sea is intrinsic to the nucleon bound state equation, originating by a non-perturbative phenomena . In this last case one can regard virtual $K-\Lambda$ or other strange meson-strange baryon pair states as fluctuations of the proton ground state [9].

Reactions in which a non-strange probe induces an associated strangeness production could certainly help in a better understanding of the strange quark content if a theoretical description of the reaction is made at fundamental level, taking into account the direct interaction with the quark substructures. In this context the electromagnetic induced reactions add the usual characteristic of a clean probe and the $(e,e'K^+)$ reaction can be regarded as a very promising tool to investigate the strangeness content of the proton.

2. Virtual photon strangeness production

The associated strangeness production process on a proton (quadri-momentum p_p ($\mathbf{p}_p,E_p$)) is obtained through the absorption of a virtual photon with momentum q ($\mathbf{q},\omega$) that gives as final state a meson K^+ with momentum p_K ($\mathbf{p}_K,E_K$) and a neutral hyperon (Λ^0 , Σ^0) with momentum p_Y ($\mathbf{p}_Y,E_Y$). Important kinematical variables are $s = W^2 = (q + p_p)^2 = (p_K + p_Y)^2$ and $t = (q - p_K)^2 = (p_Y - p_p)^2$ being the latter the squared momentum transfer from the photon to the hyperon and the former the squared invariant mass of the initial and final state. The W dependence of the production cross section gives account of the threshold behavior and of the resonant or non-resonant characteristic of the process. The t dependence characterizes completely, if interaction effects for the kaon in the final state can be neglected, the proton to hyperon transition form factors while a dependence on both $\mathbf{q}$ and $\mathbf{p}_K$ must be restored if a distorted wave for the kaon has to be used as could be the case for the production on a proton bound in a nucleus.

Electroproduction cross section for the reaction in which in the final state the scattered electron and the produced kaon are detected is given [10] by :

$$\frac{d^4\sigma}{dq^2 \, ds \, dt \, d\alpha} = \Gamma \, 2\pi \, \frac{d^2\sigma}{dt \, d\alpha} \tag{1}$$

where Γ is the virtual photon flux and the single photon production is

$$2\pi \frac{d^2\sigma}{dt\,d\alpha} = \frac{d\sigma_T}{dt} + \varepsilon \frac{d\sigma_L}{dt} + \varepsilon \frac{d\sigma_{TT}}{dt} \cos 2\alpha + \sqrt{2\varepsilon(\varepsilon+1)}\ \frac{d\sigma_{LT}}{dt} \cos \alpha \qquad (2)$$

with ε being the virtual photon polarization parameter and α the angle between the scattering electron plane and the kaon production plane with the tri-momentum transfer q belonging to both planes. Γ and ε depend only on the electron vertex kinematics. The different terms in (2), that can be separated by measuring the ε and α dependence of the cross section, are :

- $\sigma_T = 1/2\ (\sigma_{\parallel} + \sigma_{\perp})$ the contribution from purely transverse photon and is the only part occurring in the photoproduction by real unpolarized photons;
- σ_L the contribution from longitudinally polarized photons;
- $\sigma_{TT} = 1/2\ (\sigma_{\parallel} - \sigma_{\perp})$ the interference term of the transversely polarized photons;
- σ_{LT} the interference of transversely and longitudinally polarized photons.

$\sigma_{\parallel}$ and $\sigma_{\perp}$ are the cross sections for photons having the electric vector parallel and perpendicular to the hyperon production plane, respectively.

More terms should be considered in the cross section if the reaction is studied with a polarized electron beam and using a polarized proton target or measuring the final hyperon polarization.

The four contributions to the unpolarized scattering and the recoil polarization as well as the polarized target response functions are combinations of bilinear products of independent helicity amplitudes, which, in the case of electroproduction of a pseudoscalar meson with initial and final baryon having $J = 1/2$, are limited to only six (four in the case of photoproduction) by the requirements of Lorentz covariance, parity conservation and gauge invariance. Then the amplitude for kaon electroproduction using the Pauli spinor space can be expressed in the CGLN [11] form :

$$F = \sum_{i=1}^{6} F_i\ \chi^+_{s_2}\ \Omega_i\ \chi_{s_1} \qquad (3)$$

where the six spin-space operators are

$$\Omega_1 = \sigma \cdot \hat{e} \qquad\qquad \Omega_2 = i\ (\sigma \cdot \hat{p}_K{}^c)\ (\sigma \cdot (\hat{e} \cdot \hat{q}^c)) \qquad \Omega_3 = (\sigma \cdot \hat{q}^c)(\hat{p}_K{}^c \cdot \hat{e})$$

$$\Omega_4 = (\sigma \cdot \hat{p}_K{}^c)(\hat{p}_K{}^c \cdot \hat{e}) \quad \Omega_5 = (\sigma \cdot \hat{q}^c)(\hat{q}^c \cdot \hat{e}) \qquad\qquad \Omega_6 = (\sigma \cdot \hat{p}_K{}^c)(\hat{q}^c \cdot \hat{e})$$

where σ is the proton spin, ε the unit photon polarization vector, the ^{c} refers to CM momenta, the $^\wedge$ indicates unit vector and the F_i are the six independent amplitudes.

Given a dynamical model of the electroproduction, the amplitudes are deduced. In principle they could be derived from QCD starting with current quarks and involving explicit contributions from strange quark pair production but we are not aware of an effort of this kind, while several representations of the process exist in the literature that use different kinds of Feynmam diagrams involving "effective" hadronic field theories such as quantum hadrodynamics. This approach makes less clear the possibility to address, with the experimental study of the process, the question of the strangeness content of the proton. The theory assumes the existence of strong coupling parameters which can be phenomenologically determined from the available photo- and electroproduction data. Some models [12] impose crossing symmetry and unitarity constraints asking the simultaneous reproduction of the crossing related reactions such as kaon photoproduction p (γ,K$^+$) Λ and radiative capture p (K$^-$,γ) Λ that are described by the same transition amplitude evaluated in different kinematic regions of the S matrix.

3. Theoretical models and available data

In 1966 Thom [13] demonstrated that a realistic description of kaon photoproduction in the 1. - 1.4 GeV interval can be obtained from five leading Feynman diagrams corresponding to s and u channel exchange baryons (p, Λ, Σ^0) and t channel exchange mesons (K and K*(892)) (see Fig. 1 (a)-(e)).

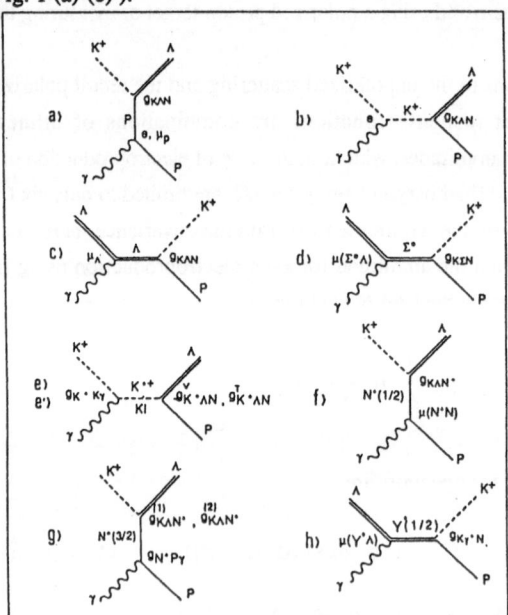

Fig.1 Feynman diagrams considered in photo- electro-production. K*, K1 are vector, pseudovector mesons; N*(1/2), N*(3/2) are nucleon resonances; Y*(1/2), I = 0,1 are hyperon resonances.

Thom's fit to data depended strongly on the strength of the KNΛ coupling constant, which he gave as -1.1> g$_{KN\Lambda}$/ (4π)$^{1/2}$> -2.6. This value is considerably smaller than the coupling deduced from purely hadronic processes and from the value expected from SU(3) symmetry. Dover and Walker [14] have deduced a value of ~-4.7 from a dispersion relation analysis of $K^{\pm}p$ forward amplitudes [15], and other determinations are -3.7 [16] and -4.13 [17]. The unbroken SU(3) symmetry gives a relation between $g_{KN\Lambda}$ and $g_{\pi NN}$ coupling constants :

$$g_{KN\Lambda} = - \frac{(1+2\alpha_F)}{\sqrt{3}} \, g_{\pi NN}$$

where $\alpha_F = F/(F+D)$ is the fraction of F-type coupling for the pseudoscalar meson nonet. Taking into account the values of $g_{\pi NN}$ and α_F together with the validity level of the known broken SU(3) symmetry, the constraints imposed by SU(3) on the coupling constant value range are -4.71÷ -4.16 [18] , -4.4 ÷ -3.0 [19].

After Thom's analysis a variety of models have been considered [12, 19-21] introducing more diagrams in the dynamic of the process taking into account excited states in the direct s channel (spin 1/2 and 3/2 nucleon resonances Fig. 1 (f-g)) and in the cross t (kaon resonances Fig. 1 e') and u (hyperon resonances Fig. 1 h) channels. Many models still give coupling constant values in disagreement with hadronic process and SU(3) estimations. This discrepancy can be at least partly resolved by inclusion of t channel K1(1270) resonance [19, 21]. The interference of strong K1 contribution with the Born term lowers $g_{KN\Lambda}$ but only at a price of very strong K1ΛN coupling. Adelseck and Saghai [19] have shown in a re-analysis of the photo-production data that, taking into account systematic errors and disregarding internally inconsistent data, a good fit is obtained using SU(3) consistent coupling constant values.

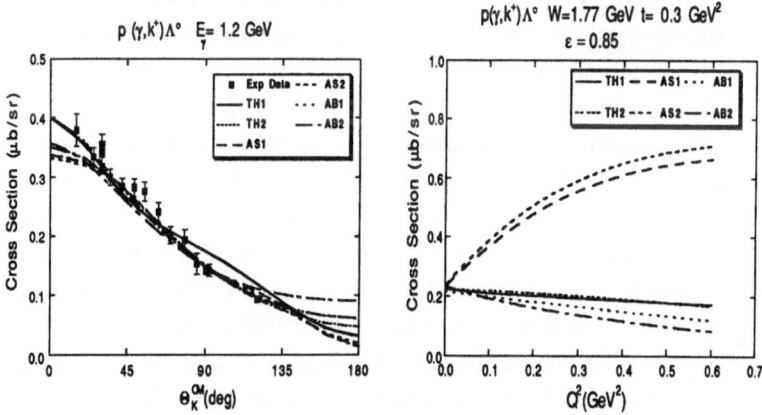

Fig.2 a) Fit of several models to photo-production data. b) Electro-production prediction of the same models. TH1,TH2 [13]; AS1,AS2 [19]; AB1,AB2 [20].

In general, phenomenological χ^2 parameter fits are not unique due to a large number of coupling constants which are usually not linearly independent. The presence of many local minima implies that different parametrizations can be found that fit the data equally well when a restricted set of kinematical conditions are considered. In Fig.2a the fit of different models to a set of data on photoproduction process is reported. All the models can be said to be consistent with the data. When in the same excitation region we consider the case of electroproduction (Fig. 2b) computations show much more capability to discriminate among different models.

Data on Λ electroproduction date back to the seventies. After the pioneering work at CEA [22], activity has been carried out at Cornell [23] and Desy [24]. The current experimental knowledge, coming from data taken with low duty cycle and low intensity beam, is unsatisfactory . Sparse data exist and in order to make comparisons with models and predictions they are extrapolated to a common W value. No systematic separation of the different contributions has been attempted. For this reasons contradictory statements are found in literature, for instance on the importance of the longitudinal part to the cross section. This item is of particular interest because if an important contribution of longitudinal component is shown, this means particular importance of t-channels kaon exchange Feynman graphs 1b and 1e-e', from the separate measurement of this component the kaon form factor can be deduced. Recently, Ji and Cotanch [25] have shown that the kaon form factor, measured in an interval region accessible at CEBAF up to a momentum transfer of 2 Gev^2, can be used to discriminate among different predictions based on quark model and among the results expected by a vector-meson dominance model of the reaction.

Much less is known on Σ^0 production. Bennhold [26] performed a least-squared fit to the available Σ^0 data with the Born terms. To improve the rather large χ^2 he allowed also for one Δ resonance to contribute to the fit. The inclusion of the $\Delta_{33}(1236)$, known to be very important in pio photo- and electro-production, reduced the χ^2 only slightly. Only the s-channel $\Delta(1700)$ with spin $3/2^-$ was able to significantly improve the fit to the data. Williams, Ji and Cotanch [27] do not include in their model the parameters associated with t-channel (K*,K1) exchanges and were able to simultaneously describe all three Λ,Σ^0 and $\Lambda(1405)$ reaction channels in kaon photoproduction p (γ,K^+) Y and the crossing related radiative capture p (K^-,γ) Y. The energy dependence of the Σ^0 cross section suggests an important resonance contribution near W = 1.9 GeV. This energy behaviour can only be reproduced, in part, by the model if the $\Delta(1620)$ $(J=1/2^-)$ and $\Delta(1910)$ $(J=1/2^+)$ isobars are included in the fit. Since isospin conservation forbids Δ contributions to Λ and $\Lambda(1405)$ production, Σ^0 production appears to offer the only opportunity to study their effects. From their model they deduce an improved value for the ratio of $\Lambda(1450)$ electromagnetic transition moments R = $\mu_{\Lambda(1405)\Lambda}/\mu_{\Lambda(1405)\Sigma^0}$ = - 0.21 $\pm$ 0.08. This small, negative ratio violates the SU(3) prediction R = + $\sqrt{3}$ severely, broken flavor SU(3) predicts an even larger ratio. This implies that the $\Lambda(1405)$ may not be the SU(3) singlet ground state and, due to absence of other Σ^* or N^* with similar mass and parity, it cannot be included in a SU(3) octet. These results support the conjecture that the $\Lambda(1405)$ is not a pure three quarks system, but is a more complicated structure. The large sensitivity of the $\Lambda(1405)$ photo- and electro-production to the $KN\Lambda(1405)$ coupling constant indicates that this parameter can be accurately determinated and that the other problems connected with this hyperon will be much better settled when more extensive data become available.

4. Topics more connected with quarks and QCD

Experiments that measured both channels Λ and Σ^0 [23, 24] showed a rapid variation with Q^2 of the Σ^0 relative to Λ yield. For photoproduction ($Q^2=0$) one finds a cross section ratio $\Sigma^0/\Lambda = 1$ whereas for increasing Q^2 the ratio falls quickly even if there is a certain inconsistency in the data as shown in Fig.3. Nachtmann [28] on one side and Cleymans and Close [29] on the other have shown that the ratio should go to 0 as the Bjorken variable $x=Q^2/2M\omega$ approaches 1 for the same reasons that make $F^{en}_2(x)/F^{ep}_2(x) = 1/4$ as $x \rightarrow 1$.

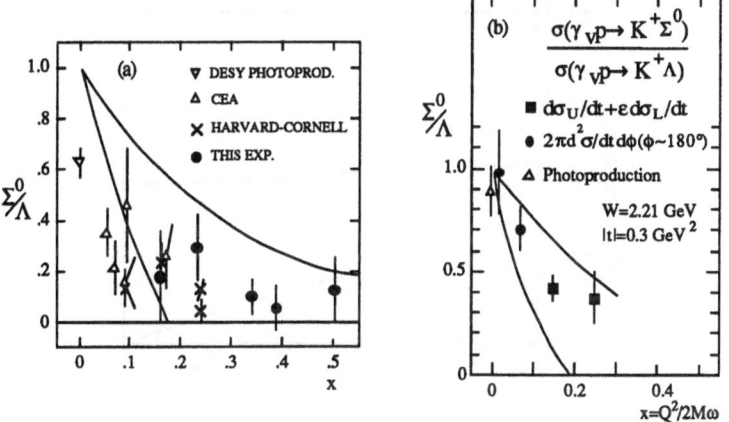

Fig.3 The ratio of the ($K^+ \Sigma^0$) to the ($K^+ \Lambda$) cross section as a function of the Bjorken parameter x. Figures (a) and (b) are taken from ref. 23c and 24b, respectively.

The Pauli principle applied to the colored quarks implies that in the proton the uu pair must be in overall s=1 spin state whereas a ud pair is equally likely to be in s=0 or s=1 state. The chromomagnetic QCD interaction raising the energy of the spin-triplet uu pair shifts the momentum distribution u(x) to higher x and makes the u flavor distribution harder than the d distribution. Then as $x \rightarrow 1$ u(x)>>d(x) and in this limit one sees only u quarks in the proton. Then a virtual photon with a Bjorken parameter near 1 sees the proton as a leading u quark which carries nearly all the momentum and the remainder which is in a state of isospin 0, for this reason the Σ^0 (I=1) production is strongly suppressed in comparision with the Λ (I=0) production. This argument should strictly apply to the transverse cross section only. In fact if the longitudinal cross section dominates, the K exchange term, that in this case is very important, could cause the vanishing of the Σ/Λ ratio due to the big difference in the coupling constants $g_{\Lambda NK} \gg g_{\Sigma NK}$. Available experiments do not separate the transverse and longitudinal cross sections and then one cannot conclude that the parton mechanism is responsible for the observed suppression; a separate determination will greatly improve our understanding of the reaction mechanism and of the proton structure. Apart the expected limit value for the ratio, an important question is also raised by the rapidity of the falling to zero of the Σ/Λ ratio. This is connected with the ratio of probabilities to produce a Σ (Λ) from an $I = 1$ ($I = 0$) state from the pair of ud quarks left as remainder of the direct interaction of the u quark with the virtual photon and could be a function of W. It should be governed by the mechanism to produce a $s\bar{s}$ pair and then to be directly connected with the strangeness content of the proton and with the way in which constituent strange quarks are formed starting from current strange quarks due to the gluon splitting or to the sea. The curves shown in Fig. 3 are the result of an old speculation by Nachtmann [28]. Theoretical and experimental investigations are urgently needed.

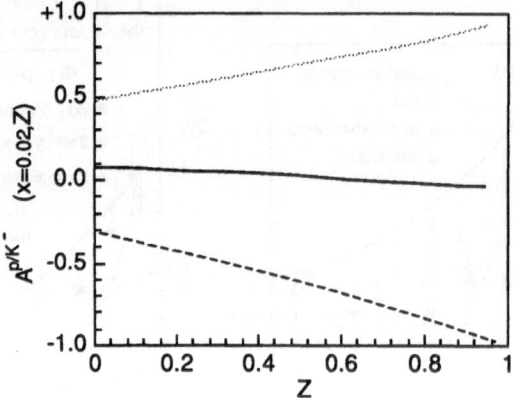

Fig. 4 Asymmetry K^- electroproduction on the proton for x=0.02 as a function of z. The dotted and dashed curves correspond to the maximum positive and negative polarization, respectively. The solid curve is the expected result for zero sea polarisazion.

Semi-exclusive leptoproduction of hadrons in deep inelastic scattering of polarized leptons off polarized nucleons has been proposed by Frankfurt et al. [30] and by Close and Milner [31] to sort out the valence and sea contributions to the quark polarization. Frankfurt and coworkers consider only negative and positive pion production in parallel/antiparallel beam-target polarization. They consider the total pion charge produced, i.e. $N(\pi^+)$-$N(\pi^-)$, and they demonstrate that the measurement of the virtual photon asymmetries for proton and neutron can give estimates of the net polarization of the u and d valence component in the nucleon. Moreover they show that the multiplicities of charged pions can be used to extract the strange sea polarization in the nucleon. Close and Milner compute virtual photon asymmetries for positive and negative pion and kaon production as a function of x and $z = E_{meson}/\omega$. When both x and z are small, the pion asymmetries are very sensitive to the sea polarization while a comparision of asymmetries for positive and negative pions provide information on the u and d polarization. They suggest that a comparision of π^+ and K^+ asymmetries could be a tool to study the relative polarization of strange and non-strange sea. Particularly interesting is the case of K^- (composed by only sea quarks s and $\bar{u}$) to probe the polarization of the sea quarks.

In Fig. 4 the K^- asymmetry for the proton is shown as a function of z. The dotted and dashed curves correspond to maximum positive and negative sea polarization, respectively. Then the kaon production semi-exclusive measurements could also play a fundamental role in future experiments devoted to the comprehension of the nucleon spin puzzle.

5. Conclusions

The interest raised in the scientific community by the problem of the nucleon strangeness content together with the near future availability of a facility under construction (CEBAF) have renewed the interest in Kaon electroproduction studies after more than a decade since the last dedicated experiments. Hyperon resonances, kaon form factor, production of strange quark and its transformation from current to constituent quark are subjects that will be addressed in experiments of the new accelerator generation. Theoretical studies that will develop computation of cross sections starting with basic degrees of freedom are urgently needed to plan future experiments, giving the right advice on the needed experimental accuracy and on the kinematical regions where different models can be discriminate and clean signature of the strange quark behaviour in the proton can be found. The study of the process on the nucleon and its comprehension is mandatory for any quantitative study on the creation and propagation of strangeness in nuclei trough the electroproduction of hypernuclei.

References

1. R.Koch, Z. Phys. C15, 161 (1982)
   J.Gasser,M.E.Sainio and A.Svarc, Nucl. Phys. B307, 779 (1988)
   J.Gasser et al. Phys. Lett. B253, 252 (1991)
2. J.F.Donoghue and C.R.Nappi, Phys. Lett. B168, 105 (1986)
3. J.Ashman et al., Phys. Lett. B206 (1988) 364; Nucl. Phys. B328, 1 (1990)
4. L.A.Ahrens et al., Phys. Rev. D35, 785 (1987)
5. G.Altarelli and C.G.Ross, Phys. Lett. B212, 391 (1988)

104

G.Preparata, P.Ratcliffe and J.Soffer, Phys. Lett. B231, 483 (1989)

R.Decker, T.Leize, M.Nowakowski, Z. Phys. C50, 305 (1991)

6. H.Genz and G.Hohler, Phys. Lett. B61, 389 (1976)

G.Hohler et al., Nucl. Phys. B114, 279 (1976)

7. G.Karl, these Proceedings

8. R.L.Jaffe, Phys. Lett. B229, 275 (1989)

9. S.J.Brodsky, Proceedings of the Topical Conference on Electromagnetic Physics with Internal Targets, 9-12 January, 1989, Stanford, R.G.Arnold Edit., World Scientific, p.3

10. E.Amaldi, S.Fubini and G.Furlan, Pion Electroproduction, Springer-Verlag, Berlin, 1979.

11. G.F.Chew, M.L.Goldberger, F.E.Low and Y.Nambu, Phys. Rev. 106, 1345 (1957)

12. R.Williams, C.R.Ji and S.R.Cotanch, Phys. Rev. D41, 1449 (1990)

13. H.Thom, Phys. Rev. 151, 1322 (1966)

14. C.B.Dover and G.E.Walker, Phys. Rep. 89, 1 (1982)

15. P.Baillon et al., Phys. Lett. B50, 383 (1974)

16. O.Dumbrajs et al., Nucl. Phys. B216, 277 (1983)

17. M.Bozoian et al., Phys. Lett. B122, 138 (1983)

18. M.M.Nagels, T.A.Rijken and J.J.de Swart, Phys. Rev. D12, 744 (1975) ; D15, 2547 (1977); D17, 768 (1978) ; D20, 1633 (1979)

19. R.A.Adelseck and B.Saghai, Phys. Rev. C42, 108 (1990)

20. R.A.Adelseck, C.Benhold and L.E.Wright, Phys. Rev. C32, 1681 (1985)

21. R.A.Adelseck and L.E.Wright, Phys. Rev. C38, 1965(1988)

22. C.N.Brown et al., Phys. Rev. Lett. 28, 1086 (1972)

23. C.J.Bebek et al., Phys. Rev. Lett. 32, 21 (1974) ; Phys. Rev. D15, 594 (1977); Phys. Rev. D15, 3082 (1977)

24. T.Azemoon et al., Nucl. Phys. B95, 77 (1975) ; P.Brauel et al., Z. Physik C3, 101 (1979)

25. C.R.Ji and S.R.Cotanch, Phys. Rev. D41, 2319 (1990)

26. C.Bennhold, Phys. Rev. C39, 1944 (1989)

27. R.A.Williams, C.R.Ji and S.R.Cotanch, Phys. Rev. C43, 452 (1991)

28. O.Nachtmann, Nucl. Phys. B74, 422 (1974)

29. J.Cleymans and F.E.Close, Nucl. Phys. B85, 429 (1975)

30. L.L.Frankfurt et al., Phys. Lett. B230, 141 (1989)

31. F.E.Close and R.G.Milner, Probing the polarized sea by inclusive leptoproduction of hadrons,Oak Ridge National Laboratory Report (1989)

Few-Body Systems, Suppl. 6, 105—116 (1992)

A NEW PARTIAL WAVE ANALYSIS
AND
THE NUCLEON-NUCLEON INTERACTION

J.J. de Swart, R.A.M. Klomp, T.A. Rijken, and V.G.J. Stoks

Institute for Theoretical Physics, University of Nijmegen, the Netherlands
Electronic mail: U634999@HNYKUN11.BITNET

Abstract

We present some recent results of the Nijmegen partial-wave analysis of all NN scattering data below $T_{lab} = 350$ MeV. We compare the predictions of various NN potential models with the NN scattering data and with the phase parameters at 50 MeV.

1. Introduction

For many years, the Nijmegen group has been investigating the baryon-baryon interaction. After the construction of several hard-core nucleon-nucleon and hyperon-nucleon potentials, this culminated in 1978 in the construction of the soft-core NN potential [1]. This was, and still is, one of the best NN potentials presently available. There is a coordinate-space version [1] and a momentum-space version [2], which are totally equivalent; they produce exactly the same phase parameters at all energies. In order to improve this potential, we needed to get both a better understanding of the underlying theory and a better knowledge of the experimental data. For that purpose we started to develop theoretical and computational tools to perform phase-shift (or rather partial-wave) analyses of the NN scattering data. The first major step, the analysis of the pp scattering data, has been published several years ago [3, 4].

At present we have obtained also fits (still preliminary) with the np scattering data and with the combined pp and np scattering data. In this talk we discuss some of the features of our way of analyzing the data. In order to present the results of our analyses, we will discuss some special topics. In Sec. 3 we will discuss our determination of the $NN\pi$ coupling constants. We consider this determination one of the nicest results of our partial-wave analyses. Because we have obtained a good knowledge of the experimental data and because we can compute the experimental observables properly, in Sec. 4 we make comparisons of various NN potential models with the NN data. This is repeated in Sec. 5, where we now compare the potential predictions with the phase parameters at 50 MeV. Our important conclusion is that our value for the mixing parameter ε_1 is in perfect agreement with modern potential predictions.

2. Partial-Wave Analysis

The first main problem in doing partial-wave analyses is: How to calculate the scattering amplitudes in terms of phase shifts (and mixing parameters)? We want to stress that this is a non-trivial problem, because the proper treatment of the electromagnetic part of the interaction and its effect on the phase shifts and the scattering amplitude is very important and at the same time rather complicated. For example, in the case of pp scattering the one-pion-exchange (OPE) amplitude has to be calculated properly, including its distortion by the Coulomb interaction. This means that the analytical expression for the OPE scattering amplitude cannot be used. Here also lie some of the differences between the various partial-wave analyses that have appeared in the literature: In the Nijmegen [4], VPI&SU [5], and Saclay [6] analyses, the higher partial waves and Coulomb distortion effects are treated differently.

The second main problem is: What is a good parametrization for the energy dependence of the phase parameters? This is important when doing multi-energy (m.e.) partial-wave analyses. The basis of any partial-wave analysis (PWA) is a good m.e. analysis. In such a m.e. PWA all experimental data at all energies are used. Such an analysis should be compared with several single-energy (s.e.) PWAs, in which the data in some energy interval are used in determining the phase parameters at one single energy. This energy usually corresponds to the middle of that energy interval; however, this is not strictly necessary. In s.e. PWAs at several energies we determine the phase parameters and their errors. These s.e. values for the phase parameters should scatter statistically around the curve representing the phase parameters as determined in the m.e. PWA.

We think that our analyses, and especially our energy dependence of the phase parameters, is of such quality that the situation sketched above is correct. Therefore, we believe that for the Nijmegen analyses the 'best' value for a particular phase parameter is the value as obtained in the m.e. analysis, rather than the value as obtained in the s.e. analysis.

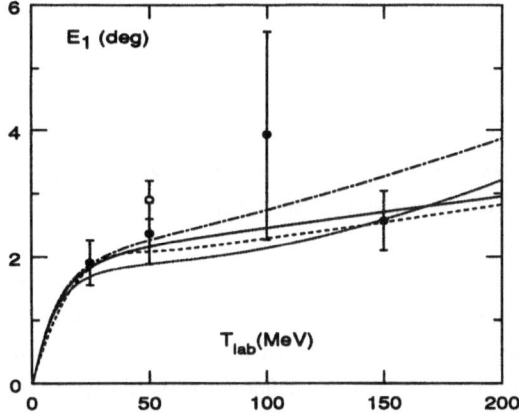

Figure 1: Mixing parameter ε_1 in degrees versus T_{lab} in MeV. •: s.e. result; ○: result Basel group [26]; solid curve: m.e. result; dash-dotted curve: Nijmegen potential; dotted curve: Paris potential; dashed curve: Bonn potential.

As an example, consider the $^3S_1 - {}^3D_1$ mixing parameter ε_1 in np scattering around 100 MeV. In a s.e. analysis at 100 MeV, ε_1 is not accurately known due to the absence of spin-correlation data. This is shown in Fig. 1, where our s.e. results are represented by the black dots. At the adjoining energies at 50 and 150 MeV the available spin-correlation data provide for an accurate determination of ε_1. These data, and the data at lower and higher energies make that ε_1 is fixed rather well in the m.e. analysis, represented by the solid line. This means that also at 100 MeV, ε_1 is in fact much more accurately determined by the data than the s.e. result would suggest. The example demonstrates that a s.e. analysis without an accompanying m.e. analysis is not very useful for making statements with regard to the accuracy with which the phase parameters are determined.

The energy dependence of the lower partial waves in our analyses is determined as follows. In our analysis we divide the interaction into two parts: a long-range part V_L which is well known and essentially model independent, and a short-range part V_S which is treated phenomenologically. The long-range part $V_L = V_{EM} + V_{NUC}$ consists of the complete electromagnetic interaction V_{EM} and the tail of the nuclear potential V_{NUC}. The electromagnetic potential contains the non-static Coulomb potential [7] (including relativistic corrections and two-photon-exchange corrections), the magnetic-moment interaction [8], and the vacuum polarization interaction [9]. The tail of the nuclear potential is dominated by the one-pion-exchange (OPE) potential, but contains also contributions of shorter range due to multi-pion exchange or the exchange of heavier bosons (like ρ, ω, η). For these shorter-range contributions we use the Nijmegen potential [1]: Nijm78. However, because we only need the tail of the potential (outside $r = 1.4$ fm), any decent potential would have sufficed here.

Using this long-range potential V_L, the radial Schrödinger equation

$$(\Delta + k^2)\psi = 2M_{\text{red}}V_L\psi \tag{1}$$

is solved for $r > b = 1.4$ fm. Relativistic effects are taken into account via the potential and by using the relativistic expression for the c.m. energy

$$E = \sqrt{m_1^2 + k^2} + \sqrt{m_2^2 + k^2} - (m_1 + m_2) \,. \tag{2}$$

The presence of the centrifugal barrier makes that the Schrödinger equation need only be solved for a small number of lower partial waves. Still, the equation has to be solved for all energies at which experimental data have been measured. The phenomenology, necessary to describe the short-range interaction is represented by a boundary condition at $r = b$. The boundary condition for each of the lower partial waves is parametrized by a square well potential of range $r = b$. The depth V_S of this potential is independent of r, but is allowed to be energy dependent and different in the different partial waves. It can be shown that V_S is an analytic function of the energy. Inclusion of a certain interaction in the long-range potential tail V_L implies that the corresponding left-hand singularities in V_S are removed. In our analyses the nearest left-hand singularity in V_S is a cut starting at $T_{lab} = -40$ MeV and is due to two-pion exchange. The nearest right-hand cut lies at $T_{lab} = 280$ MeV and is due to inelasticities of the pion production threshold. Fortunately, at energies below 400 MeV inelasticities are still very small, so in our analyses we extend the energy range to be considered to 350 MeV. Because V_S is an analytic function of k^2, regular in the cut plane between $T_{lab} = -40$ MeV and $T_{lab} = 280$ MeV, we can parametrize it conveniently as a power series in k^2. This analytic parametrization guarantees a pretty good energy dependence of our phase parameters.

A major problem in doing np analyses is that the np data base is not rich and accurate enough to determine both $I = 0$ and $I = 1$ partial waves. Therefore, in our analysis the $I = 0$ lower partial waves are searched for, whereas the np $I = 1$ partial waves (except the $^1S_0$ np phase shift) are obtained from the corresponding pp partial waves, after correcting them for Coulomb distortion and mass difference effects, and charge-independence breaking of the pion-nucleon coupling constants [10]. At present (June 1991) our data base contains 4208 NN scattering data (1766 pp scattering data and 2442 np scattering data). A large number of data have been removed from the data base because they are of poor quality (more than 3 standard deviations off). We need 51 parameters to parametrize the lower partial waves up to $J = 4$ and we reach $\chi^2_{min} = 4186.3$ which is less than 1 per data point.

From the fact that we need about 50 parameters to obtain a satisfactory description of the scattering data, we tentatively conclude that any potential will need about the same number of parameters for a good description of the NN scattering data. This explains why recent 'good' potential models (such as Nijm78 [1], Paris80 [11], Bonn87 [12]) only arrive at a χ^2/N_{data} in the order of 2: they use only about 13 parameters. In this context it is interesting to remark that the latest Nijmegen Reidlike pp potential (see Sec. 4), which fits the pp data as well as (and even slightly better than) our m.e. PWA, has exactly the same number of parameters as this PWA.

3. Determination of the $NN\pi$ Coupling Constants

One of the important results of our new partial-wave analysis is the accurate determinations of the various $NN\pi$ coupling constants at the pion pole. Let us sketch the present situation.

In the construction of the various Nijmegen NN potentials [13, 1], the $NN\pi$ coupling constant was determined by fitting to the NN data of pre-1969 using the Livermore phase-shift analyses [14]. In 1975 the rather successful hard-core potential model D used $f^2 = 0.074$, while the soft-core potential Nijm78 used the value $f^2 = 0.077$ at the pion pole. A few years later, in a phase-shift analysis of the low-energy pp scattering data, the tensor combination of the triplet P waves indicated that the $pp\pi^0$ coupling constant should be small. At that time, a value of $f_p^2 \approx 0.075$ was suggested by us [15]. Again, some years later, in a preliminary PWA of the pp scattering data below $T_{\text{lab}} = 350$ MeV [16], we found $f_p^2 = 0.0725(6)$. This preliminary version did not contain the magnetic-moment interaction [8] and it used a much smaller data base than presently available. The newer, updated value [4] is $f_p^2 = 0.0749(7)$. These values are significantly smaller than the at that time accepted value for the charged-pion coupling constant $f_c^2 = 0.079(1)$, as determined from πN scattering [17].

In 1987 it was clear to us that there was a large discrepancy between the value for the $pp\pi^0$ coupling constant as determined from the pp scattering data [16] and the value for the charged-pion coupling constant as determined from the πN scattering data. Because there was no obvious reason to doubt either one of these determinations, it was concluded [16] that there apparently is a large breaking of charge independence in the coupling constants. However, subsequent theoretical model calculations have not been able to explain such a large breaking. The differences were always found to be rather small and in most models the charged-pion coupling was found to be smaller than the $pp\pi^0$ coupling (see, e.g., Refs. [18, 19]). If we are to believe the theoretical model calculations which rule out a large charge-independence breaking, we can only come to the conclusion that the determination of at least one of these two coupling constants should be incorrect.

We are confident of our value for f_p^2 extracted from the pp scattering data. We therefore believe that the previously accepted high value for f_c^2 as determined in πN scattering can no longer be taken for granted. Recent determinations of this coupling constant in the VPI&SU analyses of the πN scattering data by Arndt and co-workers [20] resulted in $f_c^2 = 0.0735(15)$. The value for f_c^2 could also be determined in an analysis of the data on the charge-exchange reaction $\bar{p}p \rightarrow \bar{n}n$ below $p_{\text{lab}} = 950$ MeV/c, which resulted in [21] $f_c^2 = 0.0751(17)$. Both results are within one standard deviation from the value for f_p^2 determined in the Nijmegen pp analysis [4], so large charge-independence-breaking effects need no longer be invoked.

This result is also supported by the preliminary result of our NN partial-wave analysis, which now also includes the np scattering data. Introducing the three relevant coupling constants

$$
\begin{aligned}
f_p^2 &\equiv f_{pp\pi^0} f_{pp\pi^0} && \text{for} && pp \to pp \,, \\
f_0^2 &\equiv -f_{nn\pi^0} f_{pp\pi^0} && \text{for} && np \to np \,, \\
2f_c^2 &\equiv f_{np\pi^-} f_{pn\pi^+} && \text{for} && np \to pn \,.
\end{aligned}
\tag{3}
$$

we find at the pion pole

$$
f_p^2 = 0.0751(6) \,, \qquad f_0^2 = 0.0752(8) \,, \qquad f_c^2 = 0.0741(5) \,,
\tag{4}
$$

which implies a value for the $nn\pi^0$ coupling constant of $f_n^2 = 0.075(2)$. Assuming that charge independence between the coupling constants holds, we have also performed a combined analysis where we use one coupling constant only, i.e., $f^2 \equiv f_p^2 = f_n^2 = f_c^2$. We then find

$$
f^2 = 0.0749(4) \,,
\tag{5}
$$

and $\chi_{\min}^2$ rises with 6.8. Comparing with the result (4) shows that there apparently is no significant charge-independence breaking in the $NN\pi$ coupling constants. This corroborates the results of various theoretical model calculations [18, 19] which find that charge-independence-breaking effects are small.

There are several ways for demonstrating that what we determine is indeed the strength of the OPE potential, where we here mention the possibility to extract the corresponding pion mass. We find $m_{\pi^0} = 135.6 \pm 1.3$ MeV, to be compared with the experimental value [22] of $m_{\pi^0} = 134.9739(6)$ MeV, and $m_{\pi^+} = 139.4 \pm 1.0$ MeV, also in excellent agreement with the experimental value [22] $m_{\pi^+} = 139.5675(4)$ MeV.

4. Quality of NN Potentials

Another result of a complete PWA is that a good knowledge of the data is available. This can then be used to see how well the various potential models fit the experimental data. The s.e. analyses at a number of different energy bins provide us with error matrices for the lower partial-wave phase parameters. Because our combined analysis of the pp and np data is not in a final stage yet, we here focus mainly on the pp analysis. The s.e. analyses provide us with ten error matrices E_n. The error matrix is the inverse of half the second derivative matrix of the χ^2 hypersurface with respect to the phase parameters. Denoting the deviation of some model phase parameters from the s.e. phase parameters in each energy bin by $\mathbf{d}$, the χ^2 of the model can now be written as a sum of the s.e. contributions $\chi_{se,n}^2$ and the contributions from the representation matrices $\chi_{rep,n}^2$, i.e.,

$$
\chi^2(\text{mod}) = \sum_n \left(\chi_{se,n}^2 + \chi_{rep,n}^2 \right) = \chi_{se}^2 + \sum_n \mathbf{d}^T E_n^{-1} \mathbf{d} \,.
\tag{6}
$$

Although the error matrices E_n give a pretty good representation of the χ^2-surface within a certain energy bin, one should note that it is not an exact representation for several reasons. First of all, the higher partial-wave phase parameters and the normalization constants are fixed at their s.e. values. Furthermore, the data have been clustered at some central energy within an energy bin using the results of the m.e. fit, and next to that we have used the approximation that the χ^2 hypersurface is quadratic in the neighborhood of the minimum. The advantage of using this approximate method is that the phase parameters of a potential model to be tested need only be calculated at a small number (10) of energies. This saves a lot of computer time, while the results are more than sufficient for their purpose.

The quality of the representation of our χ^2-surface was tested in two ways. First, we used our m.e. phase parameters as model phase parameters and calculated the corresponding χ^2-contribution. The difference between $\chi^2(\text{mod})$ given by Eq. (6) and the $\chi^2_{\min}$ reached in our m.e. analysis is only 0.35. This means that the χ^2 as calculated directly on the data and the χ^2 calculated via Eq. (6) only differ by 0.02%. This shows that the approximation that the χ^2 hypersurface of the s.e. analyses is quadratic up to the $\chi^2_{\min}$ of the m.e. analysis, is actually very good. As a second test we used the Nijm78 soft-core potential [1] to compare the χ^2-value obtained using Eq. (6) with the χ^2-value obtained from a direct comparison with the data. We are now farther away from the minimum χ^2. The difference is now about 2%, which is satisfactory. It allows us to use Eq (6) to make statements with regard to the quality of some potential model.

We have compared a number of different NN potential models which have appeared in the literature with the pp scattering data. However, in order to make a fair comparison possible, we only consider the 3–350 MeV region. The reason is that the $^1S_0$ phase shift values at the interference minimum (0.38254 MeV) and at 1.0 MeV are very accurately known. So if the $^1S_0$ of a potential model is a little bit off, the χ^2 contribution can be enormous. For example, the Paris potential [11] gives a χ^2 contribution of more than 4500 on these two energy bins alone, whereas at the other energy bins between 3 and 350 MeV the quality of the potential is very satisfactory.

Still, even in the 3–350 MeV region the quality of most potential models is very poor. This is partly due to the following. In our analysis (and also in the VPI&SU analysis [5]), there is a difference between the pp $^1S_0$ and the np $^1S_0$ phase shifts of about 2°. At low energies the difference is even larger (about 5° at 10 MeV). Such a difference cannot be obtained using a nuclear potential model where one only includes the electromagnetic interaction. Indeed, adding the electromagnetic interaction to the Argonne v_{14} potential [23] which was fitted to the np data, we arrive at a χ^2/N_{data} of more than 7 for the pp data. This large value is for a part due to an incorrect pp $^1S_0$ phase shift. When we give the Argonne v_{14} potential perfect pp $^1S_0$ phase shifts, the χ^2/N_{data} on the pp data drops to about 4.3. For the Nijm78 [1] and Paris80 [11] potentials such a situation does not apply, since these models were explicitly fitted to the np as well as the pp scattering data.

Therefore, in the following we will focus on 4 potential models which were constructed to explicitly fit the pp data: the Nijm78 soft-core potential [1], the parametrized Paris80 potential [11], an update of the Bonn potential especially fitted to the pp scattering data [24] (denoted by Bonn89), and a Reidlike Nijmegen potential [10] NijmRdl91, where

Table 1: χ^2-values at the 8 single energies between 3 and 350 MeV of the four potential models mentioned in the text and the pp partial-wave analysis. For the total χ^2 one has to add the $\chi^2_{se} = 1510.5$ of the s.e. analyses as in Eq. (6).

energy	Nijm78	Paris80	Bonn89	NijmRdl91	PWA91
5	52.2	18.4	8.6	15.1	17.7
10	76.4	33.2	46.0	13.8	18.8
25	66.2	12.0	20.0	3.2	0.9
50	531.2	347.7	333.2	9.7	16.4
100	131.1	42.0	56.0	23.0	20.4
150	206.1	419.5	306.0	20.6	21.8
215	198.0	170.3	312.0	11.9	15.2
320	425.5	572.1	518.9	7.1	3.3
3–350	1686.7	1615.2	1600.6	104.3	114.5
χ^2/N_{data}	2.0	2.0	2.0	1.0	1.0

each partial wave is fitted separately to the pp scattering data. The Reidlike potential is constructed in order to have a phenomenological potential model which reproduces the phase parameters of our partial-wave analysis. It provides a very good representation of this analysis. The results are shown in Table 1. We note that the old Nijm78 and Paris80 potentials, and the new Bonn89 potential are roughly of the same quality: $\chi^2/N_{data} \approx 2$. The new NijmRdl91 potential fits the data much better and is as good as (and even a little bit better than) the m.e. analysis: $\chi^2/N_{data} = 1.0$.

At this point it is perhaps good to clarify a question which was raised at the Elba Conference. There somebody tried to imply that there is something wrong with the Nijmegen data base. We have compared our data base and our predictions with the VPI&SU analyses using the SAID 1989 solution. This concerns NN data in the 8–325 MeV energy range. The data bases contain

$$1113 \ pp \text{ data and } 2265 \ np \text{ data } = 3378 \ NN \text{ data in SAID 89,}$$
$$1382 \ pp \text{ data and } 2274 \ np \text{ data } = 3656 \ NN \text{ data in Nijmegen PWA.}$$

From these numbers it appears definitely not true that we reject more data than VPI&SU. We have compared the Paris80 potential directly (not using Eq. (6)) with these two data sets. The χ^2/N_{data} of the Paris80 potential on the interval 8–325 MeV yields

$$pp \text{ data} : \chi^2/N_{data} = 2.27 \text{ in SAID 89 and } 2.15 \text{ in Nijmegen PWA;}$$
$$np \text{ data} : \chi^2/N_{data} = 3.31 \text{ in SAID 89 and } 3.28 \text{ in Nijmegen PWA;}$$
$$NN \text{ data} : \chi^2/N_{data} = 2.96 \text{ in SAID 89 and } 2.85 \text{ in Nijmegen PWA.}$$

The differences between these two comparisons with the data are: SAID 89 contains phase parameters slightly different from the phase parameters calculated by us using the Paris80 potential, the inclusion of the Coulomb interaction in the pp analyses is definitely a source of difference, different $NN\pi$ coupling constants are perhaps used, etc. We make the observation that for np the analyses agree very well, while also for pp the agreement

Table 2: The difference $\Delta\chi^2$ (see text) in the 3–350 MeV energy range of various potential models using all potential phase shifts, or using one particular phase shift only.

model	all phases	only one phase shift				
		$^1S_0$	^3P	$^1D_2$	^3F	sum
Nijm78	1572.2	302.0	462.8	403.6	442.1	1610.5
Paris80	1500.7	137.2	835.0	637.0	459.6	2068.8
Bonn89	1486.1	630.5	743.4	341.4	193.8	1909.1
NijmRdl91	–10.2	5.6	0.0	–12.0	1.1	–5.3

is reasonable. We would like to stress here that in the Paris80 potential it is possible to calculate the pp phase parameters as well as the np phase parameters. When one has only np phase parameyters available and one tries to compare with the pp data, one should not complain when the results are shown to be incorrect.

It is also very instructive to see how the different partial waves contribute to the total χ^2. For that purpose we start with the m.e. phase shifts and substitute the $^1S_0$ phase shift of the different potential models. We then calculate the difference $\Delta\chi^2$ between this χ^2 and the χ^2 of the m.e. analysis. This is also done for the $^1D_2$, the triplet P and the triplet F phase shifts. These four separate contributions can then be summed and compared with the χ^2 as obtained when we take all the potential phase shifts together. The results are presented in Table 2.

The agreement between the sum of the $\Delta\chi^2$-contributions substituting the potential model phase shifts one at a time, and the $\Delta\chi^2$-contribution using all potential model phase shifts simultaneously is satisfactory for the old Nijm78 potential and the new NijmRdl91 potential. For the Nijm78 potential the $\Delta\chi^2$-contributions are about the same for each of the separate contributions. On the other hand, for the Paris80 potential the $^1S_0$ phase shift is rather good, whereas the ^3P (and less the $^1D_2$) phase shifts are not. Similarly, for the Bonn89 potential both the $^1S_0$ and the ^3P phase shifts are not very good. Moreover, for both the Paris80 and Bonn89 potentials, the last column in Table 2 (the sum of the separate contributions) is substantially higher than the second column (using all potential phase shifts together). This means that the correlation between the various phase shifts in these models is very important. This is not a rather nice feature.

5. Analyses at 50 MeV

Recently, a very accurate pp analyzing power experiment at 50.04 MeV [25] and a measurement of the spin-correlation parameter A_{zz} in np scattering at 67.5 MeV [26] were reported. These experiments, together with the other pp and np scattering data already present in the 50 MeV region, provide us with a fairly complete set of NN scattering data around 50 MeV. This makes that the phase parameters at 50 MeV can now be determined rather accurately. It will be interesting to see how the results of the

Table 3: *pp* phase shifts in degrees at 50 MeV of the partial-wave analysis and of various potential models. The numbers between square brackets denote the difference with the partial-wave analysis in standard deviations.

phase	PWA	Nijm78	Paris80	Bonn89	NijmRdl91
$^1S_0$	39.14±0.09	39.58[5]	38.75[4]	38.25[10]	38.82[4]
$^1D_2$	1.70±0.01	1.63[7]	1.80[10]	1.68[2]	1.70[0]
$^3P_0$	11.36±0.12	11.80[4]	11.81[4]	12.66[11]	11.43[1]
$^3P_1$	−8.26±0.04	−8.36[2]	−8.41[4]	−8.34[2]	−8.28[0]
$^3P_2$	5.84±0.02	5.78[3]	5.72[2]	5.56[14]	5.80[2]

Table 4: *np* $I = 0$ phase parameters in degrees at 50 MeV of the partial-wave analysis and of various potential models. The numbers between square brackets denote the difference with the partial-wave analysis in standard deviations.

phase	PWA	Nijm78	Paris80	Bonn87	NijmRdl91
$^1P_1$	−9.83±0.24	−8.64[5]	−10.94[5]	−10.48[3]	−9.78[0]
$^3S_1$	62.70±0.50	60.40[5]	62.30[1]	62.20[1]	62.50[1]
$^3D_1$	−6.43±0.08	−6.60[2]	−6.77[4]	−6.98[7]	−6.44[0]
ε_1	2.16±0.48	2.27[0]	1.89[1]	2.08[0]	2.15[0]

various potential models compare with these phase parameters.

In Table 3 we present the *pp* phase shifts at 50 MeV. The first column gives the phase shifts as determined in the Nijmegen partial-wave analysis. In the following columns we give the *pp* phase shifts of the different potential models where in square brackets we give the difference with the PWA in standard deviations. For example, the $^3P_0$ phase shift in the PWA is $11.36 \pm 0.12°$. The Nijm78 potential predicts $11.80°$ for this phase shift. The difference with the PWA value is $0.44°$ which amounts to 4 standard deviations. Important to note is that the new Bonn89 potential has a $^3P_0$ phase shift which is 11 standard deviations too large and a $^3P_2$ phase shift which is 14 standard deviations too low. These are indications for the fact that this new Bonn89 *pp* potential has a tensor potential which is too strong, whereas its spin-orbit potential is too weak. We would like to point to the very good performance of the NijmRdl91 potential for these $I = 1$ phase parameters.

In Table 4 we present some of the $I = 0$ phase parameters at 50 MeV. The numbers between square brackets again denote the difference with the PWA in standard deviations. For the Bonn potential we have now given the results of the full Bonn87 potential [12]. From Table 4 we can draw the conclusion that the NijmRdl91 potential again gives an excellent fit, while the other potentials give roughly similar predictions. Important is our conclusion that the ε_1 mixing parameter at 50 MeV is $\varepsilon_1 = 2.16 \pm 0.48°$. This is substantially lower than the result of the analysis by the Basel group [26] of $\varepsilon_1 = 2.9 \pm 0.3°$, shown as an open circle in Fig. 1. We want to stress here again that the data base for both analyses are similar. Only the conclusions differ. Our value for ε_1 is in perfect agreement with the modern *NN* potentials.

6. Conclusions

The Nijmegen partial-wave analyses of all NN scattering data below $T_{lab} = 350$ MeV has a $\chi^2/N_{data} = 1.00$. This is a very satisfactory result. Because we have $N_{df} = 3850$ degrees of freedom we would expect $\langle \chi^2_{min} \rangle = 3850 \pm 90$. We find $\chi^2_{min} = 4171$. This means that there is still room for improvement of our analysis of $\Delta\chi^2 \approx 320$, provided that the data form a **statistical** data set. This possibility for improvement can be used as a powerful tool to check theoretical improvements.

The analysis can also be used to test the quality of potential models. We can calculate the potential phase shifts and make a direct comparison with the data. But we can also include the tail of a potential model in the long-range part V_L which enters the analysis via the Schrödinger equation (1). Performing a m.e. analysis using this potential tail then provides information on its quality.

In our analysis we use the heavier-boson exchanges of some potential model (in the present analysis we use the Nijmegen soft-core potential [1]) to give a non-OPE contribution to the intermediate partial waves with $5 \leq J \leq 8$. This also provides us with a test for the quality of the non-OPE part of a particular potential model.

Our way of analyzing the data by using a potential tail allows us to determine the $NN\pi$ coupling constants. Perhaps in a later stage we will also be able to determine, e.g., the ω and ρ coupling constants.

Finally, we can study charge-independence-breaking effects between the pp and np $^1S_0$ phase shifts and the triplet P waves.

Part of this work was included in the research program of the Stichting voor Fundamenteel Onderzoek der Materie (FOM) with financial support from the Nederlandse Organisatie voor Wetenschappelijk Onderzoek (NWO).

References

[1] M.M. Nagels, T.A. Rijken, and J.J. de Swart, Phys. Rev. D **17**, 768 (1978).

[2] T.A. Rijken, R.A.M. Klomp, and J.J. de Swart, Nijmegen Report THEF-NYM-91.05.

[3] J.R. Bergervoet, P.C. van Campen, W.A. van der Sanden, and J.J. de Swart, Phys. Rev. C **38**, 15 (1988).

[4] J.R. Bergervoet, P.C. van Campen, R.A.M. Klomp, J.-L. de Kok, T.A. Rijken, V.G.J. Stoks, and J.J. de Swart, Phys. Rev. C **41**, 1435 (1990).

[5] R.A. Arndt, L.D. Roper, R.A. Bryan, R.B. Clark, B.J. VerWest, and P. Signell, Phys. Rev. D **28**, 97 (1983); R.A. Arndt, J.S. Hyslop III, and L.D. Roper, Phys. Rev. D **35**, 128 (1987).

[6] J. Bystricky, C. Lechanoine-Leluc, and F. Lehar, J. Phys. (Paris) **48**, 199 (1987).

[7] G.J.M. Austen and J.J. de Swart, Phys. Rev. Lett. **50**, 2039 (1983).

[8] V.G.J. Stoks and J.J. de Swart, Phys. Rev. C **42**, 1235 (1990).

[9] L. Durand III, Phys. Rev. **108**, 1597 (1957).

[10] R.A.M. Klomp, V.G.J. Stoks, and J.J. de Swart, in preparation.

[11] M. Lacombe, B. Loiseau, J.M. Richard, R. Vinh Mau, J. Côté, P. Pirès, and R. de Tourreil, Phys. Rev. C **21**, 861 (1980).

[12] R. Machleidt, K. Holinde, and Ch. Elster, Phys. Rep. **149**, 1 (1987).

[13] M.M. Nagels, T.A. Rijken, and J.J. de Swart, Phys. Rev. D **12**, 744 (1975); *ibid.* **15**, 2547 (1977).

[14] M.H. MacGregor, R.A. Arndt, and R.M. Wright, Phys. Rev. **182**, 1714 (1969).

[15] J.J. de Swart, W.A. van der Sanden, and W. Derks, Nucl. Phys. **A416**, 299c (1984).

[16] J.R. Bergervoet, P.C. van Campen, T.A. Rijken, and J.J. de Swart, Phys. Rev. Lett. **59**, 2255 (1987).

[17] R. Koch and E. Pietarinen, Nucl. Phys. **A336**, 331 (1980); O. Dumbrajs, R. Koch, H. Pilkuhn, G.C. Oades, H. Behrens, J.J. de Swart, and P. Kroll, Nucl. Phys. **B216**, 277 (1983).

[18] L.K. Morrison, Ann. Phys. (NY) **50**, 6 (1968); E.M. Henley and Z.-Y. Zhang, Nucl. Phys. **A472**, 759 (1987).

[19] Th.A. Rijken, V.G.J. Stoks, R.A.M. Klomp, J.-L. de Kok, and J.J. de Swart, Nucl. Phys. **A508**, 173c (1990); Prof. Fan Wang and R. Timmermans, private communication.

[20] R.A. Arndt, Z. Li, L.D. Roper, and R.L. Workman, Phys. Rev. Lett. **65**, 157 (1990); R.A. Arndt and R.L. Workman, Phys. Rev. C **43**, 2436 (1991); R.A. Arndt, Z. Li, L.D. Roper, and R.L. Workman, Phys. Rev. D **44**, 289 (1991).

[21] R.G.E. Timmermans, T.A. Rijken, and J.J. de Swart, Phys. Rev. Lett. **67**, 1074 (1991).

[22] M. Aguilar-Benitez *et al.*, Phys. Lett. **B239**, 1 (1990).

[23] R.B. Wiringa, R.A. Smith, and T.L. Ainsworth, Phys. Rev. C **29**, 1207 (1984).

[24] J. Haidenbauer and K. Holinde, Phys. Rev. C **40**, 2465 (1989).

[25] J. Smyrski, St. Kistryn, J. Lang, J. Liechti, H. Lüscher, Th. Maier, R. Müller, M. Simonius, J. Sromicki, F. Foroughi, and W. Haeberli, Nucl. Phys. **A514**, 319 (1989).

[26] M. Hammans, C. Brogli-Gysin, S. Burzynski, J. Campbell, P. Haffter, R. Henneck, W. Lorenzon, M.A. Pickar, I. Sick, J.A. Konter, S. Mango, and B. van den Brandt, Phys. Rev. Lett. **66**, 2293 (1991).

Few-Body Systems, Suppl. 6, 117—127 (1992)

UNCERTAINTIES OF PHASE SHIFT ANALYSES AND THE NUCLEON-NUCLEON INTERACTION

H. Leeb and D. Leidinger

Institut für Kernphysik, Technische Universität Wien,

Wiedner Hauptstraße 8-10/142, A-1040 Wien, Austria

Abstract

Based on Darboux transforms inverse scattering methods are successfully applied to determine the nucleon-nucleon potential for uncoupled partial waves from the corresponding phase shifts. A complete nonparametric error analyses is performed in order to determine reliable uncertainties for the nucleon-nucleon potential due to the incomplete knowledge of the phase shift data.

1. Introduction

The investigation of the nucleon-nucleon interaction is a longstanding question of nuclear physics. Although of fundamental importance we are still today not able to give its complete description in terms of a basic theory like QCD. Our present understanding is mainly guided by the boson exchange picture originally introduced by Yukawa [1] which has been refined considerably in the last decades. From the QCD point of view this boson description is a semiphenomenological one. Today there exist several well established nucleon-nucleon potentials [2-4], which reproduce the nucleon-nucleon phases shifts reasonably well. These potentials are extensively used to evaluate observables in few- and many body systems. However, a direct comparison reveals significant differences between these potentials. Therefore the important question arises as to what extent available phase shift analyses really determine the nucleon-nucleon potential.

Empirical nucleon-nucleon potentials, which are based on a non-relativistic treatment, are obtained from experimental data via phase shift analyses. Today several comprehensive phase shift analyses of nucleon-nucleon data are available [5-7]. The evaluation of the nucleon-nucleon potential from phase shifts is an inverse scattering problem at fixed angular momentum. Although the mathematical basis of this inverse scattering problem of non-relativistic quantum mechanics was already given in the early sixties by Gel´fand, Levitan [8], and Marchenko [9] only few applications are concerned with the nucleon-nucleon interaction [10-12]. Nucleon-nucleon potentials from boson exchange models as well as phenomenological potential parametrisation have been favored. The reason for this is certainly the inherent complexity of inversion methods which deter many scientists from their application. Apart from early calculations [10,11] with a poor data base only recently Kirst et al.[12] have parametrized the experimental S-matrix by rational expressions and used them in inversion procedures of Gelfand, Levitan and Marchenko type. Thus they determine successfully energy independent nucleon-nucleon potentials for the uncoupled partial waves. The same potentials can be obtained from Darboux transforms which allow to construct exactly solvable quantum models. In this case one has the additional advantage that the numerical calculation becomes not only simpler but also more stable because the potential is given almost analytically.

In this paper for the first time we apply the inverse scattering methods to the nucleon-nucleon phase shifts of uncoupled partial waves in the formulation of Darboux transforms. In particular we consider the Arndt- and Saclay-phase shift analyses as given in the program SAID [13] for the $^1S_0$ and $^1P_1$ channels. The main aim of the present paper, however, is the evaluation of uncertainties in the nucleon-nucleon potential generated by the uncertainties of phase shift analyses. Modern phase shift analyses have reached a level of sophistication that this question is indeed justified. There exist well established techniques to determine errors within parametric statistics. However, because of the high nonlinearity of the inverse scattering problem the results are doubtful, specifically no probability statement can be given. The use of inversion procedures allows us to propose a full non-parametric statistical analysis which yields probability distributions for the potential. We have performed such an analysis for the $^1S_0$ channel using single energy errors of the phase shift. Although the question of uncertainties is of fundamental interest with respect to indications on subnucleonic degrees of freedom we are not aware of any systematic study of this kind.

2. Darboux transforms

Exactly solvable models based on Darboux transforms are very useful for the inversion of scattering data [14,15]. In the case of the inverse scattering problem at fixed angular momentum the well known Bargmann potentials [16] are specific examples of this class of transformations [17]. The Bargmann potentials are characterized by a rational S-matrix which is well suited to describe the k-dependence of realistic S-matrices. However, a rational form with many poles and zeros is required to approximate the experimental nucleon-nucleon S-matrix in the whole energy range satisfactorily. The inversion of such a rational S-matrix is easily managed by a matrix generalisation of Darboux transforms [18] which is briefly sketched in the following.

We start our considerations with the regular solution $\varphi_0(k,r)$ and the Jost solution $f_0(k,r)$ of the free radial Schrödinger equation at the angular momentum quantum number ℓ ($\lambda=\ell+1/2$). The double iterated Darboux transform of an arbitrary solution $\eta_0(k,r)$ of the free Schrödinger equation is given by

$$\eta_2(k,r) := \eta_0(k,r) - X^T Y^{-1} w_0 , \tag{1}$$

where Y is the matrix,

$$Y := \begin{pmatrix} \dfrac{W[\varphi_0(\beta_1,r),f_0(\alpha_1,r)]}{\beta_1^2-\alpha_1^2} & \cdots & \dfrac{W[\varphi_0(\beta_1,r),f_0(\alpha_N,r)]}{\beta_1^2-\alpha_N^2} \\ \cdots & \cdots & \cdots \\ \dfrac{W[\varphi_0(\beta_N,r),f_0(\alpha_1,r)]}{\beta_N^2-\alpha_1^2} & \cdots & \dfrac{W[\varphi_0(\beta_N,r),f_0(\alpha_N,r)]}{\beta_N^2-\alpha_N^2} \end{pmatrix} \tag{2}$$

given in terms of the Wronski determinants

$$W[\varphi_0(\beta_i,r),f_0(\alpha_j,r)] := \varphi_0(\beta_i,r)\left[\frac{d}{dr} f_0(\alpha_j,r)\right] - \left[\frac{d}{dr}\varphi_0(\beta_i,r)\right]f_0(\alpha_j,r) . \tag{3}$$

The quantities X and w_0 are vectors of the form

$$X^T := \left(\frac{W[\eta_0(k,r),f_0(\alpha_1,r)]}{k^2-\alpha_1^2}, \ ..., \ \frac{W[\eta_0(k,r),f_0(\alpha_1,r)]}{k^2-\alpha_1^2} \right) \tag{4}$$

$$w_0^T := \left(\varphi_0(\beta_1,r), \ ... \ , \varphi_0(\beta_N,r) \right) . \tag{5}$$

It is straightforward to show that the transformed function $\eta_2(k,r)$ is a solution of the radial Schrödinger equation at the same angular momentum ℓ with the potential,

$$V(r) = -2 \frac{\hbar^2}{2m} \frac{d^2}{dr^2} \ln \det Y . \tag{6}$$

Because of the specific choice of the transformation (1-5) the potential (6) is completely characterized by the parameter set $\{(\alpha_i, \beta_i) \, , \, i=1,2, \ldots , N \}$. Furthermore it must be emphasized that the required solutions $\varphi_0(\beta_i,r)$ and $f_0(\alpha_i,r)$ are spherical Bessel- and Hankel-functions, respectively. Therefore, for a given set of (α,β)-values the potential (6) is in principle known analytically at all radii.

In this exact scheme the set of (α,β)-values is directly related to the poles and zeros of the S-matrix associated with the potential (6). An easy derivation yields

$$S(k) = \prod_{m=1}^{N} \frac{(k - \beta_m) \, (k + \alpha_m)}{(k + \beta_m) \, (k - \alpha_m)} \quad , \quad \mathrm{Im}\,\beta_m < 0 \, , \; \mathrm{Im}\,\alpha_m < 0 \; . \tag{7}$$

This simple rational form is well suited to describe nucleon-nucleon S-matrices [12,19] deduced from experimental data.

Finally it should be remarked that the potential (6) is mathematically equivalent to those obtained from Marchenko- or Gel´fand, Levitan-type procedures [12]. However, our method via the Darboux transforms is not only simpler but has also the additional advantage that it does not suffer from formulation dependent numerical inaccuracies neither in the origin nor in the asymptotic region.

3. Inversion of Nucleon-Nucleon Phase Shifts

We use the exact solvable Bargmann type model for the inversion of nucleon-nucleon phase shift data in uncoupled partial waves. For this purpose a rational approximation of the k-dependence of the corresponding S-matrix is required (7). In principle such an approximation can be determined by fitting α- and β-values of (7) to the considered S-matrix. However, because of the oscillating nature of the S-matrix many poles and zeros are required for a satisfactory description thus leading to ill-posedness in a direct fit. In particular the symmetry properties of a nucleon-nucleon S-matrix can hardly be fulfilled. Therefore we follow the procedure of Kirst et al. [12] and approximate in a first step the k-dependence of the phase shifts δ_ℓ by the rational form,

$$\delta_\ell(k) = \frac{k^{2\ell} \left(a_0 - a_s k + \sum_{i=2}^{\bar{N}} a_{2i-1} \, k^{2i-1} \right)}{1 + \sum_{i=1}^{N/2} b_{2i} \, k^{2i}} \quad , \quad \bar{N} = \frac{N}{2} - \ell \; . \tag{8}$$

For s-waves a_0 corresponds to the phase shift at k=0 and a_s is the scattering length. Because of the smoothness of $\delta_\ell(k)$ excellent approximations can be abtained with polynomials of low order N (even). Subsequently the required rational representation of the S-matrix is obtained from a symmetric Padé approximate to e^z. Thus the symmetry properties of the nucleon-nucleon S-matrix are automatically fulfilled in every order of the approximation. The poles and zeros of these representations of the S-matrix are the α- and β-values, respectively, and characterize the potential (6) uniquely. For more details of this procedure we refer to ref. 12.

In a first test of the inverse scattering scheme we have reconstructed the Reid-Soft-Core potential [20] for uncoupled partial waves. In particular we have evaluated from the potentials the $^1S_0$, $^3P_1$ and $^1P_1$ phase shifts at different enrgies by the Numerov method. Using the calculated phase shifts at 56 energies between E_{Lab}=0.1 MeV and E_{Lab}=2000 MeV excellent fits ($\Delta\delta$<0.05 degree) could be obtained with (8) using relatively low order representations, N=14 ($^1S_0$), N=12 ($^3P_1$), and N=10 ($^1P_1$). Similar to ref. 12 the Reid-Soft-Core potentials are satisfactorily reproduced by our inversion procedure.

Our main task is the application of the inversion scheme to experimental data. Today high quality phase shift analyses [5-7] are available, which are based on complete and up to date nucleon-nucleon data sets. As examples we use the Saclay and the Arndt phase shift data sets given in the interactive program SAID [13]. In particular we determine the potentials for the

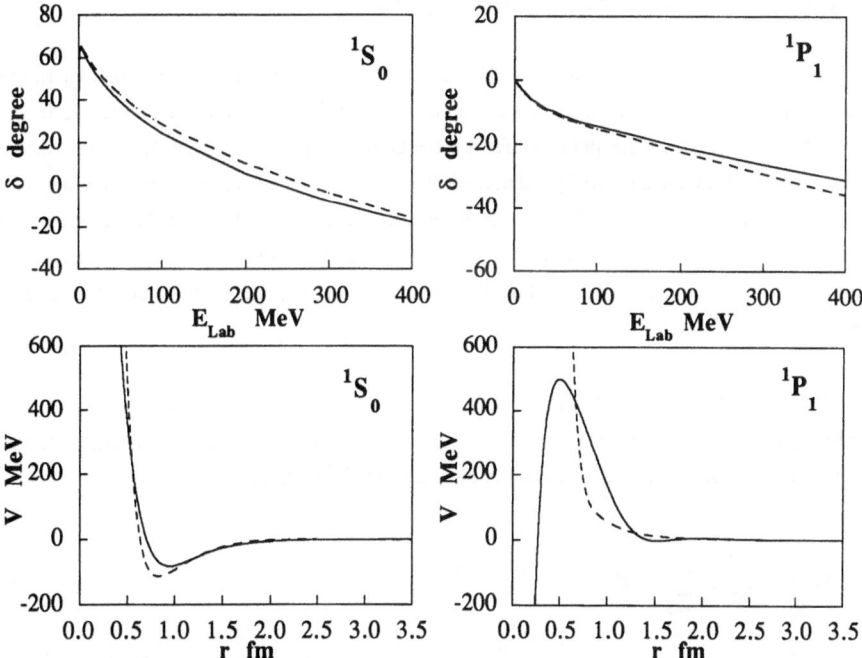

Fig. 1. Comparison of the phase shift analyses of Arndt (solid line) and the Saclay group (dashed line) and their associated energy independent potentials.

$^1S_0$ and $^1P_1$ partial waves with the inversion scheme. A comparison of the different phase shift data and their associated potentials is displayed in Fig. 1. Although there is a slight difference between the $^1S_0$ phase shifts of the two analyses the associated potentials are essentially the same for $r \geq 1.2$ fm. For larger radii, $r \geq 2.7$ fm, the inverted potentials are in full agreement with the one-pion exchange term. At smaller distances, however, there are significant differences particularly visible in the depth and position of the potential minimum. To a large extent these differences at small radii are generated by the fact that the phase shifts are fixed only up to $E_{Lab} \sim 400$ MeV. The differences between the two phase shift analyses are even greater in the $^1P_1$ partial wave due to the centrifugal barrier. Here even the sign of the potential at small distances is badly determined. A similar unstable behaviour has also been found by Kirst et al. [12].

3. Uncertainties

The previous calculations clearly demonstrate our limited knowledge of the nucleon-nucleon potential. Although these differences were mainly caused by the unknown high energy behaviour ($E_{Lab} > 400$ MeV) there arises the important question to what extent the nucleon-nucleon potential is really determined by the phase shift data at present. Today this question is of particular importance in connection with investigations which study the role of subnucleonic degrees of freedom in few- and many-nucleon systems.

Empiric nucleon-nucleon potentials are determined from phase shift analyses either by fitting parameters of a potential ansatz [20] or more elegantly by inversion techniques as in the previous chapter. Since the phase shifts are deduced from experimental data with certain measurement errors it is also obvious that apart from ambiguities the phase shift values are only known with certain error bars. In most analyses this uncertainty is ignored. In fact the phase shift errors seem to be much smaller than the differences between modern meson theoretical approaches. Nevertheless they certainly restrict our knowledge of the nucleon-nucleon potential.

The concept of uncertainty is well defined in statistics when repeated measurements vary randomly about its expectation value. These measurements of a quantity x have a distribution and the standard deviation Δx is a common measure of the uncertainty of x,

$$\Delta x := \sqrt{\frac{1}{N} \sum_{i=1}^{N} (x_i - \bar{x})^2} \ , \tag{9}$$

where $\bar{x}$ is the mean value of x from N repeated measurements. If the distribution of x is normal the standard deviation is the half-width of an interval that contains 68% of all observations. Error estimates based on standard deviations are only useful if the probability content of the associated interval is known. For distributions that are not normal there is no a priori relation between Δx and the probability content. The assumption of normal distributions is justified for the nucleon-nucleon cross section data, but the distributions are certainly distorted for the deduced phase shifts and potentials because of the strong non-linear relations between these quantities. Therefore the probability distributions for these quantities must be assumed (parametric statistics) or determined by techniques of non-parametric statistics. In the first case, however, no conclusions about the probability content can be drawn [15].

It is a useful feature of the exactly solvable models of Bargmann-type that nearly every reasonable k-dependence of the S-matrix can be described by the rational form (7) and subsequently lead to an almost analytically given potential. This is the key feature allowing a non-parametric error analysis for the nucleon-nucleon interaction with minimal effort. In the following we want to introduce a procedure to determine reliable statistical uncertainties in the nucleon-nucleon potential. In principal our considerations should start from the cross section data which have a known probability distribution. However, such an error analysis is rather tedious. Since we want to demonstrate the principle of the procedure we start our analysis for a partial wave with a set of phase shift values $D=\{\delta(E_1),\delta(E_2), ... ,\delta(E_N)\}$, where E_i denotes different energies. Since in this stage we had no access to the detailed statistical results of phase shift analyses we have assumed the phase shift values at different energies to be independent and normally distributed. In the next step K phase shift data sets $D_1, D_2, ... ,D_K$ have been constructed by choosing randomly each phase shift $\delta(E_j)$ within their normal distribution. Applying the inversion scheme described in chapter 2 and 3 to each phase shift data set D_j, $j=1, ... , K$ yields immediately the associated potential V_j, $j=1, ... , K$. These potentials represent a statistical ensemble which can be used to determine the probability distributions of the potential values at each radius or more sophisticated quantities, e.g. correlation coefficients.

In practice we have applied this method to the $^1S_0$ phase shift analysis of the Saclay group [13]. Unfortunately, no consistent error bars have been available for this phase shift analysis. Therefore we used those obtained in analyses at single energies [13] which are summarized in Table 1. With this uncertainties we have constructed 216 randomly distributed data sets D_j. For

Table 1 Uncertainties of the $^1S_0$ phase shifts obtained in phase shift analyses at single energies.

E_{Lab} MeV	$\Delta\delta$ degree	E_{Lab} MeV	$\Delta\delta$ degree	E_{Lab} MeV	$\Delta\delta$ degree
25	0.26	450	0.30	750	0.49
50	0.11	500	0.24	800	0.41
100	0.33	550	0.38	850	0.80
200	0.36	600	0.32	900	0.76
300	0.27	650	0.49	950	0.91
400	0.28	700	0.64	999	0.92

each data set the associated potential has been determined by our inversion procedure. These potentials represent a statistical ensemble which allows the determination of uncertainties. In particular we consider the probability distributions for the potential at different radii. Some characteristic histograms for these distribution are displayed in Fig. 2. As expected the distributions are rather sharp in the asymptotic region while they get considerably broader in the interior reflecting the uncertainty in the high energy behaviour of the phase shifts. Because of the deviations from a normal distribution there is no direct relation between the standard deviation and the probability. Nevertheless our analysis allows the determination of intervals with a given probability. The arrows in Fig. 2 indicate the borders of the interval which contains 68% of the potential data, while double arrows indicate the interval with 95% content. The probability content of the indicated interval correspond to that of one or two standard deviations of a normal distribution, respectively. In order to get a better insight in the size of the uncertainties we have displayed in Fig. 3 the error bands obtained from the intervals with

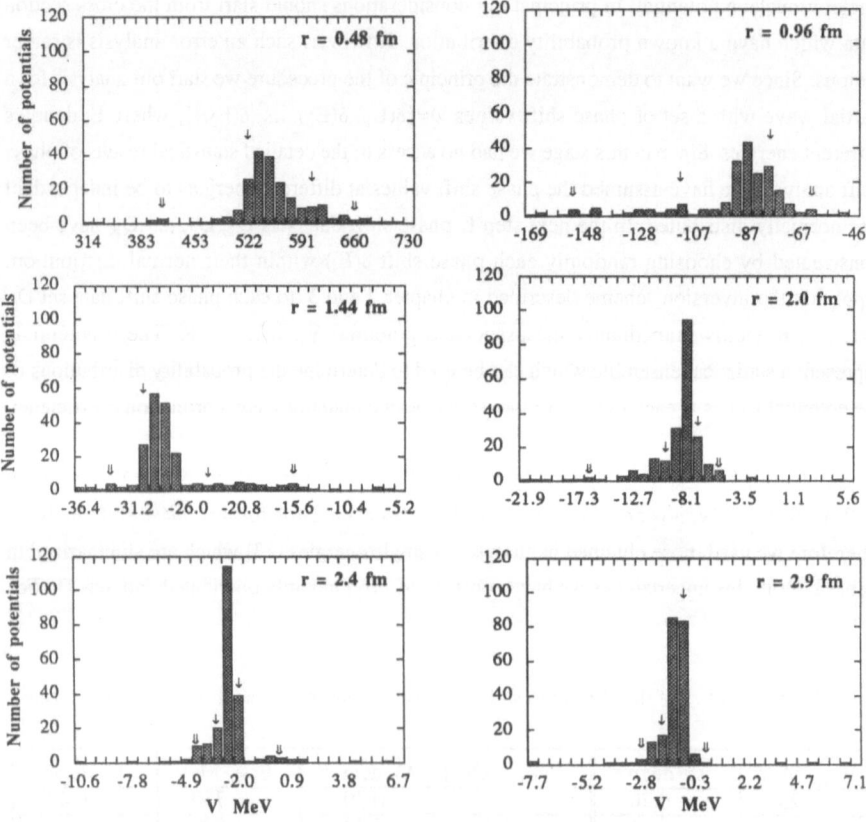

Fig. 2. Probability distributions for the $^1S_0$ potentials at different radii obtained in the non-parametric analysis of the Saclay phases with the error bars of Table 2.

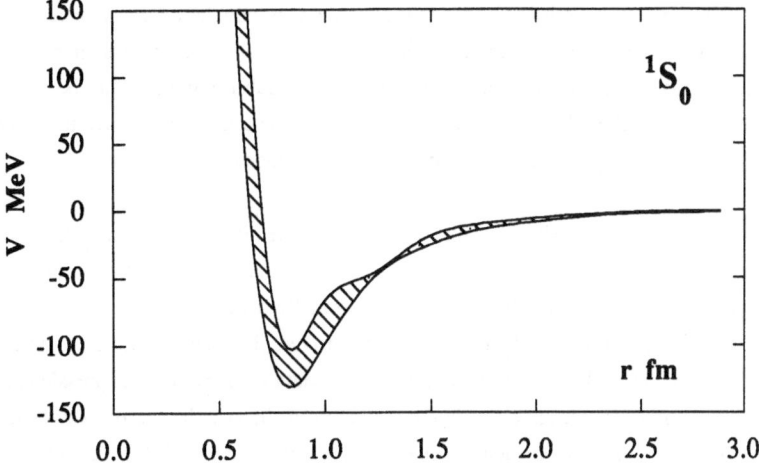

Fig. 3. Error bands with 68% content for the $^1S_0$ potential obtained from the Saclay phases

68% content. It is obvious that the potential is rather well determined for $r \geq 1.3$ fm. However, in the region, 0.8 fm $\leq r \leq 1.3$ fm, where the concept of a nucleon-nucleon potential retains its validity, our analysis indicate rather huge uncertainties in the potential. Finally, it should be remarked that our assumption of complete independence of the input phase shifts spoils slightly the distributions because it leads in few cases to asymptotically oscillating potentials which are unrealistic.

For completeness we should also mention that there is an uncertainty due to the unknown high energy behaviour of the phase shifts. This limitation is a principal one because of the opening of inelastic channels and the breakdown of the non-relativistic description. However, this will effect the potential only at small distances ($r \leq 0.5$ fm).

4. Summary and Conclusions

We have successfully applied the inversion scheme of Bargmann-type to the uncoupled partial waves of the nucleon-nucleon scattering process. In particular we have evaluated energy-independent $^1S_0$- and $^1P_1$-potentials associated with the phase shift analyses of the Saclay group and of Arndt et al. [13]. The new compact formulation via Darboux transforms is not only simpler but also numerically well behaved because the potential is given in nearly analytical form.

The main goal of this paper, however, is a systematic study of the uncertainties in the nucleon-nucleon potential. We have demonstrated that exploiting the advantages of the inversion scheme a non-parametric statistical analysis is feasible. Our calculation of the uncertainties of the $^1S_0$ Saclay phase shifts leads to an ensemble of potentials with the proper probabiblity distribution. Thus reliable error bands with a definite probability content can be given. Our results are also an excellent basis for a systematic study of the effects of the uncertainties on the properties of few-body systems. Investigations in this directions are underway.

For a complete reliable analysis of the nucleon-nucleon potential essential improvements must still be achieved. First of all there is a need of consistent statistical information of the phase shift analyses. Our assumption of independent and normally distributed phase shifts at different energies is a very crude one and can easily be improved. The best would be a start of the analysis from the original nucleon-nucleon scattering data using rational expressions for the phase shifts. Another important point is the inclusion of the coupled partial waves into the analysis. There is no principal difficulty because the mathematical basis for the corresponding inversion procedure has been developed many years ago [21] and there are also some simple applications [10, 11]. However, for a statistical analysis inversion schemes for coupled partial waves, which are eassy to apply, are still required.

The work has been supported by Fonds zur Förderung der Wissenschaftlichen Forschung in Österreich (Proj. Nr. P6608)

References:

1. Yukawa, H.: Proc. Phys.-Math. Soc. Japan **17**, 48 (1935)
2. Nagels, M. M., Rijken, T. A., and de Swart, J. J.: Phys. Rev. **D 17**, 768 (1978)
3. Lacombe, M., Loiseau B., Richard, J.M., and Vinh Mau, R.: Phys. Rev. **C 21**, 861 (1980)
4. Machleidt, R., Holinde, K., and Elster, Ch: Phys. Rep. **149**, 1 (1987)
5. Arndt, R. A., Roper, L. D., Bryan, R. A., Clark, R. B., VerWest, B. J., and Signell, P.: Phys. Rev. **D 28**, 97 (1983); Arndt, R. A., Hyslop III, J. S., and L. D. Roper, *ibid* **35**, 128 (1987)
6. Bystricky, J., Lechanoine-Leluc, C., and Lehar, F.: J. Phys. (Paris) **48**, 199 (1987)
7. Bergervoet, J. R., van Campen, P. C., Rijken, T. A., and de Swart, J. J.: Phys. Rev. **C 38**, 15 (1988); Bergervoet, J. R., van Campen, P. C., Klomp, R. A. M., de Kok, J.-L., Rijken, T. A., Stoks, V. G. J., de Swart, J. J: Phys. Rev. **C 41**, 1435 (1990)
8. Gel'fand, I. M., and Levitan, B. M.: Am. Math. Soc. Transl. **1**, 253 (1955); B. M. Levitan, B. M.: Generalized Translation Operators and Some of the Applications, Davey, New York, 1964

9. Agranovitsch, Z. S., and Marchenko, V. A.: The Inverse Problem of Scattering Theory, Gordon, New York, 1963

10. Newton, R. G., and Fulton, T.: Phys. Rev. **107**, 1103 (1957)

11. Benn, J., and Scharf, G.: Nucl. Phys. **A183**, 319 (1972)

12. Kirst, Th., Amos, K., Berge, L., Coz, M., and von Geramb, H. V.; Phys. Rev. C **40**, 912 (1989)

13. Arndt, R. A., and Roper, L. D.: Scattering Analysis Interactive Dial in SAID, Virginia Polytechnic Institute, Blacksburry, 1988

14. Naidoo, K., Fiedeldey, H., Sofianos, S. A., and Lipperheide, R.: Nucl. Phys. **A419**, 13 (1984)

15. Leeb, H., Fiedeldey, H., Lipperheide, R.: Phys. Rev. C **32**, 1223 (1985)

16. Bargmann,V.: Rev. Mod. Phys. **21**, 488 (1949)

17. Schnizer, W. A., and Leeb, H.: Inverse Methods in Action; Springer, Berlin-Heidelberg, 1990, page 455

18. Schnizer, W. A., PhD thesis, University of Graz, Austria, 1990

19. Hartt, K., and Yidana, P. V. A.: Phys. Rev. C **31**, 1105 (1985)

20. Reid, R. V.: Ann. Phys. (N.Y.) **50**, 411 (1968)

21. Newton, R. G.: Phys. Rev. **100**, 412 (1955)

Few-Body Systems, Suppl. 6, 128—135 (1992)

Few-
Body
Systems
© by Springer-Verlag 1992

S – D Transition in the Nucleon-Nucleon System

Ingo Sick
Universität Basel, CH-4056 Basel, Switzerland

1   Introduction

Important properties of the fundamental nucleon-nucleon (N-N) interaction, especially for the isoscalar component, are still poorly known. This concerns in particular the $S - D$ transition amplitude. The accurate knowledge of this quantity is important for the calculation of some of the central quantities of nuclear physics:

(i) For a nucleon-nucleon potential that explains the deuteron binding energy, the binding energy of the $A = 3, 4$ systems and nuclear matter strongly depend on the S–D transition; a stronger tensor force leads to a reduced binding energy. Of particular importance is the question whether these binding energies can be explained without introducing a three-body force. Only for a very weak tensor force can the binding energy be explained without invoking a three-body force.

(ii) The probability of $L = 2$ states in light nuclei depends directly on the strength of the S–D transition. Without accurate knowledge of these D-states, an understanding of light nuclei — test cases for the description of nuclei in terms of nucleons, as the Schrödinger equation can be exactly solved — remains incomplete.

(iii) These $L = 2$ states have also important consequences for an understanding of non-nucleonic degrees of freedom. In the observables most sensitive to these non-nucleonic degrees of freedom — magnetic isovector form factors of light nuclei — S–D transitions and meson exchange currents have effects of the same size, but opposite sign. To determine the mesonic effects, one needs to accurately know the S–D transition contribution.

The most direct way to investigate the $T = 0$ N-N interaction is an accurate measurement of observables in neutron-proton scattering. Phase shift analyses of the scattering data produce phases that contain the physical information. For the $S - D$ transition the quantity of interest is the mixing parameter ϵ_1 that describes transitions from the triplet-S to the triplet-D state. Phase shift analyses of the experimental set of data show that in the energy region relevant for nuclei, ϵ_1 is very poorly determined ($\pm 50\%$). This results basically for two reasons:

• Only observables involving the measurement of *two* spins of the incoming/outgoing nucleons are sensitive to ϵ_1; as such experiments are very difficult, only very few data on such observables are available.

• In most observables sensitive to ϵ_1, this parameter is strongly correlated with other (comparatively uninteresting) phases such as $^1P_1$ which are not sufficiently constrained by other independent information.

Since 1982 the Basel group has been carrying out a systematical program to measure these two types of quantities at an energy of $\sim 70 MeV$. This energy range is of particular interest as the knowledge of ϵ_1 there has the most important consequences for the properties of nuclei mentioned above. At significantly lower energies, ϵ_1 is determined by the slope at zero energy imposed by the $\pi - N$ coupling constant, the deuteron asymptotic S/D-state ratio, and the s-phase; the predictions of different N-N potentials are very close. At energies significantly higher than $70 MeV$, test calculations on the A=3 system have shown that the D-state probability and binding energy are no longer sensitive to the value of ϵ_1 [1]; this reflects the fact that nuclear systems are mainly sensitive to properties of the N-N interaction at momenta below the Fermi momentum k_F.

A study of the sensitivities of various observables has been carried out by Binstock and Bryan [2]. They show that the only observables sensitive to ϵ_1 are the ones involving the observation of at least 2 spins in either initial or final states. The most sensitive observable is the quantity A_{zz} measured by scattering of longitudinally polarized neutrons from longitudinally polarized protons. The observable $\Delta\sigma_l$, the change of the total cross section in polarized neutron–polarized proton scattering, was shown by Bugg [3] to be very sensitive as well.

To determine ϵ_1 one has to measure also those quantities that are strongly correlated with ϵ_1 , particularly the $^1P_1$ phase. From the angular distribution, $^1P_1$ can be obtained if the D-phases are known. The vector analyzing power A_y measured in $\vec{n} - p$ scattering is particularly sensitive to the $^3D_1$ and $^3D_3$ phases.

The series of experiments the Basel group is performing concerns the 2-spin observables A_{zz} and $\Delta\sigma_l$, the analysing power A_y, and the angular distribution $\sigma(\theta)$. In the present paper we briefly describe our measurements and give first results on ϵ_1 .

2   Experiments

The polarized neutrons needed for our experiments are produced using the neutron-beam facility we have built up at the Paul Scherrer Institut (PSI). This facility is based on the $d(\vec{p}, \vec{n})2p$ reaction at $0°$. This source is singled out by a number of features:
– The two protons are mostly left in the nearly-bound singlet-s state, thus leading to a quasi-monoenergetic neutron beam; the width of the peak in energy is $\sim 2MeV$, and the low-energy tail is very low. Such a 'monoenergetic' neutron beam leads to very clean data, simplifying greatly the background subtractions necessary due to $e.g.$ elements other than hydrogen in the polarized targets.
– The neutron spin is flipped upon flip of the proton spin in the atomic beam-type polarized source; rapid spin flip leads to small systematic errors.
– the neutron beam is carefully collimated, such that the intensity falls at the edge of the beam by 3 orders of magnitude within 2-3 mm; this allows the use of the small polarized targets now available.
– The good time structure of the cyclotron beam (pulses of FWHM 0.7 ns) allows an accurate measurement of the neutron energy via time - of - flight (TOF).
– A high neutron intensity ($5 \cdot 10^5/s, cm^2$ at the target) can be achieved as the polarized proton intensity reaches more than $5\mu A$.
– Any beam energy up to 70MeV is available
– With the combination of solenoid and dipole magnets installed, any spin direction of the neutron (longitudinal, sideways vertical, sideways horizontal) can be obtained; this allows the measurement of those observables that are the most sensitive.

Figure 1 shows the layout of the facility; a more detailed description, and the measurements done to calibrate the neutron polarization, are given in [4, 5]. At present the absolute polarization of the neutron beam is known to ±4%, attempts to significantly reduce this systematical error are underway.

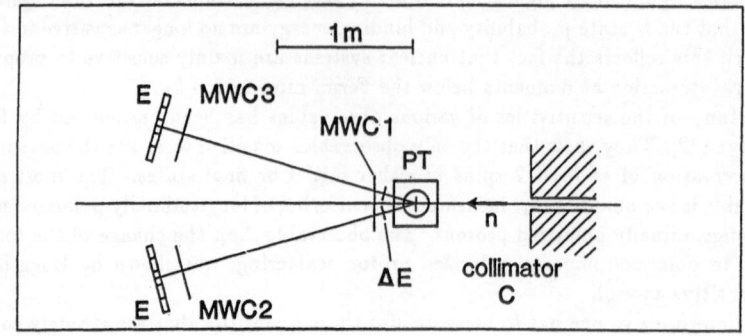

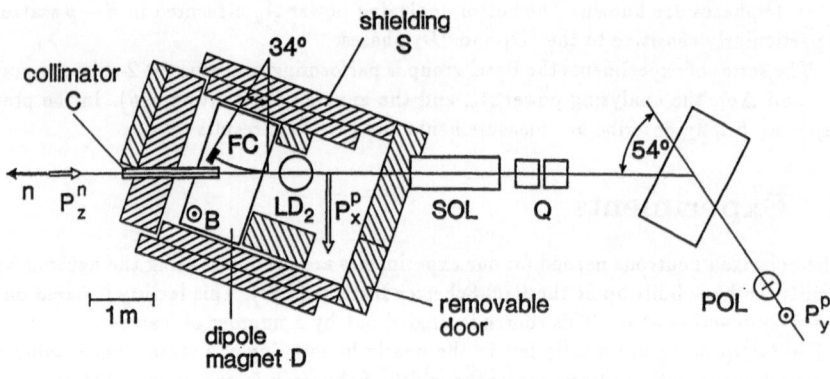

Figure 1: Neutron source, with proton polarimeter POL, spin precession solenoid SOL, liquid deuterium target LD$_2$, dipole magnet B, Faraday cup FC, and collimator C. The insert shows the setup used for the measurement of A_{zz} with polarized proton target PT, thin and thick plastic counters $\Delta E, E$, and multi wire proportional chambers MWPC.

In figure 1 we also show the setup for one particular experiment, the A_{zz} measurement at backward neutron angles, where elastic $\vec{n} - \vec{p}$ scattering was investigated by observing the recoiling protons at forward angle.

Depending on the observables measured, the various experiments involve liquid hydrogen targets, active hydrogen targets in the form of plastic scintillators or the polarized butanol target provided by the PSI polarized target group of S.Mango. This target consists of a slab of butanol, 2 – 25mm in thickness depending on the observable. A slab rather than the standard beads is used in order to avoid contributions from the cooling medium, ^3He. The target thickness is measured to 1% accuracy using X-ray absorption.

For detectors we use, depending on the observable, thin or thick plasic counters, thin plastic veto counters and multi wire proportional chambers ($\sim$ 800 channels read out by a PCOS3 system). The quantities measured are the various time of flight differences, the TOF difference to the cyclotron RF, the amplitudes in all the $\Delta E, E$ detectors, and the MWPC information. From this data one can determine the neutron energy, the energy and particle type of the detected particles, the trajectory of the recoiling proton and the coordinates of its origin in the target.

A special setup is used for the $\Delta \sigma_l$ experiment [6]. Here we employ a polarized target 2.5 cm thick. To detect the transmitted neutrons, we use a special flux monitor consisting of 20 individual units. Recoil protons, produced in the scintillator plastic or the CH_2 plates that separate the plastics, are detected via coincidences between at least 2 consecutive plastics. By using many ΔE counters rather than one E counter, the sensitivity to gain shifts is greatly reduced. This approach allows us to measure the high rates of transmitted· neutrons needed to determine the change in the transmission rate with the desired error of 10^{-5}.

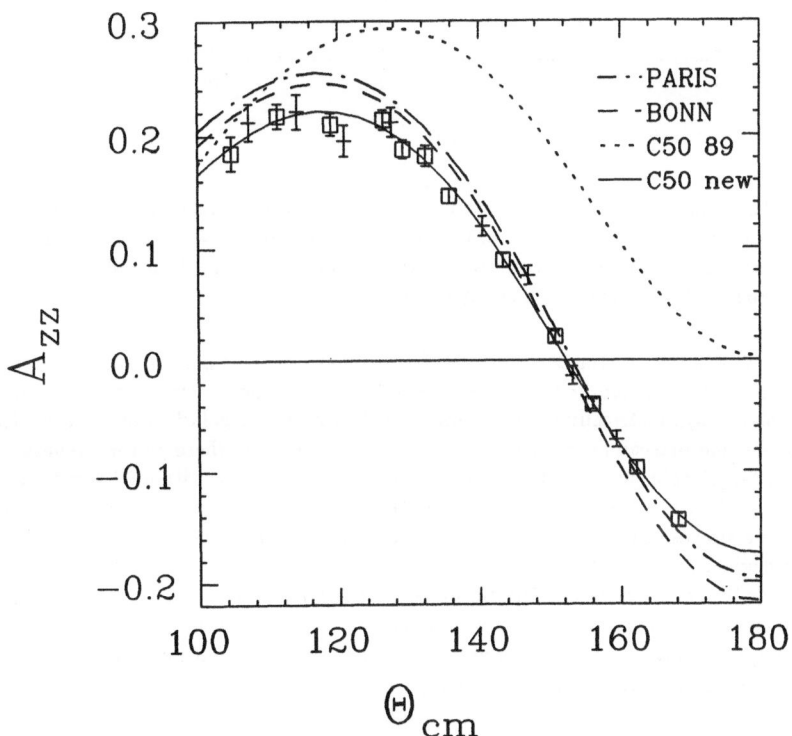

Figure 2: A_{zz} at 67.5 MeV laboratory energy, compared to the prediction of the Bonn (dashed) and Paris (dash-dotted) potentials. Our new phase shift fit is given by the solid line, the C50 SP89 fit of Arndt by the dotted line.

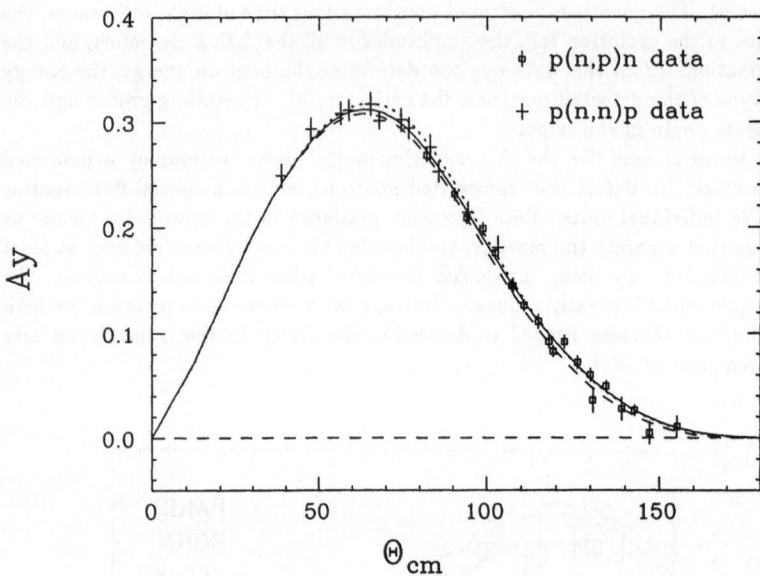

Figure 3: Analyzing power A_y as a function of the center of mass angle at 67.5MeV. The results were obtained via observation of the recoil protons (squares), or detection of the scattered neutrons (crosses). The curves are the predictions of the Bonn (dashed) and Paris (dotted) potentials, and the new phase shift fit.

In order to suppress systematical errors, the neutron (incident proton) spin is flipped every few seconds. In addition, we systematically detect the reaction products with detectors placed symmetrically to the beam; by calculating the polarization observables from super-rations one can further reduce the errors. For the case where we use a polarized target, the target spin direction is also reversed regularly. For longitudinally polarized neutrons, the data are taken with both polarities of the solenoid field. By taking the appropriate averages of the different runs, experimental asymmetries are suppressed to a very high degree.

With this neutron facility and the various experimental setups we have measured the various observables listed above. Here we only show a few examples.

In figure 2 we show the results of the A_{zz} experiment, in figure 3 we show the results for the measurement of A_y at the same energy.

3 Phase shift analysis

Our results, together with the results for $d\sigma/d\Omega$ and A_{yy} by the Karlsruhe group [7][8], are added to the world data base (omitting the erroneous [1] Harwell σ_T and $d\sigma/d\Omega$ [11],[12] data) and analyzed by means of Arndt's phase shift analysis code [13]. We perform a *single-energy* phase shift analysis in the range 32 to 68 MeV to avoid the bias introduced by an energy dependence taken from models. The phases are assumed to be linear over the energy range studied. In order to test this assumption and to study the energy dependence of the phases this range is divided in a second step into a lower (32 to 50.5 MeV) and an upper (49.5 to 68 MeV) range. For all phases the expected energy dependences are observed.

We vary all phases with L$\leq$2 (except $^3D_3$), including $^1F_3$, the mixing parameters $\epsilon_1, \epsilon_2, \epsilon_3$, and a free normalization parameter for every experiment. The other partial waves are taken from the BONN potential; using other parametrizations like the PARIS or NIJMWEGEN [14] potential does not change the results. Similarly, varying the energy dependence of the phases or fixing some of the less significant phases on the BONN predictions has negligible effect. The $^1S_0$ pp phase is taken from the PARIS potential; the charge splittings for the higher T=1 phases are also determined via χ^2 minimization and are found to be similar to the ones of the most recent complete PSA [15]. The final total χ^2 (np and pp data) for the complete energy range is 506 for 488 degrees of freedom. Fig.2 includes the new fit as well as the 1989 50 MeV single energy fit [13] without the new data. For better presentation in the figure, the new fit curve has been multiplied by the renormalization factor 0.941 required by the analysis. This is within the 6% combined normalization error of beam and target polarization.

In general the predictions of the most recent meson theoretical models, the BONN [16] and the PARIS [17] potentials, are close to our solution. The most significant difference occurs for ϵ_1, presented in figure 4. The error bar on the new value of ϵ_1 represents the diagonal element of the error matrix which corresponds to the parameter latitude given by the usual '$\chi^2_{min} + 1$' criterion for simultaneous variation of all other parameters.

The impact of the new data is shown by the striking reduction, by a factor of 3, of the uncertainties in ϵ_1 and $^1P_1$. The improvement is mainly due to the addition of the A_{zz} and $\Delta\sigma_l$ data. The value of $\epsilon_1(2.72^\circ \pm 0.25^\circ)$ is significantly higher than the predictions of the potential models and of Arndt's 1987 analysis [15]. One should bear in mind that this value is closely linked to the value of $^1P_1$. More positive values of $^1P_1$ — determined by cross section data — require larger values of ϵ_1 in order to fit A_{zz}. Although the new value of $^1P_1(-9.4^\circ \pm 0.2^\circ)$ is much more negative than the recent value of Ref.[15] $(-4.1^\circ \pm 0.6^\circ)$ and essentially removes the notorious discontinuity at 50 MeV caused by the Harwell data, it is still more positive than the Bonn (-10.5°) and Paris (-10.9°) predictions. Our new value of ϵ_1 is in line with a recent PSA [18] which includes new data above 200 MeV; there, the data set is also precise enough to yield an accurate value of

---

[1] The Harwell σ_T data are lower by 3 to 4% compared to three other, more recent data sets. See Ref. [9] and references therein. The cross section angular distributions are distorted since at backward angles the energy dependence of the proton detection efficiency was neglected while at forward angles the calibration of the neutron detector efficiency was based on an incorrect determination of the neutron transmission through lead. The authors admit to a discrepancy below 65 MeV when comparing the angle integrated cross section to the Harwell σ_T values. Renormalizing the data near 90° to the low Harwell σ_T resulted in yet another distortion. Altogether, inclusion of the Harwell cross section data results in small values of $|^1P_1|$, as observed first by Ref.[10].

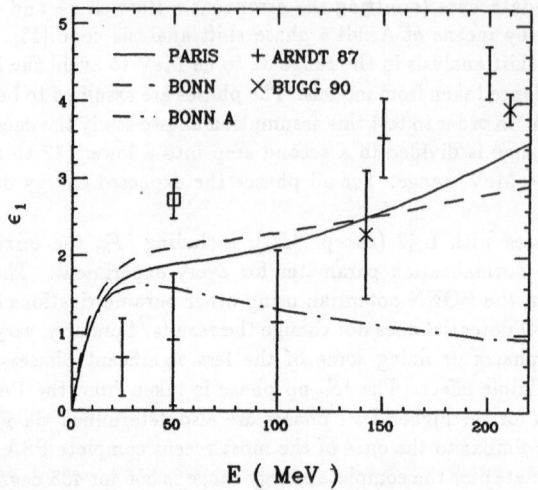

E (MeV)

Figure 4: ϵ_1 as a function of energy. Curves represent the predictions of the BONN (dashed), PARIS (solid) and BONN A (dash-dot) potentials. Single energy PSA results are given by the square (this work), horizontal bars (Ref.[15]) and crosses (Ref. [18]).

ϵ_1 , and a value of ϵ_1 larger than predicted by the various potentials is found.

The value we find for ϵ_1 is significantly higher than that given by modern N-N potentials. The discrepancy is largest with respect to the OBE approximation BONN A [16] (see Fig.4) which is characterized by a weak tensor force; the other BONN potentials as well as the PARIS potential have a stronger tensor force. The present results on ϵ_1 indicate that an even stronger tensor force is required.

References

[1] R. Machleidt. *12th Conf. on Few Body Problems in Physics*, Vancouver, Canada, Contrib. Papers:G 51, 1989.

[2] J. Binstock and R. Bryan. *Phys. Rev.*, D 9:2528, 1974.

[3] D.V. Bugg, J.A Edgington, W.R. Gibson, N. Wright, N.M. Stewart, A.S. Clough, D. Axen, G.A. Ludgate, C.J. Oram, L.P. Robertson, J.R. Richardson, and C. Amsler. *Phys. Rev. C*, 21:1004, 1980.

[4] R. Henneck, C. Gysin, M. Hammans, J. Jourdan, W. Lorenzon, M. A. Pickar, I. Sick, S. Burzynski, and T. Stammbach. *Nucl. Instr. and Meth*, A259:329, 1987.

[5] M. A. Pickar, S. Burzynski, C. Gysin, M. Hammans, R. Henneck, W. Lorenzon, I. Sick, A. Berdoz, and F. Foroughi. *Phys. Rev.*, C 42:20, 1990.

[6] P. Haffter, J. Campbell, C. Brogli-Gysin, R. Henneck, J. Jourdan, M. Hammans, G. Masson, , M. A. Pickar, I. Sick, J.A. Konter, S. Mango, and B. van den Brandt. *to be publ.*

[7] G. Fink, P. Doll, T.D. Ford, R. Garrett, W. Heeringa, K. Hofmann, H.O. Klages, and H. Krupp. *Nucl. Phys. A*, 518:561, 1991.

[8] P. Doll et. al. *12th Int. Conf. on Few Body Problems in Physics*, Vancouver, B.C.:Contr. Papers C16, 1989.

[9] P.W. Lisowski, R.E. Shamu, G.F. Auchampaugh, N.S.P. King, M.S. Moore, G.L. Morgan, and T.S. Singleton. *Phys. Rev. Lett.*, 49:255, 1982.

[10] R. A. Arndt, J. Binstock, and R. Bryan. *Phys. Rev.*, D 8:1397, 1973.

[11] P.H. Bowen, J.P. Scanlon, G.H. Stafford, J.J. Thresher, and P.E. Hodgson. *Nucl. Phys.*, 22:640, 1961.

[12] J.P. Scanlon, G.H. Stafford, J.J. Thresher, P.H. Bowen, and A. Langsford. *Nucl. Phys.*, 41:401, 1963.

[13] R. A. Arndt. *private communication*, 1990.

[14] M.M. Nagels, T.A. Rijken, and J.J. de Swart. *Phys. Rev. D*, 17:768, 1978.

[15] R. A. Arndt, J.S. Hyslop, and L.D. Roper. *Phys. Rev. D*, 35:128, 1987.

[16] R. Machleidt. *Adv. Nucl. Phys.*, 19:189, 1989.

[17] M. Lacombe, B. Loiseau, J.M. Arndt Richard, R. Vinh Mau, J. Cote, P. Pires, and R. deTourreil. *Phys. Rev.*, C21:861, 1980.

[18] D. V. Bugg. *Phys. Rev. C*, 41:2708, 1990.

Few-Body Systems, Suppl. 6, 136—147 (1992)

NEW RESULTS ON THE πNN SYSTEM

E.T. Boschitz

Institut für Experimentelle Kernphysik, Universität Karlsruhe

7500 Karlsruhe 1, Federal Republik of Germany

Abstract

New experimental results are presented on the major πd reaction channels, i.e. elastic scattering, breakup, charge exchange and absorption. The data are compared to recent calculations based on relativistic Faddeev theory. Some open theoretical problems are discussed and an outlook on future experiments is given.

The systematic experimental investigation of the πNN system is very important in intermediate energy physics because its various reaction channels can, in principle, be calculated exactly using the relativistic Faddeev formalism, and the known properties of the pion-nucleon and nucleon-nucleon subsystems. During the past few years, remarkable theoretical progress on this system has been achieved [1]. Several theory groups have refined their calculations to the extent that a detailed comparison, even with sensitive polarization observables, on reactions with low cross sections, is possible. Moreover, it has been recognized that, for a stringent test, the comparison between theory and experiment must be extended to data from as many reaction channels as possible. Neglecting electromagnetic channels the following five pionic channels are coupled:

$$NN \xleftrightarrow{\ 1\ } NN \xleftrightarrow{\ 2\ } \pi d \xrightarrow{\ 4\ } \pi d$$
$$\underset{\pi NN}{\overset{3\qquad 5}{\searrow\ \swarrow}}$$

The progress on the elastic and inelastic NN channels has been reviewed recently [2], therefore I restrict myself to new experimental and theoretical results on the reaction channels 2, 4, 5.

The πd elastic scattering reaction

The πd elastic scattering reaction has been of particular interest among the different πd reaction channels because this two-body reaction is relatively simple to deal with experimentally and theoretically. πd elastic scattering can be described by four complex helicity amplitudes which means that only seven independent observables are needed for a model independent amplitude reconstruction. Theoretically, πd elastic scattering is well described at forward angles by the impulse approximation, and only at large angles multiple scattering and pion absorption effects become important.

The experimental and theoretical progress until 1989 has been reviewed [3]. New data on iT_{11} at 50 MeV were reported by Stevenson et al. [4]. This experiment addressed the so-called "P_{11}-problem". Namely, in the course of developing a unified theory of the NN-πNN system the pion elastic scattering channel had to be coupled to the pion absorption channel via the P_{11} πN amplitude. Field theoretical considerations [5] led to a model in which this amplitude was split into a "pole" and a "non-pole" part, both of which were large, but the sum had to be small because of the on-shell behaviour of the P_{11} πN amplitude up to 300 MeV. It turned out that this new improved version of the Faddeev approach disagreed with tensor polarization data while the standard three body theory which neglected pion absorption or kept "pole" and "non-pole" term small, agreed quite well. This "P_{11} puzzle" became an interesting problem. As pointed out by Afnan and McLeod [6] it arises because in πd scattering, in the intermediate state the "pole" term is Pauli blocked for certain partial waves and the tensor polarization turned out to be very sensitive to the exact way in which the P_{11} πN t-matrix is split. A possible solution of the P_{11} problem was suggested by Jennings [7]. He observed that, in the impulse part of the πd amplitude, the Pauli exchange term to the contribution coming from the P_{11} pole part appears to be compensated to a large extent by a certain diagram with a four particle intermediate state: the initial pion is absorbed by one nucleon before the second nucleon emits the final pion. Jennings and Rinat [8] then added this contribution to the amplitude from the NN-πNN model to show that, in fact, this observation was adequate. In a more precise calculation including the self-energy of the pion for the propagating two-nucleon state (with no pion) it has been shown by Mizutani et al. [9] that the degree of the near cancellation is not as good as claimed by Jennings, and there is an imaginary part which is larger than the reduced real part. But the fact remains that the four-particle process is important and not to be neglected. Therefore, what one apparently has learnt from this P_{11} problem is the incompleteness of the NN-πNN model once contained within the two- and three- particle states: it should also include four particle $\pi\pi$NN states. In Fig. 1, the

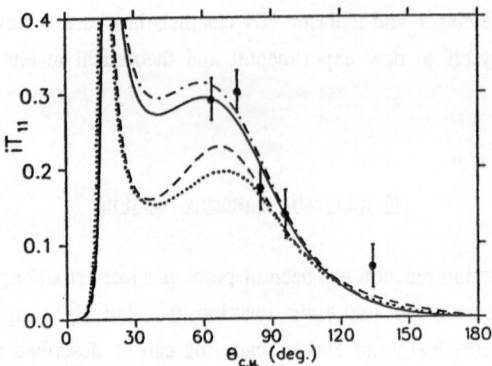

Fig. 1. iT_{11} measurements from Stevenson et al. (Ref. [4]) compared with theoretical calculations. The solid curve is the full NN-πNN prediction of Blankleider; the dashed curve is this model without any P_{11} contributions; the dash-dotted curve is the full calculation with the addition of the Jennings terms; and the dotted curve is the three-body theory of Garcilazo.

experimental results of Stevenson et al. are compared with Blankleiders full model (NN-πNN theory) with and without the "Jennings term", and the standard three body theory (Garcilazo or Blankleider without the P_{11} amplitude). Clearly, the NN-πNN theory is favoured but the accuracy of the data is not good enough to show the importance of the "Jennings term".

New data on iT_{11} and $\tau_{22} = T_{22} + T_{20}/\sqrt{6}$ at forward angles and for a number of energies in the Δ(3,3) region have been obtained by Weßler et al. [10] (Fig. 2). These measurements are part of a program to obtain seven independent observables at several energies and over the full angular range, as far as possible. At the same time precise iT_{11} data may inspire further theoretical investigations on the possibility of extracting information on the NΔ interaction. Ferreira et al. [11] tried to explain systematic discrepancies between cross section and vector analyzing power data, and "reliable" theoretical predictions in terms of a residual NΔ interaction which is not contained in the standard three-body description obtained from the relativistic Faddeev theory. However, the results from Ferreira et al. were obtained in calculations where the phenomenological NΔ interaction was added in Born term to the background few-body amplitudes. Recently, Alexandrou and Blankleider [12], and Garcilazo [13] investigated the effect of including such an NΔ interaction to all orders, within a unitary few body calculation of the πNN system. It was found that the higher order NΔ interaction terms have as much influence on the πd observables as the lowest order term, and, in fact, tend to cancel the effects obtained by adding the NΔ interaction in Born term. This, unfortunately, invalidates the conclusions of Ferreira et al., and leaves us with the question of the origin of the large discrepancies between theory and experiment.

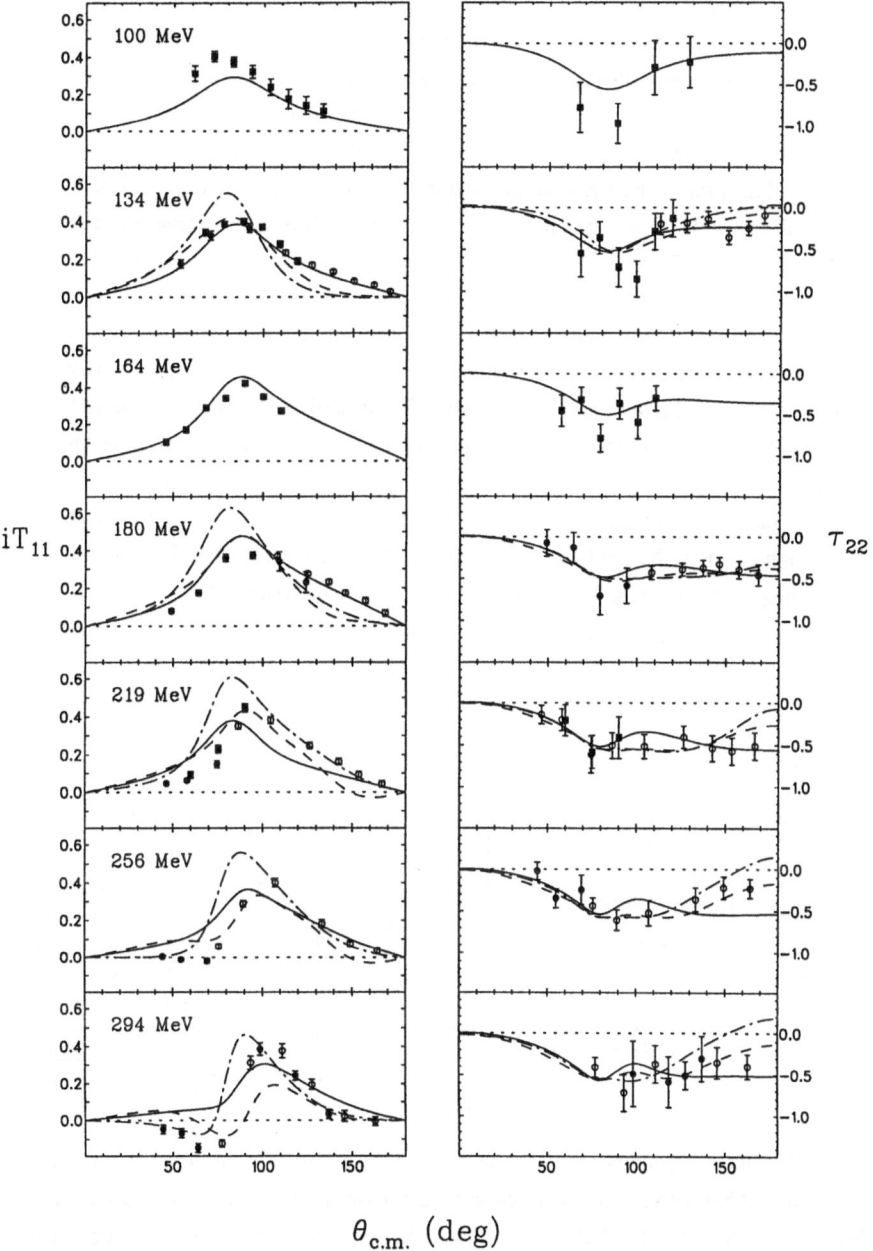

Fig. 2. Measurements of iT_{11} and τ_{22} from Wessler et al. (Ref. [10]). The full dots are recent measurements, the open circles earlier measurements by the same group. The curves are theoretical predictons from Garcilazo (solid curve), from Blankleider (dashed curve) and from the Lyon group (dot-dashed curve).

The determination of charge symmetry breaking effects from $\pi^{\pm}d$ elastic scattering is a challenging task. Experimentally, the measurement of absolute cross sections on the level of a few percent is very difficult. Theoretically, there is the question of how to treat the external Coulomb corrections (pure Coulomb, Coulomb-nuclear interference, and finite deuteron size effects), and internal Coulomb perturbations (mass differences between the neutron and the proton, mass and width differences among the intermediate Δ isobars, and the Δ's self-energy). These topics have been discussed by Rinat and Alexander [14] and by Fröhlich et al. [15]. Fröhlich et al. concluded that the reliability of the different Coulomb treatments must be carefully investigated before one may extract information on charge symmetry breaking. They demonstrated that the vector analyzing power iT_{11} is more sensitive to different Coulomb prescriptions than either $d\sigma/d\Omega$ or the tensor polarization t_{20}. For this reason Tacik et al. [16] measured iT_{11} in $\pi^{\pm}d$ elastic scattering at 180 MeV (where earlier charge symmetry breaking effects, found in total cross-section asymmetry measurements [17], passed through zero). In spite of the rather poor statistical accuracy of the ΔiT_{11} data the most refined Coulomb treatment of Fröhlich et al. is favoured. This information can now be used for the interpretation of recent absolute differential cross-section measurements by Kohler et al. [18]. Statistical and systematic uncertainties in the cross sections were typically ±3%. The results are shown in Fig. 3. The data of Kohler et

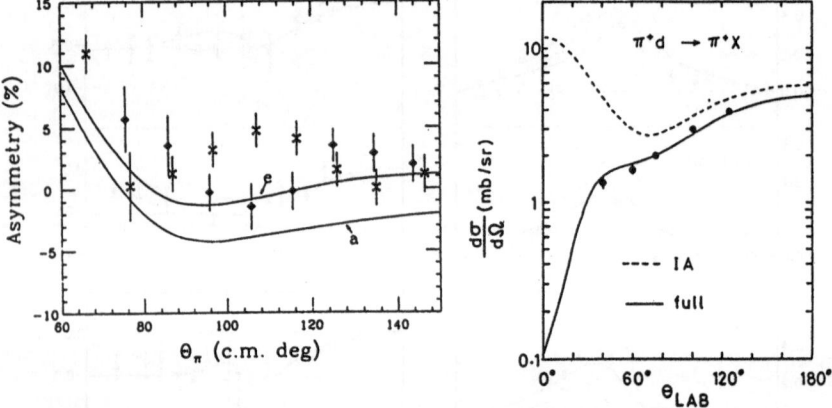

Fig. 3. Charge asymmetry for πd elastic scattering at 65.0 MeV. Crosses are earlier data by Balestri et al. (Ref. [19]), the diamonds new data from Kohler et al. (Ref. [18]). The two curves are calculations by Fröhlich et al. (Ref. [15]) where (a) is the simplest, and (e) the most comprehensive treatment of Coulomb effects.

Fig. 4. Differential cross section of the inclusive reaction $\pi^+d \rightarrow \pi^+X$ at 96.5 MeV. The experimental points are from Khandaker et al. (Ref. [20]). The curves are the predictions from Garcilazo, the solid curve corresponds to the full Faddeev calculation, the dashed one is the result of the impulse approximation.

al. contradict the "bump" structure in the angular distribution of the asymmetry found by Balestri et al. [19] and they are in reasonable agreement with the curve (e), which corresponds the most refined Coulomb treatment of Fröhlich et al., which was also favoured by the ΔiT_{11} measurement. The calculations by Fröhlich et al. do not include charge-symmetry-breaking mechanisms for the hadronic forces such as the mass differences of the Δ isobars.

The πd breakup reaction

The πd breakup reaction is the dominate πd reaction channel. Recently, the single and double differential cross sections for the inclusive $\pi^+d \to \pi^+X$ reaction were measured by Khandaker et al. [20] and calculated by Garcilazo [21]. As can be seen in Fig. 4 and Fig. 5 there is excellent agreement between the experimental data and the theoretical predictions. Obviously, triple differential cross sections obtained in kinematically complete experiments will provide a more stringent test of the theory. Apart from very early data, such measurements were performed by Hoftiezer et al. [22], Gyles et al. [23], List et al. [24], Pancella et al. [25], and recently by Mathie et al. [26]. In the experiments of Ref. [22] and [24] the pion and proton of the breakup reaction were detected by time-of-flight and a magnetic spectrometer, respectively, in the other experiments time-of-flight was used for the detection of both particles. Due to some acceptance problems of the magnetic spectrometer in Ref. [22], and underestimated corrections of out of the scattering plane events [23], there were inconsistencies between the various data sets. It was shown in Ref. [25] that after applying the proper corrections to the raw data there is good consistency now between the data of Ref. [23], [24], and [26].

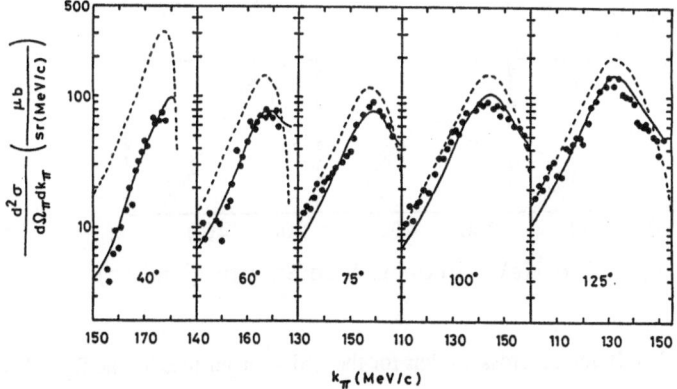

Fig. 5. Double differential cross section for the same reaction as in Fig. 4. The labelings of the data and the curves are the same as in Fig. 4.

Relativistic Faddeev calculations of the $\pi^+d \rightarrow \pi^+pn$ reaction at the kinematics of Ref. [22] was performed by Matsuyama [27] and Garcilazo [28]. The principle differences between both calculations are specific approximations and the treatment of the P_{11} πN partial wave. Matsuyama studied the contribution of the various reaction mechanisms such as the impulse process, pion multiple scattering and the NN final state interaction for several kinematical cases. He found that for small neutron recoil kinematics the impulse approximation is dominant and determines the gross structure of the cross section, but the higher-order processes also contribute in some cases. Although the calculation reproduced the experimental data of Ref. [22] within a factor of 2 the agreement was not satisfactory. For large neutron recoil kinematics the impulse term was found to be relatively small and the NN FSI contributed significantly. Garcilazo on the other obtained good agreement with the data of Ref. [22] for proton momenta above 400 MeV/c. In most kinematical cases there were severe discrepancies between theory and the data below 400 MeV/c proton momentum, apparently due to experimental problems.

It is well known that relativistic Faddeev calculations for the πd breakup reaction are technically very involved. Therefore it is very important to test such a theory not only

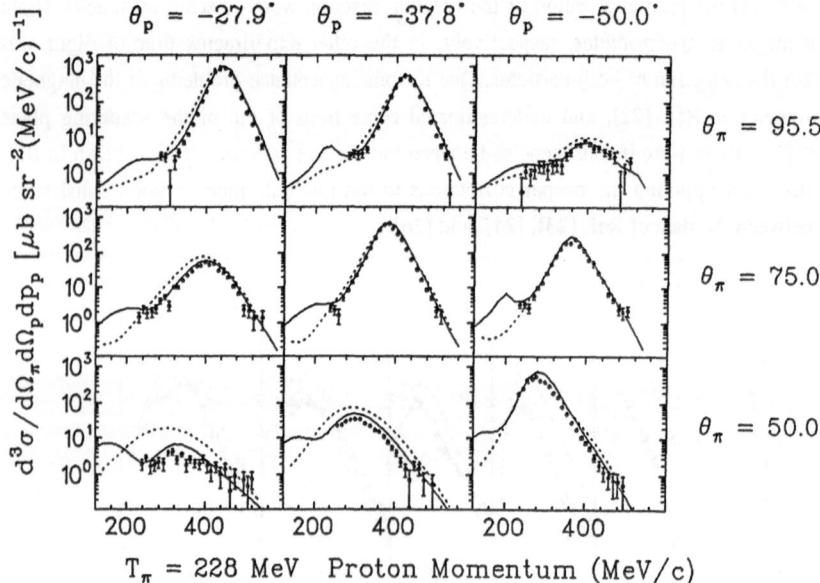

Fig. 6. Triple differential cross section for the $\pi^+d \rightarrow \pi^+pn$ reaction at $T_\pi = 294$ MeV for nine different pion-proton angle pairs plotted as a function of proton momentum. The data are from Gyles et al. (Ref. [23]). The solid curves are recent calculations by Garcilazo (Ref. [28]), the dotted curves are impulse approximation calculations by Blankleider and Matsuyama (Ref. [29]).

against "consistent" experimental data, but also against a different theoretical approach. Therefore I would like to show a comparison between the predictions from the relativistic Faddeev calculation (Garcilazo [28]) and the results of two independent impulse approximation calculations (Blankleider and Matsuyama [29]). The latter two calculations have been checked to agree within 10%. In Fig. 6, the solid curve presents the Faddeev calculation, the dashed curve presents the impulse approximation. In the quasi free scattering region, where both calculations should agree, excellent consistency is observed. It is also interesting to see how rapidly the impulse approximation deviates from the relativistic Faddeev calculation and the data as one moves away from the quasi free scattering kinematics. This contradicts some widely spread prejudice that the πd breakup reaction is the least interesting πNN reaction channel because it is dominated by quasi free scattering. This is only true for a narrow kinematically regime. Away from it the full Faddeev calculation is required to reproduce the data.

The πd charge exchange reaction

The πd charge exchange reaction was measured recently by Moinester et al. [30] and by Tacik et al. [31]. Moinester et al. measured the single differential cross section for the π^-d $\rightarrow$ π^0X reaction using the LAMPF π^0 spectrometer. These data agree very well with the predictions from Garcilazo. Tacik et al. performed a kinematically complete measurement of the π^+d $\rightarrow$ π^0pp reaction at T_π = 228 and 294 MeV. The two outgoing protons were detected in coincidence with plastic scintillator detectors, and their momenta determined by time-of-flight, at several angle pairs in regions of phase space far away from quasi free kinematics. The data were compared with predictions from the relativistic Faddeev calculation of Garcilazo, with the impulse approximation and a simple phase space calculation. The results, obtained at 294 MeV, are shown in Fig. 7. The solid lines represent the full three-body Faddeev calculation, the dashed lines the impulse approximation and the dotted line the three body phase space. There is remarkably good agreement between the data and the Faddeev calculation, considering that the calculations constitute predictions and have not been scaled or adjusted to fit the data. For many angle combinations at larger proton momenta the data are quite well described by the phase space calculation, indicating a constant matrix element. There are exceptions, however. For the combination (60°, -60°), for example, the decrease of the cross sections with increasing proton momentum is described much better by the full calculation. The most significant differences between the full calculation and the phase space occur at the lowest proton momenta, where the phase space is always decreasing while the full calculation is generally rising. This rise must be attributed to the increased importance of the quasifree process, π^+n $\rightarrow$ π^0p$_2$, with p$_1$ acting as a spectator at low momenta. It is interesting to note that for the two angle combinations (20°, -60°) and (20°, -45°) a similar situation occurs for large p$_1$ momenta. That is, the quasifree process π^+n $\rightarrow$ π^0p$_1$ with p$_2$ acting as a

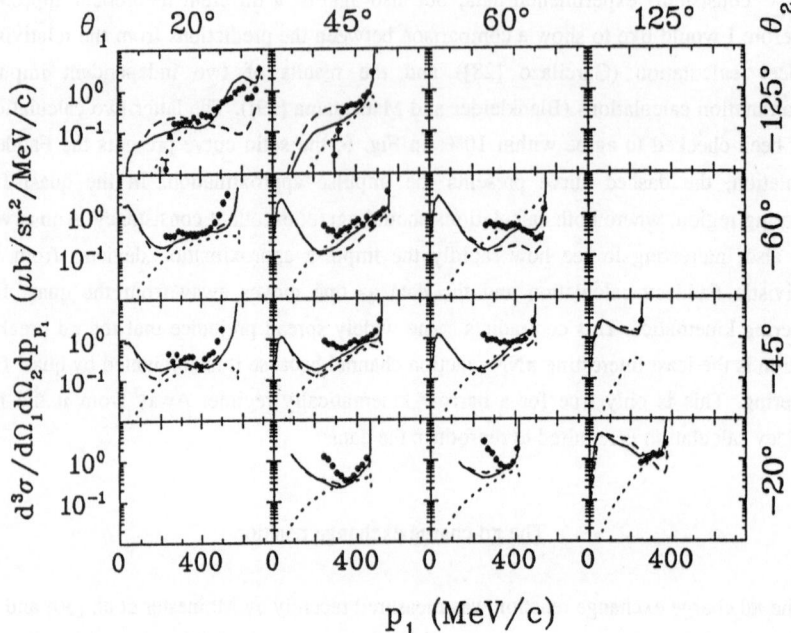

Fig. 7. Triple differential cross sections for the $\pi^+d \rightarrow \pi^0pp$ reaction at $T_\pi = 294$ MeV for twelve different proton-proton pairs as a fuction of the momentum of one proton. The data are from Tacik et al. (Ref. [31]). The solid and the dashed curves are the predictions from Garcilazo (the solid curve corresponds to the full Faddeev calculation, the dashed one to the impulse approximation). The dotted curves are the result of a phase space calculation by Tacik.

spectator becomes important. It is evident from Fig. 7, that in general the cross sections increase more quickly than the full calculation in these momentum ranges. It is interesting to speculate at this point whether the underprediction of the data is caused by an underestimation of the strength of the quasifree process, or whether it is caused by some interference between the quasifree and higher-order contributions. In order to shed some light on this question I show, as the dashed line in Fig. 7, the results of an impulse approximation calculation, that is, the calculation with only the quasifree contribution, ignoring all the higher-order terms. For many angle combinations, most notably (20°, -125°), (45°, -125°), (60°, -60°), and 125°, -20°), there are significant differences between the full calculation and the impulse approximation, especially at larger proton momenta. In these regions, the full calculation provides a better description of the data, indicating the importance of higher-order processes. At lower values of proton momenta, however, closer to quasifree kinematics, there is little difference between the full calculation and the impulse approximation. It is thus still not clear why the low momentum data are underpredicted. This may be due to (i) the deuteron wave function used in the calculation,

(ii) neglecting higher-order contributions such as the residual $N\Delta$ interaction (arising, for example, from the exchange of a ρ meson or from a direct $\pi\Delta\Delta$ vertex), or (iii) effects of the $(\pi, 2\pi)$ reaction channel which is only partially included in the present calculation through the π-nucleon input.

The πd absorption reaction

The πd absorption reaction has been reviewed in detail in Ref. [2] (as the inverse reaction). However, there are two new measurements using a polarized deuteron target. Feltham et al. [32] at TRIUMF have measured spin-transfer observables for the $\pi\vec{d} \to \vec{p}p$ reaction at a number of pion energies spanning the $\Delta(3,3)$ resonance (105 MeV $< T_\pi <$ 255 MeV). These parameters correspond to K_{SL} and K_{SS} of the $p\vec{p} \to \vec{d}\pi$ reaction for incident proton energies ranging from 500 - 800 MeV. These data can provide an important constraint on the determination of the partial wave amplitudes describing this fundamental reaction.

The second measurement was the study of the $\pi^+\vec{d} \to pp$ reaction at $T_\pi = 450$ MeV. It was performed at the Leningrad Nuclear Physics Institute (LNPI) by Bashanov et al. [33]. As described in Ref. [2] there have been considerable theoretical problems in describing this fundamental pion absorption reaction, in particular at higher energies. The energy T_π = 450 MeV is interesting because in this energy region an "anomalous" behaviour of the angular distribution of the differential cross section as well as the analyzing power A_{y0} was observed. For this reason the very sensitive observable iT_{11} was measured at several energies around $T_\pi = 450$ MeV. Much larger values of iT_{11} were found than was observed at lower energies by Smith et al. [34]. At present, there are no reliable theoretical predictions but it appears that it will be difficult to account for the anomalous energy dependence of iT_{11} which may be caused by the opening of the $NN \to N^*N$ channel.

In summary, further progress has been achieved on the different πd reaction channels. The πd breakup channels appear to be the easiest to be described by the theory. New data from TRIUMF on iT_{11} [35] may provide a better test. For the πd elastic channel spin transfer measurements are in preparation at PSI. With those, a model independent amplitude reconstruction should be possible. Theoretically, there remain the "P_{11}" and "ΔN" problems. A new comprehensive theoretical study treating the nucleon and the delta on the same footing is on the way [36]. The biggest theoretical headache presents the pion absorption reaction where some spin observables are extremely sensitive to almost all parts of the theory. It appears that, with plenty of data around, the understanding of the πNN system becomes mainly a theoretical task.

References

1. Garcilazo, H., Mizutani, T.: πNN Systems. Singapore: World Scientific Press 1990
2. Lehar, F.: Review Talk. 7th Int. Conference on Polarization Phenomena in Nuclear Physics, Paris, July 1990. Published by Les Editions de Physique, Les Ulis Cedex A, 1990.
   Boschitz, E.T.: ibid.
   Blankleider, B.: Review Talk. IUCF Topical Conference "Particle Production Near Threshold", Sept./Okt. 1990, Nashville, Indiana.
3. Boschitz, E.T.: Review Talk. Proceedings of the Third International Symposium on Pion-Nucleon and Nucleon-Nucleon Physics Gatchina, April 1989. Academy of Sciences of the USSR. Edited by B.P. Konstantinov, Leningrad Nuclear Physics Institute 1989.
4. Stevenson, N.R., Schubank, R.B., Shin, Y.M., Amaudruz, P., Delheij, P.P.J., Healey, D.C., Jennings, B.K., Ottewell, D.F., Sheffer, G., Smith, G.R., Wait, G.D., Brack, J.T., Feltham, A., Hanna, M., Johnson, R.R., Ronzon, F.M., Sossi, V., Vetterli, D., Weber, P., Grion, N., Rui, R., Kohler, M., Ristinen, R.A., Mathie, E.L., Tacik, R., Yeomans, M., Gossett, C.A., Wagner, G.J.: Phys.Rev.Lett. **65**, 1987(1990)
5. Blankleider, B., Afnan, I.R.: Phys. Rev. **C 24**, 1572(1981)
   Mizutani, T.: Phys. Rev. **C 24**, 2633(1981)
6. Afnan, I.R., McLeod, R.J.: Phys. Rev. **C 31**, 1821(1985)
7. Jennings, B.K.: Phys. Lett. **B 205**, 187(1988)
8. Jennings, B.K., Rinat, A.S.: Nucl. Phys. **A 485**, 421(1988)
9. Mizutani, T., Fayard, C., Lamot, G.H., Saghai, B.: Phys. Rev. **C 40**, 2763(1989)
10. Weßler, M., Boschitz, E.T., van den Brandt, B., Chaumette, P., Deregel, J., Durand, G., Efimovyhk, V., Fabre, J., Gill, D., Gyles, W., Konter, J.A., Kovaljov, A., Mango, S., Meier, R., Prokofiev, A., Ritt, S., Tacik, R., Thiel, W., Wait, G.: to be published
11. Ferreira, E., Dosch, H.G.: Phys. Rev. **C 40**, 1750(1989)
    and references therein.
12. Alexandrov, C., Blankleider, B.: Phys. Rev. **C 42**, 517(1990)
13. Garcilazo, H.: Phys. Rev. **C 42**, 2334(1990)
14. Rinat, A.S., Alexander, Y.: Nucl. Phys. **A 404**, 476(1983)
15. Fröhlich, J., Saghai, B., Fayard, C., Lamot, G.H.: Nucl. Phys. **A 435**, 738(1985)
16. Tacik, R., Boschitz, E.T., Gyles, W., Weßler, M., Mango, S., van den Brandt, B., Konter, J.A., Gill, D.R., Weber, P.: Phys. Rev. **C 42**, 1841(1990)
17. Pedroni, E., Gabathuler, K., Domingo, J.J, Hirt, W., Schwaller, P., Arvieux, J., Ingram, C.H.Q., Gretillat, P., Piffaretti, J., Tanner, N.W., Wilkin, C.: Nucl. Phys. **A 300**, 321(1978)

18. Kohler, M.D., Brack, J.T., Clausen, B., Kraushaar, J.J., Kriss, B.J., Ristinen, R.A., Vaziri, K., Smith, G.R., Ottewell, D.F., Sevior, M.E., Trelle, R.P., Stevenson, N.R.: Phys. Rev. C **44**, 15(1991)

19. Balestri, B., Fournier, G., Gerard, A., Miller, J., Morgenstern, J., Picard, J., Saghai, B., Vernin, P., Bertin, P.Y., Coupat, B., Lingeman, E.W.A., Seth, K.K.: Nucl. Phys. A **392**, 217(1983)

20. Khandaker, M.A., Doss, M., Halpern, I., Murakami, T., Storm, D.W., Tieger, D.R., Burger, W.J.: Phys. Rev. C **44**, 24(1991)

21. Garcilazo,H.: Phys. Rev. Lett. **65**, 293(1990)

22. Hoftiezer, J.H., Baker, S.D., Clement, J.M., Dragoset, W.H., Duck, I.M., Felder, R.D., Judd, D.M., Mutchler, G.S., Pepin, G.P., Phillips, G.C., Umland, E.A., Allred, J.C., Hungerford III, E.V., Mayes, B.W., Pinsky, L.S., Williams, T.M., Furic, M., von Witsch, W.: Phys. Rev. C **23**, 407(1981)

23. Gyles, W., Boschitz, E.T., Garcilazo, H., List, W., Mathie, E.L., Ottermann, C.R., Smith, G.R., Tacik, R., Johnson, R.R.: Phys. Rev. C **33**, 583(1986)

24. List, W., Boschitz, E.T., Garcilazo, H., Gyles, W., Ottermann, C.R., Tacik, R., Mango, S., Konter, J.A., van den Brandt, B., Smith, G.R.: Phys. Rev. C **37**, 1594(1988)

25. Pancella, P.V., Mutchler, G.S., Baker, S.D., Kruk, J.W., Duck, I.M., Corcoran, M.D., Phillips, G.C., Clement, J.M, Buchanan, J.A., Mayes, B.W., Pinsky, L.S., von Witsch, W., Andrade, E., Garcilazo, H., Laget, J.M.: Phys. Rev. C **38**, 2716(1988)

26. Mathie, E.L., Pafilis, V., Huber, G.M., Lolos, G.J., Naqvi, S.I.H., Papandreou, Z., Yeomans, D.M., Sevior, M., Trelle, R.P., Ottewell, D., Smith, G.R., Healey, D., Garcilazo, H.: Phys. Rev. C **41**, 193(1990)

27. Matsuyama, A.: Nucl. Phys. A **379**, 415(1982)

28. Garcilazo, H.: Phys. Rev. C **35**, 1828(1987)

29. Blankleider, B., Matsuyama, A.: private communication

30. Moinester, M.A., Erell, A., Piasetsky, E., Alster, J., Bowman, J.D., Baer, H.W., Cooper, M.D., Irom, F., Sennhausser, U., Ziock, H., Leach, M.: XI International Conference on Particle and Nuclei, Kyoto, 1987, Abstract Book I (unpublished)

31. Tacik, R., Boschitz, E.T., Gyles, W., Ottermann, C.R., Weßler, M., Wiedner, U., Garcilazo, H., Johnson, R.R.: Phys. Rev. C **42**, 1846(1990)

32. Feltham, A., Trelle, R.P., Jones, G., Gill, D., Healey, D., Lolos, G.J., Mathie, E.L., Olszewski, R., Ottewell, D., Papandreou, Z., Pavan, M., Rui, R., Sevior, M., Smith, G.R., Sossi, V., Wait, G., Walden, P., Weber, P.: Phys. Rev. Lett. **20**, 2573(1991)

33. Prokofiev, A.N.: private communication

34. Smith, G.R., Mathie, E.L., Boschitz, E.T., Ottermann, C.R., Gyles, W., List, W., Mango, S., Konter, J.A., van den Brandt, B., Olszewski, R., Johnson, R.R.: Phys. Rev. C **30**, 980(1984)

35. Mathie, E.L.: private communication

36. Blankleider, B.: private communication

Few-Body Systems, Suppl. 6, 148—163 (1992)

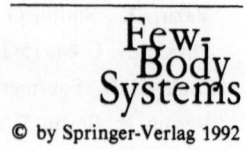

© by Springer-Verlag 1992

MULTIQUARK EXOTICS

I.M.Narodetskii

Institute for Theoretical and Experimental Physics,

Moscow, 117259 the USSR

The talk will survey a few of the recent advaces in study of multiquark confined states and their connection with hadron-hadron scattering. We also briefly discuss some relevant theoretical results obtained within the nonperturbative QCD.

1. Introduction

The calculation of the spectrum of multiquark states has been a topic of investigation for quark model theorists for many years starting from the pioneering papers by Jaffe [1] and Aerts et al. [2]. The multiquark hadrons have recently became the subject of renewed interest. There are three notable reasons for this interest. First, multiquark systems are a nice testing ground for QCD motivated hadron models. Multiquark hadrons have been studied within the framework of the various quark models, inqluding mean-field type approximation [3], pairwise color-electric force of the $\lambda_i \lambda_j$ type [4-7], flux-tube quark models [8], etc. Second is the question of the nature of some of the meson resonances. In particular, the resonance AX observed at ASTERIX and Crystal Barrel detctor in $\bar{p}p$ annihilation at rest [9] caused some suggestions about $q^2\bar{q}^2$ nature of this state. Another candidates for $q^2\bar{q}^2$ states are C(1480) which was observed in $\phi\pi^0$ system, and $\rho(1405)$ found in the amplitude analysis of $\eta\pi^0$ [10]. Finally, and closely related to both preceeding points, there is a question of the nature of the short-range hadron-hadron interaction [11].

In this talk, we shall discuss some recent works along this line [12-16]. References on other related works can be found in these references. Before discussing the formalism, it is instructive to start with a brief look at the main ideas of the potential description of multiquark states.

2. Multiquark Hadrons as Bound States of Constituent quarks.

In the quark potential model quarks as the constituents of hadrons are interpreted as quasiparticles to be used in calculations based on effective Hamiltonians. The Hamiltonian formalism implies that the quark motion is slow relative to the characteristic scale of the vacuum fields, so the nonlocal effects in the interaction of quarks with the gluonic fields can be disregarded. To get some quantitative estimates[1] it is convenient to look at the gauge invariant Green function describing the propagation of a color singlet $q\bar{q}$ system in the vacuum background fields. In the Feynman-Schwinger representation it is given by [15]

$$G(x,\bar{x};y,\bar{y}) = \int_0^\infty ds \int_0^\infty d\bar{s} \int \mathcal{D}z \int \mathcal{D}\bar{z} \, \exp(-K) \, \langle W(C) \rangle, \qquad (1)$$

where

$$K = m_q^2 s + m_{\bar{q}}^2 \bar{s} + \frac{1}{4} \int_0^s \dot{z}_\mu^2(\sigma)d\sigma + \frac{1}{4} \int_0^{\bar{s}} \bar{z}_\mu^2(\bar{\sigma})d\bar{\sigma}, \qquad (2)$$

and

---

[1]Here we follows discussion of ref. [14].

$$\langle W(C) \rangle = \frac{1}{n_c} \, Sp \, \langle P\exp(ig\!\int\! A_\mu dz_\mu)\rangle \qquad (3)$$

is the Wilson loop integral over the closed contour C consisting of a quark path $z_\mu(\sigma)$ and an antiquark path $\bar{z}_\mu(\bar{\sigma})$ ($z_\mu(0) = x_\mu$, $z_\mu(s) = y_\mu$, $\bar{x}_\mu(0) = \bar{x}_\mu$, $\bar{z}_\mu(\bar{s}) = \bar{y}_\mu$) and initial and final links connecting points $(x,\bar{x})$ and $(y,\bar{y})$. The path integral is effectively performed over a bunch of trajectories around the classical one. The later has a typical time scale T_q - a period of quark motion around the classical orbit. For heavy quarks, their motion is governed by the color Coulomb law, and T_q is a period of quark motion over the Coulomb orbit,

$$T_q \approx \frac{4\pi n^3}{m_q(\frac{4}{3}\,\alpha_s)^2} , \qquad (4)$$

where $\alpha_s(r)$ is the effective QCD running constant. This gives

$$T_q \approx 6.7 \text{ fm for } c\bar{c}, \quad T_q \approx 3.5 \text{ fm for } b\bar{b}. \qquad (5)$$

For light quarks orbiting in the linear potential, the classical period of the motion is

$$T_q \approx 2 \, \{m_q(M-2m_q)\}^{1/2}/\sigma, \qquad (6)$$

where $m_q \approx 0.3$ GeV is the constituent mass of a quark, M is the mass of a meson, and σ is the string tension. For M $\approx$ 1 GeV one has $T_q \approx 1$ fm. These values should be compared with the space-time gluonic scale T_g which enter all the correlation functions of the gluonic fields through $\langle W(C)\rangle$. The Monte-Carlo calculations suggest that $T_q \approx 0.3$ GeV [17]. In this case $T_q \gg T_g$, and the $q\bar{q}$ interaction reduces to the local one[2].

These considerations partly justify the use of the constituent quark models where the phenomenological potential and the perturbative spin-spin interaction are employed as interquark

---

[2]Possible exclusion is toponium (if $m_t > 100$ GeV), and accuracy should be checked for the lowest S-wave mesons.

potential. In this model, any multiquark state can be constructed in exactly the same way as ordinary $q\bar{q}$ mesons or qqq baryons. The wave function describing a system of n quarks and m antiquarks (n+m ≥ 4) satisfies the Schrödinger equation

$$(H_0 + V) \; \psi = E\psi. \tag{7}$$

The only requirement on the potential energy operator V is that the total Hamiltonian $H = H_0 + V$ would have a continuous spectrum of color-singlet hadrons which do not interact at large distances because of their color neutrality. However, in the quark constituent model there is no obvious relation between the $q\bar{q}$ potential and the interaction governing the multiquark systems. For baryons it is usually assumed [18]

$$V(qqq) = 1/2 \sum_{i<j} V_{ij}, \tag{8}$$

where V_{ij} is the $q\bar{q}$ potential. The simplest generalization of this potential is given by [19]

$$V = -3/16 \sum_{i<j} (\lambda_i \lambda_j) \; V_{ij}. \tag{9}$$

Here λ_i are the SU(3) color generators for the i-th quark normalized as $\lambda_i^a \lambda_i^a = 16/3$.

The validity of this model for the description of multiquark hadrons has been challenged [20], because the model leads to the spurious long range Van der Waals forces between separated hadrons which contradict experiment. The reason for the occurence these unwanted faetures is its failure to incorporate the property, fundamental to QCD, of *local gauge invariance* properly. Nevertheless, one may still consider the quark potential model as a useful approximation at small interquark distances, although the precise range of this approximation is difficult to estimate. In what follows, for the sake of simplicity and clarity, we shall restrict our discussion within the framework of the additive quark force model (9) and postpone a discussion of more realistic multiquark interactions until Section 5.

3. Multiquark exotics. Confined multiquark states.

The notion of confined multiquark states is more or less evident in the bag model, where confinement is simulated by the boundary condition given by the bag pressure. In the quark potential model the situaion is different because the physical spectrum of The Hamiltonian H has no discrete eigenvalues corresponding to confined states. Nevertheless, the multiquark configurations constructed on a restricted basis of states are always confined. Only states including the complete basis of color, spin, isospin and coordinate functions have a chance to dissociate into colorless hadrons.

To illustrate this point we first consider 4-quark states – the simplest non-trivial example of multiquark system which can exist in both meson-meson and confined phases. In the $q^2\bar{q}^2$ system, there are two independent color wave functions, $|c_T\rangle = |3-\bar{3}\rangle$ and $|c_M\rangle = |6-\bar{6}\rangle$, where $\bar{3}$ and 6 are (anti)triplet and sextet representations for diquarks while 3 and $\bar{6}$ are the corresponding representations for antidiquarks. It follows from eq. (9) that pure T- and M-states are always confined in all relative coordinates. They are defined by the Schrödinger equation

$$(H_0 + V_c) \, \Psi_\nu = E_\nu \Psi_\nu, \tag{10}$$

where E_ν are the (discrete) masses of the confined states, Ψ_ν are the corresponding wave functions, and V_c is given by

$$V_c = \langle c_T|V|c_T \rangle = 1/2 \, (V_{12} + V_{34}) + 1/4 \, (V_{13} + V_{24} + V_{14} + V_{23}),$$

and

$$V_c = \langle c_M|V|c_M \rangle = -1/4 \, (V_{12} + V_{34}) + 5/8 \, (V_{13} + V_{24} + V_{14} + V_{23})$$

for T- and M-states, respectively.

In the q^6 system the situation is a bit more complicated.

Here there exist five independent color-singlet wave functions belonging to the representation f = {222} of $SU(3)_c$. The kernel (9) as an operator in color space may also be written as a sum of ten representations, nine of which correspond to the symmetry $[42]_c$ and the remainder to the symmetry $[6]_c$. In particular, the symmetric kernel which is appropriate to define the six-quark primitives in the quark potential model is given by

$$V_c = 1/5 \; SpV = 1/5 \sum_{i<j} V_{ij}.$$

We emphasize that existence of multiquark confined states in the quark potential model is based on very general properties of the interaction (9) as an operator in color space and is valid for any multiquark system: $q^4\bar{q}$, $q^3\bar{q}^3$, q^6, q^9, etc, although the techical difficulties grows rapidly as the number of quarks is increased. In the flux tube model [8], where the above arguments can not be applied, multiquark confined states arise by imposing appropriate boundary conditions on the multiquark wave function.

More delicate is the problem of the interpretation of these states and their connection with hadron-hadron scattering. Since most of the confined states lie above the disintegration threshold they will be unstable against fission into hadronic channels and may dissociate through the superallowed decay that is a zeroth-order process in the coupling constant α_s. The presence of the open hadron channels raises an important problem of whether or not the multiquark states can manifest themselves as conventional resonances. In principle they could be relatively narrow if their decay into hadrons is suppressed for some dynamical reasons, otherwise they may totally disappear in the continuum. In the latter case the multiquark hadrons play a similar role as doorways states in strong interactions. The problem of the interplay between the properties of multiquark states and particular features of hadronic interactions will be discussed in the next section.

4. Hadron Interactions in the QPM.

The standard treatment of the scattering problem in the constituent quark model uses the RGM Ansatz for the multiquark function Ψ:

$$\Psi = \Psi_{RGM} = \mathscr{A} \left\{ \sum \Phi_{AB}(\zeta_A, \zeta_B) \chi_{AB}(r_{AB}) \right\}, \tag{11}$$

where A,B are the color singlet hadrons, $\Phi_{AB}(\zeta_A, \zeta_B)$ is the product of their intrinsic wave functions, χ_{AB} describes the relative motion of A and B and $\mathscr{A}$ is the antisymmetrizer for quarks and antiquarks.

In principle the expansion (11) may be considered to be complete if all contributions of the excited states of clusters A and B are taken into account. In practice, however, the commonly used approximation is to neglect the intrinsic excitations of each clusters limiting (for the case of the NN sistem) to the expansion of the antisymmetrized product of $(0s)^3$ three-quark clusters in a relative S-wave[3]. To overcome this approximation we need a more flexible theoretical Ansatz to treat a possible effect of excited quark clusters.

Keeping these considerations in mind we can try to generalize eq. (11) by introducing a kind of interpolating Ansatz [12]

$$\Psi = \Psi_{RGM} + \sum_\nu a_\nu(E)\Psi_\nu, \tag{12}$$

where Ψ_ν have been defined in eq. (10) and $a_\nu(E)$ are the variational parameters to be defined below. Because Ψ in eq. (12) should be considered as a trial function, there is no problem with quark double counting, though Ψ_{RGM} and Ψ_ν are not orthogonal. Note

---

[3]Recall that in this approximation the interaction between the quark clusters looks like a short range repulsion. To find a quantitative description of the attractive NN force one has to add to quark and gluon exchange the pion exchange between quarks and also the phenomenological σ-meson exchange NN potential [21].

that admixture of the hidden color states in the form of the multiquark bag only, removes the unphysical color Van der Waals forces, since these configurations are restricted to short distances.

In order to eliminate the quark variables we project the multiquark Shrödinger equation onto Ψ_ν and onto the state

$$\langle r | h_{AB} \rangle = \Phi_{AB}(\zeta_A, \zeta_B)\delta(r-r').$$

This procedure is well known in nuclear cluster calculations [22]. In this way, we get a system of coupled equations for χ_{AB} and $a_\nu(E)$. Upon eliminating a_ν one gets the effective Schrödinger equation for χ_{AB} ($\equiv \chi_\alpha$) that is given in symbolic shorthand by

$$H_0\chi_\alpha + \sum_\alpha (K_{\alpha\alpha'} + V_{\alpha\alpha'})\chi_{\alpha'} = (E-m_\alpha)\chi_\alpha, \tag{13}$$

with $m_\alpha = m_A + m_B$. The kernels in eq. (13) have an obvious physical meaning and correspond to two different interaction mechanisms relevant to the description of the short-range hadron-hadron interactions (see Fig.1). The first one, $K_{\alpha\alpha'}$, is what is usually referred to as the quark-gluon exchange

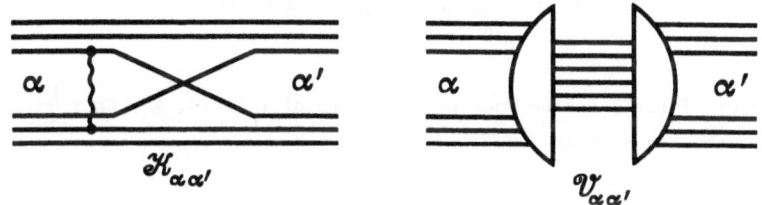

Fig.1

contribution. It is a nonlocal kernel due to the effect of antisymmetrization given in eq. (11) . The second kernel, $V_{\alpha\alpha'}$, is the hadronic force due to transitions into intermediate states of multiquark bags. We shall refer to this kernel as the Quark Optical Potential (QOP). It has the form

$$V_{\alpha\alpha'}(r,r';E) = \sum_{\nu} \frac{f_{\alpha\nu}(r,E)f_{\alpha'\nu}(r',E)}{E-E_{\nu}}, \tag{14}$$

where

$$f_{\alpha\nu}(r,E) = N \left[\lambda_{\alpha\nu}(r) + g_{\alpha\nu}(E_{\nu} - E)y_{\alpha\nu}(r)\right], \tag{15}$$

with

$$\lambda_{\alpha\nu}(r) = \langle\Psi_{\nu}| V - V_{c}|\Phi_{\alpha}\rangle, \quad \text{and} \quad g_{\alpha\nu}y_{\alpha\nu}(r) = \langle\Psi_{\nu}|\Phi_{\alpha}\rangle, \tag{16}$$

and N is the combinatorial factor. In eq. (16) the matrix element containing color, spin and isospin is written explicitly in terms of the fractional parentage coefficient $g_{\alpha\nu}$. The remaining part, $y_{\alpha\nu}$, is the overlap integral of the coordinate parts of Ψ_{ν} and Φ_{α}.

It is convenient, for illustrative purposes, to consider an extremely simplified situation when it is possible to neglect the kernel K in eq. (13). In this case one can easily solve the Schrödinger equation for the scattering amplitude $A(E)$. The partial wave amplitude is given by

$$A(E) = \sum_{\mu\nu} f_{\mu}(k,E)\tau_{\mu\nu}(E)f_{\nu}(k,E), \tag{17}$$

where $f_{\mu}(k,E)$ is the Fourier transform of $f_{\mu}(k,E)$, $E = 2m + \frac{k^2}{m}$, m is the hadron mass, and

$$\tau_{\mu\nu}^{-1}(E) = (E-E_{\nu})\delta_{\mu\nu} - \Delta_{\mu\nu}(E), \quad \Delta_{\mu\nu}(E) = \langle f_{\mu}|G_{0}|f_{\nu}\rangle, \tag{18}$$

with $G_{0}(E)$ being the Green's function of two non-interacting hadrons.

Eqs. (17,18) show explicitly how the coupling of hadrons to a confined multiquark states causes a pole in the S-matrix. Keeping for simplicity only one multiquark state one has a nonlinear equation that determines the position of the S-matrix pole E_{R} in the complex energy-plane

$$E_R - E_1 - \Delta_{11}(E_R) = 0. \qquad (19)$$

The difference between E_R and E_1 comprises the so-called hadronic shift of a multiquark confined state due to its coupling to open hadron channels. Generally, eq. (19) may have either real or complex solutions. The real solutions correspond to the multiquark states bound relative the dissociation threshold, while the comlex poles can manifest themselves as resonances in scattering channel. If the imaginary part of the pole position is small, then hadron coupling to the confined state generates a strong energy dependence in the S-matrix, otherwise the role of the confined state is merely to provide a microscopic picture of the short range mechanism. One can easily show that in the $\lambda_i\lambda_j$ model with spin-spin perturbative interaction the form-factor $f(k,E)$ is proportional to δm, where $\delta m = \Delta m(Nq) - 2\Delta m(hadrons)$ is the difference of the color magnetic energies. Therefore

$$f(c^{2-2}\!c^2) : f(s^2 s^{2-2}) : f(o^2 o^{2-2}) = (m_{J/\Psi} - m_{\eta_c}) : (m_\phi - m_{s\bar{s}}) : (m_\rho - m_\pi) \approx 1:2:5$$

Since $\Gamma_R = kf(k_R)^2$, we conclude that the multiquark resonances are seen more clearly in the heavy quark systems. In the light quark system relatively narrow resonances can occur only at the disintegration threshold.

In order to make these ideas more concrete, we shall now apply them for several multiquark systems.

5. Examples

In fig. 2 effect of superallowed rearrangement of the quintet $c^{2-2}c^2$ state with spin S=2 and $l_{qq} = 1_{\overline{qq}} = 1 = 0$[4] is shown for the full Cornell model of ref. [23] for two values of the charmed quark constituent mass. The clear resonance structures in the spectrum $\sigma(E) = 4\pi m^2 |A(E)|^2$ are showing up shifted relative to the masses M_4 of confined states by 120-150 MeV. For the case $m_q = 1.84$

---

[4]Here l_{qq} ($1_{\overline{qq}}$) is the qq ($\overline{q}\overline{q}$) relative angular momentum and l is the orbital momentum of diquark relative to antidiquark.

158

GeV one gets a resonance with mass 6.65 GeV and width ≈ 200 MeV. For m_q = 1.65 GeV one gets even a narrower resonance with width ≈ 100 MeV and mass 6.57 GeV [13].

The similar analysis of the AX resonance supplied by the $^3P_0$ qq pair creation supports the existence of resonance structure in tensor isoscalar channel in the vicinity of the $\rho\rho$ threshold. The found exotic 4q state seems to be of the mesonic molecule type rather than the Jaffe multiquark bag [24].

We shall now turn to a consideration of the formal technique used in the description of NN scattering [12]. The most essential point of our analysis is the following theorem:

The terms of QOP with $x_\nu \geq 1$ produce (for the energy $E \leq E_\nu$ effective repulsion while those with $x_\nu < 1$ correspond to effective attraction in the NN system.

Here x_ν^2 = 10 $g_{ST}^2 x_\nu^2 z_\nu^2$, where g_{ST} are the fractional parentage coeffitients and z_ν are the spectroscopic values of the 6q

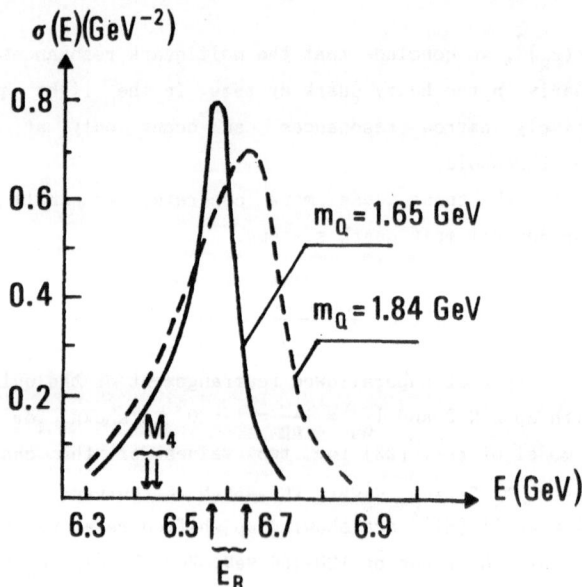

Fig.2

confined states. A simple estimation in the chromoharmonic model yields $z_1 \approx 1$, $z_2 = 0.213,\ldots$ Recalling the values $g_{10} = g_{01} = 1/3$ one gets

$$x_1^{\ 2} = 10/9 z_1^{\ 2} \approx 1.1, \qquad x_\nu^{\ 2} \ll 1 \ (\nu \geq 2)$$

Thus we are led to the important conclusion that the origin of a possible attraction in the nucleon-nucleon QOP is the admixture of excited multiquark states in the NN wave function.

To find out whether this effect is sufficient to obtain a net attraction in the NN system or it is masked by a large repulsive term due to the admixture of the 6q ground state it is convenient to introduce the notion of eigenvalues $\mu_\nu(k)$ for the kernel $G_0 V$. Recall that the position of a bound state with the energy $-\alpha_0^{\ 2}/m$ is

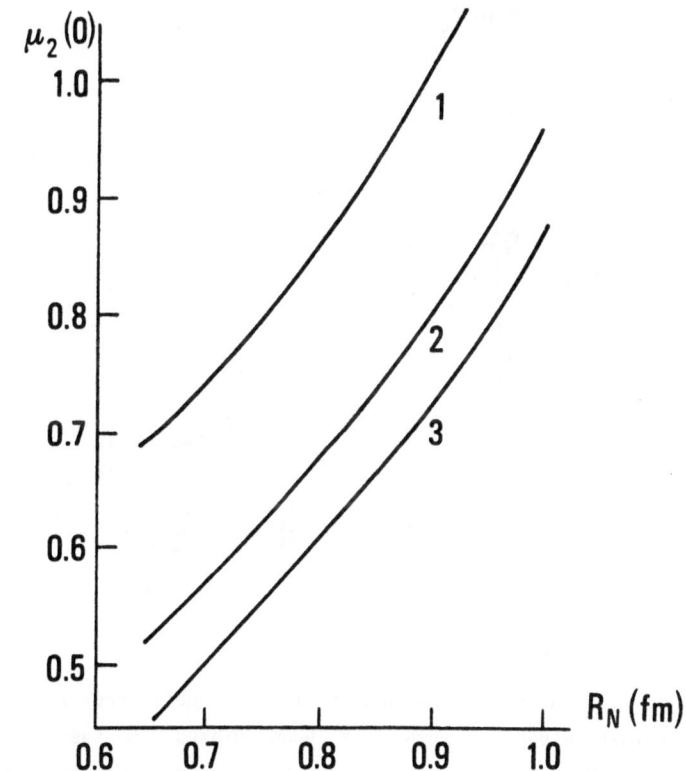

Fig.3. $\mu_2(0)$ for different m_q and r_N. Curves 1 - 3 correspond

defined by the equation $\mu_\nu(i\alpha_0) = 1$. A close consideration [12] shows that for the two-level QOP the quantity $\mu_2(0)$ can serve as a signal for binding the 6q system. To visualize the dependence of μ_2 on the parameters of the quark potential model we now consider a number of models differing in values of m_q and the charge radius of the proton R_N.

For every model the parameters of the qq interaction can be fixed by the physical masses N and Δ. The result is summarized in Fig. 3. For the values commonly used in the quark potential model ($m_q \approx 300$ MeV, $R_N \approx 0.6$ fm) the two-level QOP produces a half of attraction needed to bind the 6q system. However, the find attraction is not sufficient to obtain the positive phase shifts

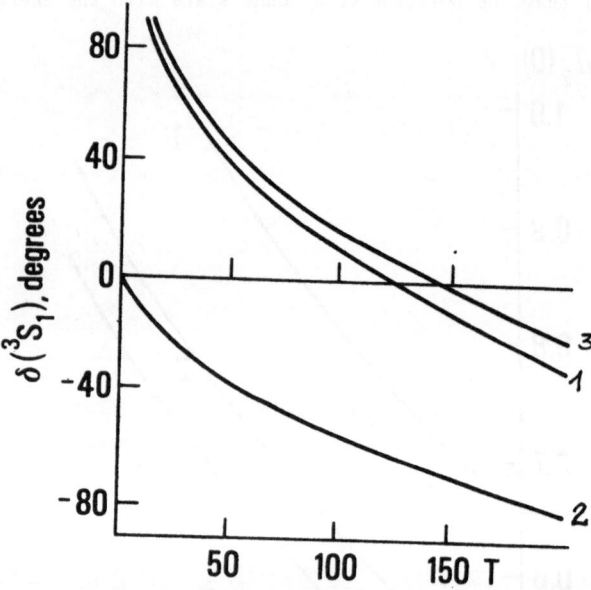

Fig.4. NN phase shifts for $m_q = 100$ MeV. 1 - QOP + RGM, 2 - QOP with $\nu = 1$, 3 - QOP with $\nu = 1,2$

at low energy. An evident explanation for this deficiency is that the explicit meson exchange has been missed in the above calculation. The meson forces are certainly not negligible but their contribution to the short-range NN interaction may be not so dominant as it is usually assumed.

Clearly there is a room in the quark potential model for an increase the strengh of the attractive NN interaction by increasing R_N and lowering the constituent quark mass m_q. In particular, for $R_N \approx 0.8$ fm one can find attraction which is strong enough to produce a weakly-bound 6q bound state but at the price of introducing unreasonably small quark masses, $m_q \approx 100$ MeV. Note that in the contribution of the RGM kern els in this case is marginal (Fig.4). In the absence of exact solution, this is a first indication that the confined 6q states play an important role in the NN scattering. Recall in this connection that the QOP method gives us the guidance in finding an optimal phenomenological parametrizatior of the core region in "realistic" NN potential, see ref. [11] for a review.

6.Conclusions. From Phenomenology to QCD.

In this talk we have outlined techniques and results relevant to a discussion of multiquark states in the context of the constituent quark model. The most interesting avenue of further research is in the direction of more realistic quark models, either derived from the phenomenology or put on a more sound theoretical basis. In particular, using the gauge-invariant representation of multiquark Green function and separating non-perturbative effects in the form of Wilson loop one can show that the confining forces have no $\lambda_i\lambda_j$ factor, but instead at large distances can be written as

$$V(r_i) = \sigma L_{min}(r_i), \qquad (20)$$

where L_{min} is the minimal length of multiquark strings. The potential of eq.(20) is identical with that derived in the flux tube quark model based upon stong-coupling QCD.A string starts from every quark i and ends on an antiquark $\bar{i}$. Three strings can end or start from an antisymmetric Y-junction. In particular, for meson $L_{min} = r$ (interquark distance) and for baryon L_{min}

162

corresponding to Y-junction is given by [25]

$$L_{min} = \frac{\sqrt{3}}{2} (r_{ij}^2 + r_k^2 + 2 |[r_{ij}, r_k]|)^{1/2}. \tag{21}$$

An analytic expressiom for two Y-junctions in the $q^2\bar{q}^2$ system has benn found in ref. [26] for a particular case of the planar geometry. Using then the non-perturbative approach developed in series of papers (see [14] for a review) one obtaines under some assumptions the proper time Hamiltonian for the multiquark hadrons. Even though it has been formulated in the literature, its effect in hadron spectroscopy has not been studied in details. This problem certainly deserves a further investigation.

References

1. R.L.Jaffe, Phys. Rev. Lett. 38, 195 (1977)

2. Th.M.Aerts, P.G.J.Mulders, and J.J. de Swart, Phys. Rev. D17, 260 (1978)

3. T.Goldman *et al.* Nucl. Phys. A481, 621 (1988)

4. J.Weinstein and N.Isgur, Phys. Rev. D27, 588 (1983), K.Maltman and S.Godfrey, Nucl. Phys. A452, 669 (1986)

5. A.M.Badalyan *et al.* Nucl. Phys. B282, 85 (1985)

6. B.Silvestre-Brac *et al.* Phys. Rev. D36, 2083 (1987)

7. Yu.S.Kalashnikova *et al.* Sov. J. Nucl. Phys. 46, 689 (1987)

8. J.Carlson and V.R.Pandharipande. Phys. Rev. D43, 1652 (1981)

9. B.May *et al.* Phys. Lett. B225, 450 (1989) K.Peters *et al.* Cristal Barrel Collaboration, in Proc. of the First Biennial Conference on Low Energy Antiproton Physics (Stockholm, 1990), p. 161

10. Particle Data Group: Phys. Lett. B239, (1990)

11. B.L.G.Bakker, and I.M.Narodetskii, *Multiquark Exotics in Hadronic Interactions.* Preprints ITEP-83,84 (1991)

12. Yu.S.Kalashnikova, I.M.Narodetskii. Int. J. Mod. Phys. A, 4, 335 (1989)

13. Yu.S.Kalashnikova, I.M.Narodetskii. Z. Phys., C43, 273 (1989)

14. Yu.A.Simonov, *A New Non-perturbative QCD Approach to Hadron Physics,* TPI-MINN-90/10-T, *February* 1990.

15. Yu.A.Simonov, Nucl. Phys. B307, 512 (1988), B324, 67 (1989)

16. A.M.Badalyan, Frascati preprint LNF-91/017(R)

17. A. Di Giacomo, private communication, September 1991

18. J.M.Richard, Phys. Lett. 100B, 515 (1981)

19. H.J.Lipkin, Phys. Lett. 45B, 267 (1973)

20. O.Greenberg, H.Lipkin, Nucl. Phys. A370, 349 (1981)

21. A.Faessler, Progr. Part. Nucl. Phys. 11, 171 (1984)

22. A.I.Baz, M.V.Zhukov, ZhETF 70, 397 (1976)

23. E.Eichten *et al.* Phys. Rev. D21, 203 (1980)

24. Yu.S.Kalashnikova, in press.

25. M.Fabre de la Ripelle, Yu.A.Simonov, Orsay preprint (1991)

26. K.Boreskov, A.Dubin, I.M.Narodetskii, in press

Few-Body Systems, Suppl. 6, 164—175 (1992)

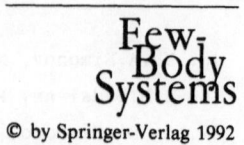

MULTIQUARK SPECTRA

IN LARGE HARMONIC OSCILLATOR BASES

R. CEULENEER and C. SEMAY*

Faculté des Sciences, Université de Mons-Hainaut, B-7000 MONS, Belgium.

Abstract

Theoretical S-wave spectra of the $q^2\bar{q}^2$ system composed of u and d quarks are calculated in harmonic oscillator bases including all the configurations up to $N\hbar\omega$ oscillator excitation energy with $N \leq 6$. In a first step the calculations are carried out using a natural generalization of the quark-antiquark interaction proposed by Bhaduri et al. and, in a second step, this interaction is supplemented with an effective potential aimed to simulate the annihilation of $q\bar{q}$ pairs. Our results are discussed in connection with bag model calculations.

1. Introduction

In the quark model of hadron spectroscopy nothing prevents the existence of systems consisting of two quarks and two antiquarks. These exotic objects, called diquonia, are generally described as systems of non-relativistic fermions interacting via pairwise forces deduced from the description of the meson and baryon spectra, so that the study of diquonia becomes, to some extent, similar to

---

* Chargé de Recherches F.N.R.S.

that of the four-nucleon problem. However, the investigation of the $q^2\bar{q}^2$ system gives rise to peculiar problems. For instance, gluons being members of a colour octet, the annihilation of a quark-antiquark pair into a gluon does not occur in colourless $q\bar{q}$ and q^3 hadrons but, in diquonia, this process must be taken into account. On the other hand it is well known that the popular colour charge dependence $\tilde{\lambda}_i \tilde{\lambda}_j$ of the interaction, which gives the correct hyperfine splitting for the usual hadrons, leads for the $q^2\bar{q}^2$ system to unobserved van der Waals' forces between $q\bar{q}$ colour singlets [1]. Of course, these peculiarities make the choice of an interparticle potential suitable for the description of multiquark systems more problematic than in the nuclear case.

In this paper we present a non-relativistic description of diquonia composed of u and d quarks in which the two-body interaction includes an annihilation potential. The corresponding four-body problem is treated variationally in large harmonic oscillator bases using a procedure similar to that used in our extended shell-model description of ^{4}He [2]; this method is outlined in Sec.2. The choice of the two-body interaction used in the calculations, and especially the potential simulating the one-gluon annihilation process, is discussed in Sec.3. Our results are commented in Sec.4 and concluding remarks are presented in Sec.5.

2. Method of calculation

2.1. Definition of the four-body basis

In terms of the individual coordinates of the particles relative to some inertial reference frame, the translationally invariant Hamiltonian for a diquonium $q_1 q_2 \bar{q}_3 \bar{q}_4$ composed of equal-mass quarks and antiquarks interacting via pairwise forces is given by

$$H = 4m + \sum_{i=1}^{4} \frac{\vec{p}_i^{\,2}}{2m} - \frac{\vec{P}^{\,2}}{8m} + \sum_{i<j}^{4} V_{ij} \tag{1}$$

with

$$\vec{P} = \sum_{i=1}^{4} \vec{p}_i \tag{2}$$

Our method of calculation consists essentially in extracting from the representation of H in large harmonic oscillator bases the eigenstates of physical interest as functions of the harmonic oscillator length parameter $b = (\frac{\hbar}{m\omega})^{\frac{1}{2}}$.

The basis states are products of three factors $|C>, |T>$ and $|S>$ denoting the colour, flavour and spin-space wave functions, respectively. The possible colour bases for a colourless diquonium are two-dimensional. In order to construct states having a given permutation symmetry the most convenient basis is

$$|3> = |\bar{3}_{12}\ 3_{34}> \quad ; \quad |6> = |6_{12}\ \bar{6}_{34}> \tag{3}$$

In state $|3>$ the quarks 1 and 2 are in the antisymmetric representation $\bar{3}$ of SU(3) and the antiquarks 3 and 4 in the antisymmetric representation 3; the state $|6>$ is the symmetric counterpart of $|3>$.

The flavour of the $q^2\bar{q}^2$ system is characterized by the isospin quantum number T. As only u and d quarks are considered, the flavour basis states can be written

$$|T> = \left[\left[\frac{1}{2}\frac{1}{2}\right]^t \left[\frac{1}{2}\frac{1}{2}\right]^{\bar{t}}\right]^T \tag{4}$$

where the square brackets are used to denote SU(2) coupling.

The spin-space basis wave functions $|S>$ are taken of the form

$$|S> = [(a_1 a_2)^j_\epsilon (a_3 a_4)^{\bar{j}}_{\bar{\epsilon}}]^{J_0} \tag{5}$$

The two-quark functions are given by

$$(a_1 a_2)^j_\epsilon = \mathcal{N}\{[\phi_{a_1}(x_1)\phi_{a_2}(x_2)]^j + \epsilon[\phi_{a_1}(x_2)\phi_{a_2}(x_1)]^j\} \tag{6}$$

where $\mathcal{N}$ is a normalization factor; the two-antiquark functions $(a_3 a_4)^{\bar{j}}_{\bar{\epsilon}}$ are given by a similar formula. The Pauli principle implies

$$\epsilon(-1)^t = \bar{\epsilon}(-1)^{\bar{t}} = \begin{cases} +1 & \text{for } |C> = |6> \\ -1 & \text{for } |C> = |3> \end{cases} \tag{7}$$

The functions $\phi_{a_i}(x_k)$, in which x_k stands for all space and spin coordinates of the kth particle, are normalized eigenstates of the three-dimensional isotropic harmonic oscillator for a single fermion. These wave functions are thus specified by the usual quantum numbers $a_i = (n_i, \ell_i, j_i = \ell_i \pm \frac{1}{2})$ and by the oscillator length parameter b. As in the flavour factor $|T>$, the square brackets denote SU(2) coupling. Bases characterized by an even (odd) parity are obtained by selecting the orbital angular momentum numbers ℓ_i so that the sum $\sum_{i=1}^4 \ell_i$ is even (odd).

2.2. Separation of the centre-of-mass motion

The multiparticle harmonic oscillator bases constructed in terms of the individual coordinates of the particles are not translationally invariant. Nevertheless, an exact separation of the centre-of-mass motion can be achieved in such bases provided they include *all* the basis states with excitation energy up to a given number of oscillator quanta [3,4]. This remarkable property is based on the fact that the independent-particle harmonic oscillator Hamiltonian, or shell-model Hamiltonian, for a system of A equal-mass particles

$$H_{SM} = \sum_{i=1}^{A} (-\frac{\hbar^2}{2m} \vec{\nabla}_{\vec{x}_i}^2 + \frac{m\omega^2 \vec{x}_i^2}{2}) \qquad (8)$$

has exactly the same form in terms of Jacobi coordinates; thus

$$H_{SM} = \sum_{i=1}^{A} (-\frac{\hbar^2}{2m} \vec{\nabla}_{\vec{y}_i}^2 + \frac{m\omega^2 \vec{y}_i^2}{2}) \qquad (9)$$

where the vectors $\vec{y}_i$ are given by

$$\vec{y}_\alpha = \frac{1}{[\alpha(\alpha+1)]^{\frac{1}{2}}} \sum_{\beta=1}^{\alpha} \vec{x}_\beta - (\frac{\alpha}{\alpha+1})^{\frac{1}{2}} \vec{x}_{\alpha+1} \quad ; \quad 1 \le \alpha \le A-1 \quad (10)$$

$$\vec{y}_A = A^{\frac{1}{2}} \vec{R}_{cm} \qquad ; \qquad \vec{R}_{cm} = \frac{1}{A} \sum_{i=1}^{A} \vec{x}_i \qquad (11)$$

The expressions (8) and (9) of H_{SM} generate two different bases of the *same* $N_0 \hbar\omega$ space, that is to say, the space including *all* the eigenstates of H_{SM} which satisfy the condition

$$\sum_{i=1}^{A} (2n_i + \ell_i) \le N_0 \qquad (12)$$

In the basis generated by expression (9) of H_{SM} each basis state is equal to a function of the $A-1$ internal coordinates $\vec{y}_\alpha$ multiplied by a harmonic oscillator function of the vector $\vec{R}_{cm}$. It is thus apparent that the diagonalization of a translationally invariant Hamiltonian in a $N_0 \hbar\omega$ space yields eigenstates which are at the same time eigenstates of the centre-of-mass Hamiltonian

$$H_{cm} = \frac{\vec{P}^2}{2mA} + \frac{mA\omega^2}{2} \vec{R}_{cm}^2 \qquad (13)$$

with eigenvalues

$$E_{cm} = (2n_{cm} + \ell_{cm} + \frac{3}{2})\hbar\omega \qquad (14)$$

As the spectrum of an operator matrix in a given vector space does not depend on the choice of a particular basis, this result holds when the diagonalization of a translationally invariant Hamiltonian is carried out using the independent-particle basis generated by expression (8) of H_{SM}. It is also apparent that the eigenstates of such a Hamiltonian describing the same internal structure of the A-particle system but corresponding to different values of n_{cm} are degenerate. Therefore it is possible to identify the eigenstates characterized by $n_{cm} = 0$, and via E_{cm} the corresponding values of ℓ_{cm}, by comparing the spectra obtained in the $N_0\hbar\omega$ and $(N_0 - 2)\hbar\omega$ spaces. This is of practical interest since in a basis coupled to zero total angular momentum (for $J_0 = 0$ in the basis states $|S >$ defined above), ℓ_{cm} is precisely equal to the intrinsic angular momentum J of the system under study. To put it in another way, the eigenstates of the energy matrices characterized by $n_{cm} = 0$ and a given value of ℓ_{cm} in a basis including all the configurations up to $N_0\hbar\omega$ of oscillator excitation and coupled to zero total angular momentum $(J_0 = 0)$ are at the same time "good states" $(E_{cm} = \frac{3}{2}\hbar\omega)$ in a basis coupled to angular momentum J, equal to ℓ_{cm}, and including all the configurations up to $N\hbar\omega$ of oscillator excitation with $N = N_0 - J$. In addition, the intrinsic parity π of these states is given by $(-1)^N$; consequently they have natural parity $(-1)^J$ or unnatural parity $(-1)^{J+1}$ according as to whether N_0 is even or odd.

The diagonalization of the model Hamiltonian in complete $N_0\hbar\omega$ bases requires the handling of large matrices. As an example we present in Table 1 the dimensions $D(N_0)$ of the $|S > |T > |C >$ basis used in our calculations for $J_0 = 0$ and $t = \bar{t} = 0$ as a function of N_0.

Table 1. Dimension $D(N_0)$ of the $|S > |T > |C >$ basis for $J_0 = 0$ and $t = \bar{t} = 0$ as a function of N_0.

N_0	0	2	4	6	8
$D(N_0)$	2	30	236	1236	4936

However, the use of the Lanczos algorithm [5] to extract the eigenstates of physical interest from huge matrices combined with a computing procedure based on the storage of numerous geometrical quantities have made our calculations feasible with reasonable economy on a modest computer (an IBM 4341). We wish also to stress that the method is completely self-checking as technical errors reveal themselves, for instance, in incorrect non-half-integer expectation values of the operator $(\hbar\omega)^{-1}H_{cm}$. These checks demonstrated that the numerical accuracy of the quantities calculated in a given $N_0\hbar\omega$ space is better than one part in 10^6.

3. The two-body interaction

The most popular phenomenological quark-antiquark potentials used in non-relativistic descriptions of ordinary mesons are of the form

$$V(r_{ij}) = f(r_{ij}) - A + Br_{ij} + g(r_{ij})\vec{s}_i\vec{s}_j \tag{15}$$

The observed mesons masses are not sufficient to determine completely the parameters appearing in (15). For instance, Weinstein and Isgur in their study of the $K\bar{K}$ molecules [6] utilize

$$f(r_{ij}) = -0.374\exp(-1.028r_{ij}^2) \; ; \; A = 1.036 \; ; \; B = 0.902 \tag{16}$$

$$g(r_{ij}) = \frac{1.168}{m_im_j}\exp(-21.73r_{ij}^2) \tag{17}$$

$$m_u = m_d = 0.375 \; ; \; m_s = 0.600 \tag{18}$$

with $V(r_{ij})$ and masses measured in GeV and $r_{ij} = |\vec{r}_i - \vec{r}_j|$ in fm, whereas Bhaduri et al. [7] obtain excellent fits for the spectra of a great variety of mesons using

$$f(r_{ij}) = -\frac{0.1027}{r_{ij}} \; ; \; A = 0.9135 \; ; \; B = 0.9410 \tag{19}$$

$$g(r_{ij}) = \frac{0.0774}{m_im_jr_{ij}}\exp(-2.2r_{ij}) \tag{20}$$

$$m_u = m_d = 0.337 \; ; \; m_s = 0.600 \; ; \; m_c = 1.870 \; ; \; m_b = 5.219 \tag{21}$$

It has also been argued that a linear potential might not be the best way to simulate confinement [8].

The colour dependence of the two-body interaction is usually obtained by multiplying $V(r_{ij})$ by the SU(3) Casimir operator $-\frac{3}{16}\tilde{\lambda}_i\tilde{\lambda}_j$ where the matrix $\tilde{\lambda}_i$ is the colour generator of the ith particle ($\tilde{\lambda}_i \to -\tilde{\lambda}_i^*$ for an antiparticle). In this way the quark-quark interaction is equal to half its quark-antiquark counterpart, which yields reasonable estimates for the baryon masses [9]. However, as already mentionned, this colour dependence leads to undesirable long-range forces between colour-singlets. In the present work it is supposed that this flaw has little effect on the calculated $q^2\bar{q}^2$ spectra and the structure of the corresponding eigenstates.

The hyperfine term in (15) originates in the direct scattering part of the one-gluon exchange interaction represented by diagram (1) in Fig. 1. This is the only one-gluon exchange contribution allowed in $q\bar{q}$ and q^3 systems. In the $q^2\bar{q}^2$ systems however, it ought to be supplemented by the annihilation process represented by diagram (2) in Fig. 1.

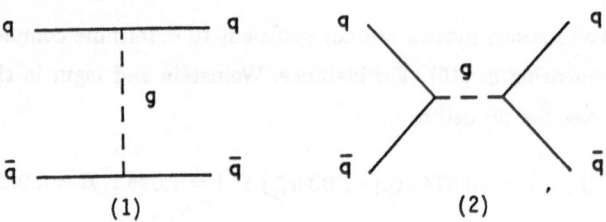

Fig. 1 (1) one-gluon scattering diagram (2) one-gluon annihilation diagram.

As the potential simulating this process cannot be determined by comparing the calculated and measured values of observables, the only way to specify such a potential is to compare the results of the non-relativistic reduction to the order v^2/c^2 of both diagrams (1) and (2). This reduction yields for diagram (1) the hyperfine potential [10,11,12]

$$V_{ij}^h = -\frac{8\pi}{3}\alpha_s \frac{\vec{\lambda}_i.\vec{\lambda}_j}{4}\frac{\vec{s}_i.\vec{s}_j}{m_i m_j}\delta^{(3)}(\vec{r}_i - \vec{r}_j) \tag{22}$$

and for diagram (2)

$$V_{ij}^a = \frac{2\pi\alpha_s}{m_i m_j}P_{ij}^g\delta^{(3)}(\vec{r}_i - \vec{r}_j) \tag{23}$$

where the projection operator P_{ij}^g selects the $q_i\bar{q}_j$ pairs with gluon quantum numbers, that is to say, colour octet, spin triplet and isospin singlet. The comparison of expression (22) of the hyperfine potential with the phenomenological form $g(r_{ij})\vec{s}_i.\vec{s}_j$ appearing in (15) shows that the delta operator in (22) has been substituted by a smeared operator proportional to $g(r_{ij})$; using the fact that the expectation value of the colour operator $\frac{\vec{\lambda}_i . \vec{\lambda}_j}{4}$ for a colour singlet is equal to $-\frac{4}{3}$ one obtains

$$\delta^{(3)}\left(\vec{r}_i - \vec{r}_j\right) \rightarrow \frac{9m_i m_j}{32\pi\alpha_s} g(r_{ij}) \tag{24}$$

The same substitution made in expression (23) of V_{ij}^a yields

$$V_{ij}^a = \frac{9}{16} g(r_{ij}) P_{ij}^g \tag{25}$$

Note that the annihilation potential removes the isospin degeneracy of the spectra calculated using potential (15).

4. Results

When the two-body interaction used in the calculation is restricted to potential (15) the isospin t and $\bar{t}$ of the q^2 and $\bar{q}^2$ pairs are good quantum numbers. In our work we have considered the bases characterized by $t = \bar{t}$; the basis states have thus the same permutation symmetry in flavour for both q^2 and $\bar{q}^2$ pairs. We have diagonalized Bhaduri's potential in $N_0 \hbar \omega$ bases of this type up to $N_0 = 8$; we were thus able to extract "good eigenstates" up to $N = 8$ for $J^\pi = 0^+$ and $N = 6$ for $J^\pi = 2^+$. The results of this calculation have been discussed in Ref. [13]. The most striking feature emerging from this study is that, among the extracted S-wave eigenstates, two states with quantum numbers $J^\pi = 0^+$ and energies close to the $\rho\rho$ threshold (1.540 GeV) have wave functions which strongly favour their decay in two ρ mesons rather than in two pions, which makes the existence of 0^+ diquonia with masses around 1.6 GeV plausible; on the other hand the lowest 2^+ states, whose decay in two pions is forbidden in the S-wave channel, are found slightly higher in energy.

The best $N = 6$ diquonia candidates are compared in Fig. 2 with those obtained by Jaffe [14] in the framework of the bag model. It is seen that for the $t = 0$ states both approaches yield quite similar results : our states at 1.524 and 1.620 GeV with quantum numbers 0^+ and 2^+ correspond respectively to the nonet bag model states at 1.450 and 1.650 GeV denoted $C^0(9^*)$ and $C^0(9)$ in Jaffe's notation. The $t = 1$ states represent a set of three degenerate states with total isospin $T = 0$, 1 and 2 which correspond to the bag model states denoted $C^0(36), C_\pi(36)$ and $E_{\pi\pi}(36)$, respectively; Fig. 2 shows that the 2^+ set is in

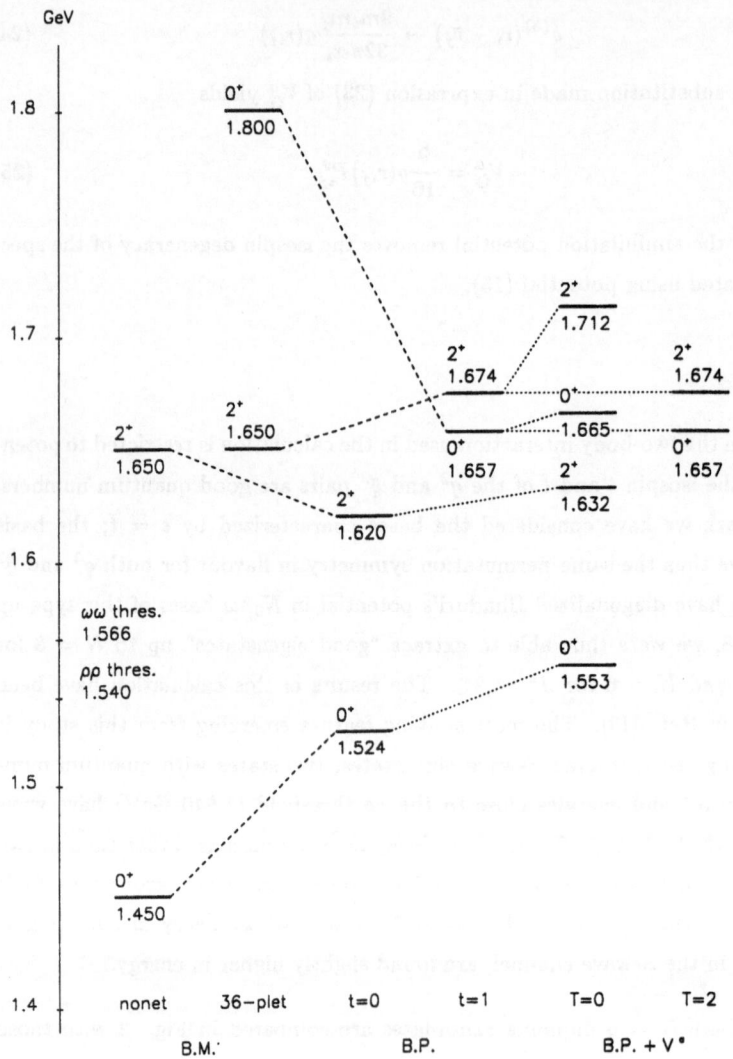

Fig. 2. Best diquonia candidates around the $\rho\rho$ and $\omega\omega$ thresholds yielded by Jaffe's bag model (B.M.), non-relativistic calculations in $N \doteq 6$ harmonic oscillator bases using only Bhaduri's potential (B.P.) and Bhaduri's potential supplemented by an annihilation term (B.P. $+ V^a$).

good agreement with its bag model counterpart. Actually the only noticeable discrepancy between the results of both approaches in this mass region is the presence in our 0^+ spectrum of a $t = 1$ state at 1.657 GeV whereas the lowest non-strange 0^+ bag model state belonging to a 36-plet is found at 1.800 GeV.

The effects of the annihilation processes on the $q^2\bar{q}^2$ spectra calculated in large harmonic oscillator bases have been investigated by Semay [11,12]. These effects were simulated by adding to Bhaduri's potential the potential V_{ij}^a defined by equations (25) and (20). As it is seen from Fig. 2, the isocalar states $(T = 0)$ are shifted both for $t = 0$ and $t = 1$ to slightly higher energies when V_{ij}^a is turned on. For the isotensor states $(T = 2)$ the intermediate isospin quantum numbers are necessarily equal to one so that the matrix elements of V_{ij}^a between such states vanish identically; accordingly the $T = 2$ members of the $t = 1$ multiplets with spin parity quantum numbers 0^+ and 2^+ are insensitive to V_{ij}^a. The effects of V_{ij}^a on the isovector states $(T = 1)$ have not been considered because such states require the introduction in the basis of states with mixed flavour symmetry $(t = 0, \bar{t} = 1$ and $t = 1, \bar{t} = 0)$. In this respect it is worth noting that the isoscalar bases must include both the $t = \bar{t} = 0$ and $t = \bar{t} = 1$ states so that the dimension of an isoscalar basis is twice that of its isotensor counterpart. Therefore the calculation of the isoscalar 2^+ eigenstates was restricted to $4\hbar\omega$ spaces; the $N = 6$ results presented in Fig. 2 concerning these states have been obtained by extrapolating $N = 0, 2$ and 4 results.

With the aim of analyzing the physical contents of the corresponding eigenstates, the wave functions have been recast in the form

$$\sum_{s,i,c} A_{s,i,c} \left[\left[q_1 \bar{q}_3 \right]^{s,i} \left[q_2 \bar{q}_4 \right]^{s,i} \right]^{J,T} |c> \tag{26}$$

where s and i are the common spin and isospin of both the quark-antiquark pairs and $|c>$ is used to denote the colour singlets $|1> \equiv |1_{13}1_{24}>$ and $|8> \equiv |8_{13}8_{24}>$; the amplitudes $A_{s,i,c}$ have been extracted from the eigenstates considered using the relations

$$|3> = \sqrt{\frac{1}{3}}|1> - \sqrt{\frac{2}{3}}|8> \; ; \; |6> = \sqrt{\frac{2}{3}}|1> + \sqrt{\frac{1}{3}}|8> \tag{27}$$

and standard shell-model techniques. For the states displayed in Fig. 2 the probabilities $P_{\pi\pi} = |A_{0,1,1}|^2$ and $P_{\eta\eta} = |A_{0,0,1}|^2$ are much smaller than $P_{\rho\rho} =$

$|A_{1,1,1}|^2$ and $P_{\omega\omega} = |A_{1,0,1}|^2$. This result, which is similar to that obtained using potential (15), shows that the introduction in the two-body interaction of a term simulating the annihilation processes does not modify fundamentally the wave functions. The new piece of information yielded by such a term is that $P_{\rho\rho}$ and $P_{\omega\omega}$ are roughly equal, indicating that diquonia composed of u and d quarks occur with comparable probabilities in the $\rho\rho$ and $\omega\omega$ channels.

5. Concluding remarks

We found that the mass spectra of the S-wave diquonia composed of u and d quarks obtained by diagonalizing a non-relativistic model Hamiltonian in large harmonic oscillator bases are comparable with bag model states. Though resonances observed in experiments such as $\bar{p}n$ annihilation and $\gamma\gamma$ reactions [15] and π^-p scattering [16] might be described as members of these spectra, it is clear that further work is necessary to establish firmly the existence of diquonia. It is of course possible to improve the accuracy of the calculated eigenstates by enlarging the model spaces used in our work; in this respect it is worth mentioning that shell-model practitioners plan to handle routinely bases coupled to a defined total angular momentum including no less than 50,000 configurations [17]. However we believe that more significant steps towards a real understanding of exotic multiquark systems would consist in improving the procedures used to mimic confinement and the methods developed to estimate the widths of the theoretical states.

5. References

1. O.W. Greenberg and H. J. Lipkin : Nucl. Phys. **A370**, 349 (1981).

2. R. Ceuleneer, P. Vandepeutte and C. Semay : Phys. Rev. **C38**, 2335 (1988).

3. J.B. French, E.C. Halbert, J.B. Mc Grory and S.S.M. Wong : Advances in Nuclear Physics, Vol. **3**, edited by M. Baranger and E. Vogt, New York 1969.

4. R. Ceuleneer and M. Gilles : Letter al Nuovo Cimento **30**, 119 (1981).

5. R.R. Whitehead, A. Watt, B.J. Cole and I. Morrison : Advances in Nuclear Physics, Vol. **9**, edited by M. Baranger and E. Vogt, New York 1977.

6. J. Weinstein and N. Isgur : Phys. Rev. **D41**, 2236 (1990).

7. R.K. Bhaduri, L.E. Cohler and Y. Nogami : Il Nuovo Cimento **65A**, 376 (1981).

8. M. Fabre de la Ripelle : Phys. Lett. **B205**, 97 (1988)

9. B. Silvestre-Brac and C. Gignoux : Phys. Rev. **D32**, 743 (1985).

10. A. Faessler, G. Lübeck and K. Shimizu : Phys. Rev. **D26**, 3280 (1982).

11. C. Semay : Thesis 1989, Université de Mons-Hainaut.

12. C. Semay : J. Phys. **G17**, 413 (1991).

13. R. Ceuleneer and C. Semay : J. Phys. **G15**, 1783 (1989).

14. R.L. Jaffe : Phys. Rev. **D15**, 267 (1977).

15. K.F. Liu and B.A. Li : Phys. Rev. Lett. **58**, 2288 (1987).

16. D. Alde et al. : Phys. Lett. **216B**, 451 (1989).

17. B.A. Brown : Nucl. Phys. **A507**, 25c (1990).

Few-Body Systems, Suppl. 6, 176—194 (1992)

Few-
Body
Systems
© by Springer-Verlag 1992

THE QUARK MODEL, DEUTERON FORMFACTORS AND NUCLEAR MAGNETIC MOMENTS

Amand Faessler

Institut für Theoretische Physik

Auf der Morgenstelle 14

Universität Tübingen

7400 Tübingen, Germany

Abstract:

Deuteron properties and nuclear magnetic moments are studied in the non-relativistic quark cluster model. The charge monopole, quadrupole and magnetic-dipole form factors and the tensor polarization of the deuteron in this microscopic meson-quark cluster model are calculated. The deuteron wave function is derived from a microscopic 6-quark Hamiltonian which, in addition to a quadratic confinement potential, includes the one-pion and the one-gluon exchange potentials between quarks. The electromagnetic current operators are constructed on the quark level, i.e., the photon is coupled directly to the quarks. Aside from the one-body impulse current, pionic and gluonic exchange current corrections are included. Due to the Pauli principle on the quark level, new quark interchange terms arise in the one-body and two-body current matrix elements, that are not present on the nucleon level. While these additional quark exchange currents are small for low momentum transfers, we find that they appreciably influence the electromagnetic structure of the deuteron beyond a momentum transfer of $q = 5fm^{-1}$. We construct the quark exchange current operators on the quark level and then eliminate the quark degrees of freedom. The photon is directly coupled to the quarks. This effective current contains new non-local and isospin dependent terms which are generated by the Pauli principle on the quark level (quark exchange between nucleons). When we evaluate these quark exchange currents in nuclei, we use harmonic oscillator wave functions as nuclear wave functions including short-range Brueckner correlations. We solve the Bethe-Goldstone equation to include this short-range correlations in our effective NN potential, which is derived from a microscopic quark Hamiltonian. We investigate also the role of these additional quark exchange currents in the magnetic moments and the elastic magnetic form factors of several closed shell±1 nuclei, such as ^{15}N, ^{17}O and ^{39}K. We specifically discuss the results for M_1 electron scattering ^{15}N and the isovector magnetic moment for the A=39 system. Quark degrees change the Schmidt value of this isovector moment by -20 %.

1. INTRODUCTION

The first idea about the nature of the nucleon-nucleon interaction came from Yukawa in 1935. He assumed that the strong interaction between two nucleons is carried by an interaction quantum, which is a particle of a medium heavy mass of about 200 MeV, the meson. After finding the π meson one thought one has found the carrier of the strong nuclear force. But the fifties saw a time where more and more mesons were found which contribute to the nucleon-nucleon interaction. One of the high points of this development was the suggestion of Gregory Breit in 1958 that the short range repulsion should be due to a vector, isoscalar meson, the omega meson (ω) of the order of 800 MeV. It was a big success of the meson exchange theory of the nucleon-nucleon interaction when this ω meson was found also experimentally.

But just the ω meson shows that this cannot be the whole story. Flavor symmetry $SU(3)$ predicts from the ρ-nucleon coupling the ω-nucleon coupling squared $g^2_{\omega NN}/(4\pi) = 4.5$. In reality one needs to reproduce the short repulsion according to the Bonn-potential values between 12 and 24. Normally flavor $SU(3)$ is only violated within 30 % or less. The need to blow up the $\omega - NN$ coupling constant by a factor 3 and more indicates that the ω meson must carry a load for which it is not prepared. After we learned that the nucleon is composed out of three valence quarks, gluons and sea quarks it is natural to look on the quark level for the nature of the short range repulsion. Indeed we will show in the next chapter that the short range repulsion of the NN interaction can be understood in the quark model by the symmetry of the 6-valence quarks.

The main purpose of this lecture is to search for quark degrees of freedom in the deuteron properties and in finite nuclei.

In the third chapter we calculate the deuteron wave function in the quark cluster model for the electromagnetic deuteron form factors of the elastic electron deuteron scattering we include the impulse term where the photon is directly interacting with the quarks and exchange currents. These exchange currents can be on the quark level due to quark exchange stemming from the antisymmetrization of the 6-valence quarks in the two nucleons. They can originate also from meson exchange currents. The specific quark degrees of freedom show up in these meson currents if also a quark pair is exchanged due to the antisymmetrization on the quark level between the two nucleons. Without quark exchange the meson exchange currents can be included on the nucleon level by suitably choosing the photon nucleon vertex function. Additional contributions come from gluon exchange currents. These are exchange currents where the photon is absorbed by a quark-antiquark pair and the antiquark is again annihilated with the quark by exchanging a gluon to some other valence quark. Again these terms show up as quark degrees of freedom which cannot be treated on the nucleon level only if at the same time a quark pair is exchanged due to antisymmetrization of the 6-valence quarks.

Finally in chapter 4 we calculate magnetic moments and magnetic form factors in finite nuclei including quark degrees of freedom. On one side we calculate contributions in the impulse approximation where the photon exchanged between the electron and the nucleus or the magnetic moment operator and the nucleus is interacting with a quark. But these terms can be included on the nucleon level by again suitably chosing the photon nucleon vertex function. Quark degrees of freedom which cannot be described effectively on the nucleon level arise by additional quark exchange between two nucleons due to the antisymmetrization of the valence quarks. In addition one gets also quark degrees of freedom involved if one has meson and gluon exchange currents between quarks of different nucleons. This leads to two body currents in the magnetic moments and the magnetic form factors stemming from quark degrees of freedom.

The quark degrees of freedom are best seen at high momentum transfer above $q = 5$ fm^{-1} for the magnetic dipole form factor of the deuteron. Especially the spin degrees of freedom of the quark exchange currents give here a large contribution.

The magnetic form factors for elastic electron scattering in nuclei away by one nucleon from double closed shells are affected by quark exchange currents but even in the maxima this contributions are so small that their effect could not be seen relative to effects coming from configuration mixing (Arima and Horie, ref. 6) and relative to the meson exchange current. A large effect from genuine quark degrees of freedom due to quark exchange currents can be seen in the isovector magnetic moments of the two nuclei with one proton and one neutron more or less as double closed shell nuclei. Specificly for ^{39}K and ^{39}Ca where the Schmidt value is relative small (due to the cancellation of the spin and the orbital part) one obtains a surprisingly large effect of 21% of genuine quark degrees of freedom (quark exchange current) relative to the Schmidt value. The quark degrees of freedom improve here the agreement with the experiment considerably.

The work presented here has been published in a series of papers together with A. Buchmann and Y. Yamauchi (ref.9,10). The magnetic form factors for elastic electron scattering on finite nuclei and the work on the isovector magnetic moments is contained in a paper with Yamauchi, Buchmann and Arima (ref. 11).

2. THE QUARK MODEL AND THE NUCLEON NUCLEON INTERACTION

At large distances the nucleon-nucleon interaction can be represented quite successfully by exchanging mesons between the centre of mass of nucleon 1 and the centre of mass of nucleon 2. At smaller distances as indicated in fig. 1 it makes no sense to exchange mesons between the centre of mass of nucleon 1 and of nucleon 2 since due to antisymmetrization it is not even known where the centre of masses of the two nucleons are, since the antisymmetrizer attaches the 6-valence quarks in each of the ten terms of the antisymmetrization to other nucleons. The difficulty of the meson exchange model at short distances might be indicated by the too large $\omega - NN$ coupling constant needed to reproduce the data as indicated in the introduction.

At short distances (below $1 fm$), where the quark contents of the two nucleons overlap, the quark degrees of freedom should play a major role in describing the nucleon nucleon interaction and especially the short range repulsion.

In the non-relativistic quark model one uses for the quark quark interaction the one gluon exchange potential (OGEP) and a linear or quadratic confinement term. In addition the 6-quark Hamiltonian of the two interacting nucleons contains the rest mass of the quarks and the kinetic energies.

$$\hat{H}_{Quark} = \sum_{i=1}^{6}[m_i + \frac{\mathbf{p}_i^2}{2m}] + \sum_{i<j=1}^{6} [V_{qq}^{OGEP}(i,j) - a\boldsymbol{\lambda}_i \cdot \boldsymbol{\lambda}_j r_{ij}^2] \tag{1}$$

$\boldsymbol{\lambda}_i$ are the eight color octet matrices acting on quark i. The scalar product between the two color matrices $\boldsymbol{\lambda}_i$ and $\boldsymbol{\lambda}_j$ is to be taken over the eight different color matrices for quark i and quark j.

The last term is the quadratic confinement term which is described microsesimicly by the exchange and the interaction of many gluons.

The one gluon exchange potential (OGEP)

$$V_{qq}^{OGEP} = \frac{g^2}{4\pi}\frac{1}{4}\boldsymbol{\lambda}_1 \cdot \boldsymbol{\lambda}_2[\frac{1}{r_{12}} + \frac{\pi}{m_q^2}(1 + \frac{2}{3}\boldsymbol{\sigma}_1 \cdot \boldsymbol{\sigma}_2)\delta(\mathbf{r}_{12}) \ + \ tensor \ + \ \mathbf{l} \cdot (\mathbf{S}_1 + \mathbf{S}_2) + \ldots] \tag{2}$$

thus contain in addition to the color Coulomb and the color magnetic interaction, which are written explicitly, tensor and two-body spin orbit contributions and additional momentum dependent terms. We did not write them down explicitly since they are quite lengthy although the tensor term plays an important role for the description of the deuteron.

In addition to the quark Hamiltonian (1) we have to include to describe the nucleon-nucleon interaction at large distances pion exchange between the different quarks and the exchange of a scalar and isoscalar σ meson. The pion-quark coupling constant is adjusted so as to reproduce the known pion nucleon coupling with the quark wave function of the nucleon. The σ-quark and by that the σ-nucleon coupling is the only quantity adjusted to the nucleon-nucleon phase shifts. The other parameters like the mass of the quark m_q, the strong gluon-quark coupling constant $g_s^2/(4\pi)$, the oscillator length b and the confinement parameter a are adjusted to three-quark data like the nucleon mass, the Δ mass the root mean charge radius from electron-proton scattering and the stability condition for the mass of the proton. The last condition says that the derivative of the proton mass as a function of the oscillator length b should have a minimum at the charge root mean square radius including also the charge distribution from the pion cloud which surrounds the nucleon (ref. 4).

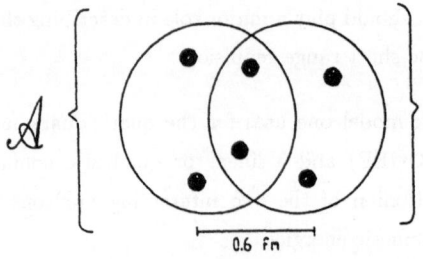

Fig. 1:

Qualitative sketch of the nucleon-nucleon interaction in the meson exchange model at large distance between two nucleons consisting out of three valence quarks and the interaction between two nucleons at smaller distances where the quark contents of the two nucleons overlap. In the last case it makes no sense to describe the nucleon-nucleon interaction by the exchange of mesons between the centre of gravity of nucleon 1 and of nucleon 2. Due to the antisymmetrization it is not even known where these centres of gravitiy are since the antisymmetrizer attributes in all different ten terms different quarks to nucleon 1 and nucleon 2. The short range part of the nucleon-nucleon interaction must be described within the quark model.

As already mentioned above the only quantity directly fitted to two nucleon data is the σ meson-nucleon coupling. $(g_\sigma^2/(4\pi) = 3.34)$. The same parameters which reproduce the nucleon-nucleon phase shifts as shown in fig. 2 and 3 describe also the hyperon nucleon interaction. That means with the same parameters we can describe the world data collection (ref.14 and 15) of the nucleon-lambda and the nucleon-sigma interaction.

The quark cluster model explains also nicely the short range repulsion of the nucleon-nucleon interaction (ref.1 and 16). Fig. 4 shows the spatial symmetries of the 6-quarks at small distances.

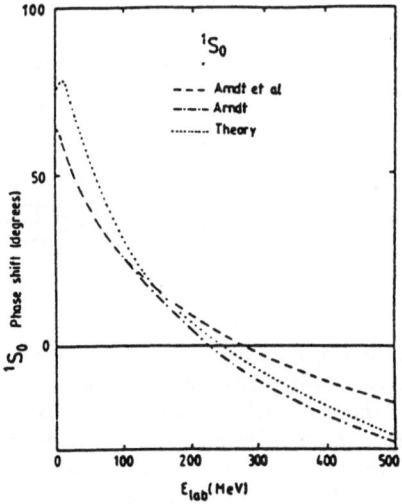

Fig. 2:

Singlet S_0 phase shift from proton-proton scattering (isospin $T = 1$). The dashed curve are the experimental phase shifts by Arndt and Mc Gregor while the short dashed line is the calculation of the quark model (ref.12).

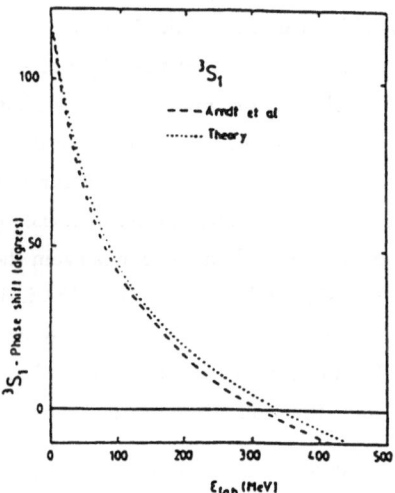

Fig. 3:

The figure shows the $^3S_1$ nucleon-nucleon phase shifts. The long dashed lines are the experimental values of Arndt and Mc Gregor (ref.13), while the short dashed line is the result of the present model calculation (ref.12).

From Fig. 4 one sees that it is more probable by the weight 8/9 to have at short distance the [42] symmetry compared with the completely symmetric orbital wave function [6] which has only the weight 1/9. Fig. 4 indicates also the lowest energy realizations of

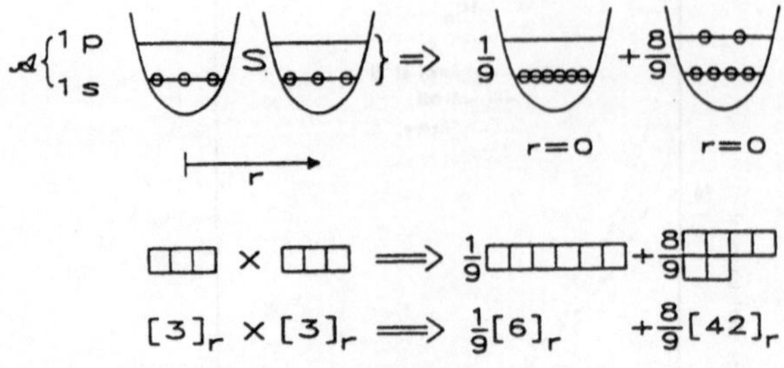

Fig. 4:

The left hand side shows two nucleons in the quark model at distance r. In each of the two nucleons all three valence quarks are in the 1s state. Group theory tells us that if the two nucleons are in a relative orbital S-state the permutation symmetry of all 6 valence quarks can either be only completely symmetric [6] or have the [42] symmetry. The square of the Clebsch-Gordan coefficients of the permutation group of 6 objects gives also the probability to find the completely symmetric spatial representation [6] or the [42] symmetry.

the orbital symmetries [6] and the mixed orbital symmetry [42]. The last configuration requests at zero distance of the two nucleons ($r = 0$) that two quarks are in the $1p$ state for the lowest energy realization of this configuration. The usual way of representing the two nucleon wave functions by 6 quarks in the $1s$ state is only at most available with the probability 1/9. It is obvious from that the [42] orbital symmetry cannot be neglected. If for a moment we neglect 1/9 compared to 8/9 we have at small distances $r \approx 0$ at least two harmonic oscillator quanta excited. Or in other words at least two quarks have to be not in the $1s$ state. For the lowest configuration they are in the $1p$ state. That means at short distances this configuration with the probability 8/9 has at least two harmonic oscillator quanta excited. If one moves again the two nucleons apart at distance r as on the left hand side of Fig. 4, one sees that inside the two nucleons one has no harmonic oscillator quanta excited. Since one has to conserve the number of harmonic oscillator quanta, the two quanta must be contained in the relative motion. If one expands the relative S wave function of the two nucleons in harmonic oscillators

$$u(r_{12}) = \alpha_1|1s> +\alpha_2|2s> +\alpha_3|3s> +\dots \quad with: \alpha_1 = 0 \ (Pauli forbidden) \quad (3)$$

one finds that the $1s$ amplitude must be zero since all parts of the wave function have to contain at least two harmonic oscillator quanta if one considers the orbital [42] symmetry. Thus the relative wave function $u(r_{12})$ is dominated at small distances by $|2s>$ and therefore has a node near the so-called hardcore radius ($r = 0.4 fm$). This node is seen in the asymptotic phase shift measured by the differential cross section. To explain the node one requests that the nucleon-nucleon interaction potential has a hard or soft core at this radius. In reality the node in the wave function is enforced by the orbital [42] symmetry.

3. DEUTERON FORM FACTORS

To describe deuteron properties one has first to reproduce the deuteron binding energy of 2.2 MeV. With the σ-nucleon coupling constant $g_\sigma^2/(4\pi) = 3.34$ one obtains a binding energy of 3.3 MeV. This discrepancy between theory and experiment seems at the first moment disappointing, after the big success for the nucleon-nucleon phase shifts and even the hyperon-nucleon scattering data. But one has to keep in mind that in the non-relativistic quark model one is not only calculating the binding energy of the deuteron but the total energy of the order of 1875 MeV. Although one is fitting the nucleon mass one should keep in mind that a discrepancy of 1 MeV is not much for such a number. But the quality of the deuteron wave function has to be better than that to study quark degrees of freedom in the deuteron form factors. Thus we readjust the σ-nucleon coupling constant and obtain $g_\sigma^2/(4\pi) = 2.68$. To calculate the deuteron we include the same diagrams as for the nucleon-nucleon interaction. A typical selection from these diagrams is shown in fig. 5. The results (ref.10,17) are shown in table 1.

$g_\sigma^2/(4\pi) = 2.68$	$P_D[\%]$	$\mu_D[\mu_N]$	$Q_D[fm^2]$
Quark Model	5.23	0.850	0.266
Exp.	—	0.857	0.286

Table 1:
Deuteron properties for the σ-nucleon coupling constant $g_\sigma^2/(4\pi) = 2.68$. The D state admixture is given in % while the magnetic moment is indicated in units of nuclear magnetons $[\mu_N]$ and the quadrupole moment Q_D in units of $[fm^2]$. The first line gives the theoretical values, while the second line lists the experimental data.

The elastic electron deuteron cross section is described by a longitudinal and a transversal form factor.

$$\left(\frac{d\sigma}{d\Omega}\right)_{eD} = \left(\frac{d\sigma}{d\Omega}\right)_{Mott}[A(q) + B(q)tg^2\frac{\theta}{2}]$$
$$A(q) = F_c^2(q) + \alpha_1 q^2 F_Q^2(q) + \alpha_2 q^2 F_M^2(q) \tag{4}$$
$$B(q) = \beta q^2(1 + q^2)F_M^2(q^2)$$

q is the momentum transfer and $F_C F_Q$ and F_M indicate the Coulomb, the quadrupole and the magnetic dipole form factors. The transversal form factor $B(q)$ is only given by the magnetic dipole form factor F_M.

We first give the results in the impulse approximation shown in fig. 7.

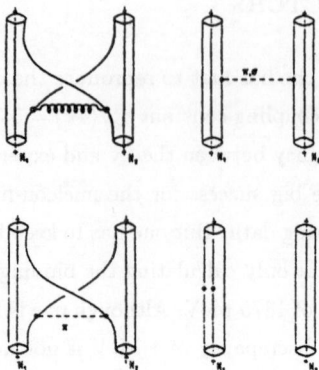

Fig.5:

Selection of diagrams included in the calculation of the deuteron wave function. One includes the one gluon exchange potential (OGEP) between all different quarks, pion exchange with and without antisymmetrization for the large distances and the exchange of the scalar and isoscalar σ meson between the quarks. The pion cloud is also included in each nucleon (intracluster pion exchange).

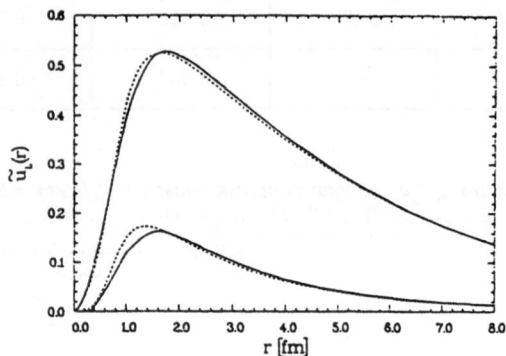

Fig. 6:

The deuteron S and D wave functions. Full curves: relative deuteron wave function in the quark cluster model; dashed curves: two-nucleon model wave functions calculated with the Paris potential (ref.18).

The diagram of fig.7a can be also described phenomenologically on the nucleon level by introducing a suitable photon nucleon vertex function. But the quark exchange diagrams (b) and (c) cannot be included phenomenologically on the nucleon level. They represent genuine quark degrees of freedom. For the Coulomb and the quadrupole form factors the quark exchange diagrams of fig. 7 do not contribute appreciably and they are not measurable at the moment. But the magnetic dipole form factor seems to have

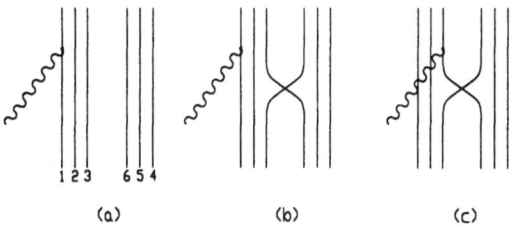

Fig. 7:

One body impulse current in the quark cluster model: (a) direct term, (b,c) quark interchange terms. The solid lines represent quarks and the wavy lines photons.

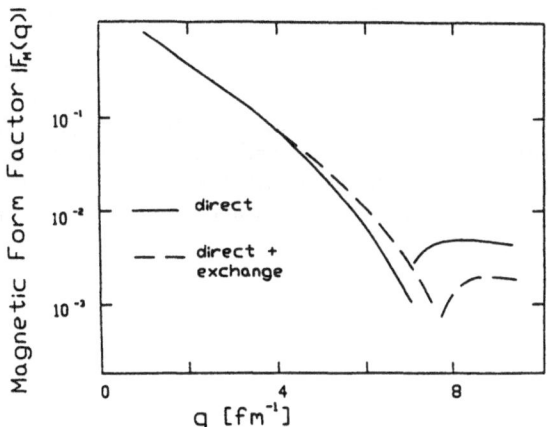

Fig. 8:

Magnetic form factor $|F_M(q)|$ as a function of the momentum transfer q for the direct impulse term (Fig. 7a; solid line) and the direct plus exchange terms (all diagrams in Fig. 7; dashed line).

a measurable effect.

Due to the large spin g-factors the spin magnetization current of the quark exchange term is contributing appreciably as shown in Fig. 9. The spin exchange term as given by diagrams in Fig. 7 b,c can flip the spin of the nucleons and therefore these terms represent large spin magnetization currents. One sees in Fig. 9 that at momentum transfers above 6 to 7 fm^{-1} the differences could easily be detected. Thus elastic electron scattering on the deuteron and especially the transversal form factor can show genuine quark degrees of freedom.

Fig. 10 and Fig. 11 give the differential cross section calculated with the impulse terms only (Fig. 7a). Only the transversal form factor is sensitive to the quark exchange

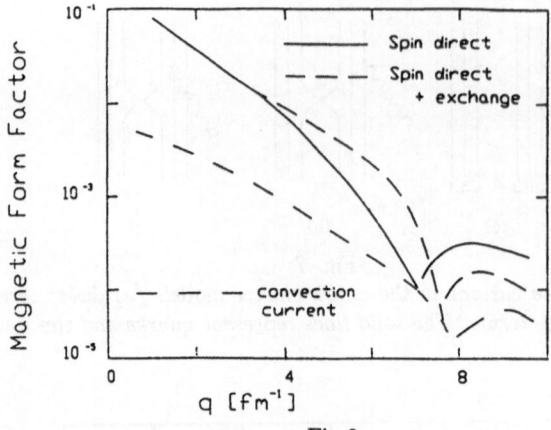

Fig.9:

Contributions to the magnetic form factor from the spin magnetization current (solid and dashed lines) and the convection current (dashed dotted line). The spin magnetization current is devided in the direct part (Fig. 7a; solid line) and the contributions from the direct and exchange parts (Fig.7b, 7c; dashed line). For the convection current contribution (dashed dotted line) the spin exchange terms do not contribute appreciably.

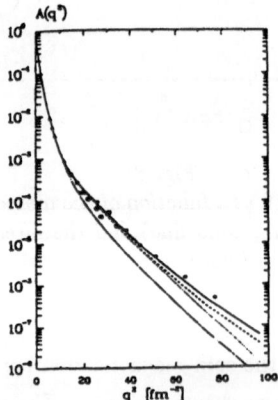

Fig.10:

Longitudinal form factor $A(q)$ from the elastic electron deuteron cross section (4) compared with the data. The theoretical calculation includes only the impulse approximation with quark exchange currents as shown in fig. 7, but no meson and gluon exchange currents (Fig. 12). This longitudinal form factor is not appreciably effected by the quark exchange diagrams of fig. 7b and 7c. The data are taken from ref. 19. The solid line gives the direct quark terms only (Fig.7a), while the dashed line includes also the quark exchange terms Fig. 7b and 7c.

diagrams (Fig. 7b and 7c).

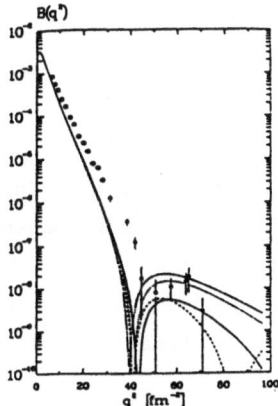

Fig. 11

Transversal form factor $B(q)$ from elastic electron-deuteron scattering (4) compared with the data. The theoretical calculation shows the impulse approximation without and with exchange currents. The data are taken from ref. 19. The solid line gives the direct quark term only (fig. 7a), while the dashed line includes also the quark exchange terms (fig. 7b and 7c).

Fig. 12 shows in addition to the impulse diagrams of Fig. 7 also the two-body pion and gluon pair exchange diagrams in the quark cluster model.

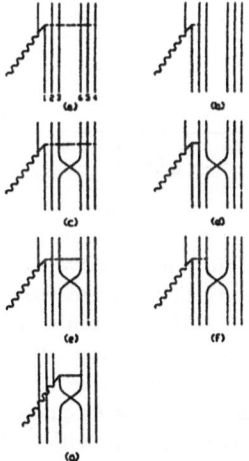

Fig.12:

The two-body pion and gluon pair currents in the quark cluster model: (b) direct inter cluster term, (c-g) quark interchange terms; the dashed line represents the pion or the gluon exchanged between quarks. For the gluon the matrix element represented by the diagram of Fig. 12a vanishes due to color selection rules (no color singlet gluons).

The two-body exchange currents indicated in Fig. 12 can be calculated in two ways. Either one uses the minimal substitution

$$\mathbf{p}_i \rightarrow \mathbf{p}_i - \frac{e_i}{c}\mathbf{A} \tag{5}$$

in the one gluon exchange potential or one calculates diagrams according to the Feynman rules. If one is consistent in the non-relativistic reduction one obtains the same results. (This twofold derivation of the two-body currents is a very useful check of the calculation.) Again the magnetic dipole form factor is most sensitive to the pion and gluon exchange currents with quark exchange (Fig. 12c-g). These results are indicated in Fig. 13.

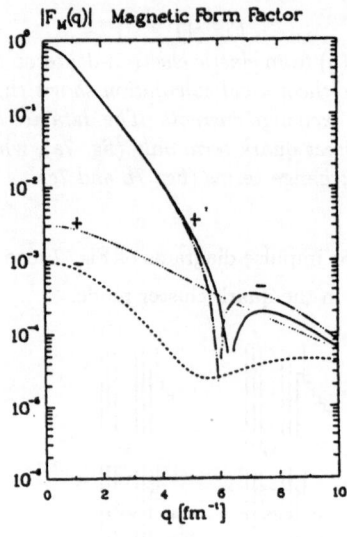

Fig.13:

Deuteron magnetic-dipole form factor as a function of the momentum transfer q. The dashed-double-dotted curve is the conventional impulse approximation; the dotted curve gives the pion pair exchange current without quark antisymmetrization. The dashed curve is the diagram with the pion exchange current with quark antisymmetrization. The gluon exchange currents give slightly smaller contributions and are not shown here. One sees that at high momentum transfers the pion exchange current with quark antisymmetrization is even larger than the usual pion exchange current without quark antisymmetrization.

4. QUARK MODEL AND THE MAGNETIC MOMENTS OF NUCLEI

To describe the magnetic moments and the elastic electron scattering of nuclei we start with the impulse approximation corresponding to Fig. 7 but now we have instead of two nucleons for the deuteron 16 nucleons for ^{16}O. But still the photon exchange between the electron and the nucleus or the magnetic moment operator $\hat{\mu}$ interacts only with one quark. The interaction is given by the product of the vector potential A_μ and the single quark current J^μ.

$$H_{int} = -A_\mu J^\mu \tag{6}$$

The one quark current operators

$$J_0 = \sum_{i=q} e_i e^{i\mathbf{q}\mathbf{r}_i}$$

$$\mathbf{J} = \sum_{i=q} \frac{e_i \hbar}{2m_i}[e^{i\mathbf{q}\mathbf{r}_i}\overset{\leftrightarrow}{\nabla}_i + \mathbf{q} \times \boldsymbol{\sigma}_i e^{i\mathbf{q}\cdot\mathbf{r}_i}] \tag{7}$$

represent the charge of the different quarks with the momentum transfer q from the photon and their currents. The one quark magnetic operator is given by:

$$\hat{\mu} = \sum_{i=q} \frac{e_i \hbar}{2m_i c}[\mathbf{l}_i + g_i \mathbf{s}_i] \tag{8}$$

in eqs. (7) and (8) the sums run over the quarks q. e_i is the charge of the different up and down quarks. m_i is the quark mass. The magnetic moment operator is later on more suitable derived from the currents.

$$\hat{\mu} = \lim_{|\mathbf{q}|\to 0} \frac{\hbar}{2ci}\vec{\nabla}_q \times \mathbf{J} \tag{9}$$

The currents (7) yield on the nucleon level one nucleon currents (the diagram of Fig. 7a) and two nucleon currents if one includes the antisymmetrization of the quarks. To make the calculation feasible we restrict ourselves always to the antisymmetrization of the quarks in two nucleons. In this way one obtains on the nucleon level atmost two-body currents. Further contributions to two-body currents one obtains due to meson and gluon exchange in the same way as in Fig. 12. The results shown here include only the two-body currents due to quark antisymmetrization the derivation of the two-body currents due to quark and gluon exchange without and with antisymmetrization on the quark level is still in progress.

The two-body nucleon currents corresponding on the quark level to Fig. 7 are again derived either by minimal substitution (5) or by using Feyman rules. With consistent non-relativistic reductions both procedures yield the same operators. Integration over the internal quark degrees of freedom yield then non-local two-body currents on the nucleon level.

$$J_\mu(\mathbf{r}, \mathbf{r}'; \mathbf{R}; \mathbf{q})$$

$$\mathbf{r} = \frac{1}{6}[\mathbf{r}_1 + \mathbf{r}_2 + \mathbf{r}_3 - \mathbf{r}_4 - \mathbf{r}_5 - \mathbf{r}_6]$$

$$\mathbf{R} = \frac{1}{6}[\mathbf{r}_1 + \mathbf{r}_2 + \mathbf{r}_3 + \mathbf{r}_4 + \mathbf{r}_5 + \mathbf{r}_6] \tag{10}$$

Here $\mathbf{r}$ and $\mathbf{r}'$ are the non-local distances between the two nucleons considered. $\mathbf{R}$ is the centre of gravity of these two nucleons. The nuclear wave function is chosen to be a shell model Slater determinant. The short range correlations between the nucleons are important and are included solving the Bethe-Goldstone equation.

$$\left[\frac{\mathbf{p}_1^2}{2m_N} + \frac{1}{2}m_N\omega^2 r_1^2 + \frac{\mathbf{p}_2^2}{2m_N} + \frac{1}{2}m_N\omega^2 r_2^2 + Q(1,2)V_{N_1 N_2}(\mathbf{r}_1, \mathbf{r}_2; \mathbf{R})\right]\psi_{\alpha\beta}(1,2) = E_{\alpha\beta}\psi_{\alpha\beta} \tag{11}$$

Here the non-local nucleon-nucleon interaction $V_{N_1 N_2}$ is obtained using for the quark-quark interaction from Hamiltonian (1) the quark wave functions shown in Fig. 4 on the left hand side for the distances r and on the right hand side for r'. All the internal quark degrees of freedom are integrated out so that the nucleon-nucleon interaction depends only non-locally on the distance of the two nucleons on the left and on the right and on the centre of gravity of the two interacting nucleons R. Q (1,2) is the Pauli operator in the harmonic oscillator space. Fig. 14 shows relative ^1S and ^3S wave functions from the solution of the Bethe-Goldstone equation.

The electron nucleus cross section and especially the transversal form factor B(q) is now determined in the following way:

$$\left(\frac{d\sigma}{d\Omega}\right)_{eN} = \left(\frac{d\sigma}{d\Omega}\right)_{Mott}\left[A(q) + B(q)tg^2\frac{\vartheta}{2}\right]$$

$$with: \quad B(q) \propto q^2(1 + q^2)|T_{JM}^{magn.}(q)|^2 \tag{12}$$

$$T_{JM}^{magn.}(q) = \frac{(-i)^J}{4\pi e}\int d\hat{q}[Y_J \times \mathbf{J}]_M^J$$

The magnetic moments are determined from the one-body operator (8) and the contributions from the two-body currents defined as in eq. (9).

$$\mu = \sum_{i=1}^{A} < Nucl..; JJ|\hat{\mu}_{iz}|Nucl.; JJ >$$

$$+ \lim_{q\to 0}\sum_{i<j} < Nucl.; JJ|\frac{\hbar}{2ci}\vec{\nabla}_q \times bfJ_{ij}|Nucl.; JJ > \tag{13}$$

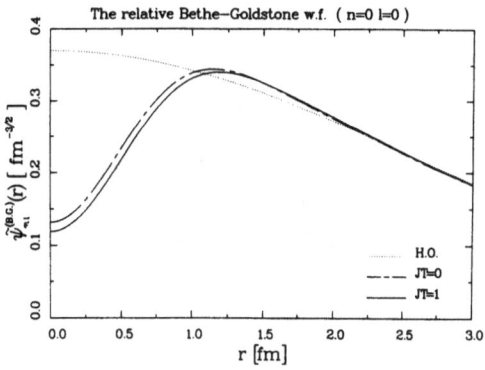

The relative Bethe–Goldstone w.f. (n=0 l=0)

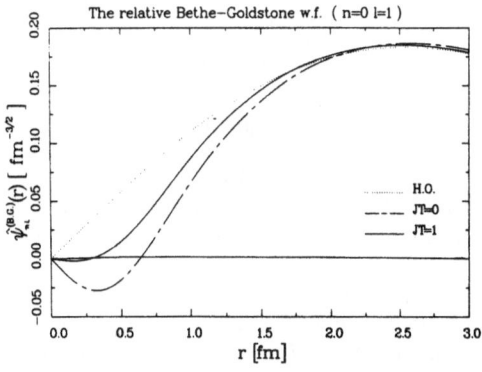

The relative Bethe–Goldstone w.f. (n=0 l=1)

Fig. 14:

Relative harmonic oscillator S-wave function (dotted line) compared with the solution of the Bethe-Goldstone equation for ³S (solid line) and ¹S (dashed-dotted line). One clearly sees the short-range correlations which stem here from quark degrees of freedom.

From Fig. 15 one sees that the quark exchange effects cannot be seen in elastic scattering on nuclei with one proton or one neutron less or more than double closed shells. The situation is different for the magnetic moments as can be seen in table 2.

For the mass $A = 39$ system ^{39}K and ^{39}Ca one obtains a large effect due to quark exchange currents (Fig. 7b and c). Although the agreement with the experimental data is by that greatly improved, we won't not so much stress the better agreement but the fact that genuine quark degrees of freedom as contained in the graphs of Fig. 7b and c can be seen in magnetic moments. The fact that just the mass 39 system is so sensitive is due to the fact that the spin and the orbital contribution for the Schmid value cancel each other by a large part and due to the fact that the quark exchange current contributions (QEC) are roughly proportional to the mass.

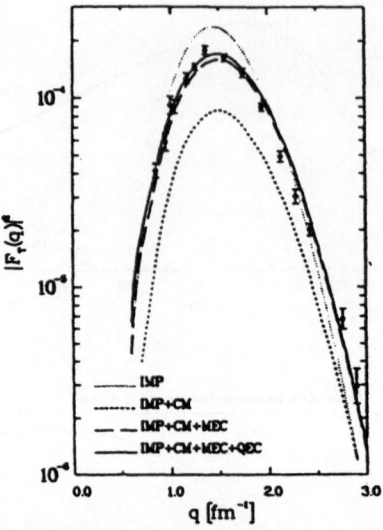

Fig.15:

Transversal form factor of ^{15}N as a function of the momentum transfer **q** the dotted line gives the impulse approximation which describes the interaction of the elasticly scattered electron with only one nucleon. The dashed line is the impulse approximation plus the configuration mixing according to Arima and Horie (Ref. 6). The long dashed line includes the meson exchange currents calculated between the nucleons. The solid line also includes quark exchange effects.

A	CM^6	MEC	QEC	EXP
15	5.5	7.2	−9.1	11.1
17	−16.9	15.5	1.6	1.3
39	42.2	−48.3	−20.9	−38.4

Table 2

Isovector magnetic moments in nuclei with one proton or one neutron more than double closed shell nuclei. The values are given in percentages of the relative change to the Schmid value. 5.5 means an increase of 5.5 % of the Schmid value due to configuration mixing as first proposed by Arima and Hori (Ref.6) the isovector magnetic moment is calculated as the difference between the magnetic moments of nuclei with one proton and one neutron more or less than double closed shell nuclei.

5. CONCLUSIONS

In this invited talk we have essentially communicated three messages:

(i) A short range repulsion of the nucleon-nucleon interaction is not due to the hard or soft core in a static nucleon-nucleon potential but is due to many body symmetries of the 6-valence-quarks of the two interacting nucleons. At short distances the spatial symmetry is of [42] nature with the probability 8/9. This requests that one has at least two harmonic oscillator quanta in the relative wave function. That means one has a node in the interaction region which produces a hard core phase shift in the differential cross section.

(ii) In the elastic electron deuteron scattering genuine quark degrees of freedom can be seen in the transversal form factor which is proportional to the magnetic dipole form factor $F_{M1}(q)$. These quark degrees of freedom show up at momentum transfers above $5fm^{-1}$ or $1GeV/c$. They are essentially due to spin magnetization currents.

(iii) In finite nuclei one finds genuine quark effects which cannot be explained on the nucleon level for the isovector magnetic moments of nuclei one nucleon away from doubly closed shells. For the mass A=39 system the quark exchange currents due to the antisymmetrization of the quark wave function contribute 20 % to the Schmid value.

I would like to thank Drs. A. Buchmann and Y. Yamauchi with whom this work has been performed. In the investigations reported in chapter 4 for finite nuclei the contributions of Prof. Akito Arima were also essential.

6. REFERENCES

1. A. Faessler, F. Fernandez, G. Lübeck, K. Shimizu, Phys.Lett.**112B** (1982)201; Nucl. Phys. **A402** (1983)555
2. M. Oka, K. Yazaki, Progr. Theor. Phys. **66** (1981)556 and 572
3. M. Oka, K. Yazaki, in Quarks and Nuclei, ed. W. Weise (World Scientific, Singapore) p. 489
4. K. Bräuer, A. Faessler, F. Fernandez, K. Shimizu, Z. Phys. **A320** (1985)609
5. K. Shimizu, Rep. Progr. Phys. **52** 1989)1
6. A. Arima, H. Horie, Progr. Theor. Phys. **11** (1954)509 and **12** (1954)623

7. K. Shimizu, Phys. Lett. **148B** (1984)418

8. F. Fernandez, E. Oset, Nucle. Phys. **A455** (1986)720

9. Y. Yamauchi, A. Buchmann, A. Faessler, Nucl. Phys. A **A494** (1989)401

10. A. Buchmann, Y. Yamauchi, A. Faessler, Nucl. Phys. **A496** (1989)621

11. Y. Yamauchi, A. Buchmann, A. Faessler, A. Arima, Nucl. Phys. **A526** (1991)495

12. A. Faessler, F. Fernandez, Phys. Lett. **124B** (1983)124

13. R. Arndt et al., Phys. Rev. **C15** (1977)1002

14. U. Straub, thesis University of Tübingen

15. A. Faessler, U. Straub, Ann. Physik **47** (1990)439; Nucl. Phys. **A483** (1988)686

16. A. Faessler, Nuclear and Particle Physics p. 238: Editors: C. J. Burden, B. A. Robson, 1990, World Scientific

17. A. Buchmann, H. Ito, Y. Yamauchi, A. Faessler, J. Phys. **G 14** (1988)1037

18. M. Lacombe et al., Phys. Lett. **B101** (1981)139

19. M. E. Elias et al., Phys. Rev. **177** (1969)2075; S. Galster et al., Nucl. Phys. **B 32** (1972)221; R. G. Arnold et al., Phys. Rev. Lett. **35** (1975)776; **58** (1987)1723

Few-Body Systems, Suppl. 6, 195—205 (1992)

DEUTERON FORM FACTORS FROM MESON-THEORETICAL NUCLEON-NUCLEON INTERACTIONS *

J. Pauschenwein, W. Plessas, and L. Mathelitsch

Institute for Theoretical Physics

University of Graz

Universitätsplatz 5, A-8010 Graz, Austria

Abstract

We review the description of the electromagnetic structure of the deuteron by realistic meson-theoretical NN interaction models. We exemplify the present situation particularly in the cases of the Paris and Bonn potentials vis-à-vis recent accurate data of the elastic and magnetic form factors as well as the tensor polarization from elastic e-d scattering. The effects of certain ingredients in the calculations, such as relativity, meson-exchange currents, and nucleon form factors are addressed. A good qualitative agreement is achieved within our most-refined calculation. Nevertheless, a detailed comparison with the whole experimental data base reveals that neither meson-exchange model considered can   really provide an accurate description over the whole momentum-transfer region up to $q^2 \lesssim 80$ fm^{-2}. Remarkable discrepancies, in particular, from electric form factor measurements occur already at relatively low q^2.

1. Introduction

Recent years have witnessed big progress in the application of the presently most advanced NN interaction models that were derived from meson-exchange dynamics. Beyond $3N$ and $4N$ bound-state calculations on binding energies and wave functions we may mention also studies of electromagnetic reactions involving ^3H, ^3He as well as ^4He and, notably, of N-d scattering. For the latter process, predictions of elastic [1] and

---

* Talk presented by W. Plessas

subsequently also of break-up [2] observables have become available for such models as the Paris [3] and Bonn [4] potentials. The approaches to solving the $3N$ Faddeev equations have been proven to be quite reliable [5], and in many places the results have been found in remarkable good agreement with a variety of N-d data. Nevertheless, certain discrepancies between theory and experiments remained. Specifically, they concern the nucleon analyzing power A_y in elastic N-d scattering [6,7] and the cross sections of particular configurations of N-d break-up [2,8]. Up till now it is not clear from where these shortcomings originate. Beyond possible NN on-shell uncertainties, likely sources are inadequate NN off-shell properties and omissions of $3N$ forces. Effects from the latter two can hardly be disentangled on the grounds of the $3N$ system as they can influence the results in the same manner. Indeed, it was proved recently on the basis of rigorous 3-body scattering theory that such an ambiguity exists, i.e. variations of the NN off-shell behaviour affect $3N$ observables just as explicit $3N$ forces [9].

In this situation it has become increasingly important to look at electromagnetic processes (involving the deuteron) that allow definitely to determine the off-shell behaviour of the NN system. Once this has been pinned down separately from the $3N$ system, one can get unambiguous evidence on the role of $3N$ forces. Only NN interaction models that have survived this test of their off-shell properties will be useful in revealing the effects of $3N$ forces.

Here, we thus consider the very performance of the Paris and several versions of the Bonn potentials in elastic e-d scattering. Not only have these models so far been considered to provide a rather accurate description of all elastic NN data but they were hitherto also most extensively applied in $3N$ studies. Our results to be discussed in the following will add further important evidences on the present performance of meson-exchange NN dynamics, as it comes specifically through the above-mentioned potential models of the Paris and the Bonn groups.

2. Elastic e-d scattering

2.1 Ingredients of the calculation

We performed our calculations of the deuteron electromagnetic structure functions in the conventional framework, i.e. first evaluating the nonrelativistic impulse approximation (IA) and then adding the relativistic Darwin-Foldy, spin-orbit, and nuclear-motion contributions (RC) as well as meson-exchange currents (MEC); for the latter we specifically included in the present study only the most important ones, i.e. the π-pair and π-retardation as well as $\rho\pi\gamma$-currents. The remaining MEC are not significant for the observables considered here, as their individual contributions largely cancel each other so that their total effect does not exceed a few percent.

In the case of the Bonn potentials, at least, the strong-interaction vertices that enter into the calculation of MEC were taken consistently with the underlying NN force model (same functional dependences and same cut-offs of form factors). For the Paris potential we relied on the usual monopole form factor with a cut-off $\Lambda_\pi = 1000$ MeV and the value $g_{\pi NN} = 13.63$ for the π-coupling constant. In all cases pseudo-scalar π-coupling was employed, and the $\rho\pi\gamma$-current was treated after ref. [10], with coupling constants $g_{\rho\pi\gamma} = +0.578$ [11] and $\kappa_\rho = 6.1$ and 6.6 for the Bonn and Paris potentials, respectively.

For the nucleons we used Dirac form factors and employed the parametrization of Höhler et al. [13]

2.2 Results for e-d observables

The results from the full calculations containing all of the above ingredients are displayed in Figs. 1-3 for the Paris potential [3] and both the Bonn OBEPR as well as full models [4]. From the semi-logarithmic plots of Fig. 1 one may conclude that the predictions of all the models are in fair agreement with each other and with the experimental data in the domain $q \lesssim 4$ fm^{-1}. Only for $q > 4$ fm^{-1} sensible discrepancies among the various potentials appear.

In the framework of the present calculation, the full Bonn potential performs best for the electric form factor $A(q^2)$ even towards rather high momentum transfers $q \lesssim 10$ fm^{-1}. Anticipating the later discussion, we note already here that this success is supported by the usage of Höhler nucleon form factors. If another parametrization was used for the nucleon form factors, the full Bonn result would deviate from the $A(q^2)$ data beyond $q \gtrsim 6$ fm^{-1} [26].

The situation is to some extent reversed for the magnetic form factor $B(q^2)$. In the region $q > 4$ fm^{-1}, the Paris and the Bonn OBEPR potentials are closer to the experimental data than is the full Bonn model. We emphasize, however, that these results are also subject to considerable changes depending on which nucleon form factors are used; see the subsequent discussion and cf. also refs. [27, 28].

The reproduction of $A(q^2)$ data appears in a different light when a more careful comparison is done by looking at percentage deviations instead of logarithmic plots. This is done in Fig. 2 for the region of rather low momentum transfers $q \leq 4$ fm^{-1}. It is surprising that neither potential cannot really describe the rather accurate experimental data. Yet, the Höhler parametrization of the nucleon form factors is the most favourable one for this domain [28, 29].

For the tensor polarization $T_{20}(q^2)$ new data have recently been published [24, 25]. All potentials are able to reproduce them within the error bars (Fig. 3). One must be aware, however, that also this observable is much sensitive to the ingredients of the calculation (nucleon form factors, ...) for $q \gtrsim 5$ fm^{-1}.

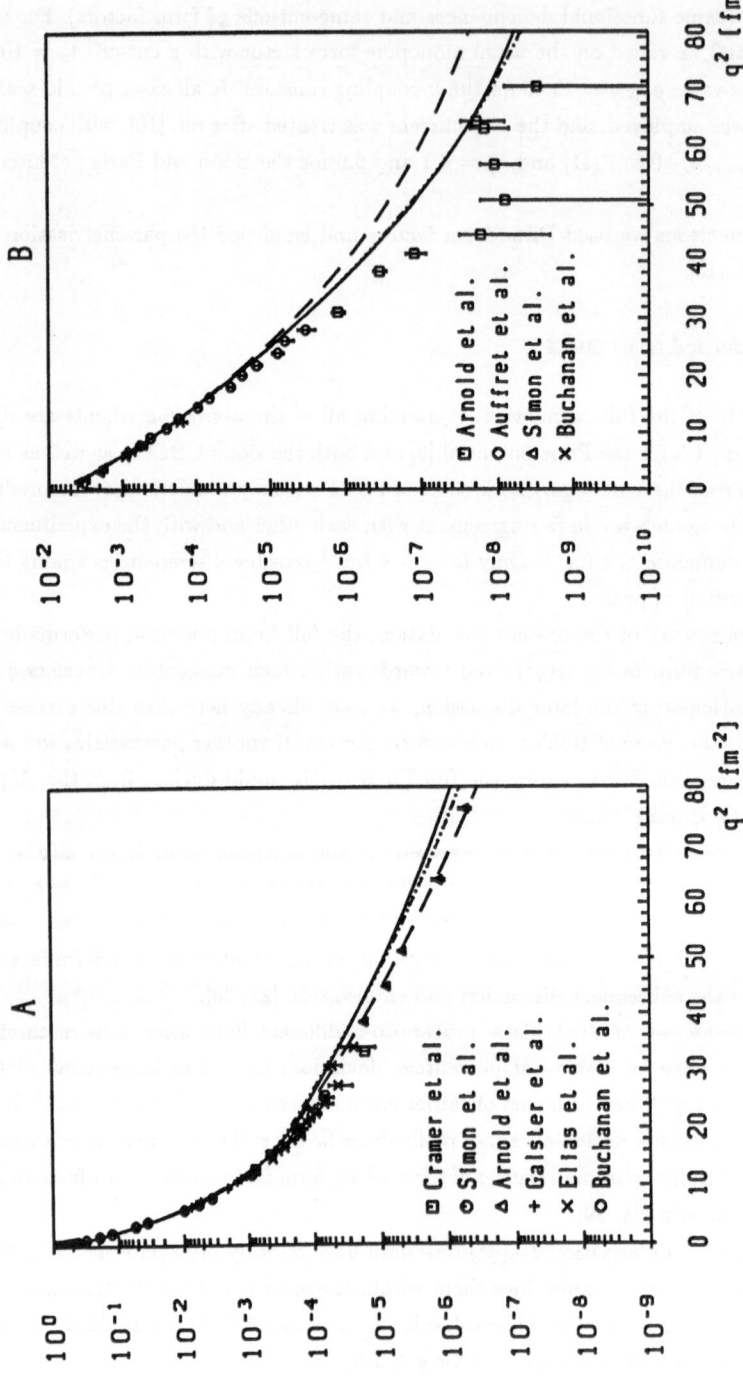

Fig.1. Deuteron electric and magnetic form factors with RC and MEC included for the Paris [3] (dash-dotted), full Bonn (dashed), and Bonn OBEPR [4] (solid) potentials. Experimental data from refs. [14-21].

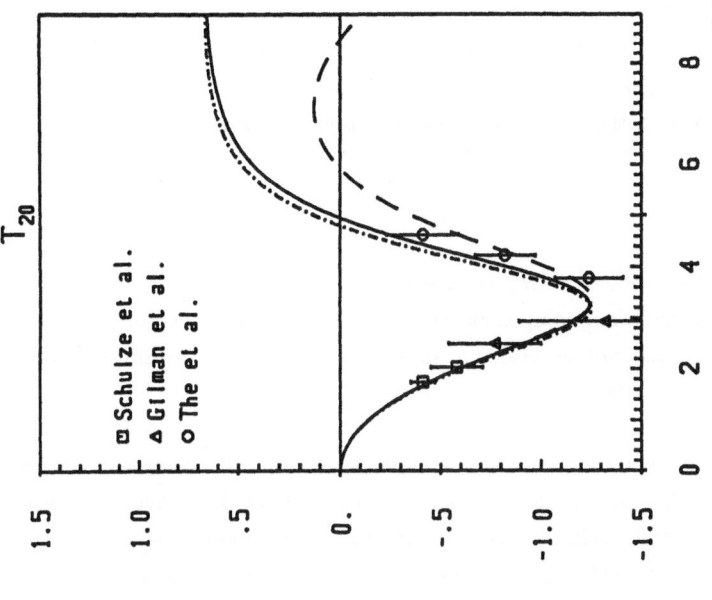

Fig. 3. Deuteron tensor polarization from the same calculation and for the same cases as in Fig. 1. Experimental data from refs. [23-25].

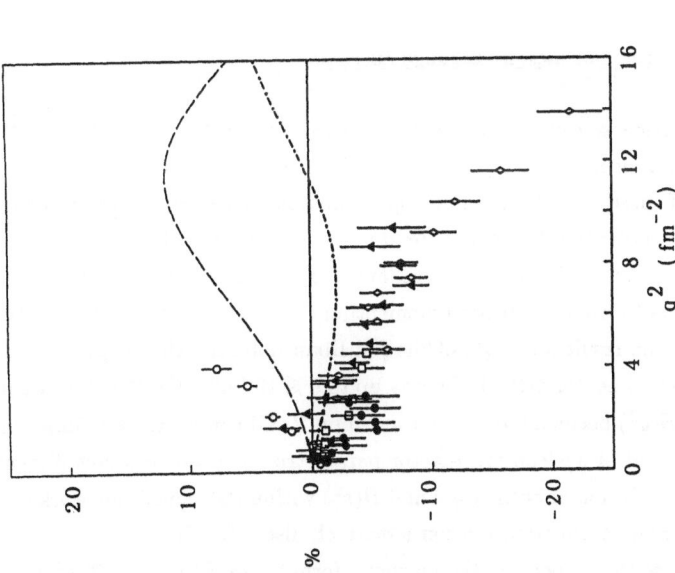

Fig. 2. Deuteron electric form factor at low momentum transfers represented as percentage deviations relative to the OBEPR result. Same calculation and notation as in Fig. 1. Experimental data from refs. [15, 18, 22].

2.3 Explicit energy dependence of the full Bonn potential

With respect to the full Bonn potential we remark that our calculation takes care of the explicit energy dependence occurring in this model. The necessary modification of the current operator [30] can effectively be accommodated through a deuteron wave function Ψ_S derived from the original one, Ψ_E, by applying the Perey transformation [31, 32]:

$$\Psi_S = \left(1 - \frac{\partial V}{\partial E}\right)^{1/2} \Psi_E. \tag{1}$$

This turns out to be a practical way to account for the explicit energy dependence (i.e. to eliminate non-nucleonic components) in the context of e-d observables [33]. The corresponding effect must not be neglected as it is of considerable magnitude; some works before did not pay attention to this fact, see, e.g., ref. [34].

Indeed, our consideration of the effect of explicit energy dependence in the NN potential has recently been confirmed by the appearance of another version of the full Bonn potential [35] whose energy dependence has been eliminated beforehand through the folded-diagram technique [26]. It turns out that the predictions of these new (energy-independent) full Bonn potential are quite close to the results of the energy-dependent full Bonn potential, once the corresponding modifications have been carried out in the latter case (cf. Fig. 4). The remaining discrepancies, especially in the magnetic form factor $B(q^2)$, might be caused by the still different treatment of the energy dependence (Perey transform vs. folded-diagram method) and, to a larger extent, also the readjustment of the meson parameters in the folded-diagram full Bonn potential [35].

2.4 Effects of relativistic and meson-exchange currents

Due to the limited space we cannot give a detailed discussion of RC and MEC here. We only note that a comparison of the dashed lines in Figs. 1 and 4 allows for an estimation of the total effect of these contributions, on top of the nonrelativistic IA result, for the case of the full Bonn potential; the global trend is similar for the other cases.

In particular, the combined effects of RC and MEC make both the electric and magnetic form factors to rise at increasing momentum transfers, cf. also ref. [27]. In our present calculation, the result for $A(q^2)$ of the full Bonn potential thus happens to lie just on top of the experimental data all the way up to $q \lesssim 10$ fm^{-1}. On the other hand, the predictions for $B(q^2)$ become too high for $q \gtrsim 5$ fm^{-1}. Of course, these findings are also subject to the kind of nucleon form-factor parametrization, but it seems difficult to achieve a good description of both $A(q^2)$ and $B(q^2)$ within the same framework over the whole range of momentum transfers considered; cf. also ref. [27].

It is our experience that especially the magnetic form factor $B(q^2)$ at $q \gtrsim 5$ fm^{-1} is governed by a subtle interplay of potential effects (in particular, deuteron wave function behaviour at $p \gtrsim 2$ fm^{-1}), of RC as well as MEC, and of the nucleon structure (mainly,

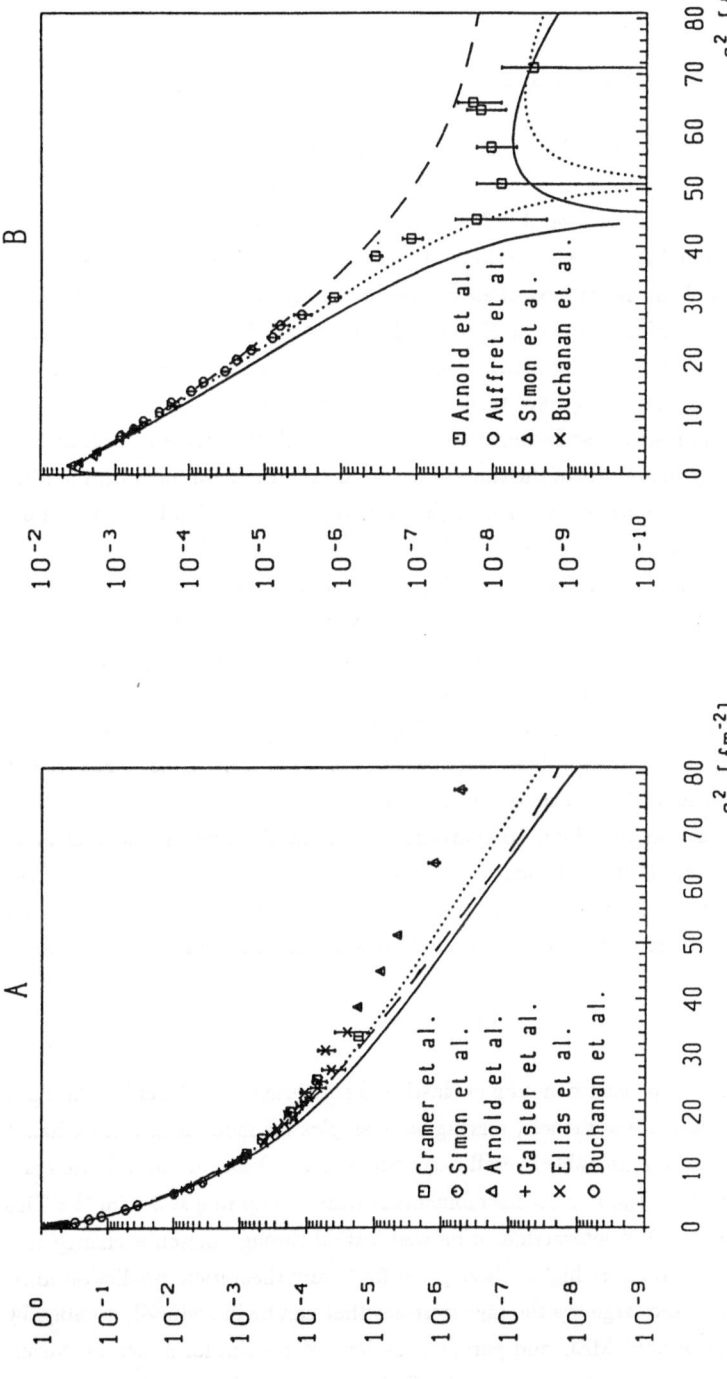

Fig. 4. Deuteron electric and magnetic form factors in IA for the energy-dependent full Bonn potential [4] without (solid) and with (dashed) energy correction after eq. (1). The dotted line represents the IA result for the energy-independent, folded-diagram, full Bonn potential [35].

neutron electric form-factor parametrization). Considerable changes can be generated by turning these ingredients thus leaving the reproduction of the experimental data at $5 \lesssim q \lesssim 9 \text{ fm}^{-1}$ open to controversy.

2.5 Influence of the nucleon structure

We have previously hinted at the significant role played by the kind of model adopted for the nucleon form factors. The corresponding effects have already been studied by many groups in the cases of several NN potentials [27, 28, 34, 37].

Out of the available parametrizations of nucleon form factors, by now only two appear as reasonable in the context of elastic e-d scattering, i.e. the one of Höhler et al. (H) [13] and the one of Iachello et al. [38] and Galster et al. [16] according to Lomon's best fit [39], here abbreviated by IJLG, as usual. Other parametrizations like, e.g., the Gari-Krümpelmann model [40] have turned out as inadequate.

All our results presented so far have been obtained with the model of Höhler et al. In Fig. 5 we display the results of the same calculation as in Figs. 1-3 but with the IJLG parametrization of the nucleon form factors in the two cases of the full Bonn potentials. While the differences between H and IJLG are not significant at low q^2, the effect of the latter model generally consists in lowering the predictions for $A(q^2)$ towards higher momentum transfers. As a consequence, e.g., the prediction of the (energy-dependent) full Bonn potential falls off below the data points, whereas the folded-diagram full Bonn potential now gets in better agreement with experiment. The same trend of lowering the higher-q^2 prediction can be observed for $B(q^2)$, at least with the energy-dependent full Bonn potential. In this case, however, the effect of using IJLG is not large enough to achieve agreement with the experimental data.

As a consequence of the above comparison, and of similiar investigations of other groups, we must state that insufficient constraints on the nucleon form factors, particularly on the neutron electric form factor, introduce considerable uncertainties in the description of the deuteron structure, especially towards higher momentum transfers.

3. Conclusions

We have discussed the present situation of elastic e-d scattering with the aid of the Paris and Bonn potentials. These models, serving as examples for modern meson-exchange NN dynamics, yield a reasonable overall description of the deuteron form factors and the tensor polarization T_{20}, at least for momentum transfers up to $q \approx 5$-6 fm^{-1}). This proves the long-range NN interaction to be well settled through meson-exchange theory. For momentum transfers higher than $q \approx 6$ fm^{-1} any theoretical prediction must be taken with care. Too large are the uncertainties that can be introduced, notably, by different (assumptions on) MEC and parametrizations of nucleon form factors. Nevertheless, it seems that in neither framework all the electric and magnetic form-factor

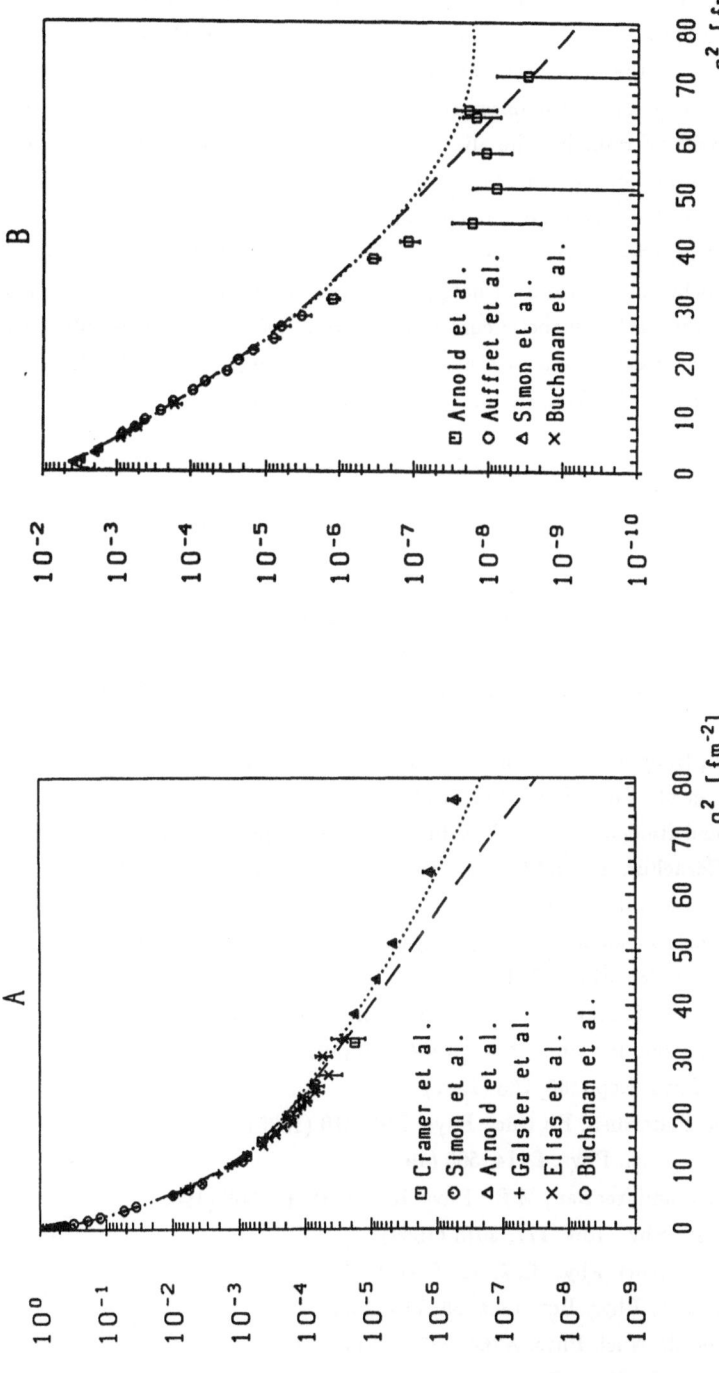

Fig. 5. Deuteron electric and magnetic form factors with RC and MEC included from the same type of calculation as described in Sec. 2.1 but with the nucleon form factors of IJLG [39] for the energy-dependent full Bonn (dashed) and the energy-independent, folded-diagram, full Bonn (dotted) potentials.

data can   be reproduced up to $q \lesssim 10$ fm^{-1}, say, with the meson-exchange NN potentials so far applied. However, as revealed by a more careful examination, some problems also exist in the electric form factor $A(q^2)$ at relatively low momentum transfers. It is worthwhile to reexamine this domain, avoiding semi-logarithmic presentation of the results, vis-à-vis the rather accurate modern data base available there.

In looking for remedies of the observed shortcomings, we wish to point out that beyond the necessity of resolving the above-mentioned intricacies it will be desirable to take into account also effects from nucleon excitation (into $\Delta(1232)$ and $N^*(1440)$). Some studies have so far been carried out in this direction, suggesting the importance of the corresponding effects [41, 42, 43], but there is no calculation considering explicit isobar effects both in a meson-exchange NN potential (valid beyond the inelastic threshold) and in the electromagnetic current-interaction. Of course, this problem will have to be resolved - possibly in a consistent way - before one can argue about further missing effects, in particular from subnuclear degrees of freedom.

References

1. Koike, Y., Haidenbauer, J., and Plessas, W.: Phys. Rev. C **35**, 396 (1987)
2. Witala, H., Cornelius, T., and Glöckle, W.: Few-Body Systems **5**, 89 (1988); Witala, H., Glöckle, W., and Cornelius, T.: ibid. **6**, 79 (1989); Phys. Rev. C **39**, 384 (1989)
3. Lacombe, M. et al.: Phys. Rev. C **21**, 861 (1980)
4. Machleidt, R., Holinde, K., and Elster, Ch.: Phys. Rep. **149**, 1 (1987)
5. Cornelius, T. et al.: Phys. Rev. C **41**, 2358 (1990)
6. Plessas, W. and Haidenbauer, J.: Few Body Systems Suppl. **2**, 185 (1987)
7. Witala, H., Cornelius, T., and Glöckle, W.: Nucl. Phys. **A491**, 157 (1989)
8. Glöckle, W., Witala, H., and Cornelius, T.: In: Selected Topics in Nuclear Structure (Proc. of the 25th Zakopane School of Physics), ed. by J. Styczen and Z. Stachura. Singapore: World Scientific, 1990
9. Polyzou, W.N. and Glöckle, W.: Few-Body Systems **9**, 97 (1990)
10. Gari, M. and Hyuga, H.: Nucl. Phys. **A264**, 409 (1976)
11. Towner, I.S. : Phys. Rep. **155**, 263 (1987)
12. G. Höhler and Pietarinen, E.: Nucl. Phys. **B95**, 210 (1975)
13. G. Höhler et al.: Nucl. Phys. **B114**, 505 (1976)
14. Buchanan, C.D. and Yearian, M.R.: Phys. Rev. Lett. **15**, 303 (1965)
15. Elias, J.E. et al.: Phys. Rev. **177**, 2075 (1969)
16. Galster, S. et al.: Nucl. Phys. **B32**, 221 (1971)
17. Arnold, R.G. et al.: Phys. Rev. Lett. **35**, 776 (1975)
18. Simon, G.G. et al.: Nucl. Phys. **A364**, 285 (1981)
19. Cramer, R. et al.: Z. Phys. C **29**, 513 (1985)

20. Auffret, R.G. et al.: Phys. Rev. Lett. **54**, 649 (1985)

21. Arnold, R.G. et al.: Phys. Rev. Lett. **58**, 1723 (1987)

22. Platchkov, S. et al.: Nucl. Phys. **A510**, 740 (1990)

23. Schulze, M.E. et al.: Phys. Rev. Lett. **52**, 597 (1984)

24. Gilman, R. et al.: Phys. Rev. Lett. **65**, 1733 (1990)

25. The, I. et al.: Phys. Rev. Lett. **67**, 173 (1991)

26. Plessas, W. in: Selected Topics in Nuclear Structure (Proc. of the 25th Zakopane School of Physics), ed. by J. Styczen and Z. Stachura. Singapore: World Scientific, 1990

27. Schiavilla, R. and Riska, D.O.: Phys. Rev. C **43**, 437 (1991)

28. Mosconi, B. and Ricci, P.: Few-Body Systems **6**, 63 (1989); ibid. **8**, 159 (1990)

29. Pauschenwein, J., Mathelitsch, L., and Plessas, W.: in preparation

30. Friar, J.L. in: Mesons and Nuclei, Vol. 2, ed. by M. Rho and D. Wilkinson. Amsterdam: North Holland, 1979; Lecture Notes in Physics **108**, 445 (1979)

31. Perey, F.G. and Buck, B.: Nucl. Phys. **32**, 353 (1962); de Forest, T., Jr.: Nucl. Phys. **A163**, 237 (1971)

32. Friar, J.L.: Ann. Phys. (N.Y.) **104** , 445 (1979)

33. Pauschenwein, J., Mathelitsch, L., and Plessas, W.: Nucl. Phys. **A508**, 253c (1990)

34. Chung, P.L. et al.: Phys. Rev. C **37**, 2000 (1988)

35. Haidenbauer, J., Holinde, K., and Johnson, M.B.: Preprint KFA-IKP(TH)-1991-26

36. Johnson, M.B., Haidenbauer, J., and Holinde, K.: Phys. Rev. C **42**, 1878 (1990)

37. Hummel, E. and Tjon, J.A.: Phys. Rev. Lett. **63**, 17 (1989); Phys. Rev. C **42**, 423 (1990)

38. Iachello, F., Jackson, A.D., and Landé, A.: Phys. Lett. **43B**, 191 (1973)

39. Lomon, E.L.: Ann. Phys. (N.Y.) **125**, 309 (1980)

40. Gari, M. and Krümpelmann, W.: Z. Phys. **A322**, 698 (1985)

41. Sitarski, W.P., Blunden, P.G., and Lomon, E.L.: Phys. Rev. C **36**, 2479 (1987); Blunden, P.G., Greenberg, W.R., and Lomon, E.L.: Phys. Rev. C **40**, 1541 (1989)

42. Dymarz, R. and Khanna, F.C.: Phys. Rev. Lett. **56**, 1448 (1986); Phys. Rev. C **41**, 2438 (1990); Nucl. Phys. **A507**, 560 (1990)

43. Obersteiner, P., Plessas, W., and Pauschenwein, J.: Few-Body Systems Suppl. , to appear

Few-Body Systems, Suppl. 6, 206—211 (1992)

Few-
Body
Systems
© by Springer-Verlag 1992

INTRODUCTION TO THE DISCUSSION SESSION ON PHOTO- AND ELECTRODISINTEGRATION OF THE DEUTERON

H. Arenhövel

Institut für Kernphysik

Johannes Gutenberg-Universität

D-W-6500 Mainz, Germany

This session is exclusively devoted to the electromagnetic break-up of the deuteron. It reflects the fact that notwithstanding its long history deuteron photo- and electro-disintegration is still an active and interesting field of research (for a recent review on photodisintegration see [1]). The purpose of this introduction is - on the one hand - to put the various contributions to this session into a general perspective and - on the other hand - to supply complimentary material of interesting recent experimental results which is of relevance to the general theme.

The special role of the electromagnetic deuteron break-up is due to (i) the simple structure of the deuteron which allows exact solutions at least in the non-relativistic domain, and (ii) the specific features of the electromagnetic probe being well known and interacting only weakly with the nuclear system. The present emphasis lies on the study of subnuclear degrees of freedom (d.o.f.) for which the two-body system offers a unique opportunity. In electromagnetic reactions like electro- and photodisintegration, these subnuclear d.o.f. are manifest through meson exchange currents (MEC) and isobar configurations (IC). At present we have strong experimental evidence for subnuclear d.o.f. in $E1$-transitions in $d(\gamma, N)N$ over a large energy region and in $M1$-transitions in $d(e, e')np$ near threshold at higher momentum transfers and in the Δ-resonance region.

I will start the discussion with the inclusive electrodisintegration. The cross section for $d(e, e')np$ is given by

$$\frac{d\sigma}{dk_2^{lab} d\Omega_e^{lab}} = 6c\left(\rho_L F_L + \rho_T F_T\right),$$ (1)

where c is determined from the electron kinematics and ρ_L and ρ_T describe essentially the longitudinal and transverse polarization of the virtual photon (see [2] for further details). The two form factors F_L and F_T contain information on the dynamic properties of the np system. They depend on the internal excitation energy E_{np} and on the momentum transfer $\vec{q}$.

While the longitudinal form factor is little affected by MEC - a consequence of Siegert's hypothesis - one finds considerable contributions from MEC and IC to the transverse form factor in various kinematic regions. A survey is shown in fig. 1. Particularly strong contributions are found close to the break-up threshold at high momentum transfer.

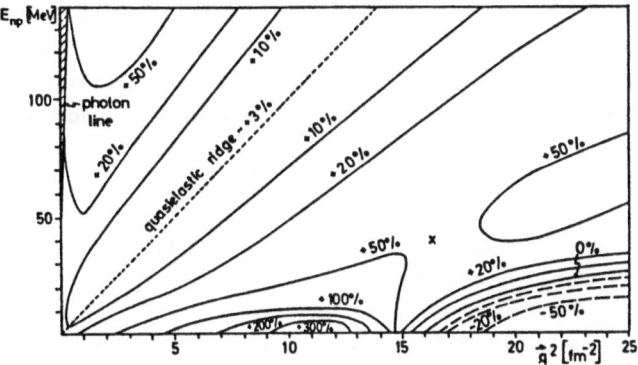

Fig.1 Contour plot of the changes (in percent) of the transverse form factor F_T by MEC and IC from [3].

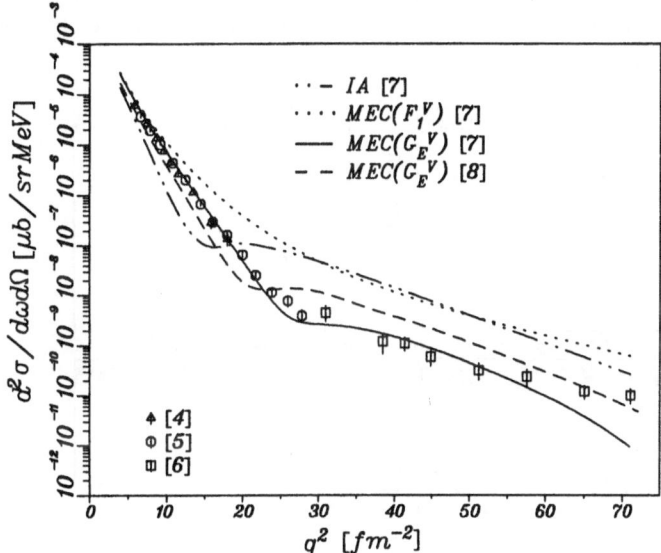

Fig.2 $d(e, e')np$ cross section at 155^0. Theoretical curves calculated with Bonn QC potential using G_E^V or F_1^V for MEC from [7] and for Argonne V_{14} potential from [8].

Let me briefly illustrate the present situation in $d(e, e')np$ near threshold in fig. 2, where experimental cross sections at backward angle and averaged over $E_{np} = 0 - 3\,MeV$ for $\vec{q}^2 \leq 30\,fm^{-2}$ and over $E_{np} = 0 - 10\,MeV$ for $\vec{q}^2 \geq 30\,fm^{-2}$ are shown together with various theoretical results. The normal theory without subnuclear d.o.f., called impulse approximation (IA), exhibits a minimum at about $\vec{q}^2 = 12\,fm^{-2}$ which is caused by a destructive interference of the transitions from the $^3S_1$- and $^3D_1$-components of the deuteron to the $^1S_0$-final scattering state. The large deviation from experimental data is obvious, but inclusion of standard MEC - mainly π- and ρ-MEC - and Δ-IC leads to a satisfactory description at not too high momentum transfers, say $\vec{q}^2 < 6\,fm^{-2}$. In the higher q-region larger uncertainties arise from electromagnetic and hadronic form factors, potential model dependence and relativistic corrections. Above about $\vec{q}^2 = 15\,fm^{-2}$ the normal and MEC contributions interfere destructively and thus small changes in different calculations can result in large differences in the cross sections. For this reason, one finds considerably different results depending on the potential model and whether one uses G_E^V or F_1^V as e.m. form factor for MEC [9]. One should also keep in mind that the SLAC data are averaged over a wider energy range (10 MeV) compared to the other data. In this high momentum transfer region, new experimental data, averaged over 3 MeV, is now available and will be presented in the first contribution by C.F. Williamson [10].

More detailed information on the dynamics of subnuclear d.o.f. can be obtained from exclusive reactions. Here, the cross section

$$\frac{d\sigma}{dk_2^{lab}d\Omega_e^{lab}d\Omega_{np}^{c.m.}} = c\left(\rho_L f_L + \rho_T f_T + \rho_{LT} f_{LT}\cos\phi + \rho_{TT} f_{TT}\cos 2\phi\right) \qquad (2)$$

is governed by four structure functions f_α ($\alpha = L, T, LT, TT$) (for details of the notation see [2]). Besides the diagonal and transverse structure functions, one has two additional longitudinal-transverse and transverse-transverse interference terms which give new information on the system. As an example, I show in fig. 3 the structure functions for a specific energy and momentum transfer.

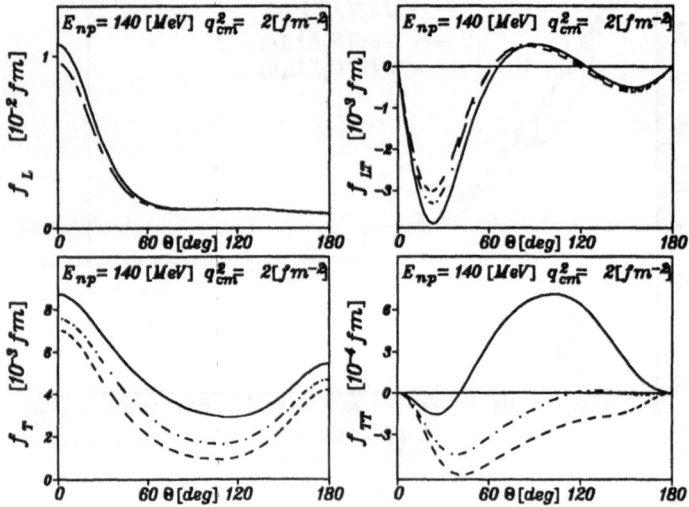

Fig.3 The four structure functions at $E_{np} = 140\,MeV$, $q^2 = 2\,fm^{-2}$ for normal theory (dashed) and with additional MEC (dash-dot) and IC (solid curve).

The longitudinal structure function f_L shows a general forward peaking which originates from the charge interaction with the proton. f_L is only little affected by IC contributions. The transverse structure function f_T is much more influenced by MEC and IC leading to a drastic increase, in particular in the minimum around 90^0. Among the two interference functions, f_{LT} shows some sensitivity to subnuclear d.o.f., most pronounced in the forward minimum, while f_{TT} is very strongly affected by contributions from MEC and IC leading even to a sign change. Even though the interference structure functions are considerably smaller than f_L and f_T, a separation of them will be rewarding for testing our theoretical understanding of subnuclear effects in greater detail.

For f_{LT} such a separation has been achieved for the first time by Tamae et al. [11]. More recently, f_{LT} has been separated for quasi-free kinematics at two places: Bonn and NIKHEF. The Bonn results are presented here by Reichelt [12]. I will complement this by showing in fig. 4 the results of NIKHEF [13, 14] for the ϕ-asymmetry

$$A_\phi = \frac{\rho_{LT} f_{LT}}{\rho_L f_L + \rho_T f_T + \rho_{TT} f_{TT}}, \tag{3}$$

which is proportional to f_{LT}. The interesting result is that relativistic effects are important

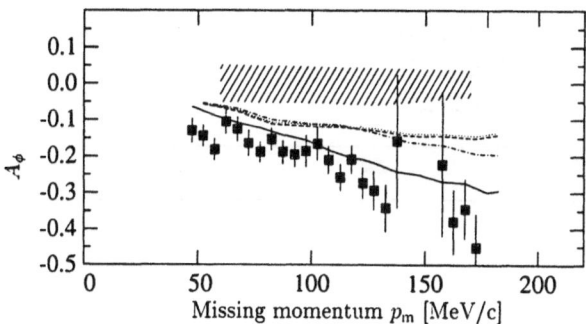

Fig.4 A_ϕ as function of the missing momentum; relativistic calculation from [15] including FSI: solid line; nonrelativistic calculations: dash-dot: PWIA, dotted: with FSI, dashed: with FSI, MEC and IC from [13].

and lead to a better description of the data. This will be discussed in the contribution of Mosconi [16] devoted to relativistic corrections. In fact, the study of relativistic effects has become another interesting subject in the electromagnetic deuteron break-up. Relativistic contributions to the current operator have already been proven to be of importance in understanding the 0^0- and 180^0-cross section in $d(\gamma, p)n$. But the existing experimental data did not allow definite conclusions. Fortunately, the situation will change with new experimental data from Mainz [19] which I show in fig. 5 with theoretical predictions using different potentials. The improved quality of the new data is very impressive, in particular for 180^0, and the agreement of the theory is quite satisfactory.

The study of polarization effects in deuteron electrodisintegration utilizing polarized electrons and/or polarized targets is a further very interesting topic which is receiving increased attention since the development of polarized beams and targets make such investigations experimentally feasible. A general survey has recently been started in order to find out to what extent the use of polarized electrons and/or polarized targets will allow a more detailed investigation of the dynamical features of the nuclear system than is possible without the use of polarization. This will be discussed in the contribution of Leidemann [18].

In view of the more fundamental framework of QCD, we are interested in the limits of the conventional theory with only nucleon, meson and isobar d.o.f. In other words, the question is: Where will we be able to see the onset of quark-gluon dynamics, i.e., where will the conventional framework clearly fail to understand certain experimental data? The last contribution of Kondratyuk is devoted to this question [20].

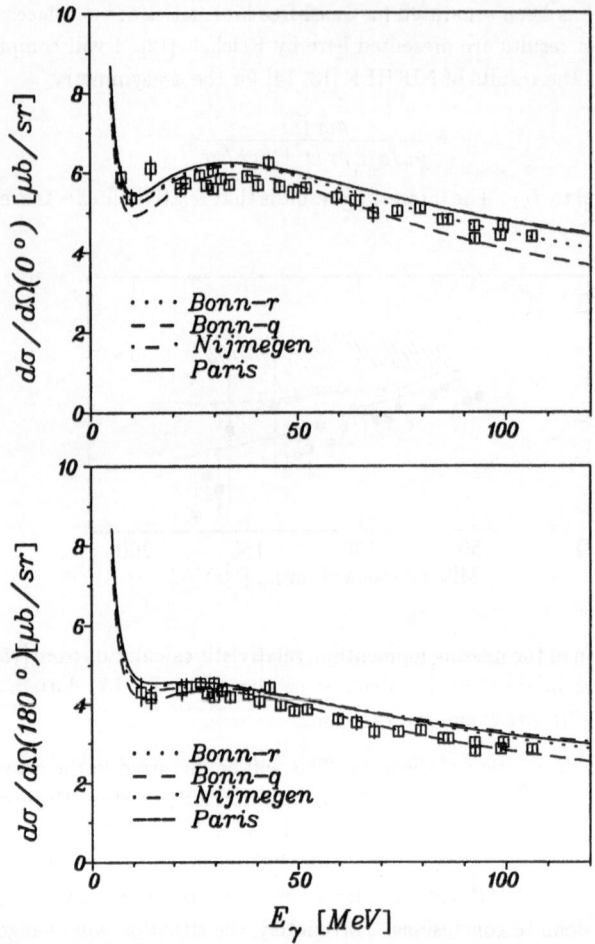

Fig.5 Differential cross section of $d(\gamma, p)n$ at 0^0 and 180^0 from [19] with theoretical predictions for Paris, Bonn-r and -q and Nijmegen potentials.

Acknowledgement: I would like to thank R. van de Vyver for allowing me to present the unpublished experimental results on the 0^0- and 180^0-cross sections of $d(\gamma, p)n$.

References

[1] Arenhövel, H. and Sanzone, M.: Few–Body Syst., Suppl. **3**, 1 (1991)

[2] Arenhövel, H., Leidemann, W. and Tomusiak, E.L.: Z. Phys. **A331**, 123 (1988); **A334**, 363(E) (1989)

[3] Fabian, W. and Arenhövel, H.: Nucl. Phys. **A314**, 253 (1979)

[4] Bernheim, M. et al.: Phys. Rev. Lett. **46**, 402 (1981)

[5] Auffret, S. et al.: Phys. Rev. Lett. **55**, 1362 (1985)

[6] Arnold, R.G. et al.: Phys. Rev. **C42**, R1 (1990)

[7] Leidemann, W., Schmitt, K.-M. and Arenhövel, H.: Phys. Rev. **C42**, R826 (1990)

[8] Schiavilla, R. and Riska, D.O.: Phys. Rev. **C43**, 437 (1991)

[9] Singh, S.K., Leidemann, W. and Arenhövel, H.: Z. Phys. **A331**, 509 (1988)

[10] Williamson, C.F.: contribution to this conference

[11] Tamae, T. et al.: Phys. Rev. Lett. **59**, 2919 (1987)

[12] Reichelt, T.: contribution to this conference

[13] van der Schaar, M.: Ph.D. thesis, Rijksuniversiteit Utrecht (1991)

[14] van der Schaar, M. et al.: to be published

[15] Hummel, E. and Tjon, J.A.: to be published

[16] Mosconi, B.: contribution to this conference

[17] Leidemann, W., Tomusiak, E.L. and Arenhövel, H.: Phys. Rev. **C43**, 1022 (1991)

[18] Leidemann, W., Tomusiak, E.L. and Arenhövel, H.: contribution to this conference

[19] Zieger, A. et al.: private communication

[20] De Sanctis, E. and Kondratyuk, L.A.: contribution to this conference

Few-Body Systems, Suppl. 6, 212—217 (1992)

Few-
Body
Systems
© by Springer-Verlag 1992

DEUTERON THRESHOLD ELECTRODISINTEGRATION

Claude F. Williamson
Department of Physics and Bates Linear Accelerator Center
MIT, Cambridge, MA, USA

Cross sections for the D(e,e')pn reaction in the region of the breakup threshold have been measured at the MIT/Bates Linac by a multi-institutional collaboration for $10 < Q^2 < 40$ fm^{-2}. These measurements extend the previous Saclay data and are in excellent agreement with them in the region of overlap. The present cross sections averaged over 3 MeV from threshold appear to be consistently smaller than recently reported data from SLAC averaged over 10 MeV. None of the measurements shows a minimum in the cross sections for $Q^2 < 40$ fm^{-2}. Some recent potential model calculations appear to be in fair agreement with the data, but the existing calculations based on hybrid quark models appear to be in poor agreement with the data.

Introduction

As the only bound two-nucleon system, the deuteron is an especially important laboratory for studying the nucleon-nucleon interaction. A very useful workbench in this sub-atomic laboratory for studying the contribution of non-nucleonic degrees of freedom to the nucleon-nucleon interaction is the threshold electrodisintegration of this system. The threshold breakup is dominated by the isovector M1 transition from the 1^+, T=0 triplet ground state to the unbound 0^+, T=1 singlet excited state. The nucleon transition amplitudes $^3S_1$ to $^1S_0$ and $^3D_1$ to $^1S_0$ cancel very strongly around $Q^2 = 15$ fm^{-2} leaving the non-nucleonic degrees of freedom as the principal mediators of the threshold breakup reaction. For $Q^2 > 20$ fm^{-2}, there is a destructive interference between the one-body nucleon amplitudes and the two-body meson exchange currents (MEC's) and isobar currents (IC's). The cross sections for these large values of Q^2 depend very sensitively on the details of this cancellation.

By measuring the continuum breakup cross sections in the vicinity of the threshold for $Q^2 > 20$ fm^{-2} one can thus study the medium and short range structure of the non-nucleonic currents contributing to the two-nucleon interaction. In particular the M1 isovector character of the dominant transition in this excitation range favors π and ρ meson exchange currents and Δ isobar currents (IC's). The importance of this nuclear reaction has long been recognized and has been the subject of numerous theoretical[1] and experimental[2] investigations.

Motivation for the Bates Experiment

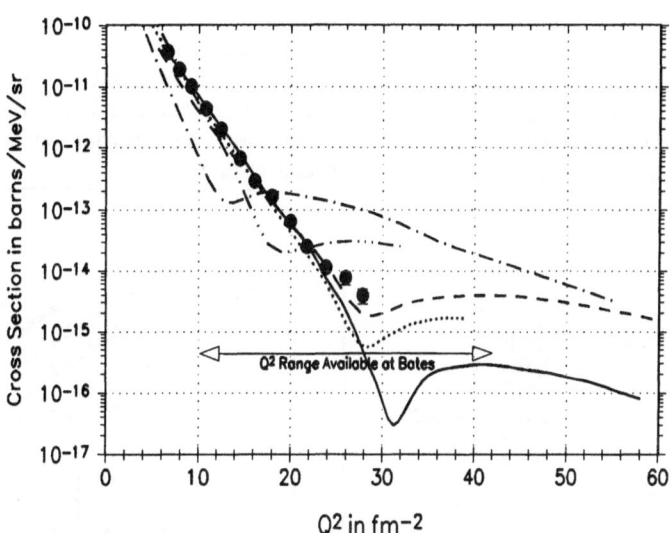

Figure 1. Experimental and theoretical situation for deuteron threshold electrodisintegration in 1986. The experimental data are taken from Ref. 3, and the theoretical curves are described in text.

The theoretical and experimental situation for the D(e,e′)pn threshold studies as it existed in 1986 is illustrated in Fig. 1. At that time a new experiment had just been reported from Saclay[3] which covered the range in Q^2 up to $Q^2 = 27.8$ fm^{-2}, and these data are shown in Fig. 1. Also shown in Fig. 1 are representative theoretical calculations chosen from those listed in Ref. 1. The dot-dash curve and the 3dot-dash curve are from Leidmann and Arenhövel using the Paris potential. The former is the one-body impulse approximation calculation with no MEC or IC contribution. The latter includes π, ρ, and Δ contributions for multipoles up to L = 4. The solid curve is from Mathiot using the Paris potential and two-body contributions from π, ρ, and Δ. The dashed curve is from Cheng and Kisslinger, and the dotted curve is from Yamauchi, et al.. Both of these two latter calculations are based on the hybrid-quark-hadron model. All data and theoretical calculations shown in Fig. 1 are averaged over the first 3 MeV above threshold.

The range of Q^2 available at Bates at a scattering angle of 160° is also shown in Fig. 1. Clearly the highest Q^2 points from Saclay were just at the edge of a kinematic region where the various theoretical predictions diverged by more than an order of magnitude. Bates possessed the capability of providing data in this Q^2 range where various theories diverged widely. It was recognized that such data would be crucial in guiding theoretical under-standing of this important process. The experiment was carried out by a collaboration from the University of Massachusetts, MIT, American University, CEN Saclay, University of Illinois and Shizuoka University.

Description of the Experiment

As can be seen from Fig. 1, the predicted cross sections lie in the range of 10^{-37} to 10^{-40} cm²/sr in the Q^2 region of interest. In order to measure such small cross sections one must maximize the available luminosity and reduce backgrounds to an absolute minimum. The target and detector systems were both designed to achieve these objectives.

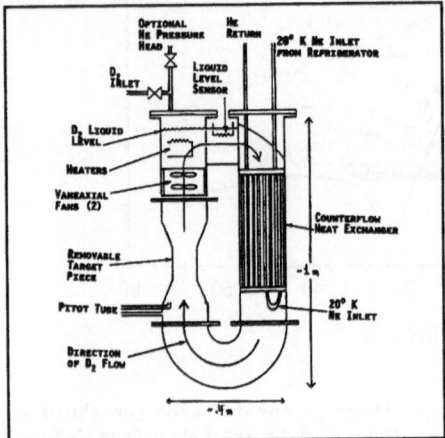

Figure 2. Vertical section view of the liquid deuterium target used in the Bates threshold electrodisintegration experiment.

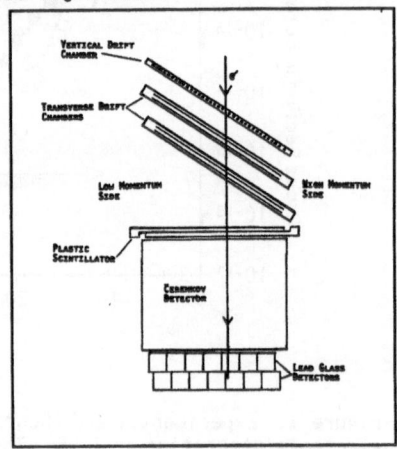

Figure 3. Focal plane detection system for the ELSSY spectrometer. The dispersion direction is in the plane of the drawing.

A vertical section view of the high power liquid deuterium target used in this experiment is shown in Fig. 2. Liquid deuterium at a temperature of 20-26 K was pumped through the loop at velocities of 1-2 m/sec transverse to the beam direction. Heat deposited by the beam and thermal leaks was ex-tracted by a 200 W cryogenic refrigerator through a heat exchanger using He as a cryogen. This target was capable of accepting average beam currents up to 50 microamperes.

A set of tungsten alloy slits approximately 20 r. l. thick defined the region of target seen by the spectrometer and blocked scattering from the end windows of the target cell. Both the position and the horizontal extent of the beam were monitored by carefully surveyed secondary emission foils to prevent beam halo or scattered beam from striking the target cell in the region of spectrometer acceptance.

The scattered electrons were momentum analyzed by ELSSY,the Bates high-resolution energy loss spectrometer. This instrument[4] and the principle of operation of its detector system[5] are described in the literature. For the present experiment the detector system was upgraded by adding wire chambers which reduced the cosmic ray background by improved angular discrimination, by adding an improved Čerenkov detector, and by adding an array of lead glass detectors to discriminate between electrons and the cosmic ray muons of momentum > 3 GeV that could trigger the Čerenkov detector. A schematic of the detector system is shown in Fig. 3. Data were taken at a laboratory scattering angle of 160° at bombarding energies ranging from 347 MeV to 903 MeV. At 903 MeV the target was also filled with liquid ^{1}H in order to assure that the line shape, bombarding energy, and absolute normalization were well understood.

Experimental Results

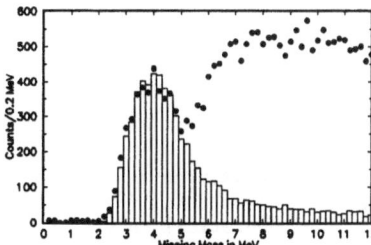

Figure 4. Spectrum of 347 MeV electrons scattered from ^{2}H at 160°. Filled circles are data, bars are from the Monte Carlo simulation.

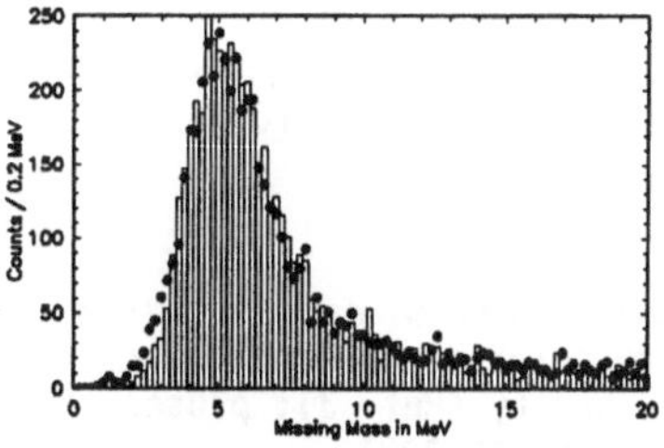

Figure 5. Spectrum of 903 MeV electrons scattered from ^{1}H at 160°. The filled circles are the data and the bars are the Monte Carlo simulation.

The measured scattered electron spectrum for ^{2}H is shown in Fig. 4 for a bombarding energy of 347 MeV and for ^{1}H at 903 MeV in Fig 5. Because of the extended target geometry the line shape must be calculated by Monte Carlo methods. The bar graphs in these figures are the full Monte Carlo calculations, and the good agreement indicates that the line shapes are well understood.

The preliminary results of this experiment[6] are shown in Fig. 6 along with the Saclay results (Ref. 3) and the recently reported SLAC results[7]. The Bates and Saclay data were taken with a missing mass resolution of 1.0 - 2.5 MeV and were averaged over the first 3 MeV above threshold. The SLAC data were taken with an energy resolution of 12 - 20 MeV and were averaged over the first 10 MeV above threshold. Also shown in Fig. 6 are representative theoretical calculations by Dymarz and Khanna[8] (solid curve); Leidmann, Schmitt, and Arenhövel[9] (dotted curve); and Schiavilla and Riska[10] (dashed curve for E_{np} = 1.5 MeV, dot-dash curve for E_{np} = 5 MeV). For comparison the hybrid quark model calculation of Cheng and Kisslinger (Ref. 1, 3dot-dash curve) is also shown in Fig. 6.

The present Bates cross sections agree very well with the previous Saclay measurements in the region of Q^2 overlap, but lie below the SLAC measurements in the overlap region. It is not clear at this time if this indicates an increasing cross section as a function of increasing E_{np} since the

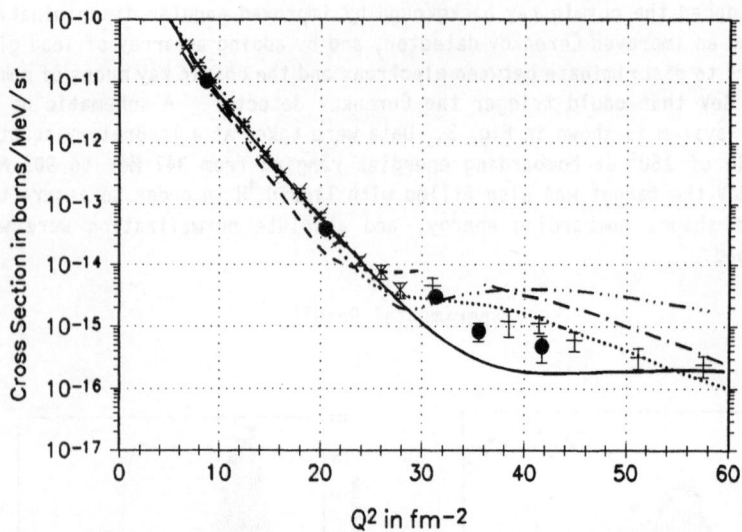

Figure 6. Recent data for D(e,e')np threshold electrodisintegration with representative theoretical calculations; solid circles, Bates (Ref. 6); X's, Saclay (Ref. 3); pluses, SLAC (Ref.7). The theoretical curves are discussed in text.

Bates data have not yet been averaged over the same E_{np} range as the SLAC data. The Bates cross sections are classified as preliminary, although it is unlikely that any changes resulting from a more complete analysis will shift the results by more than the quoted error bars.

Discussion and Conclusions

There are a number of serious unresolved questions which plague all existing theoretical calculations. There is uncertainty in how to represent the medium and short range parts of the two-body meson and isobar currents. There is uncertainty in the nucleon form factors for $Q^2 > 20$ fm^{-2}, especially in the neutron electric form factor. There is uncertainty in the importance of relativistic effects at momentum transfers approaching the rest mass of the deuteron. There is uncertainty in the choice of the correct form factors for the two-body vertices since the non-relativistic reduction of the problem is not presently understood. These problems are explored in the various theoretical studies referenced above, and it invariably happens that for $Q^2 > 20$ fm^{-2} there is significant sensitivity to the choices for the various uncertain elements.

The above should not be taken as a disparagement of the theoretical efforts that have been made. As can be seen from Figs. 1 and 6 the data indicate that for $Q^2 > 30$ fm^{-2} as much as 80% of the one-body amplitude is being canceled by the two-body currents. Obtaining theoretical agreement with the data requires a very accurate calculation of the numerous contributing amplitudes. In view of the uncertainties noted above, the agreement shown in Fig. 6 to within factors of 2 to 4 of the theoretical calculations with the data should be considered an interim success.

High quality data are now becoming, or will become available from other experiments on the deuteron and other few-nucleon systems. Examples of these experiments are tensor polarization in elastic scattering, magnetic elastic scattering, quasi-elastic scattering, exclusive experiments in which a nucleon or a nucleon and a pion are detected in coincidence with the electron, and finally experiments that exploit the spin degrees of freedom with polarized electrons and polarized targets. One can also look forward to the advent of CEBAF whose high-intensity electron beams of energy up to 4 GeV will allow extending many of these experiments to $Q^2 > 100$ fm^{-2}. These data will present both a challenge and an opportunity to theoretical understanding of the non-nucleonic degrees of freedom in few-nucleon systems.

References

1. V. Z. Jankus, Phys. Rev. 102, 1586 (1956).
   R. J. Adler, Phys. Rev. 169, 1192 (1968).
   J. Hockert, et al., Nucl. Phys. A217, 14 (1973).
   J. A. Lock and L. L. Foldy, Ann. Phys. 93, 276 (1974).
   W. Fabian and A. Arenhövel, Nucl. Phys. A258, 461 (1976).
   B. Sommer, Nucl. Phys. A308, 263 (1978).
   W. Fabian and A. Arenhövel, Nucl. Phys. A314, 253 (1979).
   L. S. Kisslinger, Phys. Lett. 111B, 307 (1982).
   W. Leidmann and H. Arenhövel, Nucl. Phys. A393, 385 (1983).
   J. F. Mathiot, Nucl. Phys. A412, 201 (1984).
   A. Buchmann, W. Leidmann, and H. Arenhövel, Nucl. Phys. A443, 726 (1985).
   Y. Yamauchi, R. Yamamoto, and M. Wakamatsu, Nucl. Phys. A457, 602 (1986).
   T.-S. Cheng and L. S. Kisslinger, Nucl. Phys. A457, 602 (1986).

2. R. E. Rand, R. F. Frosch, C. E. Littig, and M. R. Yearian, Phys. Rev. Lett. 18, 469 (1967).
   D. Ganichot, D. Grosstête, and D. B. Isabelle, Nucl. Phys. A178, 545 (1972).
   G. G. Simon, et al., Nucl. Phys. A324, 277 (1979).
   M. Bernheim, et al., Phys. Rev. Lett. 46, 402 (1981).

3. S. Auffret, et al., Phys. Rev. Lett. 55, 1362 (1985).

4. W. Bertozzi, et al., Nucl. Inst. and Methods 162, 211 (1979).

5. W. Bertozzi, et al., Nucl. Inst. and Methods 141, 457 (1977).

6. K. S. Lee, et al., Submitted to Phys. Rev. Lett.

7. R. G. Arnold, et al., Phys. Rev. C42, R1 (1990).

8. R. Dymarz and F. C. Khanna, Phys. Rev. C41, 828 (1990).

9. W. Leidmann, K.-M. Schmitt and H. Arenhövel, Phys. Rev. C42, R826 (1990).

10. R. Schiavilla and D. O. Riska, Phys. Rev. C43, 437 (1991).

Few-Body Systems, Suppl. 6, 218—222 (1992)

L/T STRUCTURE OF THE ELECTRODISINTEGRATION OF THE DEUTERON

U. Dittmayer, F. Frommberger, W. Förster, R. Gothe, G. Happe,
C. Hüffer, F. Kalleicher, G. Kranefeld, N. Leiendecker,
G. Pfeiffer, H. Putsch, H. Reicke, T. Reichelt, B. Schoch,
D. Wehrmeister, M. Wilhelm

Physikalisches Institut der Universität Bonn
Nussallee 12, 5300 Bonn 1, Germany

When regarding the interaction of an electromagnetic probe with the deuteron, two classes of experiments can easily be distinguished:

In the first class the neutron is regarded as the target and the proton is supposed to act merely as a spectator. Examples for these experiments are the photoproduction of pions on the deuteron to separate the different isospin contributions of the production amplitudes or - to give a more recent example - the scattering of polarized electrons, where the quasi-elastically scattered neutron is observed in coincidence in a neutron polarimeter. The measured neutron polarization is supposed to be a direct measure of the neutron electric form factor [1].

In any way the deuteronphysics is an unwanted, though unavoidable ingredient, which should be kept small. This requires an adequate theoretical description of the deuteron.

In the second class of experiments we investigate the deuteron itself in order to understand the most fundamental bound nuclear system and to build the basis for such theoretical work.

The gross features of the deuteron being well known, now new experimental techniques allow the search for small effects like non-nuclear degrees of freedom, relativistic corrections or final state interactions. Such new techniques are for instance the operation of electron beams with a high duty cycle and/or a substantial degree of polarization, the polarimetry of outgoing nucleons or the set-up of polarized targets, especially those, which can sustain an electron beam.

Recent experiments [2] and also theory [3, 4] have shown the importance of separating the different structure functions in the deuteron electrodisintegration. Expecially the f01 structure function is remarkably model dependent. It is due to a longitudinal/transvers interference and is apparent experimentally as a left/right asymmetry in e'p coincidence rates. Here left/right means left or right with respect to the virtual photon in the scattering plane.

At the ELSA facility in Bonn data have been taken with a set-up, which allowed the left/right measurement being made simultaneously. By this method many sources of systematic errors like target thickness or beam charge will cancel (see fig. 1).

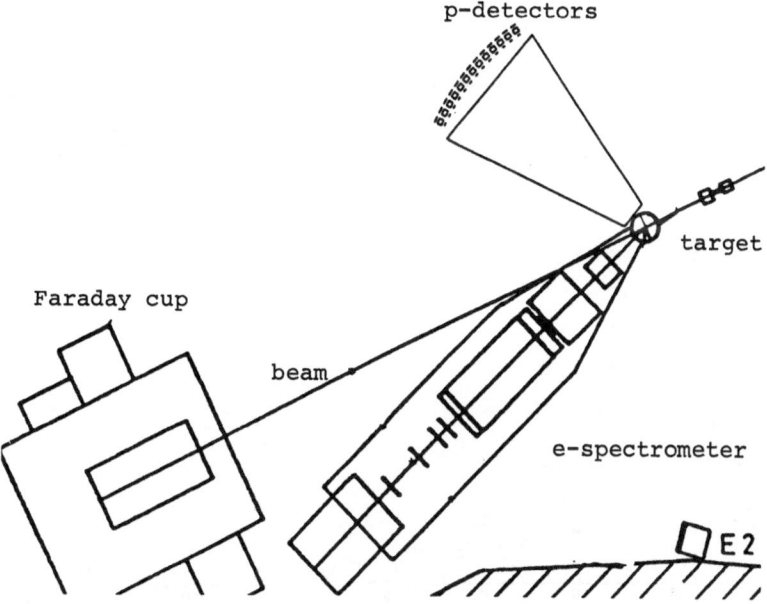

Fig. 1 Apparatus for measuring the f01 structure function

We measured under the following experimental conditions:

beam: E = 1.2 GeV, i = 30 nA, duty cycle = 30 % typical

target: 10 cm LD2

e':   standard Bonn 1.8 msr magnetic spectrometer
Θ = 21 degrees, momentum = 1.105 GeV

p: 13 modules of TOF-telescope-detectors (5*10 cm each) in 4 m distance; no shielding required

The proton detectors were positioned such that the central modul sits under parallel kinematics and the other 6 detector pairs are spaced 5 degrees apart in the hadron CMS. In this way we cover the range from 0 to 30 degrees, where the figure-of-merit (asymmetry**2 * cross section) is largest.

We had a very clean proton signal. Randoms amount to less than 0.1 % and empty target contributions to less than 1 %.

Due to the large momentum acceptance of the electron spectrometer of ± 12 % also events outside the quasi-elastic peak are recorded. The scaling variable x varies between 0.5 and 1.5; this will allow the determination of f01 also for x not close to one.

In a first step of the data analysis a 10 MeV cut in the E' spectrum was made. This ensures fixed q, ω kinematics on the quasi-elastic peak. With this cut applied, the spectrum of missing momenta (fig. 2) shows the dip near the origin due to phase space and the fast fall-off expected from the deuteron wavefunction. The six detector pairs are clearly separated.

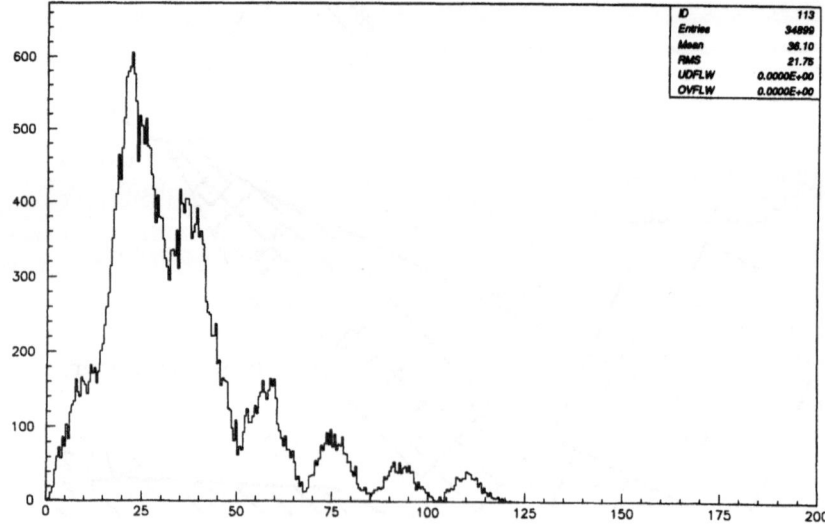

Fig. 2 Missing momentum spectrum for quasi-elastic events

A crucial question is how well the asymmetry, from which f01 finally is deduced, can be determined. Therefore a careful discussion of systematic errors is necessary. Apparatus relate asymmetries, which may lead to systematic errors, can come mainly from four sources:

1.   Small misalignments in the positioning of the proton detectors
2.   Finite acceptance of the apparatur like target length, solid angles etc.
3.   Limited resolution of the electron momentum of about 1 MeV
4.   Accuracy of the beam energy of about 1 MeV.

1. and 2. give rise to an apparatus related asymmetry which, however, can be calculated with sufficient acuracy. The whole set-up (angles, distances etc.) has been carefully measured and in the calculation enters only the shape of the cross-section, which is practically model-independent. 3. and 4. may pose a problem, because a shift in beam energy or electron momentum of 1 MeV moves the q vector by about 2 mrd. This will result in an asymmetry of 3 %, which, however, is independent of angle. It is worthwhile to note, that a deviation of beam energy and electron momentum of the same amount in the same direction leaves the q vector essentially unchanged.

Fig. 3 shows the apparatus related asymmetry due to a deviation in beam energy of 1 MeV (f01 switched off, full curve) and the resulting asymmetry (f01 switched on, dotted curve). f01-values are taken from [3].

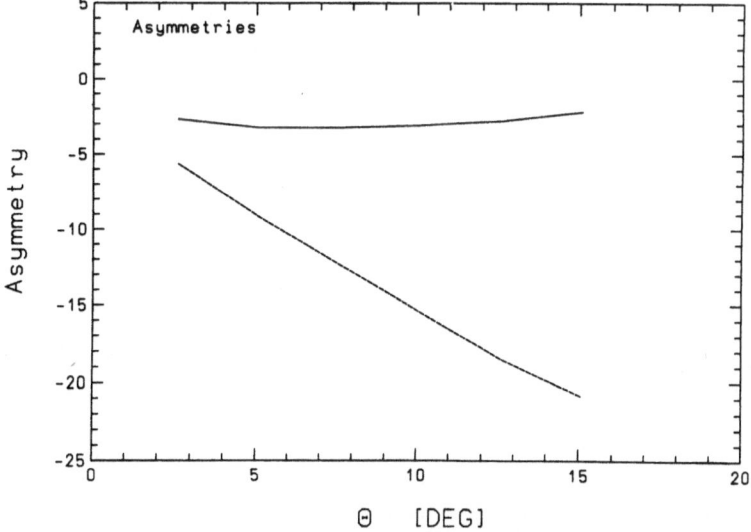

Fig. 3 Asymmetries vs. proton CMS angle

The angular dependence of both asymmetries is completely different. This important fact will allow us to disentangel the different contributions and make a precise determination of f01 possible.

Conclusion:

The f01 structure function of deuteron electrodisintegration has been measured at the Bonn ELSA facility for q**2 = 4.5/fm**2. The high duty cycle of ELSA allowed the use of non-magnetic proton spectrometers, which made a simultaneous measurement for the whole interesting angular interval possible. These detectors are easy to build and operate, they are cheap and do not require shielding. They are practically not limited in size: so - given the excellent performance - an extension from a linear array to a two dimensional one is feasible without problems. This will allow the detection also of the out-of-plane protons, which are essential for the determination of the 4. and 5. structure function.

References:

1.   The first experiment of this kind has been done at MIT/BATES where data taking was finished in June 1991

2.   M. van der Shaar: A Study of Deuteron (e,e'p) Response Functions; Dissertation, Utrecht 1991

3.   W. Fabian, H. Arenhoevel: NP A314 (1979) 253

4.   B. Mosconi, P. Ricci: NP A517 (1990) 483

Few-Body Systems, Suppl. 6, 223—228 (1992)

DEUTERON ELECTRODISINTEGRATION
WITH RELATIVISTIC CORRECTIONS

B. Mosconi$^{(+,*)}$, J. Pauschenwein$^{(**)}$ and P. Ricci$^{(*)}$

(+) *Dipartimento di Fisica, Università di Firenze, Firenze, Italy*

(*) *Istituto Nazionale di Fisica Nucleare, Sezione di Firenze, Firenze, Italy*

(**) *Institut für Theoretische Physik, Universität Graz, Graz, Austria*

Abstract. Exclusive deuteron electrodisintegration is studied in the quasi elastic region with emphasis on the effect of nucleonic and pionic relativistic corrections (RC). We consider the longitudinal-transverse structure function, where RC give sizeable effect in agreement with the experiment, and the neutron polarization in the $d(\vec{e}, e'\vec{n})p$ reaction, where the magnitude of the RC effect depends on the choice of F_1 or G_E as nucleon charge form factor. It is also shown that calculations with F_1 or G_E give rather similar results when RC are included.

1. Introduction

Some measurements of the exclusive cross section for deuteron electrodisintegration with determination of its independent structure functions (SF) have already been performed [1-3]. The availability of experimental data on the individual SF allows a more deep check of our understanding of the electromagnetic (EM) interactions in the two-body system. In particular, the interference SF (f_{LT}, f_{TT}) because of their sensitivity to small variations in the nuclear charge (ρ) and current (**j**) densities are expected to give valuable information on the role of non-nucleonic degrees of freedom (meson-exchange (MEC) and isobar-excitation (IC) currents) in the EM interactions and on the need for relativistic corrections (RC). Instead, the longitudinal f_L and the transverse f_T SF are dominated by the non relativistic (NR) impulse approximation (IA). A further step forward will be taken with measurements of spin observables which offer, in principle, the opportunity of selecting an appropriate one for testing specific points of the theory. For example, the neutron polarization in the polarization transfer reaction which we shall discuss, and the beam-target vector asymmetry have been picked out [4] because of their sensitivity to the different models of the electric form factor of the neutron (G_E^n). The extensive calculations of the Mainz group are in the standard or IA+MEC+IC theory [5] which is very successful for the EM interactions in the few-body nuclei [6] up to $Q^2 = 1(GeV/c)^2$. Our aim is to improve such theory taking into account the nucleonic and pionic (in PV coupling theory) RC which have proved to be essential for explaining the data on the differential cross section in deuteron photodisintegration [7]. Let us recall that our group [8] first has shown that such RC were able to resolve the long standing discrepancy in the energy spectrum of the forward cross section measured at Mainz [9] in 1976. In a previous paper [10] to which we refer for our notations, we have studied the relativistic effect on the SF at the quasi-elastic peak and found that it is sizeable in f_{LT}, in f'_{LT} (the fifth SF arising for polarized electrons) and also in f_L. We assumed the Dirac F_1 form factor (FF) as nucleon charge density in [10]. As is well known, a controversy exists in the literature about the use of F_1 or the Sachs G_E FF. We shall discuss this problem in Sect.2 where we also list the nucleonic and pionic RC considered in our calculations. In Sect.3 we present our theoretical results, compare them with the available data and state some conclusions.

2. Theory

The problem of the nucleon charge FF (F_1 or G_E) has been usually resolved in favour of G_E because it can be interpreted as the Fourier transform of the nucleon charge density distribution in the Breit frame of the elastic eN scattering. Clearly, such argument is not particularly significant. Looking at the problem from the relativistic point of view, the two possible choices derive from the Dirac and the Sachs forms of the relativistic γNN-vertex

$$
\begin{aligned}
\Gamma_\mu^D &= F_1\gamma_\mu + iF_2\frac{\sigma_{\mu\nu}q^\nu}{2M} \quad, \\
\Gamma_\mu^S &= \frac{1}{1+\tau}\left(G_E\frac{P_\mu}{2M} + iG_M\frac{\epsilon_{\mu\nu\alpha\beta}}{4M^2}P^\nu q^\alpha\gamma^\beta\gamma^5\right) \quad,
\end{aligned}
\tag{1}
$$

where $P_\mu = p_\mu + p'_\mu, \tau = -(q_\alpha^2/4M^2)$ and the two sets of EMFF are related by

$$
G_E = F_1 - \tau F_2 \quad, \quad G_M = F_1 + F_2 \quad.
\tag{2}
$$

Including only lowest order corrections in the NR expansion, Γ_μ^D and Γ_μ^S give the charge density operator in the two-component spinor space as

$$
\begin{aligned}
\rho^D &= F_1 - (F_1 + 2F_2)\frac{q^2 + i\sigma \times \mathbf{P}\cdot\mathbf{q}}{8M^2} \quad, \\
\rho^S &= G_E - G_E\frac{q^2}{8M^2} - i(2G_M - G_E)\frac{\sigma \times \mathbf{P}\cdot\mathbf{q}}{8M^2} \quad.
\end{aligned}
\tag{3}
$$

Thus, the NR charge FF is predicted to be F_1 starting from Γ_μ^D and G_E starting from Γ_μ^S. Because of Eq.2 the use of G_E amounts to include in IA calculations a sizeable part of the Darwin-Foldy (DF) correction as derived from Γ_μ^D. While the spin-orbit terms are the same apart from differences of higher order, the DF terms differ substantially because of the F_2-part in the Dirac form. On the other hand, these relativistic corrections almost completely cancel the NR differences in ρ^D and ρ^S. Similar results hold for the current density. We want to recall that both forms are valid for the on-mass-shell γNN-vertex and thus, they are not completely adequate for nucleons bound in nuclei where the half-off-shell form should be used. Assuming that the on-shell form of the vertex is adequate, Γ_μ^D seems more founded than Γ_μ^S. It follows by minimal substitution from the Dirac Hamiltonian and at least for structureless nucleons, it satisfies the Ward-Takahashi identity . Also, it does not seem possible [11] to formulate an effective chiral invariant Lagrangian predicting Γ_μ^S. However, these arguments are not definite. For example, one may think to a unitary transform of the Dirac Hamiltonian from which Γ_μ^S arises by minimal coupling [12]. At the same time one should have new spinors and a new nucleon propagator to be used in the evaluation of MEC and of pionic RC. For the moment such possibility has not been fully explored so that a more consistent scheme of the EM nuclear interactions including MEC and pionic RC exists for Γ_μ^D . On the other hand, the choice of G_E over F_1, leaving unchanged the structure of MEC and RC , simply amounts to differences of higher order in the NR expansion [12]. Thus, considering the NR theory an effective approximation to a truly covariant theory we shall take a practical point of view and look at the numerical predictions in the two cases. Concerning this point, a comment is worth doing about the result of Gross and Riska [13] who apparently succeeded in obtaining MEC with arbitrary EM and also hadronic FF. As noted in [12,14], this result comes out because of the use of prescriptions for off-shell extension of the γNN-vertex and for the inclusion of the hadronic FF in the meson propagator which are very particular.

As for RC, we have considered the nucleonic and pionic RC so effective in deuteron photodisintegration . In detail they are : i) the Darwin-Foldy and spin-orbit terms discussed above. They contribute to both ρ and $\mathbf{j}$ but the DF contribution to $\mathbf{j}$ does not modify the transverse amplitude being proportional to $\mathbf{q}$; ii) the nucleonic corrections to ρ which effectively take into account the distortions of the wave functions due to the nuclear motion [15]. Besides the kinematic terms explicitly given for the deuteron in [8b], the dynamic terms corresponding to the static one-pion-exchange potential are also included [16]; iii) the pionic corrections to ρ as derived by Friar [16] using the Foldy-Wouthuysen reduction method. The PV pion-nucleon coupling theory has been considered with the choice $\mu = 1$ and $\nu = \frac{1}{2}$ for the parameters introduced by Friar. We recall that $\nu = \frac{1}{2}$ means no retardation term in the corresponding one-pion-exchange potential and that $\mu = 1$ uniquely defines the PV coupling in the quasi-potential formalism developed by Jaus and Woolcock [17].The small effect of the pionic RC to ρ justifies a posteriori the neglect of the analogous contributions to $\mathbf{j}$.

3. Results and Conclusions

We consider the deuteron exclusive electrodisintegration at the quasi-elastic peak where the kinematics of the process is completely defined by the value of the four-momentum transfer squared $Q^2 = -q_\alpha^2$. The calculations have been performed with the Paris potential [18] and the Höhler model [19] of EMFF. For the technical problem caused by the very slow convergence of the multipole expansion of the IA amplitude we refer to [5]. The same problem of bad convergence affects also the transition amplitude induced by the nucleonic RC and can be similarly cured [10].

As said in the Introduction, we have predicted in [10] a considerable RC effect in f_{LT}, which has been verified in the NIKHEF experiment [3]. In the left side of Fig.1 we report the angular distribution of f_{LT} as given in [10], for $Q^2 = 5fm^{-2}$ and $Q^2 = 10fm^{-2}$. The three curves correspond to the IA theory, the IA+MEC+IC theory, and to the full theory. We note that while the meson exhange effect is negligible, RC strongly modify f_{LT} at forward and backward angles. The deep forward minimum undergoes a further 15 -30 % lowering and a clear peak develops close to 180° over the almost vanishing NR predictions. These results correspond to calculations with F_1 but the changes are very little when G_E is instead used. Thus, this relativistic effect is independent of the choice of either F_1 or G_E as nucleon charge FF. In the right side of Fig.1 the asymmetry A_Φ, which is proportional to f_{LT} being defined as the ratio of the difference over the sum of the cross section at $\Phi = 0°$ and at $\Phi = 180°$, is compared to the experimental points taken at NIKHEF [3] at $Q^2 = 0.21(GeV/c)^2$. The values of the missing momentum correspond to the forward angle minimum. Clearly, both the NR calculations give values of A_Φ sistematically less negative than the data, while the inclusion of RC allows us to get close to them. The remaining discrepancy should not be significant in view of the systematic uncertainty [3] on the experimental points.

Now we turn to the x component (according to the Madison convention) of the neutron polarization $P_x^{\prime n}$ in the polarization transfer $d(\vec{e}, e'\vec{n})p$ reaction. The kinematical conditions (quasi elastic peak in coplanar geometry at $Q^2 = 12fm^{-2}$) are those considered in [4] where it has been shown that $P_x^{\prime n}$ is independent of the NN potential model and strongly sensitive to G_E^n for forward emitted neutrons. In Fig.2 we report our results in two approximations (IA+MEC+IC theory and full theory) as a function of the neutron polar angle ϑ_{cm}^n (and for $\Phi^n = 0°$). The relativistic effect for forward emitted neutrons is drastic in calculations with F_1 (left side of Fig.2) leading to a change of sign of $P_x^{\prime n}$. Instead , RC induce a small lowering of the NR curve in calculations with G_E (right side of Fig.2). It is worth noting that here again the final predictions are close each other, the RC effect compensating for the large differences in the NR results. The sensitivity of $P_x^{\prime n}$ to different choices of G_E^n remains that discussed in the IA+MEC+IC theory in [4].Thus, it should help discriminate among the existing models of G_E^n.

In conclusion, calculations including RC are almost independent of the choice of

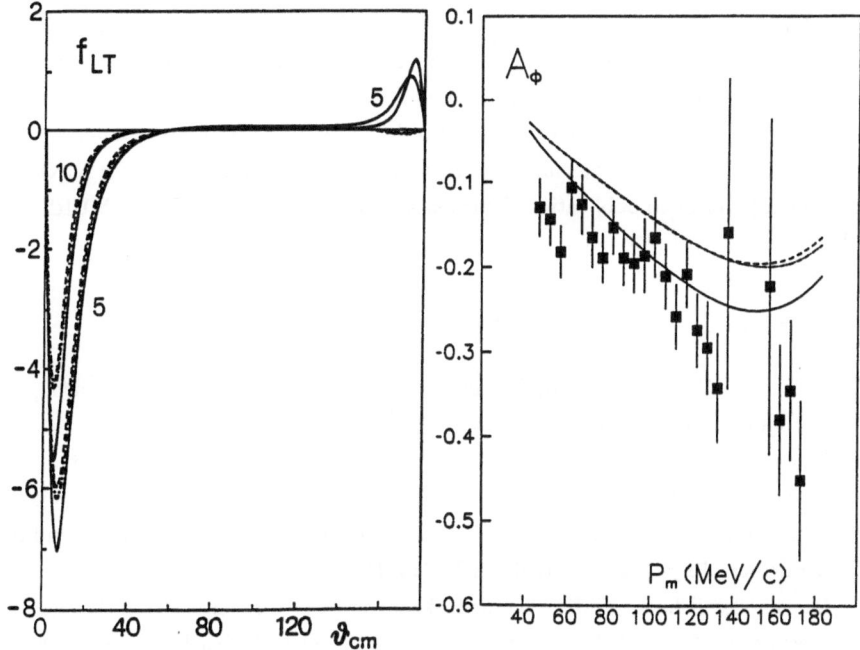

Fig.1. Angular distribution of f_{LT} (left) at $Q^2 = 5fm^{-2}$ and $Q^2 = 10fm^{-2}$, and asymmetry A_Φ as a function of the missing momentum (right) at $Q^2 = 0.21(GeV/c)^2$. Data are from [3]. Dashed line, IA theory; dot-dashed line, IA+MEC+IC theory; solid line, full theory.

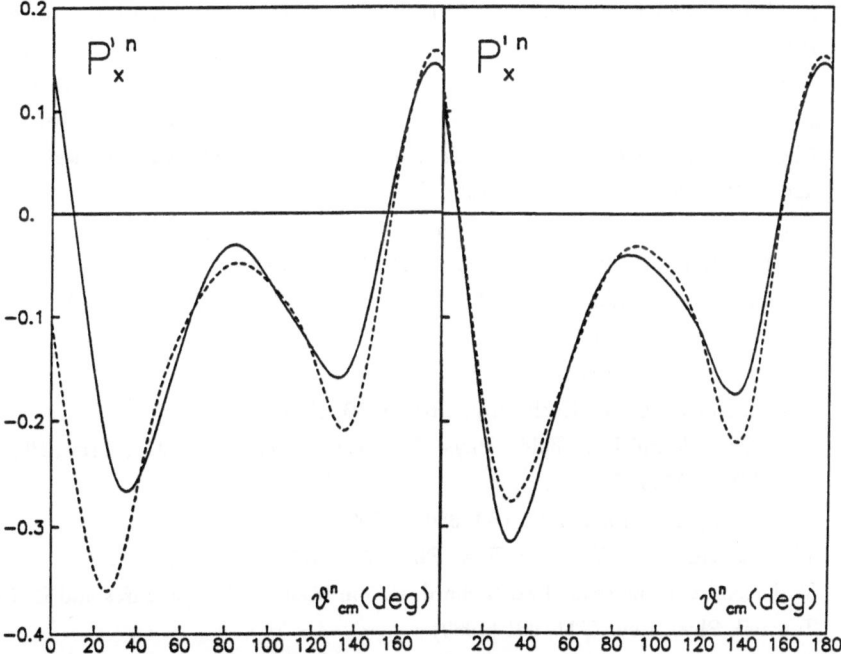

Fig.2. Angular distribution of $P_x'^n$ at $Q^2 = 12fm^{-2}$.Calculations with F_1 (left) and with G_E (right). Dashed line, IA+MEC+IC theory; solid line, full theory.

F_1 or G_E (a point not yet settled in the theory) in the quasi-elastic region . The reasons are that the nucleonic degrees of freedom dominate here the EM transitions and that the Dirac and Sachs forms of the nucleonic ρ and $\mathbf{j}$, strongly different in the NR limit, become very similar when RC of order $(v/c)^2$ are taken into account.

We thank G. van der Steenhoven for providing us with the data of the NIKHEF experiment [3] prior to publication. This research was partly supported by MURST of Italy.

References

[1]   T. Tamae, H. Kawahara, A. Tanaka, K. Namai, M. Sugawara, Y. Kawazoe, H. Tsubota and H. Miyase, Phys. Rev. Lett. **59**, 2919 (1987)

[2]   M. van der Schaar, H. Arenhövel, Th. S. Bauer, H.P. Blok, H. J. Bulten, M. Daman, R. Ent, E. Hummel, E.Jans,G.J.Kramer, J.B.J.M. Lanen, L. Lapikás, J.H. Mitchell, G. van der Steenhoven, J.A. Tjon, P.K.A. de Witt Huberts and A. Zondervan, Phys. Rev. Lett. **66**, 2855 (1991)

[3]   M. van der Schaar ,Ph.D. Thesis, Rijksuniversiteit Utrecht, 1991 ; G. van der Steenhoven, Proc. of the 5th Symp. on Mesons and Light Nuclei, Prague, 1991

[4]   H. Arenhövel, W. Leideman and E. L. Tomusiak, Z. Phys. **A331**, 123 (1988)

[5]   W. Fabian and H. Arenhövel, Nucl. Phys. **A314**, 253 (1979)

[6]   J.-F. Mathiot , Phys. Rep. **173**, 63 (1989)

[7]   B. Mosconi and P. Ricci, Proc. of the 5th Symp. on Mesons and Light Nuclei, Prague, 1991

[8]   A. Cambi, B. Mosconi and P. Ricci, a) Phys. Rev. Lett. **48**, 462 (1982); b) J. Phys. **G10**, L11 (1984) ; c) Proc. of the 1st Workshop on Perspectives in Nuclear Physics at Intermediate Energies (Trieste, 19830 Eds. S.Boffi, C. Ciofi degli Atti and M.M. Giannini (World Scientific, Singapore, 1984) p.139

[9]   R.J. Hughes, A. Zieger, H. Wäffler and B. Ziegler, Nucl. Phys. **A267**, 329 (1976)

[10]   B. Mosconi and P. Ricci , Nucl. Phys. **A517**, 483 (1990)

[11]   J. Adam Jr., E. Truhlik and D. Adamova, Nucl. Phys. **A492**, 556 (1990)

[12]   S.K. Singh, W. Leidemann and H. Arenhövel, Z. Phys. **A331**, 509 (1988)

[13]   F. Gross and D.O. Riska, Phys. Rev. **C36**, 1928 (1987)

[14]   H.W.L. Naus and J.H. Koch, Phys. Rev. **C39**, 1907 (1989)

[15]   R. A. Krajcik and L. L. Foldy, Phys. Rev. **D10**, 1777 (1974); J. L. Friar, Phys. Rev. **C12**, 696 (1975)

[16]   J. L. Friar , Ann. Phys.(NY) **104**, 380 (1977)

[17]   W. Jaus and W. S. Woolcock, Helv. Phys. Acta **57**, 644 (1984)

[18]   M. Lacombe, B. Loiseau, J.M. Richard, R. Vin Mau, J. Côté, J. Pirés and R. De Tourreil, Phys. Rev. **C21**, 861 (1980)

[19]   G. Höhler, E. Pietarinen, I. Sabba-Stefanescu, E. Borkowski, G.G. Simon, V.H. Walther and R.D. Wending, Nucl. Phys. **B114**, 505 (1976)

Few-Body Systems, Suppl. 6, 229—235 (1992)

© by Springer-Verlag 1992

DEUTERON PHOTODISINTEGRATION AT INTERMEDIATE ENERGY AND QUARK MODELS

E. De Sanctis,[1] A. Kaidalov,[2] L.A. Kondratyuk,[2] P. Rossi,[3] N. Bianchi,[1] P. Levi Sandri,[1] V. Muccifora,[1] E. Polli,[1] A.R. Reolon[1]

1. INFN-Laboratori Nazionali di Frascati, P.O. Box 13, I-00044 Frascati, Italy
2. Institute of Theoretical and Experimental Physics, Moscow 117259, USSR
3. INFN-Sezione Sanitá, Viale Regina Elena 299, I-00161 Roma, Italy

Abstract: We have examined the behaviour of the forward and backward data for the γd→pn reaction at intermediate energy in the framework of the quark-gluon model and Regge phenomenology. Our model reproduces rather well the experimental values.

In this paper we examine the data of the reaction γd→pn available at intermediate energies with the aim of determining whether this process is more economically described in terms of the quark degrees of freedom, rather than nucleon and meson degrees of freedom. In particular, we will try to establish whether the deuteron photodisintegration amplitude obeys to the predictions of the quark-gluon model and Regge phenomenology.

We will first look at the deuteron photodisintegration with the high-energy physicist eye, and then we will try to extrapolate this point of view to the intermediate energy region.

At high energy, it is usefull to distinguish two different kinematical regions: *i*) high momentum transfers [$t \gg 1$ (GeV/c)^2] and *ii*) small momentum transfers [$t \leq 1$ (GeV/c)^2].

In the region *i*), according to quark-dimensional counting rules, the cross section of the reaction at constant centre-of-mass angle should be a very fast decreasing function of energy:[1]

$$\frac{d\sigma}{dt}(\vartheta_{c.m.} \approx 90°) \sim s^{-11} \, , \tag{1}$$

where s and t are the usual Mandelstam variables. The only data available at high energies, specifically those recently taken at SLAC[2] at E_γ=(0.8-1.6) GeV (see Fig.1), suggest this behaviour according to the simple constituent-counting rule, but this is not a conclusive proof that quarks are needed in an explanation. First, the data cover a limited energy and angular region, corresponding to $|t| = 1\div2$ $(GeV/c)^2$. Second, conventional nuclear theories can scale as s^{-11}.

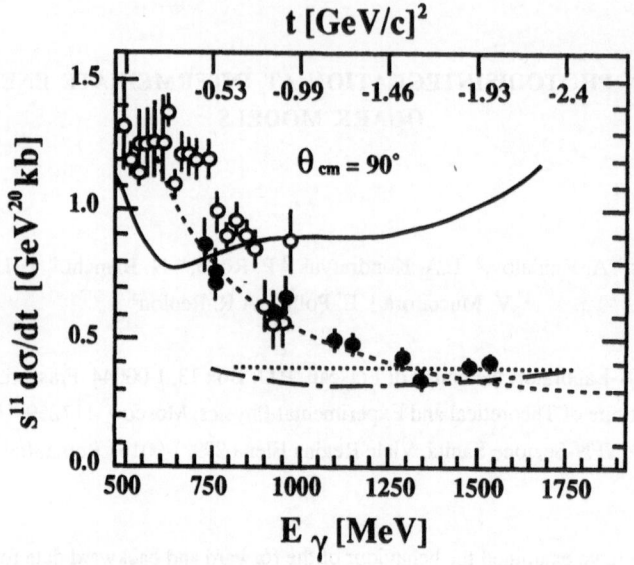

Fig. 1 Comparison of high energy $\gamma d \to pn$ data at 90° to quark model predictions: solid circles Ref. [2], open circles Ref. [3]. The dotted line indicates an energy dependence of s^{-11}; the dashed curve indicates the dependence predicted by the factorization model of Brodsky and Hiller;[1] the solid curve is the prediction of a meson-nucleon model; the dot-dashed curve is our prediction [Eq.(8)] (adapted from Ref. [2]).

In the region $ii)$, that is at sufficiently high energy and small t or u, the photodisintegration amplitude is dominated by the exchange of 3 valence quarks in t- or u-channel (see Fig. 2a) with any number of gluons exchanged between them. In the framework of $1/N_c$ expansion in QCD, this is the consequence of the dominance of the planar-quark-gluon graph. This expansion was first considered by t'Hooft[4] who proposed to analyze the properties of non-abelian quantum field theory in the large N_c limit. Then, the behaviour of different quark-gluon graphs according to their topology was discussed by Rossi and Veneziano.[5] To describe different binary reactions at high energy, Kaidalov[6] has proposed the so-called quark-gluon-model (QGM). This model is based on the propeties of $1/N_f$ expansion in QCD and can be considered as a microscopic model for Regge phenomenolgy, which, in its turn, is based on fundamental properties of scattering amplitudes such as analiticity, unitarity and crossing-symmetry.

In the Regge language the dominant contribution of 3-quark-exchange correponds to the fermion Regge pole [see in Fig. 2b where the wavy line describes the exchange of a Reggeon, which is an assembly of 3 quarks plus many gluons with angular momentum $\alpha(t)$]. The analysis of binary hadronic reactions[5-6] shows that this picture works very well at high energies and $|t|$ or $|u| \leq 1(GeV/c)^2$. However, due to the duality property of scattering amplitudes,[5] this approach can work also in the intermediate energy region.[7] If in the direct s channel the resonance behaviour is essential, the duality property ensures rather good interpolation of the amplitude in average by its Regge asymptotics. In Ref. [9] it has been shown that this approach can describe the reactions $pp \to \pi^+d$ and $\bar{p}d \to \pi^-p$ in the full energy range, starting almost from the threshold.

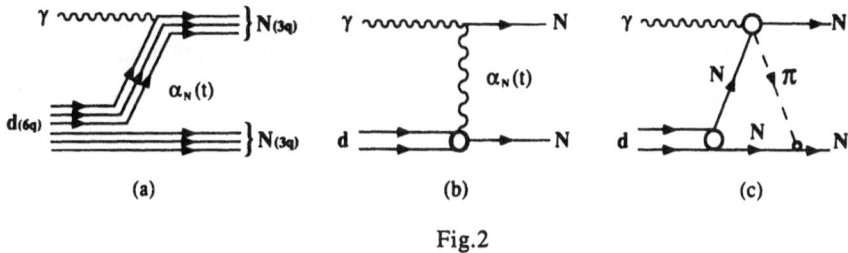

Fig.2

Let us note that at large energies the diagram of Fig. 2b includes the contribution of the graph of Fig. 2c. The amplitude $\gamma N \to \pi N$ in Fig. 2c can be described by the Regge pole exchange and the triangle can be included into the vertex $d \to N+Reggeon$. This statement can be valid at sufficiently high energy and small t. In the resonance region, $E_\gamma \leq 1$ GeV, when the local correspondance between the Regge and resonance amplitude is absent, the contribution of the diagram in Fig. 2c can not be incorporated completely into the amplitude corresponding to the diagram in Fig. 2b. This statement is also not true for large t when the amplitude $\gamma N \to \pi N$ can not be described by the Regge exchange. The vertex $d \to N+Reggeon$ can not be described only by nucleon degrees of freedom in deuteron and contains essential contributions from non-nucleonic components and, in particular, from 6-quark bag admixture in the deuteron wave function.[9]

Let us parametrize the cross section in the form:

$$\frac{d\sigma_R}{dt} = \frac{1}{64\pi s} \frac{1}{P_{c.m.}^2} (|T(s,t)|^2 + \frac{1}{R} |T(s,u)|^2), \tag{2}$$

where $P_{c.m.}$ is the photon momentum in the centre-of-mass system, $T(s,t)$ and $T(s,u)$ are the photodisintegration amplitudes and R is the forward-to-backward ratio of the cross section values. The use of Eq. (2) can be justified for $|t| \leq 1$ (GeV/c)^2 when the first term is dominant or for $|u| \leq 1$ (GeV/c)^2 when the second term is dominant. The energy behaviour of T(s,t) for fixed t, which corresponds to the fermion Regge pole exchange, can be written as:[7]

$$T(s,t) \approx F(t) \left[\frac{s}{s_0}\right]^{\alpha_N(t)} exp\{-i\frac{\pi}{2}(\alpha_N(t) - \frac{1}{2})\}, \tag{3}$$

where $\alpha_N(t)$ is the trajectory of the N Regge-pole, F(t) is the residue of the pole, s_0 is equal to the square of the deuteron mass, m_d^2, and the factor in the brackets is the phase factor. [T(s,u) is given by Eq. (3) substituting t with u]. The baryon Regge trajectory deduced from the data on πN backward scattering is known to have some non linearity:[8]

$$\alpha(t) = \alpha(0) + \alpha'(0) t + \frac{1}{2}\alpha''t^2 , \tag{4}$$

where $\alpha'(0) = 0.9$ GeV^{-2}, $\alpha'' = 0.25$ GeV^{-4}, and the intercept for the nucleon Regge trajectory N_α (which is relevant in this case) is $\alpha_N(0) = -0.5$. Therefore the energy behaviour of the cross section for small t and high photon energy is predicted to be:

$$\frac{d\sigma_R}{dt} \sim \frac{|T(s,t)|^2}{s^2} \sim \left[\frac{s}{s_0}\right]^{2\alpha_N(t)-2} , \tag{5}$$

which is much more flat dependent on s as compared with the region i) of large $|t|$ or $|u| \sim s \gg m^2$. In particular, for example, at t=0, one has: $d\sigma_R/dt \sim s^{-3}$.

The dependence of the residue F(t) on t can be taken from ref. [9]:

$$F(t) = B \left[\frac{1}{m_N^2-t} exp(R_1^2 t) + C exp(R_2^2 t) \right] , \tag{6}$$

where the first term in the square brackets takes into account the nucleon pole in the t-channel and the second term is related to the contribution of non nucleon degrees of freedom in deuteron. In ref.[9] Eq. (6), with B = 9.09 GeV^2, $R_1^2 = 3$ GeV^{-2}, $R_2^2 = -0.1$ GeV^{-2}, and C = 0.7 GeV^{-2}, was used to describe data on the reactions pp→π$^+$d and $\bar{p}$d→π$^-$p. In our case, the coupling of photons to the current generated by the quark charges should vanish at scattering angle ϑ=0, when the transverse motion of quarks is neglected. To take into account the relative suppression of this coupling at ϑ=0 as compared with the γ coupling to the quark magnetic moments we assumed B=(C$_1$ + C$_2$sin^{2}ϑ). Moreover, we took a different value for the parameter R$_1$: the coupling of pions to nucleons is not local as compared to photons, then we expected that in our case R_1^2 should be smaller than in the reaction π$^+$d→pp. We chose $R_1^2 = 1$ GeV^{-2}.

The forward-to-backward ratio R was discussed using this approach in ref.[10]. In the naive quark-model, this ratio should be related to the charges of u- and d-quarks as:

$$R = \frac{(d\sigma/d\Omega)_{0°}}{(d\sigma/d\Omega)_{180°}} = \frac{2z_u^2 + z_d^2}{2z_d^2 + z_u^2} = 1.5 , \qquad (7)$$

while in the quark gluon string model, which takes into account the difference for distributions of u-and d-quarks in nucleon, this ratio should increase from 1.5 at $E_\gamma \sim 0.2$ GeV up to 4 at $E_\gamma \to \infty$.

Strictly speaking Eq. (2) should be used at sufficiently high energies, say $E_\gamma > 1$ GeV; nevertheless, as it was said above, this approach might be valid already at lower energy, provided that the Δ-resonance tail contribution is taken into account. Then, we tried to reproduce the scarse data available at energy $E_\gamma \geq 400$ MeV by the expression :

$$\frac{d\sigma}{dt} = \frac{d\sigma_R}{dt} + \frac{d\sigma_\Delta}{dt} , \qquad (8)$$

where $d\sigma_R/dt$ is the Regge prediction as given by Eq. (2), and $d\sigma_\Delta/dt$ is the Δ-resonance tail contribution. We parametrized this tail according to the graph of Fig. 2c calculated in the infinite momentum frame, using the assumptions of the reduced QCD amplitude approach.[1] Using eq. (4) from ref. [11] we wrote the cross-section corresponding to Fig. 2c in the following form:*

$$\frac{d\sigma_\Delta}{dt} = \frac{1}{64\pi s} \frac{1}{P^2_{c.m.}} (| T^{res}_\Delta (s,t) |^2 + \frac{1}{R_{res}} | T^{res}_\Delta (s,u) |^2), \qquad (9)$$

where: $|T^{res}_\Delta (s,t)|^2 = 4 |A^{res}_{\gamma p \to \pi^0 p}(s_1,t)|^2 D^2(q_1^2) \Delta_d^2$, $T^{res}_\Delta (s,u)$ has a similar expression, and the interference between the two contributions is neglected. Here, the factor 4 takes into account the contributions of π^0 and π^- mesons in the intermediate state of Fig. 2c; the forward-to-backward ratio R_{res} is in this case equal to 1; $A^{res}_{\gamma p \to \pi^0 p}(s_1,t)$ is the amplitude of the reaction $\gamma p \to \pi^0 p$, averaged over the angular distributions; the function $D^2(q_1^2)$ takes into account the pion propagator and the form factor in the vertex πNN; and Δ_d is the deuteron structure factor[11] which, in the reduced amplitude approach, has the meaning of the distribution amplitude.

The estimates based on the hybrid model of deuteron with a realistic value[11] [$(0.3 \div 0.7)\%$] of the admixture of 6q-bag in deuteron give $\Delta_d^2 = (1.5 \div 2.5) \cdot 10^{-5}$ GeV^2. These values are also in agreement with experimental data on the probability of the Pontecorvo reaction $\bar{p}d \to \pi^- p$ at rest. Therefore, we took: $\Delta_d^2 = 2 \cdot 10^{-5}$ GeV^2.

The function $D^2(q_1^2)$ has the form:

---

* In what follows we denote with the suffix 1 all variables relevant to the reaction $\gamma N \to \pi N$.

$$D^2(q_1^2) = \frac{4m^2 f_\pi^2 \; \tilde{q}_1^2 F_\pi^2(q_1^2)}{m_\pi^2 (q_1^2 - m_\pi^2)^2} , \tag{10}$$

being:

$$q_1^2 = (p_n - \frac{p_d}{2})^2; \quad \tilde{q}_1^2 = 4m^2 [\frac{p}{E_p + m} - \frac{\frac{p_d}{2}}{E(\frac{p_d}{2}) + m}]^2; \quad F_\pi^2(q_1^2) = [1 - \frac{q_1^2 - m_\pi^2}{\Lambda^2}]^{-2}; \quad \text{with } \Lambda^2 = 1.4$$

GeV^2 and $f_\pi^2/4\pi = 0.08$.

We parametrized the amplitude $A^{res}_{\gamma p \to \pi^0 p}(s_1, t)$ in the form:

$$|A^{res}_{\gamma p \to \pi^0 p}(s_1, t)|^2 = 64\pi^2 s_1 \left(\frac{p_\gamma^1}{p_\pi^1}\right)_{c.m.} C_3 \, \delta \, \sigma_0 \, A(\vartheta) \frac{m_\Delta^2 \Gamma_{\pi N} \Gamma_\Delta}{(s_1 - m_\Delta^2)^2 + m_\Delta^2 \Gamma_{\pi N}^2} \tag{11}$$

where $\sigma_0 = 380$ μb/sr; δ is a damping factor introduced to suppress the resonance contribution at high energies, far from the resonance:

$$\delta = \frac{E_0^2 + \left[E_\gamma^{res}\right]^2}{E_0^2 + E_\gamma^2}; \quad A(\vartheta) = \frac{1 + b_1 \cos^2\vartheta}{4\pi}; \quad \text{and } \Gamma_{\pi N} = \Gamma_\Delta \left(\frac{P_\pi^1}{P_\pi^{res}}\right)^3 \frac{1 + (P_\pi^{res} a)^2}{1 + (P_\pi^1 a)^2};$$

C_3 and b_1 are free parameters, $a = 0.2$ GeV^{-1}, $P_\pi^{res} = 0.227$ GeV, and $\Gamma_\Delta = 0.14$ GeV. In distinction from Eq. (2), the formula (9) can be used also for large angles.

In Fig. 3 we compare all the experimental data available at $E_\gamma \geq 0.35$ GeV and at forward and backward angles with the results obtained with Eq. (8) for the following values of parameters: $C_1 = 8.4$ μb$^{1/2}$GeV^3, $C_2 = 7.98$ μb$^{1/2}$GeV^3, $C_3 = 0.8$, $b_1 = -0.3$, $R = 1.5$. As it is seen, our calculation reproduces rather well the experimental values. For the used set of parameters the contribution of Regge term $d\sigma_R/dt$ is dominant in the whole considered photon energy region. Nevertheless at $E_\gamma = 0.4$ GeV the contribution of $d\sigma_\Delta/dt$ is not very small and reaches 30-40%.

As it was mentioned before, we expect that our approach is valid at high energy and $|t| \leq 1$ (GeV/c)^2 or $|u| \leq 1$ (GeV/c)^2. Nevertheless, we compared our model, without adjusting any parameter, also with the SLAC data[2] at $\vartheta_{c.m.} = 90°$, which correspond to $|t| = 1 \div 2$ (GeV/c)^2. As it is seen in Fig. 1(dot-dashed curve), we found a reasonable agreement between our prediction and the data at $E_\gamma \geq 1.25$ GeV. Moreover, it is worth mentioning that our model gives also a very flat prediction for $s^{11} d\sigma/dt$ at $E_\gamma = (0.7 \div 1.75)$ GeV. This suggests that the Regge tails can give important contributions at all angles when the energy is not very high. Of course, while the energy increases, the Reggeon contributions at $\vartheta_{c.m.} = $const decreases exponentially like $exp(-2\alpha'(0) \, s \, lns)$. In conclusion, it is clear the need of good data on the deuteron photodisintegration in broader energy and angular intervals in order to determine which model describe better the mechanism of the process.

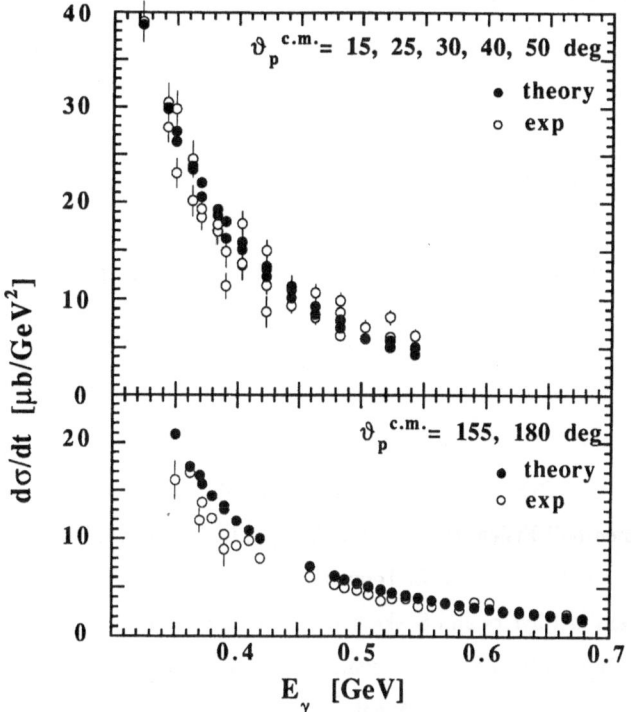

Fig. 3 Comparison of forward and backward data γd→pn above 350 MeV to our prediction [Data points from K.Baba et al. Phys. Rev. C28, 286 (1983); J. Arends et al., Nucl. Phys. A412, 509 (1984); K.H. Althoff et al., Z. Phys. C21, 149 (1983)].

References

[1] S. Brodsky, G. Farrar, Phys. Rev. **D 11**, 1309 (1975); and   S. Brodsky, J. Hiller Phys. Rev. **C 28**, 475 (1985).

[2] I. Napolitano et al. Phys. Rev. Lett. **61**, 2530 (1988),

[3] R. Ching and C.Schaerf, Phys. Rev. **141**, 1320 (1966).

[4] G. t'Hooft, Nucl. Phys. **B72**, 461 (1974).

[5] G.C. Rossi, G. Veneziano, Nucl. Phys. **B 123**, 507 (1977).

[6] A.B. Kaidalov, Z. Phys., **C 12**, 63 (1982) .

[7] A.B. Kaidalov, Sov. Physics Uspekhi **14**, 600 (1972); and E.M. Levin, Sov. Physics Uspekhy **16**, 600 (1973).

[8] V.A. Lyubimov. Sov. Phys. Uspekhy 20, 691 (1975).

[9] A.B.Kaidalov Sov. J. Nucl. Phys., to be published, ITEP Internal Report **91-7**.

[10] E. De Sanctis, A. Kaidalov, L. Kondratyuk. Phys. ReV. **C. 42**, 1764 (1990).

[11] L. Kondratyuk and C. Guaraldo, Phys.Lett., **B 256** (1991) 6.

Few-Body Systems, Suppl. 6, 236—241 (1992)

POLARIZATION OBSERVABLES IN DEUTERON ELECTRODISINTEGRATION

W. Leidemann

INFN, gruppo collegato di Trento,

Dipartimento di Fisica, Università di Trento, I-38050 Povo, Italy

E.L. Tomusiak

Department of Physics and Saskatchewan Accelerator Laboratory,

University of Saskatchewan, Saskatoon, Canada

H. Arenhövel

Institut für Kernphysik, Johannes Gutenberg-Universität Mainz,

D-6500 Mainz, Germany

Abstract: The electrodisintegration of the deuteron with polarized beam and target is investigated. The additional polarization form factors (inclusive reaction) and structure functions (exclusive reaction) are discussed. The sensitivity of these form factors and structure functions to the potential model, to meson and isobar degrees of freedom, and to electromagnetic form factors is studied in different kinematical regions.

Deuteron electrodisintegration with beam and/or target polarization offers a large variety of new observables. They contain much more detailed information on the electromagnetic deuteron response than do non polarization observables. Thus one may hope that their study will improve our knowledge of the underlying nuclear dynamics. The following discussion of polarization effects in deuteron inclusive and exclusive cross sections addresses this question. In particular it is shown how the various observables are affected by the NN interaction model and by subnuclear degrees of freedom.

The non-relativistic cross section for the reaction $\vec{d}(\vec{e}, e'N)N$ is derived in Ref. 1. The expression for the cross section contains as many as 41 structure functions and thus is quite complicated. Ref. 1 should be consulted for a detailed description of all terms in the exclusive cross section, since here we only give an abbreviated form

$$\frac{d\sigma}{dk_2^{lab}d\Omega_e^{lab}d\Omega_{np}^{cm}} = S_0 + hS(h) + P_1^d S(P_1^d) + P_2^d S(P_2^d) + hP_1^d S(h, P_1^d) + hP_2^d S(h, P_2^d), \quad (1)$$

where h, P_1^d and P_2^d are the degree of longitudinal electron polarization and the deuteron vector and tensor polarization, respectively. The subdivision of the cross section into the different parts reflects the dependences on the various kinds of polarizations. The unpolarized cross section is denoted by S_0, while the further notation is selfexplanatory. Any of the 41 structure functions enters only into one of six different parts: $f_{L/T/LT/TT}$ in S_0, f_{LT}' in $S(h)$, $f_{L/T}^{1M}$ and $f_{LT/TT}^{1M}$ ($M = 0, \pm 1$) in $S(P_1^d)$, $f_{L/T}^{2|M|}$ and $f_{LT/TT}^{2M}$ ($M = 0, \pm 1, \pm 2$) in $S(P_2^d)$, $f_T'^{1|M|}$ and $f_{LT}'^{1M}$ ($M = 0, \pm 1$) in $S(h, P_1^d)$, and $f_T'^{2M}$ ($M = 1, 2$) and $f_{LT}'^{2M}$ ($M = 0, \pm 1, \pm 2$) in $S(h, P_2^d)$. Further ingredients of the cross section can be best illustrated by giving one of its terms in greater detail, e.g.,

$$S(h, P_1^d) = \frac{\alpha}{6\pi^2} \frac{k_1^{lab}}{k_2^{lab} q_\nu^4} \left(\rho_T' \sum_{M=0}^{1} f_T'^{1M} cos(M(\phi - \phi_d)) d_{M0}^1(\theta_d) \right. \quad (2)$$

$$\left. + \rho_{LT}' \sum_{M=-1}^{1} f_{LT}'^{1M} cos(M(\phi - \phi_d) + \phi) d_{M0}^1(\theta_d) \right).$$

Lab frame momenta of the initial and final electron are denoted by k_1^{lab} and k_2^{lab}, respectively, while q_ν^2 is the four momentum transfer squared ($q = k_1 - k_2$). The structure functions are calculated in the final np cm system to which also $\Omega_{np}^{cm} = (\theta, \phi)$, the spherical angles of the relative np momentum, refer. The angles θ_d and ϕ_d describe the direction of the orientation axis $\hat{d}$ of the polarized deuteron target with respect to the coordinate system associated with the momentum transfer $\hat{q}$, while $\hat{d}$ is the axis with respect to which the deuteron density matrix is diagonal. The quantities $\rho_{T/LT}'$ together with $\rho_{L/T/LT/TT}$ – the latter enter in the cross section terms without electron polarization – describe the virtual photon density matrix. Note that our definition of the ρ's includes also the kinematical effects of the boost from the lab to the cm system.

The structure functions $f_{\mu'\mu}^{IM}$ ($\mu'\mu = L, T, LT, TT$) and $f_{\mu'\mu}'^{IM}$ ($\mu'\mu = T, LT$) are proportional to either the real or imaginary parts of the quantities

$$v_{\mu'\mu IM}(\theta) = \hat{I}\sqrt{3} \sum_{m_d m_d'} (-)^{1-m_d} \begin{pmatrix} 1 & 1 & I \\ m_d & -m_d & -M \end{pmatrix} \sum_{sm_s} t_{sm_s \mu'm_d'}^*(\theta) t_{sm_s \mu m_d}(\theta), \quad (3)$$

where $t_{sm, \mu m_d}(\theta)$ is the reduced transition matrix for the process $d+e \rightarrow np+e'$. Note that the structure functions depend on θ, on E_{np}, the np final state cm energy, and on $\vec{q}_{cm}$, the cm three-momentum transfer.

The inclusive cross section is given in detail in Ref. 2. Here we restrict ourselves to an abbreviated expression with a notation analogous to that of eq. (1)

$$\frac{d\sigma}{dk_2^{lab}d\Omega_e^{lab}} = \sigma_0 + P_1^d\sigma(P_1^d) + P_2^d\sigma(P_2^d) + hP_1^d\sigma(h, P_1^d) + hP_2^d\sigma(h, P_2^d). \qquad (4)$$

Again we give in more detail that part of the cross section which depends both on electron polarization h and on deuteron vector polarization P_1^d

$$\sigma(h, P_1^d) = \frac{\alpha}{\pi^2} \frac{k_1^{lab}}{k_2^{lab}q_\nu^4} (\rho_T' F_T'^{10} d_{00}^1(\theta_d) + \rho_{LT}' F_{LT}'^{1-1} cos(\phi_d) d_{-10}^1(\theta_d)). \qquad (5)$$

Comparing the last relation with eq. (2) one sees that only two of the five structure functions of $S(h, P_1^d)$ do not vanish in the integration over Ω_{np}^{cm}. Similar reductions occur for all the other cross section terms and thus only 10 of all the 41 structure functions survive the integration and lead to the following form factors: $F_{L/T}$ in σ_0, F_{LT}^{1-1} in $\sigma(P_1^d)$, $F_{L/T}^{20}$, F_{LT}^{2-1}, and F_{LT}^{2-2} in $\sigma(P_2^d)$, $F_{LT}'^{1-1}$ and $F_T'^{10}$ in $\sigma(h, P_1^d)$, and $F_{LT}'^{2-1}$ in $\sigma(h, P_2^d)$. Note that F_{LT}^{1-1} and $F_{LT}'^{2-1}$ vanish below pion threshold.

By a proper variation of the polarization parameters h and P_l^d one can separate the various beam, target, and beam-target asymmetries A_i for the exclusive reaction ($A_e = S(h)/S_0$, $A_d^{V/T} = S(P_{1/2}^d)/S_0$, $A_{ed}^{V/T} = S(h, P_{1/2}^d)/S_0$) and corresponding asymmetries α_i for the inclusive one. All asymmetries are functions of the deuteron orientation angles θ_d and ϕ_d and the A_i also of the angles θ and ϕ. One can utilize the variables θ_d, ϕ_d and ϕ for the further separation of the various form factors [2] and structure functions [3]. In this way almost all the polarization form factors can be determined by just one asymmetry measurement. The exceptions are F_L^{20} and F_T^{20}. They can only be separated by applying further Rosenbluth techniques. The separation of the various structure functions of the exclusive cross section is comparatively more complicated due to their large number and one asymmetry measurement is sufficient for the determination of only a few of them. Particularly complicated is the separation of the 16 structure functions of $S(P_2^d)$. In some of these cases even six asymmetry measurements are required for a determination.

Before discussing the results for the various form factors and structure functions we shortly describe the theoretical framework of our calculation (for more details see Refs. 1 and 2). For the t-matrix elements we evaluate explicitly all electromagnetic multipoles up to the order $L = 6$, while for the higher multipoles we take the Born approximation for the final state. For the deuteron and np scattering wave functions we use Paris, Bonn (r-space version), Nijmegen, and Argonne V_{14} and V_{28} potentials. The latter explicitly includes Δ-degrees of freedom within a coupled channel (CC) approach. Above pion threshold V_{28} is modified for the $^1D_2$ channel in order to give a better description of this channel. For the other potential models we use the impulse approximation (IA) for the calculation of the isobar channels. In the current operator we include meson exchange (MEC) and isobar current (IC) contributions.

The possible determination of the neutron electric form factor G_{En} in polarized electron scattering off few-body nuclei has been emphasized in the first theoretical calculations. They have indeed led to very promising results (see e.g. Ref. 1). Then, with Ref. 2 we have initiated a more general study of polarized deuteron electrodisintegration, where the role of the electromagnetic nucleon form factors is not the dominating topic, since we also aim at finding interesting potential model and subnuclear current effects. We start the discussion by reporting the most important results of Ref. 2. With respect to Δ-degrees of freedom we find various interesting aspects. The form factors F_T, $F_T'^{10}$, F_T^{20}, and F_{TT}^{2-2} all exhibit quite considerable IC contributions in the Δ-resonance region at lower momentum transfer. $F_T'^{10}$ shows comparatively the strongest influence for this kinematics. The two transverse form factors F_T^{20} and F_{TT}^{2-2} are also affected in other kinematic regions. The latter one even shows rather strong IC contributions for quasifree kinematics. However, these two form factors are about an order of magnitude smaller than F_T and $F_T'^{10}$. Furthermore, the purely longitudinal and longitudinal-transverse form factors also show influences from Δ-degrees of freedom. We only mention here F_L, F_L^{20}, and F_{LT}^{2-1}, which show notably different results if one compares the IA with the CC calculation. As regards the MEC we do not find such a variety of effects as due to the Δ-resonance. The strongest MEC influences are present at the deuteron breakup threshold. Because of the dominance of the $M1$ transition to the $^1S_0$ partial wave $F_T'^{10}$, F_T^{20}, and F_{TT}^{2-2} are affected similarly as the well-known F_T. One further interesting result is the rather strong G_{En} dependence of $F_{LT}'^{1-1}$ in the quasifree region, while at the same time potential and interaction current effects remain negligibly small.

The structure functions of the exclusive deuteron electrodisintegration exhibit very interesting effects. Since a detailed discussion would be quite involved we will only discuss some results for a few structure functions, but we would like to mention that a

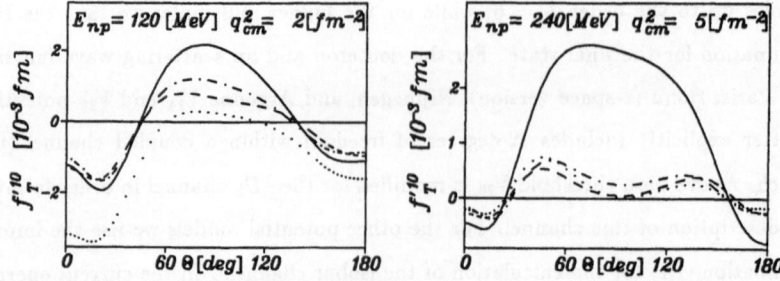

Fig. 1. Influence of FSI, MEC, and IC on $f_T'^{10}$ (Paris potential, $G_{En}=0$). Normal part (N) consisting of one-body currents plus Siegert operators with inclusion of FSI (dashed curve), Born approximation as N but without FSI (dotted curve), N with additional MEC (dash-dotted curve) and further addition of IC (full curve).

more complete study will be soon published [3]. Here we choose cases where one or two asymmetry measurements are sufficient for an experimental separation. Fig. 1 shows for $f_T'^{10}$ the effects of final-state interaction (FSI) and of MEC and IC at $E_{np} = 120$ MeV and $q_{cm}^2 = 2$ fm^{-2} and at $E_{np} = 240$ MeV and $q_{cm}^2 = 5$ fm^{-2}. One sees that $f_T'^{10}$ is strongly affected by IC, in particular for the kinematics in the Δ-resonance region (E_{np} = 240 MeV). Also MEC effects are present, but in comparison to the IC ones they are less important. The potential model dependence is small and therefore not shown, but the CC result with V_{28} is still missing and will probably be somewhat different from the IA results. Fig. 2 shows for the form factors f_{TT}^{10}, f_{LT}^{10}, and f_{LT}^{1-1} besides the effects of FSI, MEC, and IC also the potential model dependence (for kinematics see the figure). One finds relatively large MEC and IC contributions for f_{LT}^{10} and f_{TT}^{10}, while f_{LT}^{1-1} is somewhat influenced by MEC and IC only at proton forward and backward angles. The dependence on the potential model is quite interesting for all the shown cases. The results for f_{TT}^{10} at $E_{np} = 120$ MeV and $q_{cm}^2 = 2$ fm^{-2} are shown for two groupings of the potential models, the Nijmegen and Bonn potentials on the one hand and the Paris and V_{14} potentials on the other hand. Also f_{LT}^{10} exhibits rather large potential model effects and again one has similar results for Nijmegen and Bonn potentials. Looking at the size of the relative effects one finds the largest potential model dependence of Fig. 2 for f_{LT}^{1-1}. It is readily seen that at proton forward angles all four potentials lead to quite different results. Unfortunately, here we cannot discuss the various effects on the structure functions in greater detail, but have to refer for this to Ref. 3. In this reference we will also study the influence of the electromagnetic form factors on the structure functions. For the cases in Figs. 1 and 2 such effects are rather small and thus were not explicitly mentioned.

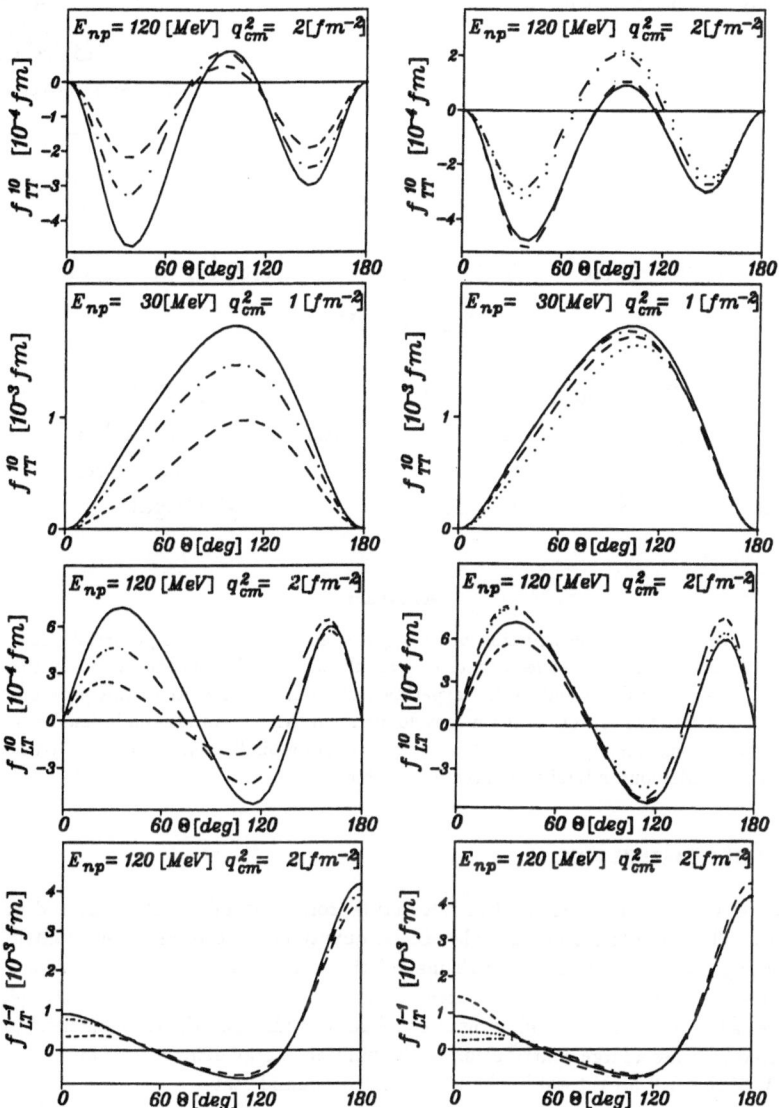

Fig. 2. The structure functions f_{TT}^{10}, f_{LT}^{10} and f_{LT}^{1-1}. Left: Influence of FSI, MEC and IC (Paris potential, $G_{En}=0$) with notation as in Fig. 1. Right: Potential model dependence for total contribution ($G_{En}=0$) with Paris (full curve), V_{14} (dashed curve), Bonn (dotted curve), and Nijmegen (dash-dotted curve) potentials.

References

1. Arenhövel, H., Leidemann, W., Tomusiak, E.L.: Z. Phys. A **331**, 123 (1988); **334**, 363(E) (1989)

2. Leidemann, W., Tomusiak, E.L., Arenhövel, H.: Phys. Rev. C **43**, 1022 (1991)

3. Arenhövel, H., Tomusiak, E.L., Leidemann, W.: to be published

Few-Body Systems, Suppl. 6, 242—253 (1992)

Few-
Body
Systems
© by Springer-Verlag 1992

Three-nucleon Potentials due to π and ρ Exchange

S. A. Coon

Physics Department

New Mexico State University

Las Cruces, NM 88003, USA

M. T. Peña

Centro de Fisica Nuclear-INIC

1699 Lisboa Codex

Portugal

Abstract

We review the construction of $\pi\pi-$, $\rho\pi-$, and $\rho\rho-$exchange potentials from off-mass-shell meson-nucleon scattering amplitudes which are constrained by the symmetries of QCD and by the experimental data. The results of this program are put into the context of recent developments in NN potentials and in few-body calculations. Finally, we present perturbative triton binding energy effects by the Tucson-Melbourne family of three-body forces.

1 Introduction

Three-body forces play a small, but interesting, role in atomic, molecular, and nuclear physics. The importance of this role can be quantified by ordering their contribution to binding energies. The Holstein-Primakoff three-electron force [1, 2, 3] plays almost no role in the inner shell binding energy of heavy atoms, the triple-dipole force [4] contributes a few per cent to the binding of nobel gas crystals, and the three-nucleon force is expected to make a large contribution to the binding energy deficiency of approximately 1 MeV of the $\sim$ 8 MeV total in the three-nucleon bound state.

Three-body operators arise when a theorist wishes to freeze out degrees of freedom in order to work in an (equal time) Hamiltonian formalism. The three-nucleon forces based on meson exchange have four distinct contributions. Three also appear in atomic and molecular systems, but the fourth contribution, due to spontaneously broken chiral symmetry, is unique to NNN forces. To begin the classification, we note that a relativistic field theory (QED) can be recast in perturbation theory into the Hamiltonian form with two-and three-body forces between non-relativistic electrons. The resulting Holstein-Primakoff three-body force (TBF) has two distinct contributions: one arises from retardation (overlapping exchange diagrams in which two photons are "in the air" at the same time) and the second is due to the necessity of projection operators onto positive energy electron states. Sequential iteration of the (constructed) two-body potential cannot produce the negative energy states present in the original QED: these frozen degrees of freedom appear as the electron-antielectron pair contribution in the TBF. The resulting three-body forces are then relativistic corrections, as they also are in nuclear physics [5]. The form of the HP potentials depend on choice of gauge, on the treatment of retardation, and upon the division of terms of order e^4 into contributions from the "irreducible" TBF and the iterated two-body potential [3]. All these ambiguities (which are also present in the three-nucleon force) could be avoided in the atomic case by employing a different calculational framework, but historically the Hamiltonian theory has been used. In contrast, a Hamiltonian formalism appears necessary for nuclear physics for some time to come.

A third contribution to TBF's is the polarization force caused by the presence of two systems distorting a nearby third. Electrons appear to have no substructure, so this polarization potential is not present in the HP three-electron force. It underlies, however, the Axilrod-Teller-Muto (ATM) three-atom force. This triple-dipole force is caused by the simultaneous distortions of the electron clouds of three noble gas atoms when the electron degrees of freedom are frozen. The ATM force and a close analogy to distortions of the nucleon (including the triple Δ NNN force [6]) have been discussed extensively in an earlier lecture in this series [7]. Here we merely remark that the strength of the ATM force is proportional to the electrical polarizability α_E of the neutral atom. The polarizability can be either measured, extracted from the low energy expansion of Compton scattering, or calculated by summing the dipole operator over the excited states of the atom. All three methods agree. This is not the case for two-meson exchange NNN forces; a sum over excited states of the nucleon does not saturate the Compton-like meson-nucleon amplitudes which determine the strength and structure of the potential.

The polarization analogy is most direct for the NNN force due to the exchange of two rho mesons (2ρ − TBF). The $\rho N \rightarrow \rho N$ amplitude is obtained from the Compton scattering process $\gamma N \rightarrow \gamma N$ by vector dominance. The excitation of a Δ-isobar in the quark model corresponds to the spin flip of a quark. This is the response of the nucleon to a *magnetic* field and is characterized by the magnetic polarizability β_M. Most models of the proton predict a large paramagnetic β_M of about 12 (units of 10^{-4}fm^3) due to Δ excitation. The latest Compton scattering experiments prove [8] the old suspicion that $\alpha_E \sim 12$ and $\beta_M \sim 2$. The large paramagnetic contribution from the Δ must be nearly cancelled by an equally large diamagnetic contribution from another source. Vector dominance then indicates that the Δ should not be the only important term in the 2ρ − TBF. A rigorous demonstration of this important fact will be described in Section 4.

Nuclear three-body forces are constructed by freezing out mesons, anti-nucleons, and nucleonic excitations such as the Δ-isobar. It is clear that the TBF between positive energy non-relativistic nucleons includes the nuclear analogies of the three types of contributions already discussed. Other contributions are needed to satisfy gauge invariance and spontaneously broken chiral symmetry. All contributions (disregarding meson retardation) can be modeled by covariant pole diagrams from a chiral Lagrangian. This was first done by Yang [9] and extended to ρ-exchange by the São Paulo group [10]. This model-building approach has many attractive features. It appears best to us, however, to consider as a single unit all the types of contributions. This can be done by working with Compton-like amplitudes which satisfy the symmetries of QCD and also can be constrained by experimental data. The Tucson-Melbourne program has obtained representations of (virtual) meson-nucleon scattering by utilizing the Ward-Takahashi identities of current algebra for the scattering of hadronic currents from nucleons. The pion is related by PCAC to the SU(2) axial vector current and the vector mesons ρ and ω are related to the SU(2) vector current by the current-field identity (which expresses the concept of vector meson dominance of the electromagnetic interaction). We now turn to a short description of the meson-nucleon amplitudes so obtained.

2 Off-mass-shell scattering amplitudes

Three-nucleon potentials of the two-meson exchange type are constructed by embedding a meson-nucleon amplitude into a NNN S-matrix. The two virtual (off-mass-shell) mesons are absorbed by the two external nucleons, a non-relativistic reduction of the three-body amplitude is made, and the matrix elements of S-1 are identified in Born approximation with the TBF, which can then be transformed into coordinate space if desired. The covariant Born pole terms are expected to dominate the amplitude with background contributions coming from the current algebra Ward identity constraints. The physical origin of the background terms is the summation of nucleon isobar and other resonance terms contributing to the processes. That this is so is verified by a successful confrontation of the theoretical amplitudes with the low energy πN experimental amplitudes (minus the nucleon Born terms) [12] and with pion photo- and electroproduction data near threshold [13, 14]. Nevertheless, the important feature of the Ward identity analysis is that the leading order background terms are obtained in a model-independent way and the explicit isobar models are required only for the construction of the next-to-leading order terms.

The four-momentum of the exchanged mesons q and q' provide a useful expansion parameter of the meson-nucleon amplitude. The amplitude $M_1 N \rightarrow M_2 N$, which we call T, then takes the form

$$T = T_B - T_{FPB} + \Delta T + q' \cdot C \cdot q \tag{1}$$

Here T_B stands for the covariant Born terms, T_{FPB} is the forward propagating (positive energy) nucleon term, ΔT is added to T_B so that to leading order $T_B + \Delta T$ satisfies the low energy theorems (Ward identity constraints), and $q' \cdot C \cdot q$ is a background term for which (isobar) models are necessary. The separation of the (purely pion and nucleon) terms of order f^2 (where f is the πNN coupling constant and the three-body S-matrix is of order f^4) into $T_B - T_{FPB}$ is just like that of the Holstein-Primakoff three-electron force. The grouping of the isobar contributions (the analogue of the ATM triple-dipole force) into $\Delta T + q' \cdot C \cdot q$ in Equation 1 enforces the largest degree of model independence of the off-shell amplitude. This is because to lowest order (all) isobar contributions are included in ΔT, and $q' \cdot C \cdot q$ is constructed to give contributions only in higher order terms.

What is the order needed for an useful representation of the amplitude? Nuclear physics folklore suggests that $q^2 \leq -10\mu^2$ (where μ is the pion mass) should be adequate. This estimate is supported by the Fourier transform of the short range cutoff in the calculated trinucleon pair correlation function and by the peak at $\vec{q}^2 \approx 9\mu^2$ in the calculated momentum distribution of the pions which act to bind light nuclei and nuclear matter [15]. On a hadronic scale (consider the mass of the nucleon M) this momentum region for the exchanged mesons is small because $\mu^2/M^2 \sim 1/50$.

For the process $\pi N \rightarrow \pi N$ the current algebra constraints take the form of soft pion theorems, to be satisfied when one or both pions are soft ($q^2 = 0$). An expansion of the (forward, isospin-even) current algebra-PCAC amplitude in the form

$$T = a + b\vec{q} \cdot \vec{q'} + c(\vec{q}^2 + \vec{q'}^2) + \mathcal{O}(q^4) \tag{2}$$

(where a,b,c subsume the physically distinct contributions of Equation 1) found that $\mu^3 b \sim \mu^3 b \sim \mu a$ (see Table 1) , so that up to second order the natural expansion parameter is q^2/μ^2. The natural expansion parameter of the higher order terms was demonstrated to be q^2/m_H^2 where m_H is a typical hadron mass [16]. In the case of $\rho N \rightarrow \pi N$ there is only one pion, and for $\rho N \rightarrow \rho N$ there are none. The absence of a multitude of soft pion theorems for the latter amplitudes means that the natural expansion parameter is q^2/m_H^2. This has three advantages for the $\rho\pi - \text{TBF}$ and the $2\rho - \text{TBF}$. The non-relativistic expansion should converge rapidly for $q^2 \leq -10\mu^2$. Vector dominance arguments should work well in this low q^2 region. The less rapid q^2/m_H^2 variation should not have the strong coordinate space singularities that the q^2/μ^2 variation of the $\pi N \to \pi N$ causes the $2\pi - \text{TBF}$ to have. In particular, the c term of Equation 2 is required by the soft pion theorems and this term makes it difficult to calculate the effect of the $2\pi - \text{TBF}$ by perturbation theory [17, 18]. One can hope that simple perturbation theory will be adequate for the ρ meson exchange forces.

3 Nucleon Born Terms and Meson Retardation

The forward propagating part of the nucleon Born term is subtracted because it corresponds to sequential iteration of a static (ie. non-retarded) one-meson-exchange two-nucleon potential and therefore does not contribute to an irreducible TBF. The remaining nucleon-antinucleon pair term is small (about 15% of the c coefficient, see Table 1) in the πN amplitude due to the "pair suppression" embodied in Adler's consistency condition. Without Adler's condition, which is enforced by approximate chiral symmetry, pair terms would dominate the off-shell amplitude and be reflected by two-pion-exchange NN potentials and on-shell πN scattering lengths much larger than nature allows. On the other hand, the ρ-analogue Kroll-Ruderman limit [19] ($q'_\rho \to 0$ and $q_\pi \to 0$) of pion electroproduction is entirely due to the pair term and is important in the $\rho\pi - \text{TBF}$. Finally the pair term is the only contribution to the exact Compton scattering amplitude from a point target in the low-energy limit [20].

As we have seen, the form of a TBF depends upon the representation used to account for retardation at the two-body force level. The subtraction procedure of the preceeding paragraph implicitly assumes no meson retardation, ie. an instantaneous NN potential. A consequence of meson retardation in the NN interaction is the dependence of OBEP upon the asymptotic initial energy E of the two nucleons, so that $V_\pi = V_\pi(E)$ for pion exchange. Then we are confronted again at the nucleon level [21, 22] with the same retardation ambiguities of the Holstein-Primakoff three-electron force [2, 3]. That is, when you subtract the nonrelativistic three-body S-matrix built upon T_{FPB} from the total irreducible S matrix, you intuitively expect the two contributions already seen in the HP force: pair contributions *and* contributions from a positive energy nucleon with two overlapping retarded pions. The former terms turn out to be of order f^4/M and the latter of order f^4 in an expansion of the entire three-body S-matrix in powers of $1/M$ [5, 21].

This retardation problem is complicated in the case of interacting pions and nucleons by the additional requirements of chiral symmetry. A complete analysis to order f^4/M was made using a phenomenological πN Lagrangian (supplemented by a chiral rotation due to Weinberg) which froze out Δ and higher resonances and heavy meson exchange [5]. A complete cancellation between all terms of order f^4 (overlapping retarded exchanges) was shown, so that the leading terms were due to various seagulls (including the nucleon pair term) of order f^4/M. Local terms from the earlier alternative expansion of $T_B - T_{FPB}$ in powers of q^2 made by the Tucson-Melbourne group [16] agrees with this result (in their limit of assuming no retardation). The f^4 cancellation follows from choosing to expand out fully the energy dependence in $V_\pi(E)$. On the other hand, if one wants to work with an energy-dependent NN potential, such as the full Bonn potential, then the sum of terms of order f^4 will not disappear and should be added to the NN potential. Indeed, Johnson et al. [22] suggest that the difference in triton binding energies between the energy-dependent full Bonn potential ($E_B = 6.73$ MeV) and the instantaneous potential obtained from it by a folded diagram technique [23] ($E_B = 7.8$ MeV) is due to the terms of order f^4 (which they call "extrinsic" three-body forces). This suggestion appears supported by the additional attractive contribution of 1 Mev estimated [24] for these terms. They conclude that "the many-body theories for energy-dependent and energy-independent potentials are inequivalent and that they are converging at different rates [energy-independent potentials converge faster]".

More recently, Weinberg started with the same chiral Lagrangian as that of Ref. 5 to construct TBF's due to pion exchange [25]. His result, to the $\mathcal{O}(f^4)$ he worked, can be shown to be equivalent to that of Ref. 5. These terms of $\mathcal{O}(f^4)$, however, disappear if one further requires an energy-independent potential NN potential. That is, Weinberg's three-nucleon force corresponds to using an energy dependent OPEP, $V_\pi(E)$.

4 (Broken) Chiral Symmetry and Gauge Invariance

The off-mass-shell amplitudes must not only agree with data when brought back on shell, but must also observe the constraints imposed by spontaneously broken chiral symmetry and invariance under choice of gauge. These constraints can be imposed at the level of the chiral lagrangian [5, 9, 10, 25]. Alternatively, one can ask that the "hadronic current"-nucleon scattering amplitudes obey the Ward identities of current algebra [11, 12, 14], and convert them to meson-nucleon amplitudes via PCAC for the pion and the current-field identity in the case of the rho. The equal time commutators of the Ward identity approach include the pion-nucleon sigma-term σ_N which is a direct measure of chiral symmetry breaking and currently a subject of intense study [26]. The σ-term is important in the a, b, and c terms of the $2\pi - \text{TBF}$ (see column ΔT of Table 1). These new developments may modify the earlier treatment [27]. On the other hand, the chiral symmetry breaking terms in the ρ-analogue of the $\gamma N \to \pi N$ amplitude are much smaller and, although calculated, have been neglected in nuclear forces. Of course, the $2\rho - \text{TBF}$ is not affected by chiral symmetry and is constrained by gauge invariance instead.

To construct the $\rho\pi - \text{TBF}$ one needs the ρ-analogue of the electroproduction of an *off-shell* pion. Gauge invariance provides a constraint on the Ward identity amplitude but the Born pole terms are not gauge invariant - the t-channel pion pole is the problem. A modification of the pion pole to make it gauge invariant was proposed in [14], but the final result (adapted for the $\rho\pi - \text{TBF}$) does not depend on that particular choice of Born term [28]. Thus the final amplitude is gauge invariant. Carrying the analysis a step further, one recognizes that the NNN amplitude resulting from the t-channel pion pole in the $\rho N \to \pi N$ amplitude could equally well be interpreted as a t-channel ρ-pole in the $\pi N \to \pi N$ amplitude. The latter pole contributes to the d-term of the $2\pi - \text{TBF}$ and should not be counted again, so it was dropped from the $\rho\pi - \text{TBF}$ [29]. The combined $2\pi - \text{TBF}$ and $\rho\pi - \text{TBF}$ are then gauge invariant.

The $\rho N \to \rho N$ amplitude of the $2\rho - \text{TBF}$ is the ρ-analogue of the $\gamma N \to \gamma N$. The latter is completely determined to leading order by gauge invariance and Lorentz covariance [20]. Bèg [31] extended the analysis to "charged photon" Compton scattering, showing that part of the low-energy limit was determined by the current commutator, rather than solely due to the nucleon Born terms. The $\rho N \to \rho N$ amplitude must be supplemented, however, by explicit Δ-isobar terms quadratic in the momenta of the exchanged rhos.

We close this discussion with the remark that TBF's due to pion and rho exchange which are based solely on a Δ-isobar model [32, 33] cannot, of course, have a well defined chiral limit nor obey the low energy theorems of chiral symmetry and gauge invariance.

5 Who needs a three-nucleon force, anyway?

Indirect evidence for an attractive TBF is the theorists result that "realistic" nonrelativistic NN force models underbind ^{3}H and ^{3}He by 0.5-1.0 MeV (out of 8.48 Mev for ^{3}H), and ^{4}He by up to 4 MeV (out of 28.30 MeV). It is plausible to hope that much of the needed binding comes from three-nucleon forces based on two-meson exchange. Experimental searches for direct evidence of TBF effects in the three-nucleon continuum are being carried out in conjunction with recently developed powerful Faddeev codes [34], but the binding energy deficiency remains the strongest evidence for TBF effects at present.

Recent Faddeev calculations of the triton binding energy with new potentials are fully consistent with the results summarized above [35, 36]. A fully converged 34-channel calculation with the new (1991) and charge dependent Nijmegen potentials by the Iowa-Los Alamos group yields 7.66 MeV [37]. The instantaneous version of the full Bonn potential obtained by folded diagrams gives $E_B = 7.86$ MeV (in a 5-channel calculation) [23]. Finally, it has been estimated (from the large P_D in its model deuteron) that the Bochum OBEP potential [38] should also "have some room left for contributions from TBF".

Calculations with Hamiltonians which combine the Tucson-Melbourne 2π − TBF with some of the older NN force models (Reid soft core, Argonne V14, and Paris) obtain an additional binding of 1.5 to 1.9 Mev [35, 36, 39], thus overbinding the triton. The results quoted use the πNN vertex function of the Tucson-Melbourne force papers. (The form factors and dependence of TBF effects on the form factor will be discussed in the next two sections). The accuracy of present day A=3 bound state Faddeev calculations with NN and NNN forces is quite good; the same Hamiltonian has eigenvalues which differ by only 10-20 keV [39] between the coordinate space calculations of the Iowa-Los Alamos group and the momentum space calculations of the Bochum and Hannover groups. Finally, the Tucson-Melbourne 2π − TBF together with the AV14 NN force overbinds the ^{4}He nucleus by 2.3 MeV [40].

It has long been expected that ρ-exchange would have an effect counter to that of π-exchange in three-body forces. Demonstrations of this partial cancellation have been seen in nuclear matter [29, 32]. A perturbative estimate of terms in the Tucson-Melbourne family of TBF's with an oversimplified triton wave function also verified this expectation [41]. More recently Sasakawa [42] has presented exact calculations with realistic NN potentials and $2\pi-$, $\rho\pi-$ and $\rho\rho-$ TBF's culled from a heterogeneous collection of three-nucleon force papers [32, 43, 44]. It was explained by Ellis et al. [29] that important cancellations (including that of the soft pion Fubini-Furlan-Rosetti limit against the nucleon pair term) in the $\rho\pi$ amplitude probably imply that the sum of bits and pieces from several papers do give a fair representation of the, in our opinion, more coherent Tucson-Melbourne forces. In any event, Sasakawa showed a near match between his chosen Hamiltonian and the experimental binding energy, noticing also a partial cancellation of the 2π − TBF by the $\rho\pi$ − TBF.

We have improved and extended our perturbative estimates of the Tucson-Melbourne triton binding energy effects. We now turn to a discussion of our results after a short discourse on the coupling constants and form factor of these forces.

6   Coupling Constants and Form Factors

The amplitude of the $2\pi - \text{TBF}$ is the sum of the nucleon pair term ($T_B - T_{FPB}$), equal time current commutators (ΔT), and ($q' \cdot C \cdot q$) the model-dependent background terms of $\mathcal{O}(q^2)$. An excellent description of the on-shell πN data at low energies is given by saturating the $q' \cdot C \cdot q$ axial-nucleon amplitude with the Δ-isobar [12, 16, 45]. (The contrapositive is that a bad description of the data is given by modeling the *entire* background amplitude by the Δ-isobar [30]). Explicit formulae for these covariant Δ poles (plus non-resonant Δ terms) are displayed in the Appendix A of Ref. 16, for example. The contributions of the nucleon pair term depends on the value of the πNN coupling constant $g \approx 13$ and that of the higher order $q' \cdot C \cdot q$ on the $\pi\Delta$N coupling constant $g^* \approx 2\mu^{-1}$. Because the two exchanged pions couple to two "outer" nucleons, the overall scale of the force is determined by g^2 for all terms except the very small nucleon pair term which is of order f^4/M, where $f \equiv (g\mu/2M) \approx 1$. Recent determinations of g are about 2% lower [46, 47] than the value $g = 13.4$ used in the calculations displayed below. This πNN coupling constant correction is then easily made, if desired.

The value of $g^* = 1.82\mu^{-1}$ was obtained by the Karlsruhe group from an analysis of πN scattering [30]. It corresponds to the dimensionless coupling constant $f^* = 0.26$ used by Martzolff, et al. [32]. With these values of g and g^* one can expand individually $T_B - T_{FPB}$, ΔT, and $q' \cdot C \cdot q$ (long known to have the structure $e\nu^2 + fq' \cdot q + \mathcal{O}(q^4)$ [48]) to display the ingredients of the model-independent coefficients of the $2\pi - \text{TBF}$. They are shown in Table 1 where the units are in terms of the charged pion mass [49]. A similar exercise can be made for the expansion parameters of the static πN amplitude

	$T_B - T_{FPB}$	ΔT	$q' \cdot C_\Delta \cdot q$	Total
μa	0	+1.13	0	+1.13
$\mu^3 b$	0	-1.10	-1.48	-2.58
$\mu^3 c$	-0.15	+1.15	0	+1.00
$\mu^3 d$	-0.15	-0.25	-0.35	-0.75

Table 1: Expansion coefficients of the πN amplitude

obtained from non-relativistic NΔ transition potentials. The non-zero terms depend on the value of the $\pi\Delta$N coupling constant chosen: $\mu^3 b = 4\mu^3 d = -1.40$ for the choice of $f^* = 0.26$ by Martzolff et al. [32] and $\mu^3 b = 4\mu^3 d = -1.83$ for the Hannover [33] choice of $f^* = 0.36$. In our opinion, not too much emphasis should be placed on these comparisons. The relationship between the covariant isobar contribution to the "weak axial current"-nucleon amplitude and the forward propagating isobar of the transition potential picture is not yet clear. In any event, the b and d coefficients of Martzolff are closer to those in Table 1, perhaps because they use a more physical coupling constant.

The form factors of the meson-baryon-baryon (MBB) couplings are tabulated and discussed in detail in Appendix A of Ellis et al. [29]. The vertex function is parameterized as a monopole $F_{MBB} = (\Lambda^2 - m^2)/(\Lambda^2 - q^2)$ normalized to unity on the meson mass shell ($q^2 = m^2$). The cutoff parameter is set to about 800 MeV or 5.8μ for the πNN form factor. The variety of theoretical and phenomenological arguments for such a value of Λ have been set out many times; of particular relevance to TBF's is that $\Lambda \approx 800$ MeV is consistent with the Goldberger-Treiman discrepancy [50], which is, along with the σ-term, another direct measure of chiral symmetry breaking. This value of Λ is reasonably consistent with OPEP terms in contemporary NN potentials: $\Lambda \approx 950$ MeV in Nijmegen potentials [51], $\Lambda \approx 980$ MeV for Argonne V14 [15], $\Lambda = 800$ MeV for both π and ρ in the Bochum potential [38], and $\Lambda = 800$ MeV in the present day evolution of the Bonn OBEP potentials [52]. The full Bonn potential with two-pion exchange and other higher order processes uses a much larger value of $\Lambda \approx 1200$ MeV (ie. , almost pointlike nucleons). The πNN form factor of the full Bonn potential has always been inconsistent with evidence from outside the NN system [50]; this difficulty is expected to go away when the additional physics of correlated $\rho - \pi$ exchange is put into the full model [53].

The growing consensus on a low mass cutoff for pion exchange highlights an important point emphasized by the Hokkaido group [54] and, in the modern context, by the São Paulo group [55]. The contact terms (those proportional to a coordinate-space δ-function and its derivatives) are spread out and gain in importance as Λ becomes smaller and the size of the nucleon grows. It has been argued that since these contact terms arise from nucleon structure they should not be included in potential models [54, 55]. The philosophy of the recent Bochum NN potential [38] turns this argument on its head. The Bochum group argues that one should first ascertain the MNN form factor and then follow its consequences to a potential. For the Bochum potential the low cutoff mass of the πNN and ρNN form factors means that *all* mesons more massive than the ρ are subsumed into contact terms. This reformulation of the traditional OBEP appears to give a satisfactory description of the NN data, although a χ^2 analysis is needed for an informed judgment. A nice feature of the Bochum NN potential is the small two-pion exchange term, cut down by four powers of a rapidly varying cutoff function and partially replaced by the contact terms. We note that Weinberg obtains similar contact terms in his NN potential from a four-fermion interaction [25].

We finish this section with brief comments on the other form factors of the Tucson-Melbourne TBF's [29]. There are indications (from neutrino-nucleon scattering) that the $\pi \Delta N$ form factor is similar to the πNN form factor; we take them to be the same. The ρNN vertex has two couplings; a direct Dirac coupling and the Pauli coupling to the anomalous magnetic moment of the nucleon. In the Tucson-Melbourne prescription, the ratio of the Pauli to Dirac coupling varies with the four-momentum of the ρ. The *two* cutoff masses $\Lambda_{Dirac} = 12\mu$ and $\Lambda_{Pauli} = 7.4\mu$ are determined from nucleon electromagnetic form factor data. This is a reasonable approach to accommodate both the vector dominance value of $\kappa_V = 3.7$ at $q^2 = 0$ and the on-mass-shell value of 6.6 [56]. Recently the Bochum group has also proposed a similar treatment of ρ-exchange for the same reasons [57]. The last needed coupling is $\rho \Delta N$; it is tied by vector dominance to the $\gamma \Delta N$ vertex which has been studied extensively both experimentally and theoretically. We follow these analyses and use an monopole form factor with cutoff mass $\Lambda_{\rho N \Delta} = 5.9\mu$ to approximate this vertex.

7 Expectation Values of TBF's in a Simple Triton Model

We now estimate perturbative contributions of the Tucson-Melbourne TBF's to the binding energy of the triton. We use a coordinate-space Schrödinger wave function obtained by solving the Faddeev equations with the Malfliet-Tjon I-III potential [58]. This potential is s-wave and has no tensor part; the triton wave function can be considered "semirealistic" at best. Nevertheless, our results may be indicative of what to expect from more sophisticated calculations. Our results are presented in Table 2; note that a minus sign means an attractive contribution. The full three-body force $W = W_1 + W_2 + W_3$, where the subscript 1 singles out particle 1 ("in the middle") to undergo the scattering, and W_2 and W_3 are obtained by cyclic permutations of W_1. The super-scripts refer to the isotopic decomposition of the meson-nucleon amplitude in the t-channel. Then W_3^+, for example, has the isospin factors $\vec{\tau}_1 \cdot \vec{\tau}_2$ and W_3^- carries $i\vec{\tau}_1 \cdot \vec{\tau}_2 \times \vec{\tau}_3$. We have calculated only those local terms which an examination of coefficients in the derived formulae indicate are the most important. We have also calculated some other terms of the rather complicated ρ–exchange potentials, but their contributions are much smaller than these displayed in Table 2.

Exchanged Mesons	Terms of the TBF	Form Factors	
		Tucson-Melbourne	Martzolff
$\pi\pi$	$a^{(+)}$	-0.088	-0.114
	$b^{(+)}$	+0.151	-0.431
	$c^{(+)}$	-0.370	-0.075
	$d^{(-)}$	-0.161	-0.218
$\pi\pi$ Total		-0.468	-0.839
$\rho\pi$	Kroll-Ruderman term$^{(-)}$	+0.215	+0.342
	$q' \cdot C_\Delta^{(+)} \cdot q$	-0.022	+0.261
	$q' \cdot C_\Delta^{(-)} \cdot q$	+0.090	+0.340
$\rho\pi$ Total		+0.283	+0.943
$\rho\rho$	pair plus Beg term$^{(-)}$	-0.003	-0.024
	$q' \cdot C_\Delta^{(+)} \cdot q$	-0.005	-0.101
	$q' \cdot C_\Delta^{(-)} \cdot q$	-0.007	-0.247
$\rho\rho$ Total		-0.015	-0.372
Total		-0.201	-0.268

Table 2: Contributions (in MeV) of three-body forces to the energy of the triton

The ρ–exchange TBF's do indeed modify the effect of the $2\pi - $ TBF. The third column of Table 2 contains the results of the Tucson-Melbourne potentials *with the original form factors*. It has been fashionable to take the derived strengths of a nuclear force seriously, but to consider the meson-nucleon form factor cutoffs as adjustable parameters. Many applications of the $2\pi - $ TBF have adopted this point of view, so we show another set of calculations which keep the strength constants of the Tucson-Melbourne TBF's but use another set of form factors. The fourth column labeled "Martzolff" is the result of choosing the much heavier cutoff masses from Ref. 31. (We convert their monopole parameterization of the product of two vertices to our convention to find $\Lambda_\pi = 10.6\mu$ and $\Lambda_\rho = 13.4\mu$ for *all* couplings to N's or Δ's.) The contributions of the $2\pi - $ TBF are, as is well known, strongly cutoff dependent, but the sum of TBF contributions is not, at least in this triton model. A closer examination of the third column in Table 2 shows a satisfying pattern ($2\pi : \rho\pi : 2\rho = -1 : +0.60 : -0.03$) of decreasing effect with increased mass of the exchanged meson. This column employs the, in our opinion, more realistic form factors. The fourth column corresponds to more pointlike nucleons. The sequence there ($2\pi : \rho\pi : 2\rho = -1 : +1.12 : -0.44$) may have a similar sum but one is uneasy about the next (unexplored) term. In any event, the addition of TBF's due to ρ—exchange does tend to counter the too strong attraction of the 2π — TBF and to stabilize the total TBF effect under variations in form factors.

8 Summary

The latest calculations of the A=3 bound state with current "realistic" two-nucleon potentials such as Argonne V14, the new (1991) Nijmegen potential, and a folded diagram version of the full Bonn potential confirm a persistent theoretical result about 0.7-0.8 MeV less than the experimental triton binding energy of 8.48 MeV. The Tucson-Melbourne (TM) two-pion exchange three-body force provides an additional binding of, for example, 1.7 MeV in conjunction with the AV14 potential. That is, "realistic" two- and three- nucleon potential models overbind the triton (and the alpha particle). Overlapping and retarded pion exchange graphs yield non-local terms which have not been numerically evaluated yet, but they are expected to be less important than the (already small) nucleon-antinucleon pair terms of the 2π-TM potential.

There are indications, however, that three-nucleon potentials of the two-meson exchange structure extended to include ρ-exchange will act against the overbinding effect of the 2π-TM potential. To be viable, these potentials must be based on meson-nucleon amplitudes which accurately reflect spontaneously broken chiral symmetry (for the pion) and which obey gauge invariance (because the rho is linked to the photon by the current-field identity). The off-mass-shell amplitudes must, as was the $\pi N \to \pi N$ amplitude, be confronted with both the low energy theorems arising from combined chiral symmetry breaking and gauge invariance and with the on-shell scattering data whenever possible. Such a program has been carried out by the Tucson-Melbourne collaboration and a consistent set of 2π-, $\rho\pi$-, and 2ρ- exchange potentials is available.

Our most recent estimates of triton binding energy effects due to these potentials have been presented. The $\rho\pi$-TM potential does give contributions of the opposite sign to those of the 2π-TM potential. The dependence of binding energy effects upon the parameters of the meson-nucleon-nucleon vertices is greatly lessened when the full Tucson-Melbourne family of three-body forces is employed.

Acknowledgments

This work was supported in part by the National Science Foundation of USA, the Institute for Nuclear Theory at University of Washington, a NATO Collaborative Research Grant, and by Instituto Nacional de Investigacao Cientifica of Portugal. This report was written in Lisbon. We would like to acknowledge useful discussions with Rob Ellis, Jim Friar, Ben Gibson, Karl Holinde, Bruce McKellar, Peter Sauer, Alfred Stadler, and Bob Wiringa.

References

[1] Primakoff, H. , Holstein, T. : Phys. Rev. 55, 1218(1938)

[2] Chanmugan, C. , Schweber, S. S. : Phys. Rev. 1A, 1369(1970)

[3] Zygelman, B. , Mittleman, M. H. : J. Phys. B19, 1891(1986)

[4] Axilrod, B. M. , and Teller, E. : J. Chem. Phys. 11, 299(1943); Muto, Y. : Proc. Phys. Math. Soc.(Japan) 17, 629(1943)

[5] Coon, S. A. , Friar, J. L. : Phys. Rev. **C34**, 1060(1986)

[6] Fujita, I. , Kawai, M. , and Tanifuji, M. : Nucl. Phys. **29**, 252(1962)

[7] Coon, S. A. : Few-Body Systems, Suppl. 1, 41(1986); see also Coon, S. A. : Proc. Int. Conf. Nucl. Phys., Inst. of Physics Conf. Series No 86 (Inst. of Physics, Bristol,1986)p.523

[8] Skopik, D. M. : this conference; Nathan, A. : private communication

[9] Yang, S. N. : Phys. Rev. **C10**, 2067(1974)

[10] reviewed by Robilotta, M. R. in Few-Body Systems, Suppl. 2, 35(1987)

[11] reviewed by Scadron, M. D. in Rep. Prog. Phys. **44**, 213(1981)

[12] Scadron, M. D. , Thebaud, L. R. : Phys. Rev. **D9**, 1544(1974)

[13] Dombey, N. , Read, B. J. : Nucl. Phys. **B60**, 64(1973)

[14] Macmullen, J. T. , Scadron, M. D. : Phys. Rev. **D20**, 1069(1979)

[15] Friman, B. L. , Pandharipande, V. R. , and Wiringa, R. B. : Phys. Rev. Lett. **51**, 763(1983)

[16] Coon, S. A. , Scadron, M. D. ,McNamee, P. C. , Barrett, B. R. , Blatt, D. W. E. , McKellar, B. H. J. : Nucl. Phys. **A317**, 242(1979)

[17] Bömelburg, A. : Phys. Rev. **C34**, 14(1986)

[18] Friar, J. L. , Gibson, B. F. , Payne, G. L. , Coon, S. A. : Few-Body Systems **5**, 13(1988)

[19] Kroll, N. , Ruderman, M. A. : Phys. Rev. **93**, 233(1954)

[20] Thirring, W. : Phil. Mag. 41, 1193(1950); Low, F. E. : Phys. Rev. **96**, 1428(1954), Maksimenko, N. V. , Shul'ga, S. G. : Sov. J. Nucl. Phys. **52**, 335(1990)

[21] Friar, J. L. : Ann. Phys. (N. Y.) **104**, 380(1977)

[22] Johnson, M. B. , Haidenbauer, J. , Holinde, K. : Phys. Rev. **C42**, 1878(1990)

[23] Haidenbauer, J. , Holinde, K. , Johnson, M. B. : preprint KFA-IKP(TH)-1991-26

[24] Pask, C. : Phys. Lett. **25B**, 78(1967)

[25] Weinberg, S. : Phys. Lett. **B251**, 288(1990); Nucl. Phys. **B363**, 3(1991)

[26] Gasser, J. ,Leutwyler, H. , Saino, M. E. : Phys. Lett. **253B**, 252(1990); ibid 260; Born, R. , Hurth, T. , Schilcher, K; Wu, Y. L. : Phys. Lett. **266B**, 463(1991)

[27] Coon, S. A. , Friar, J. L. , McKellar, B. H. J. : work in progress

[28] Ellis, R. G. , McKellar, B. H. J. : Phys. Rev. **D28**, 86(1983)

[29] Ellis, R. G. , Coon, S. A. , McKellar, B. H. J. : Nucl. Phys. **A438**, 631(1985); and in preparation

[30] Höhler, G. , Jakob, H. , Strauss, R. : Nucl. Phys. **B39**, 237(1972)

[31] Bég, M. A. B. : Phys. Rev.**150**, 1276(1966)

[32] Martzolff, M. , Loiseau, B. , Grangé, P. : Phys. Lett. **92B**, 46(1980)

[33] Hajduck, Ch., Sauer, P. U. , Yang, S. N. : Nucl. Phys. **A405**, 605(1983)

[34] Witała, H. , Glöckle, W. : this conference

[35] Chen, C. R. , Payne, G. L. , Friar, J. L. , Gibson, B. F. : Phys. Rev. **C31**, 2266(1985); **C33**, 1740(1986)

[36] Ishikawa, S. , Sasakawa, T. : Few-Body Systems **1**, 3(1986); ibid 143

[37] Friar, J. L. : invited talk at Nuclear Hamiltonian Workshop, Argonne, IL, Aug. 5-9, 1991

[38] Deister, S. , Gári, M. F. , Krümpelmann, W. , Mahlke, M. : Few-body Systems **10**, 1(1991)

[39] Stadler, A. , Glöckle, W. , Sauer, P. U. : Phys. Rev. **C44** (November 1991)

[40] Wiringa, R. B. : Phys. Rev. **C43**, 1585(1991)

[41] Coon, S. A. , Peña, M. T. , Ellis, R. G. : Phys. Rev. **C30**, 1366(1984)

[42] Sasakawa, T. : this conference

[43] Robilotta, M. R. , Coelho, H. T. : Nucl. Phys. **A460**, 695(1986)

[44] Robilotta, M. R. , Isidro Filho, M. P. : Nucl. Phys. **A414**, 394(1984)

[45] Wilde, B. H. , Coon, S. A. , Scadron, M. D. : Phys. Rev. **D18**, 4489(1978)

[46] Bergervoet, J. R. , et al. : Phys. Rev.**C41**, 1435(1990); de Swart, J. J. : this conference

[47] Arndt, R. A. et al. : Phys. Rev. Lett. **65**, 157(1990)

[48] Cheng, T. , Dashen, R. : Phys. Rev. Lett. **26**, 594(1971)

[49] The row of Table 1 labeled μ^3b corrects a misstatement made by one of us in Refs. 7 and 18.

[50] Coon, S. A. , Scadron, M. D. : Phys. Rev. **C23**, 1150(1981); ibid **C42**, 2256(1990)

[51] Nagels, M. M., Rijken, T. A. , de Swart, J. J. : Phys. Rev. **D17**, 768(1978)

[52] Holinde, K. , Thomas, A. W. : Phys. Rev. **C42**, 1195(1990); Haidenbauer, J. , Holinde, K. , Thomas, A. W. : Adelaide University preprint (1991)

[53] Holinde, K. : private communication

[54] Sato, M. , Akaishi, Y. , Tanaka, H. : Prog. of Theo. Phys. Supp. No 56, 930(1974)

[55] Robilotta, M. R. , Isidro Filho, M. P. : Nucl. Phys. **A451**, 581(1986)

[56] Höhler, G. , Pietarinen, E. : Nucl. Phys. **B95**, 210(1975)

[57] Gari, M. , Krümpelmann, W. : Bochum preprint, 1991

[58] Friar, J. L. , Gibson, B. F. , Payne, G. L. : Z. Phys. **A301**, 309(1981)

Few-Body Systems, Suppl. 6, 254—264 (1992)

THREE-BODY FORCE EFFECTS IN THREE-NUCLEON SYSTEMS
What was achieved by the study of three-nucleon bound states ?

Tatuya Sasakawa
University of Library and Information Science
Tsukuba 305
Japan

ABSTRACT

Calculated low energy physical quantities, such as the D/S ratio of the asymptotic normalization constants of ^{3}H and ^{3}He, the E1 photo absorption cross section, correlate linearly with the calculated binding energies.   The $p(\vec{d}, \gamma)^3$He analyzing power indicates that AV14   is only favorable realistic 2NP. For AV14 as 2NP and 3NP with Λ_π= 0.81GeV and Λ_ρ= 1.13GeV,   we   get   BE   of ^{3}H(^{3}He)=8.485(7.725)MeV,[experimentally, 8.482(7.718)MeV,], taking account of CIB,CSB and Coulomb effects in the 52 channel Faddeev equation. We also get (GT)/$\sqrt{}$3 = 0.955 (Exp. 0.961)   for the triton beta-decay by consistent calculations of the wave functions of ^{3}H and^{3}He, as well as the exchange current with above Λ_π and Λ_ρ.

1. INTRODUCTION

Stimulated by Faddeev theory proposed in 1960, many people have tried to solve the Faddeev equation. By 1983, when FEW-BODY X Karlsruhe Conference was held, people realized that the Faddeev 5-channel calculations with any realistic two nucleon potential (2NP) does not yield the triton binding energy of beyond 8MeV, against the experimental value of 8.482 MeV.

After we came back home from the Conference, we solved the 18 channel Faddeev equation with various 2NP and calculated the three-nucleon-potential(3NP) effect by first order perturbation theory. For 3NP, we took the Tucson-Melbourne potential (TM)[1]. We obtained the following conclusions [2],

(1) The three-body force effect has been thought to be small because two largest matrix elements cancel each other.

(2) In spite of the fact that each of first order matrix elements for higher partial waves is small, if we take all of these matrix elements from 18 channels into account, we get the triton binding energy that exceeds 8MeV.

(3) The favorable value of πNN cutoff mass of πNN form factor involved in TM-3NP is about 700 MeV.

Beginning from the checking up of our first order matrix elements by Bochum and Los Alamos-Iowa groups, the competitions between these groups and our group was initiated to take account of as many channels as possible to obtain a convergent result. We may say that in 1984 the study of three-nucleon systems plunged into the second epoch.

2. METHODS FOR SOLVING FADDEEV EQUATION

By 1984, various groups have made the computer code for solving the Faddeev equation. The idea are classified into three kind.

(1)Coordinate space method

This method was initiated by the Grenoble group[3,4] and later used by the Los Alamos-Iowa group [5,6]. In this method, we use a pair of Jacobi coordinates, one for the interacting pair and another one for the spectator, or the hyper-radius and hyper-angle, and solve a partial differential equation with a suitable boundary condition. The hyperspherical coordinate method developed by Fabre de la Ripelle may be classified in this category.

(2)momentum space method

This method has been used by the Purdue [7], Hannover [8], Bochum [9] and many other groups. In many cases, a separable expansion (or approximation) for nuclear interaction is utilized.

(3)Mixed coordinate and momentum space method

This method was considered and has been used by the Sendai group. First we expand the Faddeev equation by the complete set of the plane wave state of the spectator expressed in momentum space, and then, we solve a set of coupled ordinary differential equations for the interacting pair of particles in the coordinate space, hence the mixed coordinate and momentum space method. However, actually the first step should not necessarily be in momentum space. In fact, when the spectator motion subjects to the Coulomb potential, we treat it in coordinate space.

For the second step, we have considered three methods.

a. Sturm function

We used the Sturm function to represent the pole term and treated the reminder as the perturbation. This perturbation iteration should converge, [10].

b. New acceleration method

When we solved the 18 channel Faddeev equation for various 2NP [2], we used this method,[11].

c. Method of continued fractions

Finally, we came to the idea of the method of continued fractions,[12]. Its remarkable speed in computation and the wide applicability make us use this method ever since we discovered this method, which may be used not only for the usual local nuclear potentials, but also non-local forces, energy dependent forces, and Coulomb potentials.

Coulomb modified Faddeev equation

We can solve ^{3}He, only by the method (3), by which we have pushed forward the consistent calculations of ^{3}H and ^{3}He. As a result, now we may say that the studies of these nuclei under the potential model with non-relativistic treatment have been exhausted. In what follows, let us discuss what was achieved by the studies of three nucleon bound states.

3. THREE-BODY FORCE EFFECTS

TM-3NP describes the $\pi\pi$ exchange process between three nucleons. This 3NP involves two πNN form factors. The form factor represents phenomenologically all complex processes that can not be described in the form of a potential.

If we denote the form factor by $F_{\pi NN}(q^2)$, and assume its form to be

$$F_{\pi NN}(q^2) = \frac{\Lambda_\pi^2 - m_\pi^2}{\Lambda_\pi^2 - q^2} \quad , \qquad (q^2 = q_0^2 - \vec{q}^2) \qquad (1)$$

we call it the monopole π NN form factor. For instance, we obtain the triton binding energy of 8.42 MeV (Exp. 8.482 MeV),taking AV14 for 2NP, and TM-3NP with the monopole form factor of $\Lambda_\pi = 700$ MeV for the 34 channel Faddeev equation.

Although there is no profound reason of favoring the monopole form factor, this is the form adopted by many people and there is no positive argument against this form. Los Alamos-Iowa group has adopted [5,6] $5.8m \simeq$ 800 MeV as Λ_π, which is obtained by the deviation from the Goldberger-Treiman relation [16]. Their result is slightly overbound compared with the experimental value. In any way, our calculated binding energies agree with those of Los Alamos-Iowa group, if we adopt the same cutoff mass in spite of the difference in the methods for solving the Faddeev equation. In the course of studying the three-nucleon binding energy, an important finding was made that the calculated values of some physical quantities are linearly correlated to the calculated values of the binding energy. This property was first found in the charge radius [6,15] and later, in the D/S ratio of the triton asymptotic normalization constant, which was considered as a physical quantity independent of the binding energy [17,18]. Recently, it was found further that the Coulomb energy of ^{3}He [26], the E1 absorption cross section [19] and the D/S ratio of the ^{3}He asymptotic normalization constant [20] obey the same rule.

On the other hand, there may be some quantities that do not obey this rule. For instance, the Gamow-Teller matrix element of the triton beta decay does not show such a linear relationship with the binding energy [21].

4. REALISTIC TWO-NUCLEON POTENTIAL

As a cosequence of a great deal of effort, various realistic two nucleon potentials have been proposed. And one of the motivations to study three-nucleon problems was to find out the best realistic potential among them. However, as stated in the introduction, all realistic 2NP failed to reproduce the triton binding energy, and needed help from 3NP. Unfortunately, since the cutoff mass in 3NP can be chosen arbitrarily, we could not select the best 2NP from these arguments.

Meanwhile, the Bonn potential was proposed and attracted people, since it can reproduce not only the two-nucleon properties, but also the binding energy of the triton. Against this remarkable property of the Bonn potential, we think as follows. Due to the small tensor force in this potential, the energy dependence of the S-D mixing parameter will not be reproduced correctly, although we need more precise experimental results before making things more clear. To adjust the two-body data such as the deuteron binding energy with small tensor force effect, the central force acting to the S-state is made larger than usual potentials. This makes the three-body binding energy increase without help from 3NP[22]. However, since the tensor force effect is small, the triton D-state probability is smaller than other potentials. For example, while it is 8.94 % for AV14 and PARIS potentials, only 6.80 % for the BONN potential [23]. As a result, the calculated charge form factor curve is shifted a little bit towards the large momentum transfer side [20]. This makes agreement with the experimental result worse.

Recently, we calculated the radiative capure of a proton from a polarized deuteron [24]. Our result shows that AV14 is only potential that reproduces the anlyzing power A_{yy}. So far, the experiment was done for only one point. However, it is possible that AV14 is only favorable realistic potential.

5. CHARGE INDEPENDENCE BREAKING, CHARGE SYMMETRY BREAKING EFFECTS

As stated in sec.2, we can obtain the wave function of ^{3}He by solving CMF. Although the Coulomb force effect itself does not tell any important information, we can not say anything about the charge independence breaking (CIB) and the charge symmetry breaking (CSB) effects unless we can solve the three-body problem with the Coulomb force, the largest effect in CSB.

As for CIB and CSB, the following two things have been known so far.

(1) The scattering lengths for the two nucleon $^1S_0$ state are

$$a_{nn} = - 18.7 \pm 0.6 \text{ fm } [25], \quad a_{pp} = - 17.1 \pm 0.2 \text{ fm},$$

$$a_{np} = - 23.73 \pm 0.010 \text{ fm}.$$

(2) The Coulomb effect in the three nucleon system is about 650 keV, which is smaller than 764 keV, the mass difference of ^3H and ^3He.

The result (1) shows that the (attractive) 2NP between two neutrons ia stronger than that between two protons, and 2NP between neutron and proton is even stronger than these two kind of forces. From this fact, we can understand why the binding energy difference of ^3H and ^3He is larger than the Coulomb energy of ^3He. The solution of CMF taking AV14 as 2NP, and with 3NP, CIB, CSB (and Coulomb for ^3He) yields the following table [27]. For 3NP, we took not only $\pi\pi$ exchange but also $\pi\rho$ and $\rho\rho$ exchanges into account [20,27]. We will discuss in more detail abut this 3NP in sec.6.

Table 1

	2NP	3NP	CIB	Coulomb	CSB	other	sum	exp
^3H	7.673	1.044	- 0.232				8.485	8.482
^3He	7.673	1.044	- 0.232	- 0.648	- 0.075	- 0.037	7.725	7.718
				- 0.760	[26]			0.764

A few remarks will be made about this table.

(1) In Ref.9, we obtained the "model independent binding energy of ^3He", putting E(^3H) = - 8.482 MeV in the equation E(^3He) = 0.9684$\cdot$E(^3H) -

0.3799 ± 0.004 MeV, which is deduced from solutions of CMF with various 2NP,3NP and Coulomb potential for various number of channels. On the other hand, we use AV14 for 2NP and do not use any experimental value in obtaining the above table, except that the rho-meson cutoff mass was chosen so that the triton binding energy is reproduced.

If we calculate the binding energy of ^{3}He simply by subtracting the Coulomb energy from the triton binding energy, we get 8.482 - 0.648 = 7.834 MeV. This value corresponds to a value obtained from a calculation based on the above table, 2NP + 3NP + CIB + Coulomb = 7.837 MeV.

(2) In Ref.26, we have shown the result of solving CMF for (various 2NP) + (TM - 3NP) + CIB + Coulomb + CSB for various number of channels. On the contrary, we have used AV14 for 2NP, and 3NP including not only $\pi\pi$ but also $\pi\rho$ and $\rho\rho$ exchanges. However, the sum of Coulomb + CSB + other effects [26] remains the same.

(3) We need at least 34 channels to get a convergent binding energy for ^{3}H. Then, we should perform at least 34 channel calculations for ^{3}He.

CIB means that the force acting between nn or pp is different from that between np. Therefore, the CIB force depends on the third component of the isospin operator. As a result, the isospin states T=1 and T=0 are mixed in the two nucleon system. Consequently, in the three-body system, the isospin T=3/2 and T=1/2 states are mixed. By this reason, we should perform 52-channel calculations if we take isospin into account for 34 spin-angular partial waves. All of our calculations in this section was performed for 52 channels.

(4) The CIB and CSB reduce the three-nucleon force effect in all of the binding energy, the D-state probability of ^{3}H and ^{3}He [21], and the (Δ)$_\pi$ contribution of the exchange current [21].

6. CUTOFF MASS OF RHO-MESON

As stated in sec.3, 3NP involves the πNN and ρNN cutoff factors. Various values for Λ_π have been proposed so far. For each of Λ_π, we determine Λ_ρ, so that its use in the three-body force may reproduce the triton binding energy. Then we investigate which one of the pair of cutoff masses Λ_π and Λ_ρ so determined can reproduce the triton beta decay Gamow-Teller matrix element. In this way, if we use the pair of Λ_π and Λ_ρ thus determined, we will be able to reproduce the physical quantities which are correlated strongly with the binding energy as well as those which are not correlated to the binding energy.

To treat ^{3}H and ^{3}He consistently, it is very important in this argument that we should not neglect the CIB effect of 0.2 MeV in the triton binding energy. Also, the $\pi\rho$ exchange three-body force effect is about 0.2 MeV.

In all of our previous articles, we have taken Λ_π = 700 MeV. Since this value reproduce the triton binding energy and also since there is a relation $\Lambda_\pi \leq \Lambda_A$ for Λ_A = 730 MeV [28], this value of Λ_π might be a suitable value. On the other hand, $\Lambda_\pi \approx 5.8m \simeq 800$ MeV($m_{\pi c}$ =(1/3)(m_{π^0} + 2$m_{\pi\pm}$)) is often used [16],[5,6]. In this case, the triton is overbound. Therefore, if we use 800 MeV for Λ_π, we should introduce a repulsive force to cancel partly the attractive effect of the $\pi\pi$ exchange 3NP. A simple model by Coon et al. [29], shows that the 3NP that involves the rho-meson exchange give rise to a repulsive effect.

We have solved the 52-channel Faddeev equations and determined so that the triton binding energy is obtained [27]. Hereby, we took AV14 as the 2NP, and included the new Brazilian $\pi\pi$ exchange [32], $\pi\rho$ and $\rho\rho$ exchange 3NP [33,34] as well as CIB effect [23].

As a result, we obtained Λ_ρ = 700 MeV for Λ_π = 730 MeV. Since this is too small, we abandon this pair of cut-off masses. If we take the parameter κ for the vector to tensor ratio in the ρNN Lagrangian to be 0.3706 (weak coupling), we obtain the triton binding energy of 8.462 MeV. However, the Λ_ρ exceeds 2.0 GeV when we take account of the CIB effect. Therefore, we dismiss the weak coupling case and keep the strong coupling case (κ = 6.1). Our result is shown in the following table.

The set (I) yields the triton binding energy if we neglect the CIB effect. It is known that the two nucleon scattering data are reproduced, if we add the form factor of the cutoff mass Λ_π'(=1.2 GeV) to the form factor with this set of cutoff masses [30]. The fact that we don't need the form factor with a large cutoff mass means that a nucleon is swelling up in the triton than in free space. This is interesting. However, we don't adapt this set of cutoff masses, since the triton is underbound.

Table 2

	Λ_π(GeV)	Λ_ρ(GeV)	B E (MeV) (without CIB)	(with CIB)	GT/$\sqrt{3}$
(I)	0.8	1.3	8.413	8.189	0.945
(II)	0.81	1.13	8.717	8.485	0.955
(III)	0.85	1.35	8.711	8.479	0.948

The πNN cutoff mass in the set (III) is given by a soliton model [32]. It is interesting that the set (III) reproduces the triton binding energy. We list in the last column of table 2 the Gamov-Teller matrix element of the triton beta-decay divided by $\sqrt{3}$. Since the experimental value is 0.961 $\pm$ 0.003, the set (II) is most favorable.

Let BR$_{\pi\pi}$ represent the new Brazilian 2π exchange 3NP, K.R. the Kroll-Ruderman term and W the sum of the three-body force operators,

$$W = BR_{\pi\pi} + (\pi\rho)_\Delta + (\rho\rho)_\Delta + K.R.$$

If we let Ψ denote the three-body wave function, $\langle\Psi|W|\Psi\rangle$ is divided in the case (II),

$$(BR)_{\pi\pi} : (\pi\rho)_\Delta : (\rho\rho)_\Delta : K.R. = -2.268:0.286:0.006:0.148$$

If we call the sum of $(\pi\rho)_\Delta$ and K.R. the $\pi\rho$ exchange three-body force in the narrow sense,

$$(\pi\pi) : (\pi\rho) : (\rho\rho) = -2.0 : 0.4 : 0.005$$

From table 1, 3NP contributes 1.044 MeV to the triton binding energy. Therefore, the contributions from various exchange processes of 3NP are

$$(\pi\pi), (\pi\rho)_\Delta, (\rho\rho)_\Delta, K.R. = -1.298, 0.164, 0.003, 0.085 \text{ (in MeV)}$$

This result shows that heavier meson effects to the three-nucleon force should be negligible.

REFERENCES

1. Coon, S.A., Scadron, M.D., McNamee, P.C., Barret,B.R., Blatt, D.W.E., McKellar, B.H.J.: Nucl.Phys. A317, 242 (1979)

2. Ishikawa, S., Sasakawa, T., Sawada, T., Ueda, T.: Phys. Rev. Lett. 53, 1877 (1984)

3. Benayoun, J.J., Gignoux, C.:Nucl. Phys. A190, 419 (1972); Laverne, A., Gignoux, C.: Nucl. Phys. A203, 597 (1973)

4. Merkriev, S.P., Gignoux, C., Laverne, A: Ann. Phys.99, 30 (1976)

5. Chen, C., Payne, G.L., Friar, J.L., Gibson, B.F.: Phys. Rev. C 31, 2266 (1985)

6. Chen, C., Payne, J.L., Friar, J.L., Gibson, B.F.: Phys. Rev. C 33, 1740 (1973)

7. Kim, Y.E., Harper, E.P., Tubis, A.: Phys. Rev. C 7, 968 (1973)

8. Hadjuk Ch., Sauer, P.U.: Nucl. Phys. A322, 329 (1979)

9. Gloeckle, W., Hasberg, G., Neghabian, A.R.: Z.Phys.305A, 217 (1982)

10. Sasakawa, T., Okuno, H., Sawada, T.: Phys. Rev. C 23, 905 (1981)

11. Horacek, J., Sasakawa, T.: Czechoslovak J. Phys. B34, 1 (1984)

12. Horacek, J., Sasakawa, T.:Phys. Rev. A 28, 2151 (1983); ibid. A 30, 2274 (1984); ibid. C 32, 70 (1985)

13. Sasakawa, T., Sawada, T.: Phys. Rev. C 22, 320 (1980)

14. Sasakawa, T., Ishikawa, S.: Few-Body Systems 1 3 (1986)

15. Ishikawa, S., Sasakawa, T.: Few-Body systems 1 143 (1986)

16. Coon, S.A., Scadron, M.D.: Phys. Rev. C 23, 1150 (1981)

17. Sasakawa, T., Ishikawa, S: In Proceedings of the Tsukuba International Workshop on Deuteron Involving Reactions and Polarization Phenomena, ed. by Y. Aoki and K. Yagi (World Scientific, Singapore,1985), 341.

18. Ishikawa, S., Sasakawa, T.: Phys. Rev. Lett. 56 317 (1986)

19. Saito, T-Y., Ikeda, K., Ishikawa, S., Sasakawa, T: To appear in Phys. Rev. C, Rapid communication.

20. Wu, Y., Ishikawa, S., Sasakawa, T.: Manuscript in preparation.

21. Saito, T-Y., Wu, Y., Ishikawa, S., Sasakawa, T.: Phys. Lett. 242B, 12 (1990)

22. Ishikawa, S., Sasakawa, T.: Phys. Rev.C 36, 2037 (1987)

23. Wu, Y.: Master Thesis, Tohoku Univ. (1989)

24. Ishikawa, S., Sasakawa, T.: Submitted to Phys. Rev. $\underline{C}$, Rapid communication.

25. Schori, O. et al.: Phys. Rev. $\underline{C}$ $\underline{35}$, 2252 (1987)

26. Wu, Y., Ishikawa, S., Sasakawa, T.: Phys. Rev. Lett. $\underline{64}$, 1875 (1990); $\underline{66}$, 242 1991

27. Sasakawa, T., Ishikawa, S., Wu, Y., Saito, T-Y.: Submitted to Phys. Rev. Lett. for publication.

28. Thomas, A.W., Holinde, K.: Phys. Rev. Lett. $\underline{63}$, 2025 (1989)

29. Coon, S.A., Pena, M.T., Ellis, R.G.: Phys. Rev. $\underline{C}$ $\underline{30}$ 1366 (1984)

30. Holinde, K., Thomas, A.W.: Phys. Rev. $\underline{C}$ $\underline{42}$ R1195 (1990)

31. Meissner U.G. et al., Phys. Rev. Lett. $\underline{57}$, 1676 (1986); Weise, W.: Prog. Theoret. Phys. Supplement $\underline{91}$, 99 (1987)

32. Robilotta, M.R., Coelho, H.T.: Nucl. Phys. $\underline{A460}$, 645 (1986)

33. Robilotta, M.R., Isidro Filho, M.P.: Nucl. Phys. $\underline{A414}$, 394 (1984)

34. Matzolff, M., Loiseau, B., Grange, P.: Phys. Lett. $\underline{92B}$, 46 (1980)

Few-Body Systems, Suppl. 6, 265—269 (1992)

Few-
Body
Systems
© by Springer-Verlag 1992

BREAK-UP AND CONTINUUM STATES OF
FEW-NUCLEON SYSTEMS

W. Plessas

Institute for Theoretical Physics

University of Graz

Universitätsplatz 5, A–8010 Graz, Austria

The last few years are marked by big advances in the investigation of few-nucleon systems. Finally one of the most important aims of few-nucleon theory could have been accomplished, namely, testing modern realistic NN potentials not only in bound states but, notably, also in continuum and break-up states of the 3-N system. Before, only limited calculations could be performed that had to advocate restricting approximations or needed to rely on simplified NN forces, such as (low-rank) separable potentials. As a consequence, evidence on the fundamental dynamics in the 3-N system was hindered, and the conclusions drawn were often not really decisive. The recent successes were achieved in two different ways:

(A) The separable representation of any given NN interaction in an on-shell- and off-shell-equivalent manner and the subsequent solution of the Faddeev equations, as a coupled system of **one**-dimensional integral equations.

(B) The direct solution of the Faddeev equations, as a coupled system of **two**-dimensional integral equations for any NN interaction.

E.g., for the Paris NN potential [1] reliable predictions for 3-N scattering observables were, for the first time, obtained along approach A by the Graz-Osaka collaboration [2]; for a more detailed outline of the methods, see also refs. [3,4]. These results were subsequently confirmed in calculations along approach B by the Bochum group [5]. Though the two methods A and B solve equations of distinct type and rely on completely different and independent computer codes, the obtained results agreed within 1% [6]. As this was achieved in the rather complicated case of the Paris NN potential, this accuracy also represents the standard by which any new method for solving the 3-N (scattering) system has to be measured.

The main **findings** of a series of recent investigations of the 3-N continuum are [7–9]:

f1) Meson-exchange potentials generally yield a rather good agreement with experiment for elastic scattering observables except for the nucleon analyzing power A_y.

f2) The same 2-N forces are left with some intriguing discrepancies in certain configurations of the break-up process.

f3) The theoretical framework seems to be reliable in the range of incident nucleon energies up to $E_N \approx 50 - 60 \, MeV$.

f4) The influence of 3-N forces is hitherto unknown quantitatively.

All of the above-mentioned calculations were carried out for n-d scattering, i.e. without Coulomb forces. So far there is only one study of p-d scattering with inclusion of the Coulomb interaction and realistic NN forces [10]. The calculation is restricted to below the break-up threshold; the predictions of the Paris potential derived from it show a nice agreement with experimental data at $E_p = 1.5 - 3.0 \, MeV$.

Studies going on at present are mainly concerned with clarifying the discrepancies leftover [11]. They possibly originate from the following **problems**

p1) an inaccurate adjustment of parametric ingredients in meson-exchange NN potentials, afflicting their on-shell properties,

p2) an insufficient determination of phenomenological pp and np phase shifts (question of charge asymmetry, for example),

p3) missing 3-N forces,

p4) uncertainties connected with the NN off-shell behaviour.

Especially the latter two points represent a delicate dilemma, as they are connected to each other; the corresponding effects cannot unambiguously be disentangled by means of the 3-N system alone [12]. First, the off-shell properties have to be pinned down, by such reactions as $e - d$, $\gamma - d$ etc. [13,14]. Only then can the influences of 3-N forces be calculated reliably and their magnitudes be established. We mention that the trinucleon bound states are concerned with the same kind of problems [15,16].

Beyond the studies trying to reveal the successes and shortcomings of modern (meson-exchange) NN dynamics on a broad front [11,14,17] with established methods, still new techniques are searched for and attempted to be developed into a competitive stage. In a contribution to this conference [18] the W-matrix approach, presented before along simplistic NN potentials [19], is made applicable to the Paris potential. In spite of first promising results, it remains to be demonstrated, however, that the same reliability can be achieved as in other separable-approximation schemes, in particular the one of approach A mentioned above [2,4,6,20].

In another contribution [21], a new method is presented for the solution of the 3-body Faddeev equations in configuration space. So far the authors have only tested its applicability in low-energy n-d scattering using the Malfliet-Tjon I–III potential. For this method too, it remains to be shown that it is able to compete with alternative approaches.

In addition to the static properties of the bound trinucleons, investigations of their wave functions have recently attracted (again) considerable attention. Like in the case of the deuteron [14], the electromagnetic form factors of both ^3H and ^3He at present cannot really be explained by the most-refined calculations employing realistic meson-exchange models (such as the Paris and Bonn potentials) for momentum transfers $q \gtrsim 4\ fm^{-1}$ [22], not even with inclusion of 3-N forces [23]. Studies of other electromagnetic processes than electron scattering are therefore welcome to provide additional tests of the trinucleon wave functions. At this conference two contributions were presented dealing with the radiative capture reaction $p + d \rightarrow {}^3He + \gamma$ [24,25]. Both calculations used different dynamical inputs and were performed at distinct energies. No direct comparison of the results is therefore possible. While Ishikawa uses trinucleon wave functions calculated with local OBEP-like or OBEP potentials (Argonne v_{14}, Bonn OBEPR, ...) and including the Brazil 3-N force [26], Fonseca and Lehman, in this instance, are limited to phenomenological separable potentials. Despite the supposedly inferior dynamical ingredients, their study is nevertheless interesting, as it demonstrates in detail the influences of NN P-waves and the effect of the deuteron D-state probability on the capture cross section and tensor analyzing power. Meson-exchange-current effects are included in both investigations via Siegert's theorem; whereas Ishikawa takes into account E1 and E2 multipole operators, Fonseca and Lehman only consider E1 transitions, claiming that higher multipoles are negligible for the process at low (photon) energies [26]. One may expect that future studies will remedy still existing shortcomings in the theoretical description of $p + d \rightarrow {}^3He + \gamma$ reaction.

The same kind of reaction, though for the inverse photodisintegration process, has been considered for the 4-N system by Sofianos et al. [27]. They have made use of an improved description of the 4-N wave functions employing their integro-differential-equation approach [28]. The corresponding results are quite similar to the ones found before with the integral-equation technique [29] and they likewise reproduce modern experimental data. By including the final-state interaction via an optical-potential treatment, even the recent low-energy data could be described. No evidence was found for a giant-resonance behaviour as suggested by resonating-group calculations and earlier experiments.

We may **summarize** this session to have clearly reflected the actual status of few-nucleon investigations, namely:

s1) The solution of the 3-N problem for both bound and scattering states has become well under control. Even the most-refined NN interactions can be applied (as long as they are not explicitly energy-dependent) in ^3H and all n-d reactions. Rigorous inclusion of Coulomb forces is limited to ^3He and p-d scattering below break-up threshold.

s2) Meson-exchange NN potentials have led to a number of improvements as compared to earlier phenomenological interactions. Some discrepancies are left that suggest, in the first instance, new attempts of fine-tuning the ingredients of meson-theoretical NN models. Taking into account present-day experience from electromagnetic processes on the deuteron, one may in addition think of including non-nucleonic, but hadronic, degrees of freedom, such as isobar excitation etc.

s3) In the same spirit, effects of 3-N forces have to be taken into account. Of particular importance is their construction (parametrization), consistent with the 2-N models employed. The entanglement of effects from 3-N forces and NN off-shell properties constitutes a principle difficulty.

s4) Beyond N-d scattering, studies providing access to the 3-N bound-state and continuum wave functions become increasingly more important. The treatment of electromagnetic processes involving the 3-N system needs further efforts in several respects (usage of more-elaborate NN forces, of additional pieces in the electromagnetic current operators etc.).

s5) The 4-N system is still confronted with major technical difficulties preventing the application of realistic NN interactions in the rigorous solution of the 4-body equations.

s6) The improvement of previous methods and the invention of new techniques has been fruitful and is still welcome. The two, mostly alternative, ways of i) finding **reliable** approximation schemes and of ii) managing **considerably enlarged** numerical computations have shown similar successes. Desirably, the competition seems to go on for the benefit of the field.

References

1. Lacombe, M. et al.: Phys. Rev. C **21**, 861 (1980).
2. Koike, Y., Haidenbauer, J., and Plessas, W.: Phys. Rev. C **35**, 396 (1987).
3. Koike, Y. and Taniguchi, Y.: Few-Body Systems **1**, 13 (1986).
4. Plessas, W. in: Few-Body Methods: Principles and Applications, ed. by T.K. Lim et al. Singapore: World Scientific, 1986; Lecture Notes in Physics **273**, 137 (1987).
5. Witala, H., Glöckle, W., and Cornelius, T.: Few-Body Systems Suppl. **2**, 555 (1987); Witala, H., Cornelius, T., and Glöckle, W.: Few-Body Systems **3**, 123 (1988); Glöckle, W.: Lecture Notes in Physics **273**, 3 (1973).
6. Cornelius, T. et al.: Phys. Rev. C **41**, 2538 (1990).
7. Plessas, W. and Haidenbauer, J.: Few-Body Systems Suppl. **2**, 185 (1987); Plessas, W. in: Proceedings of the 4th Workshop on Perspectives in Nuclear Physics at Intermediate Energies, ed. by S. Boffi et al. Singapore: World Scientific, 1989.

8. Witala, H., Cornelius , T., and Glöckle, W.: Few-Body Systems **5**, 89 (1988); Witala, H., Glöckle, W., and Cornelius, T.: Few-Body Systems **6**, 79 (1989); Phys. Rev. C **39**, 384 (1989); Nucl. Phys. **A491**, 157 (1989); ibid. **A496**, 446 (1989); Glöckle, W., Witala, H., and Cornelius, T.: Nucl. Phys. **A508**, 115c (1990); in: Selected Topics in Nuclear Structure, ed. by J. Styczen and Z. Stachura. Singapore: World Scientific, 1990; Witala, H., Glöckle, W., and Kamada, H.: Phys. Rev. C **43**, 1619 (1991); Witala, H. and Glöckle, W.: Nucl. Phys. **A528**, 48 (1991).

9. Howell, C.R. et al.: Few-Body Systems **2**, 19 (1987); Phys. Rev. Lett. **61**, 1565 (1988); Rauprich, G. et al.: Few-Body Systems **5**, 67 (1988); Stephan, M. et al.: Phys. Rev. C **39**, 2133 (1989); Strate, J. et al.: Nucl. Phys. **A501**, 51 (1989); Cub, J. et al.: Few-Body Systems **6** , 151 (1989); Rauprich, G. et al.: Nucl. Phys. A, to appear.

10. Berthold, G., Stadler, A., and Zankel, H.: Phys. Rev. Lett. **61**, 1077 (1988); Phys. Rev. C **41**, 1365 (1990).

11. Witala, G. and Glöckle W.: These Proceedings.

12. Polyzou, W.N. and Glöckle, W.: Few-Body Systems **9**, 97 (1990).

13. Plessas, W. in: Selected Topics in Nuclear Structure ed. by J. Styczen and Z. Stachura. Singapore: World Scientific, 1990.

14. Pauschenwein, J., Mathelitsch, L., and Plessas, W.: These Proceedings.

15. Sasakawa, T. and Ishikawa, S.: Few-Body Systems **1**, 3 (1986); Ishikawa, S. and Sasakawa, T.: ibid. **1**, 143 (1986).

16. Gibson, B.F. and McKellar, B.H.J.: Few-Body Systems **3**, 143 (1989)

17. Sasakawa, T.: These Proceedings.

18. Sandhas, W. et al.: These Proceedings.

19. Bartnik, E.A., Haberzettl, H., and Sandhas, W.: Phys. Rev. C **34**, 1520 (1986); Haberzettl, H.: Phys. Rev. C **40** 1147 (1989).

20. Parke, W.C. et al.: Few-Body Systems, to appear.

21. Carbonell, J., Gignoux, C., and Merkuriev, S.P.: These Proceedings.

22. Henning, H., Adam Jr., J., and Sauer, P.U.: Few-Body Systems Suppl., to appear.

23. Schiavilla, R., Pandharipande, V.R., and Riska, D.O.: Phys. Rev. C **40**, 2294 (1989); ibid. **41**, 309 (1990); Schiavilla, R. and Riska, D.O.: Phys. Lett. **B244**, 373 (1990).

24. Fonseca, A.C. and Lehman, D.R.: These Proceedings.

25. Ishikawa, S.: These Proceedings.

26. Gibson, B.F.: Lecture Notes in Physics **273**, 548 (1987).

27. Sofianos, S.A. et al.: These Proceedings.

28. Fabre de la Ripelle, M., Fiedeldey, H., and Sofianos, S.A.: Few-Body Systems Suppl. **2**, 488 (1987); Few-Body Systems **6**, 157 (1989); Oehm, W. et al.: Phys. Rev. C **42**, 2322 (1990); ibid. **43**, 25 (1991); ibid. **44**, 81 (1991).

29. Casel, A. and Sandhas, W.: Czech. J. Phys. **B36**, 301 (1986).

Few-Body Systems, Suppl. 6, 270—275 (1992)

Few-
Body
Systems

THREE-NUCLEON BREAK-UP PROCESSES

T. N. Frank, H. Haberzettl[†], and W. Sandhas

Physikalisches Institut der Universität Bonn, D-5300 Bonn, Germany

[†]Department of Physics, George Washington University, Washington D.C., USA

Neutron-deuteron break-up results obtained within the framework of the AGS theory are presented. As interactions we use the original and a charge-dependent modification of the Paris potential. It is shown that comparing idealized single-configuration calculations with experimental data may lead to erroneous conclusions. A reliable analysis of a given experimental situation requires about 400 neighbouring configurations in order to simulate the experimental energy and angle resolutions. In view of the resulting huge demand on computational resources, the simplifying yet still very accurate W-matrix method is seen to be an algorithm particularly well-suited for such realistic analyses.

1. AGS formalism and conventional solution methods

The transition operators $U_{\beta\alpha}$ corresponding to the rearrangement collisions $\alpha + (\beta\gamma) \longrightarrow \beta + (\gamma\alpha)$ are determined in the AGS theory by the set of integral equations

$$U_{\beta\alpha} = (1-\delta_{\beta\alpha}) \, G_0^{-1} + \sum_{\gamma \neq \beta} T_\gamma \, G_0 \, U_{\gamma\alpha} \, . \qquad (1)$$

For the break-up operator $U_{0\alpha}$ associated with the process $\alpha + (\beta\gamma) \longrightarrow \alpha + \beta + \gamma$ we have

$$U_{0\alpha} = G_0^{-1} + \sum_\gamma T_\gamma \, G_0 \, U_{\gamma\alpha} \, , \qquad (2)$$

a relation formally contained in eq. (1) by putting $\beta=0$.

Let us recall some properties of this approach [1]. Equation (1) exhibits the typical Faddeev coupling scheme, and the two-body input enters the kernel of eq. (1) via the subsystem transition operators T_γ rather than the underlying potentials V_γ. In contrast to the original Faddeev equations, the above relations, however, are well-defined in momentum space which made them the appropriate basis of almost all recent high-accuracy three-nucleon calculations. Of particular importance is the fact that the break-up operators are given by the *purely algebraic* relation (2); in the original Faddeev theory this situation is dealt with by means of *integral equations* for three-fragment scattering states. In momentum space, after partial wave decomposition, eq. (1) represents a set of two-dimensional integral equations, and the algebraic relation (2) contains a single integration over the magnitude of the momentum variable in subsystem γ.

In principle eq. (1) can be solved by employing a direct two-dimensional discretization [2]. The dimension of the resulting linear equations, however, is so high that with today's computing facilities they could be solved only in terms of Padé approximants. Moreover, a rather small - according to our experience too small - number of meshpoints had to be chosen. Without alternative reference calculations the accuracy of this approach, therefore, remains an open question, at least in the break-up case.

Almost all other approaches are based on separable expansions of the subsystem amplitudes T_γ in the kernel of eq. (1), a procedure suggested by the fact that the dominant contributions of this amplitude (bound state and resonance poles) are indeed of separable form. This "natural" discretization technique, thus, exploits one of the characteristic properties of Faddeev-type equations, the occurrence of the subsytem amplitudes in their kernel. A rather efficient method of this type, based on the EST expansion, was developed by the Graz-Koike collaboration [3]. It has led to first fairly reliable theoretical n-d results and, not surprising in view of the above arguments, appears to be a factor of ten less time consuming than the purely numerically motivated discretization technique used subsequently in [2]. The Graz-Koike approach, however, has not yet been applied to the break-up situation.

2. The W-matrix method

The W-matrix approach [4] represents an attempt to further reduce the numerical effort by explicitly taking into account even more of the subsystem information. Following this concept we gain another factor of 10, despite the fact that we employ a much higher (sufficiently high) number of meshpoints [5].

We recall that the W-matrix method is based on a splitting

$$T_\gamma = |W_\gamma\rangle \, \Delta_\gamma \, \langle W_\gamma| + R_\gamma \tag{3}$$

constructed such that the separable part contains the whole pole and cut structure of T_γ and satisfies two-body unitarity. Quite essential in our context is that R_γ vanishes half-on-shell. This implies that but the separable part of eq. (3) enters the on-shell restriction of eq. (2). For the break-up amplitude, describing the transition from an initial two-fragment channel $\alpha = (\alpha, \beta\gamma)$ into a three-particle configuration with relative momenta p', q', we therefore have the *exact* on-shell representation

$$\langle p', q'| \, U_{0\alpha} \, G_0 \, |W_\alpha\rangle |q_\alpha\rangle = \sum_\gamma \langle p_\gamma|W_\gamma\rangle \, \Delta_\gamma \langle q_\gamma| \langle W_\gamma| \, G_0 \, U_{\gamma\alpha} \, G_0 \, |W_\alpha\rangle |q_\alpha\rangle. \tag{4}$$

Making use of the explicit form of Δ_γ and denoting the break-up and rearrangement amplitudes by $\mathcal{T}_{0\alpha}$ and $\mathcal{T}_{\gamma\alpha}$, respectively, this equation goes over into

$$\mathcal{T}_{0\alpha}(q', p'; q_\alpha) = \sum_\gamma \frac{1}{F_\gamma(p'_\gamma)} \, \mathcal{T}_{\gamma\alpha}(q'_\gamma, q_\alpha) \,. \tag{5}$$

The on-shell break-up amplitude, thus, is represented as a superposition of the half-on-shell rearrangement amplitudes $\mathcal{T}_{\gamma\alpha}$ multiplied with the inverse of the Jost functions $F_\gamma(p'_\gamma)$ or, in more general cases, of Jost matrices. Let us once more emphasize that this remarkable, extremely simple result is a consequence of characteristic aspects of the AGS theory and the W-matrix technique: the algebraic relation (2) and the half-on-shell properties of the representation (3).

Inserting the separable part of the W-matrix representation (3) also in eq. (1), we end up with the one-dimensional set of integral equations

$$\mathcal{T}_{\beta\alpha}(q'_\beta, q_\alpha) = V_{\beta\alpha}(q'_\beta, q_\alpha) + \sum_\gamma \int q'^2_\gamma \, dq'_\gamma \, V_{\beta\gamma}(q'_\beta, q'_\gamma) \, \Delta_\gamma(q'_\gamma) \, \mathcal{T}_{\gamma\alpha}(q'_\gamma, q_\alpha), \tag{6}$$

providing us with the rearrangement amplitudes entering the r.h.s. of eq. (5). This is, of course, an approximate relation. However, adjusting a parameter governing the splitting (3), allows us to minimize the remainder R_γ. Moreover, as was tested by varying the two-body input, the summation in eq. (5) effectively smears out the uncertainties of the solutions of eq. (6). The break-up results, in fact, turned out to be very insensitive to these variations in particular in configurations dominated by final state interaction (FSI). These investigations showed that even in the worst cases the accuracy of our calculations is still about 2-3%.

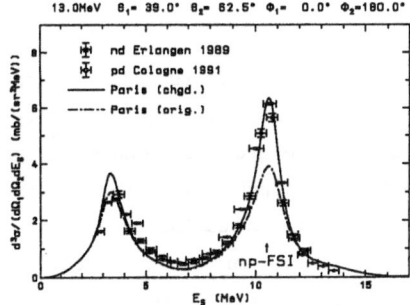

Fig.1 FSI configuration.

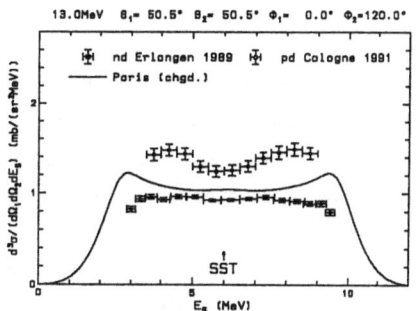

Fig.2 Space-star configuration.

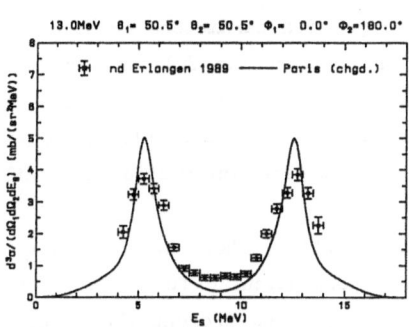

Fig.3 Coplanar configuration corresponding to SST.

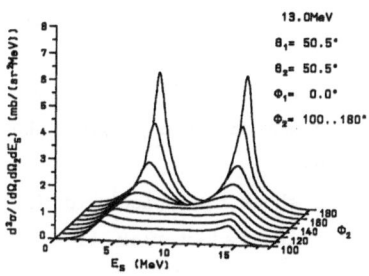

Fig.4 Steady transition from SST to coplanar configuration.

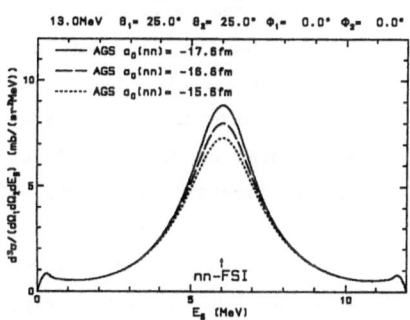

Fig.5 Variation of a_0(nn).

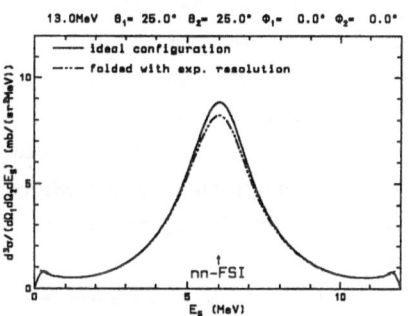

Fig.6 Effect of the experimental resolution.

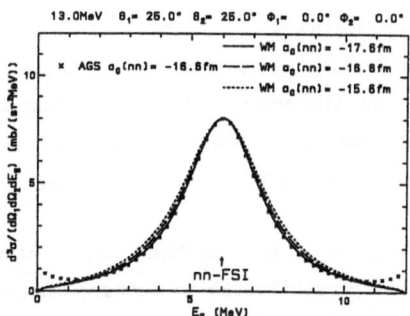

Fig.7 Comparison with Watson-Migdal model.

3. Results

For a realistic description of break-up cross sections, the charge dependence of the two-body interaction has to be included. Experimentally, the only obvious manifestation of charge dependence are differences of the scattering lengths $a_0(nn) = a_0(pp) = -17.6$ fm and $a_0(np) = -23.7$ fm, respectively. Fig. 1 shows a characteristic break-up situation treated with the original charge-independent (dash-dotted line), and with a modified charge-dependent version (solid line) of the Paris potential. This modification is obtained by replacing the $^1S_0$ contributions of the Paris potential by expressions of Malfliet-Tjon type with parameters adjusted to $a_0(pp)$ and $a_0(np)$. [The consistency of this modification follows from the fact that choosing in the np-interaction the pp-parameters we reproduce the Paris potential results]. The large differences between the resulting curves obviously necessitate employing the charge-dependent interaction for all calculations.

Fig. 2 shows the space-star (SST) configuration, where in the CMS all three particle trajectories lie in a plane perpendicular to the incident beams with angles of 120° between them. In fig. 3 we present the results obtained when changing the angle between the two neutrons from 120° to 180°, i.e., when turning the whole situation into a coplanar one, a configuration with rather pronounced FSI-peaks. It is quite instructive to study the transition from the flat curve of fig. 2 to the highly structured shape of fig. 3, a transition shown in steps of 10° in fig. 4. Bearing in mind that the uncertainty in the angle resolution may be as high as ±20°, we infer from these results that in the SST configuration experimentally one picks up quite a bit of the spurs of the FSI peaks. In FSI configurations, by contrast, a flattening of the curves due to the incorporation of lower neighbouring data is to be expected. This explains, at least qualitatively, the discrepancies between the shape of our idealized single-configuration calculations and the experimental curves of the Erlangen group [6] shown in figs. 2 and 3 (fig. 2 contains also the data of [7]). For a quantitative understanding of this situation it is, therefore, necessary to average a set of neighbouring curves over the experimental energy and angle distributions. According to our experience a reliable analysis of a given kinematical situation requires taking into account about 400 neighbouring configurations.

A careful investigation of the nn final state peak in nd scattering represents one of the few possibilities to measure the nn scattering length $a_0(nn)$. For this purpose a full three-body calculation appears crucial. Fig. 5 shows the variation of the nn-FSI peak for different values of $a_0(nn)$. In fig. 6 we demonstrate that folding with the experimental resolution yields a variation of the same magnitude which makes the relevance of the latter step particularly evident. An analysis of the data of the Erlangen group is in progress and will be reported in the near future [8].

For completeness we recall that in the conventional Watson-Migdal (WM) analysis the peaks are normalized to the same height leaving only their shapes to distinguish between different scattering lengths. Comparison of fig. 7 with fig. 5 exhibits the reduced sensitivity of the WM-procedure rather clearly. The cross section obtained for $a_0(nn)=-16.6\,fm$ within the full three-body AGS theory (crosses in fig.7), in fact, agrees almost completely with the $-17.6\,fm$ WM curve. An error of about $1\,fm$, therefore, is to be expected when analyzing FSI data by means of the WM model.

References

[1] E.O. Alt, P. Grassberger, and W. Sandhas, Nucl. Phys. **B2**, 167 (1967); W.Sandhas, in Elementary Particle Physics, ed.: P. Urban (Acta Physica Austriaca, Suppl. IX, 57 (1972)).

[2] H. Witała, contributions to this conference and references quoted therein.

[3] J. Haidenbauer and W. Plessas, Phys. Rev. C27, 63 (1983); J. Haidenbauer and Y. Koike, Phys. Rev. C34, 1187 (1986).

[4] E.A. Bartnik, H. Haberzettl, and W. Sandhas, Phys. Rev. C34, 1520 (1986); H. Haberzettl, Phys. Rev. C40, 1147 (1989); E.A. Bartnik, H. Haberzettl, Th. Januschke, U. Kerwath, and W. Sandhas,Phys. Rev. C36, 1678 (1987); T.N. Frank, H. Haberzettl, Th. Januschke, U. Kerwath, and W. Sandhas, Phys. Rev. C38, 1112 (1988); Th. Januschke, PhD Thesis Universität Bonn (1991); Th.Januschke, T.N.Frank, W.Sandhas, and H.Haberzettl, submitted to Phys. Rev. C.

[5] T.N. Frank, PhD Thesis Universität Bonn (1991); T.N. Frank, Th. Januschke, W. Sandhas, and H. Haberzettl, submitted to Phys. Rev. C.

[6] J. Strate, K. Geissdörfer, R. Lin, W. Bielmeier, J. Cub, A. Ebneth, E. Finckh, H. Friess, G. Fuchs, K. Gebhardt, and S. Schindler, Nucl. Phys. **A501**, 51 (1989).

[7] G. Rauprich et al., to appear in Nucl. Phys. A.

[8] E. Finckh, K. Gebhardt, W. Jäger, C. Jeitner, A. Schaller, J. Strate, T.N. Frank, Th. Januschke, W. Sandhas, and H. Haberzettl, to be published. Preliminary results were communicated at the Spring Meeting of the Deutsche Physikalische Gesellschaft, Darmstadt, Germany, 1991.

Few-Body Systems, Suppl. 6, 276—278 (1992)

Few-
Body
Systems
© by Springer-Verlag 1992

PHOTODISINTEGRATION OF ^{4}He

S. A. Sofianos, H. Fiedeldey, and W. Sandhas[†]

Department of Physics, University of South Africa, Pretoria, South Africa
[†]Physikalisches Institut, Universität Bonn, D-5300 Bonn 1, Germany

The photodisintegration of ^{4}He is calculated by employing recent integrodifferential equation techniques and an optical potential treatment of final state interaction. The results obtained reproduce the characteristic low-energy behaviour exhibited within the integral equation approach, and are in excellent agreement with latest experimental data also at higher energies.

Generalizing the four-nucleon AGS formalism, exact integral equations have been derived by Casel and Sandhas for the ^{4}He photodisintegration amplitude [1]. Applications of this formalism showed a considerable flattening [2] of the apparent giant resonance seen both in early measurements and subsequent resonating group (RGM) calculations [3]. Over the years a similar flattening was found also experimentally, so that recent data and the integral equation predictions are now in good agreement. Despite many still unavoidable approximations, the integral equation approach thus appears to be a rather reliable access to the problem. This method, on the other hand, is quite demanding and could not yet be applied for energies above the three-body break-up threshold. We, therefore, suggest and study a model based on two simplifications.

In calculating the three- and four-nucleon states $|\psi_{III}\rangle$ and $|\psi_{IV}\rangle$ in the Born amplitude

$$B(\vec{q}) = 2 \langle \vec{q} | \langle \psi_{III} | H_{em} | \psi_{IV} \rangle \tag{1}$$

the integrodifferential equation approach (IDEA), rather than the integral equation technique, is employed. While not fully exact, the IDEA is known to provide highly accurate bound state solutions [4]. As anticipated by this experience, the corresponding photodisintegration results (———) shown in fig. 2 coincide almost completely with the integral equation results (-----). Both curves, moreover, agree very well with the experimental data [5], which demonstrates in addition that at such energies the Born amplitude represents already a good approximation.

At low energies, the final state interaction (FSI), of course, has to be taken into account. In principle this is accomplished by going over in eq. (1) from the channel states $|\vec{q}\rangle|\psi_{III}\rangle$ to the corresponding scattering states $|\vec{q}; \psi_{III}\rangle^{(-)}$. To determine such four-body states requires the solution of quite complicated four-body (integral) equations. The essential simplification of our model consists in taking this effect into account via an optical potential treatment of the relative movement of the outgoing clusters. In other words, we use again an expression of the form (1), the plane waves $|\vec{q}\rangle$ being replaced by optical model scattering states $|\vec{q}\rangle^{(-)}$,

$$M(\vec{q}) = 2 {}^{(-)}\langle \vec{q} | \langle \psi_{III} | H_{em} | \psi_{IV} \rangle . \tag{2}$$

For the optical potential we have chosen the conventional form employed successfully in [6], but with new parameters being adjusted to the experimental p-wave phase shifts of ref. [7] which are consistent with the theoretical calculations of [8].

The corresponding photodisintegration results presented in fig. 1 show a low energy behaviour rather similar to the one obtained in the integral equation approach. They, moreover, are in excellent agreement with the latest data [9] also above the break-up threshold. In other words, we end up with a consistent picture of elastic scattering and photodisintegration.

Let us point out that in these investigations we have found a rather strong sensitivity on the optical potential parameters and, thus, on the phase shifts used in fitting them. With RGM phase shifts [10], we got, e.g., a giant resonance behaviour as in the direct RGM calculation of photodisintegration [3]. In other words, while internally consistent the RGM approach disagrees with recent experimental data (flat behaviour rather than a giant resonance). Beside its considerable technical simplifications, our model, therefore, provides a sensitive tool for judging the quality of optical potentials.

278

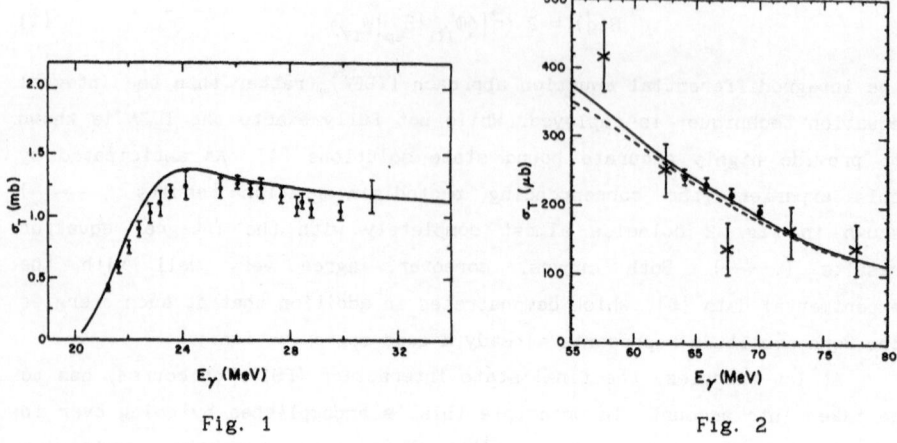

Fig. 1   Fig. 2

References

[1] A. Casel and W. Sandhas, Czech. J. of Phys. **B36**, 301 (1986) and refs. quoted therein.

[2] Going over from the simple separable interactions used in [1] to the Malfliet-Tjon (MT I + III) potential yields an even stronger flattening: S.A. Sofianos, H.Fiedeldey, H. Haberzettl, and W. Sandhas, Int. Workshop on Microscopic Methods in Few-Body Systems, Kalinin, 1988.

[3] B. Wachter, T. Mertelmeier, and H.M. Hofmann, Phys.Rev. **C38**, 1139 (1988).

[4] W. Oehm, S.A. Sofianos, H.Fiedeldey, and M. Fabre de la Ripelle, Phys. Rev. **C43**, 25 (1991).

[5] A.N. Gorbunov, Proc. of the P.N. Lebedev Institute, **71**, 1 (1976); R.T. Jones, Ph.D. thesis, Virginia Polytech Inst. and State Univ., Blacksburg, Virginia (1988).

[6] B.S. Podmore and H.S. Sherif, in Few-Body Problems in Nuclear and Particle Physics, ed.: R.J. Slobodrian et al., Laval Univ. Press, Quebec, 1975.

[7] T.A. Tombrello, Phys.Rev. **B138**, 40 (1965).

[8] J.A. Tjon, Phys. Lett. **63B**, 391 (1976); A.C. Fonseca, Few-Body Systems I, 69 (1986).

[9] G. Feldman, M.J. Balbes, L.H. Kramer, J.Z. Williams, H.R. Weller, and D.R. Tilley, Phys.Rev. **C42**, R1167 (1990).

[10] I. Reichstein, D.R. Thompson, and Y.C. Tang, Phys.Rev. **C3**, 2139 (1971).

Few-Body Systems, Suppl. 6, 279—284 (1992)

NN P-WAVE EFFECTS IN POLARIZED DEUTERON CAPTURE ON HYDROGEN: AN EXACT FADDEEV CALCULATION

A.C. Fonseca
Centro Fisica Nuclear
Av. Gama Pinto, 2
P-1699 Lisboa, Portugal

and

D.R. Lehman
Center for Nuclear Studies
Department of Physics
The George Washington University
Washington, D.C. 20052, USA

An exact Faddeev calculation of ^{1}H($\vec{d}$, γ)^{3}He has been carried out including NN P-wave interactions. It is found that the contributions from the NN P-wave interactions to the ^{3}He D-state have no more than a 3% effect on the tensor analyzing powers (TAP) at $E_d = 19.6$ MeV. However, the NN P-wave interactions in the initial-state continuum have a large effect (20% to 60% for $19.6 \leq E_d \leq 95$ MeV) on the magnitude of the TAP.

Motivated by existing measurements of tensor analyzing powers for radiative capture of polarized deuterons on protons [1-3] and the possibility of new measurements [4], plus the contrasting results of effective two-body model calculations [5-8] and the exact results of J. Torre [2], we have embarked on a program of trying to provide a comprehensive theoretical investigation of the observables in $\gamma + {}^3\mathrm{He}(^3\mathrm{H}) \leftrightarrow$ p(n) + d, with special emphasis on the polarization observables, e.g., for ^{1}H($\vec{d}$,γ)^{3}He. Our framework is that of <u>exact</u> three-body dynamics in both the initial and final states. In this talk, we report on the present stage of our program, where the NN interaction is obtained from separable potentials that reproduce well the low-energy NN data [9,10] in the $^1S_0$, $^3S_1$-$^3D_1$, $^1P_1$, $^3P_0$, $^3P_1$, and $^3P_2$ partial waves. The electromagnetic operator is limited to E1. We do not see either one of these simplifications as severely limiting our ability to extract the key dynamical understanding that is sought at this point. Ultimately, we plan to add the E2 operator, and to use realistic two-nucleon interactions like the Paris interaction through use of separable-expansion techniques like the EST method [11].

In more detail, we are concerned with the low-energy photonuclear reactions involving the A = 3 nuclei: $\gamma + {}^3\mathrm{He}(^3\mathrm{H}) \leftrightarrow$ p(n) + d. The amplitude of interest is

$$\mathcal{A} = \langle\, \Psi_{\vec{p}}^{(-)} \mid H_\gamma \mid \Psi_{3_\mathrm{He}} \,\rangle \,, \tag{1}$$

where the electromagnetic operator, H_γ, is treated perturbatively. The initial and final states are exact eigenstates of the three-nucleon Hamiltonian (no Coulomb), i.e.,

$$\text{H } \Psi_{3\text{He}} = -\text{B } \Psi_{3\text{He}} \tag{2}$$

and

$$\text{H } \Psi_{\vec{p}}^{(-)} = \left(\frac{3p^2}{4M} - \epsilon_d\right) \Psi_{\vec{p}}^{(-)} , \tag{3}$$

where M is the mass of a nucleon, $\vec{p}$ is the relative momentum of the proton and deuteron, and ϵ_d (B) is the binding energy of the deuteron (trinucleon).

Within the above framework, low-energy means that the incident photon has energy ≤ 50 MeV, or equivalently, incident deuteron energies ≤ 130 MeV. Therefore, the electromagnetic operator may be expanded into multipoles. It is known from the experimental angular distributions of $d\sigma(\theta)/d\Omega$ that the transition is primarily E1. E2 radiation is present in that it causes an E1-E2 interference effect in the angular distribution, i.e., an asymmetry about the center-of-mass angle $\theta = 90^0$, but its contribution to the total cross section $\sigma(E_\gamma)$ for $E_\gamma \leq 50$ MeV is negligible [12,5,13]. Specifically, the long-wavelength, Siegert form, of the E1 operator that we use is

$$\text{H}_\gamma = \frac{e}{2} \sum_{i=1}^{3} (\hat{\epsilon} \cdot \vec{r}_i) \tau_z^i , \tag{4}$$

where $\hat{\epsilon}$ is the polarization vector of the photon, $\vec{r}_i$ is the i-th nucleon center-of-mass coordinate, and τ_z^i is the i-th particle, z-component, isospin Pauli matrix.

The amplitude $\mathcal{A}$ is calculated by means of the formalism given by Gibson and Lehman (GL) [14]. In outline, $\mathcal{A}$ is written in terms of the three-body transition operators which connect particle-plus-correlated-pair states:

$$\mathcal{A} = B_t(z,\vec{p}) + \sum_{\nu=s}^{t} \int d^3p' \langle \vec{p} | X_{t\nu}(z) | \vec{p}' \rangle \tau_\nu \left(z - \frac{3p'^2}{4M}\right) B_\nu(z,\vec{p}') . \tag{5}$$

The first term in Eq. (5) is the photonuclear Born amplitude that corresponds to the transition amplitude from the ground state of ^{3}He to a free p + d state, where the relative motion is described by a plane wave with momentum $\vec{p}$. The second term is the rescattering contribution that involves an integral over the half-shell three-body amplitude X for the scattering of p + ν $\rightarrow$ p + d with $z = \frac{3p^2}{4M} - \epsilon_d + i\epsilon$ (ν being any correlated NN pair consistent with the underlying interactions), the intermediate ν-pair propagator τ, and the off-shell photonuclear amplitude for the transition from the ground state of ^{3}He to a free p + ν state. The present work departs from that of GL in that now the NN tensor force is included; specifically, the NN interaction is present in the $^1S_0$, $^3S_1$-$^3D_1$, $^1P_1$, $^3P_0$, $^3P_1$, and $^3P_2$ partial waves. This work also goes beyond that of Hendry and Phillips (HP) [15], where they calculated trinucleon photodisintegration with the NN tensor force present, but used a phenomenological ground-state wave function which has an incomplete D-state component in addition to the dominant S-state component. In the present work the ground state is an exact eigenstate of the same Hamiltonian as the continuum state. Our ground state has all possible L-S coupled components present: $^2S_{1/2}$, $^2P_{1/2}$, $^4P_{1/2}$, $^4D_{1/2}$. Therefore, the E1 operator connects the four L-S components of the trinucleon ground state to all possible allowed proton-deuteron final states with total angular momentum $J = \frac{1}{2}^{-}$ and $\frac{3}{2}^{-}$.

As indicated above, we consider NN interactions in the above-mentioned partial waves represented by one-term separable potentials fitted to low-energy NN observables. For the triplet interaction, we consider three sets of parameters [9] leading to 4%, 5.5%, and 7% D-state probability in the deuteron. The corresponding values for the triton binding energy, percentage D-state probability, and D- to S-wave asymptotic-normalization-constant ratio are given in Table 1, together with the asymptotic norm ratios for the deuteron.

Table 1

Properties of the deuteron and trinucleon ground-state wave functions.

P_d (%)	$\eta_d = \dfrac{C_D}{C_S}$	B (MeV)	Percentage D-state	$\eta_t = \dfrac{C_D}{C_S}$
		No P-waves		
4.0	0.0291	8.568	5.10	0.0506
5.5	0.0268	8.038	7.17	0.0432
7.0	0.0256	7.589	9.10	0.0385
		P-waves present		
4.0	0.0291	8.365	4.96	0.0489
5.5	0.0268	7.864	6.98	0.0420
7.0	0.0256	7.490	8.93	0.0377

Our aim is three-fold: (a) To study the sensitivity of photonuclear observables to the underlying NN tensor force, and in particular, to uncover the effect of the NN P-waves; (b) To study the effect of three-nucleon dynamics on the tensor observables (Phys. Lett., in press); (c) To make progress towards understanding three-nucleon photonuclear data from a microscopic viewpoint in order to shed light on the structure of the trinucleons. Previous work [6,7] on the photodisintegration of ^{3}He and the inverse reaction, seem to indicate that the observables are sensitive to the underlying NN tensor force or , in other words, to the D-state component of the ground state. This sensitivity is often related to the D to S asymptotic norm ratio in the trinucleon at fixed binding energy [8], or to the percentage D-state in the deuteron at fixed three-body binding and RMS radius [15]. These studies always imply a model for the trinucleon ground state wave function that is not an eigenstate of a given 3N Hamiltonian.

Although the percentage D-state (P_d) in the deuteron is not an observable, it is nevertheless a convenient parameter to characterize the NN tensor force. Therefore, in Fig. 1, we show the results of our exact calculations of A_{yy} at $E_d = 95$ MeV for increasing P_d. The two sets of curves correspond to calculations with and without the NN P-wave interactions. It is immediately evident that the presence of the NN P-wave interactions leads to a significantly different magnitude for A_{yy} relative to the case without the P-wave contributions. At this energy, the P-wave effect is on the order 60%, while varying P_d between 4 and 7% only affects A_{yy} by 20%. Although not shown here, we also find that the contribution of the NN P-waves is energy

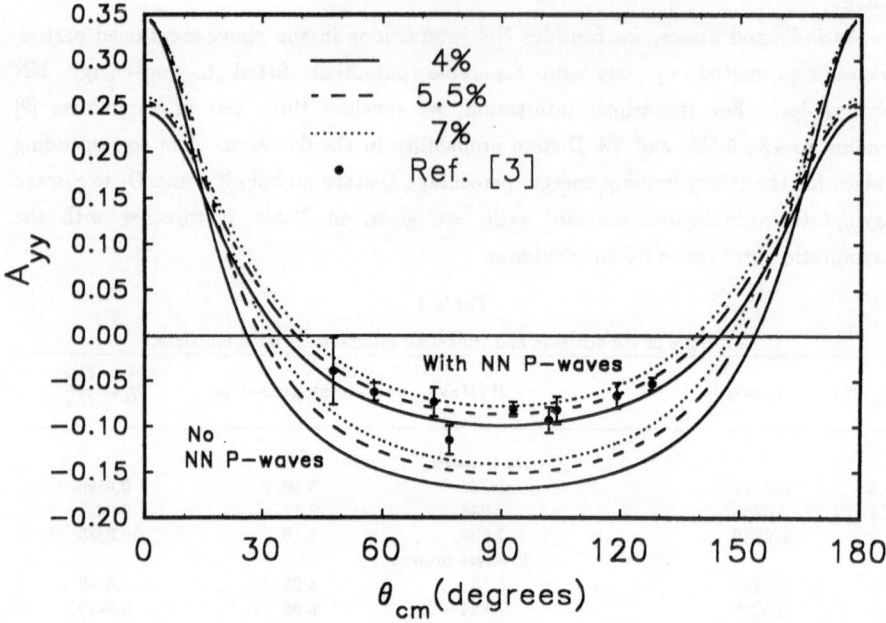

Fig. 1. Effect of the NN P-wave interactions on the Tensor Analyzing Power A_{yy} for varying P_d.

dependent and rises with E_d. Nevertheless, contrary to the claim of J. Jourdan et al., the effect of the NN P-waves may not be attributed to the D-state components of the $\Psi_{3_{He}}$ that are generated by their presence. Comparing lines 2 and 3 in Table 2, we show that by truncating all the components of $\Psi_{3_{He}}$ that are generated by the NN-waves, but keeping the initial-state unchanged, one gets basically the same results as when these components are present. On the other hand, comparing lines 1 and 2, one

Table 2

Effects of including NN P-waves on observables at $E_d = 19.6$ MeV ($P_d = 4\%$).

	$\frac{d\sigma}{d\Omega}(90^0)$ (μb/sr)		$T_{20}(90^0)$	
	Born	Full	Born	Full
NN P-waves absent	80.22	102.86	-0.119	-0.053
NN P-waves added	82.50	99.64	-0.117	-0.042
Same as directly above, but NN P-wave generated components of $\Psi_{3_{He}}$ truncated	82.81	99.84	-0.116	-0.041

clearly sees that by neglecting NN P-wave rescattering a change of 20% takes place in A_{yy}. We may further add that the contribution of the NN P-wave generated components of $\Psi_{3_{He}}$ to the capture observables is negligible in the whole energy range from threshold up to ~95 MeV. The changes coming from the NN P-waves derive mostly from their presence in the three-body continuum state which leads to additional contributions from the ground state even in the absence of NN P-wave generated components.

In Figs. 2 and 3, we show how A_{yy} (capture) and $d\sigma/d\Omega$ (photodisintegration) at 90^0 in the center-of-mass vary with energy for different P_d ranging from 4 to 7%. While the photodisintegration cross section is sensitive to P_d at the low-energy peak, the change in A_{yy} due to the variation of the tensor force rises with E_d. Although all the curves in these two figures represent calculations without NN P-wave interactions, we do not expect that their inclusion will change this pattern.

To summarize, one may conclude from Figs. 1 and 2 that sensitivity to the percent D-state probability in the deuteron is small and energy dependent. As shown in Table 1, different values of P_d correspond to a wide range of values for η_d and η_t, as well as for the percent D-state in the triton; therefore, it seems that one may not be able to draw an energy-independent relationship between tensor observables and the above mentioned properties associated with the D-states. This may be attributed to the dominance of initial-state rescattering (ISR) [18], which is responsible for significant effects on the tensor analyzing powers even at the higher energies considered, e.g., $E_d \sim 90$ MeV. For example, at $E_d = 74$ MeV, the ISR hardly affects the differential cross section, but is responsible for a change in sign of T_{20}. Further discussion of the effects of ISR, plus a presentation of how well previously used approximations handle the ISR compared to the exact results, can be found in Ref. [18].

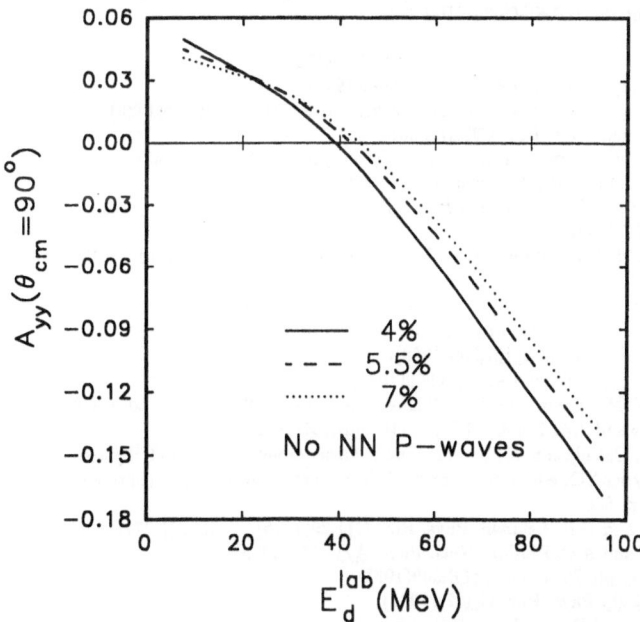

Fig. 2. Tensor Analyzing Power A_{yy} at 90^0 versus laboratory deuteron energy.

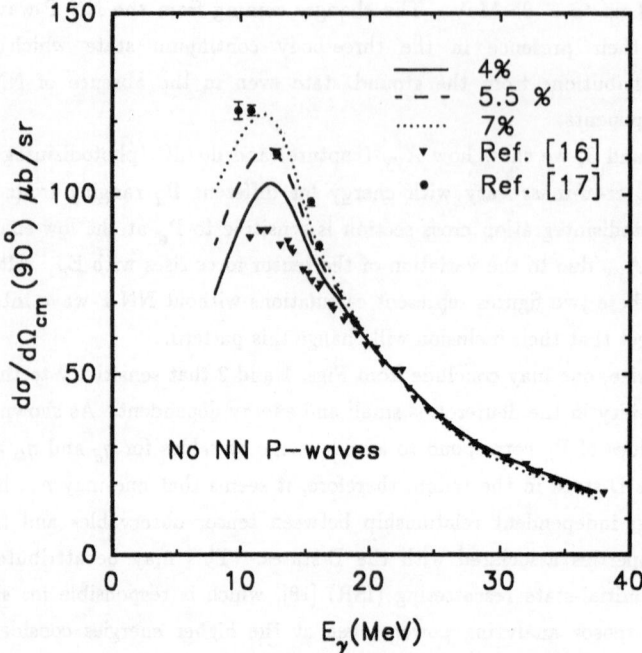

Fig. 3. Differential cross section at 90° for photodisintegration of ³He versus photon energy.

The authors extend their thanks to William C. Parke for his assistance in calculating some of the values in Table 1. This collaboration was facilitated by NATO International Scientific Exchange Research Grant No. 326/87. The work of DRL is supported in part by the U.S. Department of Energy under grant No. DE-FG05-86-ER40270.

REFERENCES

[1] M.C. Vetterli, et al., Phys. Rev. Lett. 54,1129(1985).

[2] J. Jourdan, et al., Nucl Phys. A453,220(1986); Phys. Lett. 162B,269(1985).

[3] W.K. Pitts, et al., Phys. Rev. C37,1(1988).

[4] W.K. Pitts, private communication; H.R. Weller, private communication.

[5] D.R. Lehman, Theoretical Status of Three-Nucleon Photonuclear Reactions, in: Lecture Notes in Physics 260, eds. B.L. Berman and B.F. Gibson (Springer-Verlag, New York, 1986) p. 287; Nucl. Phys. A463,117c(1987).

[6] H.R. Weller and D.R. Lehman, Ann. Rev. Nucl. Part. Sci. 38,563(1988); D.R. Lehman, Colloque de Physique 51,C6-47(1990).

[7] A.M. Eiró and F.D. Santos, Jour. of Phys. G 16,1139(1990).

[8] A. Arriaga and F.D. Santos, Phys. Rev. C29,1045(1984).

[9] A.C. Phillips, Nucl. Phys. A107,209(1968).

[10] F.D. Correll, et al., Phys. Rev. C23,960(1981).

[11] D.J. Ernst, C.M. Shakin, and R.M. Thaler, Phys. Rev. C8,46(1973); ibid, C9,1780 (1974).

[12] I.M. Barbour and J.A. Hendry, Phys. Lett. 38B,151(1972).

[13] B.F. Gibson, Calculation of Electromagnetic Observables in Few-Body Systems, in: Lecture Notes in Physics 273, eds. L.S. Ferreira, A.C. Fonseca, and L. Streit (Springer-Verlag, New York, 1987) p. 548.

[14] B.F. Gibson and D.R. Lehman, Phys. Rev. C11,29(1975); ibid, C13,477(1976).

[15] J.A. Hendry and A.C. Phillips, Nucl. Phys. A211,533(1973).

[16] G. Ticcioni, et al., Phys. Lett. 46B,369(1973).

[17] S.E. King, et al., Phys. Rev. C30,21(1984).

[18] A.C. Fonseca and D.R. Lehman, Phys. Lett., in press (1991).

Few-Body Systems, Suppl. 6, 285—290 (1992)

Few-
Body
Systems

P - D RADIATIVE CAPTURE WITH REALISTIC THREE-NUCLEON WAVE FUNCTIONS

S. Ishikawa

Institut für Theoretische Physik II, Ruhr-Universität Bochum

4630 Bochum, Germany

and

Department of Physics, Tohoku University

980 Sendai, Japan

Recent study on the proton-deuteron radiative capture reaction at low energies ($E_\gamma \cong 10$ MeV) is reported. For initial- and final three-nucleon states, solutions of the Faddeev equations are used with various combinations of available realistic nucleon-nucleon potentials and the two-pion exchange three-nucleon potential.

1. Introduction

Three-nucleon system has been playing an important role in nuclear physics as a good testing ground for nuclear interaction models. Recent progress in theoretical treatment of the three-body problem allows us to solve the three-nucleon Faddeev equations accurately not only for bound states, but also for continuum states with available realistic N-N potentials even with three-nucleon potentials. This enables us to perform consistent calculations for reactions such as breakup of three-nucleon bound state by photon, electron or muon, and their inverse capture reactions, in which we need both bound- and continuum states wave functions. I will report our recent calculations of the proton-deuteron radiative capture reaction ($p + d \rightarrow {}^3He + \gamma$) at low energies ($E_\gamma \cong 10$ MeV)[1].

2. Formalism

A transition operator for the p-d radiative capture reaction is constructed following the same procedure as Refs.[2, 3], in which we use the Siegert's theorem and the long wavelength approximation. By making use of the Siegert's theorem, we can include effects of meson exchange currents implicitly. For the initial- and final three-nucleon states, we solved the Faddeev equations numerically. This is consistent with the use of the Siegert's theorem which demands that the nuclear Hamiltonian for the initial- and final states should be the same. In our method, we express the Faddeev equation as a set of coupled integral equations in the coordinate representations, and solve it by an iterative method called as the Method of Continued Fractions[4, 5, 6]. We take into account E1 and E2 multipole operators, which are sufficient for our energy region. Since spin-parity (J_0^π) of the final state (^{3}He) is $\frac{1}{2}^+$, we need to calculate p-d states with J_0^π being $\frac{1}{2}^-, \frac{3}{2}^+, \frac{3}{2}^-$ and $\frac{5}{2}^+$. This contrasts with calculations of the nucleon-deuteron elastic or three-body breakup reactions, in which we need at least to calculate up to $\frac{15}{2}$ of J_0 to get convergent results.

3. Numerical Results

In the present calculation, we used seven kinds of realistic N-N potentials (2NP): the Argonne V14 model (AV_{14})[7], the super soft core potential models of de Tourreil-Sprung (model A,B and C) (dTS-A, -B and -C)[8] and de Tourreil-Rouben-Sprung (dTRS)[9], the soft-core potential of Reid (RSC)[10], and the r-space OBEP version of Bonn potential (BONN)[11]. Among them, first six potentials are known to underbind the triton by the order of 1 MeV compared to the experimental value of 8.48 MeV. In order to explain these differences, we should introduce a three-nucleon potential (3NP) to nuclear Hamiltonian. For 3NP, we used two-pion exchange model with a parameterization of Brazilian group[12]. The monopole-shape is taken for πNN vertex form factor with cut-off mass $\Lambda_\pi = 700$ MeV (BR_{700}). Three-body states with the maximun value of the total angular momentum of two-body subsystem being 2 are included in our calculations both for bound- and continuum states.

Below, we show our numerical results for various observables of the p-d radiative capture reaction from two points of view: effects of initial state interaction (ISI) to the Born approximation, and effects of different choice of nuclear interaction models (2NP and 3NP).

Total cross section

In Fig. 1, we plot the calculated values of the total cross section of the p-d capture reaction (σ_{tot}) at $E_p^{Lab} = 14.6$ MeV for the Born approximation (crosses) and the full calculation with ISI (full circles) against the calculated values of binding energy of three-nucleon bound state (B_3). Experimental value of σ_{tot}[13] is also plotted. The calculated values of σ_{tot} are correlated with those of B_3. Two straight lines in Fig. 1 are obtained

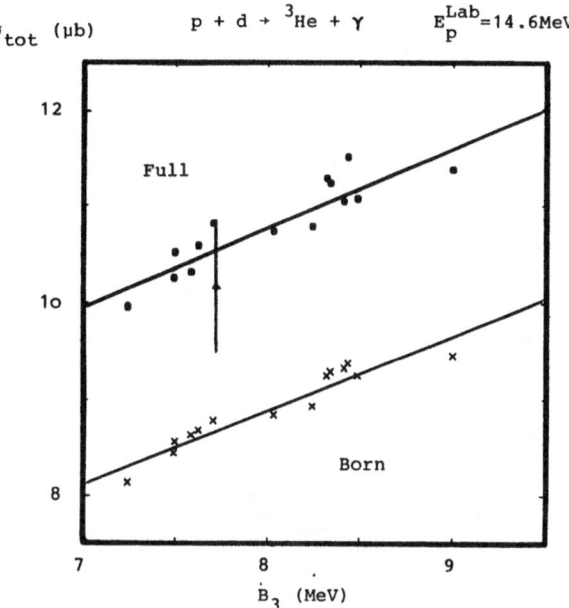

Figure 1: Total cross section for p-d radiative capture reaction at $E_p^{Lab} = 14.6$ MeV vs. trinucleon binding energy.

by the least square fitting. From this figure, we can see that ISI increases σ_{tot} by about 20%, and if we take the experimental ^{3}He binding energy (7.72 MeV), we can get σ_{tot} which is close to the experimental value.

The details of the partial wave contributions on σ_{tot} are listed in Table 1 for AV$_{14}$. This table shows that 98% of σ_{tot} comes from E1 ($J_0^\pi = \frac{1}{2}^-$, and $\frac{3}{2}^-$). While the effect of ISI to E1 is about 20%, one on E2 is negligibly small. Consequently, weight of E1 on σ_{tot} increases relatively by ISI.

Angular distributions

Usually, experimental angular distributions of the reaction p + d → ^{3}He + γ are fitted by a Legendre series;

$$\sigma(\theta) = A_0\{1 + \sum_{k=1}^{4} a_k P_k(cos\theta)\},\tag{1}$$

where θ is the angle of the outgoing photon.

King et al.[14] reported that the coefficient a_2 is sensitive to D-state probability in the ^{3}He wave function by effective two-body direct capture calculations with various separable interactions. We found that, against to this, not only calculated values of a_2 but also those of a_1, a_3 and a_4 are insensitive to different choice of 2NP and existence of 3NP. Further, we found that calculated values of a_k's are in good agreement with

Table 1: Total cross section for p-d radiative capture reaction for each partial wave at $E_p^{Lab} = 14.6$ MeV for AV_{14}.

J_0^π	Multipolarity	σ_{tot} (μb) Full	Born
$\frac{1}{2}^-$	E1	3.35	2.94
$\frac{3}{2}^+$	E2	0.07	0.07
$\frac{3}{2}^-$	E1	6.79	5.51
$\frac{5}{2}^+$	E2	0.11	0.11
	E1	10.13	8.45
	E2	0.18	0.18
	Sum	10.31	8.63

Table 2: The Legengre coefficients and the fore-aft asymmetry for AV_{14} at $E_p^{Lab} = 14.6$ MeV.

a_k	Full	Born	Exp.
a_1	0.343	0.378	0.33(1)
a_2	-0.914	-0.937	-0.89(1)
a_3	-0.337	-0.373	-0.29(2)
a_4	-0.029	-0.035	-0.08(1)
a_s	0.324	0.357	0.29(1)

available experimental values at energies around $E_p^{Lab} = 10$ MeV. As an example, we show our results of a_k's for AV_{14} at $E_p^{Lab} = 14.6$ MeV in Table 2 together with the experimental values[15, 16].

Experimental angular distributions are almost symmetry around $\theta = 90°$. Asymmetry is caused by an interference between E1 and E2. Thus, so called the fore-aft asymmetry (a_s) indicates the effect of E2. This parameter is defined as;

$$a_s \equiv \frac{\sigma(\theta_0) - \sigma(\pi - \theta_0)}{\sigma(\theta_0) + \sigma(\pi - \theta_0)}, \qquad (2)$$

where $\theta_0 (= \frac{1}{\sqrt{3}} rad. \cong 55°)$ is given by $P_2(\cos \theta_0) = 0$. The result of a_s for AV_{14} at $E_p^{Lab} = 14.6$ MeV is also presented in Table 2, which shows ISI acts to reduce the effect of E2.

Tensor analyzing power

Tensor analyzing powers of the reaction $p + \vec{d} \to {}^3He + \gamma$ were thought to be sensitive to the existing of D-state. Jourdan et al.[17] measured $A_{yy}(\theta)$ for $\theta_{Lab} = 90°$ at $E_p^{Lab} = 14.6$ MeV and 22.65 MeV. In the Ref.[17], a consistent calculation with three-nucleon wave functions solved for RSC has been performed for the first time. However, as they

Table 3: The tensor analyzing power $A_{yy}(\theta_{Lab}=90°)$ at $E_p^{Lab} = 14.6$ MeV.

| 2NP | $P_D(^2H)$ | 2NP | | 2NP+BR$_{700}$ | |
| | | Full | Born | Full | Born |
	(%)	(%)	(%)	(%)	(%)
AV$_{14}$	6.08	2.32	7.37	2.57	7.63
dTRS	5.92	2.29	7.18	2.39	7.35
dTS-A	4.43	1.98	7.27	2.13	7.55
dTS-B	4.25	2.01	6.70	2.03	6.77
dTS-C	5.45	2.28	6.75	2.24	6.80
RSC	6.47	2.16	7.20	2.41	7.48
BONN	4.81	2.38	7.45	2.27	7.46

use only RSC, it is worth while to perform calculations also for other potentials and also including three-nucleon potential.

In Table 3, we show the calculated values of A_{yy} for $\theta_{Lab}=90°$ at $E_p^{Lab} = 14.6$ MeV, together with the probability of the D-state in the deuteron ($P_D(^2H)$) for 2NP to be used. The ISI effect makes $A_{yy}(\theta)$ decrease by about 70%, that is consistent with the result of Ref.[17].

We can see that all 2NP's yield 15-30 % smaller values than the experimental value $(2.82 \pm 0.16$ %)[17]. It should be noted that BONN, which has smaller value of $P_D(^2H)$, yields the largest value for A_{yy} among the 2NP's that we used. This means that A_{yy} can not be determined simply by $P_D(^2H)$.

Introduction of 3NP makes calculated values of A_{yy} increase for almost all 2NP's except BONN. The increase of A_{yy} due to the inclusion of 3NP is large for 2NP with strong tensor component.

Among all models which we used, AV$_{14}$ with BR$_{700}$-3NP turns to be the best choice to reproduce the experimental value of $A_{yy}(\theta_{Lab}=90°)$ at $E_p^{Lab} = 14.6$ MeV. Clearly, we need more data at other energies and other angles to draw a definite conclusion.

4. Conclusion

We calculated some observables of the proton-deuteron radiative capture reaction at low energies with realistic three-nucleon wave functions. The results well reproduced the experimental values of the total cross section and angular distribution.

Available 2NP's give the tensor analyzing power A_{yy} to be smaller by 15-30% compared to the experimental value. Introduction of 3NP improves this for some potentials. We found that the 3NP effects on A_{yy} depend on the tensor character of 2NP. Thus, we conclude that tensor analyzing power could provide important information on nuclear interactions.

The calculation was supported in part by the Grant-in-Aid of the Ministry of Education, Science and Culture of Japan, Research Center for Nuclear Physics, Osaka University and the Cyclotron Radioisotope Center, Tohoku University.

References

[1] Ishikawa, S., Sasakawa, T.: submitted to Phys. Rev. C.

[2] Arenhövel, H.: From Collective States to Quarks in Nuclei, ed. by Arenhövel H., Saruis, A.M., Lecture Notes in Physics **137**, 136 (1981).

[3] Ciofi degli Atti, C.: Proc. of the International School of Intermediate Energy Nuclear Physics, Verona, Italy, July 1981 (World Scientific, Singapore, 1982), p.65.

[4] Sasakawa, T., Ishikawa, S.: Few-Body Syst. **1**, 3 (1986).

[5] Ishikawa, S., Sasakawa, T.: Few-Body Syst. **1**, 143 (1986).

[6] Ishikawa, S., Sasakawa, T.: a contribution (D47) to 12th International Conference of Few Body Problems in Physics, Vancouver, B.C., Canada, July 2-8, 1989.

[7] Wiringa, R.B., Smith, R.A., Ainsworth, T.L.: Phys. Rev. C. **29**, 1207 (1984).

[8] de Tourreil, R., Sprung, D.W.L.: Nucl. Phys. **A201**, 193 (1973).

[9] de Tourreil, R., Rouben, B., Sprung, D.W.L.: Nucl. Phys. **242**, 445 (1975).

[10] Reid, Jr., R.V.: Ann. Phys. (N.Y.) **50**, 411 (1968).

[11] Machleidt, R., Holinde, K., Elster, Ch.: Phys. Rep. **149**, 1 (1987).

[12] Robilotta, M.R., Coelho, H.T.: Nucl. Phys. **A460**, 645 (1986).

[13] Belt, B.D., Bingham, C.R., Halbert, M.L., van der Woude, A.: Phys. Rev. Lett. **24**, 1121 (1970).

[14] King, S.E., Roberson, N.R., Weller, H.R., Tilley, D.R.: Phys. Rev. Lett. **51**, 877 (1983).

[15] van der Woude, A., Halbert, M.L., Bingham, C.R., Belt, B.D.: Phys. Rev. Lett. **26**, 909 (1971).

[16] Skopik, D.M., Asai, J., Beck, D.H., Dielschneider, T.P., Pywell, R.E., Retzlaff, G.A.: Phys. Rev. C. **28**, 52 (1983).

[17] Jourdan, J., Baumgarter, M., Burzynski, S., Egelhof, P., Henneck, R., Klein, A., Pickar, M.A., Plattner, G.R., Ramsay, W.D., Roser, H.W., Sick, I., Torre, J.: Phys. Lett. **162B**, 269 (1985); Nucl. Phys. **A453**, 220 (1986).

Few-Body Systems, Suppl. 6, 291—297 (1992)

RECENT ADVANCES IN THREE-NUCLEON CONTINUUM STUDIES.

H.Witała[a] and W.Glöckle

Institut für Theoretische Physik II, Ruhr Universität Bochum, D-4630 Bochum, Germany

[a] Institute of Physics, Jagellonian University, PL-30059 Cracow, Poland

Abstract: Recent advances in the study of nucleon-deuteron elastic scattering and breakup processes are reviewed. Rigorous solutions of the three-nucleon equations using meson-theoretical two-nucleon interactions are compared to experimental pd and nd data. The possiblility for charge-independence and charge-symmetry breaking in the ^3P nucleon-nucleon forces is discussed together with the problem of the magnitude of the $^3S_1$-$^3D_1$ tensor force.

1. INTRODUCTION

The three-nucleon (3N) system plays an important role in testing our knowledge of the basic nuclear dynamics. One reason is the fact that the theoretical formulation is well founded [1] and that the three-body equations can be solved in a numerically precise way using modern computers. In this way the dynamical input can be tested unambiguously by comparing theoretical 3N observables to experimental data. In addition new high quality experimental results are becoming available which allow to test the 3N Hamiltonian based on a specific nucleon-nucleon (NN) interaction more precisely.

There is a long history of trials to formulate and solve the three-body equations. For a recent review we refer to [2] and references therein. With the advent of supercomputers we succeeded in solving the 3N Faddeev type scattering equations for any two-nucleon (2N) force in a numerically precise way [3]. In this article we concentrate on such numerically precise solutions of the 3N equations with the Paris [4] and Bonn [5] meson-theoretical 2N interactions and discuss some unique features that can possibly be used to disentangle subtle characteristics of the NN interaction which are very difficult to obtain by studying the 2N systems only. For details of the theoretical formulation and numerical performance we refer to [3,6].

2. RESULTS FOR THREE-NUCLEON SCATTERING OBSERVABLES

We now describe shortly results for 3N scattering observables. Our calculations are based on the free 2N interactions of the Paris and Bonn potentials. The existing elastic Nd scattering cross-section data are well discribed by these meson-exchange based NN interactions. In Figs. 1 and 2 we present the quality of that description for the 14.1 MeV nd elastic scattering angular distribution and for the nd total cross section. It should be stressed that high precision nd elastic scattering cross-section data are a rarity.

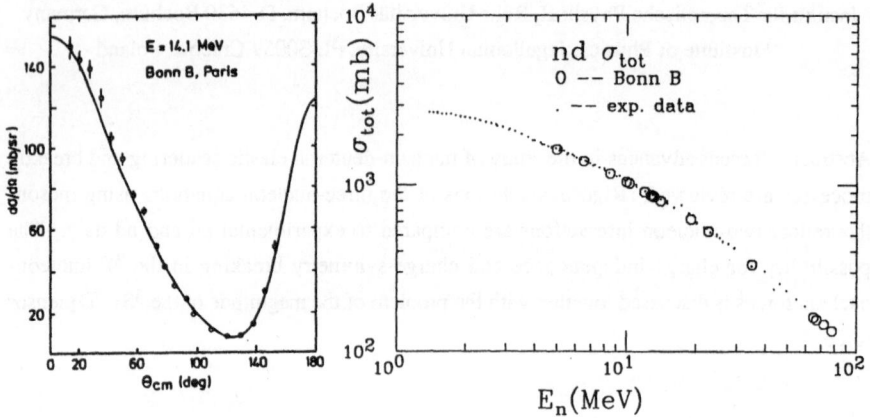

Fig. 1 Comparison between nd elastic-cross-section data [18] and predictions using the Paris and Bonn B dynamics.

Fig. 2 Total cross-section data for the nd interaction (points, [19]) compared with the Bonn B potential predictions (circles).

The tensor analyzing powers T_{20} and T_{22} at 22.7 MeV are shown in Fig. 3. Note the spectacular agreement between experiment and theory. However, the neutron and deuteron vector analyzing powers pose interesting problems. While the nucleon analyzing power A_y agrees very well with the data at nucleon laboratory energies of 50 MeV and above, significant disagreement exists at lower energies between theoretical 3N calculations based on the Paris and Bonn potentials and the experimental data. A possible key to the understanding of this discrepancy is the known sensitivity of A_y to the ^3P 2N forces [7,8]. We have shown in [9] that very different choices of ^3P 2N phase shifts describe the analyzing powers in the 2N pp and np systems similarly well. Therefore, in view of this ambiguity, it may not be too suprising that 2N potentials that are fitted to one out of several possible choices of ^3P

phase shifts fail to describe the A_y data in the 3N system. We took advantage of this freedom and demonstrated in [9] that it is possible to simultaneously describe the A_y data in the pp and np systems together with the A_y data in the 3N pd system by modifying the 2N interaction in the ^3P states. Practically, this modification was achieved by simply multiplying the Bonn B potential in the states $^3P_0, ^3P_1$ and $^3P_2\text{-}^3F_2$ by energy independent strength parameters λ. This approach introduces charge-independence breaking in these NN interactions. The nd analyzing power data which are slightly higher in the maximum than the pd data can also be very well reproduced by assuming an additional charge-symmetry breaking of the ^3P forces. It should be emphasized that the choice of the λ parameters used to modify the ^3P nn interactions is more uncertain than that for the pp system, since two nucleon nn data do not exist. We obtained in our simultaneous analysis of 2N and 3N data a sizeable charge-independence and charge symmetry-breaking only in the $^3P_0$ phase shift while the $^3P_1$ and $^3P_2\text{-}^3F_2$ phase shifts were found to be essentially charge independent. The magnitude of the effects is shown in Fig. 4 where the commonly used parameterizations [11] of the ^3P phase shifts are shown as a function of nucleon energy.

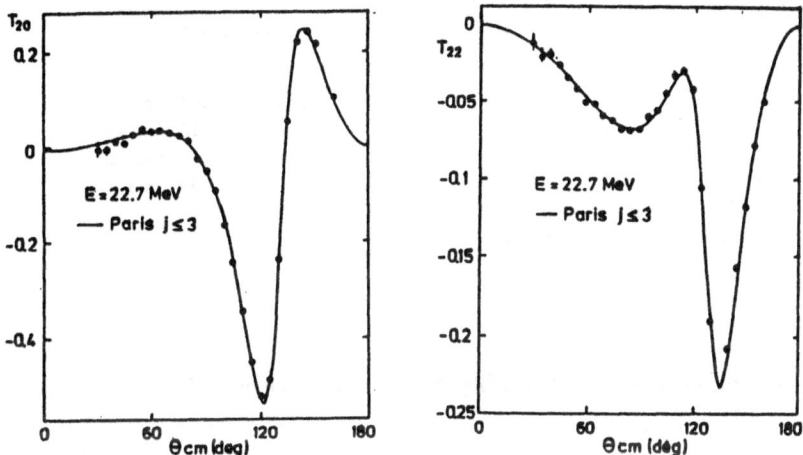

Fig. 3 Comparison between pd tensor analyzing power data [20] and Paris potential predictions.

Since we neglect the Coulomb interaction totally in our 3N calculations we assumed that the entire difference in the maxima of the pd and nd analyzing power distributions is due to charge-symmetry breaking effects. There exits a plausible argument which is based on the picture that the incident proton slows down in the Coulomb field of the deuteron causing the nuclear interaction to take place at a slightly lower energy than in the case of nd scattering. If this argument is correct then charge symmetry breaking effects in the ^3P NN interactions are probably not as large as found in our analysis [10]. However, with decreasing bombarding energy an interesting effect can be seen (see Fig. 5). While at E_N=8.5 MeV the nd and pd A_y data support our values for the λ parameters obtained at higher energies our nd prediction at

$E_N=5$ MeV overestimates the data. Simultaneously the difference between the nd and pd A_y data decreases and our pd prediction agrees with both of them rather well. These observations may indicate the smallness of Coulomb effects. Furthermore, it indicates an energy dependence of the charge-symmetry breaking interaction. In view of this result it would be extremely interesting to measure the nd analyzing powers at even lower energies.

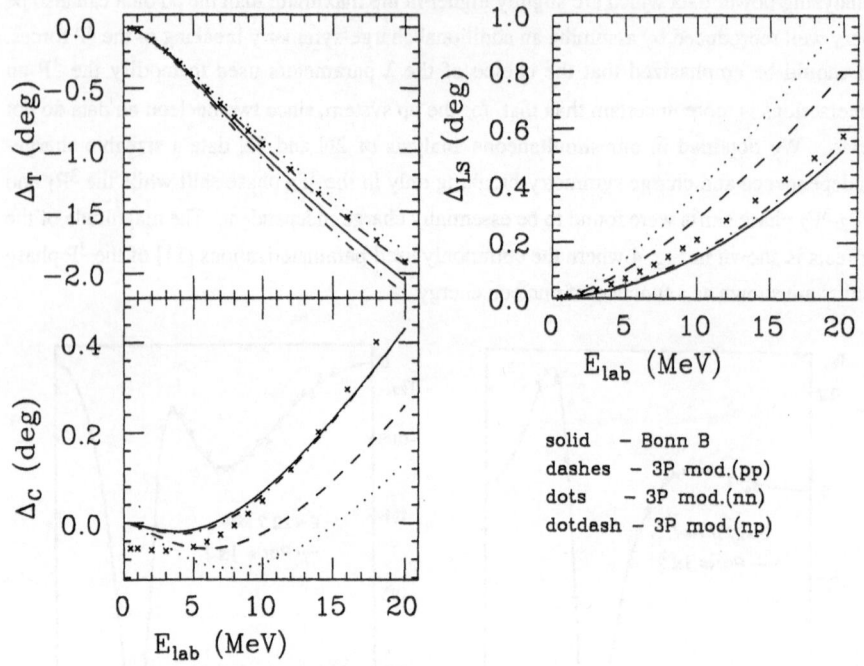

Fig. 4   Comparison of $^3P_0$, $^3P_1$ and $^3P_2$ phase shifts as given by the Bonn B potential, those of Ref. [11] and modifications proposed for the pp, nn and np systems. Crosses are results from the pp phase-shift analysis [11]. In all cases Coulomb corrections are included.

Such a possible explanation of the A_y puzzle appears very interesting considering the well-known difficulties in determining charge-independence and/or charge-symmetry breaking effects in ^3P states from the study of 2N data alone [10]. It should be pointed out that the 3N system offers a unique possibility to test the proposed modifications of the ^3P forces. Besides the vector analyzing powers, nucleon-deuteron spin-transfer coefficients also depend strongly on the ^3P force components [2]. The most promising observables are $K_z^{y'z'}$ and $K_x^{x'y'}$ whose sensitivity to ^3P force modifications not only differs from that for A_y but also from each other [9].

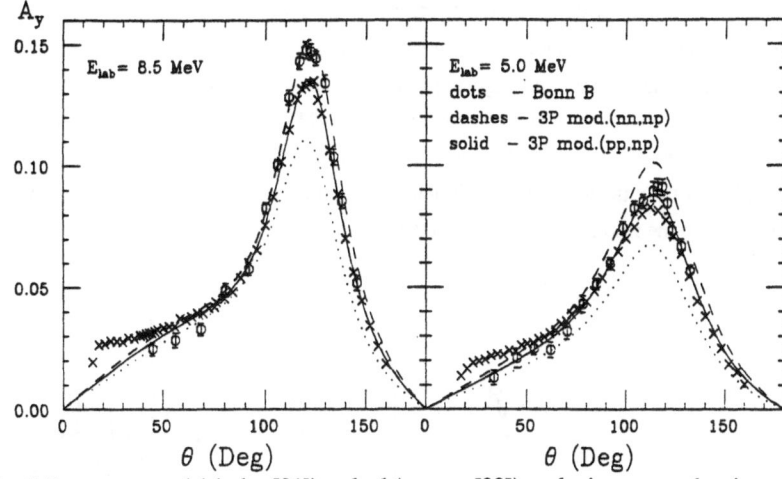

Fig. 5 Low energy nd (circles [21]) and pd (crosses [22]) analyzing power data in comparison with the Bonn B potential predictions and ^{3}P modified nd and pd dynamics as proposed in Ref. [9].

Another area where the 3N systems can contribute significantly is the determination of the NN tensor force at low energies. A measure of this force is given by the mixing parameter ε_1 of the $^3S_1$-$^3D_1$ partial waves. An accurate determination of ε_1 is in principle possible but requires complicated 2N measurements with at least two polarized nucleons. In addition such observables are also sensitive to other NN phase-shift parameters and that complicates the determination of ε_1 [10]. Among the many spin-observables we studied we found that the Nd elasting scattering nucleon-nucleon polarization transfer parameter $K_y^{y'}$ is very sensitive to the choice of the $^3S_1$-$^3D_1$ tensor force [12]. In Fig. 6 we show that the difference between the various curves is mainly due to differences in the tensor force. It is evident that the $K_y^{y'}$ data at 22.7 MeV favor the weaker tensor force of Bonn A. A recent pd measurement at 19.0 MeV leads exactly to the same picture, that Bonn A with the weaker tensor force is clearly favored and leads to a very good agreement with the data [24]. Such a weaker tensor force reduces the discrepancy between the theoretical and experimental triton binding energy to a few hundred keV [13] in comparison to about 1 MeV [14] obtained with the Paris potential and its larger tensor force.

The simple explanation of the A_y problem using modified 2N on-shell properties only and including an isospin breaking in the ^{3}P forces is very appealing. However a more exotic picture, such as 3N forces (3NF) may alter the explaination given above. In general, the usefulness of the 3N system as a tool for studying subtle characteristics of the NN interaction relies heavily on the smallnes of 3NF effects in Nd elastic scattering observables. A possibly rich source of information about this problem is the nd breakup process for which the first model calculations [15] including 3NF have shown non negligeable effects in some kinematically complete configurations. Especially interesting are configurations which are insensitive to details of the 2N forces. This is, of course, a most welcome feature for identifying clearly 3NF effects. Such a candidate is the space-star configuration [16], where in the c.m. system the final nucleons have equal energies in the plane perpendicular to the beam direction. The collinear configuration (in the c.m. system one nucleon is at rest) was also proposed as a candidate for studying 3NF effects [17]. Although the nd data recently obtained by different groups for the space star configuration have a tendency to lie above theory (Fig. 7) they do not agree sufficiently well among each other to draw a definite conclusion about the importance of 3NF effects.

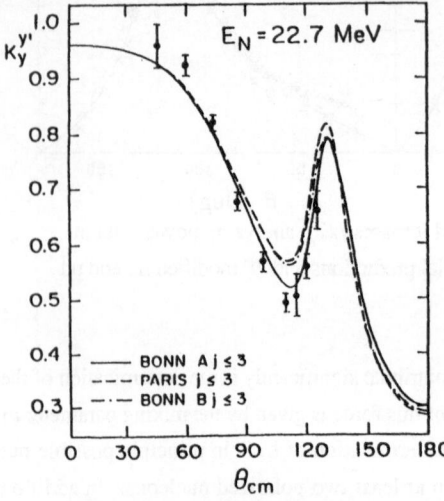

Fig. 6 Nucleon spin transfer coefficient $K_y^{y'}$ at 22.7 MeV. Circles are pd data from Ref. [11].

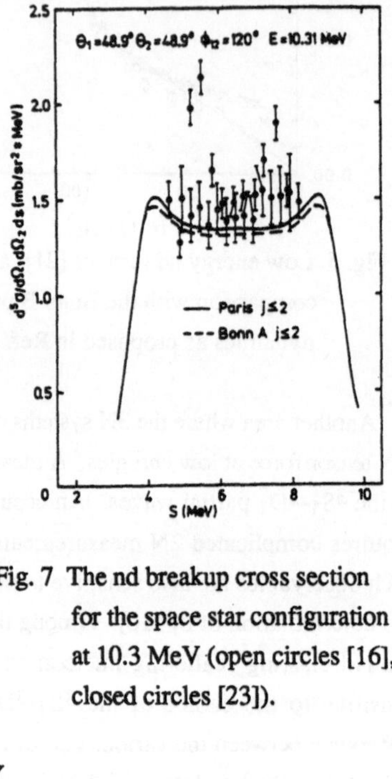

Fig. 7 The nd breakup cross section for the space star configuration at 10.3 MeV (open circles [16], closed circles [23]).

3. SUMMARY

We have shown that the new tool of rigorous 3N continuum calculations allows one to test the basic question whether or not unperturbed 2N forces describe 3N scattering quantitatively. Some results obtained with the meson theoretical NN interactions show very good agreement with the data while vector analyzing powers disagree. The reason for the latter can be traced back very likely to unsettled on-shell properties of the ^{3}P forces. Our combined analysis of 2N and 3N data shows that it is possible to describe these data satisfactorily by introducing isospin breaking effects in the ^{3}P forces. Independent information on ^{3}P forces can be provided by measuring some nucleon to deuteron spin-transfer coefficients. Experimental data for the nucleon spin-transfer coefficient $K_y^{y'}$ at 22.7 MeV clearly favor the weaker tensor force of the Bonn A potential, which also provides an improvement in the theoretical triton binding energy. Study of specific breakup configurations which are insensitive to details of the 2N dynamics would help to remove uncertainties with regard to possible 3NF effects. It will be one of the most important tasks for both experimenters and theoreticians to scrutinize such configurations in order to establish the magnitude of 3N force effects and to demonstrate the further reliability of the 3N system as a tool for studying subtle characteristics of the NN interaction.

Acknowledgement: The numerical work has been performed on the CRAY Y-MP of the "Höchstleistungsrechenzentrum" in Jülich, Germany and the North Carolina Supercomputation Center (NCSC) in Durham, USA.

References:

1. Faddeev, L.D.: Sov. Phys. JETP **12**, 1014 (1961).
2. Glöckle, W., Witała, H., Cornelius, Th.: Nucl. Phys. **A508**, 115c (1990); Glöckle, W., Witała, H., and Cornelius, Th.: Proc. 25th Zakopane School on Physics, Vol.2, ed.: Styczen, J. and Stachura, Z. (World Scientific, Singapore, 1990) p.300.
3. Witała, H., Glöckle, W., and Cornelius, Th.: Few-Body Systems Suppl. **2**, 55 (1987) Few-Body Systems **3**, 123 (1988).
4. Lacombe, M., *et al.*: Phys. Rev. **C 21**, 861 (1980).
5. Machleidt, R., Holinde, H., and Elster, Ch.: Phys.Rep. **149**, 1 (1987); Machleidt, R.: Adv.Nucl. Phys. **19**, 189 (1989).
6. Glöckle, W.: The Quantum Mechanical Few-Body Problem (Springer Verlag 1983); Lecture Notes in Physics **273**, 3 (1987).
7. Howell, C.R., *et al.*: Few-Body Systems **2**, 19 (1987).
8. Witała, H., Glöckle, W., Cornelius, Th.: Nucl. Phys. **A491**, 157 (1989).
9. Witała, H., and Glöckle, W.: Nucl. Phys. **A528**, 48 (1991).
10. Tornow, W.: Int. Conf. on Spin and Isospin in Nuclear Interactions, Telluride, Colorado USA, 11-15 March 1991, to be published in Adv. Nucl.Phys.
11. Bergervoet, J.R., *et al.*: Phys.Rev. **C 38**, 15 (1988).
12. Clajus, M. *et al.*: Phys.Lett. **B245**, 333 (1990).
13. Brandenburg, R.A., *et al.*: Phys.Rev. **C37**, 781 (1988).
14. Friar, J.L., Gibson, B.F., and Payne, G.L.: Ann.Rev.Nucl.Part.Sci. **34**, 403 (1984).
15. Meier, W., and Glöckle, W.: Phys.Lett. **138B**, 329 (1984).
16. Stephan, M., *et al.*: Phys.Rev. **C39**, 2133 (1989).
17. Witała, H., Cornelius, Th., Glöckle, W.: Few-Body Systems **5**, 89 (1989).
18. Berick, A.C., Riddle, R.A., and York, C.M.: Phys.Rev. **174**, 1105 (1968).
19. Computer index of neutron data (CINDA), International Atomic Agency, Vienna, 1988.
20. Grüebler, W., *et al.*: Nucl.Phys. **A398**, 445 (1983).
21. Tornow, W., *et al.*: Phys.Lett. **257B**, 273 (1991).
22. Sagara, K.: private communication.
23. Finckh, E.: private communication.
24. Paetz gen. Schieck, H.: private communication.

Few-Body Systems, Suppl. 6, 298—303 (1992)

LOW ENERGY SCATTERING ON TWO-BODY BOUND STATE IN CARTESIAN COORDINATES

J. Carbonell,[1] C. Gignoux [1] and S. P. Merkuriev [2]

[1] Institut des Sciences Nucléaires, 38026 Grenoble Cedex, France

[2] Computational and Mathematical Physics Department, Leningrad State University, USSR

Abstract:

Faddeev equations for a low energy elastic scattering of a particle on a two body-bound state are solved in cartesian Jacobi coordinates. The method is applied successfully to calculate n-d scattering length with realistic interactions.

1.Intrdoduction

The main difficulty in solving the three body elastic scattering in configuration space is the proper implementation of the asymptotic behaviour. Even in the low energy limit, where the problems related to the break-up threshold are absent, the very few works devoted to this task [1,2], has been obliged to overcame this difficulty either by introducing some additional parameter or by solving modified Faddeev equations.

The common root of such a trouble is the inadequacy of polar coordinates (ρ,θ) to describe the asymptotic behaviour :

$$u_b(x) \, e^{i(qy+\delta)} \tag{1}$$

where $u_b(x)$ is the (2,3) two body bound state wavefunction and q^2 the relative 1-(2,3) kinetic energy. The choice of a polar grid was imposed by practical reasons, leading to a banded linear system and allowing so reliable numerical calculations. It has been widely adopted in solving bound state as well as scattering problems. However, due to the finite size of the grid, equation (1) is not a solution of the asymptotic hamiltonian in polar coordinates. First because the bound state wavefunction $u_b(x)$ cannot be obtained as a numerical solution in the $\rho=\rho_{max}$ region, second because equation (1) is only physically acceptable in the neighbourhood of ($\rho=\rho_{max}$, $\theta=0$).

In order to avoid these difficulties, we have solved Faddeev equation in cartesian x-y Jacobi coordinates. By doing so, the asymptotic behaviour (1) is naturally obtained as a product on separable x-y coordinates into which Faddeev equation splits. This factorisation leads to a very stable local results.

2. Faddeev equations in a cartesian grid

We illustrate the method in the simplest case of three S-wave interacting identical particles at zero energy. In this case, Faddeev equations result into a set of integro-differential equations coupling the different components φ_α ($\alpha=1,...,n_c$) of the unique Faddeev amplitude φ. They will be written in cartesian Jacobi coordinates under its usual form :

$$\left[-\varepsilon_b + \Delta_x + \Delta_y - V_\alpha(x) \right] \varphi_\alpha(x,y) = V_\alpha(x) \int_{-1}^{+1} du \; \frac{xy}{x'(u)y'(u)} \sum_{\beta=1}^{n_c} C_{\alpha\beta} \; \varphi_\beta(x'(u),y'(u)) \quad (2)$$

where $x'^2(u)=\frac{1}{4}(x^2-2\sqrt{3}xyu+3y^2)$, $y'^2(u)=\frac{1}{4}(y^2+2\sqrt{3}xyu+3x^2)$, ε_b the two body cluster binding energy and $C_{\alpha\beta}$ is a $n_c x n_c$ matrix coupling the different components.

Equation (2) has been solved by using finite difference method in a rectangular x-y grid. The integration step for each coordinate is kept constant in the strong inner part of the potential and doubled several times in the small interaction region. A one dimensional grid G_q (q=x,y), defined by a set of (N+2) points $\{q_0, q_1, ..., q_N, q_{N+1}\}$, is characterised by the initial integration step, the set of points in wich it is doubled, N and its maximum value $q_{max} = q_{N+1}$:

$$G_q = \{h_q; q_{i_1}, q_{i_2}, ..., q_{i_d}; q_{N+1}; N\} \quad (3)$$

The integral term has been approximated by a N_g points Gauss quadrature formula plus linear interpolation of the integrand in the grid points. Stability is reached with $N_g=8$.

The boundary conditions for the elastic scattering are implemented in the following way:

$$\varphi_\alpha(0, y) = \varphi_\alpha(x, 0) = \varphi_\alpha(x_m, y) = 0 \quad (4)$$

$$\alpha=1,.., n_c$$

$$\varphi_\alpha(x, y_m) = u_b(x) \, \delta_{1\alpha} \quad (5)$$

The main advantage of the cartesian method is the natural way in which the asymptotic behaviour is obtained. Indeed, when the integral term in (2) is negligible, x and y variables separate. The solution compatible with (4) is then given by the product $u_b(x) \, g(y)$, where:

$$\left[-\varepsilon_b + \Delta_x - V_\alpha(x) \right] u_b(x) = 0 \quad (6)$$

$$\Delta_y \ g(y) = 0 \tag{7}$$

and the scattering length, a, can be easily extracted from the linear behaviour of g:

$$g(y) = A \ (y - a) , \tag{8}$$

Let us emphasize however that the accuracy in obtaining g(y) is intimately related to the accuracy in satisfying equation (6). The efficiency of the method lies in the fact that this accuracy can be done arbitrarily small provided that the two body wavefunction, implemented in the boundary condition (5) is an exact numerical solution of (6), i.e. obtained by the same x-grid used for solving the full three body problem. As we will see in the following section, the use of any other solution for $u_b(x)$, although more precise, can completely false the result for g.

3. Results

The results presented in this section concern the neutron deuteron (n-d) zero energy scattering. The nucleon-nucleon interaction we have used is the MT I-III potential [2]. We shall discuss three cases: the symmetric (S) and quartet case (Q) with n_c=1 and the doublet case (D) for which n_c=2. The corresponding $C_{\alpha\beta}$ values are:

$$C(S) = 2 \qquad C(Q) = -1 \qquad C(D) = \begin{bmatrix} 1/2 & 3/2 \\ 3/2 & 1/2 \end{bmatrix} \tag{9}$$

We have used essentially three different grids, G0-G2, the parameters of which are given in Table 1. Some variants denoted by primed (GN') and double primed (GN") labels, have been also used. They differ from the unprimed grids by the addition of 5 and 10 points, i.e. respectively 6 and 12 fm.

	h	q_{i_1}	q_{i_2}	q_{i_3}	q_{i_4}	q_{i_5}	q_{i_6}	q_{max}	N_q
G0	0.10	2.60	3.20	4.40	6.80	11.60		37.20	45
G1	0.05	1.05	2.05	3.05	4.25	6.65	9.85	32.25	50
G2	0.05	2.05	2.55	3.15	4.35	6.75	11.55	37.15	65

Table 1

The method provides, on the same foot, the solution of the two and three body bound state problem if one replaces equation (4) by the usual exponentially decreasing boundary conditions. The deuteron, B_d, and three body, B_t, binding energies (in MeV) are given in Table 2.

Deuteron binding energy and wavefunction, obtained with n_c=1 and $C_{\alpha\beta}$=0, are independent of the y-grid, once removed the trivial kinetic energy in the y degree of freedom. Due to the strong triplet potential, the three body symmetric case has a nearthreshold first excited bound state. The quartet model has no bound states. The doublet case corresponds to triton.

The binding energies on Table 2 has to be compared to the "exact" values B_d=2.231 MeV and B_t=8.53 MeV obtained by independent and more precise methods. Our "best grid", G2, reproduce these values better than 1%.

GX	GY	B_d	B_t (S)		B_t (D)	a(S)	a(Q)	a(D)
G0	G0	2.256	19.6	2.32	8.61	25.1	*	*
	G0'					25.2	6.48	0.545
	G0"					25.2	6.48	0.546
G1	G1	2.210	19.4	2.25	8.47	24.8	*	*
	G1'					24.8	6.49	0.519
	G1"					24.8	6.48	0.520
G2	G2	2.223	19.5	2.28	8.51	25.2	*	*
	G2'					25.2	6.51	0.525
	G2"					25.2	6.51	0.525

Table 2

The scattering length is calculated by a local analysis of the amplitude $\varphi(x,y)$ in the asymptotic region. According to (8), the function $\varphi_x(y)=\varphi(x,y)$ must be linear for all x in the grid. The linear extrapolation of φ_x outside the asymptotic region provides a family of rays, $\overline{\varphi}_x(y)$, focusing in the same point ($\overline{\varphi}_x$=0, y=a). Any astigmatism is a non convergence test. The numerical results are presented in Table 2. The following general remarks concerning their obtention are in order:

(i) Setting $C_{\alpha\beta}$=0 in equation (2), the regularity at the origin of g(y) imposed by the boundary conditions (4), requires a=0. Due to the fact that a is obtained in the asymptotic region, this requirement will be fullfilled if and only if equation (7) generates a very precise linear solution in all the grid. A small error in equation (6) will introduce a $q^2 \neq 0$ value in (7) and propagate a wrong solution. When the coherency between the input boundary conditions (5) and the numerical solution of (6) is ensured, we get indeed a value a~10^{-7}. But a 10^{-3} difference in ε_b is enough to generate a "numerical" scattering length of the order of unity.

(ii) In Figure 1 the functions $\varphi_x(y)$ are plotted for some values of x. In the S case, Figure 1.a, the scattering length is directly given by the intersection of $\varphi_x(y)$, whereas in the D case, Figure 1.b, is obtained by their linear interpolation $\overline{\varphi}_x(y)$. The range of the effective potential, created by the inhomogeneus term in equation (2), is manifested by the non linear behaviour in $\varphi_x(y)$. Due to our definition of Jacobi coordinates, a factor $\frac{\sqrt{3}}{2}$ on the y coordinate is needed to get the physical scattering lengths given in Table 2.

(iii) Results indicated by * in Table 2, exhibit an x dependency on the calculated scattering length. This dependency is due to the integral term, which, still sizeable in the vicinity of y_{max}, generates a solution different from its asymptotic limit $u_b(x)g(y)$. For all the others values, obtained by increasing y_{max}, the x stability is better than 10^{-4}.

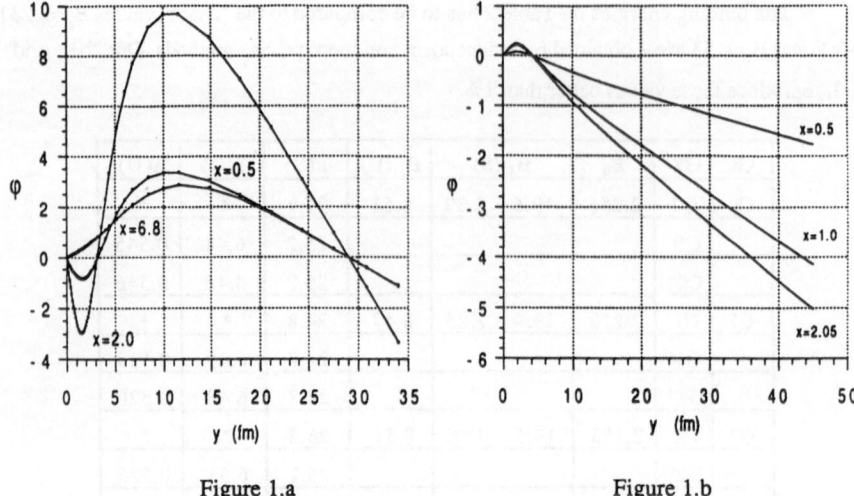

Figure 1.a   Figure 1.b

(iv) We have tested the orthogonality of the scattering Faddeev amplitude φ_s to the three body bound state wavefuntion, ϕ_b, calculated with the same method. In practice this condition, $<\phi_b/\varphi_s>=0$, requires a normalisation of φ_s and a numerical criterion for zero. It has been replaced by the following equivalent condition:

$$R = \left| \frac{\langle\phi_b/\varphi_s\rangle}{\langle\phi_b/\varphi_s^B\rangle} \right|^2 \ll 1$$

where φ_s^B is the Born amplitude. In the relevant cases (S and D), we found $R \leq 1.2 \ 10^{-3}$.

(v) The agreement between the local and integral results is better than 0.3% in all the values in Table 2.

(vi) For a given x-y grid, it has been always possible to obtain stable results as a function of x_{max} and y_{max} above $x_{max} \approx y_{max} \approx 35$ fm. This stability, better than 10^{-4} , is not to be taken as the accuracy of the method which is determined by the inner part of the grid. The comparison between the results obtained by G0 and G2 grids, which differs by a factor two in this region, give us an estimation of the accuracy.

Results provided by our best grids, G2-G2", and the stability with respect to wide variations of the grid parameters in Table 2 lead us to propose the values a(S)=25.2±0.4 fm, a(Q)=6.51±0.03 fm and a(D)=0.525±0.02 fm, with overestimated error bars.

Comparing these results to those proposed in [2], (26.03±0.01, 6.442±0.005 and 0.70±0.01 fm respectively), we remark only small deviations for the S and Q cases but a sizeable disagreement in the doublet one. Although the precision of our results is certainly less than the one exhibit in [2], we believe that the disagreement cannot be attributed to numerical inaccuracies.

The scattering length calculations has been completed by studying the low energy limit of phase-shifts. The convergence is excellent in all cases and the agreement between the $q^2 \to 0$ limit and the zero energy calculations is better than four digits.

4. CONCLUSON

We have presented a new method to calculate low energy scattering parameters. The method is independent of those used in the previous works [1, 2]. It is based in solving Faddev equations in the Jacobi cartesian coordinates. This set of coordinates seems to be the natural choice to implement properly the three body elastic scattering boundary conditions. They allows reliable numerical calculation of Faddeev amplitude if the band structure of the linear system which appears in the polar grid is preserved.

The results obtained, compared to the preceding calculations, show the validity of this approach in solving the n-d low energy scattering. However the numerical methods used in our calculations, finite differences and linear interpolation in the integral term, should be improved to reach the stability nowadays accessible in in the few body problem.

REFERENCES

[1]   J.J. Benayoun, C. Gignoux, J. Chauvin, Phys. Rev. C 23, 5 , 1854 (1981)
[2]   G.L. Payne, J.L. Friar, B.F. Gibson, Phys. Rev. C 26 , 4, 1385 (1982)

Few-Body Systems, Suppl. 6, 304—313 (1992)

p-d CAPTURE REACTIONS IN MUONIC MOLECULES

J. L. Friar

Theoretical Division, Los Alamos National Laboratory

Los Alamos, New Mexico 87545 U.S.A.

ABSTRACT

Capture reactions for very low-energy n-d and p-d systems are calculated and compared with experiment, as are low-energy n-d and p-d scattering. We find excellent agreement for the n-d scattering lengths, but poor agreement for the p-d case, which we believe is a problem with the experimental extrapolation. The n-d radiative capture is sensitive to details of the meson-exchange currents, but reasonable models agree with the data. The latter models are in good agreement with experiment when extended to the p-d case. Our large quartet capture rate resolves a long-standing anomaly. The E0 capture matrix element recently obtained from a reanalysis of internal conversion in muonic molecules is in excellent agreement with our predictions. This matrix element is very clean theoretically and provides the best test of the calculations.

1. Introduction

1.1 Overview

Complete or "exact" solutions of the Schrödinger equation with realistic potentials have been obtained during the previous six years for a variety of problems: the ground states of ^{3}He and ^{3}H [1] (including a three-nucleon force, and a Coulomb interaction in the former case), the ground state of ^{4}He [2] and low-lying continuum of ^{5}He [3], above-breakup n-d scattering[4], and, more recently, zero-energy n-d and p-d scattering[5] (including three-nucleon forces, and a Coulomb interaction for the latter case). The purpose of these calculations is to probe the limits of our understanding of the binding and interactions of the few-nucleon systems. The goal is to establish whether our current level of experimental and theoretical sophistication can unambiguously detect the presence of three-nucleon forces in these systems.

In pursuing this goal, a wide variety of few-nucleon scattering, breakup, and capture reactions have been calculated, as well as the properties of the ground states. To date, most calculated properties agree well with experiment, if models are adjusted in some way to produce the correct ground-state binding energies. Typically, models containing only two-nucleon forces will underbind the triton by 0.6-1.0 MeV and the alpha particle by roughly four times that amount. While this binding defect is only a few percent of the total potential energy, the binding energy sets the size scale, and an underbound model will be too large in size. Consequently, size-dependent properties such as the charge radius and the Coulomb energy will be too large and too small, respectively.

Adding a "realistic" three-nucleon force based on two-pion exchange provides additional binding in roughly the correct amount. There is typically a severe short-range sensitivity to these forces, which cannot be well-constrained theoretically, and which precludes an *ab initio* prediction of binding energies. Nevertheless, the addition of a reasonable three-nucleon force to the Hamiltonian can produce proper binding for the three- and four-nucleon systems, and this allows us to extrapolate calculated observables to the correct binding energy (or size scale). We will see the importance of this later. To date, the primary effect of three-nucleon forces has been to set this size scale, and, unfortunately, not to change observables in a way that is incommensurate with a modified two-body force.

1.2 Low-energy Reactions

In this talk we report on calculations of low-energy scattering and reactions involving three nucleons, which utilize complete or "exact" solutions of the Schrödinger equation. Our goal is to describe μ-catalyzed p-d fusion reactions at (essentially) zero energy, but in order to benchmark this process we must also examine the n-d capture, as well as n-d and p-d scattering, which are two of our building blocks.

The quantity that characterizes very low-energy scattering is the scattering length, a. Neutron scattering lengths can be obtained by measuring (or calculating) scattering at thermal energies, which are effectively zero on the nuclear scale. They can also be obtained by calculating the usual effective-range function, $k \cot(\delta)$, at finite energy, and extrapolating this quantity to zero energy: $-1/a$. Capture reactions at thermal energies can be directly measured. Reactions involving positively charged particles at zero energy are much more difficult to obtain. Direct laboratory measurement is inhibited by a dominating Coulomb barrier which keeps the particles apart, and dramatically suppresses reaction rates and, even worse, the amount of nuclear scattering compared to pure Coulomb scattering. Laboratory scattering measurements are typically performed at fairly high energies, and the Coulomb-modified effective-range function obtained from them is subsequently extrapolated to zero energy. This methodology presupposes that the nuclear amplitudes are rather smooth functions of energy.

1.3 Muonic Molecules

The one exception to this extrapolation scheme is provided by fusion in muonic molecules. If negative muons are captured by two positively charged ions, they can form a molecule, which eliminates the problems described above. The molecular binding keeps the particles together, and the (negative) muon charge between the ions decreases the Coulomb barrier, which greatly increases transition rates. If these rates are faster than the normal muon β-decay rate, charged-particle reactions in the molecules can be directly measured. Moreover, the size scale of these molecules is roughly a muonic Bohr radius (~ 250 fm), and the relative kinetic energy of the ions can be estimated using the uncertainty principle to be several hundred electron volts, which is negligible on the nuclear scale. Thus, muonic molecules are potentially an ideal laboratory for low-energy charged-particle reactions, analogous to that provided by thermal neutrons.

There are possible difficulties with this scenario. The familiar product of beam current with target thickness, which occurs in laboratory beam-target experimental rates, is replaced by the probability of molecular coalescence, determined by the square of the molecular (*i.e.*, non-nuclear) wave function, which must be accurately calculated. In addition, a wide variety of competing processes must be disentangled, and this requires a sophisticated understanding of the atomic and molecular physics involved. Fortunately, both of these problems have been largely resolved in recent years[6].

2. n-d and p-d Scattering Lengths

Calculating the n-d and p-d scattering lengths is relatively straightforward. The Los Alamos/Iowa collaboration prefers to solve the Faddeev partial differential equation at zero energy, subject to the appropriate boundary conditions for elastic scattering (breakup being energetically impossible)[5]. The only difficulty in principle is the set of weak, long-range, Coulomb-induced (multipole) polarization potentials, which will modify the usual Coulomb boundary conditions for p-d scattering, or even the long-range magnetic force in the n-d case, which has a similar effect. These long-range forces are important in principle, but negligible in practice[7], if one follows customary procedures. A similar situation would result if the usual Coulomb interaction were a factor of 10^6 smaller: implementing non-Coulomb boundary conditions in the usual way, although wrong in principle, would introduce negligible error.

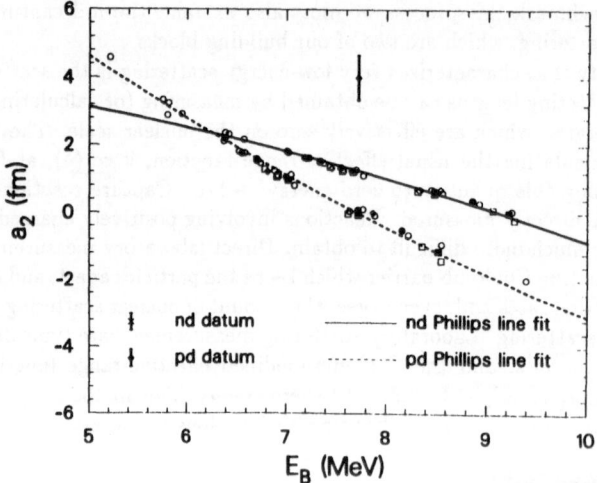

Fig. 1. n-d and p-d doublet scattering lengths plotted *vs* corresponding trinucleon binding energy.

The results of the Los Alamos/Iowa collaboration for s-wave doublet scattering are shown in Fig. 1, where the doublet scattering lengths for a variety of potential models are plotted against the ^{3}H (for n-d) or ^{3}He (for p-d) binding energy of that model. That is, with the exception of two experimental points with error bars, every point on the plot represents a theoretical calculation. This type of presentation is called a Phillips plot[8]. The filled squares and circles correspond to complete (34-channel) calculations. The circles represent models containing the Reid Soft Core N-N potential[9], while the squares represent models based on the Argonne V_{14} potential[10]. Approximately half of the points contain either the Tucson-Melbourne[11] or Brazilian[12] three-nucleon force. Almost invisible is the n-d experimental datum[13], which falls on the solid curve. The latter is a simple fit to the various theoretical points. On the other hand, the p-d datum[14-16] is far from the dashed curve, which is a fit to the p-d (Coulomb-modified) scattering-length points.

The quartet results have a similar disposition. The latter channel is insensitive to everything but the $^3S_1$-$^3D_1$ N-N force, and does not require any extrapolation. The average for the models we solved is 6.34(4) fm, while the experimental value is 6.35(2) fm; excellent agreement is obtained. In this example and later ones, we will quote a subjective "theoretical error bar", which merely reflects the spread in the theoretical values. The calculated p-d scattering length, on the other hand, is 13.6(1) fm, while the experimental value is about 11.5 fm [16]. This disagreement reflects the same problem we saw in the doublet case.

The experimental n-d scattering lengths can be directly obtained from low-energy neutron scattering, while the p-d scattering lengths must be obtained by extrapolating the Coulomb-modified effective-range function to zero energy. Unfortunately, the latter must be done from rather large energies. In order to check the extrapolation we calculated[17] the s-wave phase shifts as a function of energy. The calculations agree with the experimental values, and also extrapolate to the result one gets by solving for a directly at zero energy. For CM energies above 200 keV the doublet effective-range function is rather flat. Unfortunately, for energies below that value (and well below the data) there is enormous curvature in that effective-range function, which is equivalent to a huge effective range, r_0. This curvature is consistent with a sub-threshold pole. Although ref.[16] was well aware of the possibility of such a pole, the data gave no such indication, because the lowest datum was nevertheless too high in energy. We believe that the problem is an experimental one and will not be resolved until experiments can be performed at lower energies. We emphasize that there is no disagreement with actual measured quantities, but only with extrapolated values.

3. n-d Capture

3.1 Overview

Given ^{3}H and (zero-energy) n-d wave functions, one can calculate capture reactions accurately, provided one has an accurate model for the transition interaction. A well-studied process is thermal n-d radiative capture, which is almost entirely magnetic dipole (M1) in nature. The average of the experimental cross sections[18] is 0.518(8) mb, which is more than a factor of 600 smaller than for the corresponding n-p capture. The reason for this suppression is found in the spin-flip nature of the M1 operator in impulse approximation. The dominant part ($\sim$90%) of the ^{3}H ground state (the S-state) is an eigenfunction of this operator and, after enforcing orthogonality of initial and final wave functions, this large component does not contribute to the transition, as shown originally by Schiff[19].

Meson-exchange currents (MEC) will play a much larger role when the impulse approximation is suppressed. Although only a 10 percent correction to the n-p rate, such currents change the n-d rate by a factor of two. Thus, the latter case is sensitive to details of the MEC. We emphasize that constructing a model of the MEC requires information not contained in the potential model. Indeed, a completely specified current requires a theory of the photon-meson-nucleon system, which exists only in part. The dominant part, fortunately, is the long-range one-pion-exchange (OPE) part, for which a reasonable model can be constructed, just as it can for the N-N force. Unfortunately, the short-range parts are much more problematic. Consequently, we restricted ourselves in these calculations to a MEC model consisting only of the OPE part, modified by a short-range cutoff (a standard monopole form factor with a mass Λm_π). Typically one finds that the shorter-range contributions (*i.e.*, other than OPE) are comparable to the sensitivity of the OPE part with respect to Λ.

3.2 Results of Calculations

We can summarize our results[20] as follows. For models which are extrapolated to the physical ^{3}H binding energy (as we did for the doublet scattering length) we find:

1. In impulse approximation the quartet capture is larger than the doublet capture;

2. The impulse and MEC parts of the doublet matrix element are comparable and constructive;

3. The quartet MEC matrix element is only about 10 percent of the impulse approximation and is destructive;

4. Without MEC the capture cross section is a factor of two too low;

5. With MEC the doublet capture dominates the quartet by roughly 3:1;

6. With MEC the experimental capture cross section is reproduced within a ± 20 percent uncertainty due to reasonable variations in Λ;

7. The variation in Λ causes a variation of 25 percent in the MEC matrix elements, which greatly affects the total doublet matrix elements, but hardly changes the quartet ones;

8. Values of Λ which give a reasonable total cross section also reproduce the results of the recent experiment[21] on the circular (photon) polarization in polarized thermal neutron capture, which is sensitive to the quartet/doublet matrix element ratio.

The strong sensitivity of the total capture rate to the short-range behavior of the MEC can be used to test and refine models of the latter[22], although this is not our motivation. In order to calculate p-d capture with as little uncertainty as possible, we adjust the value of Λ to produce reasonable agreement with the n-d total capture cross section, which reproduces the polarized capture experiment as well. This gives us a reasonable (although semi-phenomenological) MEC model which can then be used with no free parameters to test our understanding of the former (p-d) reaction. We will display our sensitivity to Λ in that reaction, which is very similar to the n-d case, although slightly less so because of the Coulomb repulsion.

4. p-d Capture

4.1 Laboratory Measurements

Laboratory measurements of p-d radiative capture have been performed in the past and extrapolated (with some assumptions) to zero energy in order to obtain the astrophysical S-factor (which is proportional to the total cross section) for the s-wave reaction: S_s . The best available experimental value[23] from this source is $S_s = 0.12(3)$ keV mb. The large uncertainty reflects the difficulty in performing these experiments at the very low laboratory energies necessary for the extrapolation. The Los Alamos/Iowa prediction[24] for this quantity is 0.108(4) keV mb, and is based on the MEC models which were adjusted to the n-d capture and then extrapolated to the physical ^{3}He binding energy.

4.2 Fusion in Muonic Molecules

Fusion of protons with deuterons was discovered more than thirty years ago in muonic molecules formed in a hydrogen bubble chamber[25]. The reaction which was first seen was internal conversion: $\mu^- + p + d \rightarrow {}^3He + \mu^- + (5.5\ MeV)$. This reaction competes with the more common radiative-capture process, $\mu^- + p + d \rightarrow ({}^3He + \mu^-) + \gamma$, where the final muon almost always resides in the 1S Bohr orbit of the residual He atom[26]. In the former reaction the muon reappears and can reinitiate the complex molecular capture process and generate more fusions; hence the name: μ-catalyzed fusion.

Because the molecular energies are very small compared to nuclear energies, these nuclear reactions are s-wave in nature, as we discussed in the introduction. This restricts the possible electromagnetic matrix elements to four: E0, E2, M1(3/2), M1(1/2). The E2 matrix element, which is interesting because of its sensitivity to the tensor force, is not believed to play a significant role, while the E0 matrix element dominates the internal conversion process (and cannot occur for radiative capture). Magnetic dipole capture from the quartet and doublet states contributes only negligibly to internal conversion because the muon is very heavy and consequently generates a small current. It dominates the radiative capture, however.

Early theoretical work focussed primarily on the atomic and molecular aspects of the fusion process, although estimates of the nuclear rates were made using the relatively primitive (and largely phenomenological) methods which were available. A very influential paper[27] argued that the quartet capture should be very small, because the spins of any two nucleons would have to be aligned (J=1), which is Pauli-forbidden for two s-wave protons. Although the very small molecular energies force the p-d system into a relative s-wave, this does not mean that the p-p system resides in a relative s-wave, because of the large zero-point motion of the neutron and proton in the deuteron. Indeed, this mechanism leads to the S'-state (mixed-symmetry s-wave component of the wave function), which can produce non-negligible quartet capture.

4.3 Wolfenstein-Gerstein Effect

During the molecular formation process in a low-temperature mixture of normal hydrogen with a weak concentration of deuterium, the muon is preferentially captured by protons, and subsequently during collisions transfers from the ground state of that atom to the deuterium, which atom has a lower ground-state energy because of its larger mass. The resulting μ^--d atom has a significant hyperfine splitting in its ground state, which exceeds the value of kT in liquid hydrogen. At sufficiently high deuterium concentrations, collisions between the μ^--d atom and deuterium nuclei can lead to transitions between the atomic hyperfine states, which will eventually equilibrate into the lowest state. Thus, although the deuterium spin is randomly distributed in the liquid, the complex set of atomic processes during the formation of the molecule can lead to a non-statistical distribution of spins which depends on the deuterium concentration. After the final molecule is formed, the quartet and doublet spin states of the p-d system will also have a non-statistical spin weighting (which favors the doublet). Because doublet fusion dominates, the fusion rate will increase as the deuterium concentration increases. This is the Wolfenstein-Gerstein (WG) effect[28], which allows the experimentalist to "dial" varying amounts of quartet and doublet spin into the initial nuclear state prior to fusion, and permits these competing M1 processes to be disentangled. This interesting and valuable phenomenon accomplishes the same thing as polarized beams or targets in the more usual beam-target experiments.

Tests of the WG effect were initially successful. Subsequent tests[29-30] using improved calculations of the atomic physics were in poorer agreement, if the no-quartet conjecture was also used. This anomaly (the WG anomaly) has persisted until recently, and led to speculation about exotic atomic hyperfine mechanisms for restoring agreement; the latter have never survived quantitative calculations[6].

4.4 Molecular Wave Functions

An important simplification in calculating the reaction rates occurs when one realizes that the fusion occurs for small nuclear (*i.e.*, p-d) separations, R < 10 fm, while the average separation in the molecule is very much greater. Indeed, the p-d-μ reaction is a four-body problem which would be beyond our current calculational abilities to do exactly. Approximating the nuclear separation in the molecule by zero, the transition rate factorizes[31] into a nuclear part and a molecular part: the molecular coalescence probability. The latter is different for the two different reactions. For the internal conversion it is given by the square of the molecular wave function evaluated at the point where all three charged particles overlap. For the radiative capture, it is given by an integral with respect to the muon's coordinate over the square of the wave function. Highly accurate calculations by the Florida group[32] exist for the latter probability. In order to obtain the former probability we performed our own calculation of the molecular wave function using Faddeev techniques, and later compared this result to one obtained from the Florida wave function. Good agreement at the one percent level was obtained.

4.5 Internal Conversion

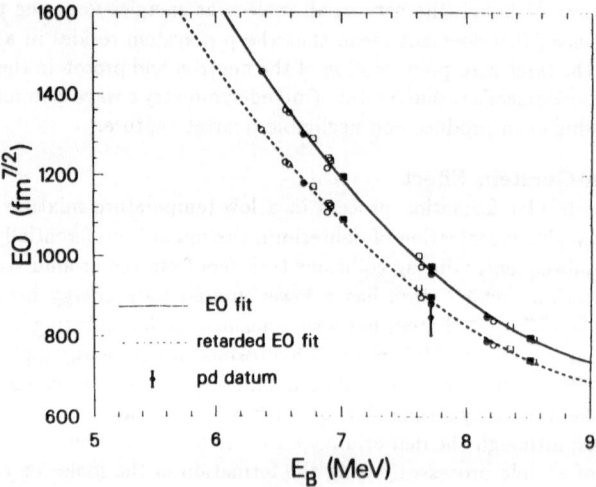

Fig. 2. E0 p-d internal conversion matrix elements plotted *vs* corresponding ^{3}He binding energy.

The internal conversion process is determined by the square of the E0 matrix element, which is just the mean-square radius of the charge distribution in the transition region, if one neglects retardation (*i.e.*, the muon's momentum) in the multipole operator. This operator is very clean theoretically and contains none of the usual exchange currents which complicate calculations of the magnetic multipoles. Our results[24] are shown in Fig. 2, after the fashion of Fig. 1. The matrix element scales roughly as E_B^{-2}. This illustrates clearly how sensitive some observables are to the binding. Our numerical results are 963(10) fm$^{7/2}$ for the unretarded matrix element and 888(10) fm$^{7/2}$ for the retarded (*i.e.*, physical) one, while the experimental matrix element extracted from the measured rate using our wave function is 844(45) fm$^{7/2}$. The experimental rate was recently obtained by reanalyzing old bubble chamber data[33]. The agreement is very good.

4.6 Radiative Capture

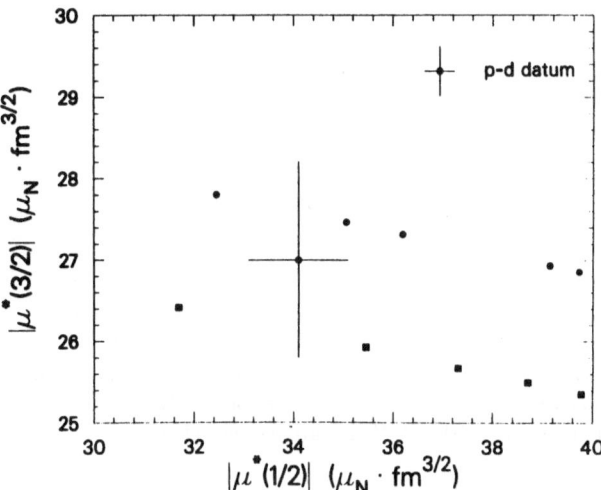

Fig. 3. p-d doublet and quartet M1 transition matrix elements plotted *vs* corresponding ^{3}He binding energy.

In addition to sensitivity to the model binding energy, the M1 matrix element is sensitive to the short-range behavior of the MEC. For purposes of display, we restrict ourselves to only two potential models, based on the RSC and AV_{14} potential models with a three-nucleon force added to produce the correct ^{3}He binding energy. For these two models we calculate[24] the quartet and doublet (reduced) matrix elements, which are displayed in Fig. 3, together with the experimental data extracted from the experimental rates[30]. As before, the circles correspond to RSC model calculations, while the squares correspond to AV_{14} results. The various values of Λ (*i.e.*, the short-range cutoff mass) for the RSC model are [5.8,7.5,8.6,17.0,99.0] from left to right, while for the AV_{14} case they are [5.8,8.6,RS,17.0,99.0]. The Riska-Schiavilla[22] prescription is denoted RS. The values of Λ fitted to n-d capture are 7.5 for the RSC case and 8.6 for the AV_{14} case. Note the sensitivity of the doublet matrix element to Λ, and the insensitivity of the quartet one. The values of Λ fitted to the n-d capture process work well for p-d. Overall, these M1 results are very satisfactory.

Table 1. Results of Ref. [24] (theory) compared to experiment. References a, b, c are Ref. [33], [30], and [23], respectively.

Process	Theory	Experiment
$\lambda_{1/2}^{\mu}[\cdot 10^6 sec^{-1}]$	0.062(2)	0.056(6)[a]
$\lambda_{3/2}^{\gamma}[\cdot 10^6 sec^{-1}]$	0.107(6)	0.11(1)[b]
$\lambda_{1/2}^{\gamma}[\cdot 10^6 sec^{-1}]$	0.37(1)	0.35(2)[b]
S_s[keV·mb]	0.108(4)	0.12(3)[c]

The calculated and experimental fusion rates are given in Table 1. The four processes are internal conversion, quartet radiative capture, doublet radiative capture, and astrophysical S-factor, respectively. The substantial quartet fusion rate confirms our previous argument that the no-quartet conjecture of ref.[27] is inadequate, and this, in fact, resolves the anomaly in the Wolfenstein-Gerstein effect.

5. Summary

We have calculated the nuclear reactions in μ-catalyzed p-d fusion. Accurate wave functions have been generated for the AV_{14} and RSC models for the first time, and these have been used to compute matrix elements. A pion-exchange-current model which reproduces thermal n-d radiative capture has been used with these wave functions to calculate radiative-capture matrix elements and rates. The latter are in good agreement with experiment. Our internal conversion rates are in very good agreement with a recent reanalysis of old bubble-chamber data. Our substantial value of the quartet capture rate resolves the anomaly in the Wolfenstein-Gerstein effect, which was based on an erroneous no-quartet conjecture.

Acknowledgements

This work was performed under the auspices of the U.S. Department of Energy. We would like to thank L. Bogdanova of ITEP, V. Markushin and L. Ponomarev of the Kurchatov Institute, and C. Petitjean of PSI for their encouragement and substantial help with the latest experimental results.

References

1. C. R. Chen, G. L. Payne, J. L. Friar, and B. F. Gibson, *Phys. Rev. C* **31**, 2266 (1985); **33**, 401 (1986); J. L. Friar, B. F. Gibson, and G. L. Payne, *ibid*,**35**, 1502 (1987).

2. J. Carlson, *Phys. Rev. C* **36**, 2026 (1987).

3. J. Carlson, private communication.

4. W. Glöckle, H. Wital̷a, and Th. Cornelius, *Nucl. Phys.* **A508**, 115c (1990).

5. C. R. Chen, G. L. Payne, J. L. Friar, and B. F. Gibson, *Phys. Rev.* *C* **44**, 50 (1991).

6. W. H. Breunlich, P. Kammell, J. S. Cohen, and M. Leon, *Annu. Rev. Nucl. Part. Sci.* **39**, 311 (1989).

7. Gy. Bencze, C. Chandler, J. L. Friar, A. G. Gibson, and G. L. Payne, *Phys. Rev.* *C* **35**, 1188 (1987).

8. A. C. Phillips, *Phys. Rev.* **142**, 984 (1966).

9. R. V. Reid, *Ann. Phys.* (NY) **50**, 441 (1968); see B. Day, *Phys. Rev.* *C* **24**, 1203 (1981) for the higher partial waves.

10. R. B. Wiringa, R. A. Smith, and T. A. Ainsworth, *Phys. Rev.* *C* **29**, 1207 (1984).

11. S. A. Coon, M. D. Scadron, P. C. McNamee, B. R. Barrett, D. W. E. Blatt, and B. H. J. McKellar, *Nucl. Phys.* **A317**, 242 (1979).

12. H. T. Coelho, T. K. Das, and M. R. Robilotta, *Phys. Rev.* *C* **28**, 1812 (1983).

13. W. Dilg, L. Koester, and W. Nistler, *Phys. Lett.* **36B**, 208 (1971).

14. W. T. H. Van Oers and K. W. Brockman, Jr., *Nucl. Phys.* **A92**, 561 (1967).

15. J. Arvieux, *Nucl. Phys.* **A221**, 253 (1974).

16. E. Huttel, W. Arnold, H. Baumgart, H. Berg, and G. Clausnitzer, *Nucl. Phys.* **A406**, 443 (1983); E. Huttel, W. Arnold, H. Berg, H. H. Kranse, J. Ulbricht, and G. Clausnitzer, *ibid*, **A406**, 435 (1986).

17. C. R. Chen, G. L. Payne, J. L. Friar, and B. F. Gibson, *Phys. Rev.* *C* **39**, 1261 (1989).

18. E. T. Jurney, P. J. Bendt, and J. C. Browne, *Phys. Rev.* *C* **25**, 2810 (1982); J. S. Merritt, J. G. V. Taylor, and A. W. Boyd, *Nucl. Sci. Eng.* **34**, 195 (1968). We have combined the results of these two latest measurements.

19. L. I. Schiff, *Phys. Rev.* **52**, 242 (1937).

20. J. L. Friar, B. F. Gibson, and G. L. Payne, *Phys. Lett.* **251B**, 11 (1990).

21. M. W. Konijnenberg, *et al.*, *Phys. Lett.* **205B**, 215 (1988).

22. R. Schiavilla, V. R. Pandharipande, and D. O. Riska, *Phys. Rev.* *C* **40**, 2294 (1989).

23. G. M. Griffiths, M. Lal, and C. D. Scarfe, *Can. J. Phys.* **41**, 724 (1963).

24. J. L. Friar, B. F. Gibson, H. C. Jean, and G. L. Payne, *Phys. Rev. Lett.* **66**, 1827 (1991).

25. L. W. Alvarez, *et al.*, *Phys. Rev.* **105**, 1127 (1957).

26. H. Bossy, *et al.*, *Phys. Rev. Lett.* **55**, 1870 (1985).

27. S. Cohen, D. L. Judd, and R. J. Riddell, Jr., *Phys. Rev.* **119**, 384 (1960).

28. S. S. Gerstein, *Sov. Phys. JETP* **13**, 488 (1961).

29. W. H. Bertl, *et al.*, *Atomkernenergie-Kerntechnik* **43**, 184 (1983).

30. C. Petitjean, *et al.*, in *Proceedings of an International Symposium on Muon-Catalyzed Fusion* $-\mu CF89$, ed. by J. D. Davies (Rutherford Appleton, 1990), p. 42; C. Petitjean, *et al.*, (unpublished).

31. L. N. Bogdanova, Yu. A. Kuperin, A. A. Kvitsinsky, V. E. Markushin, S. P. Merkuriev, and L. I. Ponomarev, *Muon Catal. Fusion* **3**, 377 (1988).

32. S. A. Alexander, P. Froelich, and H. J. Monkhorst, *Phys. Rev. A* **41**, 2854 (1990); S. A. Alexander, private communication.

33. L. N. Bogdanova and V. E. Markushin (unpublished).

Few-Body Systems, Suppl. 6, 314—325 (1992)

Few-
Body
Systems
© by Springer-Verlag 1992

MUON CATALYZED FUSION (μCF) AS A METHOD FOR STUDYING FEW NUCLEON SYSTEMS

L.N.Bogdanova

Institute for Theoretical and Experimental Physics
Moscow 117259, USSR

Abstract

The theoretical analysis of the methods to obtain information about the properties of few nucleon systems from the muon catalysis studies is presented. The kinetics of the μCF processes in the hydrogen isotope mixture is considered and the observables in μCF are related to the characteristics of the fusion reactions between hydrogen isotopes, e.g., the astrophysical S-factor in the $E = 0$ limit.

The kinetics of μCF in HD mixture is discussed and the doublet and quartet pd spin state contributions to $M1$ pd radiative capture at rest is obtained from the analysis of gamma-ray yields in $pd\mu \to \mu^3 He + \gamma$ reaction. The non-radiative fusion channel $pd\mu \to \mu + {}^3He$ is analyzed and the $E0$ transition matrix element is determined from the data on the absolute muon yields.

The possibility to extract the $M1$ pt radiative capture constant and the $E0$ matrix element of ${}^4He^*(0^+)$ excitation is explained.

It is shown that the properties of the $dd\mu$ molecule and the specific features of its formation mechanism enable one to separate the $P-$ and S-wave $dd \to pt$, n^3He reaction constants.

The spectra of reaction products for $tt\mu \to {}^4He + \mu + 2n$, $\mu^4He + 2n$ are shown to be sensitive to fusion mechanism and the possibility to extract reduced nuclear widths for ${}^6He^*$ decay is pointed out.

The perspectives and unsolved problems are outlined.

1 Introduction

Muon catalyzed fusion reactions between hydrogen isotopes are the final stage of the chain of the processes which occur when a muon gets to the hydrogen isotope (e.g. HD) mixture [1]

$$\mu \;\rightarrow\; \mu p$$
$$\searrow\; \downarrow$$
$$\mu d \;\rightarrow\; p d \mu \;\rightarrow\; \mu^3\mathrm{He} + \gamma$$
$$\searrow\; \mu + {}^3\mathrm{He}$$

Primarily a muon is captured forming a mesic atom, then the muon is transferred to a heavier isotope, next a mesic molecule is formed and finally the fusion reaction $p d\mu \rightarrow \mu^3 He + \gamma$ finishes the cycle. During the past decade, experiments in which the fusion reactions are catalyzed by muons have added a wealth of new information about the nuclear fusion reactions at energies just below the lowest energies accessible by conventional beam-target experiments. In addition to the fact that these experiments probe fusion reactions off the two body energy shell due to the presence of the muon, they introduce a degree of selectivity (by partial waves, e.g.) that is difficult to achieve at these energies by any other means. One of the many interesting questions raised by these measurements is the extent to which extrapolations based on the cross sections and polarizations measured at higher energies agree with the rates obtained from the muon catalyzed fusion (μCF) experiments. The comparisons confront data obtained using widely different experimental techniques, test the energy dependence of nuclear reaction calculations. In some cases the agreement is satisfactory, while in other discrepancies remain unresolved.

The inclusion of short range nuclear interaction effects in the three-body system bound by long range Coulomb forces is usually done with the so called factorization relations [2]. For the rate λ_L^{Jv} of fusion from the given mesic molecular state (Jv) and the state with orbital angular momentum L of relative nuclear motion it reads:

$$\lambda_L^{Jv} = K_L \rho_L^{Jv}. \tag{1}$$

Here K^L is the L-wave nuclear reaction constant:

$$K_L = \lim_{v \to 0} \left\{ \begin{array}{ll} v \sigma_0 C_0^{-2} & L = 0 \\ v \sigma_1 C_1^{-2}(vM)^{-2}/9 & L = 1 \end{array} \right. \tag{2}$$

v and M are relative velocity and reduced mass of nuclei, $C_0^2 = 2\pi\eta/(e^{2\pi\eta} - 1)$ and $C_1^2 = C_0^2(1 + \eta^2)/9$ are Gamov factors for S- and P-waves, $\eta = \alpha c/v$. The factor ρ_L^{Jv} is the probability density of nuclei being in the coalescence point in the given mesic molecule rotational-vibrational state (Jv) and is a characteristics of the three-body Coulomb wave function [3].

The progress achieved in the solving of the Coulomb three-body problem allows to extract the nuclear reaction constant reliably from the fusion rates λ_L^{Jv} measured in μCF experiments. The validity of the factorization relation (1) itself has also been a subject of special interest [2, 4] and has been proved.

However, to relate the fusion rates λ^{Jv} in the mesic molecules with the observables in the μCF experiments which are the yields and time spectra of the fusion products one has to consider all the chain of the mesic atomic and mesic molecular processes preceeding fusion, i.e. to consider μCF kinetics.

In Section 2 we consider the muon catalyzed radiative pd capture as a source of specially prepared initial spin states of fusing nuclei and present the analysis of the γ-ray yields from the reaction $pd\mu \rightarrow \mu^3He + \gamma$ in terms of doublet and quartet pd spin state contributions. The absolute muon yields from the reaction $pd\mu \rightarrow \mu + {}^3He$ are analyzed and $E0$ transition matrix element is determined from the μCF data available.

Section 3 develops an analogous program to determine the radiative and non-radiative pt fusion matrix elements from the $pt\mu$ fusion studies.

Section 4 summarizes the results already available on dd μ-catalyzed fusion and the perspectives for further progress are outlined. Section 5 demonstrates the sensitivity of μCF observables to nuclear structural properties on the example of tt fusion. In Summary we list the unsolved problems and outline the perspectives.

2 Nuclear Fusion Reactions in $pd\mu$ Mesic Molecule

2.1 Overview

The formation and fusion of $pd\mu$ mesic molecules was in the late 50's — early 60's the pioneering object of first μCF investigations. Originally it was observed in bubble chambers [5, 6, 7, 8], later in cloud chambers [9] and with counter experiments [10, 11, 12, 13]. The radiative capture in the molecule competes with the more rare internal conversion process $pd\mu \rightarrow \mu + {}^3He$ which is just the reaction discovered by Alvarez et al. [5] and identified as a muon catalyzed process. The fusion takes place from $J = 0$ (ground) state of the $pd\mu$ system and is essentially S-wave fusion. By that time (and till recently) the quartet fusion was assumed to be negligible, since the parallel spins for all the nucleons in S-waves should be Pauli-blocked.

Taking into account that the mesic molecule ground state is split into four hyperfine structure substates due to the interaction with the muon spin and the fusion proceeds from each state independently:

$$(pd\mu)_J \xrightarrow{\lambda_J} \begin{cases} \lambda_J^\mu & \mu + {}^3He \\ \lambda_J^\gamma & \gamma + \mu^3He \end{cases} \tag{3}$$

Gerstein and Zeldovich [14] predicted the enhancement of the fusion yield with the increase of the deuterium concentration C_d known as Gerstein-Wolfenstein effect [15], which was observed in the experiments [11, 13] to be in agreement with [14]. This effect is due to the transitions between the hyperfine structure states (h.f.s.) $F = 3/2, 1/2$ of the μd atom in collisions

$$\mu d(F = 3/2) + d \rightarrow \mu d(F = 1/2) + d' \tag{4}$$

which are followed by the $pd\mu$ formation and change the distribution over the h.f.s. $pd\mu$ states J, from which the fusion occurs, for the more favourable for doublet pd fusion.

However, the revision of the original estimate, which used the improved spin-flip rate λ_d [16, 17] as well as the result of the direct measurements of λ_d from PSI experiment [18], resulted in an essentially higher prediction for the yield enchancement. To resolve this puzzle it was suggested in [13] that the additional spin flip process $\mu d(3/2) + p \rightarrow \mu d(1/2) + p'$ could play a role, but the rate required was in a drastic disagreement with the theory.

Another long known puzzle in **pd** muon catalyzed fusion concerns the **pdμ** radiative fusion rate λ^γ. The fusion rates λ_J of the reactions (3) from various states J are given by formulas

$$\lambda_J = \lambda_J^\gamma + \lambda_J^\mu \tag{5}$$

$$\lambda_J^{\gamma,\mu} = \rho(w_J^{1/2} K_{1/2}^{\gamma,\mu} + w_J^{3/2} K_{3/2}^{\gamma,\mu}) \tag{6}$$

where ρ is the probability of nuclei being at zero internuclear distance in **pdμ** state $(Jv) = (00)$, $K_S^{\gamma,\mu}$ are the S-wave reaction constants for radiative and nonradiative **pd** fusion from the pure spin states $S = 1/2, 3/2$ and w_J^S are the probabilities of the spin S nuclear configurations in the h.f.s. states J [19]. Under the assumption of the doublet **pd** fusion dominance one gets the theoretical value [14, 20] (with the reaction constant $K_{1/2}^\gamma = 3K_{exp}^\gamma = (2.5 \pm 0.6) \cdot 10^{-22} cm^3 s^{-1}$ from the experiment [21]) about a factor of three larger than the values given by the analysis of the μCF kinetics in $H - D$ mixture [11, 12, 13].

In order to find the solutions to these problems we have studied the μCF kinetics in $H - D$ mixture taking into account that both spin configurations $S = 1/2$ and $S = 3/2$ contribute to the **pd** fusion. We demonstrate that the quartet **pd** fusion gives an important contribution to the observable effects and that all the fusion rates can be obtained from the analysis of the experimental data on the fusion yields as functions of deuterium fraction C_d.

2.2   The kinetics of the muon catalyzed fusion in $H - D$ mixture

The schematics of the μCF kinetics in $H - D$ mixture is shown in Fig.1.

The γ-ray and muonic yields from the reactions $pd\mu \rightarrow \mu^3 He + \gamma, \mu + {}^3He$ have been calculated as functions of deuterium fraction C_d in the hydrogen-deuterium mixture and the fusion rates [22]. The ratio of the γ-ray yields Y^γ measured at different C_d is shown in Fig.2 as the function of the quartet-to-doublet ratio, $X = \lambda_{3/2}^\gamma : \lambda_{1/2}^\gamma$, of the radiative fusion rates λ_J^γ for different spin states $(S = 1/2, 3/2)$ of **pd** system, the 'spin averaged rate' $\lambda^\gamma = 1/3\lambda_{1/2}^\gamma + 2/3\lambda_{3/2}^\gamma = \rho K_{exp}^\gamma$ being kept fixed. With the nonvanishing contribution from the $S = 3/2$ state $(X \approx 0.3)$ one can obtain a fair agreement with the SIN experiment [13] for the relative γ-ray yields without anomalous spin-flip cross section of $\mu d + p$ collisions.

The analysis of the PSI data on the muon catalyzed **pd** fusion [23] gives the values $\lambda_{1/2}^\gamma = 0.35(2) \cdot 10^6 s^{-1}$ and $\lambda_{3/2}^\gamma = 0.11(1) \cdot 10^6 s^{-1}$. The radiative fusion rates thus obtained are in a reasonable agreement with the theoretical calculations [24] which also reproduce the astrophysical factor for the reaction $p + d \rightarrow \gamma + {}^3He$ and the data on the radiative nd capture [25].

These results show that a muon catalyzed fusion can be considered as an alternative laboratory to study few nucleon interaction. The information they provide completes the results already available from polarized beam measurements at higher energies. However, recent capture measurements made with polarized deuterons and protons at energies about $2MeV$ above threshold [26] indicate that the M1 transition strength is predominantly determined by the quartet state contribution and is only about one-tenth that of total, implying a significant energy dependence of the EM transition matrix elements in this region. Additional polarized-beam measurements at energies below $2MeV$ would be quite valuable for defining this energy dependence in anticipation of the availability of Faddeev calculations at higher energies within the next two-to-three years.

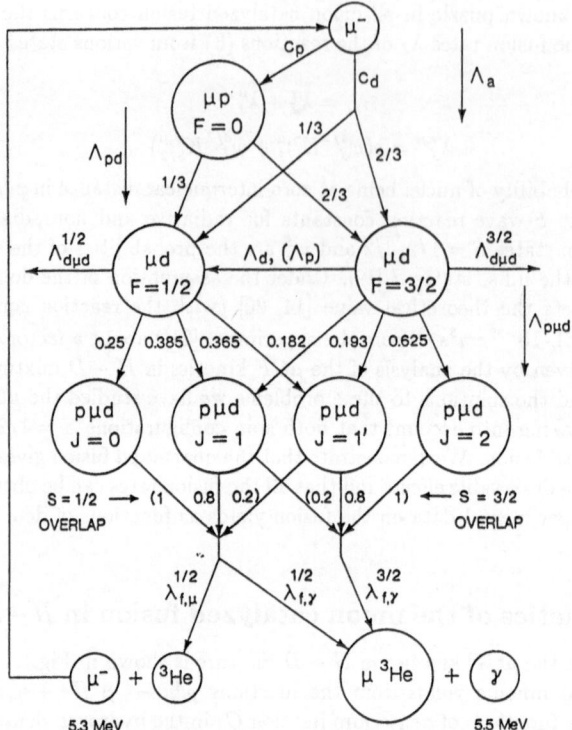

Figure 1: $pd\mu$ kinetics scheme showing the hyperfine structures, transitions and reaction channels. The rates indicated with capital Λ are due to collisions and depend therefore on density and concentration of the colliding nuclei.

2.3 The nonradiative pd fusion

The nonradiative pd fusion rate measurements can serve as a probe of the charge distribution in three nucleon system, especially sensitive to the continuum state wave function structure, thus being of interest for the modern theoretical calculations with realistic NN interaction. The fusion rate $\lambda_{1/2}^{\mu}$ is given by the formula [20]

$$\lambda_{1/2}^{\mu} = 16\pi\alpha^2 mqA_0(q^2) \mid \Psi(0,0) \mid^2 \tag{7}$$

where m and $q = 33 MeV/c$ are the reduced mass and relative momentum of $\mu - {}^3He$ system, function $A_k(q^2)$ is defined by the $E0$ term in the multipole expansion for the transition current

$$< {}^3He \mid J_0(q) \mid pd; k, S = 1/2 >= eq^2 A_k(q^2) + ... \tag{8}$$

with the initial state Coulomb interaction switched off, and $\Psi(r, R)$ is the $pd\mu$ wave function for $(Jv) = (00)$ state (R is the relative coordinate of nuclei, r is the muon coordinate).

We have calculated the muon yields from the reaction $pd\mu \rightarrow \mu + {}^3He$ as the function of deuterium fraction C_d. The results obtained with the radiative fusion rates $\lambda_{1/2}^{\gamma} = 0.35 \cdot 10^6 s^{-1}$ and $\lambda_{3/2}^{\gamma} = 0.11 \cdot 10^6 s^{-1}$ from the analysis of the γ-ray yields [23] are shown in Fig.3.

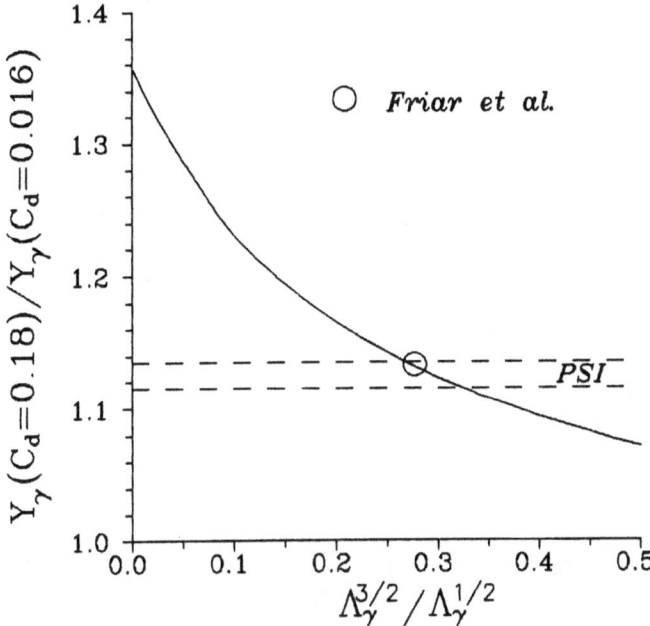

Figure 2: The ratio of γ-ray yields at $C_d = 18.1\%$ and $C_d = 1.6\%$ v.s. the quartet/doublet ratio $X = \lambda^\gamma_{3/2}/\lambda^\gamma_{1/2}$ at $\lambda^\gamma = 1/3\lambda^\gamma_{1/2} + 2/3\lambda^\gamma_{3/2} = 0.2 \cdot 10^6 s^{-1}$, — the calculation with the rates obtained in [24]. The band shown by dashed lines corresponds to the experiment [23].

Considering the rate $\lambda^\mu_{1/2}$ as a free parameter we have found that the theoretical absolute muon yields (solid line in Fig.3) agree with those measured in experiments [5, 6, 7, 8], provided the non-radiative fusion rate is $\lambda^\mu_{1/2} = (0.056 \pm 0.006) \cdot 10^6 s^{-1}$ (the error is statistical), the result having been obtained with the $pd\mu$ formation rate $\lambda_{pd\mu} = 5.6 \cdot 10^6 s^{-1}$. It is worthwhile to point out that the hypothesis of the anomalous spin-flip rate in $\mu d + p$ collisions [13] fails to describe the muon yield.

2.4   Conclusions

1. Both pd nuclear spin configurations $S = 1/2$ and $S = 3/2$ are important in μCF in $pd\mu$ mesic molecule.

2. The experimental dependence of the relative gamma yield on the deuterium fraction [13, 23] can be well reproduced by the present kinetics calculations at the doublet-to-quartet radiative fusion rates ratio $\lambda^\gamma_{1/2} : \lambda^\gamma_{3/2} \approx 3$, the spin averaged reaction constant corresponding to the in-flight measurement [21]. This result is in agreement with modern theoretical calculations of the radiative pd capture at low energies [24].

3. The nonradiative fusion rate has been obtained from the analysis of the bubble chamber experiments [5, 6, 7, 8] $\lambda^\mu_{1/2} = (0.056 \pm 0.006) \cdot 10^6 s^{-1}$. This value is in a good agreement with the recent theoretical result $\lambda^\mu_{1/2} = 0.062(2) \cdot 10^6 s^{-1}$ [24] obtained with the $E0$ matrix element and the $pd\mu$ wave function calculated in the framework of Faddeev approach.

320

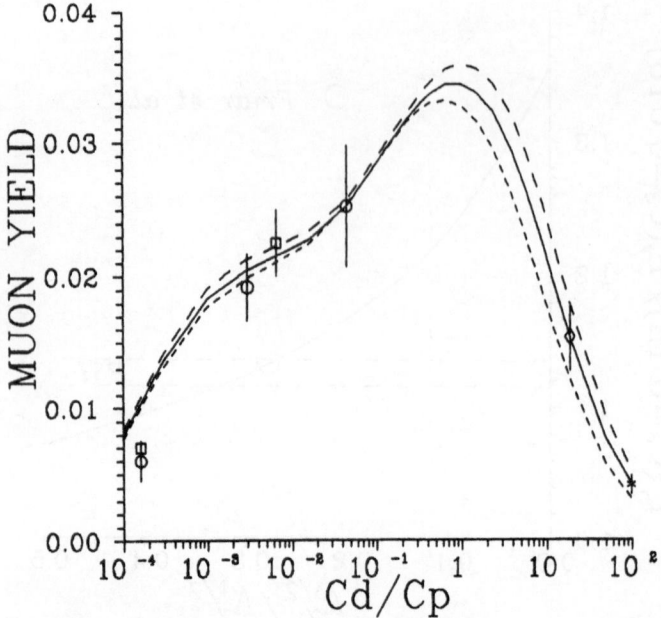

Figure 3: The muon yield Y_μ vs. the ratio $C_d : C_p$ calculated at different $pd\mu$ formation rates $\lambda_{pd\mu} = 5.6\mu s^{-1}$ (solid line) and $4.5\mu s^{-1}$ (dashed line) with $\lambda^\mu_{1/2} = 0.056\mu s^{-1}$. The experimental points: (**o**) – Alvarez et al. [5], (**◊**) – Fetkovich et al. [6], (**□**)– Schiff [7], (**×**) – Doede [8].

4. The gamma fusion spectra of a good quality are the valuable source of information about the low energy *pd* fusion and can be used for the measurements of the *S*-wave reaction constants for the *pd* radiative capture from pure spin states.

3   Muon Catalyzed *pt* Fusion and Structure of $A = 4$ System

Displaying some features similar to the *pdμ* system, *ptμ* mesic molecule opens interesting possibilities to study some specific properties of a few-nucleon system due to the presence of exotic fusion channels. The first experimental results [27] show that there is a serious hope to have in the nearest future a bulk of new data, which should be understood from the point of view of nuclear theory.

The fusion processes in *ptμ* mesic molecule are listed below:

$$pt\mu \rightarrow {}^4He + \mu + \gamma \qquad (9)$$

$$pt\mu \rightarrow \mu {}^4He(> 1S) + \gamma \qquad (10)$$

$$pt\mu \rightarrow {}^4He + \mu \qquad (11)$$

$$pt\mu \rightarrow {}^4He + \mu + e^+ + e^- \qquad (12)$$

Besides the main fusion channels [(9) and (10)] which proceed by γ-emission (with the muon either free of bound to the ^4He nucleus) there are two other fusion processes: (11) emission of an energetic (conversion) muon and (12) emission of an electron-positron pair. These processes are expected to be rare compared to γ-emission, but up to now only theoretical estimates exist on the branching ratio [28].

In $pt\mu$ mesic molecule fusion occurs from the ground state $(Jv) = (00)$ which is split into three sub-levels corresponding to the total spin of $pt\mu$ system $F = 3/2, 1/2, 1/2$ [19]. The fusion goes from each hyperfine structure state independently.

Reactions (11) and (12) are the only processes possible if the total nuclear spin S of the p and t in the $pt\mu$ molecule is $S = 0$, while the radiative processes (9) and (10) are possible only from the triplet $S = 1$ spin states. One of h.f. molecular states $F = 1/2$ is a nearly pure singlet one $S = 0$ and fusion from this state by γ emission is suppressed.

Similarly to the $pd\mu$ case considered before, the population of the molecular states depends on the hyperfine populations a_0 $(F = 0)$ and a_1 $(F = 1)$ of the μt atom from which $pt\mu$ molecule is formated. It influences the yields of reactions to radiative Y_γ and nonradiative Y_μ channels. The competition between spin-flip and molecular formation determines atomic h.f. populations at the moment of molecular formation. The spin-flip rate being dependent on tritium concentration, the ratio $R = Y^\mu/Y^\gamma$ is a function of C_t.

After the complete analysis of μCF kinetics in HT mixture is performed the $pt\mu$ fusion rates to (9)-(12) reaction channels will be obtained. This opens a unique possibility to extract the reaction constant for the pt radiative capture, where the low energy in flight data are rather scarce [29] and the concern is the determination of the relatively small $M1$ and $E2$ matrix elements. They are most effectively separated only by doing extremely complicated analyzing power measurements with polarized protons [30].

The comparison of the pt radiative capture constant and that for the mirror reaction of thermal n^3He capture [31] can test the isovector character of $M1$ matrix element [32] implied by calculations with mesonic exchange currents [33].

The radiationless muon-catalyzed pt fusion [34] is also of great interest in order to obtain the E0 transition amplitude which can be directly related to the quantity deduced either from $^4He(ee')^4He^*$ experiment [35] or from microscopic calculations [36] aimed to determine the nature of ^4He low lying excitations.

4   Muon Catalyzed $d + d$ Reactions

The selectivity of μCF processes in $dd\mu$ mesic molecule [4, 37] resulted in the direct measurement of the n - to - p ratio for the P-wave part of the charge-symmetric reactions:

$$d + d \rightarrow n + {}^3He \tag{13}$$

$$d + d \rightarrow p + {}^3H \tag{14}$$

In accordance with the experiments for these reactions done at low energies with polarized beams [38], the μCF experiment [39] yielded a surprising result: the neutron branch is enchanced about 40% with respect to the proton branch. This enhancement is much larger than that given by the integrated reaction cross section at low energies, and is in the opposite direction. It has been interpreted by some as evidence for charge-symmetry breaking of nuclear forces.

However, an essentially charge-independent R-matrix calculation of all the four-body reactions [40] accounts for these anomalous experimental results, but the absolute fusion rates given by the R-matrix analysis appear to be above 20% higher than the value extracted from μCF experiment [39].

The even-spin (S-wave) transitions of the $d + d$ reactions have branching ratios close to unity that favour slightly the $p - {}^3H$ branch. This explains quantatively why a strong temperature dependence is seen for the branching ratio in μCF [41] as the temperature is lowered below $20°C$.

The reason for this effect is in the competition between the resonant and nonresonant $dd\mu$ molecule formation mechanism which becomes noticeable in this temperature region. While the former produces $dd\mu$ molecule in the excited $(Jv) = (11), (10)$ states, the latter results in the formation of $J = 0$ states, and they favour, respectively P- and S-wave dd fusion. The analysis of dd μ- catalyzed fusion kinetics can, in principle, provide an independent determination of S- and P-wave reaction constants. This is a unique possibility, especially in the context of the recent work by a group from Munster, where the indication are that screening by atomic electrons may be producing measurable effects on the cross sections for the interactions of "bare" nuclei [42]. Recent calculations [43] suggest that screening corrections begin to be important at the lower end of the energy reached in the in-flight experiments [42]. Hence in the extrapolation of the electron screened cross-sections one should take into account the point at zero energy and bare nuclei from the μCF.

5   Nuclear Fusion in $tt\mu$ Mesic Molecule

As in case of $d + d$, the muon catalyzed fusion for reactions

$$tt\mu \rightarrow {}^4He + 2n + \mu \tag{15}$$

$$tt\mu \rightarrow \mu {}^4He + 2n \tag{16}$$

selects out the P-wave transitions, since the fusion occurs from states $J = 1, v = 0, 1$ of the $tt\mu$ mesic molecule [4]. The experimental study of the in-flight reaction $t + t \rightarrow {}^4He + 2n$ is exceedingly difficult task because of the tritium in the initial state and the three-body character of the final state [44].

For this reason some theoretical considerations are used to calculate the $tt\mu$ fusion rates [4]. Of the theoretical schemes suggested, one looks natural that reaction goes through the compound ${}^6He^*$ nucleus. The peculiarity of this scheme is the presence of strong $n - {}^4He$ correlations, related to the formation of virtual ${}^5He^*$ state. The reaction constant and, consequently, the fusion rate prove to be sensitive to model parameters and can be used to extract the reduced widths of the low-lying P-wave states in $A = 5, 6$ systems.

The sticking probability, i.e. the branching ratio to channel (16), also proves to be sensitive to nuclear reaction model. The analysis of neutron spectra for the muon catalyzed fusion in liquid and gaseous tritium measured in experiment [45] found the presence of strong n^4He correlation in the final state of reaction. The measured value of the sticking coefficient $\omega_t = (14 \pm 3)\%$ is in agreement with the resonance mechanism of tt P-wave fusion. Note, that other extreme assumption, i.e. nn final state correlations would result in a value $\omega_t = 5\%$ [4].

Resuming we express the hope that the theoretical analysis of excitations in $A = 5, 6$ neutron-rich systems will have a strong support from the μCF experiments.

6 Summary

The muon catalyzed fusion reactions between hydrogen isotopes which are actively studied now provide a unique information about few nucleon systems. The example of *pd* capture displays the sensivity of the μCF which enables one to test the existing models of nuclear structure, such as the non-nucleonic degrees of freedom in the nucleus (mesonic exchange currents) [46].

The experimental program on μCF which is in progress makes actual some problems both for nuclear and μCF theory.

For the *pt* fusion the microscopic calculation of the radiative capture and the radiationless processes (muon conversion and e^+e^- pair production) should be performed.

The analysis of *ptμ* kinetics should be completed to reliably extract the relevant fusion rates from μCF experiments.

The *ddμ* kinetics applied to μCF data will separate the S- and P-wave reaction constants for *dd* fusion mirror channels, which gives a good reference point for the electron screening effects in fusion reaction, being of importance in the analysis of stellar processes.

The *ttμ* branch of μCF is expected to give insight into nuclear models of low-lying excitations in lightest neutron-rich nuclei $A = 5, 6$.

Acknowledgements

The author is grateful for important discussions with J.L.Friar and G.M.Hale of Los Alamos National Laboratory and with V.E.Markushin and L.I.Ponomarev of Kurchatov Atomic Energy Institute.

References

[1] L.I.Ponomarev, Contemporary Physics **31**(1990) 219; S.S.Gerstein, Yu.V.Petrov and L.I.Ponomarev, Uspekhi Fiz.Nauk **160** (1990) 3 [Sov. Phys. Uspekhi (1990)].

[2] L.N.Bogdanova, V.E.Markushin, V.S.Melezhik and L.I.Ponomarev, Sov. Journ. Nucl. Phys. **34** (1981) 662.

[3] S.A.Alexander, P.Froelich, and H.J.Monkhorst, Phys. Rev. **A41** (1990) 2854; Errata (1991).

[4] L.N.Bogdanova, Muon Catalyzed Fusion **2** (1988) 359.

[5] L.W.Alvarez et al., Phys. Rev. **105** (1957) 1127.

[6] J.G.Fetkovich et al., Phys. Rev. Lett. **4** (1960) 570.

[7] M.Schiff, Nuovo Cimento **22** (1961) 66.

[8] J.H.Doede, Phys. Rev. **132** (1963) 1782.

[9] V.P.Dzhelepov et al., Sov. Phys. JETP **23** (1966) 820.

[10] A.Ashmore et al., Proc. Roy. Soc. **71** (1958) 161.

324

[11] E.J.Bleser et al., Phys. Rev. **132** (1963) 2679.

[12] V.M.Bystritsky et al., Sov. Phys. JETP **43** (1976) 606; **44** (1976) 881.

[13] W.Bertl et al., Atomkernenergie **43** (1983) 184.

[14] Ya.B.Zeldovich and S.S.Gerstein. Sov. Phys. Uspekhi **3** (1961) 593.

[15] S.S.Gerstein, Sov. Phys. JETP **13** (1961) 488.

[16] A.V.Matveenko and L.I.Ponomarev, Sov. Phys. JETP **32** (1971) 871.

[17] V.S.Melezhik and J.Wozniak, Phys. Lett. **A116** (1986) 370; L.Bracci, C.Chiccoli, P.Pasini, G.Fiorentini, J.Wozniak and V.S.Melezhik, Phys. Lett. **A165** (1989) 459.

[18] P.Kammel, W.H.Breunlich, M.Cargnelli, H.G.Mahler, J.Zmeskal, W.Bertl and C.Petitjean, Phys. Rev. **A28** (1983) 2611.

[19] D.D.Bakalov, S.I.Vinitsky and V.S.Melezhik, Sov. Phys. JETP **52** (1980) 820.

[20] L.N.Bogdanova, Yu.A.Kuperin, A.A.Kvitsinsky, V.E. Markushin, S.P.Merkuriev, L.I.Ponomarev, Muon Catalyzed Fusion **3** (1988) 377.

[21] G.M.Griffiths, M.Lal wnd C.D.Scarfe, Can. J. Phys. **41** (1963) 724.

[22] L.N.Bogdanova and V.E.Markushin, Muon Catalyzed Fusion **5** (1990) 189.

[23] C.Petityean, K.Lou, P.Ackerbauer et al., Muon Catalyzed Fusion **5** (1990) 191.

[24] J.L.Friar, B.F.Gibson, H.J.Jean and G.L.Payne, Phys. Rev. Lett. **66** (1991) 1827.

[25] J.L.Friar, B.F.Gibson, and G.L.Payne, Phys. Lett. **B251** (1990) 11.

[26] M.Vetterli et al., Phys. Rev. **C38** (1988) 2503.

[27] F.J.Hartmann, H.Bossy, H.Daniel et al., Muon Catalyzed Fusion **2** (1988) 53; P.Baumann, H.Daniel, S.Grunewald et al., in Proceedings of an International Symposium on Muon Catalyzed Fusion – μCF89, ed. by J.D.Davies (Rutherford Appleton, 1990), p. 38.

[28] L.N.Bogdanova and V.E.Markushin, Muon Catal. Fusion **4** (1989) 103.

[29] J.E.Perry and S.J.Bame, Phys. Rev. **99** (1955) 1368.

[30] D.J.Wagenaar et al. Phys. Rev. **C39** (1989) 352.

[31] V.P.Alfimenkov et al., Sov. Journ. Nucl. Phys. **31** (1980) 10; M.Suffert and R.Berthollet, Nucl. Phys. **A318** (1979) 54; F.L.H.Wolfs et al., Phys. Rev. Lett. **63** (1989) 2721; R.Welverman et al., Nucl. Phys. **A526** (1991) 265.

[32] J.D.Jackson, private communication.

[33] I.S.Towner and F.C.Khanna, Nucl. Phys. **A356** (1981) 445; J.Carlson, D.O.Risca, R.Schiavilla and R.B.Wiringa, Phys. Rev. **C42** (1990) 830.

[34] L.N.Bogdanova and V.E.Markushin, Muon Catalyzed Fusion **4** (1989) 103, Nucl. Phys. **A508** (1990) 29 c.

[35] R.F.Frosch et al., Nucl. Phys. **A110** (1968) 657;
Th.Walcher, Phys. Lett. **31B** (1970) 442;
S.Fiarman and W.E.Meyerhof, Nucl. Phys. **A206** (1973) 1;
G.Kobschall et al., Phys. Rev. **130** (1963) 1987.

[36] V.S.Vasilevsky and I.Yu.Ryibkin, Sov. Journ. Nucl. Phys. **46** (1987) 220; **48** (1988) 217.

[37] L.N.Bogdanova et al., Phys. Lett. **115B** (1982) 171; Errata **167B** (1986) 485.

[38] B.P.Adyasevich, V.G.Antonenko, and V.E.Bragin, Sov. Journ. Nucl. Phys. **33** (1981) 173; JETP Lett. **40** (1984) 112.

[39] D.V.Balin et al., Phys. Lett. **141B** (1984) 173; JETP Lett. **40** (1984) 112.

[40] G.M.Hale, Muon Catalyzed Fusion **5** (1990) 227.

[41] D.V.Balin, V.N.Baturin, Yu.A.Chestnov et al., Muon Catalyzed Fusion **5** (1990) 163.

[42] A.Krauss, H.W.Becker, H.P.Trautvetter and C.Rolfs, Nucl. Phys. **A465** (1987) 150.

[43] L.Bracci, G.Fiorentini, V.S.Melezhik et al., Nucl. Phys. **A513** (1990) 316; Phys. Lett. **153A** (1991) 456.

[44] H.M.Agnew et al., Phys. Rev. **84** (1951) 862;
V.I.Serov et al., Atomnaya Energia **42** (1977) 59.

[45] W.H.Breunlich, M.Carnelli, P.Kammel et al., Muon Catalyzed Fusion **1** (1987) 67.

[46] J.L.Friar, this volume.

Few-Body Systems, Suppl. 6, 326—331 (1992)

EFFECT OF THE NUCLEAR $d - t$ RESONANCE ON MUON STICKING IN μ-CATALYZED FUSION

J. Révai

Central Research Institute for Physics, H-1525 Budapest, P.O.B. 49, Hungary

A.L. Zubarev and L. Ya. Higer

Department of Physics, Tashkent State University, 700095 Tashkent, U.S.S.R.

V. B. Belyaev

Joint Institute for Nuclear Research, 141980 Dubna, U.S.S.R.

Abstract

The generally accepted validity of the sudden approximation for the calculation of the sticking coefficient ω_s^0 is questioned. Physically this doubt is motivated by the fact, that due to the ${}^5\text{He}^{3/2+}$ resonance, the nuclear interaction time ($\sim 10^{-20}s$) is non-negligeable compared to the muon orbiting time ($\sim 10^{-19}s$); thus the "propagation" of the muon during the nuclear process can not be excluded.

Calculations are based on a formally exact, coupled two-channel three-body formulation of the fusion process in the $dt\mu^-$ system [1,2]. After a careful definition of the sticking coefficient within this framework the sudden formula for ω_s^0 is derived pointing out the chain of approximations, leading to it. The effect of the nuclear resonance is incorporated in a modified expression for the transition amplitude (and ω_s^0), in which the simple overlap of the initial and final muon wave functions is replaced by a (free) propagation between them. The characteristic momentum (or inverse range) of this propagation is determined by the difference of the total (three-body) energy of the system and the resonance energy of the heavy-particle subsystem.

Our numerical calculations give an ω_s^0, which is roughly a factor 2 smaller, than the sudden value and is rather sensitive to the nuclear resonance parameters: changing them whithin the experimental errors results in a 10-15% variation of ω_s^0.

The sudden approximation (SA)

The number of μ-catalyzed fusion reactions

$$(dt\mu) \quad \Bigg\langle \begin{array}{l} \xrightarrow{\omega_s^0} \quad (\alpha\mu) + n + 17.6\ MeV \\[2mm] \xrightarrow{1-\omega_s^0} \quad \alpha + \mu + n + 17.6\ MeV \end{array} \tag{1}$$

a muon can produce during its life time depends crucially on the value of the sticking coefficient ω_s^0, the probability of the first reaction in (1), when the muon in the final state is bound to the α-paricle and thus is lost for generating further reactions.

The most commonly accepted way of calculating ω_s^0 is the so-called sudden approximation, according to which the nuclear reaction from the point of view of the muon can be regarded as a sudden change of its Hamiltonian. In this case the corresponding transition amplitudes are given by a simple overlap of the muon wave function before and after the reaction. The initial wave function is identified with the normalized zero internuclear distance limit of the mesomolecular ground state wave function $\Phi_0(\mathbf{R}, \mathbf{r})$, while the final states are represented as Coulombic bound or scattering states of the muon with respect to the recoiling α-particle.

However, as it is well known, the low energy nuclear $d-t$ interaction is dominated by the $^5\mathrm{He}^{3/2+}$resonance, which has a life time τ_N of the order of $10^{-20}s$ ($\Gamma \sim 100$ keV) which is not obviously negligible compared to characteristic μ-mesic orbiting times ($\tau_\mu \sim 10^{-19}s$). Thus the validity of the basic assumption of the SA ($\tau_N \ll \tau_\mu$) can be questioned.

Since it is not easy to define corrections to the SA, we started from the other side: tried to derive it from an exact formalism and discuss the approximations leading to it. For the description of the fusion reaction (1) a coupled two-channel three-body model was adopted with the Hamiltonian

$$H = \begin{pmatrix} H_1 & V_{12}^S \\ V_{21}^S & H_2 \end{pmatrix} \tag{2}$$

where H_1 describes the system in the $dt\mu$-channel

$$H_1 = T_{\mathbf{R}} + T_{\mathbf{r}} + V_{dt}^C + V_{d\mu}^C + V_{t\mu}^C + V_{11}^S = H_1^0 + V_1^C + V_{11}^S, \tag{3}$$

while H_2 acts in the $\alpha n\mu$-channel:

$$H_2 = T_{\mathbf{R}'} + T_{\mathbf{r}'} + V_{\alpha\mu}^C + V_{22}^S = H_2^0 + V_2^C + V_{22}^S, \tag{4}$$

In Eqs.(3)-(4) $\mathbf{R}(\mathbf{R}')$ are the vectors connecting the heavy particles, $\mathbf{r}(\mathbf{r}')$ are the Jacobian coordinates of the muon; V^S is the strong interaction matrix responsible for the nuclear reaction. In our approach [1,2] the reaction (1) is considered as an electromagnetic transition (treated perturbatively) between stationary two-component states of the Hamiltonian (2). Thus the strong interaction rather "forms" the sates instead of leading to transitions betwen them.

Our first approximation concerns the initial ($J = 1, n = 1$) "capture" state $|\Psi_i>$: due to the centrifugal barrier the effect   of the short range V^S can be neglected and we can identify it with the purely Coulombic (11) exited state of the $dt\mu$ mesomolecule:

$$|\Psi_i>= \begin{pmatrix} |\Phi_{11}^C > \\ 0 \end{pmatrix}$$

The final states $|\Psi_f^{\nu,E}>$ are $J = 0$ two-component, three-body states, labelled by the asymptotoc quantum numbers ν in the $\alpha n\mu$-channel:

$$\nu = (nl) \quad \text{or} \quad (\epsilon l)$$

the n and ϵ referring to bound or scattering n-α Coulombic states acompanying the neutron plane wave. This labeling leads to a precise definition of the sticking coefficient ω_s^0 and the accuracy of its calculated value depends on the accuracy of solving the coupled three-body equations for the final state.

From the formally exact equations for $|\Psi_f>$ one can derive the SA formula for the sticking coefficient ω_s^0 going through the following chain of approximations (for the details see [2]):

(a) In the present formalism the Coulombic ground state $\Phi_0(\mathbf{R}, \mathbf{r})$ of the $td\mu$ mesomo-lecule is missing. If we believe, that, according to the traditional picture, this state in a sense dominates the $td\mu$-channel, we can express this idea in this approach by approximating the $dt\mu$-channel Coulomb Green's function by a single term in its spectral representation:

$$G_1^C = (E - H_1^0 - V_1^C)^{-1} \approx \frac{|\Phi_0 >< \Phi_0|}{E - E_0}$$

(b) A less important, rather technical approximation is to neglect the neutron plane wave distortion by the strong n-α interaction in the $\alpha n\mu$-channel: $V_{22}^S \approx 0$.

(c) Our final states are in the continuous spectrum of H, therefore the calculation of total transition probabilities needed for ω_s^0 involves integration over the possible final state energies. The transition probability has a sharp peak around the $dt\mu$ ground state energy E_0 and the next approximation is to consider in the energy integrals only the contribution of this peak, replacing the energy variable in the slowly varying parts of the probabilities by E_0.

(d) Zero range nuclear interaction is assumed.

(e) According to the general arument about the smallness of characteristic μ-mesic energies compared to the reaction Q-value (17.6 MeV), the dependence of the final neutron momentum on the μ final state label ν is neglected. Thus the comleteness of $(\alpha\mu)$ states can be used when calculating the total transition probability, which is the denominator of ω_s^0.

The set of approximations (a)-(e) leads from the exact definition of ω_s^0 to the SA formula. We stess here, that this is just the list of the necessary approximations, their validity and accuracy remains to be checked.

A possible improvement of the SA

In order to demonstrate the effect of the resonant character of the nuclear $d-t$ interaction upon the sticking coefficient ω_s^0— without dropping the approximations (a)-(e) — one can start from a slightly modified dynamical equation for the final state $|\Psi^{\nu,E}>$, in which instead of the strong interaction matrix V^S the corresponding scattering operator T^S is introduced. The operator T^S takes into account multiple rescattering between the two channels and carries the energy dependence corresponding to the $^5\mathrm{He}^{3/2+}$ resonance. The same approximations (a)-(e) applied now to this equation yield a modified expression for the sticking coefficient, in which the simple overlap integral of the initial and final muon wave functions

$$\int \phi_f^*(\mathbf{r})\phi_i(\mathbf{r})d\mathbf{r}$$

entering the SA is replaced by a "propagation" between these states

$$\int \phi_f^*(\mathbf{r})\frac{e^{i\kappa|\mathbf{r}-\mathbf{r}'|}}{|\mathbf{r}-\mathbf{r}'|}\phi_i(\mathbf{r}')d\mathbf{r}d\mathbf{r}'$$

and the characteristic momentum (or inverse range) κ of the propagator is given by

$$\kappa = \sqrt{2m_\mu(E_0 - E_R + i\Gamma/2)} \qquad Im\,\kappa > 0$$

where E_R and Γ are the position and width of the $^5\mathrm{He}^{3/2+}$ resonance. For the experimental values of E_R and Γ we get a κ, which is commensurable with the muonic momenta in ϕ_i and ϕ_f and therefore a sensible effect can be expected.

Indeed, calculations performed along this line confirm this expectation: the resulting ω_s^0 strongly differ from the SA values. The results of calculations are shown in Table 1

Table 1. Values of the sticking coefficient ω_s^0 for different resonance parameters E_R and Γ. The sudden value (without strong interaction) is $\omega_s^0 = 0.895$

E_R keV	Γ keV	ω_s^0 %	E_R keV	Γ keV	ω_s^0 %	E_R keV	Γ keV	ω_s^0 %
64	70	0.469	64	200	0.597	200	100	0.653
64	90	0.486	64	500	0.774	500	100	0.769
64	100	0.496	64	750	0.828	1000	100	0.826
64	120	0.516	64	1000	0.854	1500	100	0.849
64	140	0.535	64	1200	0.866	2000	100	0.860
54	100	0.480	64	1500	0.876	3000	100	0.872
60	100	0.490	64	2000	0.885	6000	100	0.884
70	100	0.506	64	5000	0.896			

and their essence can be summarized as follows:

- For the variation of the nuclear resonance parameters within the experimental errors ω_s^0 is roughly a factor 2 smaller than the SA value and changes about 10-15 %;

- if the resonance parameters are moved away from E_0 on the real axis, the propagator resembles a δ-function and ω_s^0 approaches the SA value.

Conclusions

The main message of this talk — according to the authors' intention — is to emphasize, that the crucial question of muon sticking in the $td\mu$ fusion cycle is still open. This statement holds for the experimental situation (for a recent review see [3]), where even the latest, still preliminary results [4,5] preserve the ambiguities which were char acteristic for the last few years.

From the theoretical side our aim was to shake the overall confidence in the SA, without claiming, however, that our modified formula is the ultimate method for calculating ω_s^0. The two-channel three-body dynamics of the $dt\mu$ fusion reaction is very complicated, it involves the interplay of two-body and three-body resonances, which are produced by forces of very different range, but which happen to be "close" on the rather non-trivial enegy surface of this problem. The simplifying arguments based on non-matching of energy or distance scales are only partly acceptable. We believe, that our approach contains a "little more" from the three-body dynamics, than the SA, but in order to make a choice among the heuristic approaches a decent two-channel three-body calculation is needed.

References

[1] V. B. Belyaev, J. Révai and A. L. Zubarev, Phys. Lett. B **219**(1989)157

[2] J. Révai, A. L. Zubarev, L. Ya. Higer and V. B. Belyaev, Phys. Rev. A **43**(1991)4611

[3] J. Rafelski, H. Rafelski, *to be published in* Advances in Atomic, Molecular and Optical Physics

[4] K. Nagamine et al., μCF '90 Vienna, 1990, Abstract book, p.36

[5] P. Ackerbauer et al., PSI Annual Report 1990, p.50

Few-Body Systems, Suppl. 6, 332—337 (1992)

© by Springer-Verlag 1992

MUONIC MOLECULES OF LIGHT NUCLEI

V. B. Belyaev[1], M. Decker[2], H. Haberzettl[3], L. J. Khaskilevitch[1], W. Sandhas[2]

[1]Laboratory for Theoretical Physics, JINR, 141980 Dubna, USSR
[2]Physikalisches Institut, Universität Bonn, D-5300 Bonn 1, Germany
[3]Physics Department, George Washington University, Washington DC, USA

We present a systematic study of the muonic molecules $(p \mu N_z)$, $(d \mu N_z)$, and $(t \mu N_z)$ where N_z are nuclei with charges $Z = 1, 3, 4$. Our calculations are based on a dual expansion scheme where, in a first step, we employ expansions into surface functions which are then, in a second step, expanded into hyperspherical harmonics. To ensure the correct asymptotic behaviour of the eigenpotentials, we amend the second step by an additive expansion into channel functions. The eigenpotentials obtained show the expected attraction for the $(d \mu t)$ system. For $Z = 3, 4$, high-lying dips are found indicating resonance behaviour. At present, the radial equation is solved in extreme adiabatic approximation. First attempts are being made towards a solution in adiabatic approximation.

The dynamics of muons in hydrogen media with N_z-admixtures is expected to be essentially determined by the formation of $(N_z \mu p)$ resonances [1]. Such a situation is of considerable interest both experimentally and theoretically. It provides us, e. g., with detailed information on the strong p-N_z interaction at very low energies (a few keV), an information relevant to astrophysical questions, e. g., the pp-cycle in the sun. First theoretical investigations of such resonances have been performed in low-order Born-Oppenheimer approximation [2]. In this contribution we present a more reliable calculation based on surface function expansions, and study systematically the dependence of the properties of several muonic molecules upon charge and mass of the nuclei involved.

Let us consider three charged particles with charges Z_i and masses m_i.

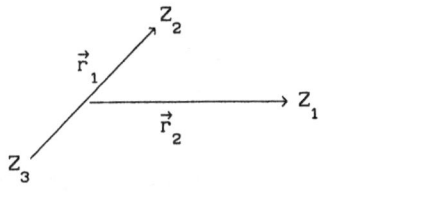

(fig. 1)

Following the notation of fig. 1, we introduce dimensionless Jacobi coordinates

$$\vec{x} = \frac{\vec{r}_1}{a} \quad , \quad \vec{y} = \frac{\vec{r}_2}{\alpha a} \quad , \tag{1}$$

where $\quad a = \dfrac{C}{m|Z_2 Z_3|} \; , \; \alpha^2 = \dfrac{m}{M} \; , \; m = \dfrac{m_2 m_3}{m_2 + m_3} \; , \; M = \dfrac{m_1(m_2 + m_3)}{m_1 + m_2 + m_3} \; , \; C = \left[\dfrac{2}{|Z_2 Z_3|}\right]^{\frac{1}{2}} \; .$

$$\tag{2}$$

Using hyperspherical variables r and ϕ,

$$|\vec{x}| = r \cos\phi \text{ and } |\vec{y}| = r \sin\phi \quad , \tag{3}$$

the total Hamiltonian of the three particles then reads

$$H = - r^{-5/2} \frac{\partial^2}{\partial r^2} r^{5/2} + H_r(\Omega) \quad , \tag{4}$$

with $\qquad H_r(\Omega) = \dfrac{\hat{\mathscr{L}}^2}{r^2} + V(r, \Omega) \quad . \tag{5}$

Here, the grand angular momentum operator $\hat{\mathscr{L}}^2$ is given by

$$\hat{\mathscr{L}}^2 = - \frac{1}{\cos\phi \sin\phi} \frac{\partial^2}{\partial\phi^2} \cos\phi \sin\phi + \frac{\vec{\ell}_1^{\,2}}{\cos^2\phi} + \frac{\vec{\ell}_2^{\,2}}{\sin^2\phi} - \frac{1}{4} \tag{6}$$

($\vec{\ell}_1$ and $\vec{\ell}_2$ are the subsystem angular momenta) and the Coulomb interaction between the three particles is of the form

$$V(r,\Omega) = \frac{C}{r}\left[\frac{Z_2 Z_3}{\cos\phi} + \frac{Z_1 Z_3}{|\alpha\sin\phi\ \vec{e}_y + \gamma\cos\phi\ \vec{e}_x|} + \frac{Z_1 Z_2}{|\alpha\sin\phi\ \vec{e}_y - \beta\cos\phi\ \vec{e}_x|}\right]$$

(7)

with
$$\beta = \frac{m_3}{m_2+m_3}\ ,\quad \gamma = \frac{m_2}{m_2+m_3}\quad \text{and}\quad \vec{e}_x = \frac{\vec{x}}{x}\ ,\quad \vec{e}_y = \frac{\vec{y}}{y}\ .$$
(8)

The solutions of the three-body Schrödinger equation

$$H\ \Psi(r,\Omega) = E\ \Psi(r,\Omega)$$

(9)

may be expanded into *surface functions* $B_n(r,\Omega)$ defined by

$$H_r(\Omega)\ B_n(r,\Omega) = U_n(r)\ B_n(r,\Omega)$$

(10)

where $U_n(r)$ denotes the eigenvalues (*eigenpotentials*) of this eigenvalue equation. In other words, we write

$$\Psi(r,\Omega) = \sum_n \frac{f_n(r)}{r^{5/2}}\ B_n(r,\Omega)$$

(11)

which immediately leads to the following coupled system of differential equations for the radial functions $f_n(r)$

$$f_n''(r) = (\ U_n(r) - E\ )\ f_n(r)$$

(12)

$$- \sum_m \left[\ 2\ \langle B_n(r)|B_m'(r)\rangle\ f_m'(r) + \langle B_n(r)|B_m''(r)\rangle\ f_m(r)\ \right]\ .$$

In this paper we restrict ourselves to the *extreme adiabatic approximation* of eq. (12):

$$f_n''(r) = \left[\ U_n(r) - E\ \right]\ f_n(r)\ .$$

(13)

In order to guarantee the correct asymptotic behaviour of the surface functions we expand them in terms of channel functions $\Phi(r,\Omega)$ which, for large r, show the clustering of the channels involved (compare [3]). At small r, an appropriate basis is given by the eigenfunctions of the grand angular momentum, the hyperspherical harmonics $Y_{[\mathscr{L}]}(\Omega)$. This suggests the representation

$$B_n(r,\Omega) = \sum_{i=1,3} \sum_{[m^i]} a^i_{n[m^i]}(r) \ \Phi^i_{[m^i]}(r,\Omega) \ + \ \sum_{[\mathscr{L}]} b_{n[\mathscr{L}]}(r) \ Y_{[\mathscr{L}]}(\Omega) \ , \qquad (14)$$

where i=1,3 indicate the two fragmentations $1 \hat{=} (1,23)$ and $3 \hat{=} (3,12)$. In the following calculations we consider only surface functions belonging to total angular momentum L = 0.

To get a feeling for the quality of this ansatz we first consider the $d\mu t$ system, for which very elaborate calculations are available [4]. Using 6 channel functions and 120 hyperspherical harmonics in the expansion (14), we find the eigenpotentials (solid lines) presented in fig. 2. They agree qualitatively rather well with the results (□ □ □) of [5], but show less attraction at intermediate distances. The corresponding $d\mu t$ energies

$$E_1 = - \ 295.54 \ eV \quad and \quad E_2 = - \ 35.80 \ eV \quad ,$$

obtained by means of eq. (13) are in reasonable agreement with the values

$$E_1 = - \ 319.14 \ eV \quad and \quad E_2 = - \ 35.83 \ eV$$

reported in [4].

When treating the $x\mu Li$ and $x\mu Be$ states, where x=p,d,t, we use 10 channel functions in the $x\mu$ subsystems in order to take into account the polarizability of the $x\mu$ atoms in a proper way. Fig. 3 shows the lowest 7 eigenpotentials of the $p\mu^6Li$ system. In figs. 4-9 we present the eigenpotentials with $x\mu$ ground-state asymptotics for various $x\mu N_z$ systems. We see that with increasing mass of particle x the size of these systems is slightly enlarged. Increasing the charge of the heavy nucleus also leads to more extended systems. The absolute values of the binding energies (in eV) given in the following table show the same tendency but more pronounced.

system	$p\mu^6Li$	$d\mu^6Li$	$t\mu^6Li$	$p\mu^7Li$	$d\mu^7Li$	$t\mu^7Li$	$p\mu^7Be$	$d\mu^7Be$
present	24.3	23.8	35.3	20.8	25.9	37.5	11.7	29.3
Ref.2	17.6	18.5	19.8	21.0	22.0	23.3	-	-

336

Finally, let us mention that by slightly increasing the attraction of the eigenpotentials we find in all cases a second bound state, which indicates that there are resonances near zero energy.

References

[1] Bystritskii, V.M., Dzhelepov, V.P., Petrukhin, V.I., Rudenko, A.I., Suvorov, V.M., Filchenkov, V.V., Kovanskii, N.N., Khomenko, B.A., Sov.Phys.JETP **57**, 728 (1983); Nagamine, K., Matsuzaki, T., Ishida, K., Hirata, Y., Watanabe, Y., Miyake, Y., Kadono, R., AIP Conf.Proc. **181**, 23 (1989); Aristov, Yu.A., Kravtsov, A.V., Popov, N.P., Solyakin, G.E., Truskova, N.F., Faifman, M.P., Sov.J.Nucl.Phys. **33**, 564 (1981); Belyaev, V.B., Haberzettl, H., Kuzmichev, V.E., Sandhas, W., Proc. μCF'90, Vienna.

[2] Kravtsov, A.V., Popov, N.P., Solyakin, G.E., Sov.J.Nucl.Phys. **35**, 876 (1982).

[3] Lin, C.D., Phys.Rev.**A 23**, 1585 (1981).

[4] Monkhorst, H.J., Proc. μCF'90, Vienna.

[5] Gusev, V.V., Puzynin, V.I., Kostrykin, V.V., Kvitsinsky, A.A., Merkuriev, S.P., Ponomarev, L.I., Few-Body Syst. **9**, 137 (1990).

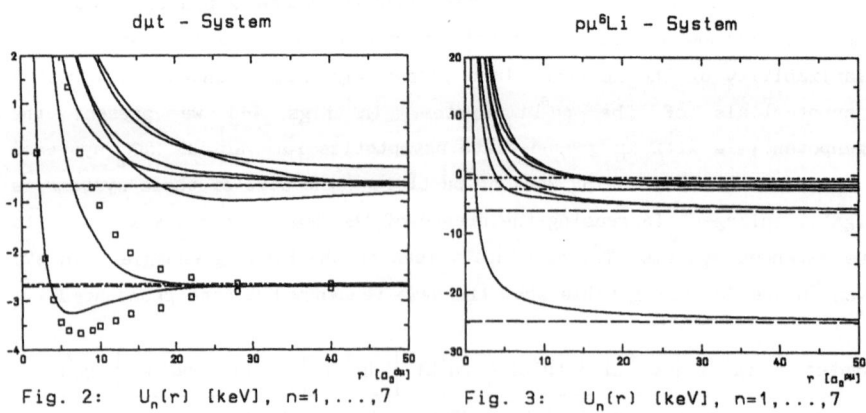

dμt - System

pμ^{6}Li - System

Fig. 2: $U_n(r)$ [keV], n=1,...,7

Fig. 3: $U_n(r)$ [keV], n=1,...,7

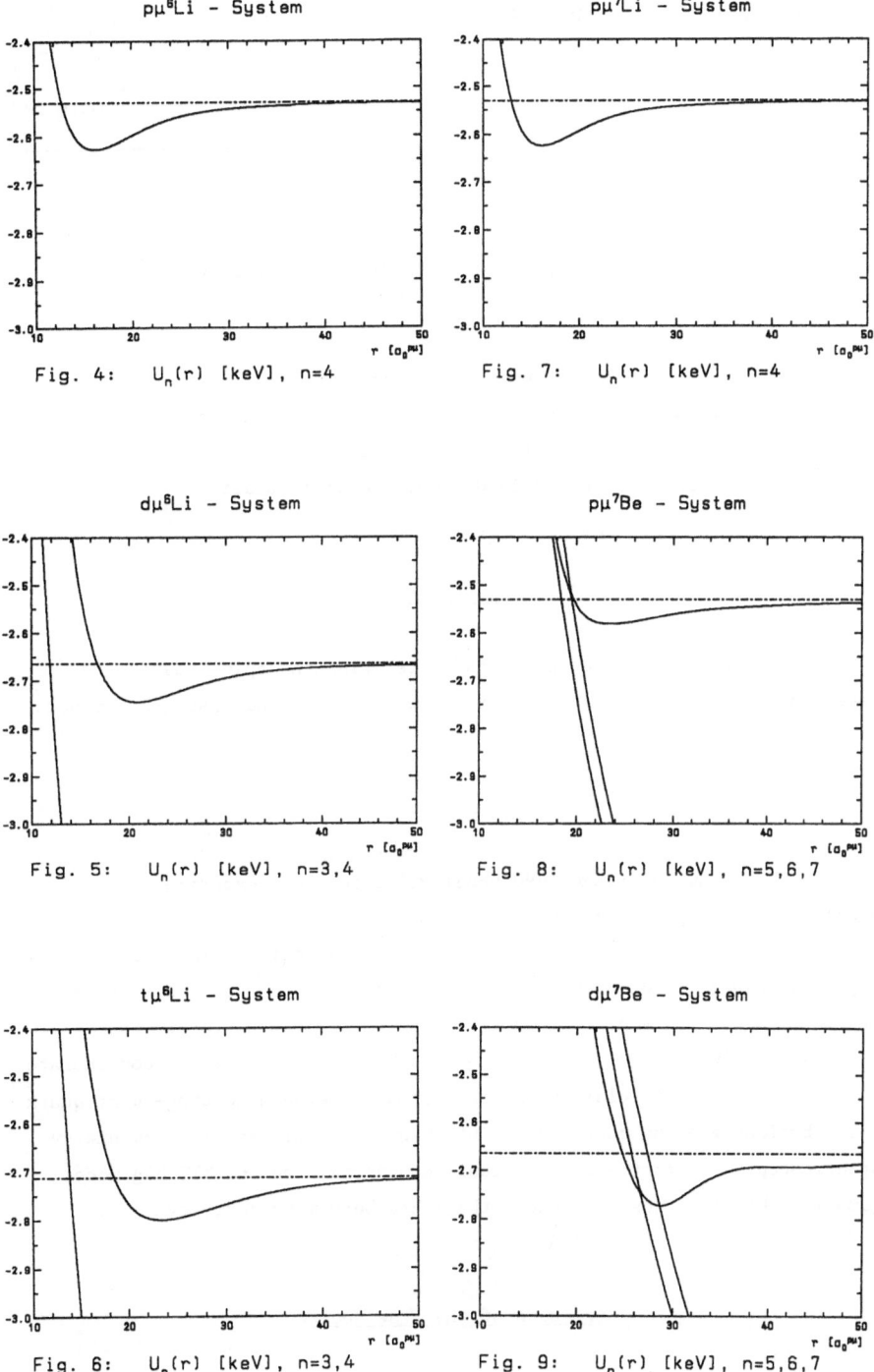

Fig. 4: $U_n(r)$ [keV], n=4

Fig. 7: $U_n(r)$ [keV], n=4

Fig. 5: $U_n(r)$ [keV], n=3,4

Fig. 8: $U_n(r)$ [keV], n=5,6,7

Fig. 6: $U_n(r)$ [keV], n=3,4

Fig. 9: $U_n(r)$ [keV], n=5,6,7

Few-Body Systems, Suppl. 6, 338—349 (1992)

THEORETICAL DESCRIPTION OF NUCLEON STRUCTURE FUNCTIONS

Anthony W. Thomas

Department of Physics and Mathematical Physics

The University of Adelaide,

GPO Box 498, Adelaide 5001, South Australia.

Abstract

We review some recent progress in relating the spin, flavor and
medium dependence of nucleon structure functions to familiar quark models.

Introduction

With the award of the 1990 Nobel Prize to those experimenters who
discovered scaling, it is timely to examine just what has been learnt from
two decades of experiments in deep-inelastic scattering. In this brief
report we cannot touch on many of those successes, such as tests of the
standard model, tests of perturbative QCD, the determination of the
running coupling constant, $\alpha_s(Q^2)$, and so on. Rather we shall concentrate
on recent attempts to clarify the relationship between leading-twist quark
distributions and the quark models which are so familiar from low energy
spectroscopy. In particular, we shall concentrate on the MIT bag model
(and its chiral extensions) where there has been most activity.

Structure of the Free Nucleon

In recent years there have been a number of attempts to relate the
quark models which have been so widely used in spectroscopic studies to

data in the deep inelastic regime.[1-9] It has been apparent for a long time that the sensible way to approach this is to use the models to calculate twist-two parton distributions at some (a priori unknown) low momentum scale (μ^2).[10] At such a scale one knows that the valence quarks carry a sizeable fraction of the nucleon momentum. One can then evolve these distributions to higher values of Q^2, where the twist-two piece dominates, using perturbative QCD.[11,12]

A specific advantage of the approach developed at the University of Adelaide [6-9] (see also a recent paper by Meyer and Mulders [13]) is that it guarantees the correct support for the calculated quark distributions. Starting from the usual expression for the twist-two quark distribution [14-16]

$$q^{(2)}(x,\mu^2) = \frac{m}{2\pi} \int_{-\infty}^{\infty} dz e^{-imxz} \langle N|\psi_+^{\dagger}(\xi^-) \psi_+(0)|N\rangle_c, \qquad 2.1$$

(where m is the mass of the nucleon, ξ^- is (z;0,0,-z) and ψ_+ is $(1+\alpha^3)\psi/2$), one inserts a complete set of energy and momentum eigenstates between the field operators. For the nucleon itself and for the intermediate states we use translationally invariant Peierls-Yoccoz states. These will be two-quark (with mass in the region of 3/4 m) and three-quark one anti-quark (with mass of order 5/4 m) states.

In the calculation of the anti-quark distribution $\bar{q}^{(2)}(x,\mu^2)$ (for which ψ and $\psi^{\dagger}$ are interchanged in equ.(2.1)) the dominant contribution is from a four-quark intermediate state (again with mass of order 5/4 m). One novel feature of this calculation is that it is quite clear that the nucleon has an intrinsic sea [7] - even in a model with just valence quarks, like the three-quark bag. Furthermore, as a result of the Pauli Exclusion Principle, this intrinsic sea will not be flavor symmetric.[7] Indeed we will find more d-$\bar{d}$ pairs in the sea. (This is because, with two spins and three colors one can insert d-quarks into five different 1s-states in a proton bag whereas there are only four states available for u-quarks.) Clearly an asymmetry such as this will have important consequences for the Gottfried sum-rule as discussed below.

Valence Quark Distributions

The dominant piece of the valence quark distribution calculated from equ.(2.1) is that involving a two-quark intermediate state. This term is controlled by two parameters, the bag radius and the mass. For the latter it is important to take into account [8] the effect of gluon exchangewhich raises the mass of a pair of quarks with spin-1 and lowers that of a

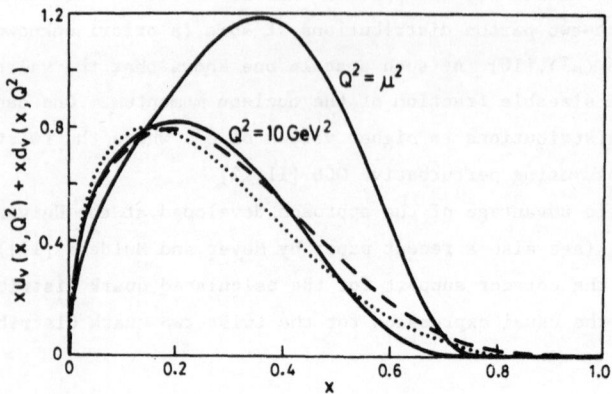

Figure 1: Valence quark distributions for the proton in the bag model (R=0.8 fm) at the bag scale μ^2 (0.25 GeV^2) and at 10 GeV^2 (solid lines).[9] The dashed and dotted lines are the Duke–Owens and MRS parameterisations, also at 10 GeV^2 [21,20]

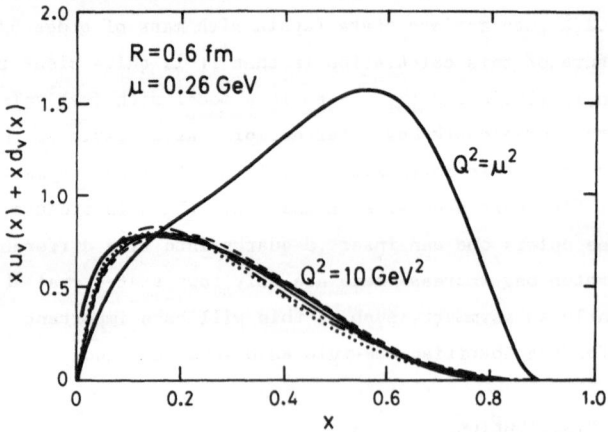

Figure 2: As for Fig. 1 but R = 0.6 fm, and we have also included the fits of EHLQ [19] and DFL [22] – from ref. [9].

spin-0 pair so that the resultant splitting is 200 MeV.

Rather than using the model for the contribution to the valence distribution from 3q-$\bar{q}$ intermediate states, we simply use a phenomenological term of similar shape (say $(1-x)^7$) with a normalisation chosen to ensure that we have three valence quarks. Under QCD evolution, this phenomenological term moves to small x so that there is no significant uncertainty for $Q^2 \gtrsim 5$ GeV^2 and $x \gtrsim 0.1$.[9] It is also worth noting that at small x we are sensitive to long-distance physics (the important values of z in (2.1) are roughly up to order $(mx)^{-1}$) which is difficult to handle in any phenomenological quark model, so it will be difficult to do any better in the near future.

In Fig. 1 we show a comparison between the valence quark distribution of the proton calculated for a bag radius of 0.8 fm and various phenomenological fits which will be loosely referred to as "data". A priori we have no way to specify the bag scale μ. Instead it is determined by seeing how far one must evolve until the agreement with "data" at 10 GeV^2 is optimal. Clearly the overall description of the "data" is rather good. Only at very large values of x ($x \gtrsim 0.7$) is there a significant difference. At such values the struck quark will have a momentum greater than 1 GeV/c and one would expect to have to include the effect of correlations. There is an additional uncertainty associated with the use of leading order QCD, which may be less reliable for higher moments and hence large x.

On the other hand, the agreement with "data" for calculations with a bag radius of 0.6 fm are essentially perfect (see Fig. 2). The improvement at large x is a consequence of the higher average momentum in the smaller cavity. Certainly it would be tempting to conclude that 0.6fm is preferred. We choose not to draw that conclusion at this stage in view of the problems just cited. Instead we are content to observe that a bag with a radius in the range 0.6 to 0.8 fm gives a very good representation of the "data". Particularly for the calculations at 0.6 fm the bag scale is rather low (e.g. 0.26 GeV in Fig. 2). For $\Lambda_{QCD} - 0.2$ GeV, as used here, this gives a rather large value of $\alpha_s(\mu^2)$. Other phenomenological studies have used similar values in perturbative calculations of QCD evolution [23], but we would be more comfortable with μ closer to 0.7 or 0.8 GeV. This does seem to be a likely, desirable consequence of including the pionic corrections needed to preserve chiralsymmetry.[24,25] While we shall not pursue this discussion, the effect of these pionic corrections on the Gottfried sum-rule will be mentioned later.

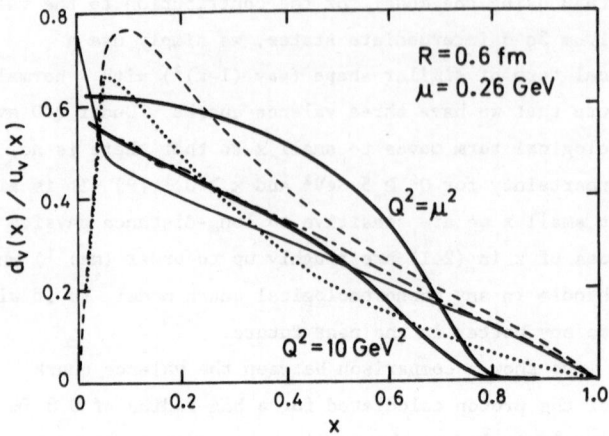

Figure 3: The bag model prediction for the d/u ratio, including the energy shift due to one gluon exchange (heavy solid curves), compared with four standard fits to world data. The parameters are as in Fig. 2 – from ref. [9].

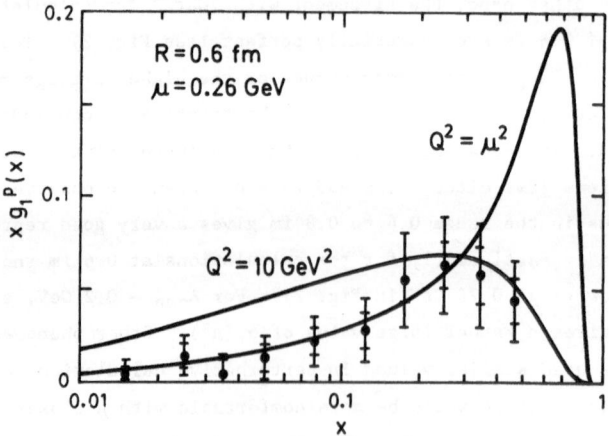

Figure 4: Prediction for the proton spin-dependent structure function in the bag model [9], in comparison with the EMC data [30].

Flavor and Spin-Dependence

Because the quark distributions measured in deep-inelastic scattering involve light-cone correlation functions, the energy of the struck quark is as important as its three-momentum. This is why, even for SU(6) wavefunctions for which the u and d quarks of all spin orientations have the same distribution of three momentum, one finds important differences in $u^{\uparrow\downarrow}$ and $d^{\uparrow\downarrow}$ by including the first-order one-gluon-exchange corrections to the energies of the intermediate (di-quark) states inserted in equ.(2.1).[8,9,26] (The lowest mass di-quark will give the hardest quark distribution.) In particular, in this unsophisticated model one can readily see that the ratio d(x)/u(x) tends to zero for x going to one. As a consequence $F_{2n/F2p} \to 1/4$ as $x \to 1$.[27] Figure 3 shows the general agreement between the "data" for d/u and our calculations. Only at very large x is there any serious discrepancy and this may also be related to the absence of short-range correlations in the bag.[28]

In Fig. 4 we see that this same, simple physics, familiar from low energy spectroscopy, also leads to a qualitative understanding of the proton spin-structure function, $g_{1p}(x)$. The crucial feature is that only a u-quark with its spin parallel to that of the proton is accompanied by a low-mass spin-singlet pair of quarks (in an SU(6) proton). As a consequence $u^{\uparrow}(x)$ is the dominant parton distribution at large x.

Of course, because we are using an SU(6) spin-flavor wavefunction, the integral of our g_{1p} agrees with the Ellis-Jaffe sum-rule [29] - unlike the data.[30] A clear indication of this problem is the quantitative disagreement at intermediate x in Fig. 4. One knows that a more sophisticated treatment of the proton wavefunction including gluonic and pionic corrections would improve the situation a little.[31-33] However it is also known that the anomaly plays a critical role in the flavor singlet distribution [35] and this is still rather controversial.[36-38] We prefer not to discuss this issue further here.

Independent of the question of the Ellis-Jaffe sum-rule, it would be extremely interesting to obtain data on the neutron spin-dependent-structure function $g_{1n}(x)$. Figure 5 shows an exciting prediction of the bag model, namely that g_{1n} should become positive at large x. However, arecent extension of that work to include pion corrections [25] has led us to question the sensitivity of this prediction to small changes in the model. This is being thoroughly investigated at the present time.

The Gottfried Sum-Rule

The Gottfried sum-rule is based on the fact that a nucleon contains

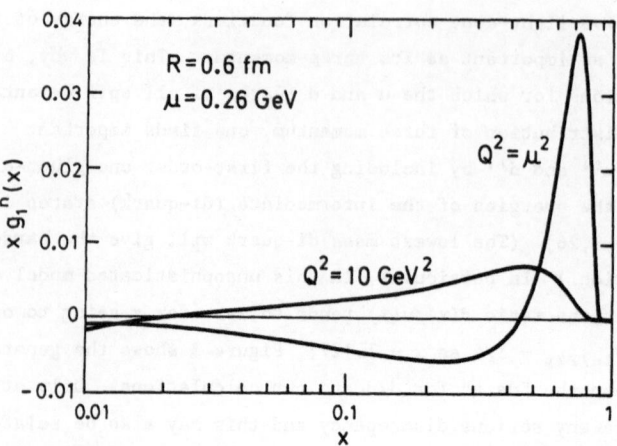

Figure 5: Prediction for the neutron spin-dependent structure function in the bag model (including gluon exchange) [9].

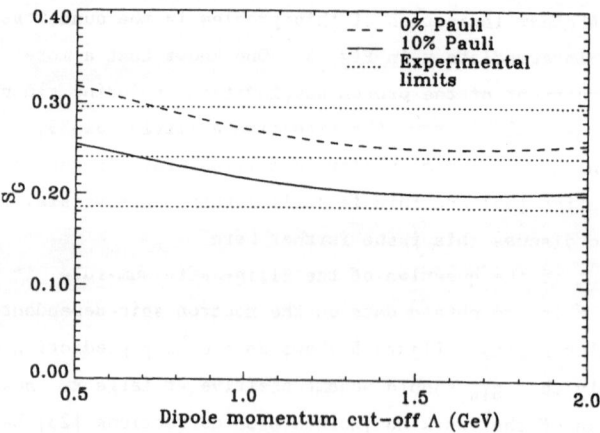

Figure 6: Prediction for the Gottfried sum rule as a function of cut-off mass at the pion vertex and the Pauli defect $(1-P_2)$ [39]. The dotted lines indicate the present experimental limits [44].

three valence quarks and on the assumption that the sea is purely
perturbative and therefore flavor symmetric. We have already mentioned
one important correction which is self-evident in our approach, namely the
fact that the intrinsic sea of the nucleon has $\bar{d} > \bar{u}$. Like the $3q-\bar{q}$
contribution to the valence quark distribution we treat this correction
phenomenologically. In particular, Melnitchouk et al.[39] took $\bar{d}-\bar{u}$ to
have the functional form $x^\alpha (1-x)^\beta$. They chose $\beta=7$ so it affected only
small x, and α was chosen to match the small x behaviour of the valence
distribution. The overall normalisation was allowed to vary between 10
and 40 percent.

Another source of flavor asymmetry in the sea is the pion cloud
required in any quark model that respects chiral symmetry. A number of
calculations of the contribution of this cloud to deep-inelastic
scattering have been made along the lines initiated by Sullivan [40-42] –
see also ref.[43] where the consequences of $\bar{d} > \bar{u}$ were noted for proton-
nucleus Drell-Yan experiments. With the recent announcement of a clear
discrepancy with the Gottfried sum-rule by NMC [44], several groups have
estimated the pionic correction.[45-47] The complete result with and
without a 10% Pauli correction is shown in Fig. 6. Clearly it is possible
to reproduce the experimental value for the sum-rule for a wide range of
πNN form-factor masses (consistent with other processes).

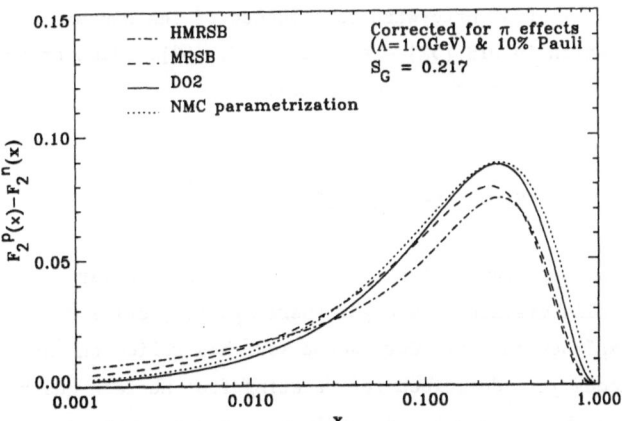

Figure 7: Comparison of the data [44,48] and
predictions for $[F_2^p - F_2^n]$ after correcting several
standard sets of structure functions for pionic
effects and the Pauli defect – from ref. [39].

However it is clearly much more informative to look at the x-dependence of $(F_{2p}-F_{2n})/x$, rather than just its integral. For the data we use the parameterisation of Ratcliffe and Soper.[48] Fig. 7 shows the best quality fit obtained by Melnitchouk et al.[39] which required a 10% Pauli correction. (The 1GeV mass for πNN and $\pi N\Delta$ dipole form-factors was preferred on other grounds.[41,42]) Clearly the quantitative agreement depends on which set of phenomenological quark distributions is used but the overall agreement is fairly good for all of them.

To conclude this discussion we mention one other fascinating development. Recent estimates of shadowing in deuterium suggest that itmay be a significant correction.[49] If so the discrepancy in the Gottfried sum-rule would be even greater and the pionic correction alone could not explain it. In view of the fact that our bag work favored a Pauli correction more like (20-30)%, it is imperative that this question be investigated thoroughly.

Semi-Inclusive Structure Functions

In a number of bubble chamber experiments with $\nu(\bar{\nu})$-beams it hasbeen possible to measure quark distributions with certain constraints - e.g. a DIS event accompanied by a low energy proton. Initially it was thought that the large-x depletion was related to the EMC effect [50], but the observation of a similar depletion in hydrogen ruled out that possibility.[51] Melnitchouk et al. have been able to demonstrate that this depletion is a straightforward consequence of the kinematics of target fragmentation in DIS [52] - see also ref. [53]. Some of those results will be presented at the conference.

Structure of the Bound Nucleon

The nuclear dependence of structure functions discovered by the European Muon Collaboration some eight years ago [54] still defies a satisfactory explanation.[55] Our recent work has relied on the successful methods just described for relating free nucleon quark distributions to the bag model. By applying the same technique to the (admittedly simple) quark bag model of nuclear matter developed by Guichon [56] we have been able to test the assumptions underlying many other calculations [57]. It seems that all calculations based on binding corrections at the nucleon level (i.e. using an impulse approximation) overestimate the effect. If correct, this means that no existing binding calculation can explain the depletion in the nuclear data at large-x. The heartening side of this finding is that the EMC data really is telling us about the energy and momentum distributions of <u>quarks</u>, not nucleons, in the nucleus.

While our early calculations were made for nuclear matter at fixed density, this has since been extended to real nuclei using the local density approximation.[58] The initial results are very promising and it is hoped to present some at the meeting.

Acknowledgements

It is a pleasure to acknowledge the contributions of the following colleagues and students who have greatly aided my attempts to understand these issues: S. Bass, F.E. Close, P.A.M. Guichon, J.T. Londergan, W. Melnitchouk, A. Michels, N.N. Nikolaev, K. Saito, A.W. Schreiber and A.I. Signal. I should also like to thank Mrs. A. Shaw for her careful preparation of the manuscript. This work was supported in part by the Australian Research Council.

References

1. Jaffe, R.L. and Ross, G.G. : Phys. Lett. 93B, 313 (1980).
2. Jaffe, R.L. : Ann. Phys. (N.Y.) 132, 32 (1981);
    Bell, J.S., Davis, A.C. and Rafelski, J. : Phys. Lett. 78B, 67 (1978).
3. Hughes, R.J. : Phys. Rev. D16, 622 (1977).
4. Celenza, L.S. and Shakin, C.M. : Phys. Rev. C27, 1561 (1983); ibid C39, 2477(E) (1989).
5. Benesh, C.J. and Miller, G.A. : Phys. Lett. B222, 476 (1989).
6. Thomas, A.W. : Prog. Part. Nucl. Phys. 20, 21 (1988); and in Vautherin, D. et al. : Hadrons and Hadronic Matter, New York: Plenum 1990, p.263.
7. Signal, A.I. and Thomas, A.W. : Phys. Rev. D40, 2832 (1989).
8. Close, F.E. and Thomas A.W. : Phys. Lett. B212, 227 (1988).
9. Schreiber, A.W., Thomas, A.W. and Londergan, J.T. : Phys. Rev. D42, 2226 (1990); Phys. Lett. B237, 120 (1990); Schreiber, A.W., Signal, A.I. and Thomas, A.W. : ADP-91-155/T95 to appear in Phys. Rev. D, (1991).
10. Cabibbo, N. and Petronzio, R. : Nucl. Phys. B126, 298 (1977).
11. Leader, E. and Predazzi, E. : An Introduction to Gauge Fields and the New Physics. Cambridge: Cambridge University Press, 1982.
12. Altarelli, G. : Phys. Rep. 81C, 1 (1982).
13. Meyer, H. and Mulders, P.J. : Nikhef preprint NIKHEF-P-12 (1991).
14. Jaffe, R.L. : Nucl. Phys. B229, 231 (1983).
15. Politzer, D.I. : Phys. Rev. D9, 416 (1974).

16. Ellis, R.K., Furmanski, W. and Petronzio, R. : Nucl. Phys. **B212**, 29 (1983).
17. De Rujula, A., Georgi, H. and Glashow, S. : Phys. Rev. **D12**, 147 (1975).
18. Chodos, A. et al. : Phys. Rev. **D10**, 2599 (1974).
19. Eichten, E. et al. : Rev. Mod. Phys. **56**, 579 (1986); ibid **58**, 1065(E) (1986).
20. Martin, A., Roberts, R.G. and Stirling, W.J. : Phys. Rev. **D37**, 1161, (1988).
21. Duke, D.W. and Owens, J.T. : Phys. Rev. **D30**, 49 (1984).
22. Diemoz, M. et al. : Zeit. Phys. **C39**, 21 (1988).
23. Glück, M., Reya, E. and Vogelsang, W. : Nucl. Phys. **B329**, 347 (1990).
24. Thomas, A.W. : Nucl. Phys. **A518**, 186 (1990).
25. Schreiber, A.W., Mulders, P.J., Thomas, A.W. and Signal, A.I. : to be published.
26. Schafer, A. : Phys. Lett. **208B**, 175 (1988).
27. Close, F.E. : An Introduction to Quarks and Partons. New York : Academic 1979.
28. Farrar, G. and Jackson, D. : Phys. Rev. Lett. **35**, 1416 (1975).
29. Ellis, J. and Jaffe, R.L. : Phys. Rev. **D9**, 1444 (1974).
30. Ashman, J., et al. : Phys. Lett. **B206**, 364 (1988); Nucl. Phys. **B328**, 1 (1990); Hughes, V., et al. : Phys. Lett. **B212**, 511 (1988).
31. Myhrer, F. and Thomas, A.W. : Phys. Rev. **D38**, 1633 (1988).
32. Hogaasen, H. and Myhrer, F. : Phys. Lett. **B214**, 123 (1988).
33. Schreiber, A.W. and Thomas, A.W. : Phys. Lett. **B215**, 141 (1988).
34. Jaffe, R.L. : Phys. Lett. **B193**, 101 (1987).
35. Efremov, A.V. and Teryaev, O.V. : Dubna preprint E 2-88-287 (1988); Altarelli, G. and Ross, G.G. : Phys. Lett. **212B**, 391 (1988); Carlitz, C., Collins, J.C. and Mueller, A.H. : Phys. Lett. **B214**, 229 (1988).
36. Jaffe, R.L. and Manohar, A.V. : Nucl. Phys. **B337**, 509 (1990).
37. Manohar, A.V. : Phys. Rev. Lett. **66**, 289 (1991).
38. Bass, S.D., Ioffe, B.L., Nikolaev, N.N. and Thomas, A.W. : ADP-91-154/T94, to appear in J. Moscow Phys. Soc. (1991).
39. Melnitchouk, W., Thomas, A.W. and Signal, A.I. : ADP-91-153/T93, to appear in Zeit. Phys. **A**, (1991).
40. Sullivan, J.D. : Phys. Rev. **D5**, 1732 (1972).
41. Thomas, A.W. : Phys. Lett. **126B**, 97 (1983).
42. Frankfurt, L.L., Mankiewicz, L. and Strikman, M.I. : Zeit. Phys. **A334**, 343 (1989).
43. Ericson, M. and Thomas, A.W. : Phys. Lett. **148B**, 191 (1984).
44. Amaudruz, P. et al. (NMC Collaboration) : Phys. Rev. Lett. **66**, 2712 (1991).
45. Kumano, S. : Phys. Rev. **D43**, 59 (1991); Kumano, S. and Londergan, J.T. : IUCF preprint IU/NTC 90-16 (1990).
46. Henley, E.M. and Miller, G.A. : Phys. Lett. **B251**, 453 (1990).
47. Signal, A.I., Schreiber, A.W. and Thomas, A.W. : Mod. Phys. Lett. **A6**, 271 (1991).
48. Preparata, G., Ratcliffe, P.G. and Soffer, J. : Phys. Rev. Lett. **66**, 687 (1991).
49. Zoller, V.R. : ITEP preprint 91058 (1991); Nikolaev, N.N. and Zakharov, V.I. : Phys. Lett. **B55**, 397 (1975).
50. Kitagaki, T., et al. (E745 Collaboration) : Phys. Lett. **B214**, 281 (1988).
51. Guy, J., et al. (BEBC Collaboration) : Phys. Lett. **B229**, 421 (1989).
52. Melnitchouk, W., Thomas, A.W. and Nikolaev, N.N. : Adelaide Preprint ADP-91-160/T99.
53. Bosveld, G.D., Dieperink, A.E.L., and Scholten, O. : Groningen preprint KVI-870 (1991).

54. Aubert, J.J., et al. (EMC Collaboration) : Phys. Lett. **B123**, 275
     (1983).
55. Bickerstaff, R.P. and Thomas, A.W. : J. Phys. **G15**, 1523 (1989);
     Frankfurt, L.L. and Strikman, M.I. : Phys. Rep. **160**, 235 (1988).
56. Guichon, P.A.M. : Phys. Lett.
57. Thomas, A.W., Michels, A., Schreiber, A.W. and Guichon, P.A.M. :
     Phys. Lett. **B233**, 43 (1989).
58. Michels, A., Saito, K. and Thomas, A.W. : to be published (1991).

Few-Body Systems, Suppl. 6, 350—355 (1992)

Few-
Body
Systems
© by Springer-Verlag 1992

THE SPIN MUON COLLABORATION EXPERIMENT

R. van Dantzig

NIKHEF, P.O. Box 41882, 1009 DB Amsterdam, The Netherlands

The method and status of CERN experiment NA47 to measure the structure function g_1 of the neutron (for the first time) and of the proton (with improved accuracy compared to earlier measurements) is outlined. First data were taken for 100 and 210 GeV polarized muons deep-inelastically scattered off a polarized deuterated butanol target.

Great interest in the spin structure of the nucleon was raised by a recent measurement of the European Muon Collaboration (EMC) at CERN [1]. The results, in conjunction with SLAC-Yale data [2], violate the Ellis-Jaffe sum rule [3], in which flavour SU(3) symmetry in the quark-parton model and an unpolarized quark-antiquark sea is assumed. The EMC results are considered [1] as a surprising indication that the proton spin is carried by gluon spin or by quark and gluon orbital angular momentum rather than by quark spin. A variety of theoretical propositions on how the nucleon spin is distributed among the quark and gluon degrees of freedom has been published. New experiments to extend and improve in precision our knowledge on the nucleon spin structure are called for. With the results one should be able to discriminate between different models and to perform important tests in the frame of QCD.

A series of new experiments [4] has been proposed. The first one is CERN experiment NA47 [5] of the Spin Muon Collaboration (SMC) [6], in which the spin structure of the neutron and the proton will be studied. In the experiment the distribution of spin in the nucleon is probed at high space-time resolution using deep inelastic muon scattering at virtual photon four-momentum squared (Q^2) in the range from 1 to 70 GeV^2. First aim is to perform measurements in the scaling regime of the principal spin-dependent structure function for the proton and the neutron, $g_1^p(x)$ and $g_1^n(x)$, in the x-range from 0.005 to 0.7, where $x \equiv Q^2/(2Mv)$ is the Bjorken scaling variable (with M the proton mass and v the virtual photon energy in the lab. system). No measurements on $g_1^n(x)$ have been made until now. From the $g_1^p(x)$ and $g_1^n(x)$ data the first moments Γ_p and Γ_n, with $\Gamma = \int_0^1 g_1(x)\,dx$, will be determined and compared with various model predictions. A test of the Bjorken sum rule for $\Gamma_p - \Gamma_n$ [7], to be achieved with about 10 % accuracy, is an important goal of the experiment. This sum rule, originally derived from current algebra [8], allows a fundamental test of QCD in the scaling limit. It reads: $\Gamma_p - \Gamma_n = (1/6) \mid g_A/g_V \mid (1 - \alpha_s / \pi)$ where g_A and g_V are the well known low energy neutron beta decay coupling constants and $\alpha_s = \alpha_s(Q^2)$ is the running coupling constant of QCD.

The g_1 structure function is deduced from the asymmetry (A) for muons with longitudinally polarized spin (helicity $\pm$ 1/2) scattered by nucleons with their spin polarized along the muon-nucleon collision axis. This muon scattering asymmetry is defined as

$$A = \frac{d\sigma^{\uparrow\downarrow} - d\sigma^{\uparrow\uparrow}}{d\sigma^{\uparrow\downarrow} + d\sigma^{\uparrow\uparrow}} \tag{1}$$

where $d\sigma^{\uparrow\uparrow(\uparrow\downarrow)}$ is the cross section for parallel (antiparallel) muon and nucleon spins. It is derived from the measured asymmetry (A_{meas}) as $A = A_{meas} / (P_\mu \, P_T \, f)$ where P_μ and P_T are the incident muon and target polarizations and f is the dilution factor arising from unpolarized nucleons in the target. The muon asymmetry A can be expressed in asymmetries, A_1 and A_2, for the exchanged virtual photon as

$$A = D (A_1 + \eta \, A_2) \tag{2}$$

where D and η are functions of the kinematical variables.

The photon asymmetry A_1 is defined as

$$A_1 (x) \equiv \frac{\sigma_{1/2} - \sigma_{3/2}}{\sigma_{1/2} + \sigma_{3/2}} \tag{3}$$

in which $\sigma_{1/2} \, (\sigma_{3/2})$ is the absorption cross section for transverse (helicity $\pm$1) virtual photons by nucleons polarized along the photon-nucleon collision axis with summed photon and nucleon spin projection of 1/2 (3/2). The asymmetry A_2 arises from interference between longitudinal and transverse photons. Since both A_2 and η are small, to a good approximation $A_1 = A/D$. The g_1 structure functions for the proton and the neutron are then obtained from:

$$g_1^{p(n)} (x) = \frac{1}{2x} \frac{A_1^{p(n)}(x) \; F_2^{p(n)}(x)}{1 + R^{p(n)}(x)} \tag{4}$$

in which $F_2(x)$ and $R(x) = \sigma_L/\sigma_T$ are spin averaged structure functions in the usual notation.

The neutron asymmetry can be derived from the proton and deuteron asymmetries A_1^p and A_1^d, while accounting for the deuteron D-state probability w_D, by the relation

$$A_1^n = \frac{1 + F_2^p/F_2^n}{1 - 1.5 \, w_D} \; A_1^d - \left(F_2^p/F_2^n\right) A_1^p \tag{5}$$

where the F_2 proton-neutron ratio has recently been measured with high precision by the New Muon Collaboration (NMC) [9].

To obtain the asymmetry A_2 and the second spin-dependent structure function g_2 [10], additional data for transverse polarization of the target nucleon are needed.

The SMC measurements make use of a newly optimised muon beam, the reinstalled EMC polarized target (beam period 1991), a strongly upgraded spectrometer and a muon beam polarimeter set-up behind the spectrometer. A new polarized target suitable for deuteron polarization and fast polarization reversal will be used in forthcoming measurements. In the following we briefly discuss the major parts of the SMC equipment .

The muon beam

The CERN muon beam line has been optimised for energies of $E_\mu \leq 220$ GeV. The muons are naturally polarized in the decay of pions (and kaons) in flight. The flux is 4×10^7 μ/burst from an SPS beam of 3×10^{12} protons on target with a spill time of 2 s and a repetition time of 14 s. About 80 % of the beam passes through the 5 cm $\emptyset$ polarized target.

The polarized target

For the first measurements, during the 1991 beam period, the EMC target [1] has been reassembled. The SMC target [11], with a new helium dilution refrigerator, is under construction. Both, the old and the new target have two sections aligned with the beam, with (longitudinal) opposite polarizations. Because of the simultaneous measurement for both spin directions, flux normalisation uncertainties essentially cancel. The length of the sections is 60 cm for the SMC target (40 cm for the EMC target), yielding a 50 % increase in luminosity as compared to the EMC experiment [1]. The target material consists of frozen beads of (deuterated) butanol with the paramagnetic dopant EHBA-Cr(V) [12].

The method of dynamic nuclear polarization (DNP) is used. The opposite polarizations are obtained by applying different microwave frequencies to the two target sections. The new requirement of the polarization of deuterons (with a dipole moment much smaller than that of the proton) over the 2 l target volume provides a technological challenge. A longitudinal magnetic field of 2.5 T with a homogeneity of 2×10^{-5} (rather than 10^{-4} for the EMC magnet) is needed to achieve high deuteron polarization (the aim is 40 %). The target polarization P_T is derived from integrated NMR signals (8 parallel measurements) with an expected accuracy of 3 %. The nucleon spins in the two target sections must be reversed frequently (and therefore rapidly) to reduce systematic errors due to a possible time dependence of the spectrometer acceptance.

Fast polarization reversal hopefully can be achieved in the new target by a transverse magnetic dipole field (up to 0.5 T) allowing the rotation of spins in the frozen-spin mode at a temperature of 50 mK. With the transverse field, it will also be possible to keep the target polarization in a direction perpendicular to the beam. This will allow an exploratory determination of the "transverse" structure function g_2 [10] from the measurement of the asymmetry A_2. Such a measurement also reduces the error in A_1 (derived from A using eq. 2) and therefore in g_1 (eq. 4).

The forward muon spectrometer

Using the SMC spectrometer, deep inelastic muon scattering events are detected and the necessary tracking information is collected. From the incoming and scattered muon the interaction vertex, determining the target section (with given polarization) where the scattering took place, is reconstructed. The scattering angle (θ) and the initial and final energy (E, E') of the scattered muon determine the kinematic variables: $Q^2 = -4\, E\, E' \sin^2(\theta/2)$ and $\nu = E - E'$.

The longitudinal vertex resolution is sufficient to distinguish between the target sections for scattering angles above 5 mrad. Since the EMC experiment [1], the spectrometer set-up and the data acquisition system have been upgraded considerably. First in 1986 by the New Muon Collaboration (NMC, CERN exp. NA37) and most recently by SMC. The present layout is shown in fig. 1. For the new experiment the large drift chamber systems in the spectrometer have been partly improved (W12), extended (W45) and replaced (W67) by arrays of streamer tubes (ST67) and drift tubes (DT67).

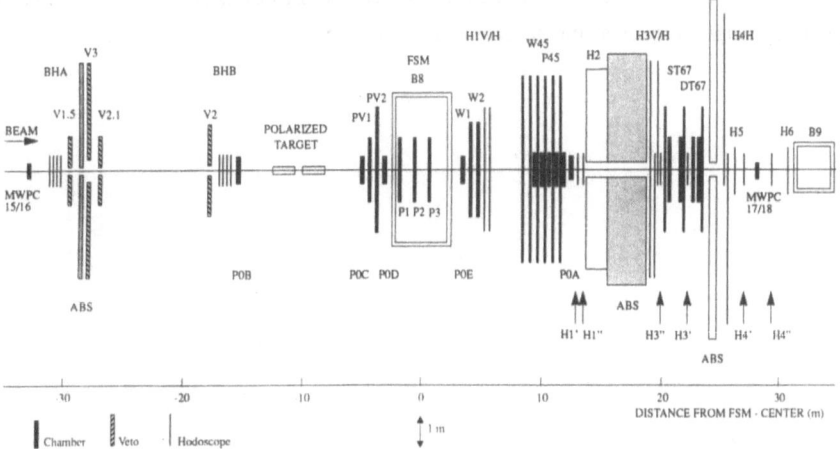

Fig. 1: The SMC spectrometer (V = veto counter, BH = beam hodoscope, P = proportional chamber, FSM = forward spectrometer magnet, W = drift chamber, H = hodoscope for triggering, ST = streamer tube array, DT = drift tube array). Note difference in longitudinal and transverse scale.

The muon beam polarimeter

The polarization, P_μ, of the incident muon beam (~ 80 %) is measured from the polarization dependence of the spectrum of the decay positrons. Fig. 2a shows the spectral shape of the positron momentum fraction $y = p_e/p_\mu$ for different muon polarizations.

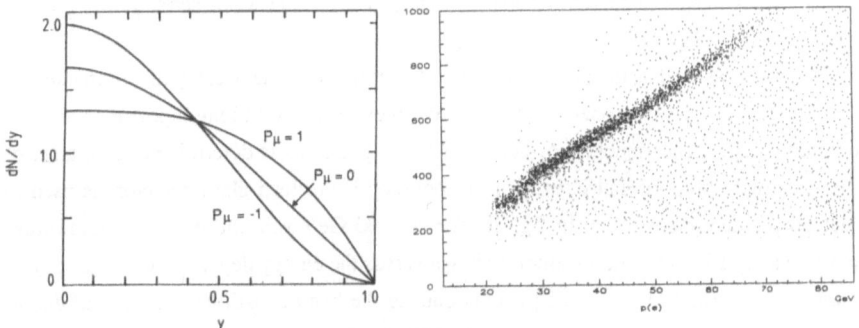

Fig. 2: a) Shape of the polarization dependent positron spectrum from decay of μ^+ in flight, expressed in the dimensionless momentum variable $y = p_e/p_\mu$, for muon polarization P_μ at the values 1, 0 and -1; b) scatterplot for polarimeter events.

This dependence arises from the CM forward-backward asymmetry in the decay of the muon with respect to its spin. The sensitivity to the muon polarization is highest near y = 0.75 measured by absolute magnitude and near y = 0.4 measured by slope. The latter measurement is somewhat less complicated because it does not require an accurate beam flux normalisation.

The polarimeter (fig. 4) has been set-up downstream of the main experiment, where the beam after having passed the spectrometer, is refocussed. Electrons and positrons are vetoed to obtain an effectively clean muon beam entering a 30 m long field free decay region.

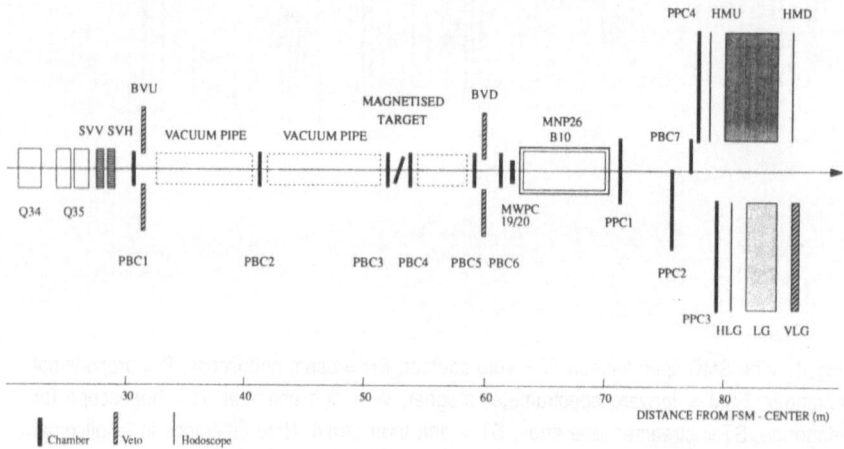

Fig. 3: Topview of the muon beam polarimeter (Q = quadrupole magnet, SV = shower veto, PB = proportional beam chamber, BV = beam halo veto, MNP26 = dipole magnet, PP = proportional chamber for scattered or decay particles, H = hodoscope, LG = lead glass detector). Detectors to the right of the beam are for $e^{\pm}$ detection, to the left for μ detection in the μ-e scattering method. Note difference in longitudinal and transverse scale.

The muons as well as the decay positrons appearing along the decay path are tracked by high rate multiwire proportional chambers before they reach a 6 m long dipole magnet. The momentum dependent trajectory of deflected decay positrons is determined using a set of larger multiwire proportional chambers. A hodoscope with lead glass detectors is used to select proper decay positrons. In fig. 2b, for the 100 GeV run, the positron momentum (approximately 100 times the y-value) is shown versus the energy deposited in the lead glass shower detector (in arbitrary units); events outside the band constitute background due to scattered electrons and muon decays inside the MNP26 magnet.

The polarimeter will be equipped with additional detectors to allow an independent determination of P_{μ} by measuring the asymmetry in the scattering of polarized muons by polarized electrons in a magnetized iron slab (the μ-e scattering method).

Status at the end of the 1991 run

In 1991 the experiment had its first beam period. The new spectrometer equipment was tested and calibrated. Global efficiencies of all detector systems were determined. The reinstalled EMC polarized target was loaded with deuterated butanol. Deuteron polarization at the level of 25 % was achieved and first data were taken during two periods of two weeks at 100 and 210 GeV beam energy, respectively. Polarization reversal was performed in less than two hours once per day by microwave repolarization (DNP). Around a million events suitable for physics analysis were acquired for polarized deuterons, which is similar to the number of events for the full series of EMC polarized proton measurements. The polarimeter and its on-line data acquisition system was tested and calibrated. About one million muon decay events within the acceptance of the analysis were measured for each period. In conclusion, the NA47 1991 run has produced a highly interesting data sample, which after a complicated analysis, may yield the first experimental information on the spin structure of the neutron.

References

[1] EMC, J. Ashman et al., Nucl. Phys. **B328**, 1 (1989) and references therein.

[2] SLAC-Yale, G. Baum et al., Phys. Rev. Lett. **51**, 1135 (1983).

[3] J. Ellis and R.L. Jaffe, Phys. Rev. **D9**, 1444 (1974) and **D10**, 1669 (1974).

[4] Contributions from G.G. Petratos and M. Dueren to this conference and
     R. Windmolders, to be published in Int. Journ. Mod. Phys. (1991)

[5] SMC, "Measurement of the Spin-dependent Structure Functions of the Neutron and the
     Proton", CERN proposal SPSC 88-47, Dec. 1988.

[6] In the Spin Muon Collaboration the following institutions participate: University of
     Bielefeld; University of California, Los Angeles; CERN; Delft University of
     Technology; University of Freiburg; GKSS; Helsinki University of Technology;
     University of Houston; JINR, Dubna; University of Mainz; University of Mons;
     University of München; Nagoya University; NIKHEF, FOM and Free University
     (Amsterdam), Northeastern University; Northwestern University; Rice University;
     Saclay DAPNIA; University of Santiago de Compostela; Tel Aviv University;
     University of Trieste; Uppsala University; University of Virginia; Warsaw University
     and Institute for Nuclear Studies;Yale University.

[7] V.W. Hughes and J. Kuti, Ann. Rev. Nucl. Part. Sci. **33**, 611 (1983).

[8] J.D. Bjorken, Phys. Rev. **148**, 1467 (1966).

[9] NMC, Phys. Lett. **B249**, 366 (1990) and NMC, to be published in Nucl. Phys. **B**.

[10] R.L. Jaffe and Xiangdong Ji, Phys. Rev. **D43**, 724 (191);
     L. Mankiewicz and Z. Ryzak, Phys. Rev. **D43**, 733 (1991).

[11] T. Niinikoski et al., Proc. 9th Symposion on High Spin Physics, Bonn, Sept. 1990,
     W. Meyer, E. Steffens and W. Thiel (Eds.), Springer-Verlag (1991), contributions on
     page 266, 272, 347, 363, 369, 378.

[12] S. Trentalange et al., same as previous reference, page 325.

Few-Body Systems, Suppl. 6, 356—361 (1992)

Few-
Body
Systems
© by Springer-Verlag 1992

FUTURE EXPERIMENTS TO MEASURE
THE SPIN STRUCTURE OF THE NUCLEON

Michael Düren

Max-Planck Institut für Kernphysik
W-6900 Heidelberg 1, Germany
Bitnet: DUE@CRUXNHD3

Abstract — Polarized lepton nucleon scattering experiments have access to six different spin structure functions and related sum rules. The spin distributions of the valence and sea quarks, of gluons and the quark's angular momentum in the nucleon are subject of investigation in the 2nd generation spin experiments SMC, E-142, HERMES and HELP. The abilities and precisions of the experiments are compared. The HERMES experiment, having the most complete experimental program, is described in more detail.

1. Introduction

The EMC/SLAC measurement [1] of the spin structure function $g_1(x)$ of the proton revealed our incomplete theoretical and experimental understanding of the internal spin structure of the nucleon. The unexpected violation of the Ellis-Jaffe sum rule [2] lead to many discussions, some even questionated the validity of the quark model and of pertubative QCD [3].

To clarify the situation, four new lepton nucleon scattering experiments have been proposed: SMC at CERN, E-142 at SLAC, HERMES at DESY and HELP at CERN [4-7]. The four experiments are based on very different experimental techniques and will provide details of the various spin structure functions of the proton and the neutron.

2. Experimentally Accessible Spin Information

The basic information of the quark spin content of the nucleon is accessible by measuring the spin dependent structure function g_1:

$$g_1(x, Q^2) = \frac{1}{2} \left[\frac{4}{9} \delta u(x) + \frac{1}{9} \delta d(x) + \frac{1}{9} \delta s(x) + \frac{4}{9} \delta \bar{u}(x) + \ldots \right]$$

where $\delta u(x) = u^+(x) - u^-(x)$ and $u^+(x)$ resp. $u^-(x)$ are the momentum distributions of u-quark with spin parallel resp. anti-parallel to the nucleon spin.

The spin-dependent part of the polarized deep inelastic cross section is given by the formula [8]

$$\frac{d^3(\sigma(\alpha) - \sigma(\pi + \alpha))}{dx\,dy\,d\Phi} = \frac{e^4}{4\pi^2 Q^2}\left[\cos\alpha\left(\left[1 - \frac{y}{2} - \frac{y^2}{4}\gamma\right]g_1(x, Q^2) - \frac{y}{2}\gamma g_2(x, Q^2)\right)\right.$$

$$\left. - \sin\alpha\cos\Phi\sqrt{\gamma\left[1 - y - \frac{y^2}{4}\gamma\right]}\left(\frac{y}{2}g_1(x, Q^2) + g_2(x, Q^2)\right)\right].$$

The kinematic variables are defined as usual: $\nu = E - E'$, $x = Q^2/2M\nu$, $y = \nu/E$ and $\gamma = Q^2/\nu^2$. α is the polar angle of the target polarization vector in respect to the beam direction and Φ is the azimutal angle in respect to the lepton scattering plane.

According to this formula the determination of $g_1(x)$ and $g_2(x)$ requires a combined asymmetry measurement off a longitudinally and a transversely polarized target. Because of the mixing of g_1 and g_2 a precise determination of g_1 from a longitudinally polarized target alone is not possible.

The determination of $g_2(x)$ has also a physical interest in its own. Besides distinguishing between several models of the nucleon $g_2(x)$ is sensitive to quark gluon correlations that appear as twist-3 operators in the operator product expansion [8].

The combination of proton and neutron spin structure functions allows to separate valence from sea contributions of the quark spin. The neutron spin structure functions $g_1^n(x)$, $g_2^n(x)$ can be obtained experimentally either from the difference of deuterium and hydrogen or from ^{3}He as the two protons spins in ^{3}He mostly cancel. Since the ^{3}He wave function has a small p-wave contribution, the neutron structure function derived from ^{3}He gets a model dependent correction.

As the deuteron is a spin-1 target there exist two higher multipole structure functions $b_1(x)$ and $\Delta(x)$. Both functions can be measured by scattering an unpolarized beam off a longitudinally tensor-polarized resp. a transversely polarized target. Both functions are expected to be small. The knowledge of the structure function $b_1(x)$ is required for a precise derivation of $g_1^n(x)$ and $g_2^n(x)$ from deuterium. $\Delta(x)$ is interesting in its own as it probes gluons which cannot be assigned to individual nucleons within the nucleus [8].

Besides the shape of the various structure functions, also their integrals allow important tests of different models and assumptions about the spin structure of the nucleon. The **Ellis Jaffe sum rule** [2], derived from SU(3)$_f$ symmetry arguments, was experimentally found to be violated [1]:

$$\int_0^1 g_1^p(x)\,dx = (0.189)_{theor} = (0.126 \pm 0.010 \pm 0.015)_{EMC/SLAC}$$

A check of the experimental result and a more precise measurement has highest priority. The integral requires high statistical precision especially at low x. The **Bjorken sum rule** [9]

$$\int_0^1 [g_1^p(x) - g_1^n(x)]\,dx = \frac{1}{6}\left|\frac{g_A}{g_V}\right| + \text{QCD corr.}$$

relates the structure functions $g_1^{p,n}$ to the axial charges g_A and g_V measured in Gamov-

Teller nuclear β-decay. A violation of the Bjorken sum rule would cast doubt on the validity of the quark model or pertubative QCD. The **Burkhardt-Cottingham sum rule** [10] $\int_0^1 g_2(x)\,dx = 0$ is less fundamental, but its violation would rule out certain classes of spin models of the nucleon [8].

The analysis of the EMC results lead to the conclusion that the integral contribution of the quarks to the proton spin $\Delta q = \int_0^1 \delta q(x)\,dx$ is small (10–20%) [1,3,8]. An analysis of the hadronic final states would allow to distinguish the contributions from valence quarks Δq_V, sea quarks Δs, gluons ΔG and from orbital angular momentum of the quarks L_z [8]:

$$\frac{1}{2}\left(\Delta u_V + \Delta d_V + \Delta s\right) + \Delta G + L_z = \frac{1}{2} \qquad \text{(nucleon spin)}$$

L_z: The angular momentum of the quarks can be deduced from the azimutal distributions of hadrons in unpolarized scattering off transversely polarized nucleons [11].

Δu, Δd, Δs: Multiplicity asymmetries of (leading) hadrons in polarized lepton scattering off longitudinally polarized nucleons reflect the Δu and Δd distributions (π^+, π^-) resp. the strange sea contributions Δs (K^-) [14].

ΔG: Both, high-$p_\perp$ jets and J/Ψ from the photon gluon fusion graph give direct information about the gluon contribution of the nucleon spin.

Table 1 summarizes which of the above items can be measured by the four proposed new spin experiments. The most complete measurement program will be done by the HERMES experiment.

3. Inventing the 'Perfect' Spin Experiment

Criteria for a precise measurement of spin asymmetries are statistical and systematic precision. The statistical precision can be quantified by a 'Figure of Merit',

$$\text{FM} = I^B\,T\,\left(p^B p^T f\right)^2$$

defined as the product of the beam current I^B, the target density T, the beam and target polarizations p^B, p^T and the fraction of polarizable nucleons in the target f. Table 1 shows a comparison of the FM for the different spin experiments. Systematic precision requires

- a frequent reversal of the polarization to cancel time dependent systematic effects,
- a good beam and target polarimeter,
- high degrees of polarization and a small dilution factor to increase the asymmetry signal versus the spin-independent cross section,
- small radiative corrections because of uncertainties in the radiative asymmetries coming from coherent contributions,
- small background and an unambiguous reconstruction and identification of the scattered lepton.

The physical relevance of the experiments is increased

- by covering a wide kinematical range that allows to verify the sum rules (low-x region) and to detect higher twist effects (Q^2-dependence),

Table 1: *Comparison of the possible measurements, the Figure of Merit and the projected statistical and systematic errors of the different future spin experiments. The figures are mainly taken from [12]. The errors for the HELP experiment are not quoted because of the small luminosity.*

measurement:	target	SMC CERN 1991-?	E-142 SLAC 1992-?	HERMES DESY $\geq$1994?	HELP CERN ?
g_1^p	H	yes	–	yes	yes
g_1^n	D	yes	–	yes	yes
	^{3}He	–	yes	yes	–
g_2^p	H	(limit)	–	yes	–?
g_2^n	D	–	–	yes	–?
	^{3}He	–	(limit)	yes	–
b_1	D	–	–	yes	–
Δ	D	–	–	yes	–
$\pi^\pm$	–	(yes)	–	yes	(yes)
L_z	–	–	–	yes	–
J/Ψ	–	(limit)	–	–	–
jets	–	–	–	–	–
Figure of Merit[1]:	H	1.0	0	6.4	0.05
	D	1.0	0	13.0	0.09
	^{3}He	0	19.2 $[0.14]^2$	2.5	0
stat.+syst. errors (%):					
$\int_0^1 g_1^p\, dx$	H	5.3+6.8	–	2.3+5.2	?
$\int_0^1 g_1^n\, dx$	D-H	19.2+15.5	–	7.6+11.2	?
	^{3}He	–	6.5+15.0	7.3+7.8	–
$\int_0^1 (g_1^p - g_1^n)\, dx$	D-H	9.0+9.6	–	3.7+7.1	?
	^{3}He-H	–	–	2.9+4.3	–

[1] *compared to SMC*   [2] *including spectrometer acceptance*

- by using p and n targets (H, D and/or ^{3}He) with longitudinal and transverse polarization,
- by the ability to measure hadron asymmetries.

Table 1 gives a comparison of the projected statistical and systematic accuracies for the determination of the sum rules. But not only a verification of sum rules is important but also a high precision measurement of the shape of the spin structure funtions that allows to distinguish between different quark models.

4. The HERMES Experiment

As the SMC and E-142 experiments are described elsewhere in this volume [13], HERMES will be described in more detail here. The HELP experiment at LEP is in many respects similar to the HERMES experiment, except that HELP uses a polarized gas jet target with a much lower luminosity [7].

HERMES intends to measure the spin structure functions by scattering polarized electrons off a polarized storage cell target at HERA. The experiment will be performed in the modified HERA East Hall and can in principle run simultaneously with the collider experiments H1 and ZEUS. Spin rotators will turn the transversely polarized HERA e^--beam into a longitudinally polarized one.

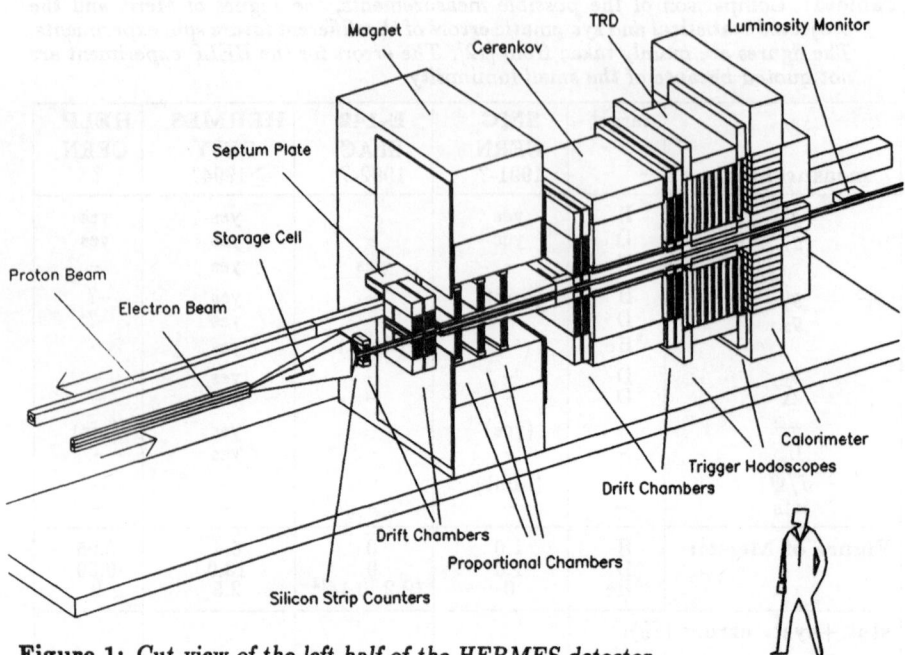

Figure 1: *Cut view of the left half of the HERMES detector.*

Polarized hydrogen and deuterium gas serves as proton and neutron target, a polarized ^{3}He target allows an independent check of the neutron structure functions. In order to achieve a luminosity as high as $L = 3.5/7/30 \cdot 10^{31} \mathrm{cm}^{-2}\mathrm{sec}^{-1}$ for H/D/^{3}He the new technique of a storage cell target will be used. The 40 cm long thin walled target cell is filled by a permanent gas flow from a source of polarized atoms. After several hundred wall bounces the atoms diffuse out of the exit holes.

The atomic beam source has been build in Heidelberg and achieves an intensity of $5.4 \cdot 10^{16} \vec{H}/\mathrm{s}$ in two HFS. This flow produces a target density of $T = 10^{14}$ atoms/cm^2 in a $24 \times 8 \times 400$ mm^3 elliptical tube at 100 Kelvin. A target polarization above 80% is achieved by Stern-Gerlach separation of atomic hydrogen and deuterium. The ^{3}He target gas is polarized by optically pumping with a polarized laser beam.

The longitudinal and transverse target polarization is realized by a longitudinal resp. transverse guiding field of $B \simeq 0.34$ T. This high magnetic field also suppresses depolarization effects by wall bounces and by the bunch field of the electron beam. The nuclear spin can be flipped within milliseconds by a HF-transition in the source. A periodical flip will minimize systematic errors in the experiment.

The major advantage of the gas target is the large fraction of polarisable nucleons $f^2 = 1.00$ compared to a dilution of $f^2 = 0.01...0.1$ in solid polarized targets.

The HERMES detector (see fig. 1) will be a forward spectrometer with a dipole magnet of 1.5 Tm. A horizontal septum-plate in the middle of the magnet will protect the HERA ring from the magnetic field. The angular acceptance is 40 mrad $< \Theta <$ 250 mrad. The kinematical range covered by the experiment is $0.02 < x < 0.8$ and $Q^2 > 1$ GeV^2. Measurements below $Q^2 = 1$ GeV^2 are also possible to study the transition between the negative asymmetries in the resonance region and the positive ones in the scaling region. To suppress regions with high radiative corrections only events below $y < 0.85$ are accepted.

Angle and momentum of the scattered electron and of hadrons are measured by silicon-strip detectors, proportional- and drift-chambers. The resolution in x will be $\sim 2-6\%$ and in $Q^2 \sim 1-2\%$ depending on the kinematical region. A calorimeter wall serves as trigger device with a threshold of $E' > 4.5$ GeV. It provides an additional energy measurements of the scattered electrons and allows in conjunction with a TRD an electron/hadron separation with a π suppression of $> 10^4$. A gas Čerenkov detector allows the identification of π's and K's to extract hadron spin asymmetries.

The HERMES experiment will provide fundamental information on the spin structure of the nucleon by the measurement of six different spin structure functions g_1^p, g_1^n, g_2^p, g_2^n, b_1 and Δ with good precision. This includes a completely independent experimental check of the EMC-results for the proton and a test of the Bjorken sum rule with a precision of 5-8%. It also promotes new experimental technologies, namely a *'polarized storage cell target inside a polarized storage ring'*.

The full approval of the experiment has been delayed until the transverse polarization of the electrons in the HERA ring is experimentally proven (1991?). The construction time of the spectrometer will be about two years. Sufficient statistics will be collected within one month of data taking for each target.

Acknowledgement

I would like to thank my collegues at Heidelberg, especially K. Rith, for efficient collaboration.

References

[1] EMC, J.Ashman et al., Phys. Lett. **206B** (1988), Nucl. Phys. **B328** (1989) 1.
[2] J. Ellis and R.L. Jaffe, Phys. Rev. **D9** (1974) 1444, **D10** (1974) 1669.
[3] review articles and references therein e.g.:
A. Manohar, Proc. Polar. Coll. Worksh., Univ. Park USA, (1990) 90;
F. Close, this volume;
B.L. Ioffe, M. Karlinger, SLAC-PUB-5235 (1990);
G. Altarelli, G.G. Ross, Phys. Lett. **212B** (1988) 391.
[4] J. Beaufays et al., CERN/SPSC 88-47, SPSC/P242 (1988).
[5] R. Arnold et al., SLAC Proposal E142, (1989).
[6] K. Coulter et al., DESY/PRC 90/1, (1990).
[7] L. Dick et al., CERN/LEPC 88-16 (1988); LEPC/89-10 (1989).
[8] R.L. Jaffe, Colloque de Physique C6 (1990) 149.
[9] J.D. Bjorken, Phys. Rev. **148** (1966) 1457, **D1** (1970) 1367.
[10] G.H. Burkhardt et al., Ann. Phys. (NY) **56** (1970) 453.
[11] T.C. Meng, Proc. XXIV Renc. de Moriond, Les Arcs (1989) 287.
[12] K. Rith, Proc. 9th Int. Symp. High Energy Spin Phys., Bonn (1990), Vol I,198.
[13] contributions of R. van Dantzig and G.G. Petratos in this volume.
[14] L.L. Frankfurt et al., Phys. Lett. **B230** (1989) 141.

Few-Body Systems, Suppl. 6, 362—367 (1992)

Few-Body Systems

MEASUREMENT OF THE NEUTRON SPIN STRUCTURE FUNCTION—TEST OF THE BJORKEN SUM RULE *

Gerassimos G. Petratos**

Stanford Linear Accelerator Center
Stanford University, Stanford, CA 94309

An experiment to measure the neutron spin-dependent structure function $g_1^n(x)$ over a range in x from 0.04 to 0.7 and with $Q^2 > 1$ (GeV/c)^2 is presented. The experiment consists of scattering a longitudinally polarized electron beam from the Stanford Linear Accelerator off a polarized ^{3}He target and detecting scattered electrons in two magnetic spectrometers. The experiment will provide a critical test of the Bjorken sum rule and valuable information in understanding the nucleon spin structure and the violation of the Ellis-Jaffe sum rule.

A few years ago the EMC group measured [1] the proton spin-dependent structure function $g_1^p(x)$ in a range over the Bjorken scaling variable x wide enough to calculate reliably $\int_0^1 g_1^p(x)dx$. The experimental result was found to disagree with the Ellis-Jaffe sum rule [2] based on quark current algebra and isospin symmetry, that assumed that the net polarization of the strange quark sea is zero. Assuming the validity of the Bjorken sum rule [3] and exact flavor SU(3) symmetry in the decays of the members of the baryon octet, the EMC data lead to two surprising conclusions: (1) that the quarks carry only a small fraction of the proton spin, consistent with zero within the experimental uncertainties and (2) that a rather large and negative contribution to the proton spin should be attributed to the strange quark sea. The Bjorken sum rule:

$$\int\limits_0^1 [g_1^p(x) - g_1^n(x)]dx = \frac{1}{6}\left|\frac{g_A}{g_V}\right|\left[1 - \frac{\alpha_s(Q^2)}{\pi}\right],\qquad(1)$$

where α_s is the QCD coupling constant, relates the nucleon spin structure functions to the ratio $|g_A/g_V|$ of the axial to vector weak coupling constants of the nucleon

---

* Work supported by US Department of Energy contract DE-AC03-76SF00515.

** Reporting on behalf of the SLAC E142 collaboration (American, Clermont-Ferrand, Harvard, Princeton, Saclay, SLAC, Stanford, Syracuse, Wisconsin).

beta decay. It is regarded as a fundamental QCD sum rule based only on quark current algebra with the standard quark charge assignments and on isospin symmetry.

The unexpected EMC result has lead to extensive discussions about the structure of the proton spin and possible explanations. Some of the questions raised include the validity of the extrapolation of $g_1^p(x)$ for $x \to 0$ and the validity of the Bjorken sum rule and perturbative QCD. The principal explanations proposed are: (a) that the nucleon spin is due to orbital angular momentum contributions from the quarks, a notion explained within the Skyrme model of the nucleon and (b) that the gluons contribute substantially to the nucleon spin. Recent reviews of many old and new theoretical developments on the subject are given in Ref. [4]. On the experimental front, the EMC result has generated a series of recently proposed or planned high precision measurements in the U.S. [5,6] and in Europe [7-9] on the neutron and proton spin-dependent structure functions using a variety of polarized targets (gas ^{3}He, NH$_3$, ND$_3$, butanol, deuterated butanol, gas hydrogen and gas deuterium).

This paper describes an approved experiment at SLAC [5] to extract the neutron spin-dependent structure function $g_1^n(x)$ by measuring the cross section asymmetries:

$$A_\parallel = \frac{\sigma(\uparrow\Uparrow) - \sigma(\uparrow\Downarrow)}{\sigma(\uparrow\Uparrow) + \sigma(\uparrow\Downarrow)} = \frac{1-\epsilon}{(1+\epsilon R)W_1} \left[(E + E'\cos\theta)MG_1 - Q^2 G_2 \right] \qquad (2)$$

$$A_\perp = \frac{\sigma(\uparrow\Rightarrow) - \sigma(\uparrow\Leftarrow)}{\sigma(\uparrow\Rightarrow) + \sigma(\uparrow\Leftarrow)} = \frac{(1-\epsilon)E'}{(1+\epsilon R)W_1} \left[(MG_1 + 2EG_2)\sin\theta \right] \qquad (3)$$

in deep inelastic scattering of polarized electrons from polarized neutrons, where $\uparrow, \downarrow$ denotes the longitudinal spin of the incoming electron (along or opposite its direction of motion) and $\Uparrow, \Downarrow$ or $\Leftarrow, \Rightarrow$ denotes the longitudinal or transverse spin of the target nucleon. The asymmetries are functions of kinematics (E and E' are the incident and scattered electron energies, θ is the scattering angle, $\epsilon = [1 + 2(1 + \nu^2/Q^2)\tan^2(\theta/2)]^{-1}$), of the unpolarized structure functions W_1 and W_2 connected via $R = (1 + \nu^2/Q^2)W_2/W_1 - 1$ and of the polarized structure functions G_1 and G_2, which in the Bjorken scaling limit of large momentum $Q^2 = 4EE'\sin^2(\theta/2)$ and energy $\nu = E - E'$ transfers, become functions only of $x = Q^2/2M\nu$: $M^2\nu G_1(x, Q^2) \to g_1(x)$, $M\nu^2 G_2(x, Q^2) \to g_2(x)$.

Polarized electrons of 22.7 GeV energy and 5μA intensity will be produced by a laser optically pumped GaAs source presently being built for the Stanford Linear Collider. This type of source, used successfully in the electron-deuteron scattering parity violation experiment [10], is capable of providing high intensity beams (in excess of 5×10^{11} electrons per 1.6 μs long pulse at 120 Hz) with an average polarization near 40%. The helicity of the beam can be reversed randomly on a pulse-to-pulse basis by reversing the circular polarization of the excitation photons. The beam polarization will be monitored during the experiment by performing Møller scattering from magnetized Permendur foils and measuring the outgoing electrons at 90° in the CM frame in a magnetic spectrometer consisting of a dipole magnet and a calorimeter.

The experiment will use a polarized ^{3}He target. The nucleon spin structure of a polarized ^{3}He target is the same as a polarized free neutron target to the extend that the ^{3}He nucleus is in its space-symmetric S state. In this state, the two proton spins are aligned antiparallel due to the Pauli exclusion principle, implying that scattering from a polarized ^{3}He nucleus represents scattering from a polarized neutron. The presence of some D state admixture in the ^{3}He ground state complicates the above picture by introducing a polarized proton component in the total polarization of the nucleus. Theoretical calculations [11] have shown that this component has a small effect in the cross section asymmetry measurements and that the theoretical uncertainty in extracting the spin structure function $g_1^n(x)$ is small.

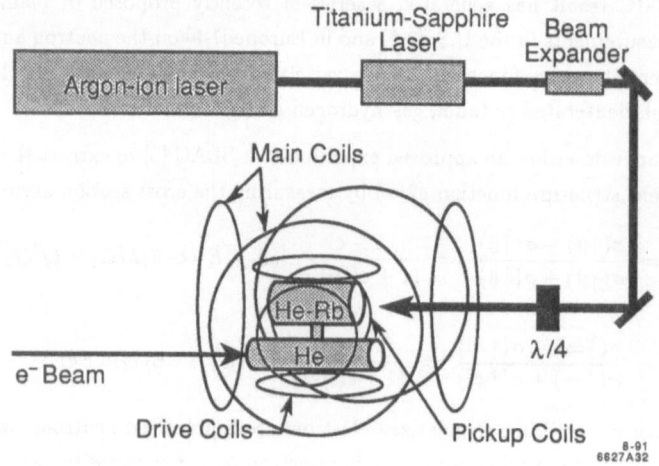

Fig. 1: The polarized ^{3}He target system.

The target is based on the technique of ^{3}He polarization by spin exchange with Rb vapor [12]. The Rb atoms are polarized via laser optical pumping by absorbing circularly polarized photons at a wavelength of 795 nm. The spin exchange from Rb to ^{3}He occurs due to the hyperfine interaction between the polarized valence electron of Rb and the ^{3}He nucleus. The ^{3}He nuclear polarization is measured by means of NMR adiabatic fast passage [12] and by observing the frequency shift caused in the electron-paramagnetic-resonance line of the Rb by the polarized ^{3}He [13].

The major elements of the target system are shown in Fig. 1. To avoid Rb depolarization by the beam, the optical pumping region is separated from the bombardment region by using a dual target cell. The main cell is a 30 cm glass tube, containing a ^{3}He density of 3×10^{20} cm^{-3} at 10 atm and a density of $\sim3\times10^{18}$ cm^{-3} of N$_2$. The N$_2$ is necessary for non-radiatively quenching the Rb excited state populated by the absorption of the laser light. The pumping cell contains several milligrams of Rb metal and is heated to $\sim 180°C$ to obtain the desired density of Rb vapor. The axis of quantization for polarization is established by the magnetic field produced by the two main Helmholtz coil sets. The drive and pickup coils are used for the ^3He polarization measurements. The lasers for optical pumping are four solid state titanium-sapphire lasers, each pumped by an argon-ion laser and producing greater than 20 watts of power.

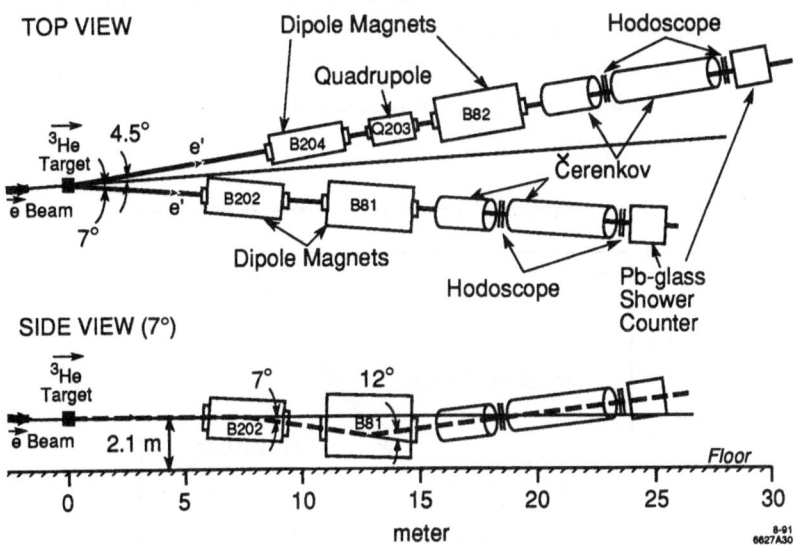

Fig. 2: The magnetic spectrometer system.

Scattered electrons of 7 to 18 GeV in energy will be detected in two magnetic spectrometers centered at 4.5° and 7° respectively, as shown in Fig. 2. Each spectrometer is based on two large aperture dipole magnets bending in opposite directions. This 'reverse' deflection design doubles the solid angle, integrated over the 7-18 GeV/c range, of the conventional design of same direction bending, used in previous polarized electron scattering experiments at SLAC [14]. The solid angle of the 4.5° arm is 0.2 msr and of the 7° arm is 0.7 msr. Proper choice of the deflection angles and the distance between the two magnets in each spectrometer allow background photons from radiative processes to reach the detectors only after having bounced twice on the spectrometer vacuum walls, resulting in an expected tolerable background. Each spectrometer is equipped with a pair of Čerenkov detectors, a pair of scintillator hodoscopes and a lead-glass shower calorimeter providing electron and pion identification with $\sim \pm 2\%$ momentum resolution, sufficient for the asymmetry measurements.

The experiment is expected to be limited by systematic uncertainties. Their percentage contributions on $\int g_1^n(x)dx$ have been estimated assuming that it has the value -0.07 as predicted by the Bjorken sum rule in conjunction with the EMC result on $\int g_1^p(x)dx$. The beam polarization measurement is inherently limited to $\sim \pm 5\%$ by a background subtraction of the signal and an uncertainty in the amount of the Møller target foil spin magnetism. The ^{3}He target polarization measurement is limited to $\sim \pm 5\%$ by the uncertainty in the density of ^{3}He in the cell and in the absolute calibration as compared to a water sample. The uncertainty in the dilution factor (the ratio of the polarizable to the total number of nucleons) is dominated by the uncertainty in the thickness of the glass cell walls and in the ^{3}He density and is $\sim \pm 5\%$. The experimental uncertainties in the measurement of the unpolarized W_2^n structure function ($\sim \pm 5\%$) and of the ratio R ($\sim \pm 2\%$) in the kinematic regime of this experiment are directly translated as uncertainties in the $g_1^n(x)$ measurement. Radiative corrections uncertainties ($\sim \pm 2\%$) are not expected to contribute in a sizable way to the asymmetry measurements.

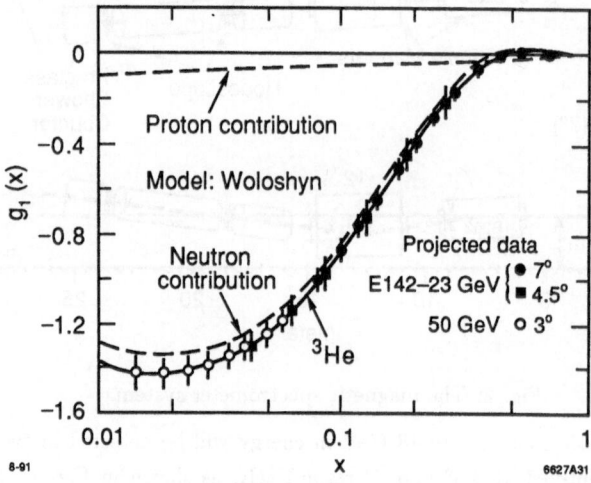

Fig. 3: Projected data for the neutron spin structure function $g_1^n(x)$.

The projected statistical accuracy of the experiment is shown for 100/10 hours of running with longitudinal/transverse target polarization in Fig. 3, assuming that $g_1^n(x)$ follows the model of Ref. [11]. The figure also shows the small correction factor relating the neutron spin structure function $g_1^n(x)$ to that of ^{3}He. The experiment will also provide a crude (statistically limited) measurement of the $g_2^n(x)$ structure function, not shown in the figure. The different contributions to the total ($\pm 15\%$) projected uncertainty on the Bjorken sum rule are shown in Table 1, where a conservative theoretical error of $\pm 20\%$ on the extraction of $\int g_1^n(x) dx$ from that of ^{3}He has been assumed.

Future measurements at SLAC will provide a better measurement of $g_2^n(x)$ as well as measurements at lower x values. Such projected measurements on $g_1^n(x)$ using a planned upgraded beam energy of 50 GeV [15] are also shown in Fig. 3. These measurements assume the same running conditions, for 3° scattering, using the new generation of polarized guns (> 70% beam polarization) being developed [16]. In summary, this experiment is expected to make significant new measurements on the

367

Table 1: Expected uncertainties on the Bjorken sum rule test.

Quantity	Type of Uncertainty	Value
$\int g_1^n(x)dx$	Statistical	$\pm.004$
$\int g_1^n(x)dx$	Systematic	$\pm.010$
$\int g_1^n(x)dx$	Total experimental	$\pm.012$
$\int g_1^n(x)dx$	Theoretical	$\pm.016$
$\int g_1^n(x)dx$	Total experimental and theoretical	$\pm.020$
$\int g_1^p(x)dx$	Experimental (EMC)	$\pm.021$
$\int [g_1^p(x) - g_1^n(x)]dx$	Combined TOTAL	$\pm.03$

neutron spin structure, and together with other proposed experiments, should provide precise tests of the Bjorken sum rule and of our understanding of the spin structure of the nucleons.

REFERENCES

[1] J. Ashman *et al.*, Nucl. Phys. **B 328**, 1 (1989).

[2] J. Ellis and R. L. Jaffe, Phys. Rev. **D 9**, 1444 (1974); **D 10**, 1669 (1974).

[3] J. D. Bjorken, Phys. Rev. **148**, 1467 (1966); **D 1**, 1376 (1970).

[4] G. Altarelli, CERN preprint CERN–TH–5675/90 (1990) ; G. G. Ross, Oxford University preprint OUTP–89–39P (1989).

[5] SLAC Proposal E142, E. W. Hughes *et al.*, (1989).

[6] J. S. McCarthy and C. Y. Prescott, private communication.

[7] CERN Proposal CERN/SPSC 88–47, V. W. Hughes *et al.*, (1988).

[8] DESY Proposal PRC 90/1, R. Milner, K. Rith *et al.*, (1990).

[9] CERN Letter of Intent CERN/LEPC 88–16, L. Dick *et al.*, (1988).

[10] C. Y. Prescott *et al.*, Phys. Lett. **B 77**, 347 (1978); **B 84**, 524 (1979).

[11] R. M. Woloshyn, Nucl. Phys. **A 496**, 749 (1989).

[12] T. E. Chupp *et al.*, , Phys. Rev. **C 36**, 2244 (1987).

[13] S. R. Schaefer *et al.*, Phys. Rev. **A 39**, 5613 (1989).

[14] G. Baum *et al.*, Phys. Rev. Lett. **51**, 1135 (1983).

[15] C. Y. Prescott, private communication.

[16] T. Maruyama *et al.*, Phys. Rev. Lett. **66**, 2376 (1991).

Few-Body Systems, Suppl. 6, 368—377 (1992)

Few-Body Systems

The Role of Sea Polarisation and Gluons in Magnetic Moments and Resonance Excitation

F E Close

Rutherford Appleton Laboratory,
Chilton, Didcot, Oxon, OX11 OQX, England.

ABSTRACT

This talk discusses the rôle of gluons and polarised sea antiquarks in the proton at low energies. Spin dependent effects, magnetic moments and the possible excitation of hybrid (gluonic) baryons are considered. The relation between a low energy electron program (such as at CEBAF) and deep inelastic polarisation is examined

INTRODUCTION

The remarkable results from EMC on deep inelastic polarised leptoproduction have generated intense interest on the spin content of the proton. In the kinematic region $x \gtrsim 0.2$ where valence quarks are dominant, the polarisation asymmetry is in excellent agreement with that predicted [1,2] based upon experience with the flavour-spin correlations of the (valence) quarks in spectroscopy. However, the integral over the polarised structure function $g_1^P(x)$ has been widely interpreted as implying that the net polarisation of quarks and antiquarks is only a small fraction of the total. One possibility is that the sea is polarised, with a negative polarisation that annuls the positive contribution of the valence quarks. Another possibility is that polarised gluons may be responsible for "hiding" the intrinsic quarks' spin polarisations. In my talk I shall consider the impact of these ideas on low energy observables and contemplate the rôle, or evidence, for gluons in the "low energy" proton.

There are three parts to this. First there is the proton of the naive constituent quark model (NCQM) - three quarks whose flavours and spins are correlated by the Pauli principle. This wavefunction describes the ratio of magnetic moments (μ_n/μ_p) well - how does this survive? Second, gluons implicitly play a rôle at $O(\alpha_s)$ in $pQCD$ through their exchange, mediating a $\vec{S}.\vec{S}$ "hyperfine" spin dependent energy shift. This splits the masses of Δ and N, may cause the transition form factor for Δ excitation to fall faster with Q^2 than does the elastic $G_M(Q^2)$ and, via Pauli, distorts the x dependence of the parton distributions $u(x)/d(x)$ at $x \to 1$ as revealed in deep inelastic scattering.

The possibility that there is polarised glue in the (deep inelastic) proton, $\Delta G \neq 0$, raises a question as to whether the static properties, such as magnetic moments, may need reconsideration. And in turn, there is a question of how much of a dynamical rôle gluons play - are there "hybrid baryons" where both quark and gluonic degrees of freedom are excited?

MAGNETIC MOMENTS

Define $\Delta q \equiv \int dx \Delta q(x)$ whose $\Delta q(x) \equiv q^\uparrow(x) - q^\downarrow(x)$ is the spin polarisation distribution of quarks in a polarised proton. Define ΔG similarly. The deep inelastic polarised experiment measures $\Delta q' \equiv \Delta q - \frac{\alpha}{2\pi} \Delta G$. The EMC results are

$$\Delta u' + \Delta \bar{u} \simeq \frac{3}{4}$$

$$\Delta d' + \Delta \bar{d} \simeq -\frac{1}{2}$$

$$\Delta s' + \Delta \bar{s} \simeq -\frac{1}{4}$$

$$\Delta q' + \Delta \bar{q} = 0.13 \pm 0.10 \pm 0.15 \simeq 0$$

First we shall compare these with the naive quark model.

The wavefunction for the VNP (very naive proton) is formed by noting that two identical quarks (the uu) will be in net $S = 1$ by Pauli (being color antisymmetric). Thus from the Clebsch's

$$p^\uparrow = \sqrt{\frac{2}{3}}(u^\uparrow u^\uparrow)d_\downarrow + \sqrt{\frac{1}{3}}(u^\uparrow u^\downarrow)d_\uparrow$$

and so the spin weighted probabilities are

$$\Delta u = \frac{4}{3}, \quad \Delta d = -\frac{1}{3} \quad ; \quad \Delta s = 0 \quad ; \quad \Delta \bar{q} = 0$$

Note that $\Delta u + \Delta d = 1$, confirming the fact that for this wavefunction the quarks carry the entire J_z of the proton, and $\Delta u - \Delta d (\equiv \frac{g_A}{g_V}) = \frac{5}{3}$. Notice also that $\Delta u = -4\Delta d$. If we ignore L_z and consider only valence quarks, the ratio of magnetic moments may be written

$$\frac{\mu_p}{\mu_n} = \frac{\frac{2}{3}\Delta u - \frac{1}{3}\Delta d}{\frac{2}{3}\Delta d - \frac{1}{3}\Delta u} = -\frac{3}{2}$$

where the empirically satisfactory result has followed by inputting the $\Delta u, \Delta d$ above. However the bad result for g_A/g_V encourages a "less naive quark model" where the above spinor representation of $p^\uparrow$ refers to the upper components of $q^\uparrow$ only (see ref 2). If the lower components introduce a "passive L_z" which dilutes the net spin by 25% then

$$\Delta u = 1 \quad \Delta d = -\frac{1}{4}; \quad \Delta u + \Delta d = \frac{3}{4}; \quad \Delta u - \Delta d = \frac{5}{4}.$$

Note that $\mu_p \mu_n = -\frac{3}{2}$ still.

Now let us return momentarily to the EMC results.

Notice the obvious, but frequently overlooked, result that these are for the sums of $\Delta q_i + \Delta \bar{q}_i$. We can rewrite these as

$$\Delta q_i + \Delta \bar{q}_i = \Delta q_v + (\Delta q + \Delta \bar{q})_{sea}$$

If all flavours of the sea are (approximately) equally polarised then $(\Delta q + \Delta \bar{q})_i^{sea} \simeq -\frac{1}{4}$ and hence $\Delta d_v = 1, \Delta d_v = -\frac{1}{4}$. Alternatively, or in addition, if there are polarised gluons in the proton then, via the anomaly, these contribute to the $\Delta q(\bar{q})$ of the EMC: denoting the latter by $\Delta q'$ then

$$\Delta q' \equiv \Delta q - \frac{\alpha}{2\pi}\Delta G.$$

So, if $\Delta s + \Delta\bar{s} - \frac{\alpha}{\pi}\Delta G \simeq -\frac{1}{4}$, the valence spin is in accord with the magnetic moments. The question of why the sea and/or glue is polarised is for the future: the valence results, and the polarisation data for $x \gtrsim 0.1$ (where valence quarks dominate and successfully predicted the data) suggest that the EMC results imply a non-trivial sea dressing a "standard" valence distribution.

If this is the case, what impact may it have on low energy properties such as magnetic moments? This has been discussed by Karl recently; I refer you to his papers and here limit myself to some comments.

The ratio of μ_p/μ_n implies only that the ratio $\Delta u/\Delta d = -4$ and says nothing about their absolute magnitudes. As noted in [3] the magnetic moments are protected from the sea (assuming that $\Delta q + \Delta\bar{q}$ is approximately flavour independent) as $\sum e_i = 0$. Indeed, the fact that electromagnetic properties and photoexcitation are well described in the constituent quark model may support this flavour independence: the probe that is linear in charge is neutral to the polarised sea, whereas deep inelastic scattering (proportional to e_i^2) is sensitive to the polarised sea.

Even if one ignores L_Z, and the fact that the EMC polarisation involves the axial rather than vector current (as in magnetic moment), the two quantities are not related unless $\Delta\bar{q} = 0$. Heuristically

$$EMC \simeq \Delta q_v + \Delta q_s + \Delta\bar{q}_s$$
$$\mu \simeq \Delta q_v + \Delta q_s - \Delta\bar{q}_s$$

Karl [4] has extended the analysis of μ_n, μ_p (in terms of $\Delta u, \Delta d$), to the whole octet. He uses octet symmetry so that

$$\mu_p \sim \mu_u\Delta u + \mu_d\Delta d + \mu_s\Delta s$$
$$\mu_n \sim \mu_u\Delta d + \mu_d\Delta u + \mu_s\Delta s$$
$$\mu_{\Xi^-} \sim \mu_u\Delta u + \mu_d\Delta d + \mu_s\Delta u$$

etc. (Strictly, the above equations hould include the $\Delta\bar{q}_i$ with opposite sign to their respective Δq_i). There are (at least) three possible ways in which $\underset{\sim}{8}$ symmetry may be used.

(A) Δq_v
(B) $\Delta q_v + \Delta q_s \equiv \Delta q$
(C) $\Delta q - \Delta\bar{q}$

Refs (4) have implicitly or explicitly assumed that $\Delta\bar{q} = 0$. As they allowed $\Delta s \neq 0$ (and their fits required this) it is clear that they have chosen to apply $\underset{\sim}{8}$ symmetry to option (B). However, it is unnatural (though not in principle incorrect) that one obtains a significant $\Delta s \neq 0$ even though $\Delta\bar{q} = 0$.

In ref [5] Karl has allowed for $\Delta\bar{q} \neq 0$ and applied $\underset{\sim}{8}$ symmetry to option (C). He makes certain assumptions to facilitate the analysis and concludes that $\Delta s + \Delta\bar{s} \simeq -0.2$. This is interesting in that it shows that, a priori, the EMC results are not necessarily in discord with magnetic moments (see also ref 6). However, it is by no means required and reasons for caution are apparent by the following example.

The requirement $\Delta s \neq 0$ is being driven, in particular, by the $\mu(\Xi^-)$ which is difficult to fit in terms of its valence quarks alone. This wants $\Delta u(\Xi^-) \neq 0$. Now the Δu is in the sea of Ξ^- and, by the supposed octet symmetry, forces $\Delta s(p) \neq 0$. But suppose, for illustration, that $m_s \to \infty$ (e.g. like a b-quark, so that $\Xi_b(bbd)$ requires $\Delta u \neq 0$). Would you accept that $\Delta u \neq 0$ in Ξ_b implied that $\Delta b \neq 0$ in the proton? This is clearly unreasonable and, although an extreme example due to the $m_s \to \infty$ assumption, it does illustrate that $m_s > m_u$ distorts the octet analysis.

The resolution of this problem is seen if we consider the octet analysis applied to option (A) above. Let the baryons be described by three valence quarks (to which $\underset{\sim}{8}$ symmetry will be applied) and a $q\bar{q}$ sea.

Suppose first that the polarised $q\bar{q}$ sea is common to all hadrons in the octet. One may be forced, by the fit, to conclude that $\Delta q_s - \Delta \bar{q}_s \neq 0$ but one cannot apportion this among the separate flavours, i.e. one cannot conclude that $\Delta s - \Delta \bar{s} \neq 0$ necessarily.

As a second example, suppose that the $s\bar{s}$ content in the sea differs from $u\bar{u}$ or $d\bar{d}$ i.e. the sea is not a flavour $\underset{\sim}{1}$, but contains $\underset{\sim}{8}$ component. The valence + sea will not in general be only $\underset{\sim}{8}$ but may have $\underset{\sim}{10}$, $\underset{\sim}{\bar{10}}$ or $\underset{\sim}{27}$ transformation properties. Thus the overall assumption of $\underset{\sim}{8}$ transformation that underwrites several recent analyses of magnetic moments may be suspect.

These are problems that merit further investigation. Future information on the polarisation of the sea [7], and its possible favour dependence, may prove important. Nonetheless, it is interesting that magnetic moments can accommodate significant Δs or $\Delta \bar{s} \neq 0$.

An alternative explanation [8] of the EMC data is that $\Delta G \neq 0$ is responsible for deviation from the naive model $\Delta u, \Delta d$. This will in general cause the correlations of quark flavours and spins to differ from that in the naive model and, thereby, destroy the magnetic moment analyses. However, this need not be so. Several years ago it was noticed [9] that a particular coherent superposition of states - $3q$ in $S = \frac{1}{2}$ and $s = \frac{3}{2}$ coupled to a gluon - preserves $\mu_n/\mu_p(\Delta u/\Delta d = -4)$. As far as ratios of static properties are concerned this wavefunction is indistinguishable from the VNP (physically, if one radiates a single gluon additively from each quark in VNP, one obtains the above "hybrid" wavefunction).

These results may suggest that we understand the static wavefunction less than we thought - the VNP may well be an oversimplified description. This suggests the following question: what rules out (or "rules in") "dynamical glue" in the low energy proton, and are there dynamical gluonic hybrids (as suggested in the MIT bag model)? If so, can we hope to identify them in $\gamma N \to N^* g$ at low energies?

GLUONIC BARYONS (very model dependent)

There is rather general expectation that the lightest glueballs should exist with masses in the 1-2 GeV region, and in turn this raises the question as to whether the gluonic degrees of freedom can be excited in a system of quarks but "independent" of these quarks. These states are known as hybrids or gluonic hadrons. Specific models predict their existence as low as 0.5 - 1 GeV above the mass of the lightest conventional quark - hadrons and so one might hope that the lightest gluonic baryons are below 2 GeV in mass, certainly accessible to excitation at low energy facilities such as CEBAF.

There are some tantalising theoretical questions about the interpretation and existence of these states. One suggestion has been that these states are dual to, or even misidentified as, the well known states that are traditionally interpreted as orbital and radial excitations in the constituent quark model. For example adding a gluon with $J^P = 1^-$ to these quarks generates the same set of J^P quantum numbers as if a quark had been excited to $L = 1$. In the latter case the spatial symmetry forces the three quarks to be in $\underset{\sim}{70}$ of SU(6) spin-flavour; in the gluonic case the colour $\underset{\sim}{8}$ of the three quarks, with Pauli, generates the $\underset{\sim}{70}$ too.

Detailed study of the spectroscopy and excitations can distinguish between these two pictures. First, the two spectroscopies are not identical: the radial excitations of the NRQM do not precisely correlate with gluonic excitations. And where they do appear to overlap, the spin dependent mass shifts at $O(\alpha_s)$ in perturbative QCD have quite different patterns. For example, consider the lightest hybrids which, according to the bag model, could be an $J^P = \frac{1}{2}^+, I = \frac{1}{2}$ state (e.g. the Roper resonance $P_{11}(1470)$). If the Roper resonance is a radially excited state then it will be partnered by a Δ state analogous to the way that $\Delta(1232)$ partners the nucleon; there is a possible candidate for such a state but its existence is by no means established. Contrast this with the hybrid spectroscopy [10] where the $\frac{1}{2}^+(Ng)$ is alone, the next heaviest being another $\frac{1}{2}^+$ and $\frac{3}{2}^+(Ng)$ but with Δg states pushed up to high masses by the $O(\alpha_s)$ mass shifts. Thus the presence or absence of the $\Delta(1630)$ would respectively eliminate or support $N(1470)$ as a hybrid.

Photo and electroproduction of the resonances also helps to probe the internal constitution of the hadrons. The $N(1470)$ and $\Delta(1630)$ will be excited by M1 radiation with characteristic strengths from proton and neutron targets if they are radial excitations of N and Δ. However, there is a selection rule that the lightest hybrid (the $\frac{1}{2}^+ Ng$) may be photoexcited from neutrons but not from protons [9]. This makes it seem unlikely that $N(1470)$ is a hybrid as it is excited from both n and p with relative amplitudes consistent with the ratio $-\frac{2}{3}$ as befits the ratio of magnetic moments. There is a possibility that the "Roper" resonance actually consists of two nearly degenerate states in which case one could be the hybrid. Photoexcitations from protons and comparison with $\pi^- p \to N^* \to n\gamma$ may help to settle this issue.

If the $N(1470) \neq Ng$ then the next candidate would be the $N(1710)$. This has no clear Δ partner and also is consistent with zero photoproduction from protons. One possibility is that this zero (if indeed it is zero) is a result of an "accidental" cancellation between competing electric and magnetic multipoles (e.g. as in a NRQM assignment); if this is so, one would expect that the cancellation does not survive as one varies Q^2. Hence it will be important to study the photo and electroproduction of the $N(1710)$: a hybrid will have a vanishing excitation amplitude (from protons) at all Q^2 whereas a NRQM excitation will be expected to show a Q^2 dependent production [11]. Recently there has been a novel development. Li [12] exploits the "hidden glue" hybrid wavefunction of ref 9 such that proton and hybrid are orthogonal states

$$
\begin{aligned}
N &= \cos\theta |q^3> &&+ \sin\theta |q^3 g; &&{}^4 8 +{}^2 8 > \\
H &= -\sin\theta |q^3> &&+ \cos\theta |q^3 g; &&{}^4 8 +{}^2 8 >
\end{aligned}
$$

This preserves the M1 transition ratios to the Roper (interpreted as hybrid by Li) and raises the possibility that the constitution of the Roper may indeed be discerned at CEBAF [13].

THE CONNECTION BETWEEN LOW ENERGY AND DEEP INELASTIC

The $O(\alpha_s)$ single gluon exchange gives spin dependent effects and, via the Pauli principle, flavour dependent effects. Its direct spin dependence elevates the mass of the $J = \frac{3}{2}\Delta(1232)$ relative to the $J = \frac{1}{2}$ nucleon. This also feeds more "energy" or, in the light cone momentum distributions a larger $< x >$, into quarks that are correlated with $S = 1$ relative to $S = 0$. The results is that as $x \to 1$ in deep inelastic polarised leptoproduction the polarised quark tends to spin parallel to the target and hence the polarisation asymmetry $A(x) \equiv g_1(x)/F_1(x) \to 1$ for both proton and neutron targets [1,2,14,15]. Such predictions are consistent with the trend of the data for protons and we eagerly await the data for neutrons.

Combining the above with the Pauli principle implies that $< x >_u > < x >_d$ and hence $u(x)$ dominates over $d(x)$ as $x \to 1$. This is qualitatively seen in data though there is still an open question as to whether $F_1^n/F_1^p(x \to 1) \to \frac{3}{7}$ or $\frac{1}{4}$. In the bag model it has proven possible to correlate these two very different regions of physics: setting the $O(\alpha_s)$ strength by the $\Delta - N$ mass splitting, the x dependence of deep inelastic scattering, both spin dependent and spin independent, is rather well described.

The relation between these two apparently disparate regimes of deep inelastic and resonance production may be made by Bloom-Gilman duality. For the unpolarised data this relates the shape of $F(x)$ and the elastic form factor. The key is that when $\gamma(Q^2)N \to W$ (where W is the mass of a final state be it inclusive or exclusive, then since $x' \simeq W^2/Q^2$ one can either measure

Inclusive: $F(x')$ where $x' = W^2$ (varies)$/Q^2$ (fixed)

Exclusive: $F(x')$ where $x' = W^2$ (fixed)$/Q^2$ (varies)

In the latter case one is essentially measuring the squared elastic or transition form factors.

Thus $F(x' \to 1) \sim (1-x')^3 \to G(Q^2 \to \infty) \sim Q^{-4}$ the well known counting rules.

These ideas appear to work in some detail. The flavour asymmetry ($u >> d$ as $x' \to 1$) appears to be related to the asymmetry in resonance excitation form factors (Δ excitation dies out faster with Q^2 than does $N^*, I = 1/2$). Stoler has shown [16] how for $Q^2 \gtrsim 3GeV^2$ the bump (resonance plus background) in the $\Delta(1232)$ region dies faster than the elastic nucleon and $I = 1/2$ dominated region around 1500 MeV.

That this duality may be quantitative may be seen by the work of Close and Thomas [15] who, in the MIT bag model, have shown that gluon exchange induced hyperfine forces in QCD can generate these asymmetries. Thus low energy excitation is complementary to deep inelastic in probing certain QCD induced phenomena. The above gluon exchange affects flavours due to the Pauli statistics correlating the constituents' flavours and spins. Its most direct manifestation is in its coupling to the spins and hence to spin-dependent phenomena such as those manifested in deep inelastic or low energy polarised scattering or resonance excitation.

One interesting feature of these experiments is that the polarization asymmetry $A(x, Q^2)$

$$A \equiv \frac{\sigma_{1/2} - \sigma_{3/2}}{\sigma_{1/2} + \sigma_{3/2}}$$

(where $\sigma_{1/2.3/2}$ are the photoproduction cross sections for transversely polarized photon and the polarized target to have net $J_Z = 1/2.3/2$) appears to maximize in the kinematic limit of Bjorken $x \to 1$, i.e. $A^p(x \to 1) \to 1$, at least for a proton target (the only target so far studied). In modern QCD we believe that the asymmetry tending to unity as $\to 1$ is due to chromomagnetic effects [15], in the wave function (one gluon exchange). If the (yet to be measured) neutron asymmetry also maximizes in this limit, then we will have important confirmation of these ideas. These chromomagnetic effects are also manifested in low-energy experiments. Chromomagnetic forces split the $\Delta(P_{33}(1232))$ and nucleon masses - the well-known hyperfine splitting of energy levels. This already shows us that QCD can make clean-cut predictions for low-energy experiments. Less well known, perhaps, is that QCD perturbations induce mixing in the quark wave functions which may be accessed most sharply by electromagnetic probes. Furthermore, if the $x \to 1$ behaviour of inclusive high-energy data relates to the low-energy exclusive behaviour, $Q^2 \to \infty$, then we may look for some characteristic differences in the Q^2 dependence of Δ/N^* form factors, G_M^p/G_M^n - the elastic magnetic form factors, and other specific final states such as $ep \to e\Lambda K/e\Sigma K$.

Spin-dependent sum rules imply that nontrivial Q^2 dependence must occur, probably in the low-energy region accessible to a milli-TeV machine. To illustrate this, consider the Drell- Hearn-Gerasimov (DHG) sum rule for real photons [17]

$$\frac{-2\pi^2\alpha\kappa^2}{M^2} = \int \frac{d\nu}{\nu}(\sigma_{1/2} - \sigma_{3/2}). \tag{1}$$

(κ is the anomalous magnetic moment of the target mass M) and the Bjorken [18] sum rule (for $Q^2 \to \infty$).

$$\frac{1}{6}\left|\frac{g_A}{g_V}\right| = \int_0^1 dx(g_1^p - g_1^n) \tag{2}$$

We can write these in a similar form if we define

$$G(\nu) \equiv \frac{m^2(\sigma_{1/2} - \sigma_{3/2})}{8\pi\sigma^2} \tag{3}$$

Thus the DHG sum rule becomes for protons

$$-\frac{1}{4}\kappa_p^2 = \int \frac{d\nu}{\nu}G^p(\nu, Q^2 = 0) \tag{4}$$

and for the proton-neutron difference

$$\int \frac{d\nu}{\nu}(G^p - G^n)(Q^2 = 0) = -\frac{1}{4}[\kappa_p^2 - \kappa_n^2] = 0.112 \tag{5}$$

The Bjorken sum rule (Eq. 2) becomes

$$\int_{Q^2/2m}^{\infty} \frac{d\nu}{\nu}(G^p - G^n)(Q^2) = \frac{1}{3}\frac{M^2}{Q^2}\frac{g_A}{g_V} \simeq \frac{0.37}{Q^2} \tag{6}$$

The EMC data for protons $\int_0^1 dxg_1^p(x, Q^2) = 0.114 \pm 0.029$ would suggest that

$$\int_{Q^2/2m}^{\infty} \frac{d\nu}{\nu}G^p(Q^2) \simeq \frac{0.2}{Q^2} \tag{7}$$

Thus, we can compare the proton-neutron difference as a function of Q^2. That the Q^2 dependence would be most interesting was noted long ago by Gilman Karliner and myself qualitatively, [19] and the above quantification follows that of Ioffe and collaborators [20].

Some comments about the significance of these equations is called for. Equation 2 applies in the scaling region. We know from unpolarized scattering that scaling is still rather good modulo logarithms at $Q^2 = 5 GeV^2$. The logarithmic breaking of scaling only changes the argument very slightly and cannot alone make correspondence with the value for real photons (eqs 4,5). The RHS of eq (6) grows as Q^2 falls and exceeds the $Q^2 = 0$ value when $Q^2 = 3 GeV^2$: non-perturbative effects must cut the integral down when $Q^2 < 3 GeV^2$. This is no surprise, but how these enter into the Q^2 dependent RHS is presently unknown.

These remarks (for the $p - n$ difference) are even more dramatic when applied to the proton target alone. As noted long ago [19] there is an interesting change of sign for the polarization asymmetry A^p. For real photons, this asymmetry will be negative over some range of energy, whereas for highly virtual photons, the asymmetry is positive over a considerable range of x as shown by the recent data from EMC, confirming the old SLAC data [21]. While one might dispute errors in the EMC integral [22], a very dramatic behaviour would have to ensure as $x \to 0$ if the integral would be negative! Indeed, if that was the resolution of the above "paradox", it would, by itself, be most interesting! If the integral is of the order of 0.1 when Q^2 is greater than 4 GeV^2, then something significant must occur as one proceeds into the nonperturbative regime of small Q^2 physics. This is not just a slight smoothing effect; here nonperturbative effects seem to have to change conditions radically! So with hints that something "has to give", combined with the fact that the integrals are weighted towards small energies which hints that the resonance region may be relevant in the sum rule, let's see what we know so far about these spin-dependent effects and then delineate the questions that arise for a milli-TeV electronic machine.

PHOTOPRODUCTION OF N^*: EMPIRICAL

Above pion production threshold are three prominent resonance bumps. The first resonance is the $P_{33}(1232)$; the second consists of $S_{11}(1550)$ and $D_{13}(1520)$, and the third is mainly $F_{15}(1690)$. Although these are the most noticeable states, a glance at the particle data tables shows that there are many other resonances within these bumps. The electromagnetic couplings of many of these are not well known. A better knowledge of some of these could have a significant impact on theory.

The spin dependence of the above prominent resonances is known, at least qualitatively. The P_{33} is dominantly photoproduced by M1 radiation and so $A = -1/2$. If M1 dominates for all Q^2 as in the NRQM, this value of A will be preserved.

The D_{13} and F_{15} are both photoproduced from protons dominantly in the helicity 3/2 mode, hence $A = -1$. So we see that the prominent states at $Q^2 = 0$ have the right sign of A to help saturate the DHG sum rule. (However, the quantitative test of this sum rule, especially the contributions above the resonance region, is not yet done.)

One of the successes of the NRQM was its prediction, [23] confirmed by data, [24] that the polarization asymmetry would change sign rapidly with Q^2. Thus, these resonances seem to do what is required to help resolve the conundrum mentioned in the previous section. However, there are interesting and significant questions outstanding. Is the Q^2 dependence of D_{13} and F_{15} the same, or is A a function of Q^2/W^2 where W is the resonance mass, or some other behaviour? If a function of $Q^2/W^2 (\simeq x')$, then at what value of x' does A change sign and how does this correlate with the possibility that A change sign as $x \to 0$ in deep inelastic data? If A changes sign at fixed $Q^2 \simeq 0.5 GeV^2$, say, does this correlate with a possible similar behaviour in the high W inelastic region? Interesting structure seems likely for all W and Q^2; we need to map it out to the best possible accuracy in order to understand the nonperturbative dynamics that are at work.

These spin dependences, and especially their Q^2 dependences, are among the sharpest probes of the nonperturbative dynamics. This is my challenge for experiment. Now I will address a parallel challenge for theory.

The above are all ideas, motivated by current excitement in the experimental community. The questions are potentially answerable at moderate Q^2, such as those accessible in Bonn and at CEBAF. Though not immediately requiring a higher energy facility their answers will stimulate yet deeper questions for that facility. There is a school that applies perturbative QCD to exclusive or resonance production as $Q^2 \to \infty$. These results imply significant changes in the helicity structure, even in the $\Delta(1232)$ region where the naive quark model does not immediately require new physics. The naive quark model must fail on rather general grounds probably within the region accessible at CEBAF/ELSA. The application of exclusive QCD probably will not be applicable at these energies. Thus there is likely a gap in the $Q^2 - W^2$ plane where no suitable theory framework yet exists and we may have to be led to that empirically.

References

[1] F.E. Close, Phys. Lett. 43B, 422 (1973); Nucl. Phys. B80, 269 (1974);
R. Carlitz and J. Kaur, Phys. Rev. Lett. 38, 673 (1977), F. Close, p209 in XIX Internat. Conf. on HEP Tokyo (1978).

[2] F. Close, "Introduction to Quarks and Partons" (Academic Press 1978).

[3] F. Close, Nucl. Phys. A497, c109 (1989);
C. Carlson & J. Milana Phys. Rev. D40, 3122 (1989).

[4] G. Karl, University of Guelph report (1991) & proceedings of this conference.
D. Choudhuri et al., Toronto preprint UTPT-91-02.

[5] G. Karl, Phys. Rev. D, (to be published).

[6] C. Avenarius, University of Oxford report, Phys. Lett. B, (to be published).

[7] F. Close & D. Sivers, Phys. Rev. Lett. 39, 1116 (1977).
F. Close & R. Milner, Phys. Rev. D, (1991) (in press).

[8] G. Altarelli & G. Ross, Phys. Lett. 212B, 391 (1988);
R. Carlitz et al., Phys. Lett. 214B, 229 (1988).

[9] F. Wagner, Proc. XVI Rencontre de Moriond (1981);
T. Barnes and F.E. Close, Phys. Lett. 128B, 277 (1983)

[10] T. Barnes and F.E. Close, Phys. Lett. 123B, 89 (1983)

[11] F.E. Close and Z.P. Li, Phys. Rev. D42, 2194 (1990)
ibid 2207 (1990)

[12] Z.P. Li, Phys. Rev. D (in press).

[13] Z.P. Li, V. Burkerdt & H. Li, Phys. Rev. (in press);
C. Carlson & N. Mukhopadyay, Phys. Rev. (in press).

[14] G. Farrar & D.R. Jackson, Phys. Rev. Lett. 35, 1416 (1975);
R. Carlitz, Phys. Lett. 58B, 345 (1975).

[15] F.E. Close, and A.W. Thomas, Phys. Lett. B212, 227 (1988)

[16] P. Stoler, Phys. Rev. Lett. 66, 1003 (1991)

[17] S. Drell and A. Hearn, Phys. Rev. Lett. 16, 908 (1966)
S.B. Gerasimov, J. Nucl. Phys. (USSR) 2, 598 (1966)

[18] J.D. Bjorken, Phys. Rev. 148, 1467 (1966); D1, 1376 (1970)

[19] F. Close, F. Gilman & I. Karliner, Phys. Rev. D6, 2533 (1972);
F. Close in Proc. of IX Rencontre de Moriond, Vol 2, ed J. Tran Thanh Van
(CNRS France 1974) p285.

[20] B. Ioffe et al, Hard Processes Vol I (North Holland, Amsterdam 1984).

[21] V. Hughes and J. Kuti, Ann. Rev. Nuc. Part. Sci. 33, 64 (1983)

[22] F. Close and R. Roberts, Phys. Rev. Lett. 60, 1471 (1988)

[23] F. Close and F. Gilman, Phys. Lett. 33B, 541 (1972).

[24] V. Burkert, Electroproduction of N* at CEBAF energies; CEBAF-R-86-011
(1986)

Few-Body Systems, Suppl. 6, 378—392 (1992)

SUMMARY OF SESSIONS ON EXPERIMENTAL FACILITIES

J. Arvieux

Laboratoire National Saturne

CE-SACLAY, F-91191 Gif-sur-Yvette Cedex, France

Th. Walcher

Institut für Kernphysik, Postfach 3980

Joh-Joachim-Becher Weg 45, D-6500 Mainz 1, Germany

Abstract

The contributions and discussions on the session "Present Status and Future Perspectives in the Experimental Activities" are summarized. In this session the points of convergence and the complementarity for the study of "Few Body Systems" with strong and electromagnetic probes were apparent. This aspect is emphasized.

I. Introduction

Today "Few Body Systems" are few nucleon systems as well as few quark systems. For the experimental investigation of these systems one needs either the spatial resolution to "see" the constituents and then conclude on their interactions, or one provides precision and uses models of the constituents and their interactions in order to understand the data. The first approach needs high energies and moderate beam qualities, whereas the second requires lower energies, but with beams of high intensity and small phase space.

At the Elba Conference examples of the second approach were presented, which can be separated into two domains. The first is the domain of the strongly interacting probes like protons, pions, kaons etc... and the second one that of electrons. Whereas for the strongly interacting probes powerful and productive facilities exist like PSI, SATURNE, TRIUMF, LEAR, etc..., the field of electron physics is hampered by the lack of state of the art accelerators. After the phasing out of the Linacs with low duty cycle, it took many years to plan and get approval for good new facilities. The field is, therefore, characterized by a rather uneven development. This was particularly apparent at Elba : the strong interaction facilities could show nice new results, whereas the electron community was presenting its facilities under construction, planned or dreamt of. However, it became also clear that results with the electron probe, once they will be available, may offer a new insight due to the easier interpretability of the electromagnetic interaction.

II. Few Body Physics with hadronic probes

At the Elba Conference overviews of few body physics done at 3 major hadronic facilities were presented : TRIUMF, SATURNE and IUCF.

TRIUMF is one of the meson factories built in the 70's to produce intense secondary beams of unstable particles, essentially pions and muons, the other two being LAMPF at Los Alamos and PSI at Villigen. After some 20 years of existence these facilities are looking for a second breath, the so-called kaon factories concept which would produce copious beams of strange mesons. Such a facility, the "KAON" factory project in Vancouver, was presented at Elba.

Strangeness is no stranger for SATURNE which, with its energy of 3 GeV, allows a copious production of mesons heavier than the pion. Especially spectacular is the production of eta mesons which has gained for SATURNE the nickname of "eta-factory". A lot of few-body experiments involve eta and heavier meson production.

IUCF represents a new type of facility betting on the very good beam qualities allowed by electron cooling. Other such facilities are in construction (COSY at Jülich) or near completion (CELSIUS at Uppsala). IUCF energy is limited to 400 MeV giving access to pion production at or just above threshold. Obviously it would be desirable to have the same beam qualities at much higher energies and a proposal for a 15 GeV Cooler Ring (LISS) at IUCF was presented to us.

II.1. TRIUMF

Elastic NN interaction

Few-body physics at hadron factories start usually with the Nucleon-Nucleon (NN) interaction. All intermediate energy accelerators have a strong NN program starting with the proton-proton (pp) interaction, due to an easier handling of proton beams and targets, then moving to neutron-proton after production of neutron beams (the neutron-neutron interaction is identical to pp by charge symmetry, except for Coulomb effects which are important only at low energies).

These programs are more than often criticized for being "systematics physics", but they study the building blocks of the nuclear interaction and that should be enough to justify them. When reading the innumerable papers in which e.g. the Paris potential (or Bonn or others) are used to calculate a variety of nuclear phenomena, one easily forgets that the validity of these models could only be ascertained by painstaking experimental studies using all resources of spin physics and subsequent phase-shift analyses.

After a few years of thorough investigation of the elastic interaction, NN studies at intermediate energies have now moved to extensive studies of inelastic channels (as we will see from results obtained at TRIUMF and SATURNE). There are still some interesting problems in the elastic channel but they now require a very high precision as it will shown from data from IUCF.

$NN \longrightarrow \pi d$. This system is the best studied of the inelastic channels due to its experimental simplicity. In most cases experiments were done in the $pp \longrightarrow \pi d$ direction using polarized proton beams interacting with (un)-polarized proton targets. The wealth of data obtained in this way does not allow an unambiguous amplitude analysis as long as the polarization of the recoiling deuteron or some polarization transfer to the deuteron spin has not been measured. By using the inverse reaction $\pi \vec{d} \longrightarrow \vec{p}p$ where pions scatter a polarized deuteron target a TRIUMF team (Feltham et al.) has measured the spin transfer coefficients K_{NN}, K_{SL} and K_{SS} where N, S and L stand for the spin direction compared to the beam direction (N = orthogonal to the scattering plane, S = sideways, L = longitudinal). As shown in fig. 1, the results of K_{SL} and K_{SS} vindicate, without being fit perfectly, the Fadeev calculations of Blankleider against a more simplified isobar model by Niskanen.

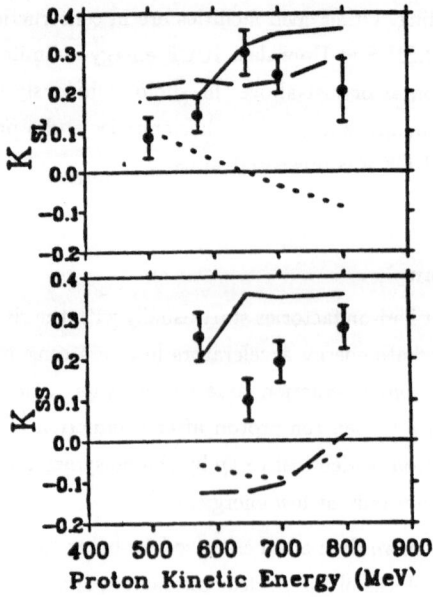

Fig. 1

The $pp \longrightarrow \pi d$ reaction is dominated, over a large energy span, by Δ-isobars in intermediate states. These isobars can only be produced by P−wave πN interactions. On the contrary, very near threshold, the interaction must be dominated by S−waves which are masked by the strong P−waves at higher energies. A very careful $np \longrightarrow d\pi^\circ$ experiment (Hutcheon et al.) has allowed to determine the ratio A_2/A_0 which measures the departure from S−waves to P−waves as a function of the pion c.m. momentum η (in units of $m_\pi c$), see fig. 2.

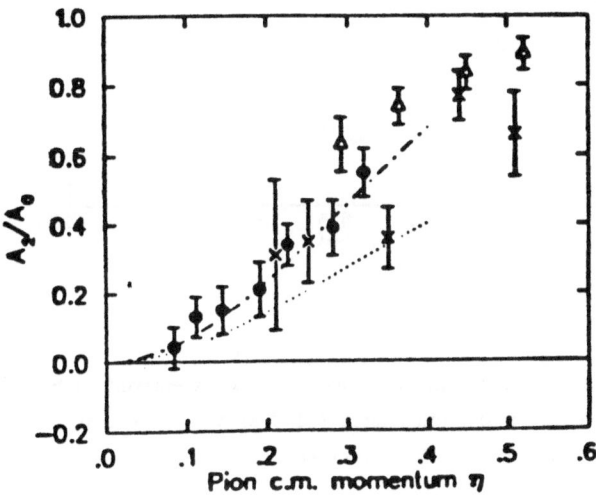

Fig. 2

NN $\longrightarrow$ NNπ. Single pion production in NN interactions depend on 3 elementary cross-sections : σ_{11}, σ_{10} and σ_{01}, where the first and second indices refer respectively to the isospin of the NN pair in the initial and final channel. Most data come from the $pp \longrightarrow pn\pi^+$ reaction whose cross-section is $(\sigma_{10} + \sigma_{11})$ dominated by the Δ^{++} $I = 1$ channel in the initial state. The other reactions are complementary : $np \longrightarrow pp\pi^-$ is proportional to $\frac{1}{2}(\sigma_{01} + \sigma_{11})$ and then it gives access to the small $I = 0$, non Δ-dominated channel. Combined with $pp \longrightarrow pp\pi^\circ$ which is pure σ_{11}, it allows to single out σ_{01}.

Total and differential $pp \longrightarrow pp\pi^\circ$ cross-sections and analyzing powers have been measured for the first time in the energy range 320-500 MeV, showing large negative analyzing powers (Stanislaus et al.). Total cross-sections have also been measured at SATURNE with the SPES 0 spectrometer (Didelez et al.) just above TRIUMF energies, from 480 MeV to 560 MeV and the small σ_{01} cross-section, could be extracted (fig. 3).

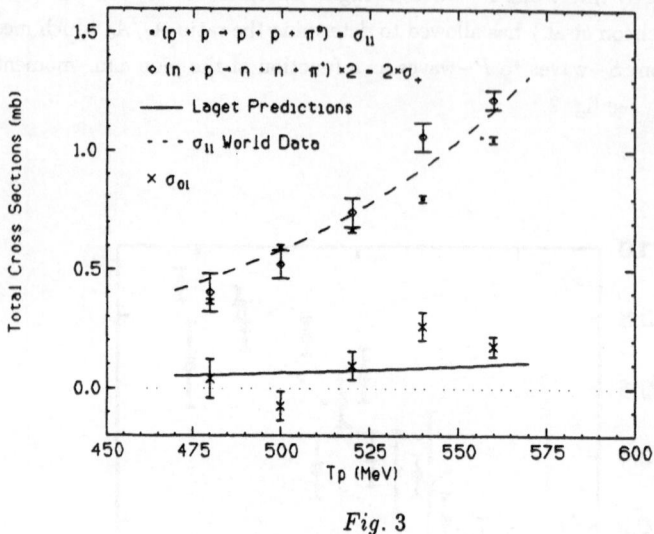

Fig. 3

The $np \longrightarrow pp\pi^-$ reaction has also been studied extensively at SATURNE up to 1134 MeV, close to the maximum energy (1150 MeV) of the free polarized neutron beam, using the Arcole detector (Y. Terrien et al.).

Polarized ^{3}He. A laser-pumped ^{3}He gas target with polarization 50-70 % over a volume of 35 cm^3 has been developed at TRIUMF (Haüsser et al.). Different experiments have been done in the elastic channel with incident protons and π^+ beams and in the break-up channels ^{3}He$(p, 2p)$ and ^{3}He(p, np) at energies ranging from 200 to 500 MeV.

The break-up $(p, 2p)$ channel is well fitted by PWIA calculations (Woloshin), see the spin correlation parameter A_{nn} in fig. 4. The agreement with the (p, pn) reaction is less outstanding (the shape of A_{nn} as a function of q is correctly predicted but the absolute value is too high and at the upper limit of the experimental error bars) but the calculations can be easily improved to get the correct normalization. A good agreement both with $(p, 2p)$ and (p, pn) would give some confidence in the use of polarized ^{3}He as a quasi-free polarized neutron target.

II.2. SATURNE

SATURNE 2 is a strong focussing synchrotron built in 1974-78 and reaching 3 GeV. It holds a world record for the intensity of polarized protons accelerated in a

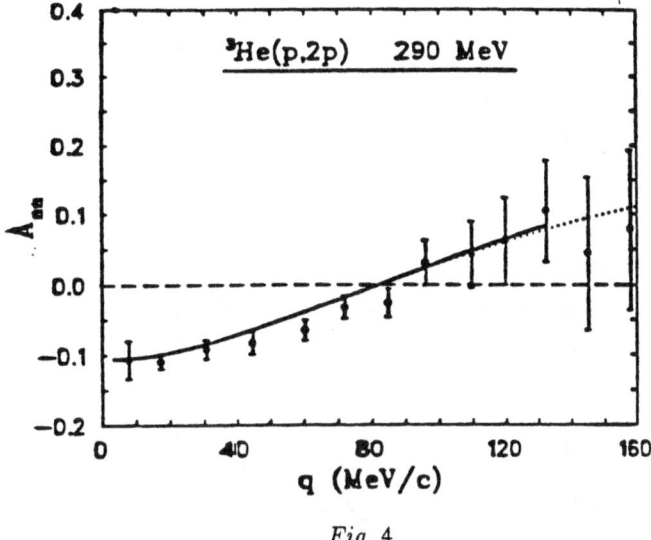

Fig. 4

synchrotron (up to 3×10^{11} per second and a polarization of 80-90 %). It is also the only synchrotron accelerating polarized deuterons up to 2.3 GeV.

Although the meson factories are all dominated by the excitation of the $\Delta(3/2,\ 3/2)$ in the intermediate states, the following baryonic excitations are also of particular interest at SATURNE :

$$
\begin{aligned}
P_{11}(1440), && J^P &= 1/2^+ \\
S_{11}(1535), && J^P &= 1/2^- \\
P_{11}(1710), && J^P &= 1/2^+
\end{aligned}
$$

Roper Resonance. The $P_{11}(1440)$ or Roper resonance is supposed to be the first radial excitation (breathing mode) of the nucleon whose quantum numbers it holds. Its mass and width are intimately related to the compressibility of the nucleon. This compressibility arises from the fact that a nucleon is a composite object made of 3 quarks surrounded by a meson cloud whose thickness (0.5 fm) is about the same as the quark core (0.5 fm).

The Roper resonance has been excited in a very elegant experiment whose idea is simple. Since the breathing mode is a $\Delta L = \Delta T = 0$ excitation, then it should be preferentially excited by a high energy $L = T = 0$ particle, namely the α-particle. Monopole excitation are strongly forward peaked and one key to the success of the experiment was the ability to reach angles close to zero degree. Finally by using the inverse kinematics, that is by sending a beam of α-particles onto a hydrogen target one can clearly separate the projectile and target excitations as shown in the spectrum of fig. 5. The angular distribution (fig. 6) has an unmistakable forward peaked monopole signature.

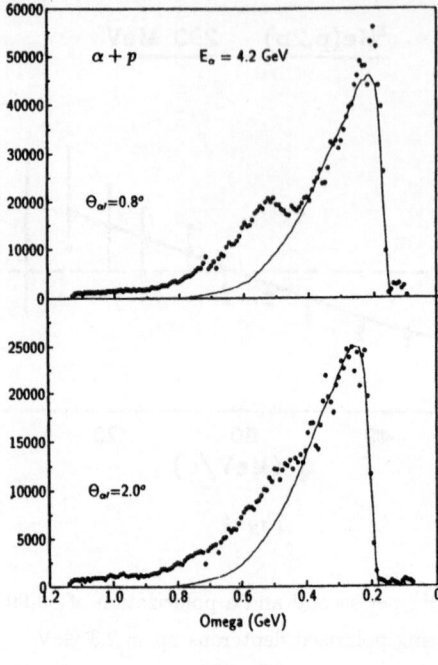

Fig. 5

S_{11} and P_{11} resonances. They have very peculiar decay modes with strong $N\eta$ branching ratios :

$$S_{11} \longrightarrow \begin{array}{ll} 35-50\ \% & N\pi \\ 45-55\ \% & N\eta \end{array}$$

$$P_{11} \longrightarrow \begin{array}{ll} 10-20\ \% & N\pi \\ 25\ \% & N\eta \\ 15\ \% & NK \end{array}$$

They are then large sources of η-mesons or conversely, by studying η-production in nuclear reactions one can deduce properties of these baryonic resonances in the vacuum (production in elementary processes) or in medium.

$pp \longrightarrow pp\eta$: Existing data on this elementary channel are very scarce. New data have been taken with the SPES 3 spectrometer between 1256 and 1450 MeV. A spectrum is shown in fig. 7. Final analysis of the data is in progress but they already indicate that ρ-exchange plays a dominant role in the S_{11} excitation (Germond and Wilkin).

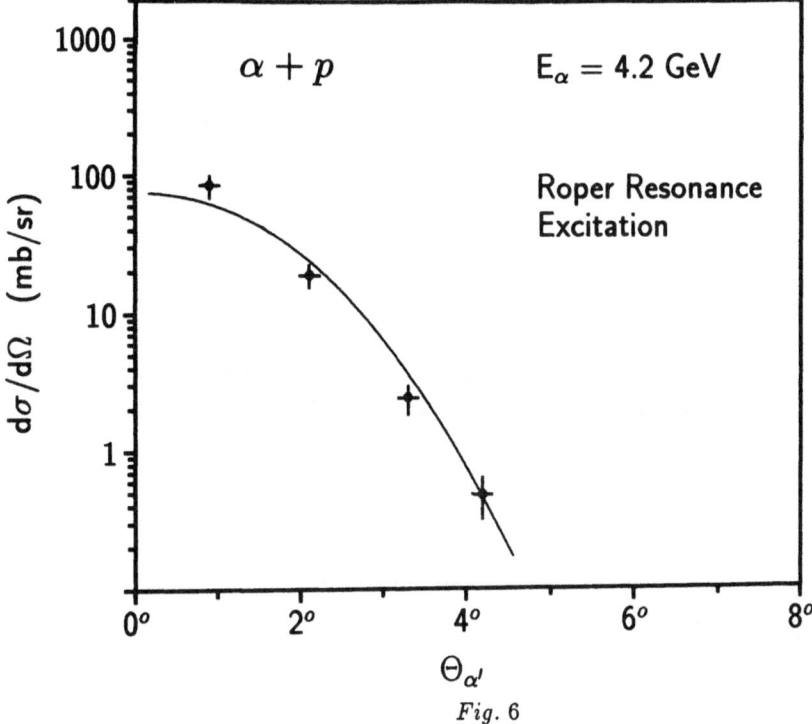

Fig. 6

pd ⟶ ³Heη : This reaction is the source of a spectacular production of η-mesons (up to 10^5 per second). The region around the ω and above has been studied recently in more details showing a clear excitation of the ω, η' and ϕ and the indication of structures at the K^+K^- and $K_0\bar{K}_0$ thresholds (fig. 8).

pp ⟶ KYN : Strangeness production at intermediate energies is a new avenue which has been opened recently with the availability of SATURNE higher energies and good energy resolution. Studies of hypernuclei have been developed at different accelerators (CERN, KEK, BNL) but the data on elementary production date from the early sixties and they are very poor. Data on $pp \longrightarrow NK^+Y$ where Y is either a $\Lambda°$ or a Σ have been obtained at SATURNE with the SPES 4 spectrometer (R. Frascaria et al.). The data (fig. 9) indicate a good agreement with calculations by J.M. Laget which take into account both π- and K-exchange. Angular distribution have also been obtained.

The next step is to use a 4π−detector to study spin effects (either in the initial state by using polarized protons or in the final state by measuring the polarization of the Λ or the Σ through their decay products). Such possibility (exp proposal 241 by D. L'Hôte et al.) would use the Arcole detector combined with an improved forward wall. Another collaboration will use a C−type magnet from CERN (exp prop. 213 by R. Bertini et al).

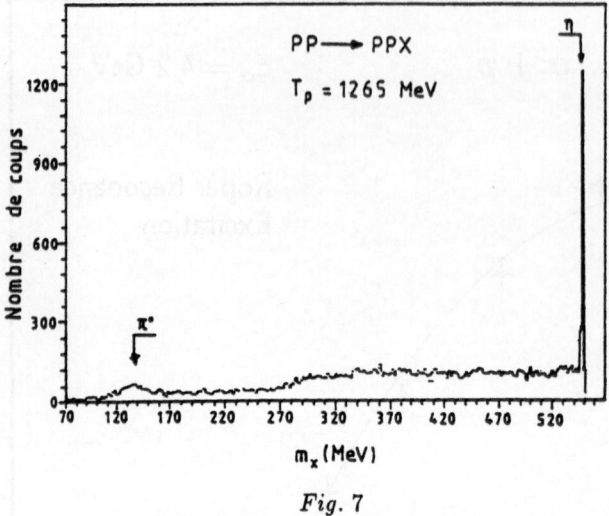

Fig. 7

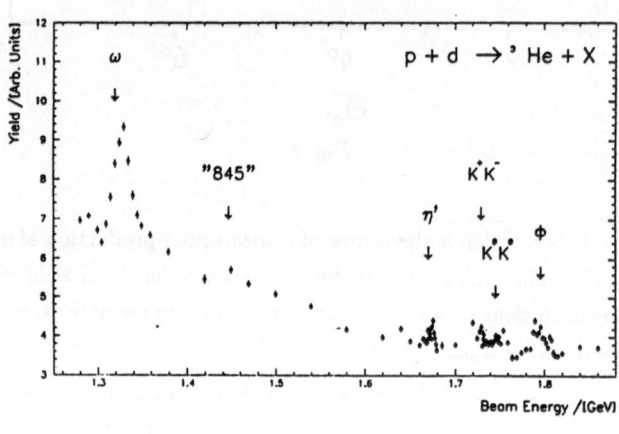

Fig. 8

II.4. IUCF

The IUCF Cooler Ring represents a new type of accelerator which benefits from novel features not readily available at existing facilities :

- extremely good momentum resolution
- small beam size (~ 0.2 mm)
- small beam divergence (2 mrad)
- variable time structure ($\sim$ CW down to 0.5 ns pulses)
- high current (≤ 10 mA)

– thin targets (1 to 100 mg × cm^{-2}).

These qualities have to be paid for by a limited luminosity ($\mathcal{L} \approx 10^{32}$ cm^{-2} s^{-1}) compared to solid targets with high intensity external beam and the filling time may become not negligible ($\sim$ 100 ms for 10^9 particles, a second or more for higher intensities) when operating in the extracted beam mode.

The Cooler Ring is shown in fig. 10 with its present experimental facilities and electron cooling.

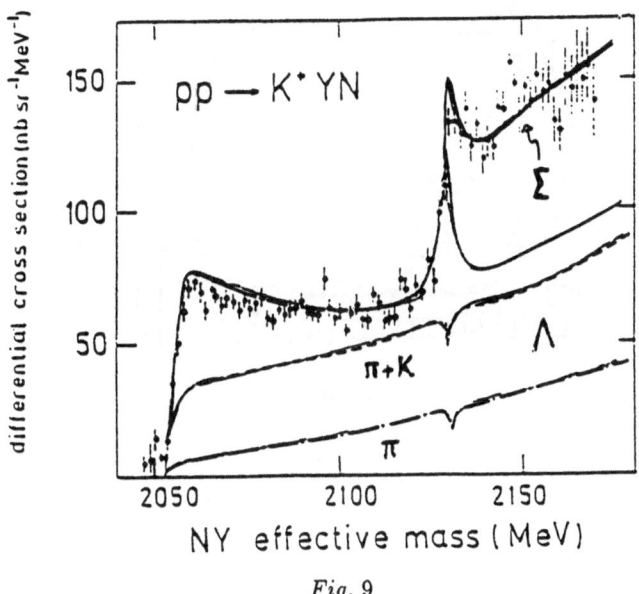

Fig. 9

Few body experiments include the study of proton-proton elastic scattering. Theory and experiment have now reached in this channel a very high degree of accuracy as shown in fig. 11 where very recent analyzing power data (K. Pitts et al.) which extend down to $\theta_{c.m.} = 5° \theta_L \gtrsim 2°$) are compared with potential model calculations and phase shift analysis.

The $pp \longrightarrow pp\pi°$ reaction has also been studied with great precision. The data (fig. 12) demonstrate the superiority of this type of Cooler for excitation functions just above threshold.

Other reactions include the study of $pp \longrightarrow \pi d$, $pp \longrightarrow pn\pi^+$, $pd \longrightarrow ^3\mathrm{He}\pi°$, $pd \longrightarrow t\pi^+$ but data are still being processed.

IUCF is now planning test experiments with a 2.2 MeV e^-cooling beam to show that the concept of a Cooler synchrotron could work at higher energies, since technolog-

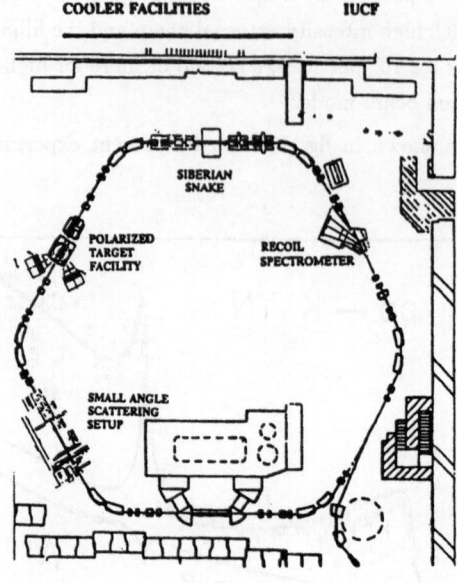

Fig. 10

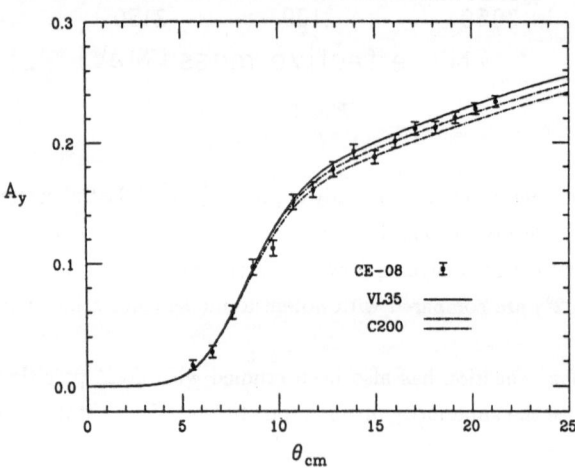

Fig. 11

ical problems will be hard to overcome (cooling a 15 GeV proton beam would require 1A of electrons at 8 MeV). Luminosities of 10^{32} cm^{-2}s^{-1} for pp scattering with polarized beam and jet target and up to 10^{33} with an unpolarized hydrogen target could be obtained. This project named LISS (for Light Ion Spin Synchrotron) could be finished by 1998 if funded rapidly.

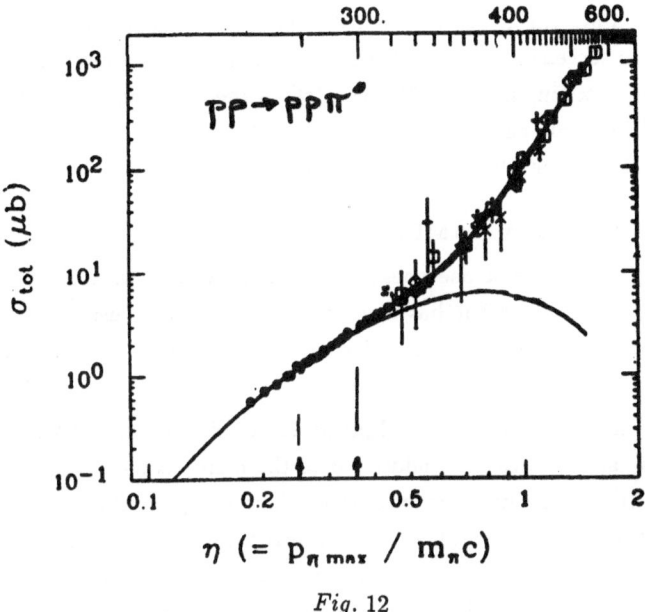

Fig. 12

III. Few Body Systems with the Electromagnetic Probe

As outlined in the introduction, this field is hoping for the new facilities under construction with will be all characterized by 100 % duty cycle. Even those who have already started, like the stretcher ring at Saskatchewan (at 300 MeV) or the Mainz Microtron stage MAMI A (at 180 MeV), talked about the interesting possibilities of their facilities in the future rather than their past results.

Three laboratories (BATES/MIT, NIKHEF and MAMI), which shall have an energy close to 1 GeV, discussed their program which has either started (MAMI) or will start in the near future.

A special role will be played by CEBAF which should begin operation in 1994 with an initial energy of 4 GeV. It will for the first time combine both high energy, i.e., good spatial resolution, and precision.

The dream of the most ambitious facility of a European project with a continuous electron beam beyond 15 GeV showed where one could go from today, provided a large number of physicists decide to join their efforts towards its realization.

III.1. Saskatchewan Accelerator Laboratory (SAL)
Pulse Stretcher Ring (PSR)

This facility is operational at 100 and 300 MeV. It has centered its program around the few nucleon system giving emphasis to photon induced reactions and a photon tagger has been successfully tested. 24 experiments have been approved. They include elastic photon scattering from ^{2}H, ^{4}He and ^{12}C in the Δ region, three body effects in the reaction ^{3}He(γ,pp)n and an investigation of ^{4}He(γ, dd). This program will be a nice continuation of that one which begun at MAMI A.

III.2. Mainz Microtron MAMI B

This facility is the first one operating at about 1 GeV with 100 % duty cycle. Since its first beam in 1990 it has started with experiments using tagged photons. The 3-spectrometer facility and the polarized electron facility are close to becoming operational.

In the spirit of the introduction all proposed experiments at **MAMI** concern "few body systems" and the reader is addressed to the more detailed description in these proceedings. However, three examples which show the possibility to investigate the borderline between the baryon/meson and the quark/gluon models shall be mentioned. The first example is the study of the form factors of nucleons bound in nuclear matter. Due to a figure of merit about 100 times larger than achievable up to now very intriguing polarization phenomena ("swollen nucleon") can be separated from the masking effects of final state interactions. A second example is the photo-production of π° and η mesons at threshold, which may allow to better understand the excitation function behaviour of these amplitudes. The last example is the scattering of polarized electrons from polarized $^3\vec{\text{He}}$. In a first step the electric form factor of the neutron will be determined with the reaction $^3\vec{\text{He}}(\vec{e}, e'n)pp$ which is complementary to the $d(\vec{e}, \vec{e}n)p$ reaction. However, in a second stage possibly three body forces will be accessible in e.g. $^3\vec{\text{He}}(\vec{e}, pp)n$ with a new sentivity.

III.3. The Amsterdam Pulse Stretcher AmPS

The beam of this facility will have very similar features at the MAMI beam. The pulsed beam of the existing Linac will be stretched to 100 % duty cycle by means of a stretcher/storage ring. This beam can be extracted and then be used in the well tried two spectrometer set-up which will be extended with non-magnetic detectors. With this equipment, high momentum components of nucleons in nuclei, medium effects on bound nucleons and nucleon-nucleon correlations will be investigated. However, the stretcher/storage ring will also be usable as a stored beam facility with internal targets. By using gas jet targets or storage bottle targets this method opens a new dimension to take advantage of the additional sensitivity of polarized experiments. In particular, it will be possible to use polarized protons and deuterons as targets which are not well suitable with extracted beams. The main thrust of the planned physics will be, however, few body systems with $A \leq 4$.

III.4. The Continuous Electron Beam Accelerator Facility CEBAF

This facility is the most advanced and ambitious accelerator in the field of electromagnetic physics. It may be well the most powerful tool for nuclear physics existing after its planned completion in 1994/95. It shall have the same beam quality as the preceeding facilities, however, with an energy of 4 GeV and consequently a resolution about five times better.

The experimental program in the non-strange sector is very similar to the Amsterdam and Mainz/program, however, the measurements will cover a much larger kinematical range and will be more significant. Additionally, the possibility to produce open strangeness adds a new dimension and precise measurements of the badly known strangeness production cross sections on nucleons and nuclei may be a key to understand the "few quark system". The planned experiments are outlined in detail in a contribution to these proceedings.

III.5. The Dream of a European Continuous Electron Beam Facility with an Energy above 15 GeV

Clearly, in order to "see" quarks directly one needs energies high enough to reach the scaling region, i.e. above about 15 GeV. This domain, together with the precision attainable with 100 % duty cycle beams, would allow a qualitatively new test of QCD at low and medium momentum transfer. It would allow to search in a convincing and systematic manner for the (missing!) link between the baryon/meson and the quark/gluon world, a realm frequently designated as "the confinement region". The problem is that at very high and low energies no experiment is addressing this region directly but needs models to extract some information. In contradiction at energies of the order of 15 GeV definite handles have been proposed as for instance :

1) The nucleus as a detector for small, bare quark baryons (colour transparency) and as a laboratory with size of the order of the range of hadronization (a few fermis).

2) Modifications of baryons and mesons induced by the nuclear binding (EMC-effect, structure functions at $x > 1$, semi inclusive deep inelastic scattering).

3) Form factors and resonances of free baryons (also $B > 1$) and mesons (spin structure functions, weak form factors, structure functions of spin=1 baryons like the deuteron, etc...).

The chances to realize such a machine depend on three prerequisites of very different character :

(i) The finding of a technical solution which is affordable. Though it is possible to build such a machine in principle, the cost of the proposed solutions are at the limits of being realizable. It is adequate to mention that the same holds for detectors.

(ii) A scientific program has to be formulated that nuclear physicists and high energy physicists alike can accept and understand. This is not too surprisingly a language problem because the proposed field is kind of "interdisciplinary".

(iii) A European community has to be formed.

IV. Conclusion

The field dealt with the few-body Conference in Elba is a frontier in physics. We don't know whether QCD is the theory of strong interactions at low momentum transfers. Since the smallest directly observable particles are baryons and mesons we have to find a theory for their description. The facilities which have been discussed in this summary will, most likely, give us the answer to this most important problem.

Few-Body Systems, Suppl. 6, 393—407 (1992)

Few-
Body
Systems
© by Springer-Verlag 1992

MAMI EXPERIMENTAL ACTIVITY

Thomas Walcher
Institut für Kernphysik
Universität Mainz
D-6500 Mainz, Germany

Abstract

The Mainz Microtron MAMI is shortly described followed by a sketch of the three main experimental facilities: the A1 collaboration realizing a 3-spectrometer set-up, the A2 collaboration realizing a photon tagger together with numerous dedicated detectors, and the A3 collaboration realizing a facility to use polarized electrons and polarized targets for the measurement of the electric form factor of the neutron $G_{E,n}$. The physics program of the first round is outlined by listing all planned experiments.

I. Introduction

The title of the conference on "Few Body Problems" has taken a double meaning. As seen from the program it covers the traditional few nucleon systems. But baryons and mesons are few quark systems, too, and now belong to this field. Beyond this more amusing observation there exists a serious relation between the two kinds of few body problems. We hope that a better understanding of the first will emerge from a study of the second system. Questions like the nucleon-nucleon force, meson exchange currents, or baryonic resonances in nuclei may have to be answered by taking quarks and gluons into account. Vice versa the polarization of baryons produced by nuclear binding may provide important insights into the structure of few quark systems at low momentum transfers.

It is in this spirit that the experimental program at the Mainz Microtron MAMI has been designed. With a maximum energy of 855 MeV the achievable spatial resolution is of the order of ~0.2 fm only and not sufficient to "see" quarks directly. However, this is a situation not strange to nuclear physicists. By far the most detailed information about the nucleus, which allowed to develop the main field model, were obtained from experiments not "seeing" the constituents of the nucleus. The precision of the measurements, as e.g. of elastic and transition form factors in electron scattering or decay studies of fission and electromagnetic radiation allowed to develop a model which was largely confirmed later on by higher energy experiments. The experimental program of MAMI can therefore best be characterized as an attempt to investigate the internal structure of nucleons and mesons by precise low q^2 measurements.

This idea has an immediate consequence for the experimental facilities: they have to be designed to reach the needed high accuracy, an accuracy unparalleled in the past at low energies ($E \lesssim 1$ GeV) with Linacs or at high energies with Linacs, Synchrotrons or secondary μ-beams ($20 \lesssim E \lesssim 400$ GeV). In section II a short summary of the most important parameters of MAMI is given. In the following three sections the set-ups of three collaborations and their first round of experiments is outlined.

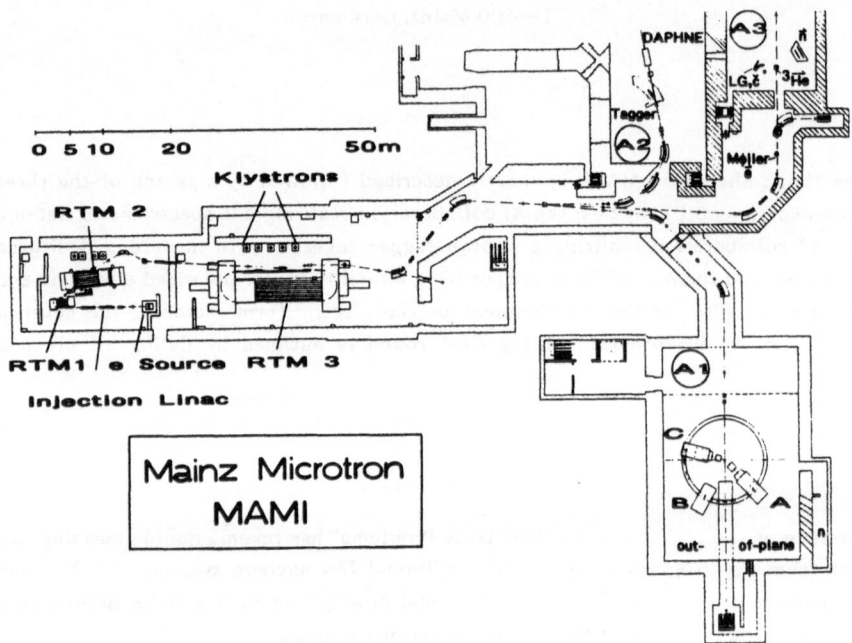

Fig. 1: Floorplan of the Mainz Microtron MAMI and its major facilities. Injector Linac and the race track microtrons RTM1 and RTM2 are in a common hall, while RTM3 fills a second hall alone. The beam handling system distributes the beam to three major experimental set-ups: A1: the 3-spectrometer set-up; A2: the tagger, here shown with one of the used detectors, the 4π detector DAPHNE; A3: the Møller polarimeter, the polarized ^{3}He target, lead glass and a Cherenkov counter as the electron arm and a plastic scintillator polarimeter for the recoil neutron.

Section III deals with the program of the collaboration A1 "Coincidence Experiments with Electrons". This program is centered around three spectrometers which are complemented by further appropriate detectors. Section IV lists the planned experiments of the collaboration A2 "Real Photons" using a large tagger spectrometer with a great number of different detectors. Section V outlines the plans of the A3 collaboration "Polarized Electrons", setting up a source for polarized electrons and targets together with detectors to measure the electric form factor of the neutron. In section VI some intentions for the future are listed.

II. The Microtron MAMI

The MAMI accelerator consists of a linear accelerator as injector and a cascade of three race track microtrons [1]. The RF accelerator structures are normal conducting with low gradient providing high reliability and easy operation. In this way a continuous wave beam is produced at medium high energies between 180 and 855 MeV at currents of up to 100 μA. The energy may be varied in steps of 15 MeV by kicking the beam off the separated tracks into the extraction channel. The energy width is about 120 keV (FWHM). This, together with the excellent phase space of 0.04 π mm mrad in vertical and 0.14 π mm mrad in horizontal direction, is a great help for building high resolution spectrometers. It may also be a prerequisite to start novel investigations of classical interactions of an electron beam with matter (see section VI).

The MAMI machine is running since summer 1990 at its full energy of 855 MeV with currents of up to 30 μA [2]. The routine operation with currents of about 100 nA for the photon work (see section IV) is limited to 60 hours per week at this time in order to allow the installation of the rest of the facility.

III. Collaboration A1: Coincidence Experiments with Electrons

The precision determination of electron scattering cross sections in coincidence with hadrons requires the use of imaging magnetic spectrometers. They provide the necessary good resolution in momentum and space to precisely define the momentum transfer and to resolve nuclear states. Furthermore, they allow an optimal suppression of background allowing the measurement of very small cross sections and in this way covering the broadest possible kinematical range.

The A1 collaboration has decided to build a set-up of three such spectrometers rotating on a large turntable around a common pivot. This arrangement is completed depending on the experiments by further magnetic and non-magnetic detectors, as e.g. a BGO-ball. Its optimization has been made with the experiments in mind described further below.

For a thorough understanding of this optimization one needs a discussion of the theoretical basis of electron-hadron coincidence cross sections, as found e.g. in Drechsel and Giannini [3]. Some detailed aspects are described in ref. [4] and [5].

Table 1: The main parameters of the 3-spectrometer set-up
(D = dipole, Q = quadrupole, S = sextupole)

spectrometer		A	B	C
magnet configuration		QSDD	clam D	QSDD
maximum momentum	MeV/c	735	870	551
kinetic energy: protons	MeV	250	340	150
kinetic energy: pions	MeV	610	740	430
momentum acceptance	%	20	15	25
solid angle	msr	28	5.6	28
extended-target acceptance	mm	50	50	50
momentum resolution		$\leq 10^{-4}$	$\leq 10^{-4}$	$\leq 10^{-4}$
angular resolution at target	mrad	≤ 3	≤ 3	≤ 3
spatial resolution at target	mm	3 - 5	≤ 1	3 - 5
range of scattering angle	degrees	18 - 160	7 - 50	18 - 160
out of plane angle	degrees	0	10	0
length of central ray	m	10.75	12.03	8.53
length of image plane	m	1.80	1.80	1.60
weight of dipoles (iron)	t	201	192	116
weight of shielding	t	100	75	85

However, three important features of the Mainz set-up shall be mentioned here:

1) Coverage of forward scattering angles

For a good separation of longitudinal and transverse cross sections as well as for reactions with large energy loss a coverage of very forward scattering angles is mandatory. Therefore, one of the spectrometers (B, see Table 1 and Fig. 2) can measure scattering angles as small as 5°.

2) Versatility of the three spectrometers

For the indicated physics one needs different spectrometers of different kinematical ranges. The small solid angle spectrometer B will normally serve as electron spectrometer in forward direction, where a large solid angle is less desirable because of high counting rates. This is also valid if it is used as a proton spectrometer. The large solid angle spectrometers A and C may be used at larger scattering angles where counting rates are low for electron-hadron coincidences. All three together are also well suited for the triple coincidence needed for the Δ resonance investigation ("Δ-spectrometer").

3) Out-of-plane measurements

A particularly important option in coincidence experiments with electrons is the measurement out-of-plane in order to determine σ_{TL} and σ_{TT} [3]. These contributions of the cross section stem from the interference of the longitudinal with the transverse and the transverse with the transverse photon amplitudes. They offer the possiblity to investigate very small L- or T-admixtures in dominant transitions of the opposite character.

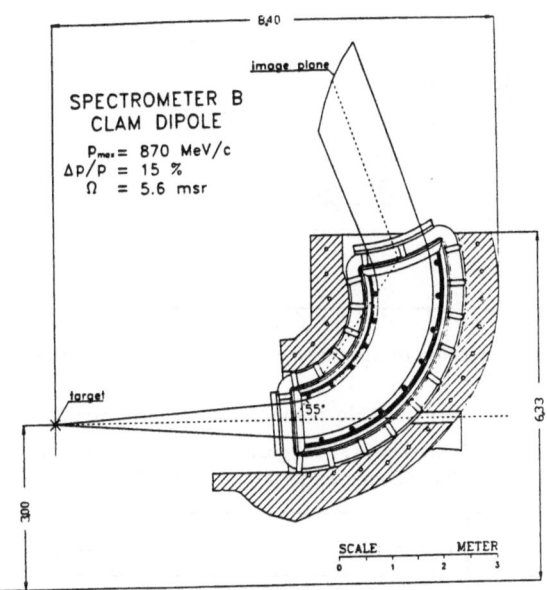

Fig. 2: Cross section of spectrometer B featuring the following peculiarities:

  • out-of-plane measurement:
    – rising of spectrometer from $0°$ to $10°$ in steps of $0.5°$
    – alternatively: out-of-plane beam line $20°$ to $40°$
  • extreme forward angle: down to $5°$

The three spectrometers are installed and aligned. They are presently being equipped with the appropriate detector systems and shielding houses (see Fig. 3). The resolution of the vertical drift chambers (VDC's) is $150\,\mu m$ which determines, together with multiple scattering in the vacuum foils, the resolutions given in Table 1.

398

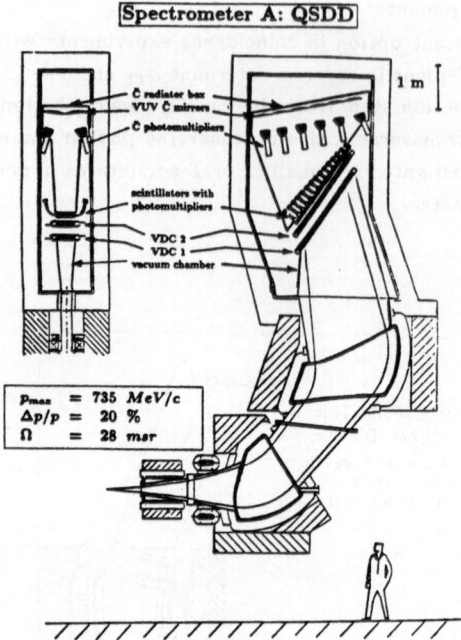

Fig. 3: Cross section of spectrometer A together with the detector assembly

The six experiments shown in Table 2 constitute the program of the first round.

Table 2: Experimental Program of Collaboration A1:
Coincidence Experiments with Electrons

	Reaction	Physics	Equipment
1.	$A(e,e'p)$ $A = {}^{12}C, {}^{16}O,$ ${}^{40,48}Ca$	separation of structure functions high-momentum components formfactor of bound nucleon	spectrometers A, B, C
2.	${}^2H(e,e'p)$	long./transv. separation high-momentum components non-nucleonic degrees of freedom	spectrometers A, B, C kryo-target
3.	${}^2H(e,e'n)$ ${}^2H(e,e'p)$	measurement of $G_{M,n}$ $\rightarrow G_{M,p}$ (as control)	spectrometer A neutron spectrometer kryo-target
4.	$A(e,e'\underline{xN})$ $\Delta^{++} = \pi^+ p$	multi-hadron final states quasi-elastic, dip and Δ region Δ-propagation in nuclei	spectrometer A or B BGO-ball [1]
5.	${}^1H(e,e'\pi^+)$	long./transv./long.-transv. structure functions $\rightarrow$ different form factors	spectrometers A, B kryo target
6.	${}^3He(e,e'\pi^{\pm})$	$\frac{\sigma_{tot}(\pi^+)}{\sigma_{tot}(\pi^-)}$ in long. channel $\rightarrow \Delta$-content in nucleus	spectrometers A, B gas target

The first two experiments A1-1 and A1-2 are devoted to a measurement of high momentum components of nucleons in nuclei. Due to the large gain of sensitivity with MAMI and the new spectrometers a considerable extension of the kinematical range will be possible. The project A1-3 will measure the neutron magnetic form factor $G_{M,n}$, a prerequisite for the determination of $G_{E,n}$ in A3-1 and A3-2. However, a precise determination of $G_{M,n}$ may reveal differences to the proton magnetic form factor $G_{M,p}$, too. This comparison will deliver an interesting constraint on nucleon models in the spirit of the introduction. The same holds for experiment A1-5, in which the axial form factor of the nucleon and a measure of the π content of the nucleon can be extracted in the framework of model assumptions. Experiments A1-4 and A1-6 will study Δ excitations in nuclei, one of the possibilities to insert a baryon different from the nucleon in the nucleus.

IV. Collaboration A2: Real Photons

The c.w. beam of MAMI provides a decisive improvement of photonuclear reactions, a field almost as old as nuclear physics itself. At electron accelerators the real photons are produced in bremsstrahl targets with an energy equal to the difference between the initial and final energy of the electron. The 100% duty factor allows to measure the energy of the final electron in coincidence with the photon induced reaction products because the accidental rate is small. This method is called "photon tagging". For its realization a broad band electron spectrometer for measuring the final energy is needed, the so-called tagger. Such a tagging facility had been already used with the 185 MeV beam of the first phase of MAMI, called MAMI A [6]. For the phase B of MAMI with 855 MeV a new large tagger has been designed and installed (see Fig. 4) and is being used for experiments. It allows the tagging of photons in the range $50 \text{ MeV} \leq E_\gamma \leq 800 \text{ MeV}$ with a resolution of $\Delta E_\gamma < 2 \text{ MeV}$. This resolution is determined by the long counter ladder and could be improved if needed.

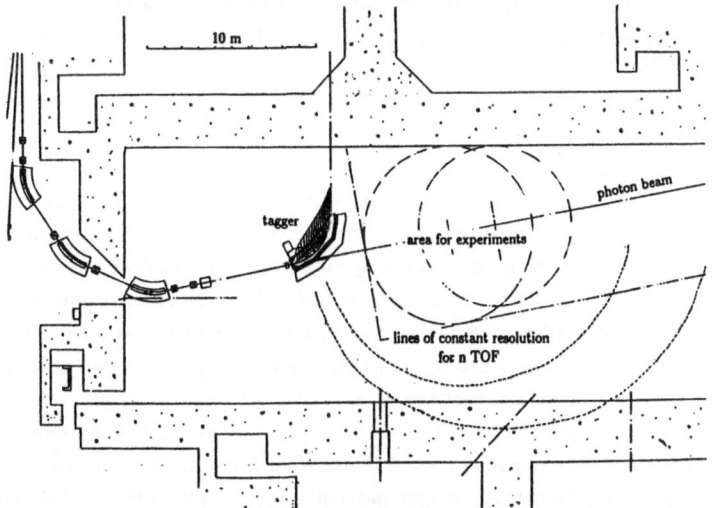

Fig. 4: The experimental area in the tagger hall

The experimental program of collaboration A2 is summarized in Table 3.

Table 3: Experimental program of Collaboration A2:
Real Photons

	Reaction	Physics	Equipment
1.	$p(\gamma, \gamma p)$ also: $^4He, ^{12}C, ^{16}O$ $p(\gamma, \pi^0 p)$	Compton scattering $\rightarrow$ proton polarizabilities change of polarizability in nuclei	$CATS$ [1] $COPP$ [2]
2.	$p(\gamma, \left\{ \begin{array}{c} \pi^0 \\ \eta \end{array} \right)$ $n(\gamma, \pi^0)$	π^0 and η production at threshold	$TAPS$ [6] ^2H-target
3.	$A(\gamma, \pi^0)A$ $A(\gamma, \eta)A$ $A = {}^{12}C, {}^{40}Ca$	mass-distribution mean free path of η in nuclei $\pi^0 \leftrightarrow \Delta(I = \frac{3}{2})$ and $N^*(I = \frac{1}{2})$ $\eta \leftrightarrow N^* \rightarrow$ medium modification of N^*	$TAPS$ [6]
4.	$^2H(\gamma, pp\pi^-)$ $^3He(\gamma, ppn)$	information on $\Delta - N$ interaction initial $\gamma + 3N$ interaction	$DAPHNE$ [3] TOF-wall[7]
5.	$A(\gamma, 2N)(A - 2)$ $A = {}^{6,7}Li, {}^{12}C$ $2N = pp, np$	photon absorption mechanism genuine two-body absorption? (Quasi-Deuteron Effect)	PIP [4] TOF-wall[7]
6.	$^{235,238}U(\gamma, fission)$	total photofission σ_γ^{fis} $= (90 - 100\,\%) \cdot \sigma_\gamma^{tot}$	$PPAC$ [5]

1. CATS = Collaboration A Two Spectrometer, 2. COPP = COincidence between Photon and Proton, 3. DAPHNE = Détecteur à grande Acceptance pour la PHysique Nucléaire Electromagnétique, 4. PIP = Pion Proton Detector, 5. PPAC = Parallel Plate Avalanche Counter, 6. TAPS = Two (or, possibly, Three) Arm Photon Spectrometer, 7. TOF = Time Of Flight (neutron detectors).

It contains again physics of the nucleon and the nucleus. The Compton scattering on the proton in A2-1 gives information on the electric and magnetic polarizability which is directly related to the internal structure via the excitation spectrum of the nucleon. A very exciting question is whether this observables are modified in the nuclear medium which will be addressed in the second step of the experiment. Project A2-2 aims to measure π^0 and η production from threshold up to higher energies. The comparison of the two cross sections may serve to further study the mesonic component of the nucleon after the threshold production of the π^0 has raised already so many questions [7]. Experiment A2-4 deals with Δ excitations in nuclei and the question of three body forces, a question particularly relevant to the traditional topic of this conference. The new precision usable by the tagging method opens a fresh look on old open problems of photonuclear reactions on the nucleus. This will be the topic of experiment A2-5. The last of the approved experiments, A2-6, will take advantage of the easy access to measure the total fission cross section and will determine in this way σ_γ^{tot}.

As indicated in Table 3, a rather large variety of detectors has been built for this program. Three of these shall be shortly described in the following:

1.) CATS (Collaboration A2 Spectrometer)

This detector arrangement (see Fig. 5) consists of two parts:

- a segmented NaJ crystal (48 cm diameter, 63 cm length) with a resolution $\Delta E/E \leq 2\%$ (FWHM) at 300 MeV and

- a half sphere approximated by 70 hexagonal BaF_2 crystals of 25 cm length.

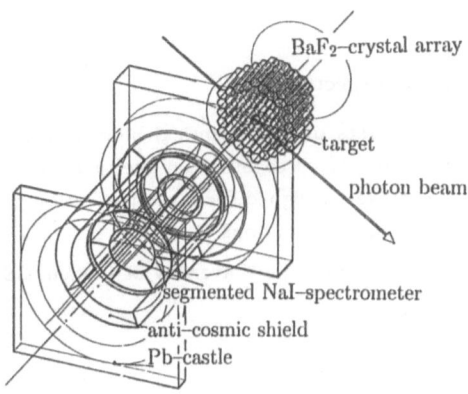

Fig. 5: CATS: a set-up of two large γ detectors for the measurement of π^o's

This detector is particularly suited for the measurement of π^o at large production angles.

This arrangement has been conceived and realized by physicists from the universities of Mainz and Bonn.

2.) DAPHNE (Détecteur à grande Acceptance pour la Physique Nucléaire Electromagnétique)

This is a cylindrical detector (see Fig. 6) which consists of a sandwich of wire chambers, scintillators and a Pb converter. It is particularly well suited for the reaction $d(\gamma,pp\pi^-)$ and $^3He(\gamma,pp)n$.

This detector has been designed and built by ALS (CEN) Saclay and the INFN, Pavia.

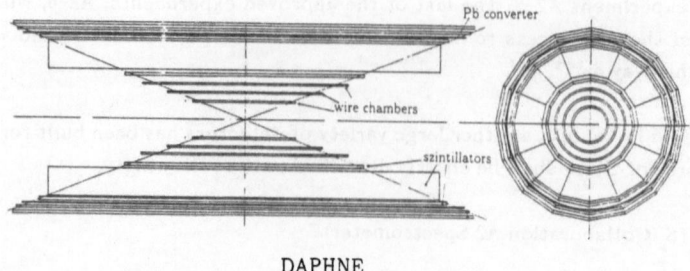

DAPHNE

Détecteur à grande Acceptance pour la Physique Nucléaire Electromagnétique

Fig. 6: DAPHNE: a cylindrical close to 4π detector for the detection
of π , p and γ's

3.)   TAPS (Two Arm Photon Spectrometer)

This spectrometer consists of four identical blocks of 88 BaF$_2$ crystals each.
Two of these blocks are placed together on one turnable arm (see Fig. 7).
This arrangement has been designed for investigation of π^o and η production in
heavy ion and photon induced reactions.

It is an instrument realized by the University of Gießen, GSI Darmstadt, GANIL
Caen and KVI Groningen.

Fig. 7: TAPS: a two arm photon spectrometer for the measurement
of π^o and η

V. Collaboration A3: Polarized Electrons for a Measurement
of the Electric Neutron Formfactor

Polarized electrons are a powerful tool to investigate some of the problems mentioned in the introduction [8]. This is basically due to the fact that new terms in the
cross section appear, which are proportional to the interference between contributions from the longitudinal (L) and transverse (T) polarization of the virtual photon
mediating the electron nucleon/nucleus interaction. A particularly interesting
application of this possibility represents the measurement of the electric form factor
of the neutron G_{En} [9], [10].

The A3 collaboration has envisaged two measurements (see Table 4).

Table 4: Experimental Program Collaboration A3:
Polarized Electrons and Electric Formfactor of the Neutron

	Reaction	Physics	Equipment
1.	$^2H(\vec{e}, e\vec{n})p$	observable $G_{E,n}$ neutron polarization	neutron polarimeter: second TOF-wall ^2H cryo target
2.	$^3\vec{He}(\vec{e}, en)$	observable $G_{E,n}$ target electron asymmetry	polarized ^3He target

Measurement of the target electron asymmetry by using polarized electrons with
a polarized neutron target. Since no pure neutron target exists, a polarized ^{3}He
target will be used. In this nucleus the two protons form a saturated spin pair
and it is estimated that its influence as spectator in quasifree kinematics of the
neutron is small. For this measurement a precise knowledge of G_{Mn} is needed,
the form factor just measured in project A1-3. A detailed discussion of the
relevant formulae is given in ref. [10].

2.) Measurement of the polarization transfer using a non-polarized deuterium target
as neutron target. As has been demonstrated by Arenhövel et al. [9] in quasifree
kinematics ($\Theta_{pn} = 180°, \vec{q} = \vec{p}_n$) practically no dependence due to the binding of
the neutron in the deuteron is left if the polarization transfer $P_n^\perp$ is measured.

For a realization of these experiments several components beyond a rather sophisticated detector set-up are needed. First a source of polarized electrons has to be provided. The collaboration A3 is using a traditional source with a GaAsP cathode, which
yields polarizations up to 40%. The source is designed to deliver 50 µA current, which
will result after preparation for injection into MAMI in 5 µA beam. The expected
lifetime is more than 100 hours. A second component is the 24 m long transport and
preparation of the beam, including adjustment of the spin direction into the accelerator. A further important part is a Møller polarimeter in front of the experiment (see
Fig. 1). All of this has been completed and the first polarized beam has been
accelerated up to the RTM1.

The detector set-up is largely common for both experiments (see Fig. 8).

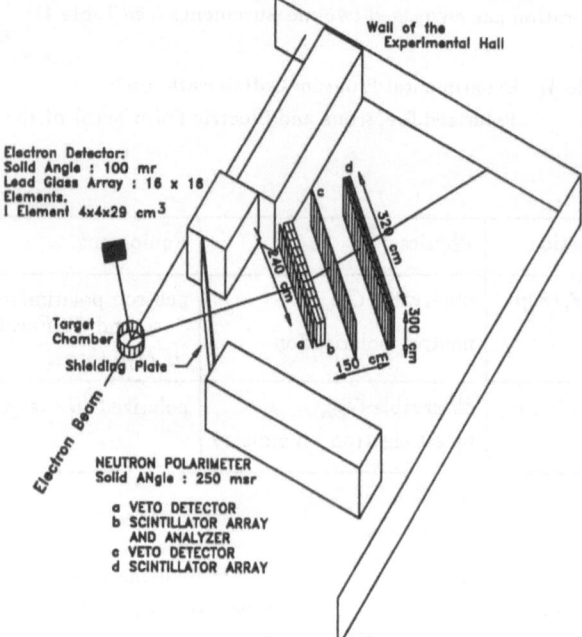

Fig. 8: Detector set-up of the A3 collaboration

The scattered electrons will be measured by means of an array of 16 · 16 lead glass
blocks of 4 · 4 · 29 cm^3, covering a solid angle of 100 msr. For the suppression of
high-energy photons an additional Cherenkov detector is provided. However, during
the first tests with the MAMI beam it turned out that this may be not needed. The
beam is so well defined and clean that little background is produced.

The neutron is detected by means of two layers of 5 cm thick plastic scintillators
with anticounters in front. They cover a solid angle of 250 msr. For the measurement
with $\vec{^3He}$ only the direction and energy of the neutron is determined using the set-up
as a hodoscope. The detection efficiency is about 10%. The $\vec{^3He}$ target is a gas target
for which a polarization of 40% at 10 µA has been measured in beam [11]. For the polarization transfer measurement on the unpolarized deuteron the two layers form a

neutron polarimeter. The efficiency is due to the requirement of double detection of the neutron only 0.1 %. However, considering that the deuteron target is a liquid target with high density of scatterers, the luminosity of the two experiments is about the same. The two experiments will enable the collaboration to determine two independent results of G_{En}, which differ only in the correction for nuclear effects.

VI. Outlook

In this paper the first round of experiments was outlined. They have been chosen with the necessity in mind to tune the machine and apparatus at the beginning. Consequently, they are not pushing the experimental difficulties to the limits. For the second phase, the following more ambitious plans exist:

1.) MAMI
   - beam splitter for two or three beams of different intensities
2.) Collaboration A1
   - a study of the reaction $p(e,e'p)\pi^0$ or η
   - triple coincidence for the study of the Δ resonance in nuclei with spectrometers A, B, C.
   - a polarimeter in the focal plane of spectrometer A.
3.) Collaboration A2
   - linearly and circularly polarized photons for
   higher accuracy in the measurement of the polarizability
4.) Collaboration A3
   - good statistics in a large q^2 range
   - improved $\vec{e}$ source
5.) Collaboration A4
   - parity violation experiments for measurement of
   weak flavour singlet form factor
6.) Collaboration X
   - X-rays by means of transition radiation and Purcell-Smith effect

Acknowledgements

The work sketched in this paper is the common effort of a great number of people. The participating physicists have organized themselves in collaborations which allowed to concentrate resources and in this way to build ambitious apparatus. These physicists came from all over the world as the following list shows:

Collaboration A1 - Spokesperson: R. Neuhausen
K.I. Blomqvist, Mainz; W.U. Boeglin, Mainz; M. Distler, Mainz; R. Edelhoff, Mainz; J. Friedrich, Mainz; D. Fritschi, Basel; M. Jones, Piscataway; J. Jourdan, Basel; M. Korn, Mainz; H. Kramer, Mainz; K.W. Krygier, Mainz; V. Kunde, Mainz; M. Kuss,

Darmstadt; J.M. Laget, Saclay; A. Liesenfeld, Mainz; G. Masson, Basel; K. Merle, Mainz; C.L. Morris, Los Alamos; R. Neuhausen, Mainz; E.A.J.M. Offermann, Mainz; R.D. Ransome, Piscataway; A. Richter, Darmstadt; A.W. Richter, Mainz; B.G. Ritchie, Tempe; S. Robinson, Basel; G. Rosner, Mainz; P. Sauer, Mainz; S. Schardt, Mainz; G. Schrieder, Darmstadt; I. Sick, Basel; Th. Veit, Mainz; Th. Walcher, Mainz;

Collaboration A2 - Spokesperson: J. Ahrens
J. Ahrens, Mainz; J. Annand, Glasgow; G. Anton, Bonn; I. Anthony, Glasgow; G. Audit, Saclay; R. Beck, Mainz; D. Brandford, Edinburgh; J. Capitani, Frascati; G. Crawford, Glasgow; N. D'Hose, Saclay; U. Dittmayer, Mainz; B. Dolbilkin, Moscow; T. Frommhold, Gießen; P. Grabmayr, Tübingen; S. Hall, Glasgow; P. Harty, Glasgow; T. Hehl, Tübingen; S. Herdade, Sao Paulo; F. Kalleicher, Mainz; J. Kellie, Glasgow; U. Kneißl, Stuttgart; R. Kontratjev, Moscow; K.-H. Krause, Mainz; B. Krusche, Gießen; W. Kühn, Gießen; M. Ludwig, Göttingen; D. MacGregor, Glasgow; C. McGeorge, Glasgow; V. Metag, Gießen; G. Miller, Glasgow; R. Novotny, Gießen; R. Owens, Glasgow; P. Pedroni, Pavia; J. Peise, Mainz; T. Pinelli, Pavia; M. Sanzone, Genua; M. Schneider, Mainz; B. Schoch, Bonn; M. Schumacher, Göttingen; A. Shotter, Edinburgh; J. Sobolowski, Mainz; F. Steiper, Gießen; H. Ströher, Gießen; G. Tamas, Saclay; G. Wagner, Tübingen; Th. Walcher, Mainz; F. Wissmann, Mainz; B. Ziegler, Mainz;

Collaboration A3 - Spokesperson: E.W. Otten
H. Andresen, Mainz; J. Annand, Glasgow; K. Aulenbacher, Mainz; P. Drescher, Mainz; M. Ertel, Mainz; D. Eyl, Mainz; H. Fischer, Mainz; A. Frey, Mainz; P. Grabmayr, Tübingen; S. Hall, Glasgow; H. Hartmann, Mainz; T. Hehl, Tübingen; W. Heil, Mainz; E. Heinen-Konschak, Mainz; J. Hoffmann, Mainz; F. Klein, Mainz; V. Kniese, Mainz; M. Leduc, Paris; R. Loos, Mainz; M. Meyerhoff, Mainz; G. Miller, Glasgow; P.J. Nacher, Paris; R. Owens, Glasgow; E. Otten, Mainz; S. Plützer, Mainz; M. Prokscha, Mainz; E. Reichert, Mainz; R. Rieger, Mainz; L. Schearer, Rolla Missouri; H. Schmieden, Mainz; K.H. Steffens, Mainz; A. Steinel, Mainz; R. Suhrkau, Mainz; Th. Walcher, Mainz; M. Welling, Mainz.

An as important prerequisite for the building of MAMI were the untiring efforts of the technical and administrational staff of the Institut für Kernphysik. This staff took for years the load of a project almost too big for the framework in which it was realized.

All these achievements became possible through the financial and organizational supports of the following institutions:
- Land Rheinland-Pfalz, Mainz
- Deutsche Forschungsgemeinschaft, Bonn-Bad Godesberg
- British Science and Engineering Research Council, U.K.
- Bundesministerium für Forschung und Technologie, Bonn
- CEA, CEN, Saclay
- INFN, Rome
- Universität, Mainz.

References:

[1] Herminghaus, H., Feder, A., Kaiser, K.H., Manz, W., von der Schmidt. H: Nucl. Instr. and Meth. 138, 1 (1976)

[2] Herminghaus, H.: Proceedings 1990 Linear Acc. Conf. Albuquerque, NM, USA

[3] Drechsel, D. and Giannini, M.M.: "Electron scattering off nuclei", Reports on Progress in Physics 52, 1083 (1989)

[4] Walcher, Th.: Proceedings of Int. School of Nucl. Phys. Erice 1989, published in Progress in Particle and Nuclear Physics 24, 189 (1990)

[5] Friedrich, J.: Proceedings 5th Int. Symp. on Mesons and Light Nuclei, Prague 1991, to be published in Few Body Systems, Suppl.

[6] Drechsel, D. and Walcher, Th. (editors): "Physics with MAMI A", University of Mainz (1988)

[7] Bernstein, A.M. and Holstein, B.R.: Comments Nucl. Part. Phys. 20.4, 197 (1991)

[8] Raskin, A.S., Donnelly, T.W.: Ann. of Phys. 191, 78 (1989)

[9] Arenhövel, H., Leidemann, W., Tomusiak, E.L.,: Z. Phys. A331, 123 (1988)
Dmitrasinovic, V. and Gross, F.: Phys. Rev. C40, 2479 (1989)

[10] Blankleider, B. and Woloshyn, R.M.: Phys. Rev. C29, 538 (1984)

[11] Heil, W.: Proc. Workshop on Future of Nuclear Physics in Europe with Polarized Electrons and Photons, Orsay 1990.

Few-Body Systems, Suppl. 6, 408—420 (1992)

NUCLEAR PHYSICS WITH 1 GeV
CONTINUOUS CURRENT ELECTRON BEAMS

P.K.A. de Witt Huberts

National Institute for Nuclear Physics and High-Energy Physics
Section K (NIKHEF-K)
P.O. Box 41882
1009 DB Amsterdam/The Netherlands
and
State University of Utrecht

Abstract

In electron-nucleus physics at ~ 1 GeV electron energy several topics are in focus; high-momentum components, nucleon-nucleon correlations, pion electro-production on the nucleon, meson fields and the properties of nucleon-resonances in the nuclear medium. Powerful tools to investigate this physics will become available with the commissioning of facilities with continuous (polarized) beams of ≈ 1 GeV electrons. Some physics ideas and planned experiments with AmPS at NIKHEF are discussed with an emphasis on few-body systems.

Introduction

Our understanding of the structure of nuclei has evolved substantially over the past decade. This progress has been achieved in particular by the use of high-luminosity beams of electrons with high-energy resolution in the energy range of (500 - 900) MeV. As an illustration consider the following examples.

- Total charge distributions at the very center of (magic) nuclei and the distribution of magnetization at the nuclear surface in odd-even nuclei have been mapped out. This data has demonstrated the limitations of a description of strongly interacting many-body systems based on independent particle motion.

- The proton spectral function $S(E_m, p_m)$ has been determined with a missing-mass resolution of ~ 100 keV by proton knockout experiments for several quantum states in a series of nuclei across the periodic table. Evidence for a substantial smearing of the Fermi surface has been found, a signature of nucleon-nucleon correlations.

- The longitudinal and transverse structure functions of ^{2}H and ^{3}He, ^{3}H measured out to large values of momentum transfer demonstrate the importance of meson exchange and isobar currents in the short-distance structure of hadronic currents in these systems.

Despite the progress achieved various physics problems remain to be addressed. I will discuss the following topics

- The longitudinal (L), transverse (T) and L-T interference structure functions for the reaction ^{2}H(e,e'p)..

- The role of charge exchange processes in the electro disintegration of ^{4}He.

- The triple coincidence (e,e'NN) data soon to be expected can in principle provide insights in the physics of nucleon-nucleon correlations. In practice theoretical developments to treat the 2 nucleon continuum final state preferably with a Fadeev approach (N + N + optical potential of (A - 2) system) are required to exploit the physics potential of two-nucleon emission experiments.

In particular the determination of interference structure functions and *polarization* observables with continuous-current electron beams of energy E_0 ~ 1 GeV will offer opportunities to achieve progress. With the commissioning of the two ~ 1 GeV continuous current electron beam facilities in Europe - MAMI-B at Mainz and the pulse stretcher and storage ring AmPS at NIKHEF/Amsterdam - this programme will start in the very near future. I will discuss a few selected and approved experiments planned with AmPS in 1992 and 1993.

Electrodisintegration of ^{2}H

Data on the separated elastic response functions $A(Q^2)$, $B(Q^2)$ of ^{2}H up to a value of 4-momentum transfer squared, $Q^2 = 60$ fm^{-2} are described well by non-relativistic (Schrödinger equation) calculations with a few hadronic degrees of freedom - nucleons (with free-space form factors) and mesons (π, ρ, ω, δ,...) to construct the electromagnetic current of ^{2}H.

This is quite surprising. Given the high recoil momenta involved special relativity should be important; e.g. Lorentz covariance (boost effects), gauge invariance and the use of the Dirac equation. Recent efforts [1,2] to describe $A(Q^2)$ and $B(Q^2)$ using the Bethe-Salpeter equation and a one boson exchange potential have indeed demonstrated the need for a relativistically covariant calculation. One might wonder in which observables of ^{2}H some effect of special relativity could show up even at relatively low momentum. As pointed out by Mosconi and Ricci [3], truncating the nucleon current operator at order (q/m) in the description of the ^{2}H(e,e'p) reaction affects the longitudinal-transverse interference structure function f_{01} ~ $2Re(J_0^*(J_+ - J_-))$ where J_0, J_+ and J_- are the three components of the nucleon current. In an experiment at NIKHEF [4,5] the structure functions f_{00}, f_{11} and f_{01} have been obtained using the expression for the (e,e'p) cross section in the one photon exchange approximation

$$d^5\sigma/d... = C\,(\rho_{00}f_{00} + \rho_{11}f_{11} + \rho_{01}f_{01}\cos\phi_{np}^{cm} + \rho_{-11}f_{-11}\cos 2\phi_{np}^{cm}) \qquad (1)$$

Measurements were performed in parallel kinematics for recoil momenta p_m in the range ($40 \leq p_m \leq 180$ MeV/c) in a range of 4-momentum transfers $Q^2 = 0.05 - 0.27$ $(GeV/c)^2$. Separated structure functions were obtained at $Q^2 = 0.21$ $(GeV/c)^2$ by a Rosenbluth separation (f_{00} and f_{11}) and f_{01} with data taken at $\phi_{np}^{cm} = \pi$ and $\phi_{np}^{cm} = 0$, while keeping C and ρ_{ij} constant. The

data are represented in terms of $R_G = (2m^2/Q^2 * f_{11}/(f_{00} - \varepsilon))^{1/2}$ with m = proton mass, $\varepsilon = q^2/2Q^2 * f_{-11}$ (estimated to be smaller than 5 % of f_{00}) and the asymmetry $A_\phi = (\rho_{01}f_{01})/(\rho_{00}f_{00} + \rho_{11}f_{11} + \rho_{-11}f_{-11})$. Data for R_G and A_ϕ are shown in Figs. 1, 2. Also theoretical results are shown, obtained with two different methods: i) a calculation by Arenhövel [6] using non-relativistic quantum theory and perturbatively adding corrections due to mesons and delta degrees of freedom to the electromagnetic interaction; ii) a calculation by Tjon and Hummel [7] with nucleons and mesons as the relevant degrees of freedom, described by a relativistic field theory.

As is anticipated for diagonal structure functions there are no significant differences between the two calculations for R_G; good agreement with the data is found and the effect of meson exchange current is calculated to be small. The comparison of theory and experiment for the asymmetry A_ϕ (proportional to the interference structure function f_{01}) is quite interesting. The relativistic (R) calculation deviates substantially from the non-relativistic (NR) one and significantly better agreement with the data is obtained. The averaged ratio of calculated values and the data amounts to 1.80 ± 0.08 (NR) and 1.18 ± 0.08 (R) where the quoted errors are statistical and the absolute systematic error of A_ϕ is ± 0.05. The relevance of a relativistically-covariant description of the electro disintegration process of ^2H even at low momentum is thus demonstrated. To obtain sufficiently accurate data for the separated structure functions is a very demanding experimental challenge requiring stable, high-luminosity beams and an accurate energy (few x 10^{-4}) and angle (~ 1 mrad) calibration. The new data of separated structure functions including the helicity-dependent one from continuous wave electron acccelerator facilities promise to yield new physics insights into the short-range (high momentum) structure of the deuteron.

L/T anomaly in $^4He(e,e'p)^3H$

In the interpretation of the proton knockout process from nuclei (A(e,e'p)A-1 with (near) quasi elastic kinematics) it is usually assumed that the virtual photon couples exclusively to a single (off-shell) proton. However, as is well known, there are at least as many neutrons in stable nuclei as there are protons. It may then very well be [8] that the virtual photon knocks out a neutron which subsequently appears via a charge-exchange reaction as a proton in the detector ((e,e'n)-(np) $\rightarrow$ (e,e'p)). The amplitude of this (two-step) process interferes with the direct (e,e'p) amplitude and could in principle lead to observable anomalies in the longitudinal (L) and transverse (T) response functions of the A(e,e'N) reaction. Another physical mechanism that may induce modifications of the (quasi)-free photon-nucleon coupling concerns the influence of mesonic degrees of freedom on the em structure of the hadronic current.

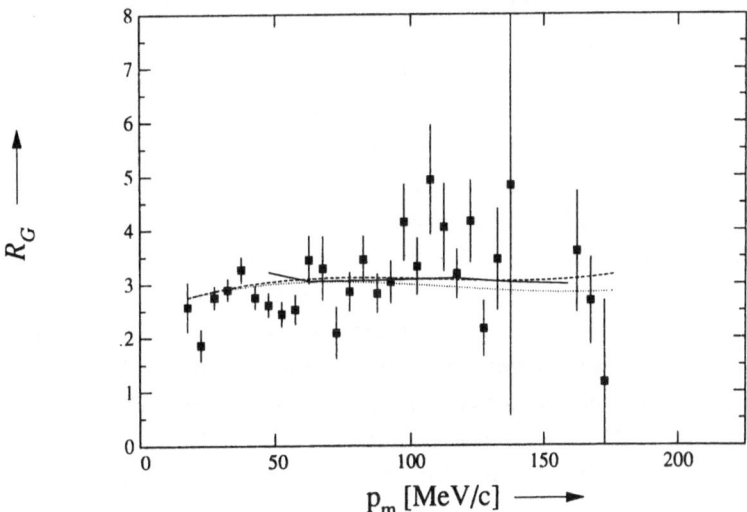

Fig. 1. The ratio R_G measured at $Q^2 = 0.21$ (GeV/c)^2. The solid (dashed) curves represent relativistic (non-relativistic) calculations. The dotted curve is a NR calculation without meson contributions.

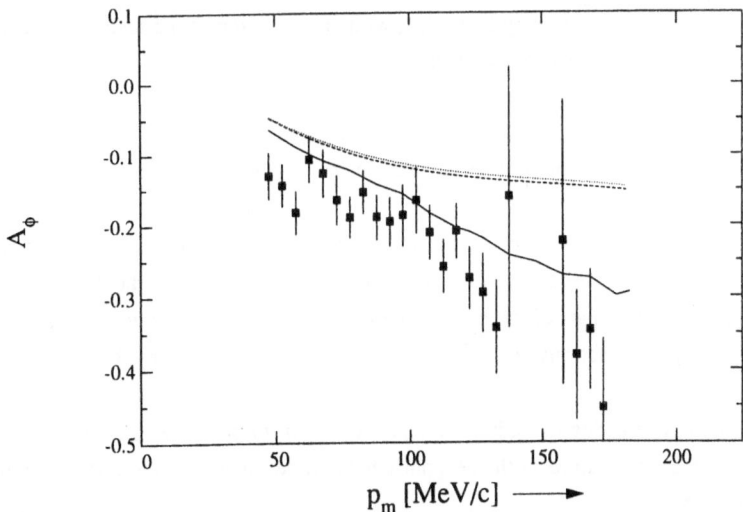

Fig. 2. The asymmetry A_ϕ measured at $Q^2 = 0.21$ (GeV/c)2. For explanation of the curves see caption of Fig. 1.

The latter effect, however, is theoretically estimated to be small in near-quasi elastic kinematics at the relatively small Q^2-values studied thusfar. What do the available data have to say about anomalous couplings? The few (L-T) separated (e,e'p) data available [9,10] for $A \geq 12$ systems indicate [11] that below the particle emission threshold in the (A-1) systyem no glaring anomalies occur. Above that threshold the transverse response appears to be substantially enhanced [11] over the longitudinal one. Thusfar a convincing explanation of the latter observation is lacking.

A significant anomaly has been found by Van den Brand [12] in the reaction $^4He(e,e'p)^3H$ connecting the ^4He and ^3H groundstates. This observation has been subsequently substantiated by (L-T) separated data from Saclay [13]. Estimates of the charge-exchange contribution to $^4He(e,e'p)^3H$ cross sections in a single-rescattering approximation [14] with correlated wavefunctions indicate the importance of this mechanism. As a result a much-improved description of the data is obtained [15] but tangible descrepancies remain. Since charge exchange (CE) in the final state has a relatively small effect on the (e,e'p) cross section it would be quite interesting to study the neutron knockout reaction (e,e'n) for which CE effects may be large. Such an experiment has been recently performed at NIKHEF-K [16] on ^4He. The difficult task of detection of the knocked out neutrons has been circumvented by detecting the recoil ^3He in a magnetic spectrometer. With 431 MeV incident electrons the kinematics were choosen such that in backward directions ^3He was detected at kinematic energy $T_\tau = 5.5$ and 7.9 MeV in a low-pressure multiwire time-projection chamber [17]. The momentum transfer was 375 MeV/c and the relative (n-^3T) kinetic energy in the center of mass system was kept constant at 70 MeV.

Two theoretical methods were employed in the interpretion of the data.
i) A coupled-channels formalism (CCIA) with the Lane potential to describe the final state interaction including the charge-exchange process.
ii) A microscopic calculation by Laget [14] with an expansion of the scattering process in a few presumably dominant amplitudes.

Data and a comparison with theory is shown in Figs. 3 and 4. The principal observation is that the charge-exchange mechanism in either theoretical approach substantially enhances the cross-section relative to the simple DWIA calculation (not including CE) and results in a correct description of the data.

Whereas the importance of the charge-exchange process has thus been demonstrated for ^4He the effect is expected to become less important for heavier nuclei due to the A^{-1} dependence of the isospin-dependent part of the Lane potential. Estimates for the role of CE processes in (e,e'p) and (e,e'n) reactions for heavier nuclei have been made by Blok and Van der Steenhoven [18].

It will be quite interesting to study the A(e,e'n) reaction (in particular the longitudinal cross section) in the near future with the continuous wave electron accelerators soon to be commissioned.

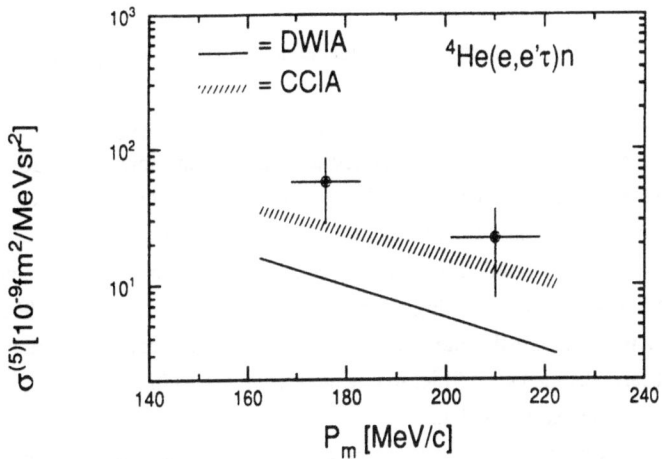

Fig. 3. Cross sections for neutron knockout from ^{4}He at two values of recoil momentum P_m. The solid line is the DWIA result (no charge exchange), the dashed area is the coupled channels calculation with charge exchange.

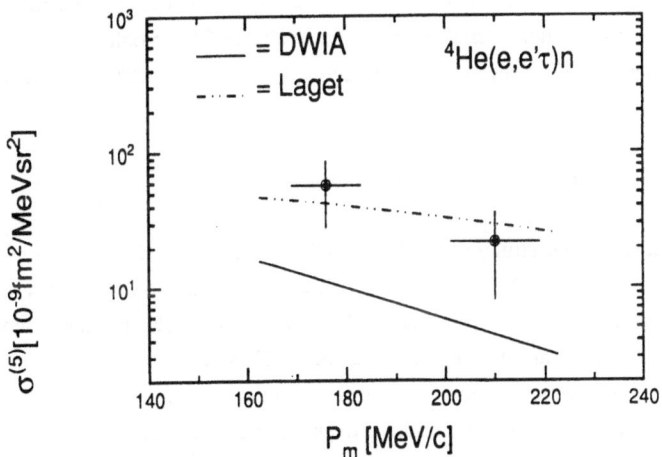

Fig. 4. Same data as in Fig. 3. The solid line is the DWIA result. The dash-dot-dot line is the result of a microscopic single-rescattering calculation by Laget [14].

Two-nucleon emission reactions A(e,e'2N)A-2

Description of nuclear systems based on fully independent-particle motion, as is assumed in selfconsistent meanfield (Hartree-Fock) approximations, however successful in their ability to describe global nuclear observables, cannot be accurate in a microscopic sense. A simple estimate based on the size of a nucleus and the radius of a nucleon shows that ~ 20 % of the volume available for independent motion of point-like nucleons is excluded. Nucleon-nucleon repulsion will lead to a smearing of the step-function occupation probability function that is hypothesized in calculations based on the Hartree-Fock approximation. The phenomenon of NN-correlations leads to observable high-momentum components in the single-nucleon spectral function. With the greatly improved signal to noise ratio provided by continuous-current electron beams the mapping of high-momentum components in heavy nuclei will soon become feasible with the (e,e'p) reaction.

It would be very interesting to hit the NN pair in close proximity with a virtual photon and study the angular and relative-energy dependence of the emitted pair of nucleons. Pioneering work on the $^{12}C(e,e'pp)$ reaction performed with a 1% duty factor beam of 500 MeV electrons has been reported recently [19]. Once continuous wave electron beams are available the virtually unexplored field of two nucleon emission processes induced by virtual and real photons will become accessible. However, in order to deduce the electromagnetic structure of short-range nucleon-nucleon currents from the (e,e'2N) cross sections one must have a reliable treatment of final state interaction processes available. This problem has been tackled by the Pavia group [20]. With several approximations adopted the (e,e'2N) cross sections can be written as

$$\text{cross section} \sim S_{fi}\, N_R(K)\, F(P)$$

where S_{fi} contains the em structure of the (γ-2N) coupling, $N_R(K)$ is the relative N-N momentum distribution and $F(P)$ represents the momentum distribution of the (NN) center of mass relative to the (A-2) systems. The effect of FSI on $F(P)$ is calculated in the approximation that the two nucleons interact with the (A-2) system via an optical potential but do not interact among themselves. In this approximation the FSI effects are not severe - $F(P)$ is typically reduced by a factor two relative to the plane wave (no FSI) approximation. However, the approximation used may be not reliable to describe the NN continuum final state process. It appears likely that the full complexity of a three-body treatment including charge exchange processes should be taken into account. This possibly can be done by treating the process of two nucleon emission in a Faddeev approach where two interacting nucleons are ejected while interacting with the (A-2) system via an isospin-dependent optical potential (Fig. 5).

It will be a challenge for theorists to tackle this problem whose solution will be indispensable for exploiting the physics potential of (e,e'NN) reactions.

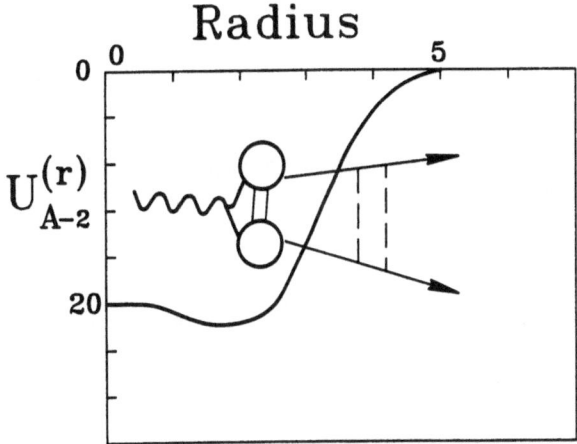

Fig. 5. An illustration of the final state interaction in (e,e'2N) with two interacting nucleons leaving the (A - 2) system represented by an (isopsin-dependent) optical potential.

The Amsterdam Pulse Stretcher (AmPS)
- a selection of scheduled experiments

Twenty-one proposals requesting a total of seven thousand hours of beam time for the initial round of experiments with AmPS have been evaluated by the Programme Advisory Committee in May 1991. These proposals were submitted by various research teams from abroad - e.g. IPN-Grenoble, Saclay, IPN-Orsay, Frascati, Lecce, Sanita-Rome, Univ. of Virginia, Univ. of Wisconsin, ETH-Zürich, IPN-Novosibirsk, often in a collaboration framework including NIKHEF staff members. The AmPS facility will be commissioned starting in April 1992. There are two mainstreams of experiments planned, i) experiments with the external beam from AmPS in stretchermode with (600 - 750) MeV energy, duty factor ≥ 30 % and (5 - 10) µA current foreseen in the initial phase of operations; ii) experiments with a stored beam in AmPS and internal targets, with a 80 mA circulating continuous current of up to 900 MeV energy electrons with optimal emittance at the internal target station, in 1993.

An overview of the AmPS facility with the experimental stations EMIN (external beam) and the Internal Target Hall (stored-beam interaction area) is shown in Fig. 6. AmPS construction is in rapid progress with the final milestone of installation foreseen in April 1992; a photograph of a curved cell installed at the first 90°-bend after the injection area is shown in Fig. 7.

The Programme Advisory Committee (PAC) had the difficult task to select the most promising experiment proposals for the initial series of experiments. In total 1600 hours of beamtime

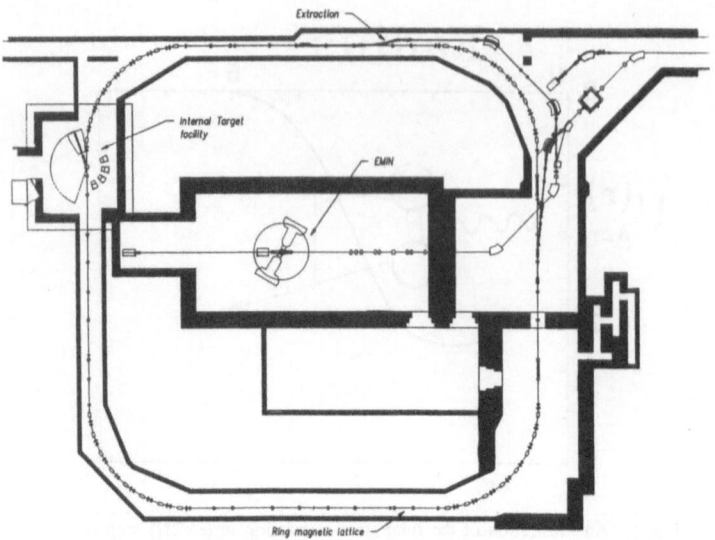

Fig. 6. The lay-out of the Amsterdam Pulse Stretcher (AmPS) with the experimental stations EMIN and ITH.

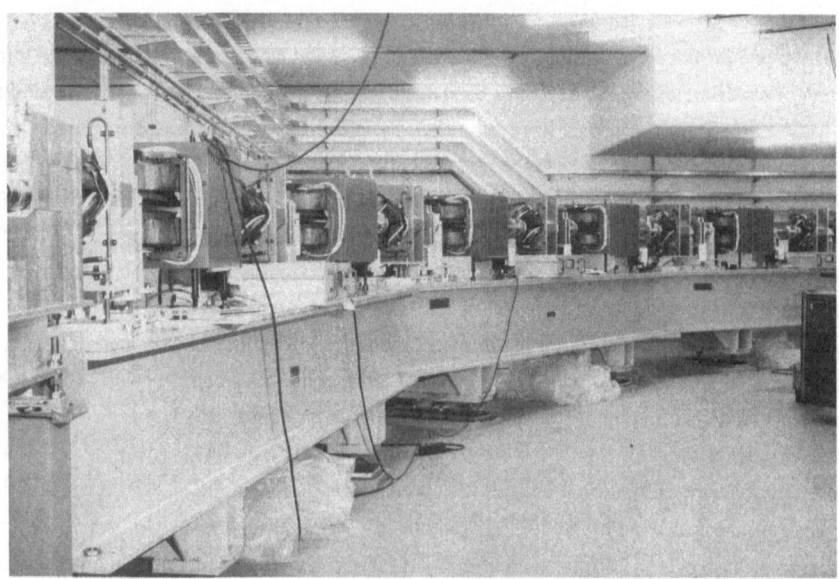

Fig. 7. Photograph of an installed 90° bend curved section of AmPS.

have been allocated to experiments with the external beam. It is interesting to note that more than 70 % of the proposed experiments focus on investigations of few-body (A ≤ 4) systems. I will not discuss the experiments in any detail here but will rather sketch the physics addressed by ranging them according to their physics topics. For more details reference is made to the original proposals or the NIKHEF Bulletin nr. 16 (September 1991).

- High momentum components. (Exp. 91-10, 91-19). (Exp. 91-19, 91-21).
- Reaction mechanism studies of (e,e'p) and (e,e'n). (Exp. 91-1, 91-2, 91-21).
- Two nucleon emission processes. (Exp. 91-2, 91-3, 91-4).
- Pion production and delta resonance physics in few-body systems (A ≤ 4). (Exp. 91-6, 91-8).
- Clusterization processes in electro disintegration reactions. Reaction (e,e'd), (e,e'α). (Exp. 91-7, 91-8).
- Relativistic effects in ^{2}H(e,e'p). (Exp. 91-15).

Internal target physics experiments

Two proposals for internal target physics have been approved with highest priority. Experiment 91-8 concerns the physics of neutral pion production and delta propagtion in ^{4}He. A windowless gasjet target will be employed. With a 80 mA circulating current of (700 - 800) MeV electrons an effective luminosity $L \approx 10^{33}$ cm^{-2}s^{-1} will be possible. Low-energy heavy recoil products of the reaction will be detected in the backward hemisphere by ultrathin gas-multiplication detectors backed up by Si microstrip detectors. The recoil detection method is particularly valuable in case that neutral pions are produced; this particular reaction channel can be identified by the type of recoiling system. NIKHEF is involved in a R & D project to develop novel gasjet technology that will enable to reach much higher densities (one to two orders of magnitude improvement) in the interaction region with the circulating beam.

Given the importance of spin-dependent response functions in electron-nucleus interactions a most interesting experiment has been approved: Experiment 91-12 will measure the target spin dependence of the (e,e'p) reaction for tensor polarized deuterium. The data will allow separation of the S- and D-wave components of the deuteron wave function. The experiment will be carried out by a collaboration of physicists from Arizona State University - Tempe, Institute of Medium Energy Physics (ETH-Zürich), Instiute of Nuclear Physics - Novosibirsk, NIKHEF, University of Mainz, NIKHEF and the University of Wisconsin-Madison. For a successful completion of this technically demanding experiment an extensive and concerted effort of various specialists is required. This will be a key experiment to assess the potential of polarized internal target physics, that appears such a promising tool in electron-hadron physics in a wide range of energies from ~ 1 GeV up to 30 GeV at e.g. HERA. The principle of the experiment is illustrated in Fig. 8. The electron beam traverses a cooled 40 mm long target cell that contains tensor polarized deuterons. The storage cell target consists of a T-shaped cell fed by an intense polarized beam from an atomic beam source (ABS) [21]. A feed ratio of ~ 4.10^{16} deuterium atoms per second is foreseen. With a suitable coating on the inside walls of

418

the cell depolarization caused by wall collisions can be minimized. The design effort aims at a luminosity $L = 10^{32}$ cm^{-2}s^{-1} with 80 mA circulating current. The scattered electron will be detected with (2 - 3)% energy resolution by three layers of CsI with a total thickness of ~ 9 radiation length. Multiwire proportional chambers will be used to provide tracking information on θ and φ angle (resolution ~ 5 mrad) and interaction-vertex position. Knocked out protons will be detected by two scintillator hodoscopes of 160 msr solid-angle each, positioned at a distance of 40 cm from the interaction zone. Optimization of the design for these multidetector hodoscopes has been pioneered by a NIKHEF-Free University collaboration [19] in the harsh environment of a 1 % duty factor beam of MEA. The calculated perpendicular asymmetry in ^{2}H(e,e'p) is plotted in Fig. 9. It is quite large and can be measured with the proposed setup discussed above for five values of the recoil center of mass angle θ_{pq} with ~ 10 % accuracy in a hundred hours of beamtime. This experiment is expected to provide important practical experience for internal target experiments and will be an important impetus for the polarized internal target physics programme with AmPS. In order to exploit the potential of spin-physics with internal targets to the full it is necessary to have longitudinally polarized electrons available. This requires implementation of a polarized electron source and spinflipper devices in the ring with substantial international participation (France, Germany, Russia). NIKHEF has submitted a project proposal (SPITFIRE) for funding by the Dutch government to implement spin physics capability at the AmPS ring.

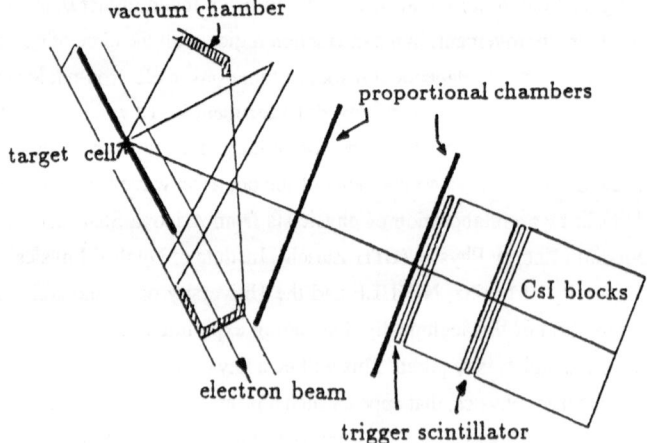

Fig. 8. Lay-out of the internal target setup to study the ^{2}H(e,e'p) reaction (experiment 91-12).

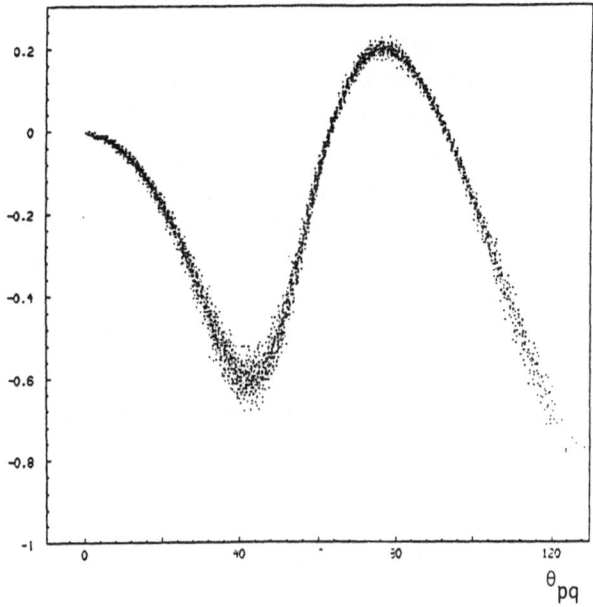

Fig. 9. Perpendicular asymmetry calculated for the ^{2}H(e,e'p) reaction as a function of the (p - q) center of mass angle θ_{pq}.

Acknowledgement

This work is part of the research programme of the National Institute for Nuclear Physics and High-Energy Physics (NIKHEF), made possible by financial support form the Foundation for Fundamental Research on Mater (FOM) and the Dutch Organization for Scientific Research (NWO).

References

1. Hummel, E., Tjon, J.A.: Phys. Rev. Lett. **63**, 1788 (1989)
2. Hummel, E., Tjon, J.A.: Phys. Rev. **C 42**, 423 (1990)
3. Mosconi, B., Ricci, P.: Nucl. Phys. **A 517**, 483 (1990)
4. Schaar, M. van der, et al.: Phys. Rev. Lett. **66**, 2855 (1991)
5. Steenhoven, G. van der: Invited talk at the International Symposium on Mesons and Light Nuclei, Sept. 1991, Prague, and to be published in Phys. Rev. Lett.
6. Arenhövel, H.: Nucl. Phys. **A 384**, 287 (1982)
7. Tjon, J.A.: Proceedings of the EPS Conference 'Hadronic Structure and Electroweak Interactions', Amsterdam, August 5 - 10, 1991; to be published in Nucl. Phys.
8. Witt Huberts, P.K.A. de: Nucl. Phys. **A 497**, 449c (1989)

9. Steenhoven, G. van der, et al.: Phys. Rev. Lett. **57**, 182 (1986)
10. Reffay-Pikeroen, D., et al.: Phys. Rev. Lett. **60**, 776 (1988)
11. Steenhoven, G. van der: Nucl. Phys. **A 527**, 17c (1991)
12. Brand, J.F.J. van den, et al.: Phys. Rev. Lett. **60**, 2006 (1988)
13. Magnon, A., et al.: Phys. Lett. **B 222**, 352 (1989)
14. Laget, J.M.: Nucl. Phys. **A 497**, 391c (1989)
15. Brand, J.F.J. van den: Phys. Rev. Lett. **66**, 409 (1991)
16. Daman, M.: PhD. Thesis, Univ. of Amsterdam, April 16, 1991 and to be published in Phys. Rev. Lett.
17. Steijger, J.J.M., et al.: N.I.M. **A 295**, 123 (1990)
18. Steenhoven, G. van der, et al.: Phys. Lett. **B 191**, 227 (1987)
19. Zondervan, A., et al.: NIKHEF Internal Report EMIN 91-01 and to be published in N.I.M.
20. Boffi, S., Proceedings of Topical Workshop on Two-nucleon Reactions, Elba IPC, September 1989
21. Singy, D., et al.: N.I.M. **A278**, 349 (1989)

Few-Body Systems, Suppl. 6, 421—434 (1992)

FEW BODY EXPERIMENTAL PROGRAM AT CEBAF

Jean Mougey
Continuous Electron Beam Accelerator Facility
Newport News, VA 23606, USA

Abstract

Nearly two thirds of the initial CEBAF physics proposals deal with experimental studies of the nucleon and few nucleon (A $\leq$ 4) systems. These studies include the structure of nucleons at high Q^2, the electro- and photodisintegration of the deuteron, and the short distance structure of ^{3}He and ^{4}He. Emphasis is given to coincidence experiments and measurements of polarization observables.

I. Introduction

Electro- and photonuclear reaction studies at low and medium energy confirmed the essential validity of a mean field description of nuclei and put quantitative constraints on this model. But they also showed that, increasing energy and momentum transfers, other processes have to be taken into account involving meson exchange currents, nucleon resonance excitations, short range N-N interactions, relativistic effects, etc··· Ultimately, one would like to understand the basic properties of nucleons and nuclei in terms of quarks and gluons with QCD as the underlying theory. At present, the most challenging task now is to search for nuclear phenomena that demonstrate unambiguously the necessity of taking the nucleon substructure and its modification in the presence of other hadrons explicitly into account.

The few-body ($\leq$ 4) systems provide the best suited "nuclear laboratory" for such studies. Covering a large range of nucleon binding energies and nuclear densities, they exhibit of all the above mentioned effects without having the complexity of heavier nuclei. Consequently, a major part of the CEBAF physics program (32 among the 47 first-round physics proposals) deals with ^{1}H, ^{2}H, ^{3}He and ^{4}He targets. An overview of the physics issues addressed in these proposals will be given, after a short description of the CEBAF experimental facilities.

II. CEBAF Experimental Facilities

The main CEBAF beam performance objectives are:

- energy	0.5 to 4 GeV	- duty cycle	100%, CW
- emittance	2×10^{-9}m	- energy spread (4σ)	10^{-4}
- current intensity	0.1 to 200 μA		

In addition, the accelerator configuration[1] allows three beams to be delivered simultaneously in the three experimental halls, with intensities which can differ by a factor up to 1000 and different but correlated energies. A polarized GaAs photocathode electron source has been developed in collaboration with the University of Illinois[2]. The source can deliver up to 100 μA beams with 49% polarization. Various techniques are under investigation to increase the polarization rate.

Complementary sets of instruments[3] have been designed for the three experimental halls, and are now under procurement. a) Hall A will have two identical, high resolution 4 GeV spectrometers (Fig. 1). Their characteristics are given in Table 1. The QQDQ design includes a large iron-dominated dipole with superconducting racetrack coils and focussing properties provided by slanted pole ends and a non-uniform field. The three superconducting quadrupoles are $\cos 2\theta$-type, current-dominated magnets with higher multipole correcting coils. The detector systems include vertical drift chambers as position detectors (~ 100 μm track resolution) and a focal plane polarimeter in the hadron arm. b) The CEBAF Large Acceptance Spectrometer (CLAS) to be installed in Hall B, together with a photon tagging facility, is a 6-coil superconducting toroidal magnet (Fig. 2) installed around the beam. The six sectors will be filled with drift chambers and particle identification detectors. The CLAS will be best suited for experiments for which broad kinematical coverage and high particle multiplicities are more important than resolution and high luminosity. c) Hall C initial equipment will consist in a High Momentum 6 GeV/c Spectrometer (HMS) and a Short Orbit (7.4m) Spectrometer (SOS). Their characteristics are given in Table 1. The moderate resolution ($\sim 10^{-3}$) should be sufficient for a substantial fraction of the few body physics program. The high momentum capability of the HMS will allow high-Q^2 physics. The SOS will be used for pion and kaon electroproduction experiments and for a variety of (e,e'p) experiments.

Liquid and high density gas cryogenic targets are being developed in all three Halls. The 1 kW cryotarget designed for Hall A will operate either with liquid H_2 and D_2 at 20K, 5 atm or with ^{3}He and ^{4}He gases at 20K (ultimately 10K) and 70 atm. With 15 cm cell length and 200 μA beam current, it corresponds to a luminosity of 5 $\times 10^{38}$cm^{-2}s^{-1}. A polarized solid state NH_3 and ND_3 target is also being developed to operate initially (but not exclusively) in Hall C. Based on recent developments at MSU[4] one expects to reach nearly 100% polarization at 1K in a 5T magnetic field. Additional instrumentation will be provided by user groups. It may include neutron detectors and polarimeters[5], large solid angle scintillator arrays[6], a possible third arm 1.3 GeV/c Multipurpose Spectrometer[7] and dedicated facilities for (e,e'K$^+$) and parity violation experiments[3].

Table 1
CEBAF Spectrometer Characteristics

	Hall A HRS (2)	Hall C HMS	Hall C SOS
Configuration	QQDQ	QQQD	QD$\bar{\text{D}}$
Optical length (m)	23.4	24.8	7.4
Momentum range (GeV/c)	$0.3 \rightarrow 4$	$0.5 \rightarrow 6$	$0.2 \rightarrow 1.5$
Momentum acceptance (%)	10	10	40
Momentum resolution (FWHM)	10^{-4}	5×10^{-4}	2×10^{-3}
Angular range	$12.5^{o} \rightarrow 165^{o}$	$12.5^{o} \rightarrow 90^{o}$	$11.6^{o} \rightarrow 168^{o}$
Positioning accuracy (mr)	0.1		
Angular resolution: horiz. (mr)	0.5	0.4 (0.8)	
(FWHM) vert. (mr)	1.0	0.9 (0.8)	
Solid angle (msr)	7	6	9
Transverse length acceptance (cm)	10	10	10
Transverse position resolution (cm)	0.1	3.4 (0.6)	

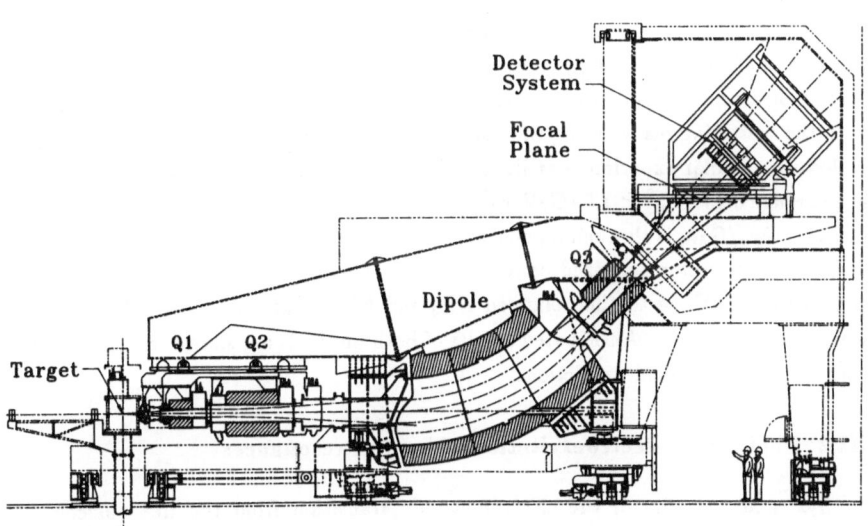

Figure 1. Hall A 4 GeV/c High Resolution Spectrometer

III. The Structure of Nucleons and Nucleon Resonances

III.1 The structure of free nucleons

The precise knowledge of $G_{EN}(Q^2)$ and $G_{MN}(Q^2)$ up to the highest possible Q^2 is of fundamental importance for the testing of microscopic models of the nucleon structure and of the γ_vNN electromagnetic coupling. It is also a necessary ingredient in the analysis of electron-nucleus data.

The usual Rosenbluth separation technique to isolate G_E and G_M is no longer adequate when $G_E^2 \ll G_M^2$ (neutron case) or at high Q^2 where the magnetic contribution dominates. Yet, it has been used recently at SLAC to measure G_{Ep} and G_{En}, up to $Q^2 = 7$ and 4 (GeV/c)^2, respectively. Preliminary results on G_{Ep} have been reported, with $\sim \pm 20\%$ uncertainties[8]. A more promising technique for determining G_{Ep} is to measure the interference term that result when longitudinally polarized electrons are incident on either a polarized proton target or when the recoil proton polarization is detected[9]. The recoil polarization technique will be used by Perdrisat et al.[10] to obtain a few percent measurement of G_{Ep} up to Q^2=4.5 (GeV/c)^2 (Fig. 3). By measuring both the transverse and longitudinal components of the proton polarization vector, the quantity G_{Ep}/G_{Mp} can be obtained without precise determination of the beam polarization and polarimeter analyzing power.

The electric form factor G_{En} of the neutron is poorly known at all Q^2 except around Q^2=0. For Q^2 up to ~ 0.7 (GeV/c)^2 G_{En} has been extracted from elastic e-d scattering[12], but the values depend on the model chosen for the deuteron wave function. As in the case of G_{Ep}, G_{En}/G_{Mn} can be determined in a model independent way by measuring the polarization component of the recoil neutron perpendicular to $\vec{q}$, in the scattering plane, using the reaction d($\vec{e},e'\vec{n}$)p at the quasielastic kinematics. Following a similar experiment at Bates, Madey et al. will determine G_{En} at CEBAF in the range $0.15 < Q^2 < 1.5$ (GeV/c)^2 with $\Delta G_{En} \sim \pm 0.015$[14]. Another method to measure G_{En}/G_{Mn} is through a polarization asymmetry measurement in the reaction $\vec{d}(\vec{e},e'n)$p on a vector polarized deuteron target. This technique will be used in Hall C by Day et al[15], with the polarized NH$_3$ and ND$_3$ targets under development at University of Virginia. Anticipated results of both methods are shown in Fig. 4 along with various theoretical predictions.

III.2 Photo- and electroexcitation of baryon resonances

Systematic studies of the electromagnetic transition from the free nucleon to resonances below ~ 2 GeV are expected to yield fundamental information on the γ_vNN* vertex as well as on the baryon substructure and the quark and gluon dynamics in confined systems. The γ_vNN* vertex is described by three amplitudes $A_{\frac{1}{2}}$, $A_{\frac{3}{2}}$ and $S_{\frac{1}{2}}$. Coincident detection of the resonance decay products, which will be possible in the CLAS detector, allow the separation of overlapping resonances and spin/isospin

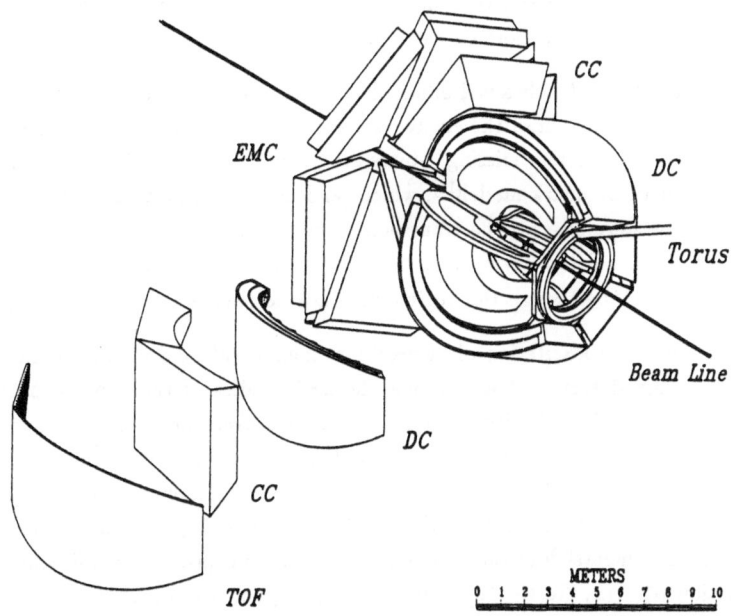

Figure 2. The CEBAF Large Acceptance Spectrometer (CLAS). Drift chambers (DC), time-of-flight counters (TOF), gas Cerenkov counters (CC) and electromagnetic calorimeters (EMC) provide particle tracking and identification. Main characteristics include: θ-range: 8° - 140°; ϕ-range (% of 2π) = 85 (50) at $\theta = 90^o(10^o)$; momentum resolution $\leq 1\%$ (FWHM); maximum luminosity $\simeq 10^{34}$ cm^{-2}s^{-1}; π/K (π/p) separation up to 2(3) GeV/c.

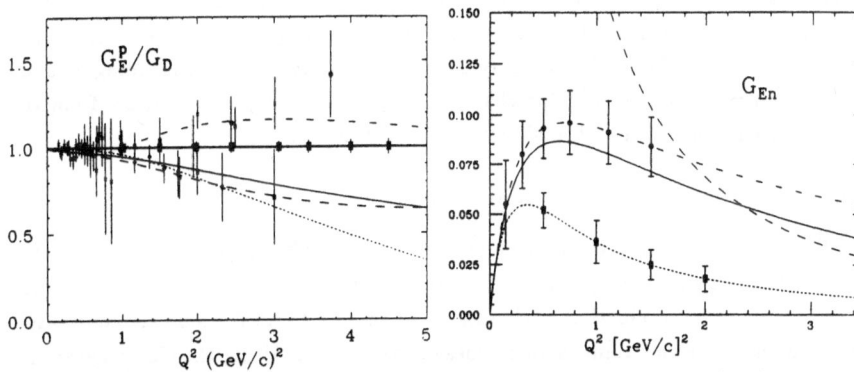

Figure 3. Data and theoretical models[11] for the proton electric form factor G_{Ep}. Solid curve: Gari and Krümpelmann; long dashes: Höhler et al.; short dashes: Iachello et al.; dot-dashed: Radyushkin. The filled circles are the projected data points from CEBAF PR-89-014[10].

Figure 4. Theoretical models [11,13] for the neutron form factor G_{En}. Solid curve: Gari and Krümpelmann; short dashes: Galster et al.; dot-dashed: G_{En}, dipole form; long-dahsed: G_{Ep}, dipole form. Open circles: expected data points, based on Galster calculation, for the CEBAF recoil polarization experiment[14]. Filled circles: expected data points, based on the dipole form, for the target polarization experiment[15].

assignments. Moreover, polarization measurements give information on the imaginary part of the amplitudes, therefore on the relative phases.

Information should be obtained about the nature of the transition (single or multiquark), the wave function of the excited state and the confinement potential. One would like also to search for "missing" states predicted by QCD motivated extensions of the non-relativistic quark model[16], which tend not to couple to the πN channel. At high Q^2, one may observe the transition from non-perturbative to perturbative regime with the predicted asymptotic behavior of the helicity amplitudes ($A_{\frac{1}{2}} \rightarrow c_1/Q^3$, $A_{\frac{3}{2}} \rightarrow c_2/Q^5$) although this regime may not be reached at CEBAF energies.

Electro- and photoexcitation of nucleon resonances will be a major program for the Hall B CLAS detector. For example, detailed studies of the $\gamma_n N \rightarrow \Delta(1232) \rightarrow N\pi$ transition with and without polarization measurements will be performed, to extract information on all multipoles including the small, poorly known contributions from the electric E_{1+} and scalar S_{1+} quadrupole amplitudes (Fig. 5). At very high Q^2, PQCD predicts $E_{1+}/M_{1+} \rightarrow 1$, while S_{1+}/M_{1+} remains ~ 0. Unpolarized measurements[18] of the reactions $p(e,e'p)\pi^o$ and $p(e,e'\pi^-)n$ will allow the determination of M_{1+}, $Re(E_{1+}M_{1+}^*)$ and $Re(S_{1+}M_{1+}^*)$ in the range $0.2 < Q^2 < 4$ $(GeV/c)^2$. Electron asymmetry will give $Im(S_{1+}M_{1+}^*)$[19], whereas a double polarization measurement (target or recoil) is needed to get $Im(E_{1+}M_{1+}^*)$. The imaginary terms vanish if the multipole are strictly in phase, which is the case if they are purely resonant.

Search for the "missing" resonances mentioned above will be undertaken in the mass range of 1500 – 2200 MeV, by looking at their decay through vector mesons[20]. Both H_2 and D_2 targets will be used for production on proton and neutron. The CLAS detector will be used to identify most of the decay products, incuding π^o's in the $\gamma^{\pm} \rightarrow \pi^{\pm}\pi^o$ and $\omega \rightarrow \pi^+\pi^-\pi^o$ channels. Expected rates per hour range from 20 – 20,000 for $\rho^o p$ and $\rho^o n$ to 1 – 200 for ωp and ωn decays.

III.3 Photo- and electroproduction of hyperons

Existing data on the elementary process $\gamma p \rightarrow K^+ Y$ ($Y = \Lambda$, Λ^*, Σ, Σ^*) are scarce and of rather poor quality. More detailed and precise experiments will bring information on the production mechanisms and the KYN, KYN* coupling constants, essential for future electromagnetic studies of hypernuclei. Using the CLAS detector, one can observe the hyperon decay and, in the case of the Λ, obtain its polarization[21]. Polarization transfer measurements using polarized electrons to produce circularly polarized photons are also planned.

Measurement of radiative decays of excited hyperon states – $\Sigma^*(1385)$, $\Lambda^*(1405)$, $\Lambda^*(1520)$ – yields information about the quark wave function of the hyperon states. Clearly, an energy tagged photon beam and a $\sim 4\pi$ solid angle coverage for photon detection are needed for these experiments[22].

Electroproduction studies of the reactions ep $\rightarrow$ eK*Λ and ep $\rightarrow$ eK$^+\Lambda$*(1520), with measurement of the hyperon decay products, has been designed to learn about the reaction mechanism and the K$^+$ and K*$^+$(892) form factors. At the same time, the electroproduction and decay of the f$_o$(975) will be measured. This state, which do not fit in the normal q$\bar{\text{q}}$ scheme, could possibly be an exotic qq$\bar{\text{q}}\bar{\text{q}}$ state or a q$\bar{\text{q}}$ - q$\bar{\text{q}}$ molecule. Using the CLAS detector at $10^{34}\text{cm}^{-2}\text{s}^{-1}$ luminosity, one expects the number of detected events per hour to be 400, 1600 and 1100 for f$_o$, Λ and Λ*(1520) respectively[23].

III.4 The nucleon weak form factors

The parity violating asymmetry A$_{\vec{e}p}$ in elastic $\vec{e}p$ scattering using a longitudinally polarized electron beam results from the interference between the amplitudes involving the electromagnetic and the neutral weak current. By varying θ_e at fixed Q^2, one can isolate terms depending linearly in F$_1^z$ and F$_2^z$. In the Standard Model and assuming strong isospin symmetry,

$$F_2^z = (\tfrac{1}{2} - \sin^2\theta_w)\, F_2^\gamma - \tfrac{1}{4}\, F_2^o$$

F$_2^o$, the flavor singlet weak form factor, contains equal contributions of all three quark flavors and is therefore sensitive to s$\bar{\text{s}}$ contributions to the proton structure. An experiment is in preparation at MIT-Bates to determine F$_2^o$ at $Q^2 \sim 0.1$ (GeV/c)^2 by measuring A$_{\vec{e}p}$ at backward angles[24]. By performing the measurement at forward angle, one can get information about G$_E^{(s)}$ (or F$_1^{(s)}$) the strangeness dependent part of the electric weak form factor, therefore about the strangeness radius of the proton[25]. The possibility to perform such an experiment at CEBAF is under investigation.

IV. Deuteron electro- and photodisintegration

IV.1 Deuteron electrodisintegration at threshold

The d(e,e')pn reaction at threshold is known to provide the cleanest evidence for isovector meson exchange currents. Quark effects have been expected to show up above $Q^2 \sim 1$(GeV/c)^2, as predicted for example by hybrid quark-hadron models[26,27]. Existing data up to $Q^2 \sim 2.7$ (GeV/c)^2 seem better reproduced by conventional models although calculations including relativistic effects are still missing. The data will be extended at CEBAF (Hall A) up to Q^2 values consistent with a cross section sensitivity of 4. $10^{-12}\mu$b sr^{-1}MeV^{-1}, a factor of 20 lower than the highest Q^2 SLAC data point, with an energy resolution better than 1.2 MeV[28].

IV.2 Coincidence d(e,e'p)n experiments

High precision systematic studies of the d(e,e'p)n reaction will be performed with the Hall A spectrometers. The Q^2-dependence of the reaction will be examined by performing longitudinal/transverse separations for protons emitted along $\vec{q}$ at $0.25 < Q^2 < 3.5$ (GeV/c^2) and p$_n$=0. The proton angular distribution will be measured up

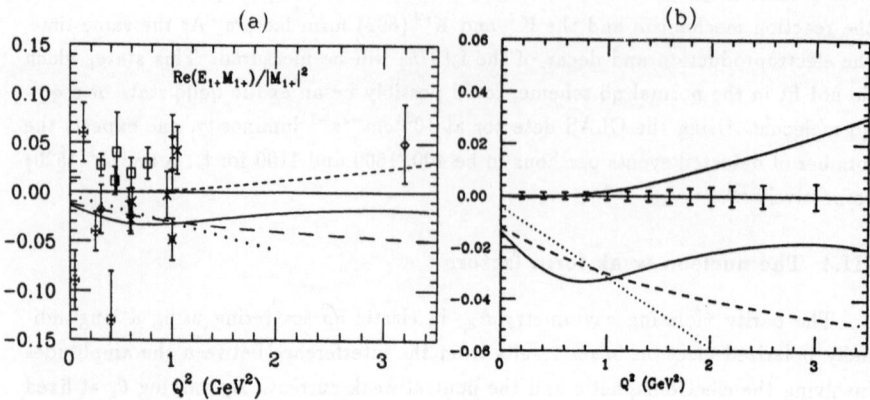

Figure 5. Interference term $\mathrm{Re}(E_{1+}M_{1+}^*)/|M_{1+}|^2$ for the $\gamma_v p \to \Delta^+(1232)$ transition: a) existing data compared with model calculations (from ref. 17); b) expected statistical errors from CEBAF PR-89-037[18].

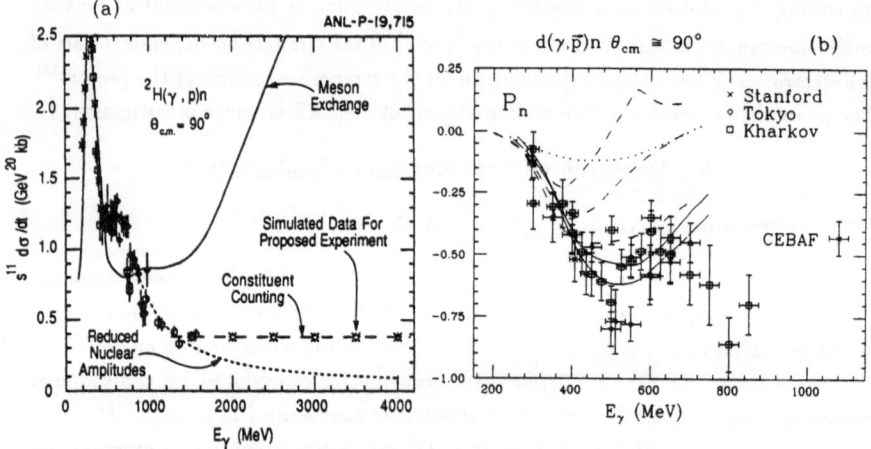

Figure 6. Deuteron photodisintegration at high energy: a) $s^{11}d\sigma/dt$ for the $d(\gamma,p)n$ reaction. Solid points are from the NE8 experiment[35], asterisks along the constituent counting line are expected data points from CEBAF experiment[36]. b) Proton polarization data[38] in the $d(\gamma, \vec{p})n$ reaction. Theoretical curves show the sensitivity to deuteron wave function[39] (dashes) and to inclusion of dibaryon resonances[40] (solid with, dotted without). Expected error bars for the CEBAF measurements[37] in the range $0.8 < E_\gamma < 1.8$ GeV are indicated at $E_\gamma = 1.08$ GeV.

to p_n=500 MeV/c at $|\vec{q}| = 1$ GeV/c. From in-plane measurements on either side of $\vec{q}$, plus a backward angle measurement, R_T, R_{LT} and $R_L + R_{TT}$ can be separated and will provide stringent tests of the microscopic models of the reaction[29].

Additional tests of the deuteron structure and reaction mechanism will be provided by measurements of all six polarization observables which can be obtained from coplanar $d(\vec{e},e'\vec{p})$ experiments[30]. Detailed non-relativistic calculations have been performed at 900 MeV incident energy by Arenhövel et al.[31] showing the sensitivity of the polarization components (corresponding to the $R_{LT'}^t$, $R_{TT'}^\ell$ and $R_{LT'}^n$ response functions respectively) to meson-exchange currents (MEC), isobar configurations (IC) and final state interactions (FSI). Similar conclusions have been reached by Shebeko et al[32]. At higher energies, relativistic calculations have been performed by Rekalo et al.[33] and are underway by Gross and Van Orden[34]. The measurements will be performed at fixed $|\vec{q}| = 1.26$ GeV/c going up to $p_n = 300$ MeV/c both sides of $\vec{q}$. By varying Q^2 at p_n=0 the outgoing proton energy will vary from 0.13 to 1.7 GeV, providing a rigorous test of the description of the two-nucleon scattering state.

IV.3 Deuteron photodisintegration at high energies

Results [35] from SLAC experiment NE8 for the $d(\gamma,p)$ cross section (Fig. 6a) appear to be consistent, above $E_\gamma \sim 1.3$ GeV, with the s^{-11} dependence expected from the constituent counting rules. It has been proposed[36] to extend these measurements at CEBAF up to E_γ=4 GeV, and also to measure the $d(\gamma, d)\pi^o$ cross section in the range $0.4 < E_\gamma < 3$ GeV. No data for this reaction have been published above 1 GeV. The constituent counting rules predict a s^{-13} dependence in the asymptotic scaling region. In another experiment[37] the induced outgoing proton polarization P_n will be measured, up to E_γ=1.8 GeV. Existing data (Fig. 6b) show that P_n is large (>50%) above ~ 300 MeV and very sensitive to reaction mechanisms, in particular to the excitation of nucleon and possibly dibaryon resonances.

IV.4 Meson photo-and electroproduction

Longitudinal pion electroproduction will be measured at CEBAF at Q^2=0.1 and 0.4 $(\text{GeV}/c)^2$ through L/T separations in $(e,e'\pi^+)$ reactions on ^{2}H, ^{3}He and ^{4}He[41] in the Δ-region and below. The goal is to study how the nucleon pion field is modified by the presence of other nucleons. Previous data[42] show a significant quenching at low Q^2. Calculations[43] predict a pion excess per nucleon of 0.09 in ^{4}He, and 0.14 in ^{208}Pb. Measurements above the nucleon resonance region and at higher Q^2 have also been proposed on a series of nuclei[44].

Using the CLAS detector, kaon production on deuterium[45] will be used to study the elementary reaction $\gamma n \to K^+\Sigma^- \to K^+\pi^-n$ in quasi free kinematics. Λn and $\Sigma^o n$ final state interactions will be studied by going away from that kinematics. Λ - Σ channel coupling effects are predicted to generate a cusp (Fig. 7) near the Σ threshold[46] in the vicinity of which two S=-1 dibaryon states have been predicted[47].

Kaon electroproduction studies on light nuclei are also planned using the Hall C spectrometers[48] and possibly the Multi Purpose Spectrometer in Hall A. Inclusive kaon rates of up to 16,000/hr are expected in the CLAS, and about 30/hr true $(e,e'K^+)$ coincidences in the HMS + SOS.

V. The Short Distance Structure of ^{3}He and ^{4}He

V.1 A=3 and A=4 elastic form factors at high Q^2

The elastic charge and magnetic form factors of ^{3}H and ^{3}He have been measured up to $Q^2 \sim 1$ $(GeV/c)^2$. Detailed "exact" calculations taking into account π and ρ-meson exchange, isobar configurations, relativistic effects and genuine three-body forces still deviate from the data in the region of the second diffraction maximum[49]. The same is true for the charge monopole form factor of ^{4}He, measured at SLAC up to $Q^2 \sim 2$ $(GeV/c)^2$. Hybrid quark-hadron models, which are still in a phenomenological stage, are also only partially successful[50]. At CEBAF the magnetic form factor of ^{3}He and the charge form factor of ^{4}He will be measured up to $Q^2 \sim 2.3$ and $Q^2 \sim 3$ $(GeV/c)^2$ respectively, corresponding to a cross section sensitivity limit of $2 \times 10^{-42} cm^{-2} sr^{-1}$ [51].

V.2 High momentum components in ^{3}He and ^{4}He

Recent $(e,e'p)$ experiments on ^{3}He and ^{4}He have demonstrated clearly the link between high momentum components and short distance nucleon-nucleon interactions[52,53]. Recoil momentum distributions have been obtained up to $p_r \sim 0.6$ GeV/c for the two- and many-body break up channels. Results for ^{4}He (Fig. 8) show that momenta larger than ~ 350 MeV/c reside mostly in the many-body break up channels. The missing energy (E_m) spectra show a broad structure in the continuum which moves towards higher E_m as p_r increases, in quantitative agreement with what would be expected from the disintegration of a nucleon pair at rest. The width reflects the center-of-mass motion of the pair. At CEBAF, it is proposed to vastly extend the range of these studies, first by precise measurements on ^{3}He and ^{4}He, allowing the separation of three unpolarized response functions at $Q^2 = 1$ $(GeV/c)^2$, for recoil momenta up to 1 GeV/c[54]. A better characterization of the contributing processes will be possible, since some of them, like MEC, affect primarily the transverse responses.

The next step is obviously to detect the partner nucleon in the pair in a triple coincidence $(e,e'2N)$ experiment. The CEBAF beam characteristics will allow the development of such experiments, very difficult to perform at lower energy (< 1 GeV) and duty factor ($< 1\%$) accelerators. Broad surveys of multi-nucleon emission processes using the CLAS have been proposed[55] with measurements of total yields, angular and energy distributions for $(e,e'N)$, $(e,e'2N)$, $(e,e'N\pi)$, $(e,e'2N\pi)\cdots$ reactions. This will allow the decomposition of the inclusive or semi-exclusive cross-section into well-identified multiparticle final states, and to obtain a first shot at more specific aspects like three-body currents, nucleon-nucleon ground state correlations and possible signatures of multiquark clusters (6q, 9q$\cdots$). More detailed studies have been

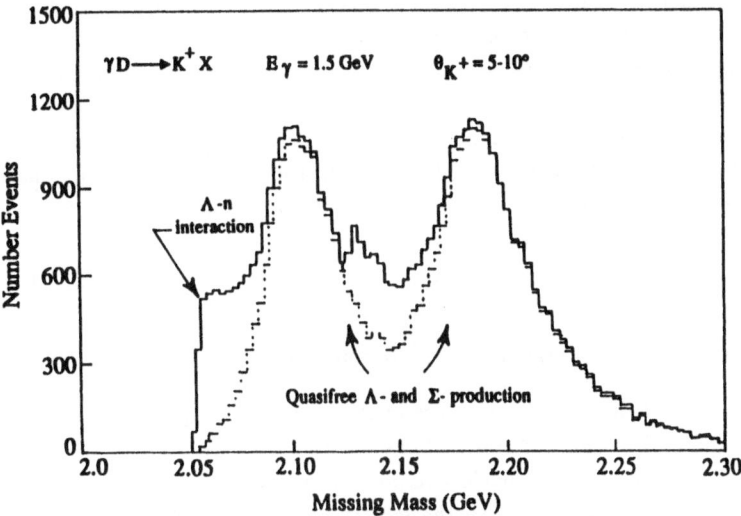

Figure 7. Monte Carlo simulation of the $\gamma d \rightarrow K^+X$ experiment using the CLAS detector[45]. The cusp near 2.13 GeV is due to the NΣ threshold. The cross section near the Λn threshold is sensitive to the Λn interaction.

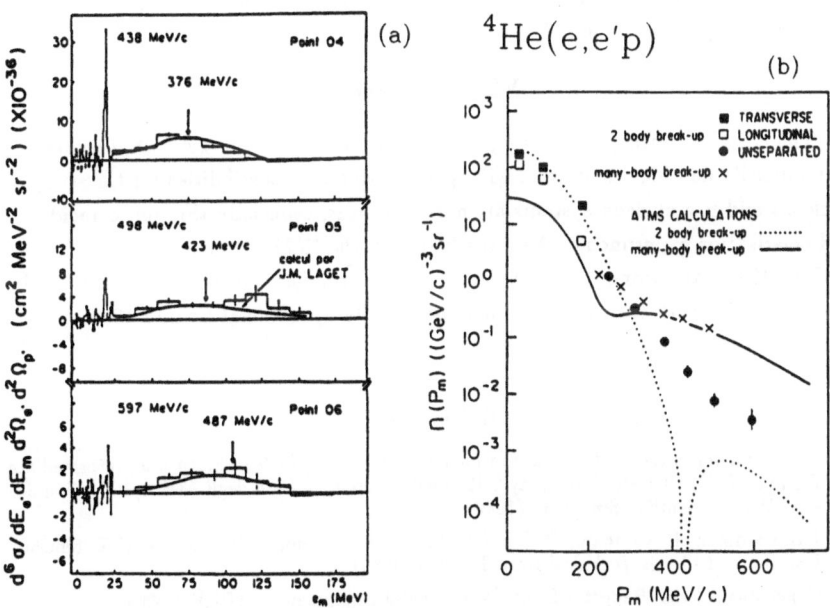

Figure 8. Experimental missing energy spectra (a) and proton momentum distributions (b) from the ^{4}He(e,e$'$p) reaction[53]. At high recoil momenta, the strength is dominantly at high missing energies in the many-body channels.

proposed using Hall A spectrometers. The ^{4}He(e,e'd) reaction cross section will be measured in the range $0.23 < Q^2 < 1.95$ (GeV/c)^2 for both d-d and d-pn break-up channels[56]. Triple arm (e,e'pn) and (e,e'2p) experiments on ^{3}He and ^{4}He have also been proposed in Hall A using existing neutron (University of Virginia) or proton (NIKHEF) scintillator arrays as second hadron detection arms[57], or possibly using the MPS[7]. Various kinematics with momentum transfers up to $Q^2=1$ (GeV/c)^2 and relative momenta in the initial pair up to 900 MeV/c have been considered. Although the (e,e'pn) reaction cross section is larger by nearly two orders of magnitude, the (e,e'2p) reaction looks more promising. One expects two-body processes to be strongly suppressed in the transverse part, which then can be used to study three-body forces[58].

V.3 Bound nucleon structure in ^{3}He and ^{4}He

The issue of the nucleon structure in the nuclear medium became one of great interest given the combined observations of the EMC effect and of the "missing" longitudinal strength in the nuclear response at moderate Q^2 values. Indeed, from the data presently available, ^{4}He is the lightest nucleus for which a strong suppression of the longitudinal strength ($\geq 20\%$ after all corrections) is observed in quasielastic (e,e'p) scattering[59]. However, the Q^2-dependence in the measured range of 0.1 – 0.4 (GeV/c)^2 appears to be consistent with the one expected from a free nucleon. At CEBAF, precise measurements will be performed[54] on both ^{3}He and ^{4}He, over a wide Q^2 range of 0.3 – 4 (GeV/c)^2 at $p_r=0$ with additional measurements at $Q^2=0.5$ and 1 (GeV/c)^2 at $p_r=\pm 300$ MeV/c.

VI. Conclusion

The CEBAF beams, with a unique combination of high energy, high duty factor, high intensity and high definition, will open exciting new possibilities for the study of nucleons and few nucleon systems at short distances, especially through coincidence and polarization experiments. A successful test of the 45 MeV injector was performed, 11 June 1991. According to the present plan, experiments will start in Hall C in the Spring of 1994, with the other two halls being fully operational by mid-1995.

References

1. H.A. Grunder, Proc. 4th Workshop on Perspectives in Nuclear Physics at Intermediate Energies, ICTP, Trieste, Italy, May 8-12, 1989, S. Boffi, C. Ciofi degli Atti and M. Giannini eds., World Scientific, Singapore (1989).

2. L.S. Cardman, B. Dunham, D. Engwall, L. Liu, N. Sereno, N. Towne and C.K. Sinclair, University of Illinois Technical Note TN-89-4 (1989).

3. Conceptual Design Report – Basic Experimental Equipment, CEBAF (1990).

4. D.G. Crabb, C.B. Higley, A.D. Krisch, R.S. Raymond, T. Roser and J.A. Steward, Phys. Rev. Lett **64**, 2627 (1990).

5. R. Madey, A.R. Baldwin, P.J. Pella, J. Schambach and R.M. Sellers, IEEE Transactions on Nuclear Science **36**, 231 (1989).

6. A. Zondervan, T.S. Bauer, R. Boontje, J.R. Calarco, W.H.A. Hesselink, E. Jans, E. Kok, R. deLeo, J.H. Mitchell, F.A. Mul, G.J.L. Nooren, A. Pellegrino, J.J.M. Stejger, J.L. Vischers, A.N.M. Zwart, Topical Workshop on Two-Nucleon Emission Reactions, EIPC, Elba, Italy, 19-23 Sept. 1989.

7. S. Frullani, F. Garibaldi, F. Ghio, M. Jodice, G.M. Urciuoli and R. DeLeo, "Multi Purpose Spectrometer", INFN-ISS 90/5 (1990).

8. P. Bosted, 4th Conf. on Intersections of Particle and Nuclear Physics, Tucson, AZ, May 23-29, 1991. NPAS-TN-91-1 (1991).

9. R.G. Arnold, C.E. Carlson and F. Gross, Phys. Rev. **23**, 363 (1981).

10. CEBAF proposal 89-014, C.F. Perdrisat and V. Punjabi spokespersons (1989).

11. G. Höhler, E. Pietarinen, I. Sabba-Stefanescu, F. Borkowski, G.G. Simon, V.H. Walther and R.D. Wendling, Nucl. Phys. **B114**, 505 (1976); F. Iachello, A.D. Jackson and A. Lande, Phys. Lett. **43B**, 191 (1973); M. Gari and W. Krümpelmann, Z. Phys. **A322**, 689 (1985); A.V. Radyushkin, Acta Phys. Polon. **B15**, 403 (1984).

12. S. Platchkov, A. Amroun, S. Auffret, J.M. Cavedon, P. Dreux, J. Duclos, B. Frois, D. Goutte, H. Hachemi, J. Martino, X.H. Phan, I. Sick, Nucl. Phys. **A510**, 740 (1990).

13. S. Galster, H. Klein, J. Moritz, K.H. Schmidt, D. Wegener and J. Bleckwenn, Nucl. Phys. **B32**, 221 (1971).

14. CEBAF proposal 89-005, R. Madey spokesperson (1989).

15. CEBAF proposal 89-018, D. Day spokesperson (1989).

16. R. Koniuk and N. Isgur, Phys. Rev. **D21**, 1888 (1980).

    F.E. Close and Zhenping Li, Phys. Rev. **D42**, 2194 (1990).

17. V. Burkert, Nucl. Phys. **B** (Proc. Suppl.)**21**, 287 (1991).

18. CEBAF proposal 89-037, V. Burkert, R. Minehart spokespersons (1989).

19. CEBAF proposal 89-042, V. Burkert, R. Minehart spokespersons (1989).

20. CEBAF proposal 89-025, J. Comfort spokesperson (1989).

21. CEBAF proposal 89-004, R. Schumacher spokesperson (1989).

22. CEBAF proposal 89-024, G.S. Mutchler spokesperson (1989).

23. CEBAF proposal 89-043, L. Dennis, H. Funsten spokesperson (1989).

24. "SAMPLE Collaboration": D.M. Beck, E. Beise, E. Belz, L. Cardman, R. Carlini, R. Carr, G. Dodson, K. Dow, M. Farkhondeh, B. Fillipone, S. Kowalski, W. Lorenzon, R. McKeown and J. Napolitano, MIT-Bates Experiment 89-06 (1989).

25. J. Napolitano, Phys. Rev. **C43**, 1473 (1991).

26. T.S. Cheng and L.S. Kisslinger, Nucl. Phys. **A457**, 602 (1986).

27. Y. Yamauchi, R. Yamamoto and W. Wakamatsu, Nucl. Phys. **A443**, 628 (1985).

28. CEBAF proposal 89-047, J. Jourdan, J. Mougey, G.G. Petratos spokespersons (1989).

29. CEBAF proposal 89-025, J.M. Finn, P. Ulmer spokespersons (1989).

30. CEBAF proposal 89-028, J.M. Finn, P. Ulmer spokespersons (1989).

31. H. Arenhövel, W. Leidemann and E.L. Tomusiak, Z. Phys **A331**, 123 (1988); erratum Z. Phys. **A334**, 363 (1989) and private communication (1989).

32. A.Yu. Korchin, Yu.P. Melnik and A.V. Shebeko, Few-Body Systems **9**, 211 (1990).

33. M.P. Rekalo, G.I. Gakh and A.P. Rekalo, J. Phys. G **15**, 1223 (1989).

34. F. Gross and J.W. Van Orden, private communication (1990).

434

35. J. Napolitano, S.J. Freedman, D.F. Geesaman, R. Gilman, M.C. Green, R.J. Holt, H.E. Jackson, R. Kowalczyk, C. Marchand, J. Nelson, B. Zeidman, D. Beck, G. Boyd, D. Collins, B.W. Fillipone, J. Jourdan, R.D. McKeown, R. Milner, D. Potterveld, R. Walker, C. Woodward, R.E. Segel, T.Y. Tung, P.E. Bosted, E.R. Kinney, Z.E. Meziani and R. Minehart, Phys. Rev. Lett. **51**, 2530 (1988).

36. CEBAF proposal 89-012, R.J. Holt spokesperson (1989).

37. CEBAF proposal 89-019, R. Gilman, R. Holt, Z.E. Meziani spokespersons (1989).

38. F.F. Liu et al., Phys. Rev. **165**, 1478 (1968); H. Ikeda et al., Nucl. Phys. **B172**, 509 (1980); A.S. Bratashevskii et al., JETP Lett. **36**, 216 (1983); V.P. Barannik et al., Nucl. Phys. **A451**, 751 (1986).

39. J.M. Laget, Nucl. Phys. **A312**, 265 (1978).

40. T. Kamae and T. Fujita, Phys. Rev. Lett. **38**, 471 (1977).

41. CEBAF proposal 89-011, H. Jackson spokesperson (1989).

42. R. Gilman, M. Bernheim, M. Brussel, J. Cheminaud, J.-F. Danel, J.P. Didelez, M.-A. Duval, G. Fournier, R. Frascaria, R.J. Holt, H.E. Jackson, J.-C. Kim, E. Kinney, J.-M. Legoff, R. Letourneau, A. Magnon, J. Morgenstern, C. Pasquier, J. Picard, D. Poizat, B. Saghai, J. Specht, P. Vernin and E. Warde, Phys. Rev. Lett. **64**, 622 (1990).

43. B.L. Friman, V.R. Pandharipande and R.B. Wiringa, Phys. Rev. Lett. **51**, 763 (1983).

44. CEBAF proposal 89-035, C.C. Chang, R. Gilman, J. O'Connell spokespersons (1989).

45. CEBAF proposal 89-045, B.A. Mecking spokesperson (1989).

46. S.R. Cotanch, Proc. Conf. on Medium and High Energy Nuclear Physics, Taipei, May 1989.

47. C.B. Dover, Nucl. Phys. **A450**, 95 (1986)
    M.P. Locher, M.E. Sainio and A. Svarc, Adv. in Nucl. Phys. **17**, 47 (1987).

48. CEBAF proposal 89-013, B. Zeidman spokesperson (1989).

49. R. Schiavilla, V.R. Pandharipande and D.O. Riska, Phys. Rev. **C41**, 309 (1990).

50. L.S. Kisslinger, Nucl. Phys. **A459**, 645 (1986).
    M.A. Maize and Y.E. Kim, Phys. Rev. **C31**, 1923 (1987).

51. CEBAF proposal 89-021, G.G. Petratos spokesperson (1989).

52. C. Marchand, M. Bernheim, P.C. Dunn, A. Gerard, J.M. Laget, A. Magnon, J. Morgenstern, J. Mougey, J. Picard, D. Reffay-Pikeroen, S. Turck-Chieze, P. Vernin, M.K. Brussel, G.P. Capitani, E. DeSanctis, S. Frullani and F. Garibaldi, Phys. Rev. Lett. **60**, 1703 (1988).

53. J.M. LeGoff, M. Berheim, M.K. Brussel, G.P. Capitani, E. DeSanctis, S. Frullani, F. Garibaldi, A. Gerard, A. Magnon, C. Marchand, Z.E. Meziani, J. Morgenstern, P. Vernin and A. Zghiche, Contributions E24 and E25, Few Body XII Conf., Vancouver, B.C., Canada, July 2-8, 1989.

54. CEBAF proposal 89-044, M. Epstein, R. Lourie, J. Mougey, A. Saha spokespersons (1989).

55. CEBAF proposal 89-027, W. Bertozzi, W. Boeglin, L. Weinstein spokespersons (1989).
    CEBAF proposal 89-031, W. Hersman, J. Lightbody, R.A. Miskimen spokespersons (1989).

56. CEBAF proposal 89-029, H.P. Blok spokesperson (1989).

57. CEBAF proposal 89-030, R.A. Lindgren, M. Epstein, G.J. Lolos, Z.E. Meziani spokespersons (1989).

58. J.M. Laget, J. Phys. **G14**, 1445 (1988).

59. A. Magnon, M. Bernheim, M.K. Brussel, G.P. Capitani, E. DeSanctis, S. Frullani, F. Garibaldi, H.E. Jackson, J.M. Legoff, C. Marchand, Z.E. Meziani, J. Morgenstern, J. Picard, D. Reffay, S. Turck-Chieze, P. Vernin and A. Zghiche, Phys. Lett. **B222**, 352 (1989); A. Magnon, private communication (1991).

Few-Body Systems, Suppl. 6, 435—445 (1992)

Few-
Body
Systems

Few Body Physics at TRIUMF and KAON

Harold W. Fearing

TRIUMF, 4004 Wesbrook Mall, Vancouver, B. C., Canada V6T 2A3

Abstract

We survey some of the interesting recent experimental and theoretical work in the few body field at TRIUMF. We discuss plans for KAON, a major extension of the TRIUMF accelerator, and describe some of the exciting few body physics which can be done at such a machine.

I. Introduction

The meson facilities, TRIUMF, LAMPF, IUCF, SIN/PSI have been on line now for twenty some years. In the early days emphasis was on the broad outlines of the field of medium energy physics, one might say, on examining the forest. At that time there were, for example, major programs surveying nucleon-nucleon scattering and pion-nucleus scattering. As the field matured, investigations focused on the individual trees of the forest, that is on specific programs and experiments designed to elucidate particular areas. For example (p,π) reactions gave information on high momentum aspects of nuclear wave functions and proton-proton bremsstrahlung put limits on the off shell behavior of the nucleon-nucleon force. Now as the meson facilities are reaching the point where major upgrades will be required to keep them on the forefront of physics, we have reached the forest floor, and are examining specific details with beautiful, precision experiments designed to pick out particular interesting bits of physics.

Thus in this talk I want to describe for you some of the "orchids in the moss", that is some of these precision and specific experiments now being carried out at TRIUMF. This will in no sense be a comprehensive survey of the experimental program, which is still very active, and covers a wide range of fields. Instead I have selected a few experimental programs which just have obtained, or shortly will obtain, results and which seem to have relevance to few body problems. The intention will be to explain why the physics is interesting, and to summarize what results have become, or shortly will become, available.

It is always important not to dwell too long in one place, but to look for new forests to explore. At TRIUMF we are planning for KAON, which stands for Kaons, Antiprotons,

Other hadrons, and Neutrinos. This will be a major new facility which will open up a vast new range of physics. Thus in the last part of this talk I want to describe the present status of our plans for KAON and to outline some of the few body physics which could be done at such a facility.

II. Few Body Experiments at TRIUMF

As it is impossible to do justice to the complete experimental program at TRIUMF, or even that portion which has relevance to few body problems, I have selected just a few topics as illustrative of some of the very interesting few body programs underway at TRIUMF.

A. The $NN\pi$ system:

Much effort has been expended in trying to understand the $NN\pi$ system. Most of this however has dealt with the "easy" reactions $pp \to d\pi$ and $pp \to pn\pi^+$, which involve only the I=1 isospin channel and are dominated by the Δ^{++}, and with measurements of only differential cross sections or analyzing powers. However several recent experiments have begun to look more involved spin correlations and at processes which involve the I=0 channel as well.

Spin transfer in $\pi\vec{d} \to \vec{p}p$: There have been a few measurements of spin transfer observables for the process $\vec{p}p \to \pi\vec{d}$ but a recent TRIUMF measurement[1] seems to be one of the first in the $\pi\vec{d} \to \vec{p}p$ direction. A polarized deuteron target was used and the polarization of the outgoing proton was measured, thus allowing determination of the coefficients K'_{LS}, K'_{SS}, and K'_{NN}[2] for a range of pion energies corresponding to proton energies of 500-800 MeV in the inverse reaction. Spin measurements such as these are interesting because they tend to sample interferences between dominant amplitudes and smaller ones which are not easily accessible directly. Thus they should have more impact on phase shift analyses and should provide more stringent tests of theoretical calculations.

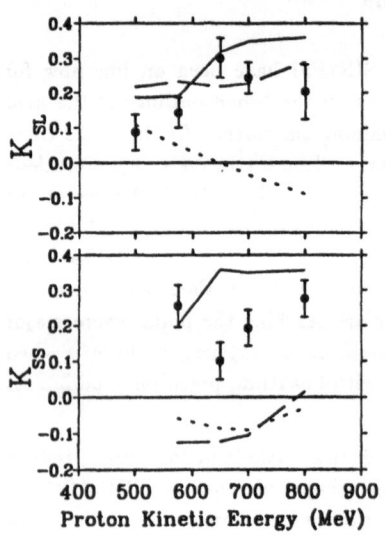

Fig. 1 shows results for K'_{SL} and K'_{SS} as a function of the equivalent proton energy in the $pp \to \pi d$ system. K'_{NN} results will be available shortly. The results agree qualitatively with the predictions of phase shift analyses[3] (solid curve) but are in strong disagreement with the calculations of K'_{SS} of Niskanen,[4] based on an isobar model, and with those of K'_{SL} and K'_{SS} of Blankleider and Afnan[5] which are based on a Faddeev formulation of the three body problem. Thus it appears that while the empirical phase shift fit is more or less reasonable, the theoretical calculations are not complete enough to reproduce these spin correlations.

Fig. 1 Spin correlation parameters K'_{SL} and K'_{SS} in $\pi d \to pp$ from Ref. 1. The solid lines are the partial wave analysis of Ref. 3. The dashed and dotted lines are respectively calculations of Refs. 4 and 5.

$NN \to d\pi$ near threshold: Most information about the $NN\pi$ system comes from $pp \to d\pi^+$, particularly in the region of the Δ resonance where the process is dominated by p-wave rescattering. Much less is known very near threshold where s-waves dominate. Such processes are interesting because there have been extensive calculations in Faddeev approaches and because one can use such calculations to relate the $NN\pi$ system to the πN system. Information about low energy πN scattering has been of particular interest lately because of its importance in relation to considerations of the strange quark content of the nucleon.

Until recently the most extensive low energy data came from old experiments on the $\pi d \to pp$ reaction. Now there has been a new measurement[6] at TRIUMF of the $np \to d\pi^0$ differential cross section using a neutron beam giving energies extending down to only a few MeV above threshold, much lower than before. The same group has also measured the analyzing power for $\bar{p}p \to d\pi^+$ at similar energies.[7] Together this data has allowed them to extract some very interesting information about the amplitudes very near threshold.

The cross section can be well parameterized as a function of the pion center of mass momentum η in the usual way as $\sigma(np \to d\pi^0) = \frac{1}{2}(\alpha\eta + \beta\eta^2)$ and the new results agree well with somewhat higher energy results from LAMPF.[8] However the value of $\alpha = 184 \pm 5\mu b$ extracted is significantly lower than the value of 275 ± 40 obtained from a fit to older data by Spuller and Measday.[9] From the angular distributions one can see that the p-waves are important almost to threshold and in any case at lower energies than predicted by the simple Watson-Bruckner theory.[10]

Finally with the addition of the $pp \to d\pi^+$ analyzing power data and the assumption of Watson's theorem, which relates the phases to the phase shifts in $pp \to pp$ and $\pi d \to \pi d$, and of charge independence, the group has been able to extract the complete amplitudes for s and p waves at energies much closer to threshold than before. These are in qualitative, but not particularly good quantitative, agreement with the Faddeev calculations of Blankleider.[11]

$np \to pp\pi^-$: For the continuum process most information comes from the $pp \to pn\pi^+$ reaction which involves only I=1 amplitudes and is dominated by the Δ^{++}. To get information about the much less well known I=0 amplitudes and about smaller pieces not dominated by the Δ^{++} an experiment[12] has been carried out using the TRIUMF polarized neutron beam to measure the cross section and analyzing powers A_N, A_S, and A_L in the reaction $np \to pp\pi^-$ at 450 MeV. These results complement similar higher energy experiments at SATURNE and LAMPF. Calculations show an interesting interplay between I=0 and I=1, and suggest that this is one of the most interesting of the $NN\pi$ reactions. The data is now being analyzed and results should be available within a year.

$pp \to pp\pi^0$: Measurements[13] have also been made on the $pp \to pp\pi^0$ process which depends only on the I=1 amplitude. Thus information on it will help separate out the I=0 pieces of the previous reaction. Total and differential cross sections have been measured for proton energies of 320-500 MeV and seem to agree reasonably well with existing data. Pion analyzing powers were also measured for the first time and tend to be large and negative.

B. Pion-proton bremsstrahlung, $\pi^+ p \to \pi^+ p\gamma$:

Pion proton bremsstrahlung should give information about the off shell properties of the πN force and about the electromagnetic properties of the Δ, in particular the magnetic moment. Most older calculations have had difficulty fitting the existing data, but two

438

recent calculations[14,15] are able to extract value of the magnetic moment using older data which is in rough agreement with SU3 or quark model predictions. A recent SIN/PSI experiment[16] measured both cross section and target asymmetry for a limited kinematic range, chosen to be sensitive to the Δ magnetic moment. Their results do not agree very well with existing theories at the higher photon energies. A new experiment[17] has been carried out at TRIUMF to measure both cross section and analyzing power over a fairly wide kinematic range. A beam of 265 MeV pions was incident on a polarized proton target and scattered pions, protons and γ's were measured. Although there have been some difficulties with the analysis, it is hoped that results for analyzing powers will be available shortly.

In parallel with the experiment a new calculation[19] is also underway at TRIUMF. Previous calculations have emphasized the importance of unitarity, and Wittman[15] made a good start toward a relativistic

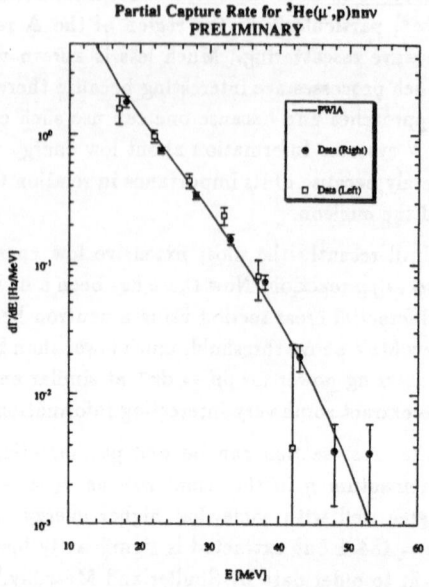

Fig. 2 Proton energy spectrum following muon capture in ^{3}He as given in Ref. 20, compared with a simple PWIA calculation.

formalism, though one which involved rather significant parameterization of background amplitudes and which was as a consequence hard to generalize to the $\pi^- p\gamma$ case. In the new calculation we start with a model of pion-nucleon elastic scattering[18] based on Feynman diagrams involving nucleon, Δ, ρ and σ intermediate states which has been shown to fit the elastic data fairly well. This is used in the Bethe-Salpeter equivalent of the two potential formalism to obtain a completely relativistic and unitary bremsstrahlung amplitude which is equivalent to an iteration of the strong diagrams with photons attached to every charged line. To actually evaluate this one has to use one of the three dimensional reductions of the BS equation and at present we are using a simple K-matrix reduction which was shown to work for the elastic case in Ref. 18. With this approximation solution of the BS equation seems feasible and will lead to a result for both π^+ and π^- bremsstrahlung similar to the π^+ result of Wittman,[15] but without need for arbitrary parameterizations of the background amplitudes.

C. Muon capture:

Muon Capture in ^{3}He: There are now preliminary results for a very interesting experiment[20] which has measured proton and deuteron spectra following muon capture in ^{3}He. The high energy tails of such spectra may be particularly sensitive to exchange currents and to high momentum components of the nuclear wave functions. In this experiment muons were stopped in a ^{3}He gas target and charged particles were detected in two identical detectors on either side of the beam. Preliminary results for the proton spectrum are shown in Fig. 2 and seem to agree with a very simple plane wave impulse approximation calculation. Early results for the deuteron spectra however seem to see excess deuterons over the simple PWIA approximation by factors ranging from two up to as much a five for deuterons above 25 MeV. Somewhat similar PWIA calculations performed at TRIUMF[21] give similar results, though off shell corrections in this calculation do increase the expected number of high energy deuterons somewhat. It will be very interesting to see how the final results compare with more detailed calculations now in progress.[21]

Radiative muon capture on the proton: There has been a major effort underway for several years to measure the rate for $\mu + p \rightarrow n + \nu + \gamma$ by stopping muons in liquid hydrogen. This gives a measurement of g_P, the induced pseudoscalar coupling of the weak interaction, free of the nuclear complications present in similar experiments on nuclei. Only a portion of the data has been analyzed so far, giving about 100 RMC events of the eventually expected 400 needed to get g_P to 10%. Preliminary indications are that the result is consistent with the Goldberger-Treiman relation.

D. Experiments with polarized ^{3}He:

There is a continuing program at TRIUMF involving scattering of protons and pions from polarized ^{3}He which has been made possible by the existence of a laser pumped ^{3}He gas target which typically reaches 50-70% polarization over a volume of up to 35 cm^3. A number of interesting results have come from this program.

p-^{3}He elastic scattering: There have been measurements[22] of elastic proton ^{3}He scattering at 500, 400, 290, and 200 MeV, with results for the cross section and for the beam and/or target polarized asymmetries A_{on}, A_{no}, A_{nn}, where the notation A_{bt} means beam or target unpolarized or polarized normal to the scattering plane. Attempts have been made to understand these data in terms of a momentum space optical potential model[23] and a relativistic DWBA[24] with marginal success. Fig. 3 shows some samples of the worst cases. Clearly the fundamental mechanism is not understood.

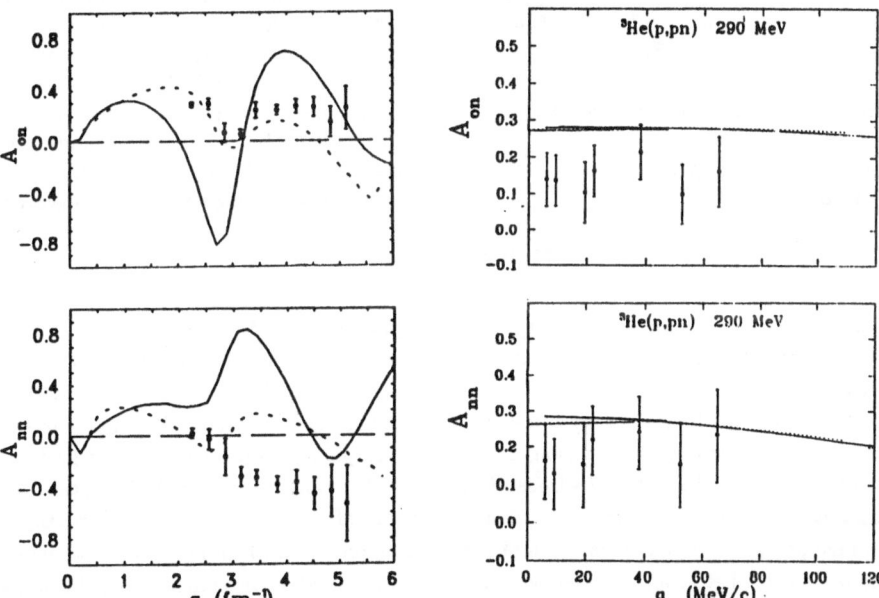

Fig. 3 Some spin observables for p - ^{3}He elastic scattering at 500 MeV.[22] Solid and dashed curves are from Refs. 23 and 24 respectively.

Fig. 4 Spin observables[25] for ^{3}He(p,pn). Curves are PWIA calculations..[26]

^{3}He(p,2p) and ^{3}He(p,pn): Measurements of the breakup channels (p,2p), and (p,pn) have also been made with polarized beam and target.[25] Here results for A_{on}, A_{no}, A_{nn} should give information on the spin momentum distributions of the nucleons in ^{3}He. Such information is very important in view of plans to use a polarized ^{3}He target to provide a polarized neutron target for measurements of the spin structure functions of the neutron. In these cases only PWIA calculations exist.[26] They seem to provide a reasonable description of the (p,2p) spin correlations at 290 MeV where the experiment has been carried out, and predict some sensitivity to the S' component of the ^{3}He wave function at low momentum transfer. However for (p,pn) the agreement is less satisfactory, particularly for A_{on} and A_{nn} (Fig. 4). Apparently final state interactions are important here, though why they should be more important for (p,pn) than for (p.2p) reactions is not clear.

π-^{3}He elastic scattering: Finally there is also nice data on elastic π^+ scattering from polarized ^{3}He with results for both differential cross section and asymmetry at 100 MeV.[27] Such data may give information on the relative importance of spin flip and non spin flip amplitudes. The most striking result is an asymmetry which approaches one at one angle as shown in Fig. 5. In this case a rather detailed calculation has been made.[28] Faddeev wave functions are used for ^{3}He and the Lipmann Schwinger equation solved, resulting in a full DWIA momentum space calculation which seems to reproduce the data, particularly the large value of A_y fairly well. One can also understand this large A_y in terms of a very simple ansatz for the amplitudes which involves only the πN elementary amplitudes.[28] Such an approach, which neglects all final and initial state rescatterings, predicts dramatically different results from the full model at higher energies however.

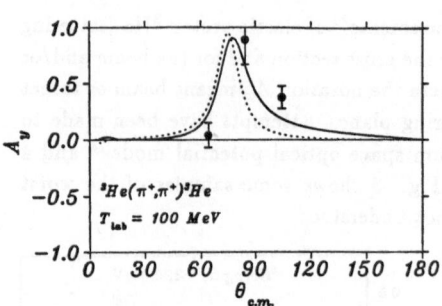

Fig. 5 Analyzing power in π-^{3}He elastic scattering as compared with full DWIA (solid) and schematic (dashed) calculations of Ref. 28.

III. KAON Proposal

We have discussed some of the things now going on at TRIUMF and now want to turn to plans for the future. These plans center on KAON, which is a proposal for a new high intensity kaon factory to be built using the present TRIUMF machine as an injector. KAON stands for Kaons, Antiprotons, Other hadrons and Neutrinos. The machine will be a multipurpose facility producing beams of kaons, pions, muons, protons, antiprotons, neutrinos and possibly other hadrons at intensities typically a factor of 100 greater than existing beams. It thus opens up a tremendous range of new physics. One can investigate strange processes. One can do much of the pion, muon and nucleon physics of the meson factories much better and at energies sufficiently high to allow investigation of the quark substructure of the hadrons. One can also use the high intensities to do high precision experiments such as rare decays and thus obtain some of the same information accessible at the very high energy machines.

The proposal for KAON envisions a complex of accelerators which would produce a primary proton beam of 30 GeV at 100 μa current and nearly 100% duty factor. The protons could be polarized or unpolarized and could be used directly for high energy proton experiments or to produce secondary beams of kaons, pions, muons, neutrinos, etc.

To achieve this result there will be five rings, alternatively dc storage rings or fast cycling synchrotrons, in addition to the present TRIUMF cyclotron which would function as an injector. The present machine currently runs routinely at 500 MeV and 100-140 μa. It is an H^- machine with proton extraction achieved by stripping the H^-. For KAON the extraction process would be modified to extract H^-. The H^- current, 100 μa at about 450 MeV, would then be stripped on injection into the Accumulator ring, the first of the five major rings. The Accumulator collects the beam from TRIUMF in preparation for injection into a Booster. The Booster operates as a synchrotron at 50 Hz and accelerates the beam to 3 GeV. The beam is then transferred to the Collector ring in preparation for the main accelerating stage. The main accelerator is a 10 Hz synchrotron which accelerates the beam to 30 GeV. Finally an Extender storage ring collects the beam and allows for slow cw extraction.

The currently favored layout is shown in Fig. 6 with the main tunnel and booster tunnel both adjacent to the present TRIUMF complex. The Accumulator and Booster would be stacked in one tunnel of radius about 35 meters. The main tunnel containing the Collector, Driver, and Extender rings would be adjacent to this and in a racetrack shape so as to allow long straight sections for extraction and collimation of the beam. The total length of the main tunnel is about 1 km.

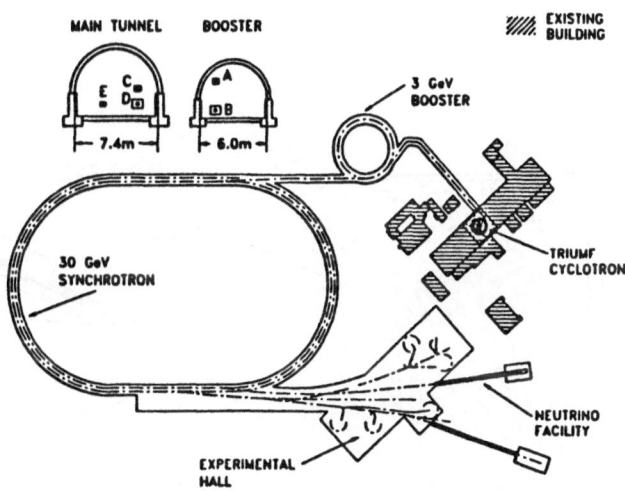

Fig. 6 Proposed layout of KAON accelerators.

The total estimated construction cost is about $700 million. The provincial government of British Columbia has from the beginning been a very strong supporter of KAON and has agreed to provide one third of the total cost. Support has been solicited from a number of other countries, as the facility will provide opportunities for physicists from all over the world. The response has been very encouraging and it appears that at least $200 million will come from other countries, probably mostly in the form of contributions of specific components. In the week following the conference the Canadian federal government announced its willingness to contribute the remaining third of the cost. Thus the major funding is in place and it appears that the project will proceed, although there are some details to work out regarding the sharing of operating costs which will be about $90 million per year. The TRIUMF staff will increase to at least 800, somewhat more than double the present size. Construction will require about five years so one can expect to have beam to experiments in 1997.

IV. Physics at KAON

A number of beam lines and experimental areas have been proposed and are being planned. Exactly which ones are built first will be determined by the types of experiments which seem most compelling. Fig. 7 shows a proposed layout of the experimental area. Kaon beams of a variety of momenta ranging from 0.40 to 20 GeV/c have been designed. Typical fluxes (shown in Table 1) are a factor of 100 greater than currently available. These beamlines can be tuned for pions as well and again give large increases in available flux. Low energy muon and pion channels can be taken off in the backward direction. There are plans for a K^0 beam and for a neutrino facility which can provide a spectrum of neutrinos peaking in the 1-2 MeV range. The primary proton beam can be used directly either in a polarized or unpolarized mode. Antiprotons are possible, as 30 GeV provides sufficient energy, though at the moment there are no definite plans for a dedicated antiproton channel or LEAR type facility. In each case the protons can be extracted from the Extender ring with nearly 100% duty factor or directly from the Driver ring which results in a pulsed structure.

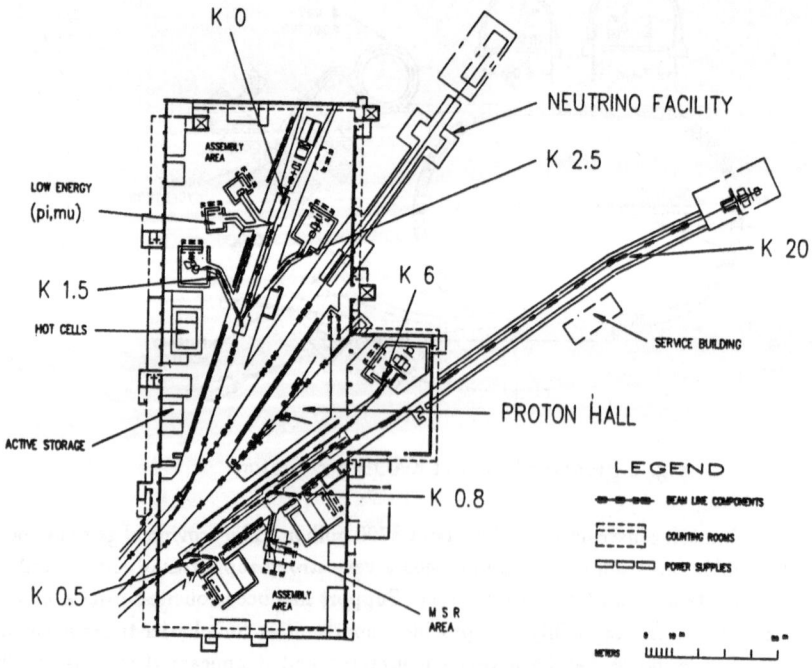

Fig. 7 Preliminary plan for the experimental areas at KAON.

Beam	Momentum (GeV/c)	Typical Flux
K0.55	0.4-0.55	
K0.8	0.55-0.8	10^7-10^8 K$^-$, K$^+$
K1.5	0.8-1.5	10^{10}-10^{11} $\pi^\pm < 2$ GeV/c
K2.5	1.25-2.5	10^9-10^{10} $\pi^\pm > 2$ GeV/c
K6	3-6	10^7-10^8 $\bar{p}$
K20	6-21	
K^0	0.5-10	9×10^7/(GeV/c) @ 2 GeV/c
ν	0-6	10^{12}/GeV-m^2 ν_μ @ 1 GeV/c
		10^9/GeV-m^2 ν_μ @ 5 GeV/c
		ν_e:$\nu_\mu \approx$ 1:100
$\mu^\pm$	stopped	Improvement by factors of 50 - 100

Table 1: Some proposed beams at KAON, their momentum range, and typical range of fluxes.

A machine such as KAON provides such a huge increase in fluxes and capabilities over present machines that the possibilities for new physics are almost endless. There have been a number of workshops to consider various possible experimental programs culminating in a summary meeting in the summer of 1990 and a multivolume proceedings outlining some of the possibilities.[29] I will mention here only a very few areas of the many that might appeal to the few body community.

High energy spin physics: Spin used to be considered an 'inessential complication' but that has proved not to be the case. In fact spin information is often crucial in unraveling details of the interactions. Thus it is planned at KAON to have a polarized proton beam ranging up to the full machine energy of 30 GeV.

For example, interesting effects have been seen both in A and in A_{nn} in high $p_\perp^2$ measurements[30] of polarized $\vec{p}+\vec{p}$ scattering at the AGS. Large polarizations have also been seen in the inclusive production of Λ and other hyperons by protons at high energies.[31] Reactions such as pp $\rightarrow \vec{H}$X, where H is a baryon could be very interesting and in principle might make possible beams of polarized hadrons and investigation of such processes as $\vec{\Lambda}$ scattering on nucleons. The one existing measurement[32] of parity violation in elastic pp scattering at high energies (6 Gev/c) shows an effect an order of magnitude larger than naively expected. All of these investigations would benefit from the higher energy and intensity of the KAON polarized proton beam.

Light hypernuclei: Normally hypernuclei are produced via a (K$^-$,π^-) reaction which allows a Λ to be produced at rest. The (π,K) reaction works also and, although it does not produce the hyperon at rest, it allows the use of π beams which have significantly greater fluxes than K beams. Improved intensities and better beam purity possible at KAON will make both of these reactions easier.

Investigation of light hypernuclei such as the hypertriton would allow one to explore properties of the $\Lambda - N$ interaction in a system which bears many similarities to the three nucleon systems which have been the staple of few body physics. Σ hypernuclei[33] are also possible and with improved kaon beams it may also be possible to produce $S = -2$ states using (K^-, K^+), double charge and double strangeness exchange reactions to produce $\Lambda\Lambda$ hypernuclei or hypernuclei involving the Ξ^-.

Few body kaon reactions: Clean kaon beams open up a number of possibilities for few particle kaon reactions. For example K-N scattering could be done with precision over the wide range of available energies and various spin correlations involving the spins of the initial or final nucleon could be measured. Radiative processes such as $Kp \to \Lambda\gamma$ and $Kp \to \Sigma\gamma$ could be examined in detail as could the reaction $Kd \to \Lambda n\gamma$ which may give information on the low energy $\Lambda - N$ force.[34] One could also use kaon reactions on the deuteron such as $d(K^-, K^0)$ or $d(K^-, K^+)$ to search for a variety of $S = 0, -1, -2$ dibaryons.

We have been able to list only a very few of the almost infinite range of possible areas which could be investigated using the new capabilities, greater intensities, and improved beams at machines such as KAON. There are many other options in, for example, hadron spectroscopy, rare K, π and μ decays, CP violation, neutrino physics, conventional medium energy physics such as π or K scattering from nuclei, and many others which we have not had space to describe.

Hopefully however we have been able to convey an indication of the flexibility of such a machine and of the vast range of interesting physics which can be investigated.

There is a tremendous sense of of excitement about the possibilities at KAON. We hope that many of world physics community will join with us in our new adventure.

This work was supported in part by a grant from the Natural Sciences and Engineering Research Council of Canada. The author would like to thank his many colleagues for providing their data prior to publication.

References

[1] A. Feltham, et al., Phys. Rev. Lett. **66**, 2573 (1991).

[2] J. J. Gammel, P. W. Keaton, Jr. and G. G. Ohlsen, *Polarization Phenomena in Nuclear Reactions* (Univ. Wisconsin press, Madison, WI, 1971), p. 411.

[3] D. V. Bugg, A. Hasan, and R. L. Shypit, Nucl. Phys. **A477**, 546 (1988).

[4] J. Niskanen, Nucl. Phys. **A298**, 417 (1978); Phys. Lett. **141B**, 301 (1984).

[5] B. Blankleider and I. R. Afnan, Phys. Rev. C **24**, 1572 (1981).

[6] D. A. Hutcheon, et al., TRIUMF preprint TRI-PP-91-10, April 1991, submitted to Nucl. Phys. A; Phys. Rev. Lett. **64**, 176 (1990).

[7] E. Korkmaz, et al., TRIUMF preprint TRI-PP-91-11, April 1991, submitted to Nucl. Phys. A.

[8] B. G. Ritchie, et al., Phys. Rev. Lett. **66**, 568 (1991).

[9] J. Spuller and D. F. Measday, Phys. Rev. D **12**, 3550 (1975).

[10] K. Watson and K. Brueckner, Phys. Rev. **83**, 1, (1951).

[11] B. Blankleider, Ph.D. thesis, Flinders University, 1980.

[12] TRIUMF Expt. 372, N. Davison, spokesman.

[13] S. Stanislaus, *et al.*. submitted to Phys. Rev. C, May, 1991

[14] L. Heller, *et al.*, Phys. Rev. C **35**, 718 (1987).

[15] R. Wittman, Phys. Rev. C **37**, 2075 (1988).

[16] C. A. Meyer, *et al.*, Phys. Rev. D **38**, 754 (1988); A. Bosshard, *et al.*, Phys. Rev. Lett. **64**, 2619 (1990).

[17] TRIUMF Expt. 446, A. Stetz and P. Kitching, spokesmen.

[18] B. C. Pearce and B. K. Jennings, Nucl. Phys. **A528**, 655 (1991).

[19] H. W. Fearing, unpublished.

[20] TRIUMF Expt. 569, S. E. Kuhn, spokesman.

[21] J. Congleton and H. W. Fearing, unpublished.

[22] O. Hausser, TRIUMF preprint TRI-PP-91-46, to be published in the Proceedings of the Telluride Conference, March 1991.

[23] R. H. Landau, M. Sagen, and G. He, Phys. Rev. C **41**, 50 (1990).

[24] L. Ray, *et al.*, Phys. Rev. C **37**, 1169 (1988).

[25] A. Rahav, *et al.*; E. J. Brash, *et al.*, to be published.

[26] R. M. Woloshyn, Nucl. Phys. **A496**,749 (1989).

[27] B. Larson, *et al.*, Phys. Rev. Lett. submitted.

[28] S. S. Kamlov, L. Tiator, and C. Bennhold, Few Body Systems (in press); C. Bennhold, *et al.*, TRIUMF preprint TRI-PP-91-56.

[29] Proceedings of a Workshop on Science at the Kaon Factory, D. R. Gill, ed., Vancouver, TRIUMF 1990.

[30] D. G. Crabb et al., Phys. Rev. Lett. **60** 2351 (1988).

[31] G. Bunce et al., Phys. Rev. Lett. **36** 1113 (1976); K. Heller et al., Phys. Rev. Lett. **41** 607 (1978); F. Nessi-Tedaldi, Proc. of the 3rd Conference on Intersections Between Particle and Nuclear Physics, Rockport, Maine, AIP Conf. Proc. **176** 1180 (1988). For a compilation of related data see T. A. DeGrand et al., Phys. Rev. D **32** 2445 (1985).

[32] N. Lockyer et al., Phys. Rev. D **30** 860 (1984).

[33] R. S. Hayano, Proc. of the 12th Int. Conf. on Few Body Problems in Physics, Nucl. Phys. **A508**, 99c (1990).

[34] R. L. Workman and Harold W. Fearing, Phys. Rev. C **41**, 1688 (1990).

Few-Body Systems, Suppl. 6, 446—459 (1992)

Few-
Body
Systems
© by Springer-Verlag 1992

NEWS FROM SATURNE IN FEW BODY SYSTEMS

R. FRASCARIA
I.P.N., 91406 ORSAY Cedex

Abstract :

A brief report on some aspects of mesonic physics in few body experiments performed at LNS (Saclay, France) is presented.

This short contribution cannot be an exhaustive report on the experimental developments at Saturne (Laboratoire National Saturne, LNS) even if it is restricted to few body systems in strong interaction. The Saturne accelerator, associated to the Mimas injector, has delivered about 4200 hours beam time with 93% efficiency for physics in 1990. The new extraction procedure allows two different experiments working simultaneously, at two different energies with a dynamic for the beam intensity of 1 to 10. This has brought forth a large amount of precise data, sometimes very new.

The main characteristics of the beams and the spectrometers of LNS are given in table 1. Other detectors, like ARCOLE, DIOGENE, PINOT and SPES3 are also permanently set up.

As it appears, a rather general tendancy is to investigate the role played by resonances other than the Δ (1236) in different reactions, the excitation of the S_{11} (1535) being the dominant fact in the results presented here. The selection of this N* excitation is obtained by either the choice of energy or momentum transfer, as well as particular channels in the final states. Due to its large branching ratio (BR) to ηN, this hadronic excitation in few body systems is connected to η-meson production, in particular near thresholds in various channels. Another important point of the production of particles near their thresholds is the specific kinematical condition - S-wave production - emphasizing the possible production of resonances or new states of matter. The last part of this short review is concerned with rare events observed in particular reactions, even if these observations lead to surprisingly higher cross sections than expected.

PARTICLE	MAXIMUM ENERGY GeV/NUCLEON	MAXIMUM INTENSITY PARTICLES PER BURST
$p, \bar{p}$	2.9	$10^{12}, 2 \times 10^{11}$ →
$d, \bar{d}$	1.15	$10^{12}, 2 \times 10^{11}$ →
^3He	1.69	10^{11}
^4He	1.15	10^{11}
$^6\vec{\text{Li}}$	1.15	7×10^8 →
$^{12}\text{C}, ^{14}\text{N}$	1.15	10^9
$^{16}\text{O}, ^{20}\text{Ne}$	1.15	1.2×10^8
^{40}Ar	0.82	10^8
^{84}Kr	0.69	2 to 6×10^6
^{127}I	0.62	PLANNED

SATURNE SPECTROMETERS					
	P/Z^{max} [GeV/c]	Angles [deg]	Solid Angle [msr]	Acceptance [%]	Dispersion cm/%
SPES 1	2	0 - 80°	3	± 4	15
SPES 2	0.75	0 - 60°	20	± 17	3
SPES 3	1.4	- 5 - 80°	10	± 40	1.4
SPES4	3.8	- 8 - 30°	0.5	± 3	7

Table 1 : The beams and the spectrometers at LNS

To begin, a first chapter reports on the free excitation of the Roper resonance N* (1440) as a way to get information on its internal structure.

EXCITATION OF THE N* (1440) RESONANCE AND THE COMPRESSIBILITY OF THE NUCLEON

The compressibility of the nucleon is related to the excitation of specific N* (1/2+) resonances having large transition matrix element to the ground state. These monopole transitions are directly related to radial oscillations of the nucleon. The Roper resonance N (1440) is a candidate for such an excitation. The E_0 transition matrix element which cannot be measured in photo-excitation is poorly known experimentally. The excitation of radial modes in hadron excitations can be selectively studied in systems which do not lead to strong excitations of other resonances, like Δ resonances or N* spin-flip modes. This is the case in α + p scattering where the direct α-particle transition operator has spin and isospin transfer equal to zero, inducing only radial excitations of the nucleon. This hypothesis is dramatically demonstrated on fig. 1, where the result of an α + p → α'+X experiment by a collaboration on SPES4 [1] is shown. The excitation spectrum taken for α particles of 7 GeV/c scattering at angles close to zero shows a strong enhancement in the tail of the α projectile excitation peaked at the pion threshold, due to the N(1440) excitation.

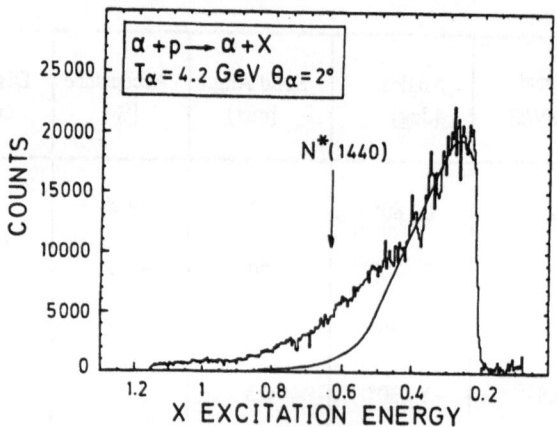

Fig. 1 : Excitation spectrum in α + p → α' + X

The angular distribution of this excitation falls off rapidly with the angle of the scattered alphas. The cross section for the transition is high, indicating a large exhaustion of the energy weighted monopole sum rule. The extraction of the compressibility depends on the nucleon radius itself and calculations are in progress to obtain constraints on the range of the compressibility modulus.

THE ROLE OF THE S_{11} (1535) IN η - MESON PRODUCTION

In a recent past the dominant role played by the S_{11} (1535) has been pointed out in the productions of π and η at backward angles in proton deuteron interactions at energies near and above η threshold [2, 3]. The interpretation of the excitation functions of these productions for very backward angles ($\theta\pi = \theta\eta = 180°$) by LAGET and LECOLLEY [4] singles out the dominance of three body mechanism where the S_{11} (1535) in intermediate states plays the dominant role. Among many others reactions where this important role can be stressed, the pp → ppη and pn → dη are good candidates. In two different calculations LAGET et al. [5] and GERMOND and WILKIN [6] have pointed out the dominance of the ρ-exchange transition amplitude NN → NS_{11} (1535). This occurs due to the large radiative width of the S_{11} resonance which implies through the vector dominance model a large coupling to the virtual ρ. On fig. 2 are plotted the few available data in comparison with the Laget calculation, showing the important role played by the ρ-exchange term.

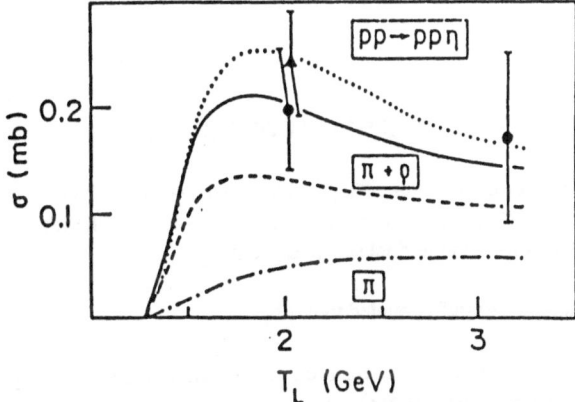

Fig. 2 : The total cross section of the pp → ppη reaction cross section is plotted against the laboratory kinetic energy of the incoming proton. The dotted curve corresponds to the $\pi + \rho$-exchange PW calculation when $\Lambda_\rho = 2.15$ GeV. The full and dashed curves correspond to the $\pi + \rho$-exchange DW calculation when $\Lambda_\rho = 2.15$ GeV and $\Lambda\rho = 2m = 1.88$ GeV, respectively. The dash-dotted curve corresponds to the π-exchange DW calculation.

The pp → ppη channel is being studied at SPES3 [7] and PINOT [8]. At SPES3 for which the angular and momentum acceptances are large the two protons are detected in coïncidence. The identification of the reaction is clear as shown on fig.3. The experimental program has covered the energy domain from $T_p = 1256$ up to 1450 MeV ; the analysis of the data is still in progress. At PINOT which is 4π detector for neutral particles (π°, η, high energy gammas) data have been taken recently in the same range of energy with a clear signature of the η. These experiments should be able to test the ρ - dominance in η - meson production.

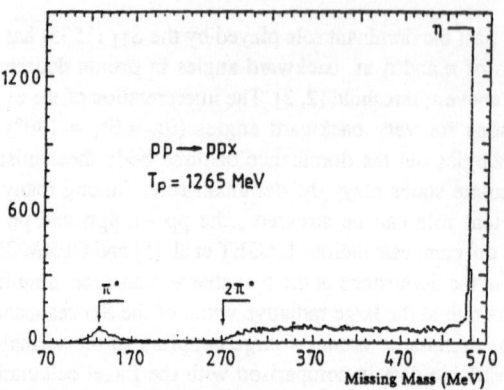

Fig. 3 : Missing mass spectrum measured in
$p + p \rightarrow p + p + X$ at $T_p = 1265$ MeV

THE N - N* (1535) INTERACTION

Single pion production elementary reactions NN $\rightarrow$ NNπ are governed by three partial cross sections : σ_{11}, σ_{10} and σ_{01} where the indexes refer to the total isospin of the nucleon pairs in the initial and the final states respectively.

The reaction pp $\rightarrow$ ppπ° is related to σ_{11} by :

(1) $\sigma(pp \rightarrow pp\pi^\circ) = \sigma_{11}$ while the reaction np $\rightarrow$ ppπ^- is the sum of σ_{01} and σ_{11} as
(2) $\sigma(np \rightarrow pp\pi^-) = 1/2 (\sigma_{01} + \sigma_{11})$.

So, from the simultaneous measurements of (1) and (2), σ_{01} can be extracted. This cross section which is pure I = 0 in the initial channel can only excite N* in intermediates states (Δ forbidden), allowing the study of the NN* interaction. This program is in progress at LNS, the reaction (1) being measured with SPES0 [9] and its modified version SPES0 2π, the reaction (2) being measured with ARCOLE [10]. This last one has ended with cross sections and asymmetry measurements from $T_p = 572$ MeV up to 1134 MeV. The asymetries are shown on fig. 4, the cross sections being not yet available. The SPES0 program on reaction (1) is still in progress, but cross sections

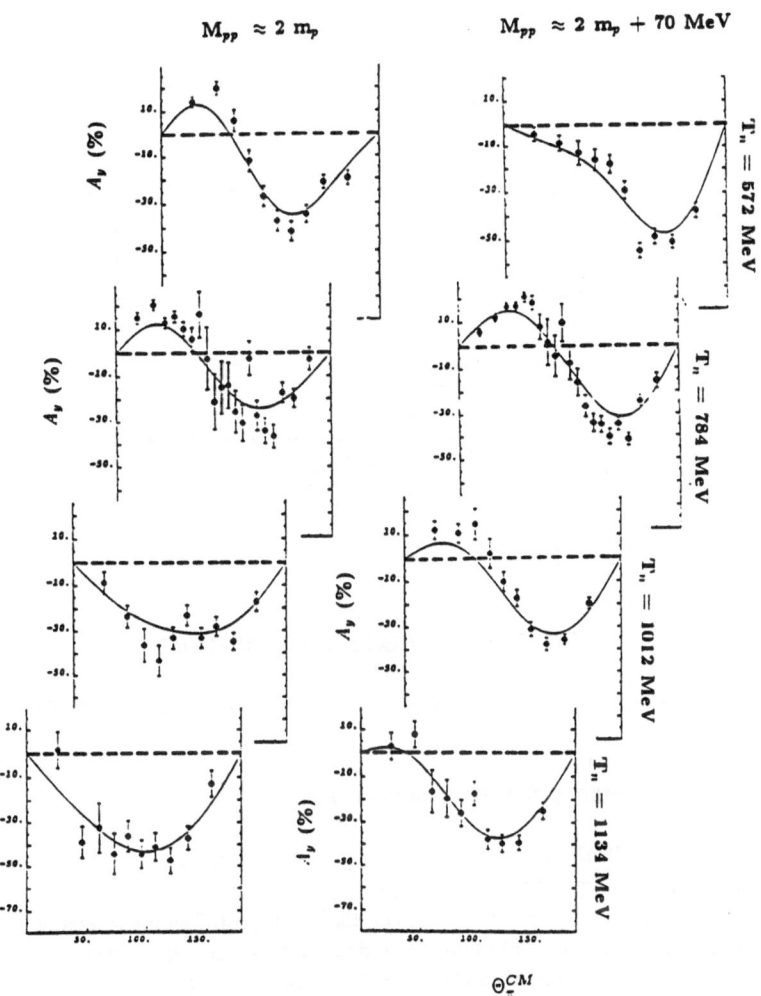

Fig. 4 : Preliminary results of beam asymmetries for four different neutron energies in the reaction np → ppπ⁻. The spectra on the left side correspond to a missing mass $M_{pp} \approx 2m_p$ and on the right side to $M_{pp} \approx 2m_p + 70$ MeV.

452

between T_p = 485 MeV and 560 MeV have already been analyzed. They are shown on fig.5 in comparison with previous np → nnπ⁺ measurements. The resulting σ_{01} cross section is also shown in comparison with a LAGET calculation [11]. This experimental program will explore in next october the energy region between 600 MeV and 1100 MeV where the S_{11} is expected to be dominant and should produce an increase of the σ_{01} cross section.

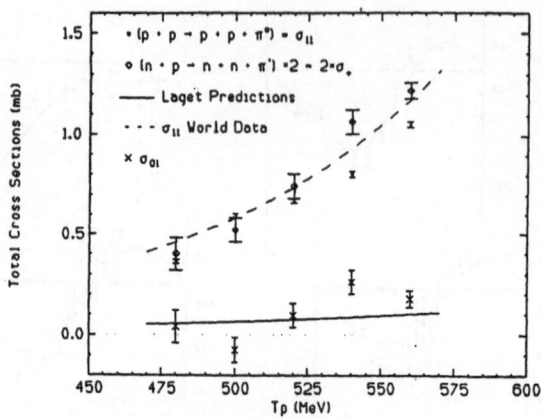

Fig. 5 : The σ_{01} cross section (crosses and fill line)

MESON PRODUCTION NEAR THRESHOLD

This program which uses the pd → ³HeX reaction has mainly two different purposes.

1 - Study of meson production mechanisms.

The pd → ³HeX reaction in the vicinity of η production (X = η) as a function of incident proton energy studied at SPES4 [ref. 12], has shown a very interesting interference pattern with the 2π or 3π channels [see fig. 6]. The interpretation of this strong cusp has been given by Wilkin [13] as due to a strong coupling between the ³He η and ³HeX channels, where X can be a J = 1⁻ 2π state or J = 0⁻ 3π state, both states having isospin 1. The results suggest that the η - ³He interaction is strongly attractive. In a new recent experiment [14], this excitation function of meson production near threshold has been pursued covering the region of ω, η' and φ - mesons. These results are shown on fig. 7. A similar behaviour to the η production is observed for the η' production. The ω-meson is also strongly produced and its $\bar{q}$ dependance - where $\bar{q}$ is the c.m. momentum of the ω - has been studied.

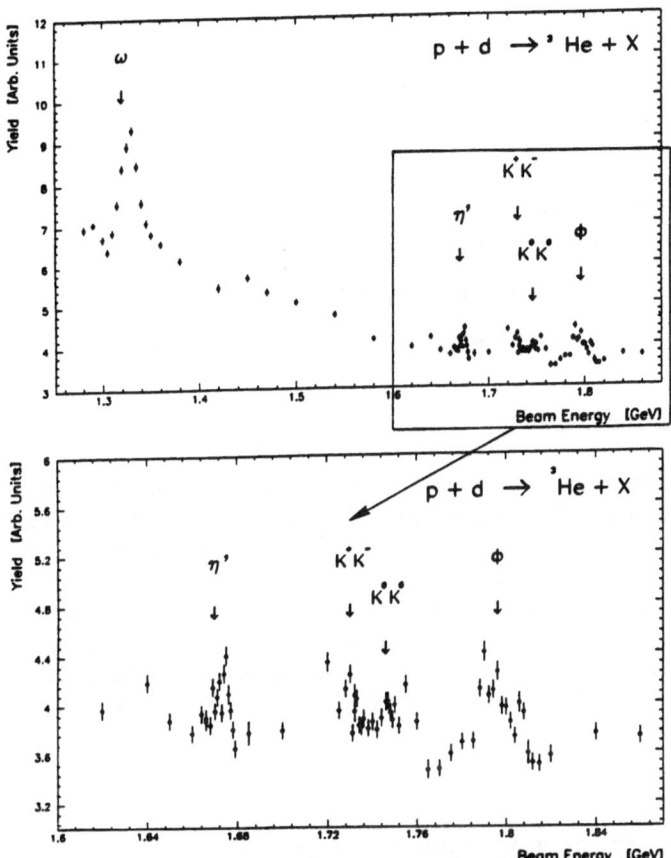

Threshold excitation function obtained at SPES 4 between 1.28 and 1.86 MeV incident proton energy in the reaction pd → ³He X. Beam energy and spectrometer momentum were adapted such a way, that the mesons are produced at rest in the c.m. system. The lower spectrum is a blow up of the upper curve.

Fig. 7 : The pd → ³HeX threshold excitation function from X = ω to X = φ

454

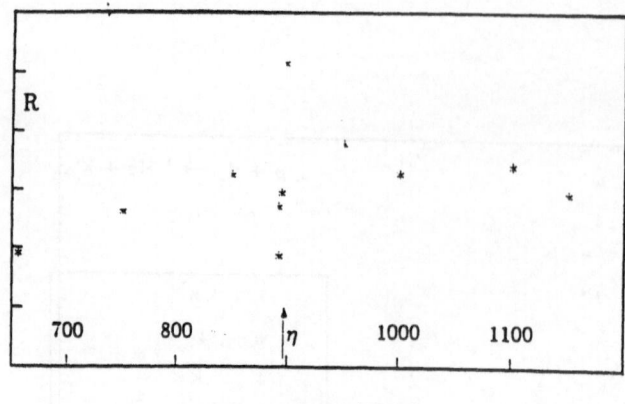

proton energy (MeV)

<u>Fig. 6</u> : The pd → ³HeX threshold excitation function of the η-meson for X

2 - Isoscalar meson structure

As far as the φ-meson is concerned a high rate production is observed as compared to its production in π⁻p → nX reaction [15]. This observation as well as the ω, η, η′ strong excitations lead to the conclusion than I = 0 meson are strongly produced near threshold in pd interactions and than the reaction pd → ³HeX is a good tool to explore the isoscalar meson spectrum. Interesting structures are observed in the vicinity of the $K\overline{K}$ threshold, where the f_0 (975) should appear. Some new information should be extracted from this experiment and its extension to more exclusive measurement (K⁺K⁻ in coïncidence) about the f_0 (975)-a_0 (980) dilemna [16], and in particular on the nature of the f_0 as a possible $K\overline{K}$ molecule.

HYPERON-NUCLEON INTERACTION

The hyperon nucleon scattering data obtained from buble chamber experiments are scarce mainly due to the difficulty which exists in producing and making interact the short-living hyperons in the same detector volume. The associated production of an hyperon-nucleon pair (YN) in few body interactions is a simple way to study the baryon baryon interaction at low relative energies. A high resolution pp → K⁺X experiment has been developed at SPES4 with a performant kaon detection [17]. A first data taking done at T = 2.3 GeV incident kinetic energy and θ = 8, 10, 12° kaon emission angles showed two remarkable features in the missing mass spectra (MM) respectively close to Λ - p and Σ - N thresholds. A theoretical analysis in terms of π and K exchange diagrams with hyperon-nucleon final state interaction (Y - N FSI) [18] was able to reproduce correctly the data. In particular a peak observed at 2131 MeV with a width of 9 MeV (FWHM) was explained as a cusp effect enhanced by a pole close to the Σ - N thresholds.

To follow the momentum and energy dependence of these structures, new data have been taken at T = 2.7 GeV and θ = 13, 16, 20 and 23° with an improved set-up and higher statistics. These results are shown in fig. 8 in comparison with the calculation in the model of ref. 18.

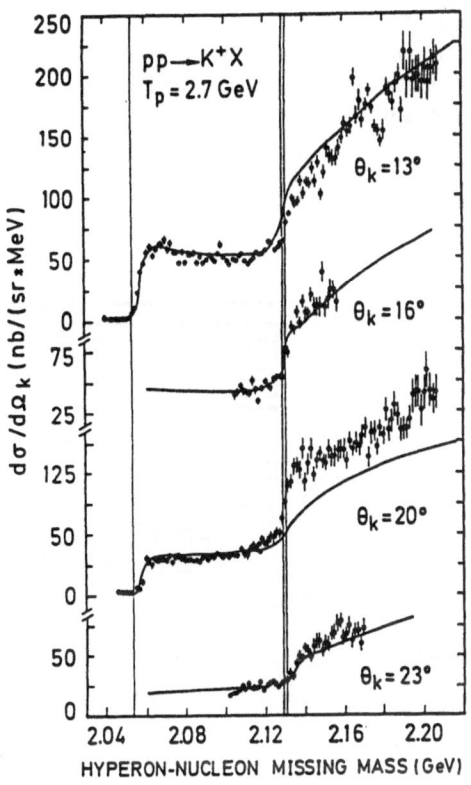

HYPERON-NUCLEON MISSING MASS (GeV)

FIGURE 8

Differences appear with the first results which are mainly :

- The previous structure at 2131 MeV seems to be not observed in these new conditions but one is seen at 2136 ± 2 MeV in the θ = 20° MM spectrum with a width of 16 MeV (FWHM).

- The theoretical model is unable to reproduce this θ = 20° spectrum above the Σ - N thresholds. The T = 2.3 GeV measurements have t values close to zero and yield very similar MM spectra. This is not the case for the new data at T = 2.7 GeV where the t values are very different and far from zero and the behaviour of the MM spectra could be strongly dependent on the 4-momentum t. A separation of the different channels is needed in further experiments. This will be realized with a new experimental area in development at LNS [19].

PROSPECTS FOR RARE EVENT MEASUREMENTS

1 - $\eta \to \mu^+ \mu^-$ decay

The large η production observed in proton deuteron interactions near threshold has lead to the development of a tagged "η-beam". This facility is now working to determine more precisely the η branching ratio in two muons $\eta \to \mu^+ \mu^-$. The η-mesons are tagged by the detection of the ^{3}H at $0°$ in pd $\to$ ^{3}He η at $0°$; a $10^5\eta$/s beam intensity is obtained. The fig. 9 shows the total cross section pd $\to$ ^{3}He η measured near threshold with SPES2 [20]. A first run with a two-arms muon detector as shown on fig. 9 has already accumulate a statistics ten times higher than the original work done by DZHELYADIN et al. [21]. The analysis of the data is in progress.

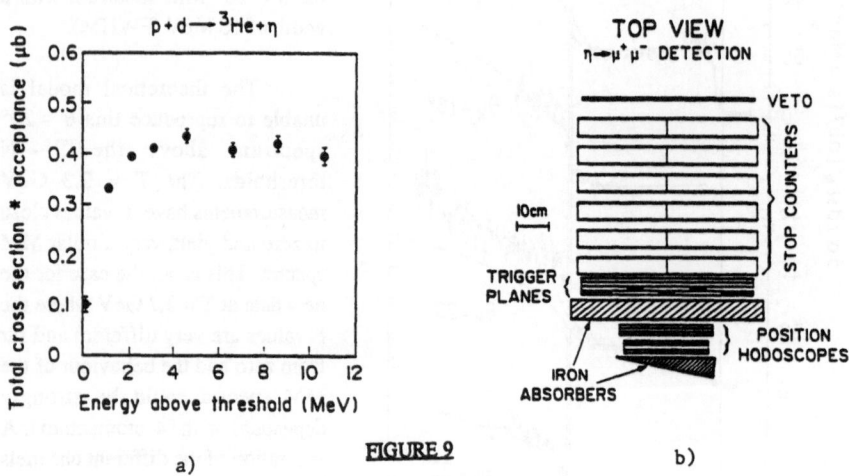

FIGURE 9

a)   b)

2 - Charge symmetry breaking in $d + d \to \alpha + \pi_0$ reactions

The search for the observation of the $d + d \to \alpha + \pi°$ reaction which is forbidden by isospin conservation in strong interaction has started a long time ago at LNS . The characterization of this reaction by the detection of the α particles with SPES4 lead to the determination of an upper limit of 19 pb . A new experiment where the α particles scattered at $\theta = 110°$c.m. and coming from the interaction of deuterons of 1.1 GeV kinetic energy are detected in SPES4 in coïncidence with either one gamma or two gammas from the π° decay by 36 lead glass detectors has succeeded in the measurement of this reaction [22]. The center of mass cross section is found to be $0.97 \pm 0.20 \pm 0.15$ pb/sn from the analysis of two different group of events [one gamma against two gammas]. In the same experiment as a check of the good operation of the set up, the dd $\rightarrow \alpha\gamma$ reaction has been also measured ; the cross section is $0.82 \pm 0.18 \pm 0.10$ pb/sr. The rather high value found for the π° production could be explained as due to the proximity of the η threshold : again the S_{11} resonance could enlarge the production rate of η-meson which by η-π mixing transforms to π_0. This important result should encourage the physics community to pursue such measurements in particular in this energy domain with an improved detector.

3 - Possible observation of $^5_\Lambda$He by means of ^{4}He(p,k$^+$)$^5_\Lambda$He

In an experiment dedicated to the observation of hypernuclei by means of (p,k$^+$) reactions on deuterium, ^{3}He and ^{4}He targets a significant signal of $^5_\Lambda$He production has been observed at SPES4 [23]. After time of flight cuts, trajectory trackings, subtraction of empty target events, a signal appears in the missing mass near the expected value corresponding to the $^5_\Lambda$He mass (only 2 MeV difference). The kinetic energy of the incident protons was 2 GeV and the angle for the detected kaons was 7°. The measured cross section is 0.4 nb/sr which is high for such large momentum transfer reaction (q = 3.6 fm^{-1}). Shimmura et al. [24] was expecting a 0.2 to 0.3 nb/sr value in a calculation with H - J potentiel including short range correlations. This experimental result has to be confirmed : if so, this (p,K$^+$) experiment would allow the investigation of high momentum components of Λ in hypernuclei, and correlatively Λ - nucleon interaction at short distance in nuclear medium. Another explanation could also imply three-body mechanisms with N* having large branching ratio into NK, in a similar way to the mechanism observed in pd $\rightarrow$ ^{3}He η reaction .

There are now convergent ways at LNS to establish reliably the validity of the meson-exchange picture in its medium range. This is achieved by studying the dynamics of light and heavy meson production in complementary few body systems. This is a good way to isolate discrepancies which could occur between precise experimental data and the meson theoretic scheme. If such deviations appear one could have the chance at the end to pull out new phenomenon due to the underlying quark structure of hadrons.

REFERENCES

[1]   IPN, Orsay ; LNS, Saclay ; I.K., Julich, U. Stockolm (INS Varsovie) , (L.I. 220, LNS)

[2]   P. BERTHET, R. FRASCARIA, J.P. DIDELEZ, Ch. F. PERDRISAT, G. PIGNAULT,
        J. BANAIGS, J. BERGER, L. GOLDZAHL, F. PLOUIN, F. FABRRI, P. PICOZZA,
        L. SATTA, M. BOIVIN and J. YONNET
        Nucl. Phys. A443 (1985) 589 - 600

[3]   J. BERGER et al., Nucl. Phys. A443 (1985) 589

[4]   J.M. LAGET and J.F. LECOLLEY, Phys. Rev. Lett. 61 (1988) 2069

[5]   J.M. LAGET et al., Phys. Lett. B 257, 3 (1991) 254

[6]   J.F. GERMOND and C. WILKIN, J. Phys. G : Nucl. Part. Phys. 15 (1989) 437

[7]   IPN (Orsay) ; CRN (Strasbourg) ; LNS (Saclay) Collaboration SPES3

[8]   D.F.S.U. and INFN, TORINO ; U. BARI, LNS (Saclay), CRN Strasbourg ;
        collaboration PINOT

[9]   IPN (Orsay) ; UCS Columbia ; SPN/DSM Saclay ; INFN Frascati, U. Bonn ;
        Collaboration SPES0

[10]   SPN/DSM, Saclay ; ARCOLE detector

[11]   J.M. LAGET et al., Phys. Lett. in press

[12]   LNS, Saclay ; Ecole polytechnique, Saclay, IPN, Orsay (L.I. 197, LNS)

[13]   C. WILKIN, workshop LNS (janvier 91)

[14]   IPN, Orsay ; LNS, Saclay ; ISKP, Bonn (L.I. 183 and 219, LNS)

[15]   D.M. BINNIE et al., Phys. Rev. D8 (1973) 2789

[16]   D. MORGAN and M.R. PENNINGTON, DTP 90/72 and Rheinfeld Workshop 1990,
        St Goar Germany (sept. 90)

[17]   R. FRASCARIA, R. SIEBERT et al., 102A (1989) 561
       R. FRASCARIA, R. SIEBERT, FEW-BODY SYSTEMS, SUPP. 2 (1987) 425
       R. SIEBERT, R. FRASCARIA et al., NUCL. PHYS. A479 (1988) 389 C

[18]   J.M. LAGET, Phys. Lett. B 259, 2 (1991) 24

[19]   SPN/DSM (Saclay) ; IPN Orsay ; LNS Saclay, Coll. Wα M. Williamsburg, Vancouver,
       U. Seoul, TRIUMF, IK, Julich, U. Bonn (LI 241, LNS)

[20]   SPN/DSM, Saclay, LNS Saclay, UCLA, Los Angeles, TRIUMF, Vancouver, GWU, BGU and
       Dubna, collaboration at LNS

[21]   R.I. DZHELYADIN et al., Phys. Lett. 97B (1980) 471

[22]   L. GOLDZAHL et al., Nucl. Phys. A (4 november 1991 issue)

[23]   IPN, Orsay ; LNS Saclay ; ISKP, Bonn (LI 182, LNS)

[24]   S. SHIMMURA et al., Nucl. Phys. A450 (1986) 147 and Prog. Théor. Phys. 76 (1986) 157

Few-Body Systems, Suppl. 6, 460—469 (1992)

© by Springer-Verlag 1992

THE FEW-BODY EXPERIMENTAL ACTIVITY AT THE SASKATCHEWAN ACCELERATOR LABORATORY[*]

D.M. Skopik

Saskatchewan Accelerator Laboratory

University of Saskatchewan, Saskatoon, Saskatchewan S7N 0W0 Canada

Abstract

The experimental program at the Saskatchewan Accelerator Laboratory (SAL) has begun to coalesce around a systematic study of few-body physics. The first round of experiments using the new tagged photon facility has started. The planned tagged photon measurements in the few-body systems, because of the small cross sections and inherently low flux available when tagging, will also make extensive use of two 4π detectors. Analysis of the initial tagged photon experiments is underway, and earlier measurements performed with a c.w. bremsstrahlung beam are now in the final stages of analysis.

Introduction

The first c.w. electron beam was extracted from the SAL pulse stretcher ring (PSR) in January, 1988. Since that time we have endeavored to improve the quality of the extracted beam and the reliability of the accelerator/ring system. Considerable progress has been made in extending the operating range of the stretcher ring; c.w. beams with energies near 100 MeV and at the highest linac energy, 300 MeV, can now be reliably extracted from the PSR. Extraction from the PSR makes use of the fact that electrons lose energy through synchrotron radiation and thus approach a 1/3 integer resonance. At the higher energies, RF power in the ring is used to slow down the extraction by alternatively storing and then spilling the beam between injection cycles. At low energies the injected beam frequency is varied to match the natural spill time. The characteristics of the accelerator-PSR system are given in Table 1. Fig. 1 shows two cycles of the extraction process where single turn injection was used. The bottom trace shows the beam current in the ring determined by a

*This work was supported by the Natural Sciences and Engineering Research Council of Canada (NSERC)

visible light spectrometer and the time distribution of the extracted current as measured by one of the photon tagger focal plane channels. The duty factor of this particular extraction set-up was ~ 60%. Multi-turn injection, which has been successfully demonstrated, will increase the available current and improve the duty factor by improving the time structure around the fill period.

Table 1: PSR Extraction Achievements

Goals	
Energy range	50-300 MeV
Maximum c.w. current	70 µA
Energy spread	±0.01 %
Vertical emittance	0.3 mm-mrad
Horizontal emittance	0.3-0.6 mm-rad
Duty factor	→ 100 %
Results To Date	
Energy range	114-293 MeV
Maximum c.w. current	8 µA
Extracted energy spread	±0.02 %
Vertical emittance	< 1 mm-mrad
Horizontal emittance	< 1 mm-mrad
Duty factor	>65 %
(Maximum stored current)	80 mA
(Minimum c.w. current)	<1 nA

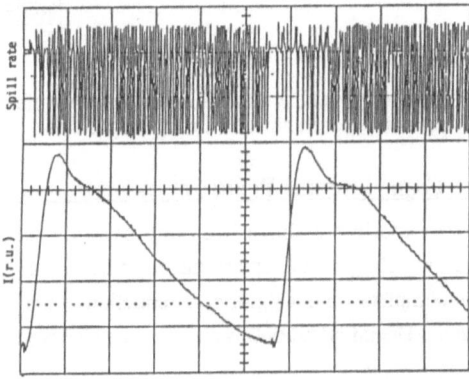

Fig. 1a. Two injection and extraction cycles. The horizontal time scale is 1.0 ms/div. One turn injection with a 300 ns beam was used.

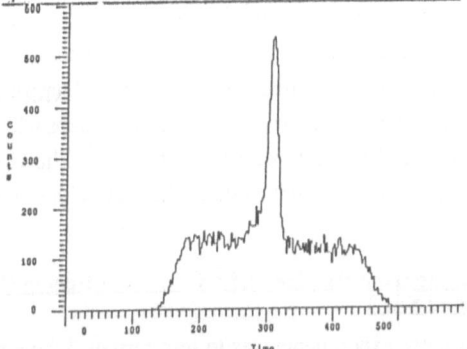

Fig. 1b. Time spectrum for the photon tagger. The TDC is started with an x-arm event, stopped with any tagged channel.

462

A plan view of the experimental areas is shown in Fig. 2. The major electron spectrometers that are part of the laboratory beam lines are a QDD and Clamshell magnet; the latter magnet is dedicated to photon tagging. Experimental area three (EA3) temporarily houses the photon tagger because its beam line was usurped by c.w. bremsstrahlung experiments. The new building addition will allow us to move the photon tagger to an enlarged EA2.

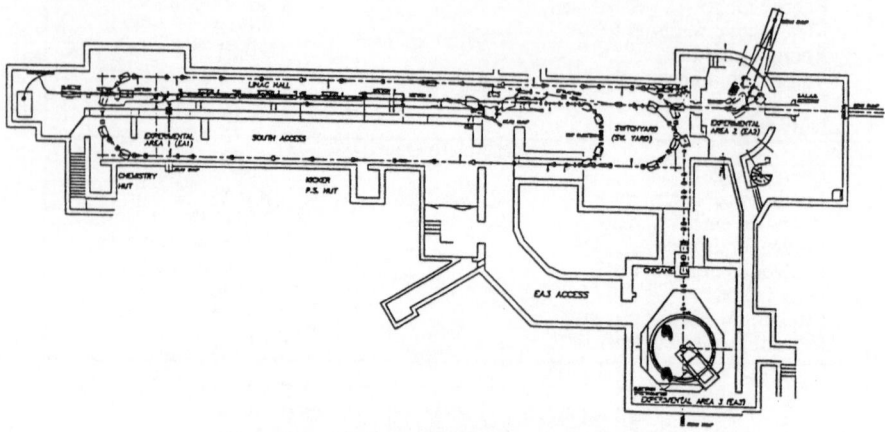

Fig. 2. Plan view of the accelerator/PSR and the experimental areas.
The photon tagger is shown in its future position.

The Experimental Program in Light Nuclei

So far, twenty four experiments have been approved by the Program Advisory Committee (PAC) for a total obligation of approximately 5,000 hours of beam time. Seven experiments have been completed and a significant reduction in the amount of committed beam time has been made. The initial proposals mainly involved experiments in light nuclei but a more diverse effort is evolving.

Central to the tagged photon efforts are large acceptance (nearly 4π) charged particle detectors. One of these was built by the University of Alberta[1] and commissioned with the tagger. It has been used to take data on deuterium and helium. Another large acceptance detector[2] used previously at the University of Illinois has been tested on our tagged photon beam line and will soon be used to measure the reaction $D(\gamma,p)n$. Other important contributions to our light nuclei program are being made by groups with large NaI detectors from Boston University and the University of Illinois. The following describes representative experiments in the few-body systems that have either been completed or are in progress.

Elastic Photon Scattering from ^{2}H, ^{4}He and ^{12}C in the Δ Region (Boston University, University of Illinois and SAL)

The motivation for these experiments was to test current Δ-hole models which have been successful in describing pion scattering. Since pion-nuclear interactions are primarily a surface effect, photon scattering which occurs throughout the nuclear volume is expected to provide sensitive tests of these models. The experiments were done with the Boston University (48×56 cm^2) NaI detector.

The resolution of this detector, 1.8%, allowed us to separate ground state from excited state scattering, and photons produced from (γ, π^0) production. For ^{12}C, the ground state was cleanly resolved except at the highest energy run at 291 MeV; for ^{4}He and ^{1}H the good resolution of the detector serves mainly to eliminate photons from π^0 decay. A typical scattered photon spectrum from ^{12}C is given in Fig. 3; it represents the best data taken to date for this reaction in this energy region.

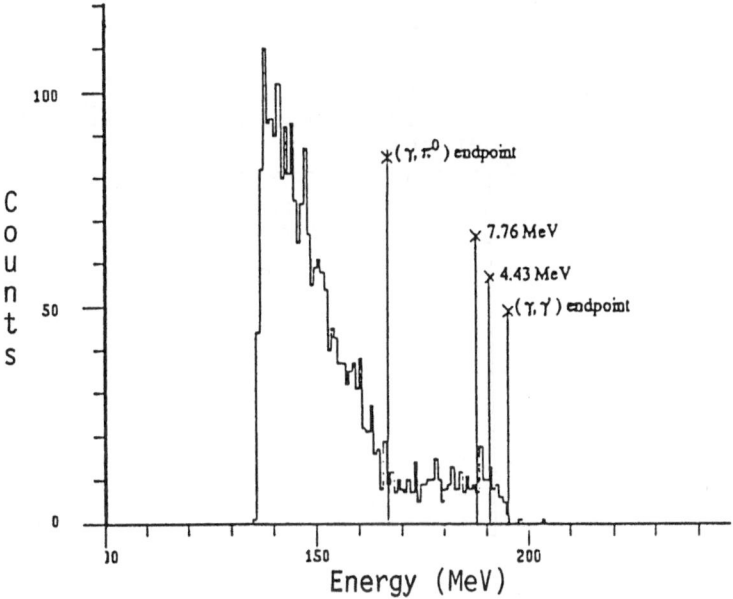

Fig. 3. Scattered photon spectrum for ^{12}C.

The ^{4}He results[3] essentially agree with the earlier Bates work[4]. The hydrogen data, taken over a range of energies from 145 to 280 MeV, are in reasonable agreement with earlier data and a comparison has been made by Petrunkin and L'vov[5] to their dispersion calculation in which they extract values of the magnetic and electric polarizabilities. These data are shown in Fig. 4.

The angular distribution for coherent scattering from ^{12}C near 200 MeV is shown in Fig. 5. The dashed line is a simple form factor model that uses electron scattering and total absorption cross sections. The solid line is the Δ-hole model of Koch et al.[6] These data demonstrate that non-delta contributions to the cross section around 200 MeV are important.

464

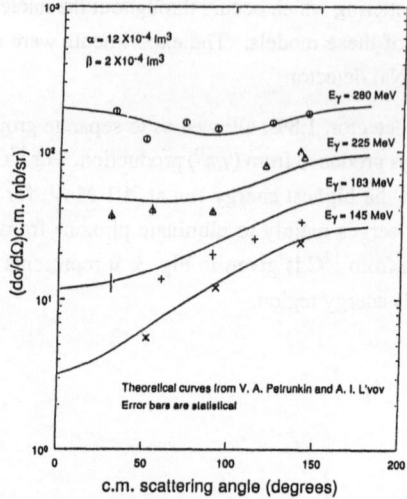

Fig. 4. Measured angular distribution for the reaction $^1H(\gamma,\gamma)^1H$ at four photon energies. The values of α and β are derived from the dispersion calculation of Petrunkin and L'vov.

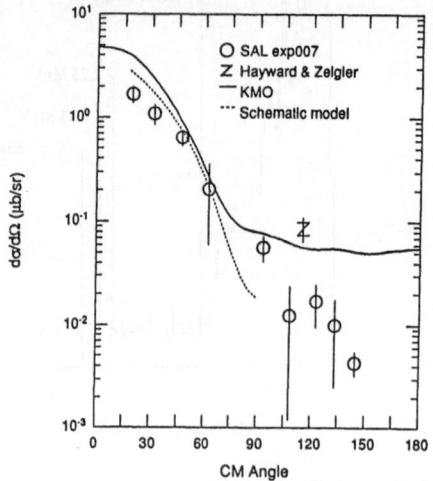

Fig. 5. Angular distribution of coherent photons scattered from ^{12}C. The photon energy was 196 MeV.

Three Body-Effects in the Reaction $^3He(\gamma,pp)n$ (SAL, NIST and the University of Maryland)

An investigation of the three-body breakup of ^3He was undertaken to look for specific evidence of three-body forces. The constraints that can be imposed upon this reaction channel are amenable to minimizing the dominant two-body mechanisms and maximizing the more subtle three-body effects in a greatly restricted reaction phase space. For the p-p-n final state, the two-body mechanisms are suppressed because:

- The p-p pair has no dipole moment to which the photon can couple.
- Charged meson exchange currents vanish.
- The formation of the Δ as an intermediate state is forbidden by spin-isospin considerations.

With these effects in mind Laget has identified kinematic regions where three-body effects are expected to dominate the cross section. These regions of phase space correspond to:

1. Minimizing the E2 absorption by forcing the p-p centre of mass breakup to be at $90°$.

2. Fixing the invariant mass of the p-p system (labelled Q) to be just below π threshold.

3. Restraining the neutron to emerge at $45°$ in the laboratory system.

The basic result of these calculations is that the cross section $\dfrac{d^3\sigma}{dp_2 d\Omega_1 d\Omega_2}$ as a function of neutron momentum is constant when only two-body forces are used, but increases when three-body forces are included. Hence by just determining the trend of the neutron momentum distribution one can make an assessment of whether or not three-body forces are present. The caveat in this calculation is that final state interactions may be extremely important. Indeed in Laget's latest calculations[8], the FSI for the three-body channel have been included and tend to reduce the effect attributed previously to the three-body force diagram which was used. Nevertheless the overall effect is still clear and our latest data will be useful for future comparisons to more sophisticated calculations.

The experiment was done with the high intensity c.w. bremsstrahlung source and four sets of $\Delta E \bullet E$ scintillator telescope arms. The telescopes were arranged so that four different left-right coincidences could give events in regions of phase space corresponding to invariant mass values of Q = 2010 and 2040 MeV. The energy and angle of each correlated p-p pair over-determines the reaction kinematics; therefore the photon energy, neutron angle and momentum could be reconstructed for each p-p event. High pressure gas cells were used with shadow bars to define the active region of the target and to eliminate the background from the cell walls. Runs with ^{1}H and ^{2}H were used for background checks and calibrations. Data shown for this experiment are from on-line analyses.

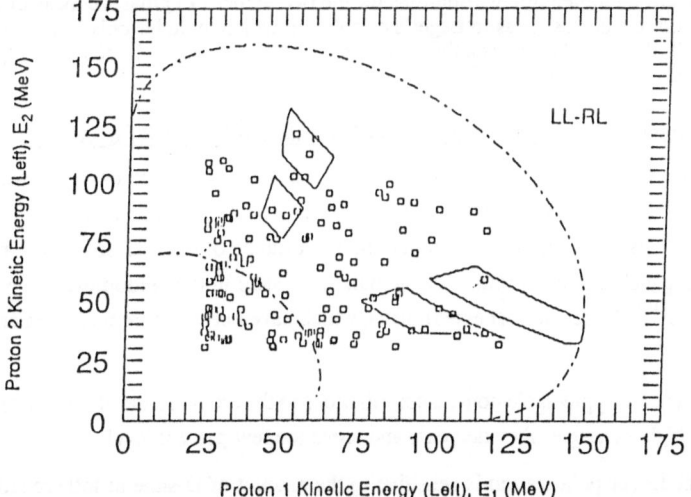

Fig. 6. A "Dalitz" type plot for the ^{3}He(γ,pp) n reaction. A similar plot with deuterium in the cell shows a lack of counts above the pion locus.

A phase space plot of one of the coincidence arms is shown in Fig. 6. The two ellipses define the extreme allowable energy and possible pion production for the three-body system. The quadrilaterals define the kinematic regions which populate $Q = 2010$ and $Q = 2040$ MeV. The background, as measured with hydrogen and deuterium in the gas cell, is negligible.

Our preliminary results for the neutron momentum distribution are shown in Fig. 7. Rather than plot a "reduced" cross section we have plotted our measured cross sections directly and removed the kinematic factor from Laget's predictions. These cross sections have been normalized by making use of Partovi's calculation[9] and a measurement of the $D(\gamma,p)n$ reaction in each of the detector telescopes. In both cases the data show an upward trend which is consistent with the requirement that three-body forces be included in the calculation. The inclusion of final state interactions lowers the cross section but does not remove the trend to increasing cross section as the neutron momentum increases.

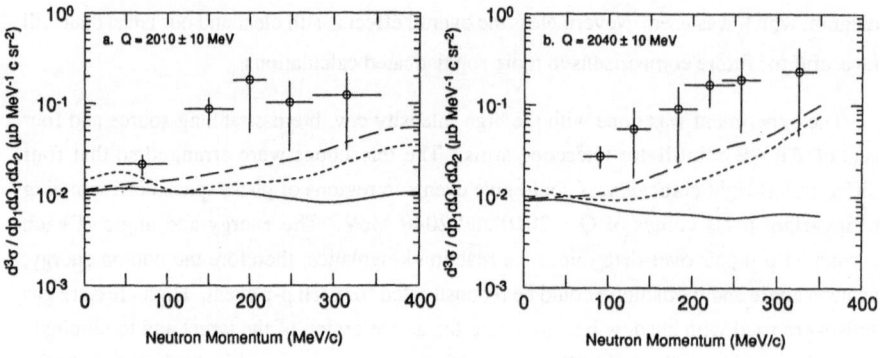

Fig. 7. Neutron momentum distributions for the two values of Q that we measured. The solid line corresponds to Laget's calculation which includes only two-body and FSI, the long dash adds a three-body force, the short dashed line reflects the addition of three-body FSI.

Measurement of the Reaction ^{4}He(γ,dd) (SAL and the University of Alberta)

Because there are identical bosons in the final state, the number of allowed matrix elements is considerably reduced for this reaction. These restrictions have been exploited at low energies in an attempt to deduce the importance of non-E2 contributions to this reaction channel and to determine the amount of D-state in the predominantly $^1S_0$ ground state of ^{4}He[10]. These newest experiments, which have been capture measurements, show that:

1. There are appreciable non-E2 contributions to the cross section at lower energies but at higher energies they decrease and are at the few percent level.

2. The tensor polarization is sensitive to the amount of D-state in the ground state of ^{4}He and for $E_d \sim 30$ MeV and the data show that there is a large capture cross section going to the D-state of ^{4}He.

3. The highest capture energy angular distribution[11] ($E_\gamma = 71$ MeV) retains the $\sin^2(2\theta)$ pattern which is indicative of capture from a ^1D continuum to the ^1S ground state. At $E_\gamma \sim 200$ MeV, the Bonn photodisintegration[12] data show a distinct departure from this angular distribution pattern.

We have started a study of this reaction using SALAD (Saskatchewan Alberta Large Acceptance Detector) in the tagged photon beam line. The nearly 4π acceptance of the detector mitigates the low cross section ($\sim 1\mu$b) for this reaction. The initial data taken with this system have been analyzed and demonstrate the feasibility of obtaining high quality results for this reaction. The SALAD detector is a cylindrical arrangement of 24 concentric $\Delta E \bullet E$ scintillators for calorimetry and two sets of wire chambers for tracking charged particles emitted from a high pressure gas cell. A front view of the detector is shown in Fig.8.

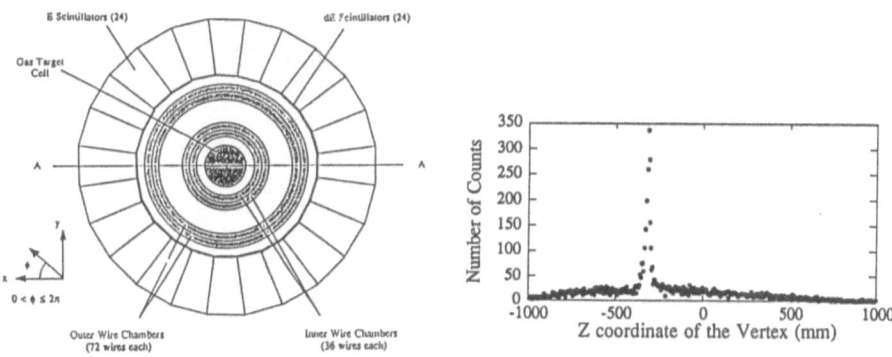

Fig. 8. Front view of the large acceptance detector, SALAD, showing the arrangement of wire chambers and scintillators.

Fig. 9. A plot of the z-position along the beam axis after imposing a 2-track trigger. The solid ^{12}C target is clearly seen in the ^2H gas cell.

A useful feature of SALAD is that solid targets can be placed along the beam axis of the detector. This feature is shown in Fig. 9 which demonstrates the detector's capability of tracking back to a solid target inside the high pressure gas cell. Besides the obvious benefit of being able to verify that the detector's tracking ability is functional, we feel sure that SALAD can be used in a diverse research program, including for example, a study of two-nucleon correlations. The layout for this particular experiment is given in Fig. 10. Eliminating the unused electron beam is important since there is minimal shielding around the detector. Cleanly dumping the primary beam has been demonstrated with the shielding arrangement shown in the figure. We have been able to accumulate time correlation spectra with signal/noise ratios that are as good as that shown in Fig. 1b and deuteron bands that can be distinguished in the $\Delta E \bullet E$ histograms, even for these long scintillators.

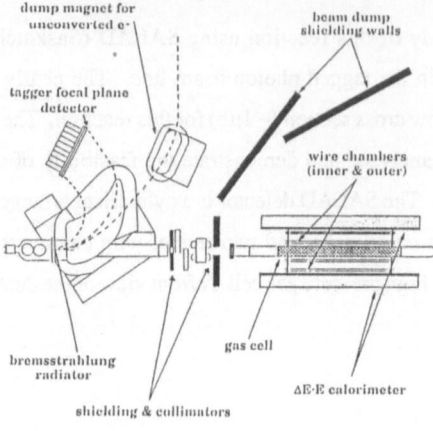

Fig. 10. Experimental layout for the ^{4}He(γ,dd) measurement.

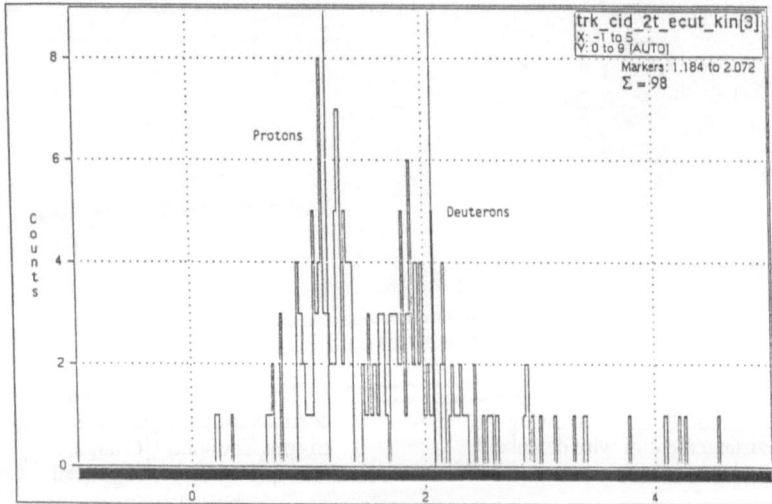

Fig. 11. Linearized p-d separation.

If the rate of energy loss is taken into account and energy and momentum conservation applied, the protons are suppressed and cleanly resolved from the deuterons. This is demonstrated in Fig. 11. As a further constraint we can also require that the two particle tracks be coplanar. Fig. 12 gives the results for re-binning the events, identified as two deuterons from the calorimetry information, but constraining them to be coplanar ($\Phi = 180 \pm 12°$). The minimum energy deuteron that we can clearly identify in SALAD is 60 MeV, which means our measurement will lie between 150 and 260 MeV. The angular range that we expect to cover is 20 to 160 degrees.

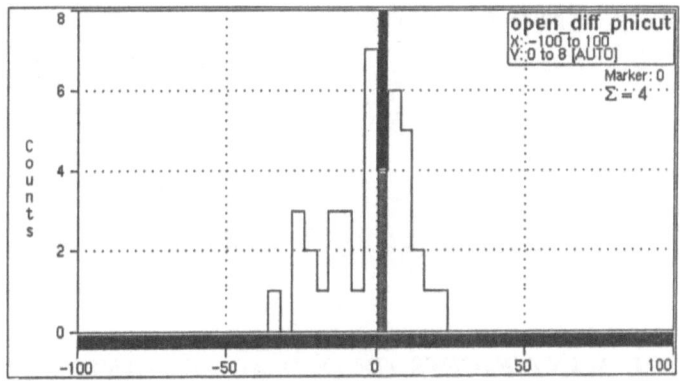

Fig. 12. Φ opening angle difference.

Outlook for Future Experiments in Light Nuclei at SAL

The few-body physics program at SAL is dominant in the future plans of the laboratory. The bulk of the experiments approved by the PAC are in few-body physics and they range from pion production on nucleons to investigating two-body breakup modes in light nuclei. This effort is made possible by the active collaboration of our many outside users who bring their expertise and their detectors to the laboratory.

Acknowledgments

Three of the experiments that I have described are the thesis projects for these Ph.D. graduate students working at SAL: Adam Sarty [^{3}He(γ,pp)n], Ru Igarashi [^{12}C(γ,γ)] and Grant O'Rielly [^{4}He(γ,dd)]. They especially, the other researchers, scientific and technical staff members who work at SAL, are gratefully acknowledged for their work which has brought us to this point.

References

[1]   P. P. Langill, M.Sc. Thesis, University of Alberta (1989).

[2]   P. D. Harty et al., Nucl. Instrum. Methods, A297, 415 (1990).

[3]   D. Delli Carpini, et al., Phys. Rev. C43, 1525 (1991).

[4]   E. J. Austin et al., Phys. Rev. Lett. 57, 972 (1986).

[5]   A. I. L'vov, private communication to H. S. Caplan, SAL.

[6]   J. H. Koch, E. J. Moniz and N. Ohtsuka, Ann. Phys. 54, 99 (1984).

[7]   J. M. Laget, Journal of Physics G, 14, 1445 (1988).

[8]   J. M. Laget, private communication.

[9]   F. Partovi, Ann. Phys. 27, 79 (1964).

[10] H. R. Weller et al., Phys. Lett. B 213, 413 (1988).

[11] W. K. Pitts et al., Phys. Rev. C 39, 1679 (1989).

[12] J. Arends, et al., Phys. Lett. B 62, 411 (1976).

Few-Body Systems, Suppl. 6, 470—477 (1992)

THE MANY FACETS OF QCD AT LOW AND MEDIUM ENERGIES: THE ROLE OF THE NUCLEUS

European Collaboration for a Continuous Wave Electron Accelerator at 15-30 GeV

J.-F. MATHIOT

*Division de Physique Théorique ***
Institut de Physique Nucléaire
F-91406 Orsay Cedex

Abstract

In the perspective of a European collaboration for a continuous wave electron accelerator in the 15-30 GeV energy range, we present the first ideas and numerical simulations on the most important and original aspects of this project. We emphasize the role that the nucleus can play in unravelling the quark dynamics at low and intermediate energies.

GENERAL SPIRIT

Detailed long-range plans for nuclear physics have recently been prepared in several European countries. They form the basis of an in-depth analysis of general trends and priorities showing that unravelling hadronic structure at the fundamental level and establishing its connection to nuclear physics are among the crucial issues.

The nature of the strong interaction is understood and there exists in Quantum ChromoDynamics (QCD) a basic theoretical framework for its description. On the experimental side however, it is difficult to imagine one single experiment to understand its most original regime, *i.e.* the non-perturbative regime of hadrons and nuclei. The many facets under which QCD appears at low and medium energies have to be understood at the same time: one is not looking for new particles. In this respect, this study is similar, in spirit, to the study of the microscopic structure of nuclei: there is not a single experiment to understand how nuclei are build up from nucleons and pions!

* Unité de Recherche des Universités Paris XI et Paris VI Associée au C.N.R.S.

Why nuclear physicists are interested by such a project? To understand this, let us recall the historical developments in the microscopic description of nuclear structure. The derivation of effective interactions in nuclei has shown that their density dependence, non-locality, and spin-orbit component are strong. With the advent of many-body techniques, phenomenological nucleon-nucleon potentials have been deduced in order to understand these features. As a first consequence, many-body forces appear to be essential to describe the saturation properties of nuclear matter. All these aspects have been synthetized with the use of nucleons and pions as relevant physical degrees of freedom in nuclei. Realistic nucleon-nucleon potentials, meson-exchange currents, and three-body forces are now understood in terms of these degrees of freedom. All these notions trace back in fact to two important properties of the underlying theory of the strong interactions: the confinement of colour degrees of freedom (quarks and gluons), and the spontaneous breackdown of chiral symetry in the physical vacuum. Such properties have received much attention during the last past years in the search for deconfinement transition and restoration of chiral symetry at high density and/or temperature. We have now to face the understanding of these crucial properties at low and medium energies.

The questions one has to solve are of course very difficult, and this may be the reason why progress in this domain is very slow. Not much can be learned from "global" experiments, *i.e.* experiments which do not give access to the appropriate scale. This may be compared with the impact that bubble chamber experiments had in the identification of processes at the scale of a few centimeters. *One is now looking for a medium to reveal quark dyna-mics at the scale of a few fermis!* This scale is precisely the size of a nucleus. *The present challenge, both for theorists and experimentalists, is to learn how one can use the nucleus as a detector!*

PHYSICS PROGRAM

The many facets of QCD can appear in two different ways, which can be characterized by the following questions:

i) How do colourless configurations of valence quarks form? This refers more generally to the study of the hadronic wave function. This aspect will be illustrated by "colour transparency" experiments.

ii) How can a (coloured) quark escape from a hadron? This aspect is illustrated by the study of hadronization and, for inclusive measurements, by deep inelastic scattering experiments.

These two aspects can be investigated from three different points of view. First of all, it is necessary to know the elementary processes. As examples related to the first aspect, we can mention the study of baryon resonances at high momentum transfer, and the study of weak neutral form factors to directly measure the strangeness content of the nucleon. Both examples will be already investigated at CEBAF, but may reveal new informations at higher momentum transfer: many important resonances appear only beyond CEBAF capabilities (the Roper resonance for instance), and the asymetries in parity-violating experiments to measure the weak neutral form factors of the nucleon increase with Q^2. The study of fragmentation functions in free space, or the determination of parton distribution functions are simple examples related to the second aspect. This latter program has been investigated for many years.

The second point of view is rather fashionable to day. It concerns the modification of hadron properties (masses and radii for instance) in the medium. The typical experiment is of course the study of parton distribution functions in nuclei (the well-known "EMC effect"). This program can be extended to the study of baryon resonances, and meson properties in the nuclear medium. It is not yet clear how to extract original informations on the structure of QCD from these experiments. They may just reveal the fact that the nucleus is not made up of free nucleons. This is already known from the study of the microscopic structure of the nucleus in terms of mesonic degrees of freedom. New experiments may concentrate on x>1 structure functions, *i.e.* in a kinematical region vorbbiden in free space.

The third, and most important point of view to our mind, is the use of the nucleus as a detector, as we already emphasized in the introduction. Two examples will be developed in the next section. We refer the reader to ref.[1] for more details.

Such an ambitious program necessitates of course quite new experimental tools. As far as the luminosity is concerned, *i.e.* the ability to measure very small cross-sections, the most important parameter for fixed target experiments is the intensity of the electron beam. With the new technological developments, the maximum ("peak") intensity that the detectors can handle could be increased by a factor of 10 to 100 compared to what has been done up to now in this energy range. For all coïncidence experiments however, the maximum intensity has to be limited in order to reduce the fortuitous over real events ratio. To increase the luminosity, one has therefore to increase the averaged intensity, *i.e.* the duty cycle: the fortuitous over real events ratio is inversely proportional to the duty cycle. It is thus essential to have a duty cycle as close to one as possible. At present, the available accelerators are of two kinds. For fixed target experiments at SLAC, ($E_e \approx 22$ GeV) the luminosity can be rather high (of the order of 10^{36}), but with a very low duty cycle (2×10^{-4}), which prevents any coïncidence experiments to use the maximum luminosity. On the other hand, experiments could be performed on storage rings (at HERA for instance with the 30 GeV electron beam), with a duty cycle close to one. However, the use of internal gaz target of rather low density, which is necessary to preserve the quality of the electron beam, limits the luminosity to about 10^{31}-10^{32}. *Combining a high duty cycle with high luminosities, one can gain a factor of 10^4 to 10^6 for coïncidence experiments.* It is easy to realize that such a jump in the luminosity really opens up a new field of research.

The development of polarized electron sources which can provide polarization higher than 80 % opens also a new domain of investigation. The study of baryon resonances, electroweak processes, and "colour transparency" for instance will gain a lot with polarization degrees of freedom. As we shall see in the next section, going to high energy is necessary to excite a new flavor degree of freedom: charm. Finally, and following the pionering work done at SLAC, such facility may provide a K_o beam of very high quality.

SELECTED TOPICS

1 -"Colour transparency" et al.

1.a) "Colour transparency"

"Colour transparency" is an original phenomenon predicted some years ago by S. Brodsky and A.Müller [2]. In the case of interest for this project, this notion refers to quasi-elastic scattering off a nucleon in the nucleus at high momentum transfer. The typical momenta involved in these experiments range from 5 to 15 [GeV/c]^2. What happens at high momentum transfer as compared for instance to quasi-elastic scattering at 0.2-1[GeV/c]² ? Two different processes are of importance here. At the electromagnetic coupling (see Fig.1) first of all, the high momentum Q transferred to the individual quark in the nucleon has to be equally splitted among the three valence quarks to end up with a nucleon in the final state. As a result of this naive argument, the transverse size of the nucleon configuration which is probed by the photon is of the order of 1/Q. The second equally important aspect is related to the subsequent propagation, in the nucleus, of the proton configuration selected by the photon. This propagation is represented by the dashed area in Fig.1. Since one demands to recover a nucleon in the detector far away from the nucleus (quasi-elastic scattering), and since the elastic cross-section is proportional to the transverse size squared of the object (dipole moment for a colourless object), the only configurations which can survive at high momentum transfer are precisely those of small transverse size. Eventhough the selection of the small component of the nucleon wave function by the photon is not efficient enough, the nucleus acts first as a *filter* of these components. The cross-section of quasi-elastic scattering in the nucleus, as compared to elastic scattering on a free nucleon, *i.e.* the factor T given by

$$T = \frac{\sigma[eA \to e'N(A-1)]}{Z\sigma[ep \to e'p']}$$

should therefore go to one when the momentum transfer increases. This is the so-called "colour transparency" effect. At small Q^2, T is of the order of 0.5, as given for instance by a Glauber type calculation. At high momentum transfer, the cross-section of the pointlike configuration ("minihadron") is small so that the nucleus is transparent and the T factor goes to one. As a function of the nucleus mass number A, one should see a decrease of T for large nucleus at moderate Q^2, while at high Q^2 the T function should be rather independent of A.

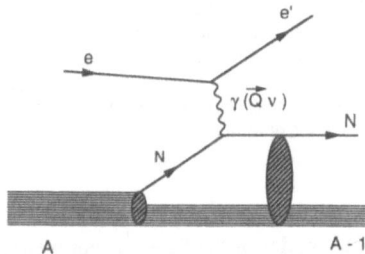

Figure 1: *Quasi elastic scattering off a nucleon in the nucleus, and subsequent propagation in the residual nucleus.*

Starting with the following decomposition of the nucleon wave function

$$|N> = Z_0 |qqq> + Z_1 |qq\bar{q}q> + Z_2 |qqqg> + ...,$$

"colour transparency" selects the small component of the wave function, *i.e.* in first approximation Z_0 |qqq> since the other components are damped by at least one power of the strong coupling constant α_s which is small at high momentum transfer. In the detector of course the nucleon has recovered its physical size back. The evolution of the nucleon wave function from its pointlike component to its physical size corresponds therefore to the transition from the well defined and calculable perturbative component to the full intrincacy of the non-perturbative nucleon wave function. The nucleus acts in this case as an *analizer* of the evolution.

This process has an important consequence for the design of the accelerator and detectors. Since we ask for quasi-elastic scattering off a nucleon in the nucleus, the energy (and momentum) resolution of the beam (and detector) should be high enough, and at least $m_\pi/2$ to be able to separate the quasi-elastic peak from the production of a single pion. Actually, it is possible to do much better! Since "colour transparency" affects the rescattering of the nucleon in the nucleus, it is essential to separate the events which correspond to no-rescattering from the ones which correspond to one rescattering. The behavior of both contributions as a function of Q^2 should reveal "colour transparency" in a very precise way. We show in Fig.2 three different simulations, corresponding to three different resolutions for the beam energy (σ_E/E) and detector resolution (σ_p/p), assuming for simplicity no Q^2 dependence of the cross-sections.

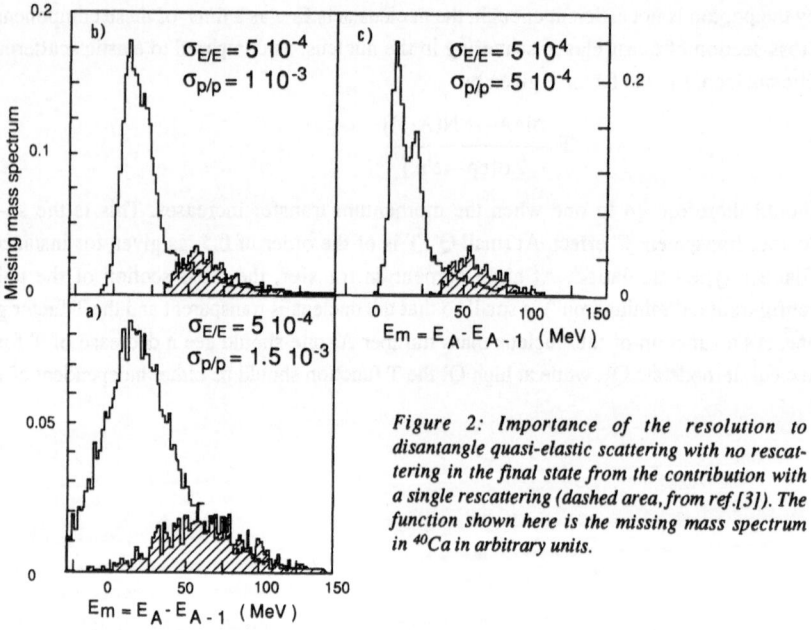

Figure 2: Importance of the resolution to disantangle quasi-elastic scattering with no rescattering in the final state from the contribution with a single rescattering (dashed area, from ref.[3]). The function shown here is the missing mass spectrum in ^{40}Ca in arbitrary units.

The first figure (a) corresponds to the characteristics of the SLAC experiment NE18 [4], while the two others correspond to typical resolutions one can think of for new detectors and machine. It is clear from these figures that a resolution of about $5\ 10^{-4}$, both in the energy of the electron beam and in the momentum of the detector, would be enough to disantangle the various processes and would give access to the full physics.

There is a whole set of experiments that can be attached to these ideas. We list here the most important ones

* Quasi-elastic scattering with a polarized beam,
* Electroproduction of π and Δ,
* Exclusive virtual Compton scattering,
* J/Ψ, Ψ' and χ electroproduction.

We will only discuss briefly the last two processes.

1.b) Exclusive Virtual Compton Scattering

Exclusive virtual Compton scattering differs from the quasi-elastic process indicated in Fig.1 only by the fact that we demand in addition a real photon in the final state, with a given energy ω. This real photon is emitted by the outgoing proton which has been hit by the virtual incident photon. In the case of virtual Compton scattering on a free nucleon, the same scaling laws as for the electromagnetic form factors hold. Therefore, the same arguments of colour transparency should also hold.

However, the main difference lies in the fact that the outgoing photon takes away an energy ω, and thus one can completely separate the kinematical conditions at the initial electromagnetic vertex (quasi-elastic kinematics), from the kinematics of the outgoing proton. There is of course a price to pay, and this is in the counting rates. This process is down by a factor α compared to quasi-elastic scattering because of the coupling to the outgoing photon. Detailed simulations are thus necessary to decide the faisability of this experiment [5].

1.c) J/Ψ, Ψ' and χ electroproduction

This part of the program illustrates the use of the charm degree of freedom in such processes. It can be related to "colour transparency" arguments by two different aspects:
      * Production mechanism and selection of pointlike components by hard gluon exchange,
      * Propagation of a $c\bar{c}$ pair in the nuclear medium and formation of a bound state.
Note that the use of spin observables can be of great help to disantangle the different production mechanisms. This process is also important to set the energy scale. We show in Fig.3 the J/Ψ production rate as a function of its transverse momentum, with two different choices of the incident electron beam energy and for a luminosity of 10^{37} nucleons/cm^2/s. While the counting rates at 15 GeV are sufficient to start doing physics, they are much more comfortable at 20 GeV, and extend to much higher transverse momentum.

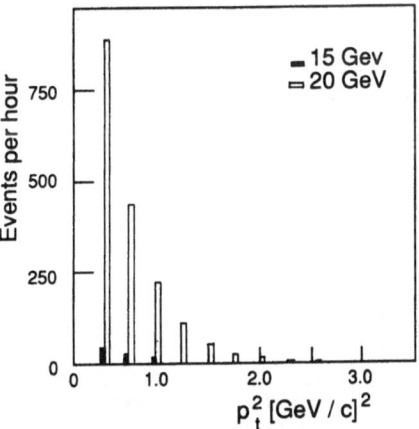

Figure 3 : Production rate of J/Ψ at 15 and 20 GeV as a function of the transverse momentum squared [7]. This illustrates the fact that near threshold the phase space varies rapidly with the energy.

2-Hadronization et al.

When a quark is kicked out from a nucleon, a colour string develops between this quark and the diquark remaining in the target nucleon. Due to the confinement of the colour electric and magnetic fields, this string has a given transverse size of the order of 1fm. The energy which

476

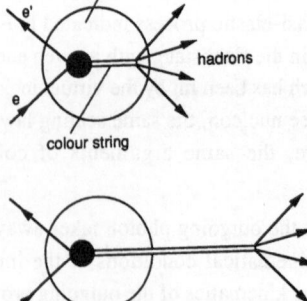

nucleus

e'

hadrons

e

colour string

Figure 4 : Cartoon indicating the use of the nucleus to access to the colour string in the hadronization process.

can be stored in the string is thus directly proportional to the length of the colour string, the scale being given by the energy density per unit length, of the order of 1 GeV/fm. The characteristic length of the flux tube, before it breaks and produces hadrons in the final state, is therefore L≈ν [fm /GeV], where ν is the energy transferred to the quark.

Up to now, one had only "global" information on the distribution of hadrons in the final state. These are the so-called fragmentation functions which give the distribution of hadrons as a function of their nature (flavor degree of freedom), and the kinematical variables $Q^2, \nu, p_T^2, y=E_h/\nu$. The properties of the colour string are almost unknown from a quantitative point of view. To have access to these properties, it is necessary to analyse the process at a scale of a few fermis. We represent in Fig.4 a cartoon indicating how the nucleus indeed can serve as a *detector* to give access to this domain. At very high energy (Fig.4 bottom), the colour string develops far away from the nucleus. The process is therefore almost tranparent to the size of the nucleus. On the other hand, if we adjust the size of the colour string to the size of the nucleus (Fig.4 top) by varying ν and A, the process can develop inside the nucleus. Looking at the attenuation of the produced hadrons as a function of ν and the size of the nucleus, one can for instance have some indication on two important properties of the

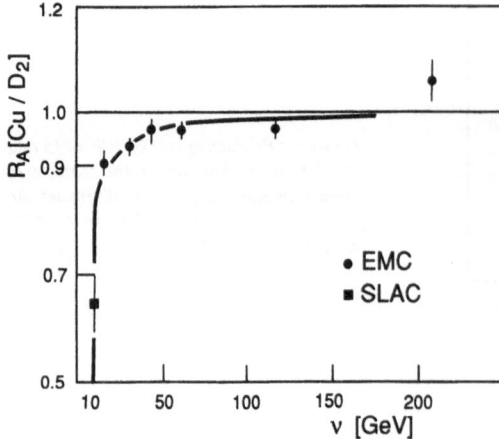

Figure 5: Attenuation of hadrons produced on a Cupper target as compared to a deuteron target. The data are taken from ref.[10]. The line is drawn to guide the eyes.

colour string: its exact length, and its cross-section in the medium, or in other words the cross-section of a quark travelling inside a nucleus [8].

These qualitative arguments are in fact confirmed experimentally [9]. We have indicated in Fig.5 the ratio of leading hadrons produced on a Cupper target with respect to a deuteron target, as a function of the transferred energy. As expected, the attenuation is very strong for small energy transfer (v<30 GeV), *i.e.* in a region where the colour string develops mainly inside the nucleus. These data are the first indications that the idea of using the nucleus as a detector can indeed be useful. However, to extract more precise information, and have now a quantitative understanding of the underlying processes, it is necessary to have precise data as a function of the various variables at our disposal: Q^2, v, p_T^2, y and A, using in addition flavor degrees of freedom (strangeness and charm production).

The design of a large solid angle detector is here necessary to detect all particles in the final state. First simulations show that about 10^7 particles are produced per second at an incident electron beam of 15 GeV, mainly protons, charged pions and photons coming from the decay of neutral pions [11]. This rate is reasonable, and almost three orders of magnitude smaller than what is expected for the LHC experimental program.

PERSPECTIVES

The idea of using the nucleus as a detector at a scale of a few fermis is certainly challenging for both theorists and experimentalists. The various aspects presented above are certainly very qualitative and need to be precise in the near future. However, the first ideas serve already as a strong motivation to go further.

REFERENCES

[1] Proceedings of the European Workshop on Hadronic Physics with Electrons beyond 10 GeV, Dourdan, October 7-12,1990, Nucl. Phys. **A532** (1991)

[2] S.J. Brodsky, in Proceedings of the Thitteenth International symposium on Multiparticle Dynamics, Eds. E.W. Kittel, W. Metzger and A. Stergion, (World Scientific, Singapore, 1982)
A. Müller, in Proceedings of the Seventeenth Rencontre de Moriond, Ed. J. Thran Than Van (Editions Frontières, Gif-sur-Yvette,1982)

[3] H. Borel et al., in ref.[1] and private communication

[4] R. Mc Keown, in ref.[1]

[5] H. Fonvieille, in progress

[6] S.J. Brodsky and P. Hoyer, in ref.[1]

[7] V. Breton and C. Martin, in ref.[1]

[8] A. Bialas, in ref.[1]

[9] N. Pavel, in ref.[1]

[10] EMC Collaboration, report at XX High En. Phys. Conf. (Munich 1988)
L.S. Osborne et al., Phys. Rev. Letters, 40 (1978) 1624

[11] P. Bertin and G. Fournier, private communication

Few-Body Systems, Suppl. 6, 478—484 (1992)

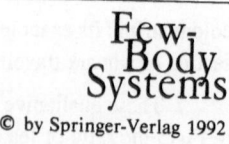

© by Springer-Verlag 1992

PRESENT AND FUTURE FEW NUCLEON STUDIES WITH COOLER BEAMS AT THE INDIANA UNIVERSITY CYCLOTRON FACILITY

John M. Cameron

Indiana University Cyclotron Facility

2401 Milo B. Sampson Lane, Bloomington, IN 47405, U.S.A.

Abstract

The advantages of electron cooled stored beams for studies of few nucleon systems are outlined. Some examples from the present IUCF program in Few body studies in Atomic and Nuclear physics are presented. The possibility of extending these techniques into the multi-GeV energy range is discussed.

Introduction

The use of electron cooling to compensate for beam heating due to nuclear targets inserted in a stored polarized positive ion beam is a new technology particularly applicable to the study of few nucleon systems. Phase space shrinking by electron cooling has already been exploited for several such studies at the Indiana University Cooler ring. The advantages of the technique include extremely good beam quality and circulating currents in the milliampere range yielding high luminosity even while using ultra thin targets. The Cooler ring also has rampable magnets and an RF accelerator cavity allowing acceleration. Some attributes are: the possibility of very narrow energy steps; detection of charged particles at very small angles (including zero degrees); use of polarized internal targets having no contaminants; heavy recoil detection; and tagging of secondary beams of pions and neutrons. The broad research program underway at the Indiana Cooler demonstrates the versatility of this type of facility.

Electron Cooling and High Luminosity with Thin Targets

The IUCF Cooler, shown in Fig. 1, is the first electron cooled storage ring to be operated with internal targets where an equilibrium is set up between heating of the beam by the target and the electron cooling [1]. Under these circumstances the

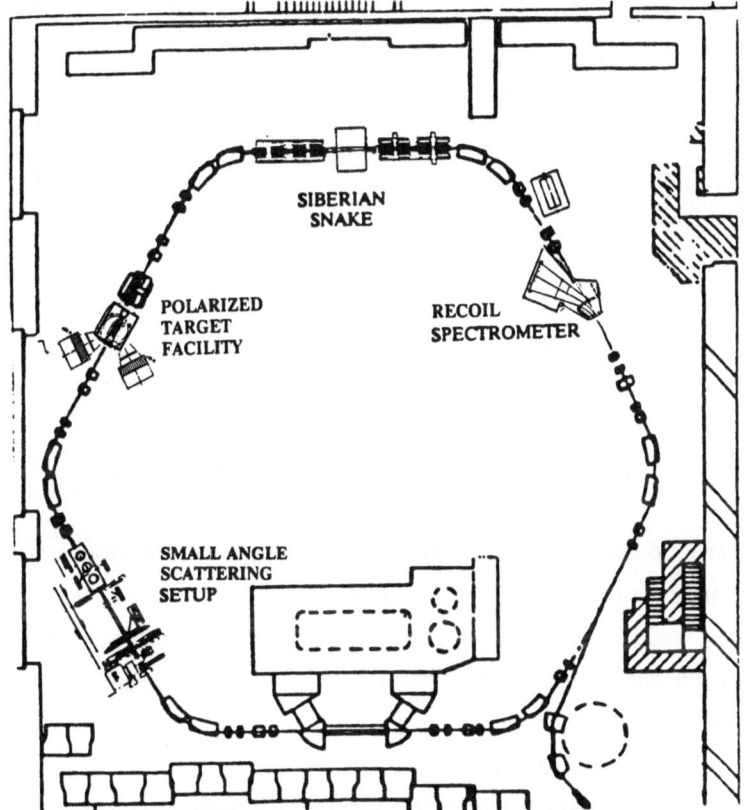

Fig. 1. The IUCF Cooler ring and present experimental facilities.

average luminosity is constrained by the accelerator aperture and the electron cooling rate. Considerable effort has gone into exploring the limits of the maximum luminosity available in the ring which will be proportional to the product of the target thickness (t) times the beam lifetime when the target is present (τ) [2]. The relationship between the target thickness and the quantity $\sigma_{loss} \approx \frac{1}{t\tau}$ for 185 MeV protons on a H_2 gas target is shown in Fig. 2. A broad minimum, corresponding to a luminosity maximum, is found centered around target thicknesses in the range 10^{14}-10^{15} atoms cm^{-2}.

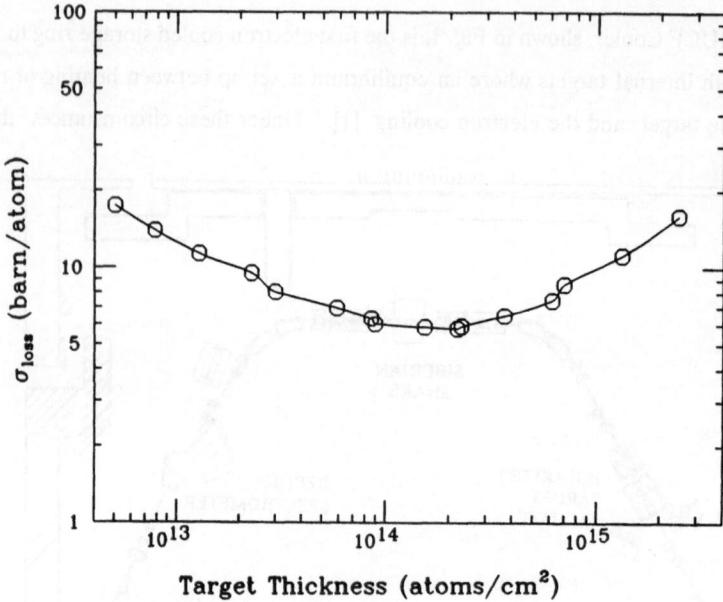

Fig. 2. The relationship between σ_{loss} and target thickness.

The electron cooling rate depends necessarily on the electron beam energy spread and current as well as on the colinearity of the proton and electron beams. Measured cooling rates are in good accord with expectations which sum the contribution due to the electron beam plus additional contributions due to the reduction of the effective angular divergence of the electron beam due to the magnetic field of the confining solenoid [3].

The most efficient method of injection for the Cooler, by stripping of electrons from ions, is not available for the fully stripped polarized positive beam and kicker injection with RF stacking is used. Both modes benefit from the use of cooling during the injection phase to reduce the emittance of the stored beam. Luminosities approaching 10^{31} cm^{-2} s^{-1} are now available with hydrogen targets and unpolarized beam, while the factor of 10 less presently available with polarized beam should be removed when a new high intensity polarized source [4] comes on line next year.

Few Nucleon Studies with the Cooler

The qualities of cooled stored beams are particularly well suited to studies of few nucleon systems and, not surprisingly, they dominate the present program at the Cooler. As it is not possible to review the whole program here, only a few examples are given.

Dielectric recombination (DR) or inverse Auger transition occurs when electron capture is accompanied by simultaneous ionic excitation. It is being studied for light atomic and molecular systems using the dense electron beam in the Cooler region as a target [5]. Energy resolution in the atom c.m. system of a fraction of an electron volt is obtained by adjusting the relative velocity of the ion and electron beams. The investigation of simple systems such as $^4He^+$ and HD^+ are of particular interest; the former because here the electron-electron interaction which mediates the DR is relatively strong compared to the electron nucleus interaction which alone dominates in heavier systems, while the latter has an important influence on the ionospheric charge density and chemistry of the earth and other planets.

Among the first nuclear physics experiments carried out with the Cooler were measurements of single charged and neutral pion production in the pp → NNπ system close to threshold [6,7]. This is perhaps the most basic problem in nuclear physics; since any consistent treatment of the NN interaction must allow for real pion emission when energetically allowed, and thus must be a three-body theory. Total cross sections and angular distributions have been measured for many energies below E_p = 325 MeV. The total cross section data for the pp → pp$\pi°$ reaction, some 10 times more precise than previous measurements, vary monotonically with energy as shown in Fig. 3. These energies are sufficiently low that only the L_{NN} ℓ_π = Ss exit channel contributes and theories involving intermediate delta formation are not applicable. Somewhat surprisingly, one finds that the cross section predicted incorporating only the Born term (or pion emission from one proton leg) in DWBA is a factor of 5 below the data and may indicate that non-resonant pion rescattering involving both nucleons is also important.

Threshold effects or cusps are seen in several reactions and it is interesting to note that no such effect is seen in the pp →pp$\pi°$ channel at the energy threshold for pp →dπ^+ or pnπ^+. Another experiment is planned to check for such effects in the pd →^{3}He$\pi°$ channel where the pd →^{3}Hπ^+ channel opens [8].

Another area where the Cooler has particular advantage is in studying spin

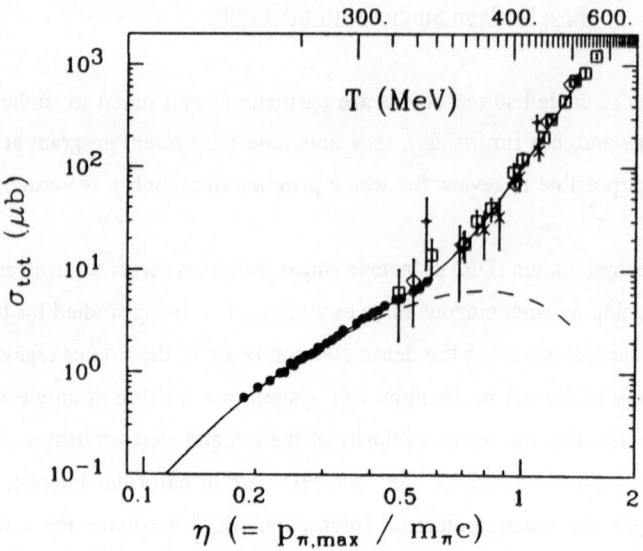

Fig. 3. Total cross section for the reaction pp →ppπ°, the new data are shown as solid circles. The dashed curve is as calculated in DWBA using only the Born term and Ss partial waves after normalization by 5.2. The effect of higher P waves for the πN system which dominate at higher η are included in the solid curve.

correlation parameters using the high circulating beam current with thin windowless polarized targets. These targets consist of polarized ions in a storage cell open to the circulating beam. Tests have shown that a tube of 8 mm internal diameter and 24 cm long can be used without adversely affecting the ring acceptance aperture. When this cell was fed with hydrogen gas at flow rates of 10^{17} atom s^{-1}, as can be obtained from an atomic beam source, the target thickness of the cell was measured to be 10^{15} atom cm^{-2} [9]. This thickness is ideal for Cooler application as described earlier. Experiments with both polarized ^{1}H and ^{3}He are being planned [9,10]. In another feasibility test [11] a polarized proton beam has been used with an unpolarized hydrogen gas target to measure asymmetries in the Coulomb-nuclear interference region at scattering angles down to 2.5° lab. The results are compared to recent phase shift predictions [12] in Fig. 4.

A Multi-GeV Stored Polarized Beam Facility

An area of nuclear physics which as yet remains unresolved is that of the short-

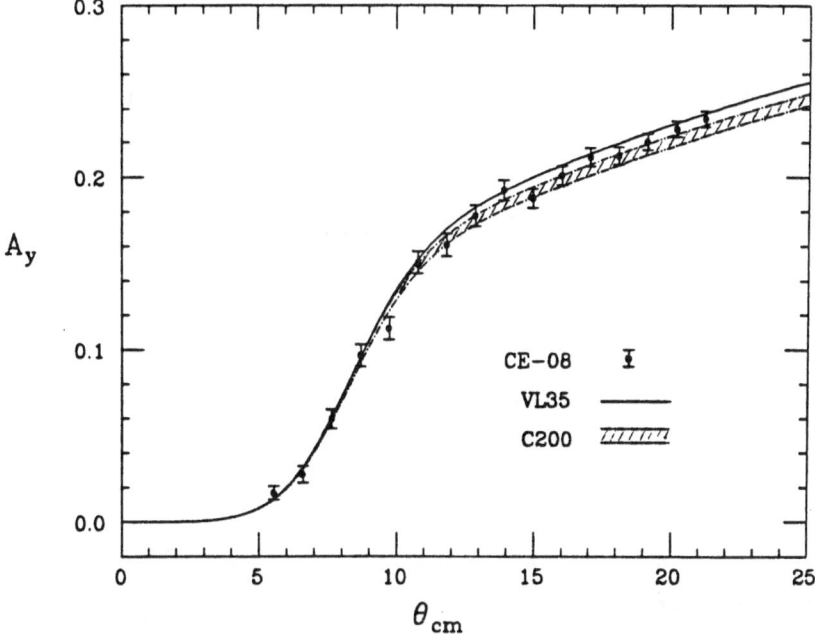

Fig. 4. The analyzing power for pp elastic scattering measured at the Cooler compared to the energy dependent (CL35) and energy independent (C200) phase shift predictions of Arndt.

range behavior of the strong interaction. This will in part be the mandate of high energy c.w. electron facilities such as CEBAF. However, complementary information can only be obtained with polarized hadronic beams. Examples of particular interest to the present audience would include detailed investigation of not only the NN interaction but also that of other hyperon-nucleon channels. Another example would be to clarify the size of the parity violating part of the NN interaction for the energy range from 1 to 15 GeV. A polarized beam of excellent quality with luminosities of 10^{33} cm^{-2} s^{-1} would allow many such experiments to be performed.

We are considering a facility which would be based on an extension of the present Cooler technology; i.e., we would use cooled beams with internal targets in a slow cycling synchrotron. Polarized beams would be made readily available by incorporating Siberian Snakes in the lattice. The machine aperture and the associated overall cost can be kept low because of the extremely small emittance of the beam injected from the Cooler. We are now embarking on a detailed evaluation phase of this opportunity, with the goal of having a complete design available in two years.

References

1. Pollock, R.E.: Ann. Rev. Nucl. and Part. Sci. 41 (1991), to be published.

2. Derenchuk, V., Pei, X., Pollock, R.E., Ross, A., Sloan, T., Sperisen, F.: IUCF Scientific and Technical Report (1991), p. 141.

3. Ellison, T.J.P.: Proc. 1991 IEEE Part. Acc. Conf. (San Francisco, 6-9 May 1991).

4. Wedekind, M., Brown, R., Derenchuk, V., Ellison, T., Friesel, D., Hicks, J., Jenner, D., Pei, A., Petri, H., Schwandt, P., Sowinski, J: IUCF Scientific and Technical Report (1991), p. 193.

5. Tanis, J.A., Clark, M.W., Dittner, P.F., Ellison, T., Forest, J.L., Foster, C.C., Graham, W.G., Haar, R.R., Jacobs, W.W., Mowat, J.R., Rinckel, T., Schneider, D., Stockli, M.P.: Proc. 17th Int. Conf. on Phys. of Elect. and Atom. Coll. Brisbane, Australia (July 1991).

6. Meyer, H.O., Ross, M.A., Pollock, R.E., Berdoz, A., Dohrmann, F., Goodwin, J.E., Minty, M.G., Pancella, P.V., Pate, S.F., v. Przewoski, B., Rinckel, T., Sperisen, F.: Phys. Rev. Lett. 65, 2846 (1990).

7. Daehnick, W.W., Dytman, S.A., Brooks, W.K., Hardie, J.G., Flammang, R.W., Bland, L., Jacobs, W.W., Rinckel, T., Pancella, P.V., Brown, J.D., Jacobsen, E.: IUCF Scientific and Technical Report (1991), p. 52.

8. Nann, H., Segel, R.E.: IUCF Experiment Proposal CE24 (1990), unpublished.

9. Haeberli, W., Pitts, W.K., Price, J.S., Ross, M.A., Meyer, H.O., Pate, S.F., Pollock, R.E., v. Przewoski, B., Rinckel, T., Sperisen, F., Sowinski, J., Pancella, P.V.: IUCF Scientific and Technical Report (1991), p. 158.

10. Sowinski, J., van den Brand, J.: IUCF Experiment Proposal CE25 (1990), unpublished.

11. Pitts, W.K., Haeberli, W., Knudson, L.D., Price, J.S., Meyer, H.O., Pancella, P.V., Pate, S.F., Pollock,, R.E., v. Przewoski, B., Rinckel, T., Sowinski, J., Sperisen, F.: IUCF Scientific and Technical Report (1991), p. 3.

12. Arndt, R.A., Hyslop, J.S., Roper, L.D.: Phys. Rev. D 35, 128 (1987).

Few-Body Systems, Suppl. 6, 485—492 (1992)

Few-
Body
Systems
© by Springer-Verlag 1992

THE SCATTERING OF POLARIZED ELECTRONS FROM POLARIZED ^{3}He: FIRST EXPERIMENTS AND IMPLICATIONS FOR THE MEASUREMENT OF G_E^n

A. M. Bernstein

Physics Department and Laboratory for Nuclear Science

Massachusetts Institute of Technology

Cambridge, MA 02139 USA

The fact that the neutron is not an elementary particle was discovered long ago when the magnetic moment was measured and found not to be equal to the Dirac value of zero. Subsequent quasi-elastic electron deuteron scattering data showed that the neutron magnetic form factor was not point like.[1] This internal structure can be explained as due to the presence of (at least) up and down quarks. If the up and down quark wave functions have the same radial dependence the charge form factor of the neutron $G_E^n(q^2) = 0$; if they are not equal then $G_E^n \neq 0$. We do not have much accurate data for G_E^n, but the measurement of the charge RMS radius from thermal neutron electron scattering[2] demonstrates that the up and down quark wave functions are not equal. This has been explained by the tensor part of the color hyperfine interaction[3,4] causing the up and down quark wave functions to differ.[4,5] A measurement of $G_E^n(q^2)$ would provide a test of nucleon models.[6]

A measurement of G_E^n is difficult not only because free neutron targets do not exist but in addition because the magnetic scattering dominates the electron-neutron cross section. The classical way to separate electric and magnetic scattering is via the Rosenbluth technique;[1] one measures the relative magnitudes of $G_M^{n\,2}$ and $G_E^{n\,2}$. Since G_E^n is very small compared to G_M^n it is not possible to make an accurate measurement. To solve this problem a "super Rosenbluth" technique[7] must be employed; this involves observing either the scattering of polarized electrons from polarized neutrons, or observing the polarization of the recoil neutrons, where there is a term in the cross section which is proportional to $G_M^n \, G_E^n$. At the present time the most practical experiments employ the $D(\vec{e}, e' \, \vec{n})$ and $^3\overrightarrow{He} \, (\vec{e}, e')$ reactions. Model calculations have been performed[8,9] and kinematic regimes have been identified in which G_E^n can be measured. Experiments for both the D^{10} and the polarized ^{3}He targets[11] have been performed at Bates during the past two years. Experiments are also being planned at Mainz.[12] In this report I shall focus on the Bates experiments with polarized ^{3}He targets.[13,14] For this case the inclusive cross section can be written a $d^2\sigma/d\Omega dE' = \Sigma + h\Delta$ where h is the electron helicity and Σ (Δ) are the spin independent (spin dependent) cross sections. In an experiment with polarized beams and targets one measures the asymmetry $A = \Delta/\Sigma$.

By necessity, in order to determine neutron properties, one becomes involved with the complexities of few body systems and the reaction mechanism for quasi-free scattering. These complexities involve the spin structure of the initial state wave function[15], and effects due to final state interactions and meson exchange currents. We need to study known quantities of the few body nuclei with super Rosenbluth techniques to ascertain that we know enough about the wave functions and the quasi-free reaction mechanism. In particular, model calculations have identified kinematic regimes for which the neutron properties are the major uncertainties and other regimes for which the few body effects dominate. These latter regimes provide a testing ground for the model calculations. For example in the quasi-free scattering of polarized electrons from polarized ^{3}He the quasi-elastic peak has the minimum model dependence and is the best place to measure the neutron form factors. When the target polarization is parallel to $\vec{q}$ (the momentum transfer) the asymmetry $A_{T'}$, due to transverse- transverse interference, is primarily sensitive to G_M^n; when the target polarization direction is perpendicular to $\vec{q}$ the asymmetry $A_{TL'}$, due to longitudinal-transverse interference, is sensitive to $G_E^n \, G_M^n$. Therefore utilizing these two directions of the target polarization allows us to test the reaction calculation and, if the test is successful, to extract a value for G_E^n. As we shall discuss below a better measurement of G_M^n is also needed.

The use of polarized ^{3}He targets is useful to the extent that the spin is carried by the neutron and that the proton spins cancel. This question has been investigated[15] using the best available nucleon-nucleon and three body interactions. The results are that for the kinematics of our experiment, $q^2 \approx -0.2 \, (\text{GeV/c})^2$, the ratio of the proton to neutron contribution is -0.138 ± 0.020 for $A_{T'}$ and 1.15 ± 0.17 for $A_{TL'}$; the errors take into account the range of values obtained using different two and three body potentials.[15] These "wave function theoretical errors" are smaller than those caused by our uncertainties in the nucleon form factors, particularly G_M^n, so they do not seem to be a problem.

Polarized ^{3}He targets were made by two techniques: both are based on optical pumping with circularly polarized resonant laser light. A CalTech-MIT group polarized the ^{3}He nucleus via the hyperfine coupling of the metastable state in the ^{3}He atom[13]; a Harvard-MIT group polarized Rb atoms and used spin exchange to polarize the ^{3}He nuclei.[14] Both targets achieved up to 30% polarization; the Rb spin-exchange target had a density of 1.1×10^{20} atoms/cm^3 which was an order of magnitude higher then the other target. The Rb spin exchange target[14] is shown schematically in Fig. 1. Both targets had a two cell geometry; the optical pumping chamber is separated from the target chamber. The two targets differ and for the sake of brevity the following description is of our spin-exchange target. The Rb is polarized using a Ti: Saphire laser with an output of $\approx$ 3W (more laser power would result in higher polarization). The polarization of ^{3}He is measured by NMR; up to 30% was achieved for the experiment (over 45% in bench tests). The time constants are $\approx$ 10 hours for the polarization to build up (or down) and $\approx$ 10 min to transfer between the two cells. The target is made of special glass to maximize the depolarization time which typically was 8 to 15 hours. The beam of up to 20 microamps had a negligible effect on the polarization (in agreement with our estimates). The experiment was typically run with 6 microamps of polarized electrons, or a luminosity of $\approx 10^{34}$ cm^{-2} sec^{-1}.

The experimental setup is shown in Fig. 2. The beam energy was 578 MeV. Two spectrometers were employed; "OHIPS" at 51° was used primarily to study the quasi-elastic peak but was also used to measure elastic scattering; and "Big Bite" at 44° was primarily used as an elastic scattering monitor. Two angles of target polarization were used. In one configuration the target polarization was reversed to check for systematic errors; at the statistical accuracy of the measurement they were not there.

The asymmetry for elastic $\vec{e}\ ^3\vec{He}$ scattering can be predicted from the magnetic and electric form factors;[7] therefore it can be used to calibrate the magnitude of P_e P_T, the product of the beam and target polarizations. This measurement was performed and the results are presented in Table I. For one measurement the ratio of the experimental to calculated asymmetry is $(43 \pm 19)\%/29\%$ and for the second case the ratio was $(13 \pm 7)\%/20\%$. The weighted average is 0.83 ± 0.31. Within this large error (due primarily to short elastic scattering runs) this result confirms the beam and target polarizations measured independently. We plan to pursue this technique in future experiments.

The results for quasi-elastic scattering are also shown in Table I. The values for the target angles θ^* relative to $\vec{q}$ are also presented. For $\theta^* \approx 0°$ the target polarization is parallel to $\vec{q}$ and one measures $A_{T'}$ which is primarily sensitive to G_M^n. For $\theta^* \approx 90°$ the target polarization is perpendicular to $\vec{q}$ and one measures $A_{TL'}$ which is sensitive to G_E^n. The results with our target[16] and with the mestability target[17] are also presented. The results for both groups are presented with both statistical and systematic errors. The results of spin exchange target have smaller statistical errors. This is due to the higher density so that better statistics were obtained with less running time. The systematic errors of both groups are comparable. The results are shown in Fig. 3 where the statistical and systematic errors have been combined in quadrature.

In Fig. 3 the calculated asymmetries have been represented with errors that take into account an estimated 10% error in G_M^n,[18] 3% errors in the proton form factors, and the uncertainties in the spin part of the three body wave functions.[15] The dominant contribution to the error in $A_{T'}$ comes from the uncertainty in G_M^n; for $A_{TL'}$ there is some contribution to the error due to the uncertainty in the spin coupling in ^{3}He. Note that these errors do not take the uncertainty of the reaction dynamics into account.

It can be seen (Fig. 3) that the two experiments are in agreement. However the results for $A_{T'}$ for our Rb spin exchange target have a somewhat smaller magnitude than the predictions of Blankleider and Woloshyn.[9] For this reason (and those given below) we do not believe that this model is sufficiently accurate to extract a value for G_E^n from the present data. We note that for $A_{TL'}$, which is sensitive to G_E^n, the model calculations[9] are in agreement with the data; these calculations used the value of G_E^n given by Galster et al.[21] It is clear from Fig. 3 that a better measurement of G_M^n is required to test the quasi-elastic reaction model and to obtain accurate data for G_E^n.

It should be noted that the Blankleider-Woloshyn calculation[9] makes many approximations including closure, neglect of FSI (final state interactions) for the nucleons, and neglect of meson exchange effects. All of these, particularly FSI, have been shown to be important for the unpolarized quasi-elastic scattering cross section.[19] The comparison of model calculations[9] for Σ (the unpolarized cross section) for data taken at Bates[20] is shown in Fig. 4. It can be seen that the peak magnitude is high by 25% for $q^2 \cong -.2$ $(GeV/c)^2$. One notes that the ratio of the calculated to the experimental cross sections approaches unity as q^2 increases. It is not clear whether the problems found in the unpolarized cross section will also be present in the asymmetries. Since there is a great deal of polarization in the final state one might conjecture that the spin dependence in the final state interactions will be considerable. This can only be settled by future calculations which we believe are essential to obtain an accurate value of G_E^n. Furthermore we plan to improve the accuracy of these measurements and to extend them[22] to higher q^2.

References

[1] Hofstadter, R., Rev. Mod. Phys. **28**, 214 (1956) and Ann. Rev. Nucl. Sci. **7**, 231 (1957).

[2] The slope of G_E^n (q^2) as $q^2 \rightarrow 0$ was measured by Krohn, V. E. *et al.* Phys. Rev. **d8**, 1305 (1973) and Koester, L. *et al.* Phys. Rev. Lett. **36**, 1021 (1976).

[3] Glashow, S. L., Physica **A96**, 27 (1979).

[4] Isgur, N., Karl, G., and Koniuk, R., Phys. Rev. **D25**, 239 (1982).

[5] Isgur, N., Karl, G. and Sprung, D. W. L., Phys. Rev. **D23**, 163 (1981) and Carlitz, R., Ellis, S. D. and Savit, R., Phys. Lett. **64B**, 85 (1976).

[6] Meissner, U. G., Phys. Rep. **161**, 213 (1988) and Theberge, S., Miller, G. A. and Thomas, A. W., Can. J. Phys. **60**, 59 (1982).

[7] Donnelly, T. W., and Raskin, A. S., Ann. Phys. **169**, 247 (1986).

[8] Arenhovel, H., Phys. Lett. **B206**, 187 (1988), Tomusiak, E. L. and Arenhovel, H., Phys. Lett. **B206**, 187 (1988).

[9] Blankleider, B. and Woloshyn, R., Phys. Rev. **C29**, 538 (1984).

[10] Bates experiment 85-05 Madey, R. and Kowalski, S., spokesmen.

[11] Bates experiment 88-02, Milner, R. G. and McKeown, R. D., spokesmen; Bates experiment 88-10, Chupp, T. E. and Bernstein, A. M., spokesmen.

[12] Otten, E., private communication.

[13] Milner, R. G., McKeown, R. D. and Woodward, C. E., Nucl. Instr. and Meth. **A274**, 56 (1989).

[14] Chupp, T. E., Loveman, R. A., Thompson, A. K., Bernstein, A. M. and Tieger, D. R., to be published and Chupp, T. E., Wagshul, M. E., Coulter, K. P., McDonald, A. B. and Happer, W., Phys. Rev.. **C36**, 2224 (1987).

[15] Friar, J. L., Gibson, B. R., Payne, G. L., Bernstein, A. M. and Chupp, T. E., Phys. Rev. **C42**, 2310 (1990).

[16] Thompson, A. K., Ph.D. Thesis, MIT, August 1991 (unpublished); Thompson, A. K., *et al.*, to be published. The group was Thompson, A. K., Chupp, T. E., Bernstein, A. M., Tieger, D. R., Dodson, G., Dow, K. A., Farkhondeh, M., Fong, W., Kim, J. Y., Loveman, R. A., Richardson, J. M., Schmieden, H., Yates, T. C. Wagshul, M. E., Zumbro, J. D., Dodge, G. E. and deAngelis, D. J.

[17] Jones-Woodward, C. E., *et al.*, Phys. Rev. **C44**, R571 (1991) and Phys. Rev. Lett. **65**, 698 (1990).

[18] Sick, I., in Nuclear Physics with Electron Scattering, in the "Liber Amicorum" meeting for C. de Vries, NIKHEF (1989).

[19] van Meijgard, E. and Tjon, J. A., Phys. Rev. Lett. **57**, 3011 (1986) and Tjon, J. A., private communication.

[20] Dow, K. *et al.* Phys. Rev. Lett. **61**, 1706 (1988).

[21] Galster S. *et al*, Nucl. Phys. **B32**, 221 (1971).

[22] Bates experiment 88-25, Bernstein, A. M. Chupp, T. E., McKeown, R. D. and Milner, R. G., spokesmen.

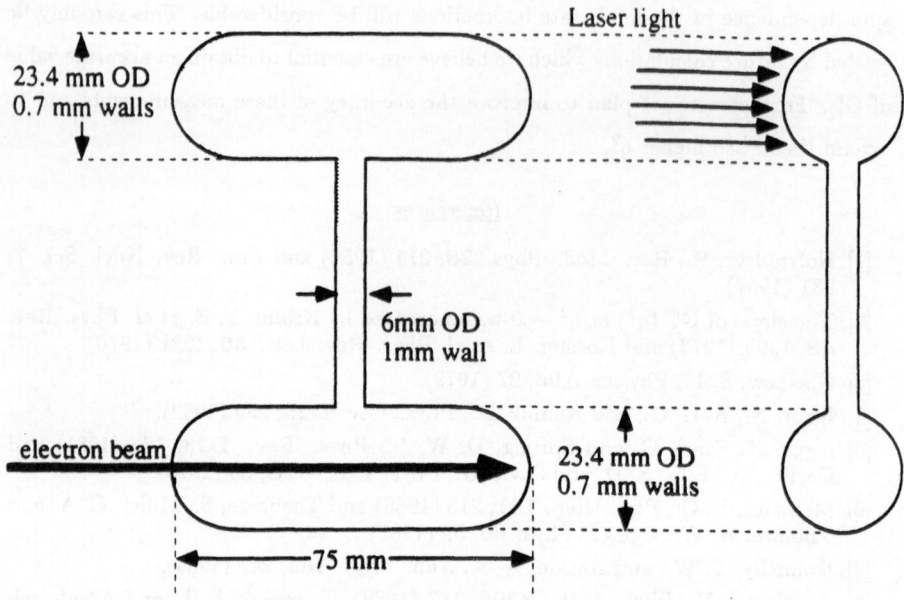

Figure 1. Schematic diagram of the Rb spin-exchange target.

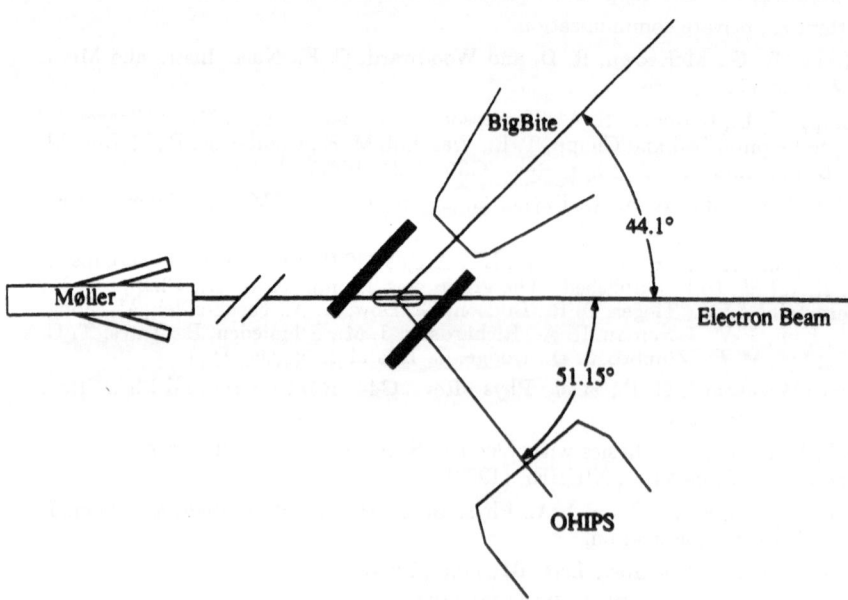

Figure 2. Schematic layout of the scattering geometry of the polarized ^{3}He target and the spectrometers. The Moller spectrometer is used to measure the polarization of the electron beam. The Helmholtz coils are shown for the $A_{TL'}$ configuration for scattering into OHIPS; the target polarization is along the magnetic field direction.

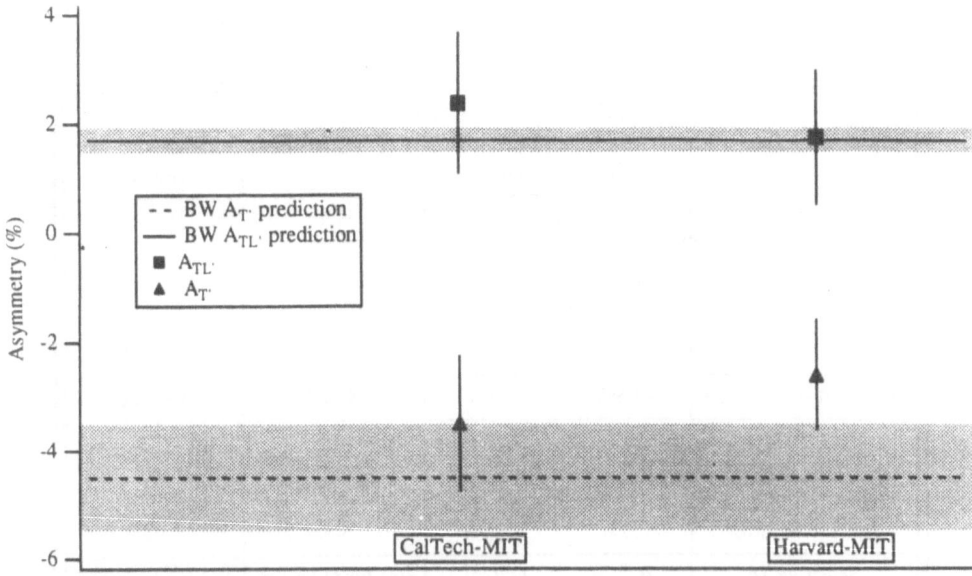

Figure 3. The measured[16,17] and calculated[9] asymmetries for quasi-elastic scattering at $q^2 = -0.2$ $(GeV/c)^2$; see text for discussion. The gray bands represent errors in the model predictions primarily uncertainties in G_M^n, and also G_E^p, G_M^p and the spin coupling in ^{3}He.

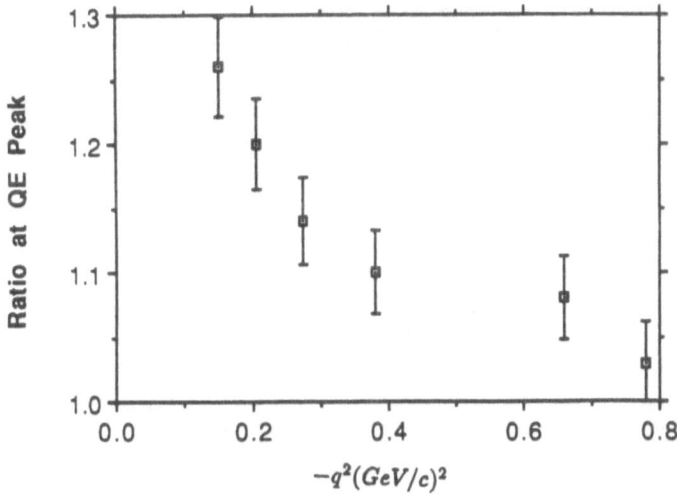

Figure 4. The ratio of the calculations of Blankleider and Woloshn[9] to the experimental data[20] at the quasi elastic peak as a function of q^2.

Table 1: Asymmetries				
Spectrometer	OHIPS	OHIPS	OHIPS	BigBite
θ	51.1°	51.1°	51.1°	44.1°
Q^2 (GeV^2/c^2)	0.20	0.20	0.23	0.18
Reaction	Quasi-Elastic	Quasi-Elastic	Elastic	Elastic
$\theta*$	3.2°	90.2°	99.1°	116.4°
Predicted Asymmetry[†]	–4.5%	1.7%	29±1%	20±0.5%
Harvard-MIT	–2.6%	1.75%	43%	13%
Statistical Error	0.9%	1.22%	19%	7%
Systematic Error	0.5%	0.35%	9%	3%
Sensitive To	$(G_M^n)^2$	$G_E^n G_M^n$	^{3}He Form Factors	^{3}He Form Factors
Caltech-MIT[§]	–3.49%	2.38% [&]	—	—
Statistical Error	1.23%	1.27%	—	—
Systematic Error	0.54%	0.44%	—	—

[†] Reference 9

[§] Reference 17

[&] Target polarization angle was slightly different in Reference 17

Few-Body Systems, Suppl. 6, 493—498 (1992)

ANOMALY IN THE TRANSVERSE RESPONSE OF
3HE(e,e'p) REACTION FOR HIGH RECOIL MOMENTUM.

J.-M. Le Goff, A. Magnon, M. Bernheim, M.K. Brussel[3], G.-P. Capitani[1], J.F. Danel, E. De Sanctis[1], S. Frullani[2], F. Garibaldi[2], F. Ghio[2], M. Jodice[2], L. Lakehal-Ayat, C. Marchand, Z.E. Meziani[4], J. Morgenstern, P. Vernin, A. Zghiche.

Service de Physique Nucléaire CEN Saclay, 91191 Gif-sur-Yvette Cedex, France
[1]Laboratori Nazionali Di Frascati, INFN Frascati, Roma, Italy
[2]Laboratorio Di Fisica, Istituto Superiore Di Sanità and INFN, sezione sanità, Roma, Italy
[3]Department of Physics, University of Illinois at Urbana Champaign, Illinois 60439, USA
[4]Department of Physics, Stanford university, Stanford, California 94305, USA

Abstract: The nucleon-nucleon interaction at short distance inside the nucleus can be studied through (e,e'p) experiments at high recoil momenta in the continuum. In order to clarify the reaction mechanism in such experiments, a separation of the Transverse (T) and longitudinal (L) components of the 3He(e,e'p) cross section was performed at the highest possible recoil momentum, 260 Mev/c. The longitudinal response is found nearly compatible with microscopic calculations. These calculations predict an enhancement of the transverse response due to reaction mechanism (a factor 1.6) but they are still about a factor 3 lower than the data.

One-body properties of nuclei have been experimentally studied rather extensively. In contrast very little information is available on two–body properties. Some contributions have been exhibited at Bates in 12C(e,e'p) experiment, which are not 1–body in character [1], but they are mainly transverse and consequently may arise from reaction mechanism. As we will see, it is possible to get information on the interaction of two close nucleons by measuring high proton momenta. The effects of reaction mechanism prevents the measurement of such rare components with the hadronic probe (e.g. (p,2p) experiments), and consequently (e,e'p) is the only way to achieve this measurement.

(e,e'p)

Before reporting on our experimental work it might be useful to give a short description of (e,e'p) experiments [2]: several hundredths Mev electrons "quasi-elastically" scatter off a given proton of the nucleus, and the outgoing proton is detected in coincidence with the scattered electron. Then the missing energy of the reaction $E_m = M_r + M_p - M_A$ is correlated to the binding energy of the proton (M_r is the mass of the undetected recoil system, M_A is the target nucleus mass, and M_p the proton mass). Furthermore the recoil momentum gives the momentum of the proton before the scattering: $\vec{P} = -\vec{P}_r$ in Plane Wave Impulse Approximation (PWIA). In the first Born approximation the six fold differential cross section for the scattering of unpolarized electrons can be written as $d^6\sigma = \Gamma\left(T + \epsilon L + \epsilon TT \cos 2\alpha + \sqrt{\epsilon(\epsilon + 1)} TL \cos \alpha\right)$, where Γ is the flux of virtual photons with a proportion ϵ of longitudinally polarized photons. When parallel kinematics are chosen (outgoing proton momentum $\vec{P'}$ parallel to momentum transfer $\vec{q}$) the interference terms TT and TL vanish. Then with a forward electron scattering angle $\theta_{e'}$ (large ϵ) and a backward one (small ϵ) one can perform an experimental separation of T and L. In PWIA, T (or L,...) is the product of T^P (or L^P,...) which encompasses the elementary off-shell electron proton cross section, times the spectral function $S^{PWIA}(E_m, P_r)$ of the bound proton, i.e. the probability to find a proton with impulsion P_r and binding energy E_m. Such factorization is only valid when the effects of Final State Interaction (FSI) and Meson Exchange Current (MEC) are neglected. Nevertheless it is convenient to present the experimental results in terms of "experimental spectral function" defined by the following equations: $T^{exp} = T^P S^{exp}_T, L^{exp} = L^P S^{exp}_L, ...$, or if the total cross section is considered without separation $d^6\sigma = \Gamma(T^P + \epsilon L^P + ...)S^{exp}$. In our analysis T^P and L^P are calculated according to the De Forest's "cc1" prescription [3]

motivations

In order to understand how N-N interaction at short distance can be studied, we can adopt the following simple picture [4]: the electron scatters on a proton belonging to a pair of close, "correlated" nucleons, while the A-2 other nucleons are spectators, the other nucleon takes all the recoil momentum. Then the mass of the recoil system (A-1 nucleons) is $M_r^2 = \left[M_{A-2} + \sqrt{M_n^2 + P_r^2}\right]^2 - P_r^2$. For a given P_r the recoil mass and consequently the missing energy $E_m = M_r + M_p - M_A$ are fixed, therefore all the continuum strength should occur at a given missing energy! Yet when the actual impulsion of the correlated pair relatively to the rest of the nucleus is considered, the continuum is expected to show a "bumpy" shape around the previously fixed missing energy.

This study was initiated at Saclay with the measurement of the ^{3}He(e,e'p)x cross section for recoil momenta up to 600 Mev/c [5]. In this experiment the continuum strength shows the aforementioned characteristic "bumpy" shape and the position of the bump shifts with increasing recoil momentum as expected. This is kinematical evidence for our simple picture, furthermore from the dynamical point of view microscopic calculations [6] predicting that the cross section is dominated by this scattering on a correlated pair, are in relatively fair agreement with the data. Then the width of the bump should reflect the momentum of the pair relatively to the rest of the nucleus, while the amplitude reflect the relative wave function of the two nucleons.

Following this experiment some questions have been raised:
– Is not ^{3}He a special case, because it is a loosely bound nucleus and when the correlated pair is taken apart the remaining system is a single nucleon and not a real nucleus?
– Do we really measure the spectral function, what is the importance of reaction mechanism effects (MEC, FSI)?

In order to answer these questions two new experiments were decided: an experiment analogous to the previous one but on ^{4}He, which is a much more dense and tightly bound nucleus, but still remains a Few-Body system for which microscopic calculations are available. We have already reported on this experiment [7], and I will mainly deal with the second one where we performed a separation of the transverse (T) and longitudinal (L) components of the ^{3}He(e,e'p) cross section at recoil momenta around 260 Mev/c. Since they involve different couplings (magnetic coupling for T, and coulomb coupling for L) the separation and the comparison of this two components must be a good test of our understanding of the reaction mechanism. It was actually proven to be so at lower recoil momenta [8]. Furthermore the longitudinal response, which couples directly to the charge distribution, should be free of MEC. This is here especially interesting, since these high recoil momenta experiments are performed in a kinematical region where sizeable effects of MEC are expected. This region is the so-called "dip" region, which in an inclusive (e,e') spectrum lies between the quasi-elastic peak (the top of this peak appears for energy transfer $\omega=\omega_{QE}=Q^2/2M_p$) and the Δ region.

Data taking and analysis

These experiments were performed with the Saclay 700 Mev electron linac (ALS) in the HE1 end-station. When measuring such rare components of the wave function one has to choose small electron scattering angles so that the cross sections remain measurable. Then for 700 Mev incident electrons the momentum transfer is relatively small (300–400 Mev/c) and since we want high recoil momentum we have high outgoing proton momentum and large energy transfer, quite larger than ω_{QE}, which means that we are necessarily in the "dip" region.

In the ^{4}He(e,e'p)x experiment the angle $\theta_{qp'}$ between the momentum transfer q and the outgoing proton momentum p' was varied from 4° to 100°. We then got six sets of kinematics for which the central recoil momentum was ranging from 250 to 600 Mev/c.

Table 1 gives the central kinematics for the separation experiment on ^{3}He. Due to very low counting rates at the backward angle we were limited to about 260 Mev/c recoil momenta. It must be noted that only central kinematics at E_m=25 Mev are rigorously aligned ($\vec{P}'$ // $\vec{q}$). In order to evaluate the subsequent background contribution of the Transverse-Longitudinal interference term (TL) a second forward kinematics was measured with a different misalignment. We were then able to perform a "pseudo" separation of the TL term, and to use its result in the separation of the T and L term.

Table 1 ^{3}He(e,e'p): q=399 Mev/c, P_r=260 Mev/c, P'=659 Mev/c, E_m=25 Mev

	E_e	$E_{e'}$	$\theta_{e'}$	$\theta_{p'}$	ϵ
forward	670	419	34°	36°	.76
backward	397	146	80°	21°	.30

The data were analyzed with a Monte-Carlo technique allowing to take into account the non-linearities of the cross section in the acceptance (the logarithmic slope of the momentum distribution is about 3%/(Mev/c), a factor 3.3 for a 40 Mev/c acceptance).

The ^{4}He experiment was performed with moderate statistical accuracy, so that systematic errors can be neglected. It was not the case for the ^{3}He experiment where a careful estimation of systematic errors had to be achieved. It involved mainly the purely experimental error (Chiefly absolute density of the cryogenic target, but also electronics dead times, variation of target density, solid angle,... :3%), the error resulting from the residual uncertainty in the reconstruction of the recoil momentum (2%), and the error due to the uncertainty on the shape of the cross section (i.e. curvature: 0.5 to 1.5%). All these components enter in the systematic error on the cross section. When the separation is performed this error is amplified, the amplification factor is variable and in spite of the relatively large $\Delta\epsilon$=0.46 it is typically a factor 5. Finally one must consider the error due to the background contribution of the TL term. This error was established by setting a 30% systematic error on the evaluated TL term, it is quite variable, always smaller than 2% for the T response, but typically 10 or 15% for L.

results and conclusion

The result of the experiment on ^{4}He have already been presented [7], I will just recall the main features. We have qualitatively the same results that on the previous experiment on ^{3}He of ref. [5]. For recoil momenta above 300 Mev/c the continuum dominates the cross section. A characteristic "bumpy" shape is observed in the continuum for P_r above 350 Mev/c as in ^{3}He but less prominent. The position of this bump shifts with increasing recoil momenta according to the already mentioned simple picture of scattering on a correlated pair. Finally we have an overall agreement between the microscopic calculation and the data.

Fig. 1 present the experimental spectral functions S^{exp} versus recoil momentum for the ^{3}He(e,e'p)d or 2-body break-up (2bbu) channel. On the left the unseparated forward and backward S^{exp} are shown, together with older data from Saclay [9] and two calculations of the momentum distribution. One is a variational calculation [10] with Argonne potential, the other one a Faddeev calculation [11] with Paris potential. The forward data are compatible both with the previous data (taken in forward kinematics) and with the theoretical prediction. The backward data exhibit a smoother slope, at P_r=250 Mev/c they nearly agree with the forward data, at P_r=290 Mev/c they clearly disagree. We therefore expect a different behavior of the T and L responses.

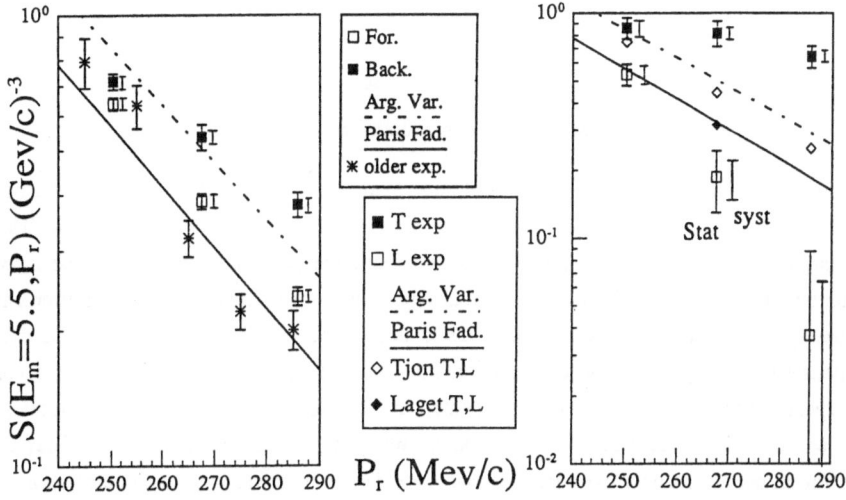

Figure 1 S^{exp} for ^{3}He(e,e'p)d (2bbu), before (left) and after T/L separation (right). Statistical errors are drawn on the points, together with systematic ones on the right.

The separated data are presented on the right of fig. 1 (note the change in vertical scale). We also display two calculations of the reaction, a Faddeev calculation [12] which uses the Malfliet-Tjon potential and neglects MEC, and a diagrammatic expansion performed by Laget [6] which is supposed to calculate them and which is using the Paris Faddeev wave function. Both calculations give very little difference between S_T and S_L (about 5%, not visible in fig. 1). The data show a considerable difference which indicates that non PWIA processes play an important role and that they are not reproduced by the current calculations.

The continuum ^{3}He(e,e'p)pn channel, which is the main topic of this experiment, is presented in fig. 2. S_T^{exp} and S_L^{exp} are plotted against the missing energy, together with Laget calculation. For such not so high recoil momentum the continuum strength does not exhibit any "bumpy" shape. The amplitude of S_T^{exp} and S_L^{exp} are very different, after integration over missing energy the ratio S_T^{exp}/S_L^{exp} is about a factor 6. Laget calculations also predict a S_T/S_L ratio larger than 1 (about 1.6) but much smaller than the experimental one. The longitudinal is not too far from the data but the transverse is really underestimating them.

498

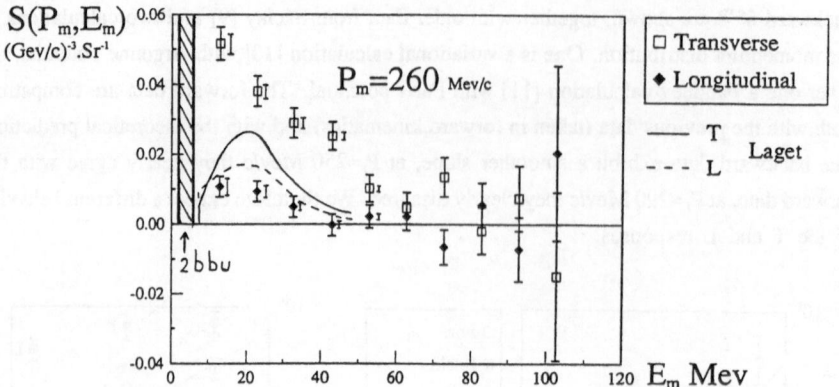

Figure 2 S^{exp} for ^{3}He(e,e'p)pn (continuum), errors as in fig. 1

Finally, we have observed on ^{4}He the same evidences for scattering on a pair of close interacting nucleons, as was observed on ^{3}He. Concerning the reaction mechanism in ^{3}He, the large discrepancy in the transverse response leave the possibility for MEC to be more important than expected. The Longitudinal is supposed to give a more reliable measurement of the spectral function, but clearly one has to understand the difference between the two responses which appears at not so high recoil momenta, before claiming to be able to measure the spectral function for high recoil momenta. In the future with several Gev accelerators one will be able to perform T/L separation for really high recoil momenta (>500 Mev/c), and for an energy transfer even smaller than ω_{QE}, i.e. in a region were the contribution of MEC should be considerably smaller. Another direction for the study of N-N interaction in the nucleus will be (e,e'2N) experiments.

1. Lourie, R.W., et al.: Phys. Rev. Lett. **56**, 2364 (1986).
2. Frullani, S., Mougey, J.: Adv. Nucl. Phys. **14**, 1 (1984).
3. De Forest, T.: Nucl. Phys. **A392**, 232 (1983).
4. Frankfurt, L., Strikman, M.: Phys. Rep. **76**, 236 (1981).
   Laget, J.M.: Nucl. Phys. **A358**, 275 (1981).
5. Marchand, C., et al.: Phys. Rev. Lett. **60**, 1703 (1988).
6. Laget, J.M.: Private communication.
7. Le Goff, J.M., et al.: International Few-Body XII conference, Vancouver 1989.
8. see for instance: Magnon, A., et al.: Phys. Lett. **B222**, 352 (1989).
9. Jans, E., et al.: Nucl. Phys. **A475**, 687 (1987).
10. Schiavilla, R., Pandharipande, V.R., Wiringa, R.B.: Nucl. Phys. **A449**, 219 (1986).
11. Hadjuck, C.H., Sauer, P.U.: Nucl. Phys. **A369**, 321 (1981).
12. Van Meijgaard, E., Tjon, J.A.: Phys. Rev. **C42**, 96 (1990).

Few-Body Systems, Suppl. 6, 499—505 (1992)

ELECTROMAGNETIC STUDIES OF FEW BODY SYSTEMS USING BLAST

F. William Hersman

University of New Hampshire

Durham, NH 03824 USA

Abstract

The MIT-Bates Linear Accelerator Center is presently installing a pulse stretcher ring. This upgrade will provide continuous, high current, polarized beams of electrons up to 1 GeV energy. A collaboration has formed to develop and install the Bates Large Acceptance Spectrometer Toroid (BLAST) at the ring internal target area. Concurrently, members of the collaboration are developing polarized and unpolarized internal targets. I will describe the facility under development, outline the broad scope of the approved program for studying few body nuclei, and focus on simulations of a few planned triple coincidence reactions.

Introduction

Elastic electron scattering has long been used to probe ground state charge and magnetization densities. Inclusive and exclusive direct knockout reactions have probed one-body densities in the quasielastic region. Recently by measuring the polarization observable T_{20} at Bates, separation of the monopole and quadrupole charge densities in deuterium was performed.[1] Inclusive and exclusive polarization dependent scattering was measured in the quasielastic regions of ^{3}He[2] and deuterium,[3] respectively, for the purpose of extracting the neutron form factor. These pioneering efforts have established spin-dependent electron scattering as an excellent tool for determining small pieces of the few body wave functions.

Information on two-body currents can be accessed through two particle knockout from nuclei. Semi-inclusive measurements, particularly of nucleon knockout in the quasielastic dip region,[4] already motivate the more challenging triple coincidence exclusive studies. Indeed, first attempts at the (e,e'2p) reaction are presently being analyzed.[5]

The Bates Large Acceptance Spectrometer Toroid (BLAST) will measure the polarization dependence of reactions and detect higher order final state coincidences to study few body nuclei. One-body structure and two-body currents will be determined simultaneously over an extensive range of kinematics. The goal of the BLAST program is a complete characterization of the full electromagnetic response accessible with 1 GeV electrons.

The facility

The MIT-Bates South Hall Ring stores circulating currents up to 80 mA of electrons with 40% polarization at energies up to 1 GeV. The large 190 meter circumference minimizes synchrotron radiation power, which could otherwise damage storage cell coatings and reduce emittance. The beam spot size at the low-β internal target location is $\sigma_{x,y} \leq 0.15$ mm, allowing narrow storage cells for polarized targets.

The source of polarized ^{3}He uses optical pumping in an RF discharge followed by metastability exchange of the atomic polarization to the nucleus.[6] Flow of 10^{17} at 50% polarization has been achieved.[7] A high intensity atomic beam source will provide polarized atoms of either hydrogen or deuterium. Two RF stages allow selection of a single nuclear m-state with polarization up to 90%.[8] New open-geometry sextupoles, based on the Heidelberg design, will allow flow of over 10^{16} atoms/sec. These sources of polarized atomic species flow into a windowless storage cell[9] to increase the target density approximately 100-fold. These targets are able to achieve densities of up to 10^{16}/cm^2. Although their density is still low, these targets approximate the ideal, being windowless, pure species with high polarization capability in a single m-state, capable of quick reversal. The axis of quantization can be oriented arbitrarily by Helmholtz coils. Nevertheless, even with the high circulating current of up to 80 mA, they allow appreciable counting rates only with solid angle acceptance on the order of 1 sr.

BLAST uses a magnetic field, trajectory tracking, time of flight, and Čerenkov detection for large acceptance momentum analysis, determination of emission direction, and particle identification. Magnetic fields of 0.2 to 0.3 T are established by eight copper coils energized to 1.4 MA–turns. Approximately 4500 hexagonal drift cells provide position information at 23 points along the curved tracks. A helium-dominated gas mixture minimizes multiple scattering for low velocity particles. Scintillators allow separation of pions from protons by comparing time-of-flight with momentum analysis. Specular reflecting gas Čerenkov detectors uniquely identify electrons and provide the event trigger. The components, electronics, and software acquisition system emphasizes off-the-shelf technology to minimize the development and construction time.

Angular coverage extends from 15° to 90° with respect to the incident beam, and $\phi = \pm 17°$ in azimuth in two opposing sectors, achieving 1 sr. Energy transfers up to 700 MeV and momentum transfers exceeding 800 MeV/c are produced with a 1 GeV

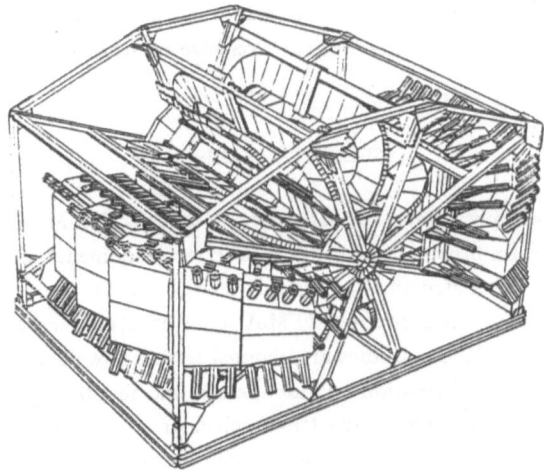

Fig. 1. The BLAST configuration showing the coils, structure, detectors, and internal target.

beam. Momentum resolution is estimated to be between 1% and 2% for electrons and 2% to 3% for protons. Angle definition is approximately 5 mr in both directions. At the design luminosity of 10^{33} e^- nucleons/cm^2 event rates of 100 Hz will be acquired.

The physics program

The BLAST physics program includes studies of the nucleon and its low lying excitations, as well as one- and two-body currents in nuclear systems. In the nucleon we plan to measure spin dependent structure functions of the transitions to the Δ and Roper. Studies of fundamental ground state and Δ excitation form factors will be conducted on both the proton and neutron bound in deuterium and ^{3}He. One-body spectral functions of protons in deuterium and helium will be measured in polarization dependent $(\vec{e},e'p)$ coincidence experiments, extracting response functions sensitive to separating the deuteron s- and d-state, and the ^{3}He s-, s'- and d-state.

Triple coincidence studies will explore the quasielastic region, the dip, and the Δ. Two nucleon knockout $(e,e'NN)$ from helium will be sensitive to the nucleon-nucleon interaction in the medium, and the role of three body forces. Triple coincidence experiments involving pions can be used to reconstruct the Δ. We will study Δ production in nuclei, damping processes, and the form of final state interactions between Δ's and nucleons. Polarization dependent triple coincidence reactions provide maximal information on the form of the current and added flexibility for isolating the spin and isospin reaction channels.

Unpolarized coincidence measurements are planned on deuterium, helium, and oxygen. Exclusive as well as inclusive polarization asymmetry measurements will be performed on the proton, deuteron, and ^{3}He. Note that all reaction channels with a particular beam and target choice are measured simultaneously. For the first time, selection of the polarization of the initial state, and completely determining the final state is possible for two and three particle exclusive coincidence experiments.

Simulations of reactions

In coincidence experiments, the capability to transfer 700 MeV of energy and 800 MeV/c of momentum provides access to nuclear recoil momenta and missing energies of similar values. Reconstructions of missing energy can be accomplished to as good as 10 MeV and missing momentum of 20 MeV/c for parallel kinematics. Several inclusive and two-fold coincidence studies are planned to measure polarization asymmetries and extract structure information on the nucleon and few body systems. In this presentation, I want to emphasize the triple coincidence capability, without and with polarization, that can provide information on complex currents in few-body nuclei.

I present below a partial list of the possible reactions that have been approved for measurement in this program.[10] All measurements on deuterium and ^{3}He will be performed with and without polarization. Several of the measurements are complementary, measuring the same final state in different kinematic regimes by detecting different particles. Others are complementary by measuring different isospin channels. Final states involving detected neutrons are included. While the BLAST time-of-flight system will provide 5% neutron detection capability, a supplementary neutron detector is separately proposed to raise the efficiency to 30%.

$$\vec{d}(\vec{e},e'p\pi^-)p \qquad {}^3\vec{\text{He}}(\vec{e},e'pp)n \qquad {}^{16}O,{}^4\text{He}(\vec{e},e'pp)$$
$$\vec{d}(\vec{e},e'pp)\pi^- \qquad {}^3\vec{\text{He}}(\vec{e},e'pn)p \qquad {}^{16}O,{}^4\text{He}(\vec{e},e'pn)$$
$$\vec{d}(\vec{e},e'pn)\pi^\circ \qquad {}^3\vec{\text{He}}(\vec{e},e'p\pi^-)pp \qquad {}^{16}O,{}^4\text{He}(\vec{e},e'p\pi^-)$$
$$\vec{d}(\vec{e},e'n\pi^+)n \qquad {}^3\vec{\text{He}}(\vec{e},e'p\pi^+)nn \qquad {}^{16}O,{}^4\text{He}(\vec{e},e'p\pi^+)$$
$$\qquad\qquad {}^3\vec{\text{He}}(\vec{e},e'n\pi^+)pn \qquad {}^{16}O,{}^4\text{He}(\vec{e},e'n\pi^+)$$
$$\qquad\qquad {}^3\vec{\text{He}}(\vec{e},e'n\pi^+)d \qquad {}^{16}O,{}^4\text{He}(\vec{e},e'\pi^+\pi^-)$$
$$\qquad\qquad {}^3\vec{\text{He}}(\vec{e},e'd\pi^+)n$$
$$\qquad\qquad {}^3\vec{\text{He}}(\vec{e},e'pp)\Delta^\circ$$
$$\qquad\qquad {}^3\vec{\text{He}}(\vec{e},e'pn)\Delta^+$$

The first Monte Carlo simulation I present is for the reaction ^{3}He(e,e'2p) using part of the theoretical calculation,[11] which includes one and two-body reaction processes on a correlated Fadeev ground state wave function for the condition of zero final neutron momentum. In this simulations only the plane wave result for the longitudinal response function was included. The theoretical input to our simulation is shown in figure 2, plotted as a function of energy transfer ω, and proton angle in the two-proton center of momentum system θ_1. V e have folded this response with a gaussian neutron momentum distribution. No dependence on momentum transfer beyond the Mott cross section was assumed.

Count rates are based on a luminosity of 10^{33} e$^-$ atoms/cm^2. The event rate at the peak of the response is approximately a few counts per second. One can assume useful

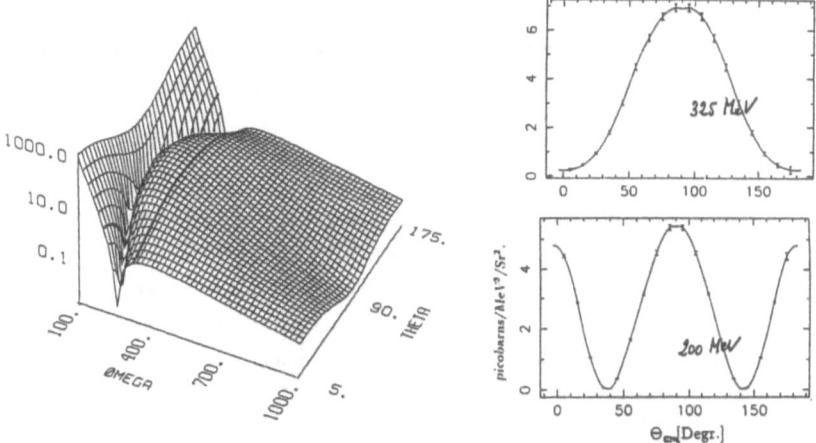

Fig. 2. Theoretical plane wave longitudinal response used for simulations of ^{3}He(e,e'pp).

Fig. 3. The simulated ^{3}He(e,e'pp) data plotted against outgoing proton angle.

determinations of cross sections at counting rates as low as a few per hour, setting a lower limit more than three orders of magnitude below the maximum rate. This criterion implies that the dependence of the cross section can be determined to energy transfers of 400 MeV, momentum transfers of 400 MeV/c, and proton-proton invariant mass of 2300 MeV.

We have estimated uncertainties of cross sections based on two-dimensional cuts of the counting rate. Figure 3 presents the dependence on proton emission angle θ_1 for different energy transfers. The cuts along the contours are indicated by bold lines on the theoretical plane wave longitudinal response. One can see the statistical precision of the measurement indicated by the error bars.

The goal of these measurements is to determine the short range structure of two nucleons bound in nuclei. Since both direct terms and two body currents can produce two body knockout, a systematic study is required. Coupling to charged meson exchange currents can initiate proton-neutron knockout while proton-proton knockout is free of this reaction mechanism. Separation of response functions can further lead to selection of the reaction. The longitudinal proton-proton knockout response function is most directly connected with the two proton correlation function. Kinematic access to a wide range of proton relative momenta with BLAST provides a comprehensive study of this reaction.

The second simulation I discuss applies to the d(e,e'pπ^-) reaction. This cross section is based on the parametrized response functions derived from a fit of the multipole matrix elements to the available Δ excitation data,[12] folded with a gaussian deuteron momentum distribution. Counting rates have been determined as a function of final state Δ momentum in the N-Δ center of momentum. An extensive range of kinematics, out to 600 MeV/c, or 1200 MeV/c of relative momentum, can be measured simultaneously with counting rates exceeding hundreds of counts per hour per 10 MeV/c bin.

The relative momentum of the Δ and nucleon is of particular interest because of the

significant presence of final state interactions. In this case the interaction is between
a Δ and a nucleon. While inclusive measurements have extracted Δ-nucleus optical
potential information, no other method is so promising for directly measuring this two
baryon interaction.

Polarization and triple coincidence

Measuring the polarization dependence of nuclear reactions considerably enhances
their selectivity. Spin dependent response functions depend on the interference between the various helicity amplitudes.[13] If a particular helicity amplitude is small, its
contribution to the unpolarized cross section can be practically impossible to extract.
On the other hand, if the small amplitude interferes with a larger amplitude leading
to a polarization asymmetry, the measurement often is possible for two reasons. First
the product of a small amplitude and a larger one is bigger than the small amplitude
squared. Secondly, asymmetries are experimentally easier to measure to high precision
than absolute cross sections.

The beam-target asymmetry R'_{LT} with the axis of quantization in the scattering
plane can enhance small components. This response function is proportional to the
interference between a longitudinal amplitude and a transverse amplitude. The longitudinal amplitude can produce direct Δ knockout in ^{3}He. This response may be closely
linked to the probability of preexisting Δ's in the ^{3}He ground state.[14]

Beam-target polarization asymmetries also allow separation of reaction channels
with different angular momentum. The transverse response function R_T is the sum
of the two photon helicities ± 1, while the beam-target polarization response function
with the axis of quantization along the momentum transfer R'_T is the difference of
the two virtual photon helicities. Comparison of these response functions can separate
the response of the polarized nucleus when one unit of angular momentum is added
or removed by the photon. In the case of Δ excitation on deuterium and s-wave
rescattering, if a unit of angular momentum is added to polarized deuterium, the final
state is spin two. If it is subtracted, the final state is a combination of spin one and two.
This reaction will be used to extract the final state interaction phase shifts between
a Δ and a nucleon in the spin two, isospin one (S=2,T=1) channel and the spin one,
isospin one (S=1,T=1) channels.

Group theoretic models based on constituent quarks with a color magnetic interaction provide a framework for exploring the baryon-baryon interaction in nuclei.[15] Six
quark states of good spin and isospin are not necessarily the combination of two three
quark eigenstates. Short range interactions in nuclei can lead to mixing of the N–N,
N–Δ, and Δ–Δ systems. Specification of the probabilities of finding Δ's in various
channels, and determination of the dynamical N–Δ interaction phase shifts could establish quarks as the relevant degrees of freedom for describing short range interactions
in nuclei.

Summary

The MIT-Bates South Hall Ring provides high continuous circulating currents of 1 GeV polarized electrons for use with internal targets. Sources of polarized protons, deuterons and ^{3}He are being developed to feed a storage cell at the internal target area to achieve luminosities approaching, and in the case of ^{3}He exceeding, 10^{33} e$^-$ atoms/cm^2. BLAST provides over one steradian acceptance of particle identification, and momentum and trajectory determination with good resolution. The combined facility represents a unique opportunity for measuring polarization asymmetries and multiparticle coincidence reactions.

A program of triple coincidence measurements on few body nuclei is described in this contribution. Coincidence cross sections involving knockout of two nucleons from helium and Δ production in helium and deuterium will be measured exclusively. Polarization asymmetries will separate helicity amplitudes and isolate specific spin channels. This new information on the baryon-baryon interaction in the N–N and N–Δ systems will characterize the response and help determine the relevant underlying degrees of freedom.

1. I. The, *et al.*, *Phys. Rev. Lett.* **67** (1991) 173.
2. C. Jones-Woodward, *et al.*, *Phys. Rev.* **C 44** (1991) R571.
3. Bates experiment #85-05, R. Meady, S. Kowalski, spokesmen.
4. R. L. Lourie, *et al.*, *Phys. Rev. Lett.* **56** (1986) 2364.
5. P. K. A. DeWitt-Huberts, this volume.
6. R. G. Milner, R. D. McKeown, and C. E. Woodward, *Nucl. Instr. Meth.* **A 257** (1987) 286.
7. K. Lee, private communication.
8. W. Haeberli, *Ann. Phys. Nucl. Sci.* **17** (1969) 373.
9. R. Gilman, *et al.*, *Phys. Rev. Lett.* **65** (1990) 1733.
10. Bates experiment #91-10, F. W. Hersman, W. Kim, and P. Galumian, spokesmen.
11. J. M. Laget, *Phys. Rev.* **C 24** (1987) 832.
12. R. C. E. Devenish and D. H. Lyth, *Phys. Rev.* **D 5** (1972) 47, and, *Nucl. Phys.* **B 43** (1972) 228.
13. A. Raskin and T. W. Donnelly, *Ann. Phys.* **191** (1989) 78.
14. R. G. Milner and T. W. Donnelly, *Phys. Rev.* **C 37** (1988) 870.
15. M. Oka and K. Yazaki, in *Quarks in Nuclei*, W. Weise, ed., World Scientific Press, Singapore (1984) 490.

Few-Body Systems, Suppl. 6, 506—511 (1992)

Few-
Body
Systems
© by Springer-Verlag 1992

STUDIES OF INCLUSIVE ELECTRON SCATTERING OFF THE LIGHTEST NUCLEI

V.D. Efros

I.V.Kurchatov Institute of Atomic Energy,

123182 Moscow, USSR

Model-independent predictions for A=3 and A=4 longitudinal (e,e') sums are shown to be of great intrinsic accuracy. In conjunction with the direct sum-rule calculations at $q>1.5$ fm^{-1} they give an unambiguous test of the single-free-nucleon approximation for the nuclear charge density. An improved version of "quasi-y-scaling" extrapolation is applied to the recent L/T separated ^{4}He(e,e') Bates data and the resulting sums are found to agree with sum-rule predictions. One way more of response extrapolation is developed and it is applied to the new low-q transversal ^{4}He(e,e') Kharkov data. Differences between the q-behavior of the experimental sums and that of sum-rule results are observed.

1. ACCURACY OF MODEL-INDEPENDENT A=3 AND A=4 SUM-RULE PREDICTIONS

1. Let R_l and R_t be the longitudinal and the transversal response functions entering doubly differential (e,e') cross section, q and ω the momentum and the energy transferred from an electron to the nucleus. We consider the inelastic sums

$$\bar{S}_{l,t}(q) = \int_{\omega_{el}^+}^{\infty} [G_E^p(Q)]^{-2} R_{l,t}(q,\omega)d\omega, \qquad (1)$$

and the corresponding total sums S_l and S_t including additional elastic $\omega=\omega_{el}$ contributions, e.g. $S_l = \bar{S}_l + (ZF_{el})^2$. Here G_E^p is the Sachs proton form factor, $Q^2=q^2-\omega^2$. The above sums may be calculated from ground-state wave functions with the help of sum rules (SR), see e.g. [1]. At sufficiently high q values, as it is believed at $q \gtrsim 2$ fm^{-1}, the SR result for S_l becomes model-independent in the frame of the traditional single-free-nucleon form for the nuclear charge density operator: $S_1 \approx Z$. Thus, this SR allows, in principle, an unambiguous test of an accuracy of the nuclear charge density form in the $2 \, \text{fm}^{-1} \lesssim q \lesssim 3 \, \text{fm}^{-1}$ range. In particular, if the free proton form factor entering Eq. (1) changes noticeably inside the nucleus this SR should be violated.

In the case of ^3He, ^3H, ^4He nuclei in Ref [2] model-independent longitudinal SR were introduced in the same frame at $q \gtrsim 0.5 \, \text{fm}^{-1}$. They are intended for reliable testing the single-free-nucleon form of nuclear charge density in this wider q range. Below we consider the ^4He case. The A=3 results are similar. The model-independent SR is [2]

$$S_1^{^4\text{He}}(q) = 2\left\{1 + [1 + 4G_E^n(q)/G_E^p(q)]\frac{F^{^4\text{He}}((8/3)^{1/2}q)}{G_E^p((8/3)^{1/2}q) + G_E^n((8/3)^{1/2}q)}\right\} \qquad (2)$$

This relation connects the experimental inelastic-plus-elastic sum with the experimental *rescaled* form factor of the elastic electron scattering. The contribution proportional to this form factor represents approximately the correlation contribution to SR. In the A=3 case such representation of correlation matrix elements was applied previously [3] in an another problem (Coulomb energy difference of A=3 nuclei).

The relation (2) includes the observable quantities only and it could have served as a direct test of the single-free-nucleon approximation for the nuclear charge density. But this relation is an approximate one and at first its intrinsic accuracy should be checked. To do this we proceed as follows. For a choice of a realistic NN force version we calculate the S_1 from the corresponding ground-state wave function using the exact SR expression. Then we calculate it with the help of Eq. (2) using F values calculated for the same NN force. The results are shown in the Table 1 for two (noncentral) NN force versions SSCa and EH. Corresponding inelastic $\bar{S}_1$ sums are represented. We see that the relation (2) is quite accurate. The reason for this is following. At moderate q values the correlation

Table 1.

q [fm^{-1}]	SSCa		EH	
	Exact	Eq. (2)	Exact	Eq. (2)
0.5	0.126	0.128	0.126	0.129
1.0	0.431	0.436	0.426	0.432
1.5	0.737	0.741	0.727	0.732
2.0	0.909	0.915	0.901	0.908

contribution to SR can be expressed in terms of elastic form factor $F((8/3)^{1/2}q)$ with great accuracy. This may be argued [2,4] proceeding from the hyperspherical expansion approach. And for higher q values this contribution becomes negligible. It is expedient to use Eq. (2) at $q \lesssim 1.5$ fm^{-1} values when one may suppose single-free-nucleon form of the charge density operator to be valid for the description of $F((8/3)^{1/2}q)$. At the same time for $q\gtrsim1.5$ fm^{-1} the correlation contribution is very small and its NN-force dependence is negligible. As a result, we have an unambiguous and accurate test of the single-free-nucleon approximation for the nuclear charge density operator in the lightest nuclei.

2. "QUASI-Y-SCALING" EXTRAPOLATION OF RESPONSES. COMPARISON OF ^{4}He(e,e') DATA WITH SUM RULES.

In order to get the sums of Eq. (1) the extrapolation beyond the range of measured ω is required. In Ref. [5] it was shown that the (unseparated) ^{4}He(e,e') spectra obtained there are well described by the modified y-scaling representation with scaling function $F(y)$ substituted by $F(y-\bar{y})$. Here $y(q,\omega)$ is the known scaling variable while $\bar{y}(q)$ is a parameter which may be interpreted as follows [6,7]. The quantity $-\bar{y}\hat{q}$ is a mean fraction of the momentum $\mathbf{q}$ transferred from electron to the residual nucleus,

$$\mathbf{p} + \mathbf{q} = \mathbf{p'} + (-\bar{y}\hat{q}) \qquad (3)$$

where $\mathbf{p}$ and $\mathbf{p'}$ are the initial and the final momenta of the knocked-out nucleon. In Ref. [7] it was proposed to use this representation for *the extrapolation* of experimental spectra. Here we apply it to *L/T* separated spectra [8] and we fix the parameter $\bar{y}$ in an improved way comparatively to [7]. (In Refs. [8,9] in the longitudinal case they tried to obtain the sums by attaching a phenomenological exponential tail to the spectra.)

An example of the corresponding fit of a spectrum from [8] is shown in Fig. 1. This fit is used for the extrapolation. The dashed curve corresponds to a pure y-scaling spectrum here. The fit was performed over

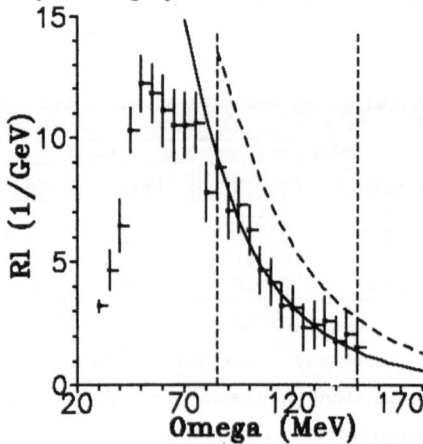

Fig. 1

the range between dashed straight lines where the speeds of the knocked-out nucleons are high enough to apply a quasi-y-scaling approach. The $\bar{y}$ value obtained equals to -53 MeV/c in this case (q=300 MeV/c) which is reasonable from the point of view of Eq. (3). The results of the comparison with the SR of the sums of Eq. (1) taken over extrapolated spectra of Ref. [8] are shown in Table 2 for the longitudinal case and in Table 3 for the transversal case. The results in the tables are normalized in such a way that the theoretical values tend to unity as q increases. The relative contributions to $\bar{S}$ sums of the spectra extrapolated tails are shown in the Table 2

q [MeV/c]	$\Delta_{tail}/(\bar{S}_l)_{exp}$	$Z^{-1}(S_l)_{exp}$	$Z^{-1}(S_l)_{theor}$
300	0.06	1.065	1.11
400	0.11	0.94	1.02
500	0.14	0.95	1.00

tables. The overall uncertainties of the experimental sums in the tables are about 10%. Taking this into account we see that full agreement between the theoretical and the experimental values exists in general. Meson exchange currents not accounted for in the SR calculation are known to increase the transversal sums. The results obtained suggest that their Table 3

q [MeV/c]	$\Delta_{tail}/(S_t)_{exp}$	$(S_t^0)^{-1}(S_t)_{exp}$	$(S_t^0)^{-1}(S_t)_{theor}$
300	0.17	0.97	0.87–0.90
400	0.20	1.11	0.95
500	0.14	1.02	0.99

contribution is rather small. Besides, there is no room for an essential nucleon swelling in ^{4}He. Other "exotics" do not manifest themselves at the considered q values either. These conclusions are based on the experimental scaling function [5] used and they are relevant to the existing accuracy of experimental spectra. Further experimental investigations are desirable in this connection.

3. ω^{-n} - EXPANSION EXTRAPOLATION OF EXPERIMENTAL RESPONSES. SUM-RULE ANALYSIS OF LOW-q ^{4}He(e,e') DATA.

A preliminary analysis of 0.875 fm$^{-1} \leq q \leq 1.5$ fm^{-1} Kharkov data [10] will be presented here. For such small q values the way of the extrapolation described above proved to be inapplicable as it was expected. We are calculating the transversal sum S_t of Eq. (1). (There are small longitudinal admixtures to the data [10] which are subtracted by using the theoretical S_l values.) In Refs. [11,12] where heavier nuclei were considered the tails of the $C\omega^{-\alpha}$ form were attached to spectra. In Ref. [11] (see also [1]) in the longitudinal case the α constant taken from the

3<α<4 interval was found by a fit to the experimental spectrum considered. The α=3 value was used as the upper bound. In Ref. [12] in the transversal case the α constant was varied at the vicinity of α=3. The power decrease of spectra as ω increases reflects the fact that there exists only finite number of momenta of the $\int_0^\infty \omega^k R d\omega$. type. While trying to determine α and C constants by a direct fit of $C\omega^{-\alpha}$ tail to the experimental spectra we were faced to some difficulties.

We calculate the spectrum tail contribution as follows. We set

$$R = \omega^{-\alpha} \sum_{n=1}^{N} C_n \omega^{-(n-1)}. \tag{4}$$

To find the α constant we consider the behavior of a *longitudinal* component R_l of a spectrum at high ω values: $R_l \simeq C_l \omega^{-\alpha}$. The point is that while extrapolating a transversal component of a spectrum the contributions of mesonic degrees of freedom not accounted for at formulating the transversal SR should be separated. We will suppose that at high ω values the contribution of *nucleonic* degrees of freedom to the transversal component of a spectrum behaves similarly to the longitudinal component: $R_t \simeq C_t \omega^{-\alpha}$ with the same α value as in the R_l case. Such an assumption stems from the similarity, up to spin variables, of the form of the operator of charge density and that of spin-current density in the single-particle approximation for these operators.

We find the constant α from the fit of the expression $C\omega^{-\alpha}$ to the experimental ^{3}He longitudinal spectra of Ref. [13] with the lowest q values. The data of Ref. [13] are the only available ones in the literature which have been measured up to high ω values with low statistical errors. We suppose that the α constant is the same for ^{3}He and for ^{4}He. (It is probably the same for other nuclei.) Different acceptable choices of ranges over which the fits were performed lead to the best-least-square α values from 3.9 to 4.1. It is natural to conjecture that α is an integer number and as a result we set "at ω⇒∞":

$$R_l(q,\omega) \simeq C_l(q)\omega^{-4}. \tag{5}$$

In Fig. 2 the description of the tail of the spectrum of Ref. [13] with the help of Eq. (5) (full line) and with the help of the expression of the form $C\omega^{-3}$ (dashed line) are shown.

The extrapolation of the spectra was performed with the help of Eq. (4) with α=4. Fitting C_n constants from Eq. (4) to the experimental spectra was performed with the least-square method. Our final results are shown in Fig. 3. Experimental (extrapolated) transversal sums are situated between full lines while theoretical results for transversal SR are situated between dashed lines. The result obtained for the highest of q values q=1.5 fm^{-1} is quite close to that obtained above for the data of Ref. [8]. We see, on the other hand, that the experimental sums and the theoretical sum rule values behave in a different way with the change of q. The results are somewhat in line with ^{3}He low q experimental results [14] which lead to a transversal sum that is noticeably lower than the SR [4]. New measurements at this low-q range are necessary as well as an experimental testing of Eq. (5).

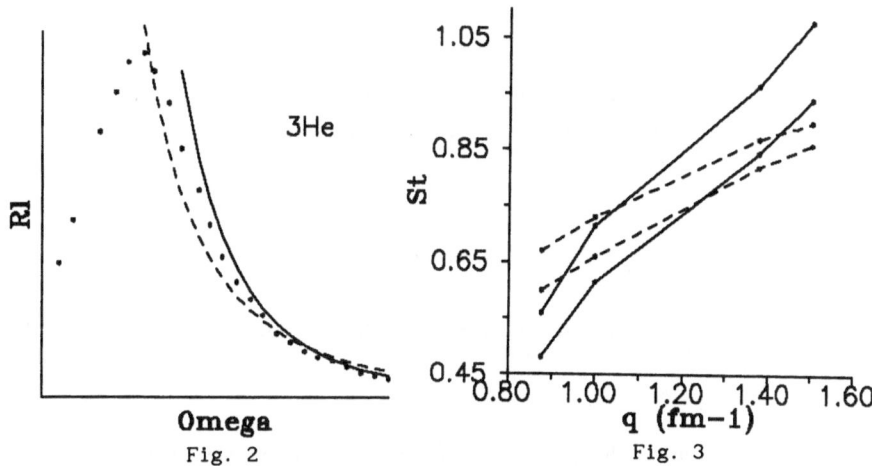

Fig. 2   Fig. 3

REFERENCES

1. Orlandini, G., Traini, M.: Repts. Progr. Phys. **54**, 257 (1991)
2. Éfros, V.D.: Sov. J. Nucl. Phys. **18**, 607 (1973)
3. Friar, J.L.: Nucl. Phys. **A 156**, 491 (1970); Fabre de la Ripelle, M.:Fizika **4**, 1 (1972)
4. Efros, V.D.: Preprint IAE-5355/2, 1991.
5. Dementij, S.V.: Sov. J. Nucl. Phys. **48**, 383 (1988)
6. Dementij, S.V.: Proc. of the Seminar VII on electromagnetic interact. of nuclei at low and intermediate enegies, p. 5. Moscow 1989.
7. Dementij, S.V., Efros,.V.D.: Z. Phys. **A 339**, 255 (1991)
8. Von Reden, K.F. et al.: Phys. Rev. **C 41**, 1084 (1990)
9. Schiavilla, R., Pandharipande, V.R., Fabrocini A.: Phys. Rev. **C 40**, 1484, (1989)
10. Buki, A.Yu. et al.: Preprint IAE-5397/2, 1991.
11. Orlandini, G., Traini, M.: Phys. Rev. **C 31**, 280 (1985)
12. Orlandini, G., Traini, M.: Phys. Rev. **C 32**, 321 (1985)
13. Marschand, C. et al.: Phys. Lett. **B 153**, 29 (1985)
14. Jones, E.C. et al.: Phys. Rev. **C 19**, 610 (1979)

Few-Body Systems, Suppl. 6, 512—525 (1992)

RELATIVISTIC APPROACHES
TO THE FEW BODY PROBLEM

L.A.Kondratyuk

Institute of Theoretical and Experimental Physics, Moscow 117259, USSR

Abstract: I discuss the applications of the Relativistic Hamiltonian Dynamics in the light -cone form to the description of few body (mainly few nucleon) systems and the relation of the Light-Cone Dynamics (LCD) to the quantum field theory and covariant equations.

When we discuss relativistic approaches we may have in mind relativistic field theory, relativistic hamiltonian dynamics, covariant equations, relativistic strings etc. In this talk I would want to discuss the relation of the Light-Cone Dynamics (LCD) to the field theory and covariant equations. Then I shall speak on the applications of LCD to the description of two-, three and four nucleon systems as well as to relativistic form-factors and nuclear deep inelastic structure functions. In particular, I want to demonstrate that the convolution formula which expresses the nuclear structure function through the integral on the quark structure functions of nucleons can be derived in the framework of LCD and this formula does not contain the so called flux factor.

1. Bethe-Salpeter equation and problem of relative time.

For two relativistic spinless particles of equal masses $m_1 = m_2 = m$ the Bethe-Salpeter equation has the following form

$$\left[\left(\frac{1}{2}P + p \right)^2 - m^2 \right] \left[\left(\frac{1}{2}P - p \right)^2 - m^2 \right] X_P(p) + \int d^4p' \, V(p,p';P) X_P(p') = 0 \tag{1}$$

where P and p are the total and relative 4-momenta of two particles, V is the interaction operator.

The Fourier component of the reduced amplitude $X_P(p)$

$$X_{P.}(x) = \int d^4 p \, e^{-ipx} \, X_{P.}(p)$$

(2)

depends on the relative time $t = t_1 - t_2$ and should not be confused with the wave function.

The normalization condition for the reduced Bethe-Salpeter amplitude has the following form [1]:

$$\int d^4 p \, d^4 p' \, X_P(p') \frac{d}{dP_\mu} \left\{ \left[\left(\frac{1}{2} P + p \right)^2 - m^2 \right] \left[\left(\frac{1}{2} P - p \right)^2 - m^2 \right] \delta \, (p' - p) + V \, (p', p; P) \right\} X_P(p) = 2 P_\mu$$

(3)

and depends on the interaction.

This is the essential complication of the Bethe-Salpeter approach.

Other problems are related to more concrete models. If for example we limit ourselves by the ladder approximation we find that such a model has not the correct static limit when m_1 becomes very large and m_2 is fixed. Moreover in the ladder approximation there are so-called anomalous solutions which do not have nonrelativistic limit and some solutions have negative norm (see e.g. ref.[1]).

Because of those problems in many applications different authors prefer to use the so-called quasipotential equations where the relative time is eliminated. However more sophisticated and consistent approach where the relative time is absent from the very beginning is the Relativistic Hamiltonian Dynamics [2 - 9] which will be discussed later. The simplest example to see how the relative time can be eliminated in Bethe-Salpeter approach would be the zero range approximation where two relativistic particles interact trough the exchange of a third one with a very heavy mass. In this case the dependence of the interaction term on p and p' can be neglected and we have the following solution of Eq.(1) for the wave function of a bound state at rest ($P_0 = M$, $P = 0$)

$$X_{0.}(p) = N \Big/ \left(p^2 + m^2 - M^2 / 4 \right)$$

(4)

where

$$X_0(p) = \int dp_0 X_P(p)$$

satisfies usual normalization condition.

2. Relativistic two-body form-factor in field theory in zero range approximation (ZRA).

As an example let us consider the form-factor of deuteron in the Breit system where the longitudinal component of momentum transfer Q is equal to zero (see e.g. ref [9])

$$F(Q) = \int \frac{P_+ \, d^4 n}{d_+ n^2 - m^2} \frac{G}{p^2 - m^2} \frac{G}{p'^2 - m^2}$$

(5)

Here G is the vertex constant for transitions d->p+n and p'+n->d'; Q=d'-d=(0,0, $\mathbf{Q_T}$) ;d, p,n , d' , p' denote particles and there 4-momenta.

It is useful to present the answer in terms of light-cone variables:

$$\alpha=n_+/d_+ \ , \ 1-\alpha=p_+/d_+=p'_+/d'_+, \ \ \mathbf{k_T}=(1-\alpha)\mathbf{n_T}-\alpha\mathbf{p} \tag{6}$$

where $n_+= (E_\mathbf{n} + n_z)/\sqrt{2}$ and the same notation for all other momenta. The elimination of the fourth component is much simpler in terms of light-cone variables as compared with instant form variables $E_\mathbf{n}$, $\mathbf{n}$.. The reason is that the inverse propagators are linear functions of n_+ or n_- but are quadratic functions of energy $E_\mathbf{n}$ It means that calculating integral over $E_\mathbf{n}$ we have to consider 6 poles while in the complex n_- -plane there are only 3 poles. The integration on n_- is equivalent to the calculation of residue in the pole $n_- =(n_T^2 +m^2)/ 2n_+$. The result is

$$F(Q^2)= \int \frac{d^2k_T d\alpha}{(2\pi)^3 \alpha \ (1-\alpha \ )} \frac{G}{\left(m_d^2 - M^2(k_T,x)\right)} \frac{G}{\left(m_d^2 - M^2(k'_T,x)\right)} \tag{7}$$

where $M^2(k_T , x)=(k_T^2+m^2)/(\alpha(1-\alpha))$ and $\mathbf{k'_T} =\mathbf{k_T}+(1-\alpha)\mathbf{Q_T}$.

The form-factor can be expressed in terms of the relativistic light-cone wave function

$$F(Q^2)= \int \frac{d^2k_T d\alpha}{(2\pi)^3} \frac{\Phi(k_T, \alpha)\Phi(k'_T, \alpha)}{2\alpha \ (1-\alpha)} \tag{8}$$

where

$$\Phi (k_T,x)= G\left(m_d^2 - M^2(k_T,x)\right)^{-1} = G\left(m_d^2 - 4(k^2 +m^2)\right)^{-1} \tag{9}$$

In the second part of Eq.(2) we have introduced the new internal variable k_3 instead of α having in mind that $M^2(k_T , x)=(k_T^2+m^2)/(\alpha(1-\alpha))=4(k^2 + m^2)$ in terms of variables k_T and k_3 the light-cone wave function has the same form as the wave function introduced by Eq.(4). Both variables $\mathbf{k}$ and $\mathbf{p}$ can be considered as relative momenta of particles in the rest frame of the bound system. However being equal in absolute values they are directed differently and are related by Wigner rotation matrix (see ref.[9]).

The wave function (9) reminds very much the nonrelativistic formula for the deuteron wave function in ZRA. However the formula (8) for the relativistic form-factor is quite different from the nonrelativistic one because of relativistic kinematics and different treatment of recoil effects. This difference can easily been seen when we shall try to introduce the coordinate space. Let us write the form-factor in the mixed representation using the locality on the transfer coordinate ρ

$$F(Q^2) = \int \frac{d^2\rho_T d\alpha\, \Phi\left(\rho_{T'}\, \alpha\right) \Phi\left(\rho_{T'}\, \alpha\right)}{(2\pi)^3 \quad \alpha \qquad (1-\alpha)} e^{iQ\rho_T(1-\alpha)}$$

We see that the definition of the transverse coordinate depends on α. This internal coupling of the variables α and ρ complicates very much the introduction of coordinate space. As a result in the coordinate space the interaction of current with internal constituencies of relativistic bound system will be nonlocal. This is related to the general difficulties of introduction of coordinate space in relativistic case. For example the introduction of the longitudinal coordinate in the Light-Cone Dynamics seems not to be useful at all [9].

3. Relativistic oscillator on a light cone.

For the relativistic oscillator model [10] the covariant function of two spinless particles with the total momentum P has the following form (compare with the reduced Bethe-Salpeter amplitude defined by Eq.(2))

$$\Psi(x_1,x_2) = \exp\left(i\, P\, (x_1+x_2)/2\right) \Phi\, (x_1 - x_2) \tag{10}$$

where the internal function $\Phi(z)$ satisfies the equation

$$(-d_\mu d_\mu + \omega^4 z^2 + m^2)\, \Phi(z) = (M^2/4)\, \Phi(z) \tag{11}$$

To eliminate the relative time it is necessary to introduce the constraint (see ref.[7])

$$(P_\mu d_\mu + \omega^2 P_\mu z_\mu)\Phi(z) = 0 \tag{12}$$

with the condition $P^2=M^2$. On the null plane $z^+ =0$ the covariant function is restricted to the light -cone function $\phi(z_T,\, z_3)$ which satisfies the following equation
$$M^2\, \phi(z_T,\, z_3) = m^2\, \phi(z_T,\, z_3) \tag{13}$$
where $z_3 = (p_- z^- + p_T z_T)/M$. The mass operator squared M^2 and the angular momentum operators l_1, l_2 and l_3 can be written in the following forms

$$M^2 = 4\,(-d^2 + \omega^4 z^2 + 2\omega^2 d_3 z_3 + m^2) \tag{14}$$

$$l_1 = -i \, (z_2 \, d_3 - z_3 \, d_2 - \omega^2 z_2 \, z_3) \,, l_2 = i \, (z_1 \, d_3 - z_3 \, d_1 - \omega^2 z_1 \, z_3),$$

$$l_3 = -i \, (z_1 \, d_2 - z_2 \, d_1) \tag{15}$$

The eigenfunctions of the operator M^2 can be written as follows

$$\phi_{JKN}(z_T, z_3) = H_J \, (\omega z_1) \, H_K \, (\omega z_2) \, H_N \, (\omega z_3) \exp(- \omega^2 z_T^2 / 2) \tag{16}$$

and are normalized with the weight $\exp(- \omega^2 z_3^2 / 2)$. It is easy to verify that the operators M^2 and l_j obey the usual commutations relations

$$[M^2 , l_j] = 0, \quad [l_j , l_k] = i \, e_{jkm} \, l_m \tag{17}$$

In this representation the angular momentum operators depend on the interaction and the mass operator squared is not rotationally invariant. As it was noticed in ref.[8] it is possible to construct more convenient representation in which the operator M^2 will be rotationally invariant and angular momentum operators will not be dependent on the interaction.

To find this representation let us perform the following "gauge" transformation

$$\phi' = V \phi , \quad M_V^2 = V \, M^2 \, V^{-1} \,, l_j V = V \, l_j \, V^{-1} \tag{18}$$

where $V = \exp(- \omega^2 z_3^2 / 2)$. This transformation changes the normalization of the wave function but does not change commutation relations. For the transformed operators we shall have the following equations

$$M_V^2 = 4 \, (- d^2 + \omega^4 z^2 + \omega^2 + m^2) , \quad l_j = -i \, e_{jkm} z_k \, d_m \tag{19}$$

It is important that in this representation we have also the standard normalization condition for the wave function ϕ'. We shall call this representation as a standard one and assume that it can be used in the Light-Cone Dynamics for any form of the interaction. In any case it is correct from the formal point of view.

4. Relativistic Hamiltonian Dynamics (RHD) and Field Theory.

The main problem of RHD is to determine the 10 generators of Poincare group in terms of dynamical variables characterizing the given system of particles. The classical as well as the quantum aspects of Poincare-invariant theories with fixed number of particles in the different forms were investigated by many authors (see, e.g., refs.[3 - 9] and references therein). In the case of the quantum problem we face two important questions : (1) wether such a problem has a formal solution and, if it exists, (2) whether this formal solution is useful in analyzing concrete physical systems.

As concerning the formal point of view, RHD is selfconsistent, i.e. for a system with a fixed number of particles one can in principle find the representation of the Poincare - group generators which is expressed through the variables of the particles. In the classical analogue of such a theory the canonical variables would not coincide with the variables of the particles and the Lorentz-invariant world lines can not be defined in the interaction region as a consequence of using collective variables (the center - of - mass and internal coordinates). Their connection to the coordinates of the particles depends on the particular interactions and it cannot always be established explicitly. In the quantum case difficulties arise in the description of the interactions of such systems with an external field. In different cases they can be bypassed using field-theoretical considerations. For example, as we have seen in Sect . 2 the analysis of the relativistic form-factor in the coordinate space in Light-Cone Dynamics is not useful at all because the longitudinal coordinate has no transparent physical meaning and the interaction of the external current with relativistic bound system appears to be nonlocal. Nevertheless the relativistic form-factor can be calculated at least in ZRA.

In quantum field theory the operator of the number of particles does not commute with the Hamiltonian and the state vector of the nucleus, for example, in different frames of references contains different admixtures of π - mesons and quark-antiquark pairs. The wave function of the bound state has its simplest form in the infinite-momentum frame (IMF) where , in particular, the creation of particles from the vacuum is absent. The advantages of the IMF for the description of bound states of hadrons were pointed out by Weinberg [11] who has stressed that there is the simple connection of the diagrams of non-covariant perturbation theory in the IMF with the unitarity condition. At the same time, in the covariant Feynman diagrams Lorentz invariance of the S-matrix in every order of perturbation theory is ensured by the mixing of contributions of intermediate states with the different number of particles and antiparticles.

To be able to have the interpretation of the results with the help of quantum field theory it is convenient to use Light-Cone Dynamics (LCD). The amplitudes written in IMF can be connected with Feynman diagrams in the IMF [11]. Nevertheless, the scheme used by Weinberg [11] and his relativistic two-particle equation are not yet equivalent to RHD in its light-cone form, since the relative time in it is not eliminated. As we have seen in Sec. 2 those schemes can be easily related in ZRA. In more general case the relation between those schemes can not be found easily. Formally, the elimination of relative time in LCD can be realized by imposing the so-called angular condition on the wave function or choosing from the very beginning of the standard representation for the internal angular momentum operators and imposing the correct commutation relations between Poincare generators.

Let us discuss the applications of LCD to the description of systems with strong interacting relativistic particles. As we know , quantum chromodynamics , as the fundamental theory of strong interactions , contains an infinite number of degrees of freedom, in contrast to RHD in any form. From this point of view LCD is to be considered as an approximate or model description of physical systems. It turns out that in many cases the approaches taking into account only a finite number of degrees of freedom in strongly interacting systems are quite good. One may refer, for example, to the successes of the constituent quark models or bag models in the description of the static properties of hadrons. LCD was also successfully applied to the description of relativistic effects in few nucleon systems (see , e.g. , refs. [12 - 18]).

The main advantage of using LCD for description of relativistic effects in nuclear physics is that the operator formalism of this scheme is the same as in the nonrelativistic quantum mechanics. As compared with the field theory the use of LCD can be interpreted as a kind of perturbative expansion not in the coupling constant but in effects of retardation and inelasticities. In some sense the use of LCD is similar to dispersion approach when only limited amount of intermediate states is taken into account. If we take into account all the possible intermediate channels then this approach will be equivalent to the field theory.

5. Three Relativistic Particles in the LCD.

The 10 generators of the Poincare group in the LCD for a system of three relativistic particles can be written as follows (see refs. [8- 9]

$$P_- = H = (P_T^2 + M^2)/2P_+, \qquad P_T = P_T , \qquad P_+ = P_+ ,$$
$$M_{+j} = -i\, P_+ d / d\, P_j , \qquad\qquad M_{+-} = i\, P_+ d / d\, P_+ ,$$
$$M_{12} = -i\, P_1 d / d\, P_2 + i\, P_2 d / d\, P_1 + \Sigma_3 , \qquad\qquad (20)$$
$$M_{-j} = i\, P_j d / d\, P_+ + i\, H\, d / d\, P_j + e^{jk} (M\, \Sigma_k + P_k \Sigma_3)/ P_+ .$$

Here $(j , k) = (1 , 2)$, P_+ and P_j are three conserved components of the total momentum , M and Σ are the mass and spin operators which depend only on the internal variables but not on P_+ and P_j.

The seven generators $P_+ , P_j , M_{12} , M_{+j}$ and M_{+-} do not depend on the interaction , since they describe transformations , which leave the light-cone (null plane) surface invariant. The other three generators P_- and M_{-j} contain the interaction. In order to define all ten generators for the case of interacting particles , it is sufficient to define the operators of the mass M and the spins Σ_k , Σ_3 and to demand that M and Σ commute with each other and three components of spin operator commute in the standard way.

The form of the mass and spin operators depends on the representation. The simplest way is to use the representation in which the spin operator does not depend on an interaction and has the same form as for free particles (see, e.g., refs. [8 - 9]). As we have seen in Sect.3 for LCD this assumption is not trivial at all.

In this representation the mass operator of the three-particle system which satisfies the cluster separability property can be written in the following form

$$M = M_0 + W = M_0 + \Sigma_{jk} W_{jk} + W_{jkl} \tag{21}$$

where the pair interactions have the form

$$W_{jk} = E_{jk} - E_{jk} = (Q_1^2 + M^2_{jk})^{1/2} - (Q_1^2 + M^2_{jk})^{1/2} \tag{22}$$

and W_{jkl} represents three body forces.

6. The Choice of Two - Body Interaction.

It is convenient to write the two-particle mass operator squared in the following form

$$M^2_{jk} = M^2_{jk} + w_{jk}$$

Then the relativistic equation for the scattering amplitude for two particles with equal masses can be written as

$$T(q', q; s) = w(q', q) + \int \frac{d^3 q'}{(2\pi)^3 E(q')} \frac{w(q', q'')T(q'', q; s)}{s - 4(q^2 + m^2)} \tag{23}$$

where q' and q are the relative momenta of particles in the initial and final states ; $s = 4 (k^2 + m^2)$. After the substitutions

$$w(q', q) = 4m (E(q')E(q))^{1/2} v(q', q) ; T(q', q; s) = 4m (E(q')E(q))^{1/2} T(q', q; s) \tag{24}$$

we find that the form of Eq.(23) becomes the same as the non-relativistic Lippman-Schwinger equation

$$t(q', q; s) = v(q', q) + \int \frac{d^3 q''}{(2\pi)^3} \frac{v(q', q'')t(q'', q; s)}{k^2/m - q''^2/m} \tag{25}$$

Therefore there are two possibilities, how to determine the relativistic two body potential $w (q', q)$:

(1) by direct reconstruction of the potential $w(q', q)$ through Eq.(23) fitting the scattering phase shifts;

(2) by the transformation (24) from the nonrelativistic phenomenological potential whose parametrization is assumed to be already known.

In both cases the two-body observables are the same as from nonrelativistic calculation with the potential $v(q', q)$ along Eq.(25). This property is valid only for particles with equal masses.

If we neglect recoil effects and effects of the Wigner rotation of the spins, then this property of the relativistic two-particle equation (23) coincides with the well-known "minimal relativity" principle [19]. Applying this principle to Eq.(23) we can use , for example, the known nonrelativistic N-N potentials to constitute the interactions for the relativistic three and four nucleon systems. This formalism was applied to the calculation of relativistic corrections (RC) to the binding energies of three and four nucleons [13 - 15], to the analysis of hadron - deuteron scattering at large angles [20] , to the calculation of RC to the magnetic moment of deuteron [16], to the relativistic calculations of deuteron form-factors [17 - 18]. In all the cases the inclusion of relativistic corrections improves the description of the experimental data.

7. Relativistic Corrections to the Binding Energies of Three and Four Nucleons.

In ref. [15] the corrections to the order $(v / c)^2$ to the binding energy of three nucleons were calculated in the framework of LCD. For the numerical calculation of RC to the binding energy of the triton it was used the parametrization of the RSC wave functions proposed in ref. [21].

The corresponding value of the nonrelativistic binding energy ε_0 for this model is known to be - 7.02 MeV. It was found that RC in the five- channel approximation (S + D - waves) $\Delta\varepsilon = - 0.5376$ MeV, in three- channel approximation (S + D -waves) $\Delta\varepsilon = - 0.5029$ MeV and in two- channel approximation (only S -waves) $\Delta\varepsilon = - 0.0998$ MeV . Therefore RC increase the binding energy and the contribution of higher partial waves seems to be very important in calculation of RC.

The composition law for the mass operator of four relativistic particles has the following form [6 , 23]

$$M = \Sigma_k (-1)^k (k-1)! \, M_k \tag{26}$$

where the mass operators of clusters are given by

$$M_2 = \Sigma_{i > j > k} (E_{ijk} + E_l) + \Sigma_{k > l, i > j} (E_{kl} + E_{ij})$$
$$M_3 = \Sigma_{k > l} (E_{kl} + E_i + E_j) ,$$
$$M_4 = \Sigma_i E_i$$

The two and three body energy operators are defined through the mass operators and total momentum operators of those clusters as in Eq. (22).

The calculation of RC to binding energy of four nucleons was made in ref. [14] using the Afnan - Tang N - N potential . and taking into account only S - waves. It was found that the relativistic corrections to different terms in the mass operator are quite noticeable. For example , RC to the kinetic energy is about - 4 MeV (giving extra binding). However , the correction to the potential energy , which is about +4.4 MeV, compensates it to large extent. It would be important to calculate this correction for more realistic potentials.

8. Current Operator in LCD and Relativistic Form-Factors of Deuteron.

The problem of the construction of the current operator in LCD was discussed in refs. [16,17, 23 , 24] . The restrictions imposed on the current operator in LCD by Poincare - invariance were originally formulated by Osborn [23]. In general case the current operator can not be closed in the sector with the fixed number of particles. However, it is possible to formulate the necessary condition which should be imposed on the current operator in terms of variables describing the given sector of particles. This condition is known as the Osborn angular condition. It is important that the "good" component of current operator J_+ is much less restricted by Poincare invariance as compared with other components.

As it was shown in ref.[24] the local interaction of relativistic constituencies can be consistent with relativistic invariance in two limiting cases : (1) at small momentum transfer q in the first two terms of expansion of the current operator in powers of q and (2) at large momentum transfers in asymptotically free theories when the interaction between constituencies vanishes. In all other cases the calculations should be based on the additional physical arguments (see, e.g. , refs. [12, 16, 17]). One of the main arguments is to use the good component of current operator which in many cases does not mix the positive and negative energy states. The comparison with the ZRA limit in field theory can also be very useful.

LCD was used for the calculation of RC to the deuteron magnetic moment in ref. [16].It was found that when RC are taken into account the discrepancies of realistic model predictions with the data become by two times smaller. For example, the RC to the deuteron magnetic moment is equal to 0.63×10^{-2} n.m. for the Paris potential and the final discrepancy μ_d (exp) - μ_d (theor) is only 0.3×10^{-2} n.m. and may be attributed to the contribution of exchange currents or to the admixture of 6q - bag.

As it was shown in refs. [17 - 18] relativistic effects improve also the description of the deuteron structure functions A (q^2) and B (q^2).

9. Convolution Formula for Nuclear Structure Function in LCD. (Is it necessary to introduce the flux factor?)

One of the important applications of LCD would be the analysis of the deep inelastic scattering of leptons on nuclei. Up to now there were no systematic use of LCD considered as a form of RHD to inelastic and deep inelastic scattering on nuclei. Frankfurt and Strikman [12] used extensively formalism of Feynman graphs considering them in IMF. For two-body problem when they impose the angular condition on the two-particle wave function this approach is equivalent to LCD. However in general case this equivalence may be absent.

An important point which I want to stress here is the difference between LCD and the approaches which treat the scattering on nucleons inside nucleus as on the virtual objects which have the masses different from the masses of free nucleons (see, e.g., refs.[26 - 28]). In LCD as in the quantum mechanics all the particles in all the intermediate states are on the mass shell. The quantity which is not conserved in the intermediate states is the energy in the instant form of RHD and P_- in LCD. Only in this case we can compare the deep inelastic structure functions of nuclei and nucleons. If we consider nucleon as a virtual particle we have to assume some relation between its structure function and the structure function of free nucleon which may be strongly model dependent.

Another thing which I want to discuss here is the derivation of the convolution formula in LCD and the problem of the so-called flux factor. Using the formalism with the virtual nucleons Frankfurt and Strikman [29] have derived the following formula relating the deep inelastic structure functions of nucleon and nucleus

$$ F_{2A}\left(x, Q^2\right) = \int d^4k \, S(k) \, \alpha F_{2N}\left(\frac{x}{\alpha}, Q^2\right) $$

(27)

where $\alpha = A \, k_+ / \, P_+$ and the spectral function of the virtual nucleon is normalized as follows

$$ 1 = \int d^4k \, S(k) \, \alpha $$

(28)

The factor α in Eqs. (27) and (28) is known as a flux factor. The normalization of $S(k)$ without this factor as in the nonrelativistic nuclear physics was strongly criticised in [29].

Using the example with the deuteron structure function I shall demonstrate that in LCD the flux factor does not appear and all the necessary conservation laws (like baryon or momentum conservation) are simply related to the normalization of the light-cone wave function.

Let us define the deuteron light-cone function

$$ < d \mid p \, n > = 2 d_+ \delta (d_+ - p_+ - n_+) \, \delta^{(2)} (d_T - p_T - n_T) (2 \pi)^3 \, \Phi(k_T, \alpha) \quad (29) $$

with the normalization condition

$$<d \mid d'> = 2d_+ \delta(d_+ - d'_+) \delta(d_T - d'_T)(2\pi)^3 \qquad (30)$$

which is equivalent to the normalization of the form-factor (8) $F(0) = 1$.

Using the definition of the nucleon structure functions

$$W_N{}^{\mu\nu} = \Sigma_N <p \mid J^*{}_\mu \mid N><N \mid J_\nu \mid p>(2\pi)^3 \delta^{(4)}(p+Q-p_N) =$$
$$= (p_\mu - Q_\mu p.Q/Q^2)(p_\nu - Q_\nu p.Q/Q^2) W_2{}^N(x,Q^2)/m^2 +$$
$$+ (\delta_{\mu\nu} - Q_\mu Q_\nu/Q^2) W_1{}^N(x,Q^2)$$

we can write the following expression for the deuteron structure functions

$$W_d{}^{\mu\nu} = <d \mid W_N{}^{\mu\nu} \mid d'> =$$

$$= \int \frac{d^2 p_T dp_+}{2p_+} \frac{d^2 p'_T dp'_+}{2p'_+} \frac{d^2 n_T dn_+}{2n_+}$$

$$\frac{1}{(2\pi)^3} 2d_+ \delta(d_+ - p_+ - n_+) \delta(d_T - p_T - n_T) 2d_+ \delta(d'_+ - p'_+ - n_+) \delta(d'_T - p'_T - n_T) W_N^{\mu\nu} =$$

$$= \int \frac{d^2 k_T d\alpha \, \Phi^2(k_T,\alpha)}{(2\pi)^3} \frac{1}{2\alpha(1-\alpha)} \frac{1}{\alpha} W_N^{\mu\nu} =$$

$$= \int d^2 k_T d\alpha \, \rho(k_T,\alpha) \frac{1}{\alpha} W_N^{\mu\nu} \qquad (31)$$

The factor $1/\alpha$ in the last equation cancels the flux factor α and we have the convolution formula in the following form

$$F_{2A}(x, Q^2) = \int d^2 k_T d\alpha \, \rho(k_T, \alpha) F_{2N}\left(\frac{x}{\alpha}, Q^2\right) \qquad (32)$$

where the normalization condition reads

$$1 = \int d^2 k_T d\alpha \, \rho(k_T, \alpha) \qquad (33)$$

Introducing instead of α the variable k_3 according to

$$\alpha = (\varepsilon(k) + k_3)/2\varepsilon(k)$$

we find

$$F_{2A}(x, Q^2) = \int d^3 k \rho(k) F_{2N}\left(\frac{x}{\alpha}, Q^2\right) \qquad (34)$$

where $\rho(k) = \Phi^2(k)/\varepsilon(k)$ or according to the principle of minimal relativity (see Sect.4) $\rho(k) = \Psi^2(k)$ with $\Psi(k)$ corresponding to some nonrelativistic phenomenological parametrization of the deuteron wave function.

Similar derivation of the convolution formula can be done for other nuclei.

As it is shown in ref. [30] the presence or absence of the flux factor is quite important for the interpretation of EMC - effect .

10. Conclusions.

1) Relativistic Hamiltonian Dynamics for fixed number of particles can be formally constructed and applied to the description of few- body physical systems;

2) RHD can be related to the field theory in the zero- range approximation;

3) The advantage of Light-Cone Dynamics is in possibility to compare the results with the Feynman diagrams in the infinite momentum frame where the contributions of the negative energy states are suppressed;

4) Few - body problem in RHD can be formulated also with the help of covariant equations with constraints which eliminate the dependence of the wave functions on the relative times;

5) Relativistic corrections improve the descriptions of the binding energy of three nucleons in realistic models as well as other observables in few -nucleon systems;

6) The formulation of the relativistic few-body problem in coordinate space seems not to be useful;

7) The convolution formula for nuclear structure functions can be derived in LCD and does not contain the so-called flux-factor;

8) One of the very interesting development in RHD would be to generalize it for many channels. In the limit of an infinite number of channels such a scheme would be equivalent to relativistic field theory.

I am very grateful to the organizers of this Conference C. Ciofi degli Atti, E. Pace, G. Salmè and S. Simula for the invitation and for the financial support.

References

1. C. Itzykson, J.-B. Zuber. Quantum Field Theory. Mc Graw-Hill Book Company, N.Y.

2. P.A.M. Dirac. Rev.Mod.Phys. 21, 392, (1949).

3. B. Bakamdjan, K.H. Thomas. Phys. Rev. 92, 1300 (1953).

4. F. Coester. Helv. Phys. Acta 38, 7 (1965).

5. L.L. Foldy, R.A. Krajcik. Phys. Rev. D12 , 1700 (1975).

6. S.N. Sokolov. Dok. Akad. Nauk. USSR 233, 575 (1977); Theor. Math. Phys. 36, 193 (1978).

7. H.Leutwyler, J. Stern. Ann. Phys. (N.Y.) 112, 490 (1979).

8 B.L.G. Bakker, L.A. Kondratyuk, M.V. Terentiev. Nucl. Phys. B158, 497 (1979).

9. L.A. Kondratyuk, M.V. Terentiev. Yad. Phys. 31, 1087 (1980).

10. R.P. Feynman, M. Kislinger, F. Ravndal.Phys. Rev. D3 , 2706 (1971).

11. S. Weinberg. Phys. Rev. 150 , 1313 (1966).

12. L.L. Frankfurt, M.I.Strikman. Phys. Rev. C76 , 215 (1981).

13. L.A. Kondratyuk, J. Vogelzang, M.S. Fanchenko. Phys. Lett. 98B, 405 (1981).

14. A.I. Veselov, L.A. Kondratyuk. Yad. Phys. 36, 343 (1982).

15. L.A. Kondratyuk, F.M. Lev, V.V.Soloviev. Few- Body Systems. 7, 55 (1989).

16. L.A. Kondratyuk, M.I.Strikman. Nucl. Phys. A426, 575 (1984).

17. I.L. Grach, L.A. Kondratyuk. Yad. Phys. 39 , 316 (1984); L.L. Frankfurt, I.L. Grach, L.A. Kondratyuk, M.I.Strikman. Phys. Rev.Lett. 62, 387 (1989).

18. P.L. Chung, F. Coester, B.D. Keister, W.N. Polyzou. Phys. Rev. C37, 2000, (1988).

19. G.Brown, A.D.Jackson, T. Kuo. Nucl. Phys. A133, 481 (1969).

20. L.A. Kondratyuk, F.M. Lev, L.V. Schevchenko. Yad. Phys. 33, 1208 (1981); 36, 377 (1982).

21. Ch. Hajduk, A.M. Green, M.E. Sainio. Nucl. Phys. A337 , 13 (1980).

22. F.M. Lev. Yad. Phys. 45, 26 (1987).

23. H. Osborn. Nucl. Phys. B38, 429, (1972).

24. B.V. Berestetsky, M.V. Terentiev.Yad. Phys. 24, 1044 (1976).

25. L.A. Kondratyuk. In : " Progress and Perspectives in Nuclear Physics at Intermediate Energy " (Ed. by S.Boffi, C; Ciofi degli Atti, M.Giannini) World Scientific, 1989.

26. S.V. Akulinichev et. al. Phys. Rev.Lett. 55, 2239 (1985).

27. B.L. Birbrair et. al. Phys. Lett. 166B, 119 (1986).

28. G.V. Dunne, A.W. Thomas. Nucl. Phys. A455, 701 (1986).

29. L.L. Frankfurt, M.I.Strikman. Phys. Lett. 183B, 254 (1987).

30. C. Ciofi degli Atti, S. Liuti. Phys. Lett. B225, 215 (1989);

Few-Body Systems, Suppl. 6, 526—531 (1992)

Few-
Body
Systems

FRONT-FORM CALCULATION OF γd→np REACTIONS AT HIGH ENERGIES

T.-S. H. Lee
Physics Division, Argonne National Laboratory
Argonne, IL 60439-4843 USA

Abstract

A front-form calculation of γd→np reaction has been performed and compared with the data at 90°.

To investigate nuclear dynamics in the high energy and/or high momentum-transfer regions, it is necessary to develop a relativistic formulation of the problem. The most well developed relativistic formulation is the Lagrangian local quantum field theory. Although significant progress has been made in recent years, rigorous theoretical methods for predicting bound states of a system of strongly interacting particles, such as the deuteron, from a given Lagrangian are still being developed. The traditional way to circumvent this difficulty is to assume that the nuclear dynamics can be effectively defined within the nonrelativistic particle quantum mechanics and the interaction potentials and operators for observables are "derived" from taking equal-time limits of a set of field theory amplitudes. This phenomenological approach has been very successful in investigating low energy nuclear physics. In this talk I shall describe a similar procedure, developed in a collaboration with Coester and Kondratyuk, to investigate the photo-disintegration of the deuteron at high energies.

Our approach is to assume that the nuclear dynamics can be defined within the relativistic particle quantum mechanics. This relatively unfamiliar theoretical framework has recently been explicitly presented in an excellent review by Keister and Polyzou [1]. To introduce this short talk, I only want to point out that this approach allows us to use the "existing" meson-exchange two-nucleon models, such as the Paris potential and coupled-channels NN⊕NΔ⊕NNπ models, to carry out a fully relativistic calculation of γd→np reaction. Our formulation follows closely that developed by Chung, Coester, Keister and Polyzou [2] in their study of e-d elastic form factors.

The form of the relativistic quantum mechanics is not unique since the energy-momentum relation $P^2 = P^\mu P_\mu = M^2$ is of a bi-linear form. Our formulation is a front-form defined by a light-like vector $n^\mu = \{1,\hat{n}\}$ with $|\hat{n}| = 1$. Any four vector is then defined with respect to this choice as

$$A^\mu : \{A^-, \vec{A}\} \quad \text{with} \quad \vec{A} : \{A^+, \vec{A}_T\} \quad \text{and} \quad \vec{A}_T \cdot \vec{n} = 0 \ .$$

The scalar product of two four vectors is defined as

$$A \cdot B = \tfrac{1}{2} \, [A^+ B^- + B^+ A^- - 2\vec{A}_T \cdot \vec{B}_T] \ .$$

It is common to choose $\hat{n} \parallel \hat{z}$ and hence in terms of the usual canonical components $A^\mu = (A^0, A^1, A^2, A^3)$ we have $A^\pm = A^0 \pm A^3$ and $\vec{A}_T = A_1 \hat{x} + A_2 \hat{y}$. We then have for a particle with mass m

$$p^\mu p_\mu = p^- p^+ - \vec{p}_T^2 = m^2 \ .$$

The front-form Hamiltonian is defined as

$$H = P^- = \frac{m^2 + \vec{p}_T^2}{p^+} \tag{1}$$

To describe a two-particle system, we first write in the absence of interactions

$$\vec{P} = \vec{p}_1 + \vec{p}_2 = (P^+, \vec{P}_T) \tag{2}$$

$$H_0 = \frac{M_0^2 + \vec{P}_T^2}{P^+} \tag{3}$$

where the mass operator M_0 can be expressed in terms of the intrinsic momentum defined by a front-form Lorentz boost defined by

$$L_F(P) \ \{p^0, p^1, p^2, p^3\} = \{M_0, 0, 0, 0\}$$

Explicitly we have

$$M_0^2 = 4 \ (m^2 + \vec{k}^2) \tag{4}$$

where

$$\vec{k}^2 = k_n^2 + \vec{k}_T^2 \ , \quad k_n = \vec{k} \cdot \vec{n} \ .$$

The components of front-form vector $\vec{k}:\{k^+,\vec{k}_T\}$ are defined by

$$k^+ = \{L_f(P)p_1\}^+ = M_0\xi$$

$$\vec{k}_T = \{L_f(P)p_1\}_T = \vec{p}_{1T} - \vec{P}_T\xi$$

$$k_n = M_0 \ (\xi - 1/2) \tag{5}$$

where $\xi = p^+{}_1/P^+$ is the momentum fraction of the first nucleon. The spin of the system is defined as

$$\vec{J} = i \ \vec{\nabla}_k \times \vec{k} + R_M \ (\xi,\vec{k}_T,m) \ \vec{s}_1 + R_M \ (1 - \xi,\vec{k}_T m) \ \vec{s}_2 \tag{6}$$

where

$$R_M(\xi,\vec{k}_T,m) = \frac{m + \xi \ M_0 - i \ \vec{\sigma} \cdot (\vec{n} \times \vec{k}_T)}{[(m + \xi \ M_0)^2 + \vec{k}_T^2]^{1/2}} \tag{7}$$

is the Melosh transformation.

The dynamics are introduced by modifying the mass operator

$$M_0^2 \rightarrow M^2 = M_0^2 + 4m \ V_{12} \tag{8}$$

As discussed in details in Refs. [1] and [2], the relativistic invariance can be achieved if we require that V_{12} is a function of only $\vec{k},\vec{\nabla}_{\vec{k}},\vec{s}_1,\vec{s}_2$ and satisfies $[V_{12} \ , \ \vec{J}] = 0$. Then the eigenfunctions of the four momentum

$$p^\mu:\{H,\vec{P}\} \quad \text{with} \quad H = \frac{M^2 + \vec{P}_T}{p^+} \ , \quad \vec{P}:\{p^+,\vec{P}_T\}$$

can be written as

$$H|\Psi_{P_d, \mu_d}\rangle = \frac{M_d^2 + \vec{P}_{dT}^2}{P_d^+} |\Psi_{P_d, \mu_d}\rangle \qquad (9.a)$$

$$\vec{P}|\Psi_{P_d, \mu_d}\rangle = \vec{P}_d|\Psi_{P_d, \mu_d}\rangle \qquad (9.b)$$

The main point is that the wave function, Eq. (9), can be directly constructed from the existing NN models. This is explicitly given in Eqs. (2.27)-(2.31) of Ref. [2] for the deuteron ground state. Extensions of these equations to calculate np scattering state is straightforward. This practical simplicity is due to the fact that the form of $M^2|\chi\rangle = E_d|\chi\rangle$ can be cast into the form of the usual nonrelativistic Schrödinger equation in momentum space, and the front-form boost transformation is only kinematic.

The amplitude of the $\gamma d \rightarrow np$ reaction is

$$T_{fi} = \epsilon_\mu \langle f|J^\mu|i\rangle$$

where ϵ^μ is the photon polarization vector and

$$\langle f|J^\mu|i\rangle = \langle \chi_{p,\mu}^{(-)}|J^\mu|\Psi_{P_d, \mu_d}\rangle = \sum_{\substack{\mu_1', \mu_2' \\ \mu_1, \mu_2}} \int d\vec{p}_1 d\vec{p}_2' d\vec{p}_1 d\vec{p}_2 \; \chi_{p\mu}^{(-)*}(\vec{p}_1'\mu_1'\vec{p}_2'\mu_2')$$

$$\langle \vec{p}_1'\mu_1'\vec{p}_2'\mu_2'|J^\mu|\vec{p}_1\mu_1\vec{p}_2\mu_2\rangle \; \Phi_{P_d, \mu_d}(\vec{p}_1, \mu_1, \vec{p}_2\mu_2) \qquad (10)$$

To proceed, we need to define the front-form matrix elements of the current operator J^μ. We choose the light-front vector n^μ such that $q^+ = 0$ and hence $\vec{q}_T = 0$ because for a real photon $q^+q^- - \vec{q}_T^2 = 0$. In the γd cm frame, this choice of front-form dynamics requires setting $\vec{p}_d = -\vec{q} \parallel \hat{n} = \hat{z}$. The current conservation then leads to the following condition

$$q_\mu \langle f|J^\mu i\rangle = q^+\langle f|J^-|i\rangle + q^- \langle f|J^+|i\rangle - \vec{q}_T \cdot \langle f|J_T|i\rangle \equiv q^-\langle f|J^+|i\rangle = 0 .$$

Hence any acceptable model of the current operator must satisfy the following condition

$$\langle f|J^+|i\rangle = 0 \qquad (11)$$

530

The total front-form momentum vector is conserved in any reaction. Since $q^+ = 0$ and $\vec{q}_T = 0$, we then have for the $\gamma d \rightarrow np$ reaction

$$\vec{q} = (q^+, \vec{q}_T) = 0 \quad \text{and} \quad \vec{q} + \vec{p}_d \equiv \vec{p}_d = \vec{p} \ . \tag{12}$$

This means that the total front-form three vectors $\vec{p}:(p^+, \vec{p}_T)$ of the final np system is equal to $\vec{p}_d$ of the initial deuteron state. Thus the impulse current matrix element takes the following "diagonal" form

$$\langle \vec{p}_1' \mu_1' \vec{p}_2' \mu_2' | J^\mu | \vec{p}_1 \mu_1 \vec{p}_2 \mu_2 \rangle$$

$$= \delta(\vec{p}_1' - \vec{p}_1) \ \delta_{\mu_1' \mu_1} \ \delta(\vec{p}_2' - \vec{p}_2) \ \langle \vec{p}_2' \mu_2' | J^{(\mu)} | \vec{p}_2 \mu_2 \rangle + (1 \leftrightarrow 2) \ . \tag{13}$$

The one-body current matrix element can be directly related to the known electromagnetic properties of a single nucleon

$$\langle \vec{p} \ \mu' | J^\mu | \vec{p} \ \mu \rangle = \bar{u}_{\vec{p}\mu'} \ \gamma^\mu u_{\vec{p}\mu} \tag{14}$$

where $u_{p\mu}$ is the front-form Dirac spinor. It can be shown for the J^+ component that

$$\langle \vec{p}\mu' | J^+ | \vec{p}\mu \rangle = 2 \ \delta_{\mu\mu'} \tag{15}$$

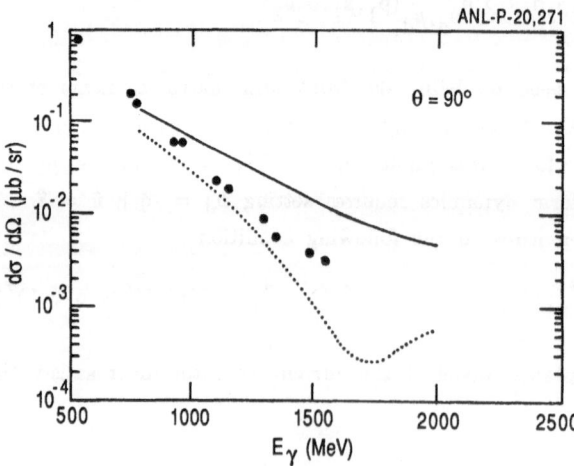

Fig. 1   The predicted differential cross section of $\gamma d \rightarrow np$ is compared with the data [4]. The dotted curve is obtained when the final state interaction is neglected.

Substituting Eqs. (13)-(16) into (10) and noting the orthogonality of the scattering and bound state wave functions, we then have

$$\langle f | J^+ | i \rangle \propto 2 \langle \chi^{(-)}_{\vec{p}\mu} | \Phi_{\vec{p}_d, \mu_d} \rangle \equiv 0 \tag{16}$$

The current conservation condition Eq. (11) is therefore satisfied.

The calculation was performed by using the Paris potential. The calculated differential cross sections at 90° are shown in Fig. 1. The dotted curve is obtained when the final state np interaction is neglected. Clearly, at GeV energies the the main contribution is due to the np final state interaction. The predicted energy-dependence of the present front-form calculation is not too different from that of our earlier coupled-channel calculation [3]. It appears that the neglect of relativistic effects is not the main reason for the failure of the meson-exchange calculations to describe the data. Note that at 2 GeV incident photon energy the final np scattering energy is nearly 4 GeV in the laboratory system. The use of the Paris potential for the final state is obviously not correct. In order to explore whether the γ d → np reaction can be described in terms of hadronic degrees of freedom, we need to develop a NN model which can account for the NN data up to about 5 GeV. It is also necessary to develop an approach to account for two-body currents in the front-form formulation.

This work is supported by the U.S. Department of Energy, Nuclear Physics Division, under contract W-31-109-ENG-38.

References

1. Keister B.D., Polyzou, W.N.: Advance in Nuclear Physics, (eds.) Negele, J.W., Vogt, E.: (Plenum Press 1991) Vol. 20, pp. 225.

2. Chung, P.L., Coester, F., Keister, B.D., Polyzou, W.N.: Phys. Rev. C 37, 2000 (1988).

3. Lee, T.-S.H.: Proceedings of the International Conference on Medium and High Energy Nuclear Physics, (eds.) Hwang, W.-Y.P., Lin, K.F., Tzeng, Y.: (World Scientific 1989) pp. 563.

4. Napolitano, J., et al., Phys. Rev. Lett. 22, 2530 (1988).

Few-Body Systems, Suppl. 6, 532—537 (1992)

RELATIVISTIC TWO BODY EQUATIONS AND NUCLEAR INTERACTIONS

M. De Sanctis, D. Prosperi

I.N.F.N. - Sezione di Roma
Dipartimento di Fisica, Università "La Sapienza", P.le A.Moro 2, 00185 Roma, Italy

Abstract

A relativistic two-body wave equation has been adapted for the study of nuclear systems. A specific One Boson Exchange model has been constructed. The introduction of the Δ-resonance degrees of freedom in our formalism is discussed.

Introduction

The present lecture is organized as follows: in Sect. 1 we will discuss the main properties of the adopted relativistic equation, in Sect. 2 we will examine a specific O.B.E. model obtained from our relativistic equation and, finally, in Sect. 3 we will briefly analyse the problems related to the introduction of the Δ-resonance in our formalism.

1. The Relativistic Equation

The study of relativistic effects in nuclear few body systems can be conveniently performed by means of a relativistic wave equation that correctly adds up the contributions of the main Feynman graphs required to describe the elementary interaction processes.

At this regard we recall that many efforts have been devoted to construct relativistic wave equations for two interacting spin 1/2 particles. However, many of these efforts give rise to formal or practical problems. In particular, the Breit equation [1] for two electrons in an external field, due to an incorrect treatment of the negative energy states does not admit normalizable bound states (continuum dissolution) [2]; the Bethe-Salpeter equation [3] in the ladder approximation, does not limit to the one-body Dirac equation when the mass of the other particle approaches infinity.

Finally, the Gross equation [4], does not exhibit an exact symmetry with respect to particle interchange, at least in the non relativistic limit.

Recently, a relativistic two-body equation, not affected by the above mentioned inconveniences, has been proposed by Mandelzweig and Wallace in the QED context [5,6]. Their procedure allows to incorporate also the series of the crossed graphs in the eikonal approximation. In this context we have examined the nuclear case [7] in which the interaction is mainly given by a field of isovector nature (the pion).

The three-dimensional Green's function, related to the graphs

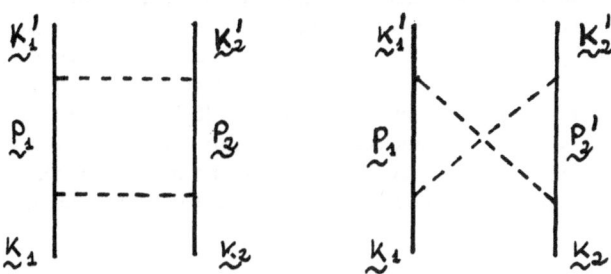

is obtained by adopting the eikonal approximation (i.e. by setting $p_2' = (p_2'^0, \vec{p}_2)$)

and the instantaneous meson exchange approximation (i.e. by neglecting the time components of the meson momenta). In consequence the Green's function, in the C.M. reference frame, has the form

$$G(E_1, \vec{p}_1; E_2, \vec{p}_2; \vec{\tau}_1 \vec{\tau}_2) = -\sum_{\alpha, \beta} \frac{(M_1^\alpha(\vec{p}_1) M_2^\beta(\vec{p}_2)) \alpha \cdot \beta}{\epsilon_1 + \epsilon_2 - \alpha E_1 - \beta E_2} \gamma_1^0 \gamma_2^0 F^{\alpha\beta} , \qquad (1)$$

where $M_i^\rho(\vec{p}_i)$ is the standard projection operator on the states of momentum $\vec{p}_i$ and energy $\rho \epsilon_i$ (with $\rho = \pm 1$ and $\epsilon_i = (\vec{p}_i^2 + m_i^2)^{1/2}$) of the $i - th$ nucleon (with $i = 1,2$). Moreover

$$F^{\alpha\beta} = \begin{cases} 1 & \text{for } \alpha = \beta \\ \frac{1}{3}(11 + 4\vec{\tau}_1 \cdot \vec{\tau}_2) & \text{for } \alpha \neq \beta \end{cases} \qquad (2)$$

By a standard use of the properties of the projection operators we can invert the G function, obtaining

$$G^{-1} = \gamma_1^0 \gamma_2^0 \Big\{ (E_1 - h_1)\hat{\rho}_2 + (E_2 - h_2)\hat{\rho}_1 + \\ + [(\epsilon_1 + \epsilon_2)(1 - \hat{\rho}_1\hat{\rho}_2) + (E_1 - E_2)(\hat{\rho}_2 - \hat{\rho}_1)]\Theta \Big\} \qquad (3)$$

with

$$\hat{\rho}_i = \frac{h_i}{\epsilon_i} , h_i = \vec{\alpha}_i \vec{p}_i + \beta_i m_i \qquad (4)$$

and the isospin operator

$$\Theta = \frac{2}{5}(\vec{\tau}_1 \vec{\tau}_2 - 2) . \qquad (5)$$

534

Note that the additional contribution proportional to Θ is absent when isoscalar mesons are considered; furthermore it is vanishing in the case of free particles and, according to the model by Sucher [2], does not give rise to the continuum dissolution disease.

Our wave equation can be written in the standard form

$$G^{-1}\psi(\vec{p}) = \int d^3 p' V(\vec{p}' - \vec{p})\psi(\vec{p}') \tag{6}$$

where V is the quasi-potential and $\vec{p}$ is the relative momentum of the nucleons. The previous equation can be used to study the bound state properties provided we identify the eigenvalue with $E = E_1 + E_2$. The energy difference $(E_1 - E_2)$ is eliminated by means of the prescription $(E_1 - E_2) = (m_1^2 - m_2^2)/E$ that ensures the obtainement of the correct nonrelativistic limit [5]. Our eq.(6), that has been derived in the C.M. of the two nucleons, can be also written in a generic reference frame. We note that the additivity of the kinetic and interaction terms suggests a simple generalization to three (or more) particles; in the configuration space it can be written

$$\left\{ \sum_{i=1}^{3}(E_i - h_i)\hat{\rho}_j\hat{\rho}_k + \sum_{i,j>i}[(\epsilon_i + \epsilon_j)\hat{\rho}_k(1 - \hat{\rho}_i\hat{\rho}_j) + \right.$$
$$\left. + (E_i - E_j)\hat{\rho}_k(\hat{\rho}_j - \hat{\rho}_i)]\Theta(\vec{\tau}_i \cdot \vec{\tau}_j) - \sum_{i,j>i}\gamma_i^o\gamma_j^o V(\vec{r}_{ij})\hat{\rho}_k \right\}\psi = 0 \quad (i \neq j \neq k) \tag{7}$$

This equation may be very useful to study the rôle of the relativistic effects in the three body nuclear systems [8], whose binding energy is still an open problem.

Also, a relativistic study of the electromagnetic interactions of the nuclear systems (such as electron scattering, photonuclear reactions photon scattering etc.) can be consistently performed by adopting the minimal substitution procedure $(\underset{\sim}{p_i} \rightarrow \underset{\sim}{p_i} - e_i A_i)$ in the kinetic terms of our equation.

We point out that the applications to the three body problem and to the interactions with an external e.m. field strictly require the absence of the continuum dissolution disease [2].

2. The O.B.E. Model

In order to test the validity of the present relativistic equation we have constructed a specific One Boson Exchange model by taking also into account the exchange of other mesons $(\rho, \omega, \sigma ...)$ besides the pion [9]. The total interaction operator V is represented by the sum of the contributions given by the different exchanged mesons, which, in turn, are constructed according to the usual prescriptions of the field theory [10]. Monopole form factors have been inserted at the vertices.

As a preliminary test of our model we have performed a non-relativistic reduction of our equation, up to the order c^{-2} according to a standard procedure [11]. We note that the term proportional to the isospin operator Θ in eq.(3), gives no contribution up to this order. As a result we obtain a wave equation with a non-relativistic Hamiltonian of the form:

$$H^{NR} = 2M + \frac{p^2}{M} - \frac{p^4}{4M^3} + V_{OBE}^{NR} + V_Q \qquad (8a)$$

where V_{OBE}^{NR} is the usual contribution that would be given by the non-relativistic reduction of the following Feynman graph,

and V_Q is a quadratic contribution of the form

$$V_Q = -\frac{1}{4M}[\vec{\sigma}_1 \vec{\sigma}_2 V_V - \sum_p \mu_p^2 V_p]^2 . \qquad (8b)$$

In the previous expression V_V and V_p represent the potentials given by the vector and pseudoscalar mesons, respectively.

We note that V_Q is sensitive to μ_p, that represents the ratio of direct to derivative coupling of the pseudoscalar fields. In particular these fields give no contribution to V_Q when pure derivative couplings are considered ($\mu_p = 0$). Furthermore, V_Q is always attractive while, if a Breit equation is adopted, one gets a quadratic term of opposite sign [11] with respect to eq.(8). As noted before, this discrepancy is due to the incorrect treatment of the intermediate negative energy states in the Breit equation [2]. As a preliminary test, we compare our interaction with the Reid soft-core (RSC) potential [12]. To this aim we have to transform the following two contributions of eq.(8a):

i) the relativistic kinetic term $-\frac{p^4}{4M^3}$, that is often inconsistently ignored,

ii) the p^2-dependent interaction terms.

As for the contribution i) we adopt the technique shown in ref.[8] by constructing an operator that does not contain the term $-\frac{p^4}{4M^3}$ but gives, up to the order $1/c^2$, the same eigenvalues and phase-shifts as $H^{NR} - 2M$. Such operator has the form

$$W = \frac{(H^{NR})^2 - 4M^2}{4M} \qquad (9)$$

For the bound state problem, the operator W gives rise to the extra term $B^2/4M$ (B: binding energy) that is negligible for most practical purposes [13]. The p^2-dependent terms are transformed by adopting the well known effective mass transformation [14].

The fitting procedure to the RSC potential has been performed by taking into account only the $^1S_0$ and the $^3S_1 - ^3D_1$ central and tensor potentials, that are reproduced quite accurately, as shown in the figure below.

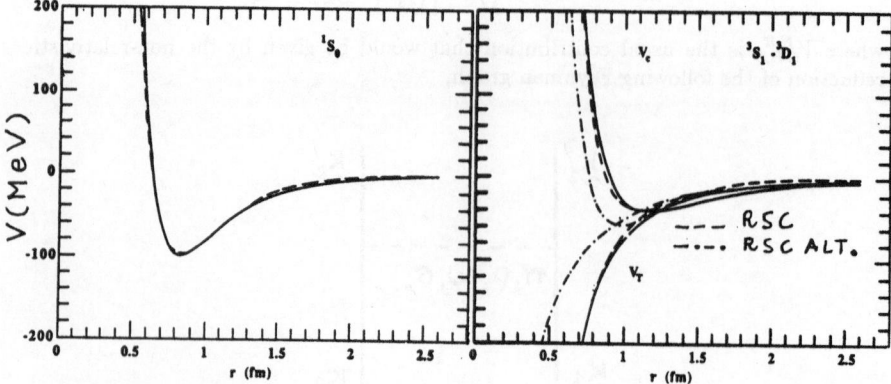

Qualitatively good fits are generally obtained for the P waves.

Some discrepancies are found for the $^1D_2$ and $^3D_2$ waves. In the cases a strongly non local effect seems to be required.

The numerical values of the fitted meson parameters are not in disagreement with the results given by other authors [4,15,16].

3. The Δ Degrees of Freedom

An interesting generalization of our model can be performed in order to study the contributions of the Δ isobar degrees of freedom, that are usually included in the quantitative descriptions of the nuclear interactions [15,17].

Our starting point is a coupled channel equation for the scattering matrix T [17]

$$T = V + VGV + \dots \tag{10}$$

where V, in this case, represents a potential matrix that mixes the four different channels of the system: NN, $N\Delta$, ΔN and $\Delta\Delta$.

Its matrix elements can be straightforwardly determined according to well established techniques [18]. However, the propagator of the spin $\frac{3}{2}$ Δ-particle cannot be directly utilized, in this context, due to an U.V. divergency.

We propose to treat the Δ particle in the spirit of the Rarita-Schwinger equation [19] by employing a subsidiary condition for the wave function Ψ. The coupled channel wave equation can be written in the form:

$$G^{-1}\Psi = V\Psi \tag{11}$$

with the subsidiary condition

$$\bar\theta\Psi = \Psi \tag{12}$$

that is necessary to extract the spin 3/2 components of the Δ field [20]. In more detail, $\Psi = (\Psi_{NN}, \Psi_{N\Delta}, \Psi_{\Delta N}, \Psi_{\Delta\Delta})$, is the four channel wave function of the system (the Lorentz indices, for the Δ case, have been omitted for brevity) V is the potential matrix defined above, G^{-1} is a diagonal matrix whose elements are of the form (see eq.(3))

$$G_{ij}^{-1} = \gamma_i^o \gamma_j^o [(E_i - h_i)\hat{\rho}_j + (E_j - h_j)\hat{\rho}_i] \qquad (13)$$

where the masses in $h_{i(j)}$ and $\hat{\rho}_{i(j)}$, are M_N or M_Δ for $i(j) = N$ or Δ, respectively. Finally $\bar{\theta}$ is a diagonal matrix whose elements are of the form

$$\bar{\theta}_{ij} = O_i O_j$$

with

$$O_{i(j)} = \begin{cases} 1 & \text{for } i(j) = N \\ \theta & \text{for } i(j) = \Delta \end{cases} \qquad (14)$$

where θ is the "on shell" ($p_\Delta^o = h_\Delta$) Behrends and Fronsdal projection operator [20].

This procedure is equivalent to the use of a non covariant propagator truncated to its positive energy contributions only [18].

We point out that in the present problem it is not possible to take consistently into account the contributions of the crossed graphs, as done in the pion case. However, we have shown that those contributions are negligible in the c^{-2} limit for the NN interaction. We assume that this property should hold also in the Δ case.

REFERENCES

1. G.Breit, *Phys.Rev.* **34** (1929) 553; **39** (1932) 616.
2. G.E.Brown and D.G.Ravenhall, *Proc.R.Soc.* **A208** (1951) 552; J.Sucher, *Phys.Rev.Lett.* **55** (1985) 1033.
3. E.E.Salpeter and H.A.Bethe, *Phys.Rev.* **84** (1951) 1232.
4. F.Gross, *Phys.Rev.* **186** (1969) 1448; **C26** (1982) 2203; **D10** (1974) 223.
5. V.B.Mandelzweig and S.J.Wallace, *Phys.Lett.* **B197** (1987) 469.
6. S.J.Wallace and V.B.Mandelzweig, *Nucl.Phys.* **A503** (1989) 673.
7. M.De Sanctis and D.Prosperi, *Il Nuovo Cimento* **104A** (1991) 921.
8. L.A.Kondratyuk, F.M.Lev and V.V.Soloviev, *Few Body Systems* **7** (1989) 55.
9. M.De Sanctis and D.Prosperi, *Nota Interna n.* **981** (1991) Dept. of Physics, University of Rome 'La Sapienza'.
10. J.D.Bjorken and S.D.Drell, *Relativistic Quantum Mechanics* (Mc Graw-Hill, New York, N.Y., 1964).
11. R.Mignani and D.Prosperi, *Nuovo Cimento* **75A** (1983) 221.
12. R.V.Reid, *Ann. of Phys.* **50** (1968) 411.
13. P.L.Chung, F.Coester, B.D.Keister, W.N.Polizou, *Phys.Rev.* **C37** (1988) 200.
14. A.M.Green, *Nucl.Phys.* **33** (1962) 218.
15. R.Machleidt, K.Holinde and Ch.Elster, *Phys.Rep.* **149** (1987) 1.
16. R.Brian and B.L.Scott, *Phys.Rev.* **177** (1969) 1435.
17. B.ter Haar and R.Malfliet, *Phys.Rep.* **149** (1987) 207.
18. E.E.van Faassen and J.A.Tjon, *Phys.Rev.* **C28** (1983) 2354; **C30** (1984) 285.
19. W.Rarita and J.Schwinger, *Phys.Rev.* **60** (1941) 61.
20. R.E.Behrends and C.Fronsdal, *Phys.Rev.* **106** (1957) 345.

Few-Body Systems, Suppl. 6, 538—543 (1992)

Few-
Body
Systems
© by Springer-Verlag 1992

NEW APPROACHES FOR BOUND AND SCATTERING STATES

J. L. Friar

Theoretical Division, Los Alamos National Laboratory
Los Alamos, New Mexico 87545 U.S.A.

1. Introduction

Since its inception, the field of few-nucleon physics has been driven by the desire to obtain accurate solutions of the Schrödinger equation for physically interesting problems. These problems include the bound and scattering states of Hamiltonians which contain "realistic" potentials. This quest has been considered so important and so difficult that a major component of our field has devoted itself as much to methods of solution as to the physics inherent in the problems. Lack of convergence of different methods in the past had many different causes, but the result was an inability to extract that physics.

This preoccupation with improving the quality of methods (both numerical and formal) has paid off. We have now achieved "exact" or complete solutions to a variety of important problems: the ground states of ^{3}He and ^{3}H [1] (including a three-nucleon force, and a Coulomb interaction in the former case), the ground state of ^{4}He [2] and low-lying continuum of ^{5}He [3], above-breakup n-d scattering[4], and, more recently, zero-energy n-d and p-d scattering[5] (including three-nucleon forces, and a Coulomb interaction in the latter case). Such solutions might be conveniently defined as those with fractional errors of less than one percent in relevant observables (or perhaps more restrictive in some cases). The realistic potentials referred to above are those containing OPEP (and thus a strong long-range tensor force), a strong short-range repulsion, and which provide at least a moderately good quantitative fit to two-nucleon scattering data. The qualitative importance of OPEP in the few-nucleon systems has only recently been recognized[6], although the importance of the tensor force has long been recognized. As much as 80 percent of the potential energy in the triton is due to OPEP[7] (in the variational sense: $<V>$). Moreover, in the recent Nijmegen N-N phase shift solutions[8], where the pion masses are *fit* rather than input parameters, they find $m_{\pi^0} = 135.6(13)$ MeV and $m_{\pi^\pm} = 139.4(10)$ MeV. The very small error bars demonstrate the importance of OPEP in the scattering data. The concomitant tensor force has proven to be a formidable obstacle to numerical solution. Techniques which work well in the absence of a tensor force often work much less well in the presence of such a force.

From an historical perspective these diverse efforts to solve the Schrödinger equation have evolved in somewhat unexpected directions, as evolutionary and revolutionary improvements have been made. The result has been that we have achieved a rather good understanding of the three- and four-nucleon systems and are beginning to develop insight into "real" nuclei with five or more nucleons (where the Pauli principle can be expected to play a pivotal role).

Improvements in the future are not only expected, they are essential. Our problems of current interest are becoming progressively more difficult and their solution will require improved methods. Moreover, as techniques improve and solutions become easier to obtain, the latter will disseminate more rapidly into the nuclear physics community.

We present below a rather personal view (*i.e.*, very incomplete and somewhat biased) of the development of the primary methods that have been used for solving the few-nucleon Schrödinger equation, all of which have achieved a significant measure of success when applied to realistic potentials, and which may be promising in the future. Much other work has contributed to this success and must necessarily go unmentioned. The new developments described in this discussion session by the various speakers have some connection with one or more of these methods.

One thing is virtually certain: the future holds surprises for us. It is not at all clear that the methods of choice today will be those at the turn of the century. Our future lies in new and improved methods.

2. Variational Techniques

The earliest technique used to solve the few-nucleon Schrödinger equation was the Rayleigh-Ritz variational principle. This method is employed because of two powerful properties:

1. The variational estimate $<H>$ is an upper bound to the ground state energy;

2. An error of order ϵ in the wave function produces a contribution of order ϵ^2 to that upper bound.

In 1935 Thomas[9] used the first of these properties to demonstrate that nucleon-nucleon forces of zero-range would lead to the collapse of the triton ground state. Subsequently, Gerjuoy and Schwinger[10] undertook 50 years ago the first variational calculation of the triton binding energy for a Hamiltonian which includes a tensor force. The latter forces have proven significantly more difficult to treat than purely central forces. Much of the work before 1970 took the variational route[11].

Two different approaches have been followed in selecting the variational trial functions. One approach uses appropriately symmetrized products of Jastrow[12] pair (two-body) correlation functions, which are physically motivated and are currently rather sophisticated[7]. The other approach is a brute-force expansion in a complete set of functions which appropriately spans the space. The variational parameters of the latter are usually chosen to be linear in order to facilitate the numerical solution. The advantage of the first approach is that it can be used for a wide variety of nuclear systems ranging from the triton and ^{4}He to ^{16}O[13] and nuclear matter[14]. Its disadvantage is that it is not yet a fully converged approach and for realistic potentials produces errors on the order of several percent in the binding energies of few-nucleon systems. Converged calculations exist for ^{3}H and ^{3}He which use the other approach[15].

The latter calculations exploit the Faddeev decomposition[16,17] of the Schrödinger wave function, Ψ. Writing

$$\Psi(\mathbf{x},\mathbf{y}) = \psi(\mathbf{x_1},\mathbf{y_1}) + \psi(\mathbf{x_2},\mathbf{y_2}) + \psi(\mathbf{x_3},\mathbf{y_3}) \equiv \psi_1 + \psi_2 + \psi_3,$$

where the function ψ is the *same* function of *different* coordinate (and spin) variables for the case of identical particles. Thus, in that case only one of these functions needs to be modeled. This representation is the essence of the Faddeev method, since it decomposes the very complicated Schrödinger wave function into simpler functions, which represent the separate (physical) rearrangement channels. It is this device which allows the boundary conditions for scattering to be implemented in a tractable and transparent way[18].

Thus, although it is the oldest weapon in the few-nucleon arsenal, the variational approach is currently very successful and holds promise for the future. We also note that other techniques (discussed below) which produce a wave function can make effective use of variational bounds to check or improve their eigenvalues.

3. Faddeev Techniques

The Faddeev approach to solving the Schrödinger equation was developed in order to resolve a problem pointed out long ago by Foldy and Tobocman[19]. Traditional scattering theory converts the Schrödinger equation to the Lippmann-Schwinger equation, whose integral form facilitates the implementation of the usual outgoing-wave boundary conditions. Nevertheless, the latter equation has no unique solution for n-d scattering above breakup threshold, because there is no way to control the amount of "initial" plane wave for the *rearrangement* channels. Specifying this component of the boundary conditions requires two additional equations which, together with the original equation, comprise the triad equations[20]. Equivalently, one can solve the three Faddeev equations (which for identical nucleons are identical equations). This clever decomposition by Faddeev isolates the rearrangement channels into the permuted Faddeev wave function components (*i.e.*, ψ_2 and ψ_3). Adding these permuted components to ψ_1 to form Ψ produces a more complex wave function with the correct rearrangement amplitudes. Although originally devised in order to satisfy the requirements of scattering theory, this scheme has proven highly effective for bound-state problems, as well. The cusps and ridges of the virtual "rearrangement channels" in the wave function are just as difficult to model numerically in the bound states as they are in the scattering states, although the boundary conditions for the former problem (finiteness) have never been in doubt.

The seminal calculations of Malfliet and Tjon[21] for realistic potentials decomposed to include only a few nucleon-nucleon partial waves ("channels") were soon followed by the calculations of the Purdue[22] and Hannover[23] groups, which involved many more partial waves. Convergence requires roughly 34 channels[1] for the trinucleon ground states, and this has become the standard for such calculations. The original converged calculations treated only strong two-nucleon forces; they have since been extended to include the Coulomb force in ^{3}He and the three-nucleon force in ^{3}H and ^{3}He. No converged Faddeev calculations exist for ^{4}He. Converged continuum calculations also exist for n-d scattering above breakup threshold[4].

4. Green's Function Monte Carlo

The powerful Green's Function Monte Carlo (GFMC) method was first applied to the few-nucleon systems in 1962[24]. In that work the integral form of the Schrödinger equation for the three- and four-nucleon bound states was solved for spin- and isospin-independent square-well, Gaussian, and exponential potentials. The form of the equation which was solved is

$$\Psi = \lambda G_0 V \Psi,$$

where Ψ is the wave function for an *assumed* binding energy, G_0 is the (non-interacting) Green's function, and V is the nuclear potential. The eigenvalue λ of this restructured problem corresponds to the strength factor of the potential required to generate the assumed binding energy. Solutions for several binding energies can be easily interpolated to produce the value $\lambda = 1$ and the corresponding binding energy.

The equation above was solved by randomly sampling the distribution of nucleon coordinates in Ψ, performing the implicit integrals, and thus generating a new estimate for Ψ/λ. Iteration of this scheme (the power method) leads to the wave function corresponding to the largest value of $|1/\lambda|$. This is usually the ground state, although cases are known[1] where the iteration is dominated by *negative* values of λ, corresponding to the strongly repulsive parts of realistic nucleon-nucleon potentials. These "ghost states", which are not solutions of the original Schrödinger equation, nevertheless control the iteration, unless special care is taken.

The advantage of the Monte Carlo procedure is that it does not suffer from increasing dimensionality (as more nucleons are added) as much as traditional procedures do. Its disadvantage is that its solution is a set of random samples, and every calculated observable has an associated statistical error. Nevertheless, the method is very powerful and its advantages greatly outweigh its disadvantages.

The extension of this procedure to spin-dependent potentials was not immediate. Spin-independent potentials generate few-nucleon wave functions which are symmetric under the interchange of the spatial coordinates of any two nucleons. These space wave functions can then be coupled to a completely antisymmetric spin-isospin wave function to generate one which satisfies the requirements of the Pauli principle. Any dependence of the potential on spin or isospin generates wave function components which are not space symmetric. A similar situation exists for heavier nuclei, where the Pauli principle requires such wave function components, which have enhanced kinetic energy and reduce the binding. Relaxing the Pauli principle might lead to states which are more deeply bound. Iteration schemes such as the one previously described want to converge to the lowest state. Even if this state is a Pauli-principle-violating "ghost state", it can affect a Monte Carlo solution, because of its statistical nature. Although one can easily enforce that principle *in the mean*, the variance (statistical error) can increase uncontrollably, rendering the solution useless. This is typical of "Fermion" problems.

Carlson[2] was the first to show that the problem was not serious for few-nucleon systems. He used the time-dependent form of the Schrödinger equation continued to imaginary time $(t \to -i\tau)$, which was first used for few-nucleon problems in 1962 [25]:

$$\Psi(\tau) = \exp(-H\tau)\Psi(0),$$

and which clearly converges for large τ to the lowest eigenvalue. This technique was originally applied to a variant of the Argonne potential[26] for the triton and the alpha particle. Subsequently[27], it was shown by direct computation using the Faddeev equations that there is no bound, symmetric triton state for Fermions corresponding to this potential, largely because OPEP is repulsive in such states. This fortuitous occurrence leads to stable iteration. The GFMC technique is currently the method of choice for the alpha particle.

5. Hyperspherical Techniques

The original impetus for this technique was supplied by Simonov[28], who recognized the geometric complexity of even the simplest light nuclei. The three-nucleon problem[18] has 3 intrinsic coordinates, usually taken to be the Jacobi coordinates: x, y, and θ, of which only one is an angle $(\cos(\theta) = \hat{x} \cdot \hat{y})$. By treating x and y as the Cartesian components of a two-dimensional vector and transforming to polar coordinates, the three intrinsic coordinates can be specified by a single length (the hyperradius) and two angles. The Schrödinger wave function of the triton can then be expanded in terms of a complete set of the angular variables and corresponding functions of the hyperradius.

Long before we had the ability to solve the Schrödinger equation for the triton, this scheme was extremely useful in providing insight into the Coulomb energy of ^{3}He, which provides the bulk of the binding energy difference of ^{3}He and ^{3}H. A truncated hyperspherical expansion[29] leads to the "magic formula", which relates the ^{3}He Coulomb energy to the charge densities of ^{3}He and ^{3}H, which can be taken from experiment. Although it is only an approximation, this relationship is accurate at the one percent level. In addition, hyperspherical variables are often used in Faddeev calculations[30], because the matrices one obtains by discretizing those equations are banded (*i.e.*, most elements are zero) in the hyperradius variable.

Extensive efforts to solve the hyperspherical equations have been only partially successful for realistic potentials, although they have been successful for other problems. Recently, large basis calculations for the Super Soft Core (C) potential[31] were reported[32], for which comparisons with other methods are possible. The reported binding energy differs from Faddeev calculations by less than 1 per cent. According to Ref. [32], the primary difficulty in achieving convergence is the (mixed-symmetry, s-wave) S'-state wave function component. The method chosen to solve the equations is also variational in nature.

An outgrowth of these methods is the integro-differential equation approach[33], which focuses on the important two-nucleon correlations in few- and many-body systems, and is equivalent to the s-wave Faddeev equations for the triton.

6. Hybrid Techniques

Some of the recent advances and new approaches have involved combinations of the somewhat arbitrary categories we have introduced (*e.g.*, a Faddeev wave function representation for a variational calculation). The methods discussed by the speakers in this session share this characteristic, and are applied to a variety of interesting problems, including purely Coulombic (*i.e.*, atomic physics) calculations. These methods therefore represent an interesting synthesis of fields, as well as techniques.

References

1. C. R. Chen, G. L. Payne, J. L. Friar, and B. F. Gibson, *Phys. Rev. C* **31**, 2266 (1985); **33**, 1740 (1986); J. L. Friar, B. F. Gibson, and G. L. Payne, *ibid*, **35**, 1502 (1987).

2. J. Carlson, *Phys. Rev. C* **36**, 2026 (1987).

3. J. Carlson, private communication.

4. W. Glöckle, H. Witała, and Th. Cornelius, *Nucl. Phys.* **A508**, 115c (1990).

5. C. R. Chen, G. L. Payne, J. L. Friar, and B. F. Gibson, *Phys. Rev. C* **44**, 50 (1991).

6. J. L. Friar, B. F. Gibson, and G. L. Payne, *Phys. Rev. C* **30**, 1084 (1984).

7. R. B. Wiringa, *Phys. Rev. C* **43**, 1585 (1991).

8. J. J. de Swart, contribution to this conference.

9. L. H. Thomas, *Phys. Rev.* **47**, 903 (1935).

10. E. Gerjuoy and J. Schwinger, *Phys. Rev.* **61**, 138 (1942).

11. L. M. Delves and A. C. Phillips, *Rev. Mod. Phys.* **41**, 497 (1969); L. M. Delves, *Adv. in Nucl. Phys.* **5**, 1 (1972).

12. R. Jastrow, *Phys. Rev.* **98**, 1479 (1955).

13. S. C. Pieper, R. B. Wiringa, and V. R. Pandharipande, *Phys. Rev. Lett.* **64**, 364 (1990).

14. R. B. Wiringa, V. Fiks, and A. Fabrocini, *Phys. Rev. C* **38**, 1010 (1988).

15. H. Kameyama, M. Kamimura, and Y. Fukushima, *Phys. Rev. C* **40**, 1 (1989).

16. L. D. Faddeev, *Zh. Eksp. Teor. Fiz.* **39**, 1459 (1960) [*Sov. Phys. - JETP* **12**, 1014 (1961)].

17. H. P. Noyes, in *Three Body Problem in Nuclear and Particle Physics*, J. S. C. McKee and P. M. Rolph, eds. (North-Holland, Amsterdam, 1970), p. 2.

18. J. L. Friar, in *Modern Topics in Electron Scattering*, B. Frois and I. Sick, eds. (World Scientific, Singapore, 1991).

19. L. L. Foldy and W. Tobocman, *Phys. Rev.* **105**, 1099 (1957).

20. W. Glöckle, *Nucl. Phys.* **A141**, 620 (1970).

21. R. A. Malfliet and J. A. Tjon, *Ann. Phys.* (N.Y.) **61**, 425 (1970).

22. R. A. Brandenburg, Y. E. Kim, and A. Tubis, *Phys. Lett.* **49B**, 205 (1974).

23. C. Hajduk and P. U. Sauer, *Nucl. Phys.* **A369**, 321 (1981).

24. M. H. Kalos, *Phys. Rev.* **128**, 1791 (1962); *Nucl. Phys.* **A126**, 609 (1969); Y. C. Tang and R. C. Herndon, *Nucl. Phys.* **A93**, 692 (1967) quote the statistical error of the former result.

25. G. A. Baker, Jr., J. L. Gammel, B. J. Hill, and J. G. Wills, *Phys. Rev.* **125**, 1754 (1962).

26. R. B. Wiringa, R. A. Smith, and T. A. Ainsworth, *Phys. Rev. C* **29**, 1207 (1984).

27. J. Carlson, J. L. Friar, and G. L. Payne, *Phys. Rev. C* **37**, 420 (1988).

28. Yu. A. Simonov, *Yad. Fiz.* **3**, 630 (1966) [*Sov. J. Nucl. Phys.* **3**, 461 (1966)].

29. M. Fabre de la Ripelle, *Fizika* **4**, 1 (1972); J. L. Friar, *Nucl. Phys.* **A156**, 43 (1970).

30. S. P. Merkuriev, C. Gignoux, and A. Laverne, *Ann. Phys.* (N.Y.) **39**, 30 (1976).

31. R. de Tourreil and D. W. L. Sprung, *Nucl. Phys.* **A201**, 193 (1973).

32. M. I. Mukhtarova, *Yad. Fiz.* **49**, 338 (1989) [*Sov. J. Nucl. Phys.* **49**, 208 (1989)].

33. M. Fabre de la Ripelle, *Few-Body Systems* **1**, 181 (1986).

Few-Body Systems, Suppl. 6, 544—549 (1992)

Few-
Body
Systems
© by Springer-Verlag 1992

NEW APPLICATIONS OF THE FADDEEV APPROACH TO THE THREE-BODY COULOMB PROBLEM

A.A. Kvitsinsky[1,2], C.-Y. Hu[3], J. Carbonell[2], C. Gignoux[2], S.P. Merkuriev[1]

[1]Department of Mathematical and Computational Physics, Institute for Physics, Leningrad University, 198904 Leningrad, USSR

[2]Institut des Sciences Nucléaires, 38026 Grenoble Cedex, France

[3]Physics Department, California State University, Long Beach, CA 90840, USA

Abstract
We present a calculation of the $e^- - (e^- e^+)$ scattering length in the framework of the bipolar expansion applied to the modified Faddeev equations (FE). Also, a new method of direct solving the three-dimensional FE for the three-body Coulomb bound state problem is described.

INTRODUCTION

This talk is a review of two recent works [1], [2] and consists of two different parts. The first one deals with the zero energy elastic scattering of e^- on the ground state of $(e^- e^+)$. This is the simplest instance of a pure Coulomb three-body scattering problem. As is known, the standard FE in the configuration space are non-compact if all the pairwise interactions are Coulombian. To bring back the compactness, they are to be modified via a cut-off procedure [3]. These modified FE were established some ten years ago but have never been used in actual calculations. We apply these equations to calculate the scattering length of the $e^- (e^- e^+)$ system. We believe this is the first Faddeev calculation of this parameter.

The second part describes a new numerical method of treating the three-body Coulomb bound state problem. It consists of direct solution of the FE in the total-angular-momentum representation. This approach was proposed recently [4] in order to avoid usual difficulty of the methods of intermediate basis expansions (f.i., the bipolar expansion) when applied to a pure Coulomb problem: slow convergence in the number of partial channels taking into account, which is due to the long-range behavior of the Coulomb interaction.

We find this new approach be pretty well performing for the bound state problem and believe it will be as well efficient in further applications to the scattering states.

1. $e^- - (e^- e^+)$ SCATTERING LENGTH

The particles are numerated by the label $\alpha = 1, 2, 3$ so that $(e^- e^- e^+) = (123)$. We shall call "the pair α" the two-body subsystem of the particles with numbers $\beta \neq \alpha$.

The modified FE for a Coulomb three-body system are of the form [3]

$$\left(H_0 + V_\alpha + \sum_{\beta \neq \alpha} V_\beta^{(0)} - E \right) \Psi_\alpha = -\hat{V}_\alpha \sum_{\beta \neq \alpha} \Psi_\beta \ , \qquad (1)$$

where Ψ_α are the Faddeev components of the total wave function, H_0 stands for the kinetic energy operator and, V_α is the potential of the Coulomb interaction in the pair α. The potentials $V_\beta^{(0)}$ and $\hat{V}_\alpha$ are constructed by the cut-off procedure [3]

$$\hat{V}_\alpha (x_\alpha, y_\alpha) = V_\alpha (x_\alpha) \zeta_\alpha (x_\alpha, y_\alpha) \ , \quad V_\alpha^{(0)} (x_\alpha, y_\alpha) = V_\alpha (x_\alpha) (1 - \zeta_\alpha (x_\alpha, y_\alpha)) \ ,$$

where x_α, y_α are the lengths of the standard mass scaled Jacobi vectors $\mathbf{x}_\alpha, \mathbf{y}_\alpha$; ζ_α is a cut-off function which separates the 2- and 3-body sectors of the configuration space. Its form may be rather arbitrary within some general requirements [3]. Roughly speaking, asymptotically ζ_α should tend fast enough to one within the 2-body sector $\Omega_\alpha : y_\alpha \geq x_\alpha^\nu$ with $\nu > 2$ and to zero outside Ω_α.

Upon projecting onto the bipolar harmonics

$$| \sigma_\alpha >= [Y_{l_\alpha} (\hat{x}_\alpha) \otimes Y_{\lambda_\alpha} (\hat{y}_\alpha)]_{LM} \ , \qquad (2)$$

Eqs. (1) are reduced to an infinite set of equations for the partial components $\Psi_{\alpha \sigma_\alpha}$:

$$(-\Delta_{\sigma_\alpha} + V_\alpha - E) \Psi_{\alpha \sigma_\alpha} (x, y) + \sum_{\sigma'_\alpha} \sum_{\beta \neq \alpha} < \sigma_\alpha | V_\beta^{(0)} | \sigma'_\alpha > \Psi_{\alpha \sigma'_\alpha}(x, y) = \qquad (3)$$

$$= -\hat{V}_\alpha \sum_{\beta \neq \alpha} \sum_{\sigma_\beta} < \sigma_\alpha | \Psi_{\beta \sigma_\beta} (x_\beta, y_\beta) | \sigma_\beta >$$

where $x \equiv x_\alpha, y \equiv y_\alpha$ and

$$\Delta_{\sigma_\alpha} = \partial_x^2 + \partial_y^2 - l_\alpha (l_\alpha + 1) x^{-2} - \lambda_\alpha (\lambda_\alpha + 1) y^{-2} \ .$$

The r.h.s. of these equations involves standard integrals over the polar angle on the $\{x, y\}$-plane with fixed hyperradius.

The presence of the screened Coulomb potential $\hat{V}_\alpha$ in the r.h.s. of Eqs. (3) provides asymptotic vanishing of the latter in the 3-body sector of the configuration space. This results in the compactness of the modified FE when applied to the scattering problem. The usual FE ($\zeta_\alpha \equiv 1$) do not possess this property.

For a system with two identical particles 1 and 2, the Faddeev components must satisfy (anti)symmetric requirements

$$\Psi_1(\mathbf{x}, \mathbf{y}) = p\Psi_2(-\mathbf{x}, \mathbf{y}) \ , \quad \Psi_3(\mathbf{x}, \mathbf{y}) = p\Psi_3(-\mathbf{x}, \mathbf{y}) \ ,$$

with parity $p = \pm 1$, so that Eqs. (3) can be reduced to equations for partial components of Ψ_1 and Ψ_3.

We solve Eqs. (3) numerically for the zero energy elastic scattering problem for the symmetric $(p = 1)$ state of the $e^- - (e^- e^+)$ system with zero total angular momentum $L = 0$. In this case the bipolar basis (2) includes equal angular momenta $l_\alpha = \lambda_\alpha$. For $\alpha = 1$ $l_1 = \lambda_1 = 0, 1, 2, \ldots$. For $\alpha = 3$ only even indices $l_3 = \lambda_3 = 0, 2, \ldots$ are involved due to the symmetry in the electron exchange.

The partial Faddeev components are subject to zero boundary conditions on the boundaries $x = 0, y = 0$. Also, they vanish asymptotically except for one channel $\sigma_1^{(0)} = \{l_1 = \lambda_1 = 0\}$ containing the zero energy asymptotic state

$$\Psi_{1\sigma_1^{(0)}}(x, y)|_{y \to \infty} \sim \varphi_0(x)(r - A) \ , \quad r = \frac{\sqrt{3}}{2} y \ , \tag{4}$$

where r is the distance between the projectile and the positronium's center of masses, φ_0 is the positronium ground state and, the constant A is the scattering length to be calculated.

For numerical solution of this problem we adopted the procedure of Ref. [5] which exploits a finite difference approximation of the FE in the cartesian coordinates on a square $x \in [0, x_{max}], y \in [0, y_{max}]$. The asymptotic condition (4) is settled on the line $y = y_{max}$ with a fairly large y_{max}. For details of the method we refer to the contribution by J. Carbonell et al. to this conference.

The cut-off function was chosen to be of the same functional form for any α:

$$\zeta_\alpha(x, y) = 2 \left[1 + \exp \left\{ (x/x_0)^\nu / (y/y_0 + 1) \right\} \right]^{-1} \ .$$

According to the compactness requirements [3] for the modified FE, the parameter ν must be > 2; x_0 should be of around the size of the 2-body bound state $\varphi_0(x)$. We used the following set of the cut-off parameters: $x_0 = 2$ a.u., $y_0 = 10$ a.u. and, $\nu = 2.3$.

Table 1 shows the results of the scattering length calculations for various sets of the bipolar partial channels taken into account, for a fixed grid with $N_x = 25, N_y = 40$ points in x, y:

$$x = 0.(0.4)4.4(0.8)12.4(1.6)20.4 \ ; \quad y = 0.(0.4)4.4(0.8)12.4(1.6)44.4 \ .$$

Table 1. Scattering length A of the $L = 0, p = +1$ state of $e^-(e^- e^+)$. N is the number of partial channels; $\{l_1\}, \{l_3\}$ are corresponding sets of angular momenta for the components Ψ_1, Ψ_3.

N	2	3	4	5	6	7
$\{l_1\}$	0	01	012	012	0123	01234
$\{l_3\}$	0	0	0	02	02	02
A (a.u.)	16.65	12.33	12.22	12.09	12.27	12.27

The convergence in N is somewhat slow, as one should expect of the bipolar expansion when applied to a pure Coulomb problem. Our $N = 7$ value $A = 12.27$ a.u. is in a good agreement with the result of Ref. [6] $A = 12.0 \pm 0.3$ a.u. obtained by the Kohn variational principle.

Of course, the limiting value of the scattering length as $N \to \infty$ must be independent of particular choice of the cut-off function. This was checked and, for fairly different cut-offs difference in $A(N)$ was found be negligible starting from $N = 4 \div 6$.

For a fixed N, the calculations exhibit a good stability with respect to grid parameters. In particular, this is how the compactness of the modified FE shows up numerically. To make that more evident, we solved also the same problem using the FE without the cut-off modifications ($\zeta_\alpha \equiv 1$) and found the results be very unstable: the scattering length varies a lot with changing x_{max}. This is due to that in this case, the contribution of the r.h.s. of Eqs. (3) does not vanish asymptotically in certain directions of the $\{x, y\}$-plane.

2. DIRECT SOLUTION OF 3-DIMENSIONAL FADDEEV EQUATIONS

By the 3-D FE we mean the equations which are obtained from Eq. (1) via separation of the Euler angles describing rotation of a three-body system as a whole. Details of deriving such equations are given in Ref. [4].

The 3-D FE can be formally written in the form (1). Now the components Ψ_α depend on three variables that fix configuration of the triangle formed by the particles. As these, we choose the hyperspherical coordinates $(\rho, \chi_\alpha, \theta_\alpha)$

$$\rho = \sqrt{x_\alpha^2 + y_\alpha^2} \, , \quad \tan \frac{\chi_\alpha}{2} = y_\alpha / x_\alpha \, , \quad \cos \theta_\alpha = (\hat{x}_\alpha, \hat{y}_\alpha) \, .$$

In these coordinates, the kinetic energy operator for $L = 0$ is given by

$$H_0 = -\rho^{-5} \partial_\rho \rho^5 \partial_\rho - 4\rho^{-2} \csc^2 \chi_\alpha \left\{ \partial_{\chi_\alpha} \sin^2 \chi_\alpha \partial_{\chi_\alpha} + \csc \theta_\alpha \partial_{\theta_\alpha} \sin \theta_\alpha \partial_{\theta_\alpha} \right\} \, ,$$

and the hyperspherical angles with different α's are related by a rotation transformation [4] involving the particle masses.

When projected onto the set of Legendre polynomials $P_{l_\alpha} (\cos \theta_\alpha)$, the 3-D FE are reduced to the partial equations (3). Compared to the latter ones, the 3-D equations have an important advantage: one can use very flexible basis expansions in all three variables $\rho, \chi_\alpha, \theta_\alpha$, such as spline expansions, which can be adjusted to cover properly all important regions of the configuration space.

To test the efficiency of such a method, we develop [2] a numerical method of solving the 3-D FE for the Coulomb bound state problem. Basically, this is as follows.

Upon scaling the components by a α-invariant factor

$$\Psi_\alpha = \rho^{-5/2} \csc \chi_\alpha \csc \theta_\alpha \, \Phi_\alpha \, ,$$

one gets new equations for Φ_α with zero boundary conditions on all the boundaries $\rho = 0; \chi_\alpha = 0, \pi; \theta_\alpha = 0, \pi$ and $\rho = \infty$ (for a bound state). We find it useful to adopt a mapping of the interval $[0, \infty)$ of ρ into the interval $[0, 1)$ of a new variable r proposed in Ref. [7]: $r = 1 - e^{-\lambda\rho}$, where λ is a free optimization parameter.

To keep the number of input parameters to minimum, we used usual form of the FE, without the cut-off modifications (i.e., $\zeta_\alpha \equiv 1$) that do not play principal role in the bound state problem. In terms of the coordinates $r, \chi = \chi_\alpha$ and $\theta = \theta_\alpha$ the equations to be solved are

$$(H_0 + V_\alpha - E) \Psi_\alpha(r, \chi, \theta) = -V_\alpha \sum_{\beta \neq \alpha} \Psi_\beta (r, \chi_{\beta\alpha}, \theta_{\beta\alpha}) \, ,$$

where $\chi_{\beta\alpha}, \theta_{\beta\alpha}$ are the angles χ_β, θ_β expressed through $\chi_\alpha, \theta_\alpha$ and

$$H_0 = -\lambda^2(1-r)^2\, \partial_r^2 + \lambda^2(1-r)\, \partial_r - 4\lambda^2\ln^{-2}(1-r) \times$$

$$\left[\partial_\chi^2 + \csc^2\chi\left(\partial_\theta^2 - \cot\theta\, \partial_\theta + \csc^2\theta \right) + \frac{1}{16} \right].$$

Next, we use a spline expansion of the components in all three variables r, χ, θ:

$$\Phi_\alpha(r, \chi, \theta) = \sum_{m=1}^{N_r}\sum_{l=1}^{N_\chi}\sum_{n=1}^{N_\theta} a_{\alpha m l n} S_m(r)S_l(\chi)S_n(\theta) ,$$

For the splines S's we take the piecewise quintic Hermit polynomial splines [8] constructing to be nonzero on two adjoint intervals. The number of splines is twice that of the intervals. Upon the orthogonal collocation procedure with two-point Gauss-quadrature points per each interval, the 3-D FE are reduced to an algebraic eigenvalue problem. This is solved by the Lanczos algorithm as described by Payne [9]. Also, we make use of a tensor representation of the algebraic problem similar to that proposed in Ref. [7].

The method above was applied to calculate the $(e^-e^-e^+)$ ground state. Uniform distribution of spline knots in all three variables r, χ, θ has been used. Table 2 presents the binding energy calculations for several grids; Ref. [10] is a variational calculation.

Table 2

$N_r N_\chi N_\theta$	$10 \times 6 \times 6$	$16 \times 18 \times 10$	$16 \times 16 \times 12$	$16 \times 18 \times 14$	[10]
$-E$ (a.u.)	0.223115	0.266880	0.2620234	0.2620217	0.26200506
λ	0.08	0.092	0.08759	0.09551	

Clearly, the method performs very efficient and pretty good results are obtained using rather coarse grids. Compared to the bipolar expansion calculations, our $16 \times 16 \times 12$ result is better than that 0.26231 of Ref. [7] obtained with 7 partial channels and 34×38 (in our notations) bicubic spline expansion.

REFERENCES

1. Carbonell, J., Gignoux, C., Kvitsinsky, A.A. : in preparation

2. Hu, C.-Y., Kvitsinsky, A.A., Merkuriev, S.P. : submitted to Phys. Rev. A.

3. Merkuriev, S.P.: Ann. Phys. **130**, 395 (1980).

4. Kostrykin, V.V., Kvitsinsky, A.A., Merkuriev, S.P.: Few-Body Systems **6**, 97 (1989).

5. Carbonell, J., Gignoux, C., Merkuriev, S.P.: contribution to this Conference.

6. Ward, S.J., Humberstone, J.W., McDowell, M.R.C.: J. Phys. **B 20**, 127 (1987).

7. Schellingerhout, N.W., Kok, L.P., Bosveld, G.D.: Phys. Rev. **A 40**, 5568 (1989).

8. Prenter, P.M.: Splines and Variational Methods. New York: Wiley 1975.

9. Payne, G.L.: in: Lecture Notes in Physics, vol. 273. Berlin: Springer 1987. PP.64-99.

10. Bhatia, A.K., Drachman, R.J.: Phys. Rev. **A 28**, 2523 (1983).

Few-Body Systems, Suppl. 6, 550—556 (1992)

Few-
Body
Systems
© by Springer-Verlag 1992

New Method for Solving Three-Dimensional Schroedinger Equation

V. S. Melezhik

Joint Institute for Nuclear Research
Dubna, Head P.O.Box 79, Moscow, USSR

Abstract

A new method is developed for solving the multidimensional Schroedinger equation without the variable separation. To solve the Schroedinger equation in a multidimensional coordinate space X , a difference grid Ω_i (i=1,2,...,N) for some of variables ,Ω, from $X = \{R, \Omega\}$ is introduced and the initial partial-differential equation is reduced to a system of N differential-difference equations in terms of one of the variables R. The arising multi-channel scattering (or eigenvalue) problem is solved by the algorithm based on a continuous analog of the Newton method. The approach has been successfully tested for several two-dimensional problems (scattering on a nonspherical potential well and "dipole" scatterer, a hydrogen atom in a homogeneous magnetic field) and for a three-dimensional problem of the helium-atom bound states.

1. Introduction

In our paper [1], an approach is proposed for solving the multidimensional Schroedinger equation without the wave function expansion over the basis in a traditional sense and computing the matrix elements. For some of variables, Ω, from $X = \{R, \Omega\}$ a difference net Ω_i is introduced (i= 1,2,...,N; the distance between nodes is characterized by the step of integration h) ; those variables are considered discrete, and one variable R remains continuous. Then the multidimensional Schroedinger equation is approximated by a system of differential-difference equations for the vector $\{\psi_i(R)\}_1^N = \psi(R, \Omega_i)$. For introducing the "difference net" procedure for the internal coordinates one can use different approaches (see, for example [2-4]). The one used here is close to the "discrete variable representation" [2]. At the further step, the scattering (or eigenvalue problem) for the system of differential-difference equations is formulated , following Ref.[5], as a system of nonlinear functional equations for the vector $z^h = \{\psi_i^h(R), t_{ij}^h\}$ (or $z^h = \{\psi_i^h(R), \varepsilon^h\}$) , where t_{ij}^h and ε^h are approximations for the searched scattering matrix t_{ij} and the eigenvalue ε of the initial multidimensional Schroedinger equation. In this approach, the problem of accuracy of the solution of the multidimensional problem reduces to the well-elaborated problem of computational mathematics on convergence of the solution $\psi^h(R, \Omega_i)$ obtained in the space $X^h = \{R, \Omega_i\}$ to the solution of the initial problem, $\psi(R, \Omega)$, in $X = \{R, \Omega\}$, instead of the convergence in the basis (see, e.g., [6]).

In the next sections we give an account of the idea of the method and discuss the results obtained in Refs.[1,7] for some two- and three-dimensional problems.

2. Method

The essence of the method developed in [1] is the following. For solving the Schroedinger equation

$$\{H(X) - \varepsilon\}\psi(X) = 0 \tag{1}$$

in the multidimensional space $X = \{R, \Omega\}$ with the Hamiltonian

$$H(X) = -\frac{1}{2M} \cdot \frac{\partial^2}{\partial R^2} + U(R, \Omega) + h_0(\Omega) \tag{2}$$

(where M is the reduced mass of the system) , a subspace Ω is extracted

$$\{h_0(\Omega) - \varepsilon_n\}\varphi_n(\Omega) = 0 \tag{3}$$

with Hamiltonian h_0 and eigenfunctions $\varphi_n(\Omega)$ which satisfy the orthogonality relations:

$$\int \varphi_n^*(\Omega) \cdot \varphi_{\acute{n}}(\Omega)d\Omega = \delta_{n\acute{n}}. \tag{4}$$

In subspace Ω the difference grid Ω_i (i=1,2,...,N) is introduced and in its nodes the values of the wave function of the system are

$$\psi(R, \Omega) \Rightarrow \psi(R, \Omega_i) = \psi_i(R) \tag{5}$$

Further , the discrete index β (β=1,2,...,∞) is introduced corresponding to the set $\{n\}$ of quantum numbers which characterize the system (3) of basis functions. After that the set of eigenfunctions $\varphi_\beta(\Omega)$, (β=1,2,...,N) of the Hamiltonian $h_0(\Omega)$ at nodal points Ω_i becomes the square matrix $\varphi_{i\beta} = \{\varphi_\beta(\Omega_i)\}$ of dimension $N \times N$. Assuming the system $\varphi_\beta(\Omega)$ to be a Chebyshev set on Ω [8] we introduce the inverse matrix $\varphi_{\beta j}^{-1}$ and represent the searched wave function $\psi(R, \Omega)$ as an expansion

$$\psi(R, \Omega) = \sum_{j=1}^{N}(\sum_{\beta=1}^{N} \varphi_\beta(\Omega)\varphi_{\beta j}^{-1})\psi_j(R). \tag{6}$$

For this expansion relation (5) is fulfilled automatically and the following relations

$$(h_0(\Omega)\psi(R, \Omega))_{\Omega=\Omega_i} = \sum_{j=1}^{N}(\sum_{\beta} \varepsilon_\beta\varphi_{i\beta}\varphi_{\beta j}^{-1})\psi_j(R) \tag{7}$$

$$(U(R, \Omega)\psi(R, \Omega))_{\Omega=\Omega_i} = \sum_{j=1}^{N} U(R, \Omega_i)(\sum_{\beta}^{N} \varphi_{i\beta}\varphi_{\beta j}^{-1})\psi_j(R) = U(R, \Omega_i)\psi_i(R) \tag{8}$$

are valid.Substituting expansion (6) into the Schroedinger equation (1) and using relations (7) and (8) we obtain the system of N differential-difference equations:

$$F_1(z) = \sum_{j=1}^{N}\{\delta_{ij}\frac{d^2}{dR^2} + 2M(\varepsilon - V_{ij}(R))\} = 0 \tag{9}$$

where $V_{ij}(R) = U(R, \Omega_i) \cdot \delta_{ij} + \sum_{\beta=1}^{N} \varepsilon_\beta \varphi_{i\beta} \varphi_{\beta j}^{-1}$.

Following the papers [1,6], we formulate the eigenvalue problem for the system of eqs. (9) as a nonlinear equation $F(z) = 0$ for the unknown eigenvalue ε and eigenfunctions $\psi_i(R)$, $z = \{\varepsilon, \psi_i(R)\}$ adding to the equation $F_1(z) = 0$ boundary conditions at $R = 0$ and $R = R_m \to \infty$ and a normalization condition for the system (9) :

$$F_2(z) = \psi_i(0) = 0, \qquad F_3(z) = \psi_i(R_m) = 0,$$

$$F_4(z) = \sum_{ij}^{N} \int \psi_i(R)\psi_j(R)dR - 1 = 0, \tag{10}$$

The scattering problem for equation (1) also can be formulated as the equation $F(z) = 0$ (9,10) for the searched $z = \{\psi_i(R), \varepsilon, t_{ij}\}$ [1,9], where t_{ij} is the reaction matrix of the scattering problem.

3. Results and Discussions

The suggested approach has been successfully tested for several two-dimensional problems in Ref. [1]. In this case

$$\Omega = x = \cos\theta \in [-1, +1], \qquad U(R, \Omega) = U(R, x),$$

$$h_0(x) = \frac{1}{2MR^2} \cdot \frac{\partial}{\partial x}(1 - x^2)\frac{\partial}{\partial x}, \qquad \varphi_\beta(x) = P_\beta(x) \tag{11}$$

where $P_\beta(x)$ are Legendre polynomials. As an example, we have solved scattering problems for a nonspherical potential well

$$U(R, x) = \begin{cases} V_0, & R \le R_0 + \gamma x^2 \\ 0, & R \ge R_0 + \gamma x^2 \end{cases} \tag{12}$$

and for a long-range nonspherical scatterer,

$$U(R, x) = \begin{cases} \infty, & R \le R_0 \\ \frac{x}{R^3}, & R \ge R_0. \end{cases} \tag{13}$$

For setting the boundary conditions $F_3(z)$ in (10) the asymptotic of equations (9) as $R \to \infty$ in the reaction-matrix representation have been used

$$\psi^{(\nu)}(R, x) = \sum_{\alpha=0}^{\infty} \{j_\alpha(kR)\delta_{\nu\alpha} + t_{\nu\alpha} \cdot n_\alpha(kR)\}\sqrt{2\alpha + 1} \cdot P_\alpha(x). \tag{14}$$

where the searched T-matrix could be used for the scattering amplitude and cross sections [1].

In Fig.1 and 2 the calculated wave functions $\psi^{(0)}(R, x)$ for the potentials (12) and (13) are presented. The calculations have been performed at the following parameters $M = 1$, $V_0 = -0.5$ and $R_0 = 1$. For solving the problem (10) the algorithm [5] with the finite-difference approximation in variable R of the order h_R^2 has been used. The calculations have been made at $h_R = 0.0125$, $R_m = 5$, $\varepsilon = 0.005$ for (12) and at $h_R = 0.1$, $R_m = 20$, $\varepsilon = 0.02$ for (13). In the Tables 1 and 2 the convergence of the method as $N \to \infty$ is demonstrated. Note here that the convergence for both considered examples was better than $\frac{1}{(N+1)!}$ [1]. Really, the quantity

$$\delta_\nu(N) = \frac{t_{\nu\nu}(N) - t_{\nu\nu}(2N)}{t_{\nu\nu}(2N) - t_{\nu\nu}(4N)} \tag{15}$$

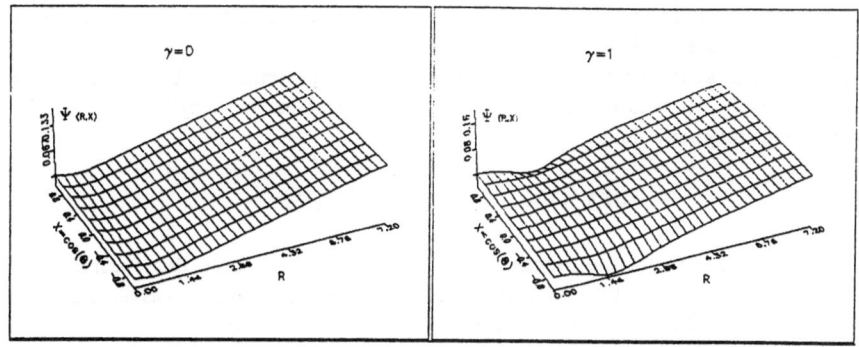

Fig.1

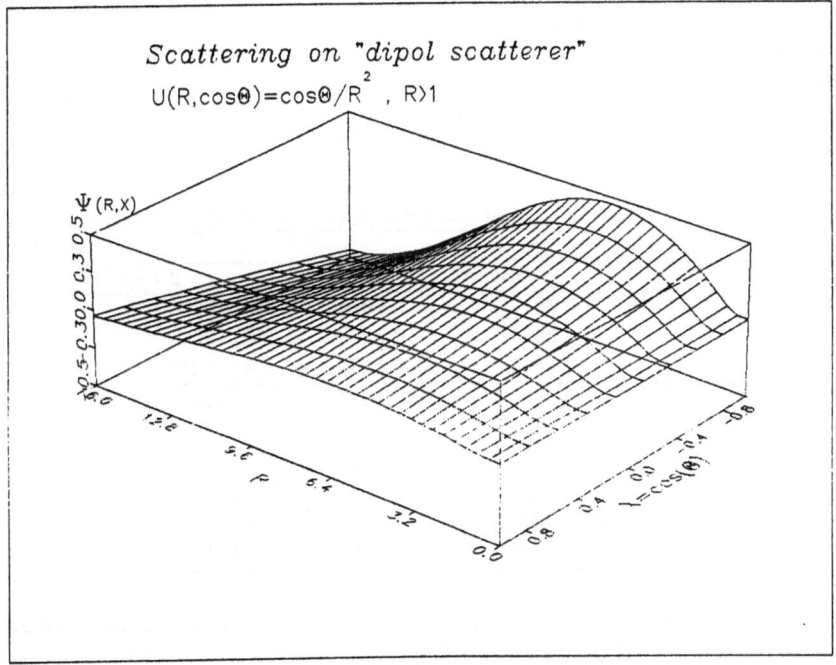

Fig.2

is estimated like $\delta^{theor}(N) \simeq \frac{(2N)!}{(N)!}$ for the $\frac{1}{(N+1)!}$ convergence of the method, and it was confirmed numerically (see Tables 1,2): $\delta_\nu^{num}(N=2) \gtrsim \delta_\nu^{theor}(N=2) = 12$. The accuracy of the calculations can also be controlled by the fulfillment of the relations $t_{\alpha\nu} = t_{\nu\alpha}$ ($\alpha \neq \nu$) (see Tables 1 and 2).

The proposed approach was used for the discrete spectrum of the Schroedinger equation too [1,7]. It was applied to a well-known problem of a hydrogen atom in a homogeneous magnetic field :

$$V(R,x) = -\frac{1}{R} + \frac{R^2\gamma^2}{8} \cdot (1-x^2) \tag{16}$$

where γ is a parameter of the magnetic field intensity. It has been demonstrated [1] that even for the strong field $\gamma = 1$ the convergence of the method was better then $\frac{1}{(N+1)!}$ for $N \geq 2$. In Fig.3 the calculated wave function is presented.

Table 1:Matrix elements $t_{\nu\alpha}$ and quantities $\delta_\nu(N)$ for nonspherical potential well (12)

		$\gamma = 1$				$\gamma = 0$
		$t_{\alpha\nu}$				$t_{\nu\nu}$
N	$\nu \backslash \alpha$	0	1	2	$\delta_\nu(2)$	
2		0.6460	10^{-10}	$-0.6714 \cdot 10^{-3}$		
4	0	0.3817	10^{-10}	$-0.4004 \cdot 10^{-3}$	529	0.05699
8		0.3812	10^{-10}	$-0.3706 \cdot 10^{-3}$		
2		10^{-10}	$0.1191 \cdot 10^{-2}$	10^{-13}		
4	1	10^{-10}	$0.5131 \cdot 10^{-3}$	10^{-12}	6.2	$0.25 \cdot 10^{-4}$
8		10^{-10}	$0.4034 \cdot 10^{-3}$	10^{-12}		
2		$-0.1678 \cdot 10^{-2}$	10^{-12}	$0.1986 \cdot 10^{-5}$		
4	2	$-0.5106 \cdot 10^{-3}$	10^{-12}	$0.8722 \cdot 10^{-5}$	4.0	$0.6 \cdot 10^{-6}$
8		$-0.3712 \cdot 10^{-3}$	10^{-12}	$0.5922 \cdot 10^{-6}$		

Table 2:Matrix elements $t_{\nu\alpha}$ and quantities $\delta_\nu(N)$ for "dipole" scatterer (13)

		$t_{\alpha\nu}$			
N	$\nu \backslash \alpha$	0	1	2	$\delta_\nu(2)$
2		0.3847	0.5916	$-0.4902 \cdot 10^{-1}$	
4	0	0.3095	0.5138	$-0.4583 \cdot 10^{-1}$	376
8		0.3093	0.5134	$-0.4581 \cdot 10^{-1}$	
2		0.5916	$-0.4253 \cdot 10^{-3}$	0.2466	
4	1	0.5137	-0.1259	0.2474	209
8		0.5134	-0.1265	0.2474	
2		-0.1226	0.6164	$-0.2522 \cdot 10^{-1}$	
4	2	$-0.4636 \cdot 10^{-1}$	0.2493	$0.9181 \cdot 10^{-2}$	
8		$-0.4581 \cdot 10^{-1}$	0.2474	$0.9150 \cdot 10^{-2}$	

The paper [7] is devoted for solving the problem of a helium atom bound state like a three- dimensional problem. The Schroedinger equation (1) for a helium-atom S-state has been written in hyperspherical coordinates

$$R^2 = r_1^2 + r_2^2 \in [0,\infty], \alpha = 2 \cdot \arctan\left(\frac{r_1}{r_2}\right) \in [0,\pi], x = \cos\theta = \frac{(\mathbf{r}_1 \cdot \mathbf{r}_2)}{r_1 r_2} \tag{17}$$

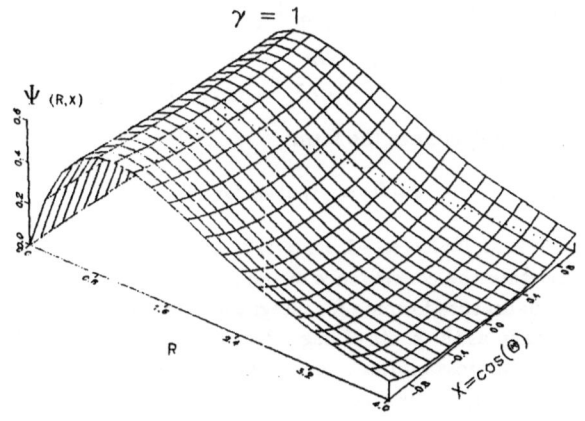

Fig.3

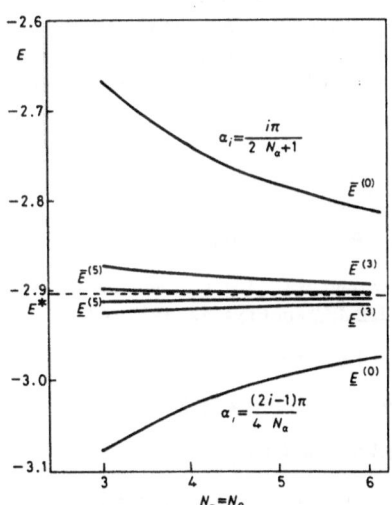

Fig.4.

where r_1 and r_2 are distances between electrons and helium nucleus. In this case

$$\Omega = \{\alpha, x\},$$

$$U(R, \alpha, x) = \frac{1}{R}\left(-\frac{2}{\sin\frac{\alpha}{2}} - \frac{2}{\cos\frac{\alpha}{2}} + \frac{1}{\sqrt{1 + x\cdot\sin\alpha}}\right) - \frac{1}{8R^2},$$

$$h_0(\alpha, x) = -\frac{2}{R^2}\cdot\left(\frac{\partial^2}{\partial\alpha^2} + \frac{1}{\sin^2\alpha}\cdot\frac{\partial}{\partial x}(1 - x^2)\frac{\partial}{\partial x}\right) \tag{18}$$

and the basis functions $\varphi_\beta(\Omega)$ have the form

$$\phi_\beta(\Omega) = P_l(x)\sin(m\alpha) \tag{19}$$

where $\beta = \{m, l\} = 1, 2, ..., N$; $l = n_\theta - 1$, $n_\theta = 1, 2, ..., N_\theta$; $m = 2n_\alpha - 1$, $n_\alpha = 1, 2, ..., N_\alpha$; $N = N_\alpha N_\theta$. To accelerate the convergence of the method by N it is natural to use the nodes of the Gauss quadrature formula for integration by coordinates x and α, i.e. grid points x_i have been taken as the roots of Legendre polynomial $P_{N_\theta + 1}(x)$ but for choosing points α_i we have the two possibilities : the nodes of the polynomials $\sin m\alpha$ or $\cos m\alpha$. This two possibilities for generation the grid Ω_i give us upper $\bar{\varepsilon}^{(0)}$ and lower $\underline{\varepsilon}^{(0)}$ approximations to the helium-atom binding energy (see Fig.4). The convergence could be improved according to the simple formula [5]

$$\underline{\varepsilon}^{(s)}, \bar{\varepsilon}^{(s)} = \frac{1}{2}(\underline{\varepsilon}^{(s-1)} + \bar{\varepsilon}^{(s-1)}) \tag{20}$$

The obtained approximation $\varepsilon = -2.9034$ at $N = N_\alpha\theta \leq 40$ is close to the results obtained by the "direct" numerical methods of finite elements $\varepsilon = -2.9032$ [9] and finite differences $\varepsilon = -2.9036$ [10]. In our opinion, two circumstances make the method attractive for solving different three-dimensional problems in nuclear and atomic physics. The first is a sufficiently simple algorithm for generation of the matrix $V_{ij}(R)$ of the system of equations (9). The second is a sufficiently fast convergence of the expansion (6) and a possibility for obtaining both upper and lower approximations to the binding energy of the system.

References

[1] V.S. Melezhik, J. Comp. Phys. **92** (1991) 67.

[2] J.V. Lill, G.A. Parker and J.C. Light, Chem.Phys.Lett.,**89** (1982)483; J.C. Light, I.P.Hamilton and J.V. Lill,J.Chem.Phys.,**82** (1985)1400.

[3] W.Yang and A.C.Peet, Chem.Phys.Lett., **153**(1988)98.

[4] R.A. Freisner, Chem.Phys.Lett.,**116**(1985)39.

[5] V.S. Melezhik, J.Comp.Phys., **65**(1986)1.

[6] S.K. Godunov and V.S. Rjabenky, *Difference Schemes*,Nauka,Moscow, 1977(In Russian).

[7] V.S. Melezhik, Nuovo Cimento **B106** (1991)537.

[8] I.S. Berezin and N.P.Zhidkov, Numerical Methods (in Russian) **v1**, Nauka (Moscow, 1959).

[9] F.S. Levin and J.Shertzer, Phys.Rev. **A32** (1985) 3285.

[10] I.L. Hawk and D.L.Hardcastl, Comput.Phys.Commun. **16** (1979) 159.

Few-Body Systems, Suppl. 6, 557—562 (1992)

Few-
Body
Systems
© by Springer-Verlag 1992

THE INTEGRODIFFERENTIAL EQUATION APPROACH AND
SOME APPLICATIONS

H. Fiedeldey

Department of Physics, University of South Africa, P O Box 392,

Pretoria, 0001, South Africa

The Integrodifferential Equation Approach (IDEA) generalizes the Faddeev equation from three to many-bodies, but retains its three-body like structure, due to the neglect of all N-particles correlations with $N \geq 3$. Employing realistic nucleon—nucleon potentials it produces results quite close to those obtained by means of the exact Faddeev-Yakubovksy equations, both treated in a three-channel approximation. This and the speed and efficiency of the IDEA are put to good use in several applications, particularly to derive an error-distribution of the three- and four-nucleon binding energies, from the corresponding distribution of potentials, obtained by Leeb et al via inversion of the experimental $^1S_0$ phase shifts with error bars. It is also applied to 2α, 3α and 4α systems to demonstrate that to a good approximation simultaneous supersymmetric transformations exist for these systems as a consequence of the occurrence of Pauli forbidden states in the 2α system.

In the Integrodifferential Equation Approach the A-body wave function is written as a sum of modified Faddeev amplitudes

$$\psi(\mathbf{x}) = \sum_{i > j = 1}^{A} \psi_{ij}(\mathbf{r}_{ij}, \mathbf{x}) \tag{1}$$

where $\mathbf{x}$ is a complete set of Jacobi coordinates, $\mathbf{r}_{ij} = \mathbf{r}_i - \mathbf{r}_j$ represents the relative coordinates, and the ψ_{ij} satisfy the modified Faddeev equation [1]

$$\left[T + \tfrac{1}{4}A(A-1)V_0(r) - E \right] \psi_{ij}(\mathbf{r}_{ij}, \mathbf{x}) = -\left[V(\mathbf{r}_{ij}) - V_0(r) \right] \sum_{k > \ell = 1}^{A} \psi_{k\ell}(\mathbf{r}_{k\ell}, \mathbf{x}) \tag{2}$$

where $r = \left\{ \sum_{i=1}^{A} x_i^2 \right\}^{\frac{1}{2}}$ is the hyperradius and $V_0(r)$ the hypercentral potential. The

Schrödinger equation is recovered when we sum over i and j. To solve this equation exactly requires its transformation into an infinite set of coupled integrodifferential equations, e.g. by partial wave expansion and truncation after convergence has been achieved. In the lowest order of the IDEA, we retain only the first term of this expansion and set $\psi_{ij}(r_{ij},x) = H_{[L_m]}(x) \, F(r_{ij},r)$ where $H_{[L_m]}(x)$ is the harmonic

polynomial of minimal degree. When all particles, as is the case for bosons, can be in a minimal configuration with $L_m = 0$, then $H_{[0]}$ is a constant. Projecting the amplitudes $F_{k\ell}(r_{k\ell},r)$ at the right hand side of (2) on the r_{ij}-space with $F_0(r_{ij},r) = r^{-(D-1)/2} \, P_0(z,r)$ where $z = 2r_{ij}^2/r^2-1$ and $D = 3(A-1)$, while $W(z)$ is the weight function, we obtain the following integrodifferential equation in two variables

$$\left\{ -\frac{\hbar^2}{m} \left[\frac{\partial^2}{\partial r^2} - \frac{\mathscr{L}(\mathscr{L}+1)}{r^2} + \frac{4}{r^2} \frac{1}{W(z)} \frac{\partial}{\partial z} (1-z^2) W(z) \frac{\partial}{\partial z} \right] + \frac{A}{2}(A-1) \, V_0(r) - E \right\} P(z,r)$$

$$= - \left[V\left(r\sqrt{\frac{1+z}{2}}\right) - V_0(r) \right] \left\{ P_0(z,r) + \int_{-1}^{+1} f_{[0]}(z,z') P_0(z',r) dz' \right\}. \qquad (3)$$

The projection function $f(z,z')$ can be written down in analytical form for A bosons in s-states, but is otherwise given by a sum which converges more rapidly with increasing A. Eq. (3) is the exact Faddeev equation for A = 3 and S-state projected potentials when we put $V_0(r) = 0$ [2]. This is called the SIDE. For local potentials acting in all partial waves it is the first term of a coupled system of equations. However due to the predominance of the hypercentral potential $V_0(r)$, the higher partial waves are partly taken into account and good results can already be obtained with a single equation for Wigner type potentials and two coupled equations for spin-dependent forces.

Previous calculations have shown that excellent results are obtained for three- and four-nucleon systems, despite the neglect of N-particle correlations for $N \geq 3$, using Malfliet-Tjon type spin-dependent and effective nucleon-nucleon forces like the Volkov and Brink-Boeker B1 potentials. For the MT I-III force we obtain for the three-nucleon binding energy E_t and the percentage S'-state $P_{S'}$, respectively, $E_t = 8.86$ MeV and $P_{S'} = 2,86$ [3] compared to converged results of $E_t = 8.88$ MeV and $P_{S'} = 2.42$ of Rosati et al. [4]. For the SIDE our results (i.e. $E_t = 8.54$ MeV and $P_{S'} = 1.95$ [3]) are essentially exact. In the case of four nucleons we get $E_\alpha = 31.29$ MeV and $P_{S'} = 1.49$ in the IDEA [3]. An indication of the contribution of higher correlations is provided by comparing the E_α = 30.68 MeV obtained by the IDEA for the MTV force to E_α = 31.3 ± 0.2 MeV [5]. Recently there has been a lively discussion on large scale extended shell model calculations for ^{4}He [6,7] and the effective central nucleon-nucleon V7 and B1 potentials, as compared to Diffusion Monte Carlo (DFMC) calculations. Ceuleneer pointed out that the effects of short-range correlations induced by the repulsive core are beyond the scope of the shell model [7]. The IDEA with binding energies of 28.7 MeV and 38.55 for the V7 and B1 potentials easily reproduces the corresponding DFMC results of 28.6 ± 0.1 MeV and 38.5 ± 0.1 MeV and for the breathing mode of the B1 potential we get 9.2 MeV. Our IDEA results were achieved on an IBM RISC 6000 model 320 Workstation. Each binding energy took only 5 s of CPU time and 4 Mbytes of operating memory [8].

For effective central nucleon-nucleon forces we have recently extended the IDEA to many-body systems (A > 4) [9]. Although this falls, strictly speaking, outside the scope of a conference on few-body physics, it is quite interesting that the IDEA remains a two-variable integrodifferential equation in two variables of the same form as the three-body Faddeev equation. Of all the many methods for the calculation of many-body nuclei, the IDEA is the only one which is a straightforward extension of the three-body Faddeev equation. Binding energies for heavier closed shell nuclei calculated by means of the IDEA compare favourably with other methods. For example for ^{12}C and the B1 potential the ground state energy predicted by the IDEA lies between 80.1 and 79.6 MeV, as compared to 82.3 MeV for the fourth order coupled cluster (FAHT-4) method and 82.9 ± 0.2 MeV for the variational Monte Carlo method [10]. For ^{16}O the IDEA yields a value between 164.7 and 164.2 MeV binding for the B1, while the Brueckner-Hartree-Fock method yields 163.7 [11]. Computing is no more laborious for these larger nuclei than for the triton and ^{4}He. The method has now also been extended to open shell nuclei and applied to ^{6}Li and ^{14}N for effective nucleon-nucleon forces [9].

Returning to few-body systems, the IDEA has also been applied to Three- and Four-Nucleon systems using realistic nucleon-nucleon forces in the so-called three-channel approximation, originally introduced for three-nucleon systems. In this approximation only even S- and D-state components are taken into account. For the three-nucleon system in the SIDE where the hypercentral potential is omitted, we recover the corresponding Faddeev integrodifferential equations. For the four-nucleon system the same set of three coupled integrodifferential equations in two variables apply, but with different more general projection functions.

The results of our calculations for the Gogny-Pires-de Toureil (GPDT), the Super-soft-core (SSC-C) and the Reid-soft-core (RSC), potentials in the three-channel approximation, are given in Table 1 for the α-particle [12]. They can be directly compared to those obtained recently by Cerba et al. [13] by solving the exact Faddeev-Yakubovsky FY integrodifferential equations in three variables, but also like ours in a three-channel approximation (see Table 1).

Our three-channel calculations are extremely fast. A converged binding energy for the RSC potential takes about 30 s on our IBM RISC 6000 Workstation.

Table 1. E_α, P_S, $P_{S'}$ and the percentage D-state P_D for the four-nucleon system in the three-channel approximation compared to the three-channel FY [13]. (Note that for the RSC potential the binding energy indicated by 18.2* is not a converged result.)

| | SIDE | | | IDEA | | | FY | |
	E_α	$P_{S'}$	P_D	E	$P_{S'}$	P_D	E_α	P_D
GPDT	25.72	0.13	4.03	26.43	0.30	4.05		
SSC	20.64	0.23	8.69	21.19	0.36	8.74	21.0	8.4
RSC	17.95	0.32	10.78				18.2*	

This property of the IDEA has been exploited to investigate the propagation of errors in the $^1S_0$ phase shift in three- and four-nucleon systems. Leeb et al. [14] have applied exact inversion techniques to determine the $^1S_0$ potential from the $^1S_0$ phase shift. Taking the experimental errors into account by means of a statistical analysis, a distribution of phase shifts is inverted to obtain a distribution of about 160 potentials, which allows one to draw conclusions of a probabilistic nature about the $^1S_0$ potential [14]. This large set of potentials is then used to calculate the corresponding distribution of three-nucleon and four-nucleon binding energies E_t and E_α, by means of the IDEA [15].

Since no error analysis for the $^3S_1$-$^3D_1$ phase shift data is available yet, we use the RSC potential for the $^3S_1$-$^3D_1$ coupled channels, but replace the $^1S_0$-RSC potential successively by one of the 160 members of our $^1S_0$-potential distribution, each time calculating E_t and E_α. In this way we obtain a set of values of E_t and E_α which appears to be normally distributed. Hence the standard deviation $\Delta E = \sqrt{<(E-<E>)^2>}$ can be directly related to the probability interval $[E - \Delta E, E + \Delta E]$ corresponding to 68% of all cases. The results we obtained were $\Delta E_t = 0.04$ MeV and $\Delta E_\alpha = 0.09$ MeV [15].

Since the tensor force contributes about 2/3 of the binding energy and the corresponding error bars are larger, particularly on the mixing parameter $\epsilon_1(k)$, we expect ΔE to be at least 3 to 4 times larger, once the experimental errors in the $^3S_1$-$^3D_1$ channels have been included in the inversion and statistical analysis.

The experimental errors in the NN interaction can therefore be expected to propagate themselves to a significant extent in the properties of three- and four-nucleon systems and are possibly even comparable to the contribution of three-body forces, etc.

Finally we consider an interesting application of supersymmetric quantum mechanics, via the IDEA, to few-body bound systems. Supersymmetric quantum mechanics has found widespread application in many fields of physics during the last few years and has for instance been discussed by Amado [16] in the context of the Efimov effect and applied by us to n+d quartet scattering [17]. The deep or shallow nature of nucleus-nucleus effective local potentials, derived either from phenomenology or microscopic theories like the Resonating Group Model (RGM), has been controversial for a long time. For the α-α interaction there exist the shallow ℓ-dependent purely phenomenological Ali-Bodmer [18] potentials and the deep Gaussian potential of Buck et al [19] $V_0(r)$, which is ℓ- and E-independent and produces an excellent fit to the α-α scattering phase shifts up to 40 MeV. In agreement with the RGM from which it is derived, this potential produces phase shifts which, if we normalise $\delta_\ell(\infty)$ to zero, have values $\delta_\ell(0) = (n^\ell_{PFS} + n^\ell_B)\pi$, where n^ℓ_{PFS} and n^ℓ_B are the so-called Pauli forbidden (PFS) and physical bound states respectively. The Ali-Bodmer potentials on the other hand satisfy $\delta_\ell(0) = n^\ell_B \pi$ in accordance with the usual form of the Levinson theorem.

Baye [20] showed that if supersymmetric (Susy) transformations are applied which remove the PFS in each partial wave, ℓ-dependent shallow potentials are obtained which maintain strict phase equivalence with the potential of Buck et al including the fact that $\delta_\ell(0) = (n^\ell_{PFS} + n^\ell_B)\pi$. These potentials however contain repulsive $1/r^2$ singularity at the origin and obey a generalised Levinson theorem. Their shapes are quite similar to the Ali-Bodmer potentials and provide a nice explanation of the controversy concerning deep and shallow potentials. This and other applications of Susy transformations have been strictly at the two-body level however.

Remarkably in the case of the α-α potential and 3α and 4α systems, we found a very close relationship between the corresponding 2-, 3- and 4-body Susy transformations. The Buck et al. potential contains two PFS in the s-state, one in the d-state and none otherwise. It is therefore necessary if we want to investigate the effect of Susy transformations on the 2α, 3α and 4α spectra, to restrict ourselves to s-projected potentials, denoted by $V_0(r)$, $V_1^{(0)}(r)$ and $V_2^{(0)}(r)$. The latter two are s-state projected potentials produced after the first and second Susy transformations respectively, each of which removes a bound state.

In Table 2 we present the results of our calculations using the IDEA for these s-projected potentials and the 3α system [21]. Similar results are obtained for the 4α-system.

We included the Coulomb potential in our calculations. It is seen that each Susy transformation removes a set of unphysical 3α and 4α states leaving the other nearly unaffected. The single remaining physical state in both 3α and 4α systems does

Table 2. The 2α and 3α binding energies for $V_0(r)$, $V_1^{(0)}(r)$ and $V_2^{(0)}(r)$ in MeV (Coulomb force included).

V_0	2α	-73.6	-25.93	$+0.092$					
	3α	-243.5	-179.6	-126.0	-78.2	-46.5	-23.8	-9.5	-2.6
$V_1^{(0)}$	3α	$-$	$-$	$-$	-77.5	-44.5	-24.3	-11.3	-2.3
$V_2^{(0)}$	3α								-0.26

not differ much for V_0, $V_1^{(0)}$ and $V_2^{(0)}$. If we omit the Coulomb interaction the picture becomes even more convincing for the 4α system, since the four 'physical' states of V_0, $V_1^{(0)}$ and $V_2^{(0)}$ are all largely unaffected by the Susy transformations. We find that the physical binding energies for $V_0(r)$ are in this case given by 8.05, 4.45, 1.82 and 0.45 MeV, while the corresponding one for $V_4^{(0)}(r)$ are 9.07, 4.35, 2.39 and 1.35 MeV respectively. The corresponding binding energies for the Ali-Bodmer potential are 7.61, 3.63, 1.99 and 0.80. This remarkable feature seems to be associated with potentials having PFS like the potential of Buck et al. In other cases the whole many-body spectrum is modified by a two-body Susy transformation.

References

1. Fabre de la Ripelle, M., Phys. Lett. **B 135**, 5 (1984)
2. Fabre de la Ripelle, M. and Fiedeldey, H., Phys. Lett. **B 171**, 325 (1986);
   Fabre de la Ripelle, M., Fiedeldey, H. and Sofianos, S.A., Phys. Rev. **C 38**, 449 (1988)
3. Oehm, W., Sofianos, S.A., Fiedeldey, H. and Fabre de la Ripelle, M. Phys. Rev. **C 42**, 2322 (1990) and **43**, 23 (1991)
4. Rosati, S. Viviani, M. Kievsky, A., Few-Body Systems **9**, 1 (1990)
5. Kameyama, H., Kamimura, M., Fukushima, Phys. Rev. **C 40**, 974 (1989)
6. Bishop, R.F., Flynn, M.F., Fosca, M.C., Buendia, E. and Guardiola, R. J. Phys. **G 16**, L61 (1990)
7. Ceuleneer, R., Semay, C., Vandeputte, P., J. Phys. **G 16**, L295 (1990)
8. Adam, R.M., Fiedeldey, H., Sofianos, S.A. and Fabre de la Ripelle, M., J. Phys. **G 16** (1991) to be published.
9. Adam, R.M., Fiedeldey, H., Sofianos, S.A. and Fabre de la Ripelle, M., (1991), submitted for publication.
10. Guardiola, R. Faessler, A., Muther, H. and Polls, A., Nucl. Phys. **A 371**, 79 (1981)
11. Bosca, M.C., Buendia, M. Guardiola, R., Phys. Lett. **B 198**, 312 (1987)
12. Oehm, W., Sofianos, S.A., Fiedeldey, H. and Fabre de la Ripelle, M. Phys. Rev. **C** (1991) to be published
13. Cerba, J. Gignoux, C. Merkuriev, S.M., Yakavlev, S.L., Institute des Sciences de Grenoble Report No ISN90.30, 1990
14. Leeb, H., Invited talk at the XIIIth European Conference on Few—Body Physics, Elba, September 1991
15. Adam, R.M., Fiedeldey, H., Sofianos, S.A. and Leeb, H. submitted for publication
16. Amado, R.D., Cannate, F., Dedonder, J.P. Phys. Rev. **A 42**, 1289 (1990)
17. Sofianos, S.A. Papastylianos, A., Fiedeldey, H. Phys. Rev. **C 42**, R506 (1990)
18. Ali, S., Bodmer, A.R., Nucl. Phys. **80**, 99 (1966)
19. Buck, B., Friedrich, H. Wheatley, C., Nucl. Phys. **A 275**, 246 (1977)
20. Baye, D. Phys. Rev. Lett. **58**, 2738 (1987)
21. Adam, R.M., Fiedeldey, H., Sofianos, S.A., submitted for publication

Few-Body Systems, Suppl. 6, 563—572 (1992)

Few-
Body
Systems
© by Springer-Verlag 1992

STUDY OF THE BOUND STATES OF FEW–NUCLEON

SYSTEMS WITH CORRELATED BASIS FUNCTIONS

S.Rosati

Dipartimento di Fisica, Università di Pisa, Piazza Torricelli, 2, 56100 Pisa, Italy

INFN, Sezione di Pisa, Piazza Torricelli, 2, 56100 Pisa, Italy

A.Kievsky and M.Viviani

INFN, Sezione di Pisa, Piazza Torricelli, 2, 56100 Pisa, Italy

ABSTRACT. The ground–state wave functions of the three– and four–body nucleon systems have been calculated by means of an expansion on a set of "correlated states"given by the product of a correlation factor with the functions of a complete basis. The correlation factor has been chosen as the product of unidimensional functions determined by an Euler variational procedure. Two different complete basis have been used, the Harmonic Oscillator (HO) and the Hyperspherical Harmonic (HH) basis. For three nucleons interacting via a realistic potential, both methods give results of a satisfactory accuracy when compared with others techniques. However, for four nucleons (interacting via a central potential) only the correlated HH expansion makes it possible to obtain accurate wave functions.

1. INTRODUCTION.

Many techniques have been used to calculate the bound–state properties of the few–nucleon systems by starting from the nonrelativistic model of structureless nucleons interacting via potentials depending on spin, isospin, tensor and spin–orbit forces. Amongst the methods proposed for the three–nucleon system, the Faddeev [1-4] and the Green Function Monte Carlo [5] (GFMC) techniques are two of the most accurate; for the four nucleon case, only the GFMC has been succesfully applied but with rather large statistical errors.

The Rayleigh–Ritz variational method was the first technique applied to treat realistic potentials, but till recently the improvement of the variational upper bound toward the exact eigenvalue was a difficult task due to the strong short–range repulsion between the particles [6-11]. The main problem is to select a set of functions suitable for describing the behaviour in the short–range repulsion region, in the intermediate–range actractive region and also in the asymptotic long–range region. A very satisfactory solution has been given only recently by expanding the variational wave function on a set of gaussian functions; the method was firstly developed for the study of muonic molecular ions [12] and successively applied to the three–nucleon problem [13].

In the approach we are pursuing [14-16], the bound–state wave function is written schematically in the form

$$\Psi = \Psi_v \sum a_k \Phi_k \ . \tag{1.1}$$

Ψ_v is a rather simple variational wave function which is able to produce a reasonable description of the system and under some respect it is similar to those used in the ATMS approach [7] and in the variational Monte Carlo [10] technique; the radial dependence of Ψ_v has been determined by an Euler type procedure. The terms Φ_k in eq. (1.1) are the functions of a complete basis, namely (i) the Harmonic Oscillator (HO) basis functions, and (ii) the Hyperspherical Harmonics (HH) functions. The terms a_k in eq. (1.1) are coefficients in the case of the HO expansion and functions of the hyperradius in the case of the HH expansion.

In section 2 the calculations of the variational wave function and its improvement by the introduction of a set of correlated basis functions is discussed for the three–nucleon system and the results obtained for two realistic potentials, the RV8 version of the Reid soft–core interaction [17] and Argonne (AV14) potential [18] are given. In section 3, a preliminary approach to the four nucleon problem is presented. The last section is devoted to a discussion of the merits and limits of the technique adopted here.

2. CORRELATED WAVE FUNCTION FOR A THREE–NUCLEON SYSTEM.

The wave function of a three–nucleon system with total angular momentum J, J_z and total isospin T, T_z can be written as a sum of three amplitudes

$$\Psi = \psi(\mathbf{x}_i, \mathbf{y}_i) + \psi(\mathbf{x}_j, \mathbf{y}_j) + \psi(\mathbf{x}_k, \mathbf{y}_k) \ , \tag{2.1}$$

where the spin-isospin dependence is implicitly understood. The Jacobi coordinates are (i, j, k cyclic)

$$\mathbf{x}_i = \frac{1}{\sqrt{2}}(\mathbf{r}_j - \mathbf{r}_k), \quad \mathbf{y}_i = \frac{1}{\sqrt{6}}(\mathbf{r}_j + \mathbf{r}_k - 2\mathbf{r}_i) \ , \tag{2.2}$$

$\mathbf{r}_i$ denoting the position of particle i. By following the notation of ref. 14, in the L–S coupling scheme the i-th amplitude with quantum numbers J, J_z , T, T_z is written as

$$\psi(\mathbf{x}_i, \mathbf{y}_i) = \sum_\alpha x_i^{l_\alpha} y_i^{L_\alpha} F_\alpha(x_i, x_j, x_k) \left\{ [Y_{l_\alpha}(\hat{x}_i) Y_{L_\alpha}(\hat{y}_i)]_{\Lambda_\alpha} [s_\alpha^{jk} s_\alpha^i]_{S_\alpha} \right\}_{J J_z} [t_\alpha^{jk} t_\alpha^i]_{T T_z} \ . \tag{2.3}$$

x_i, y_i are the moduli of the Jacobi coordinate and s (t) denotes a spin (isospin) function; each α–channel is specified by the angular momenta l_α, L_α and Λ_α and the spin (isospin) s_α^{jk} (t_α^{jk}) and s_α^i (t_α^i) of the pair j, k and the third particle i. l_α and L_α are coupled to give Λ_α , s_α^{jk} (t_α^{jk}) and s_α^i (t_α^i) are coupled to give S_α (T). The amplitude $\psi(\mathbf{x}_i, \mathbf{y}_i)$ changes sign under the exchange of the particles j and k to ensure the antisymmetry of the wave function. As a consequence, if $F_\alpha(x_i, x_j, x_k)$ is even (odd) when $j \leftrightarrow k$, the number $\ell_\alpha + s_\alpha^{jk} + t_\alpha^{jk}$ must be odd (even).

The Faddeev expansion in the configuration representation is obtained when the radial functions $F_\alpha(x_i, x_j, x_k)$ are replaced by the two–body amplitudes $\Phi_\alpha(x_i, y_i)$ and the requirement that the corresponding wave function satisfies the Schroedinger equation produces a set of two–dimensional integro–differential equations for the functions $\Phi_\alpha(x_i, y_i)$: these equations have been accurately solved [1-4] and the convergence is reached when all the important channels have been taken into account in the expansion (2.3).

As an alternative variational approach, the radial amplitude $F_\alpha(x_i, x_j, x_k)$ can be chosen to be a product of one–dimensional functions

$$F_\alpha(x_i, x_j, x_k) = f_\alpha(x_i)g_\alpha(x_j)g_\alpha(x_j) . \tag{2.4}$$

In the case of purely central potentials, such a product form for each amplitude allows for a rather accurate description of the structure of the system [14,16,19]; hence it is worthwhile to explore the flexibility of such a choice when using realistic interactions.

The optimum variational choice of the functions f_α, g_α is determined by the condition that the functional derivatives of the average value of the hamiltonian H with respect to them be zero, namely

$$\langle \frac{\delta}{\delta u_m} \Psi \mid H - E \mid \Psi \rangle = 0 , \tag{2.5}$$

where $u_m(r)$ stands either for $f_\alpha(r)$ or $g_\alpha(r)$. The Euler equations satisfied by the functions $u_m(r)$ can be written in the form

$$\sum_m \Big(A_{mn}(r)\frac{d^2}{dr^2} + B_{mn}(r)\frac{d}{dr} + C_{mn}(r) - EN_{mn}(r)\Big)u_m(r) = 0 , \tag{2.6}$$

where the eigenvalue E is the total energy of the system. The procedure to calculate the coefficients A_{mn}, B_{mn}, C_{mn}, N_{mn} and to solve the system of equations (2.6) has been sketched in ref. 16. In the present paper, the method is extended to treat also potential containing quadratic angular momentum and quadratic spin–orbit operators.

The channels included in the calculation are specified in table I, together with the corresponding form of the radial functions F_α. The four channels ($\alpha = 5 \div 8$) with $\ell_\alpha = L_\alpha = 2$ have been taken to have the same radial dependence but with the amplitudes $c_5 \div c_8$ being variational parameters; an analogous situation holds for the four channels ($\alpha = 9 \div 12$) with $\ell_\alpha = L_\alpha = 1$. Such a choice is useful to keep the number of functions to be calculated small but to a good extent it is also justified by the Faddeev theory results [4,13] in the jj coupling.

In order to test the Euler approach, we have performed the calculations for a system of three nucleons interacting via the (central) spin–dependent MT I–III potential [20]. Only the channels 1, 2 and 5 listed in table I have been considered and thus, six unidimensional function f_α, g_α must be calculated by solving the eq. (2.6). The values obtained for the binding energy, average kinetic energy and percentage of the mixed symmetry state, are 8.87 MeV, 31.30 MeV and 2.47, respectively and they compare well with the corresponding values 8.88 MeV, 31.32 MeV and 2.42 obtained with the more sophisticated method presented in ref. 15.

α	ℓ_α	L_α	Λ_α	$s_\alpha^{(jk)}$	S_α	$t_\alpha^{(jk)}$	F_α
1	0	0	0	1	1/2	0	$f_1(x_i)g_1(x_j)g_1(x_k)$
2	0	0	0	0	1/2	1	$f_2(x_i)g_2(x_j)g_2(x_k)$
3	2	0	2	1	3/2	0	$f_3(x_i)g_3(x_j)g_3(x_k)$
4	0	2	2	1	3/2	0	$f_4(x_i)g_4(x_j)g_4(x_k)$
5	2	2	0	1	1/2	0	$c_5\, f_5(x_i)g_5(x_j)g_5(x_k)$
6	2	2	2	1	3/2	0	$c_6\, f_5(x_i)g_5(x_j)g_5(x_k)$
7	2	2	1	1	1/2	0	$c_7\, f_5(x_i)g_5(x_j)g_5(x_k)$
8	2	2	1	1	3/2	0	$c_8\, f_5(x_i)g_5(x_j)g_5(x_k)$
9	1	1	0	1	1/2	1	$c_9\, f_9(x_i)g_9(x_j)g_9(x_k)$
10	1	1	2	1	3/2	1	$c_{10} f_9(x_i)g_9(x_j)g_9(x_k)$
11	1	1	1	1	1/2	1	$c_{11} f_9(x_i)g_9(x_j)g_9(x_k)$
12	1	1	1	1	3/2	1	$c_{12} f_9(x_i)g_9(x_j)g_9(x_k)$

Table I. *Quantum numbers and radial functions for the various $\alpha = 1 \div 12$ channels included in the variational wave function (2.3,4). The coefficients c_α, $\alpha = 5 \div 12$, are trial parameters.*

The values of the binding energy B, the average kinetic energy, the percentage of the various states and the mass radius calculated in correspondence of realistic potentials with all the twelve channels listed in table I taken into account, are presented in table II. The dependence of B on the number of channels taken into account is similar for both the potentials considered, which means that our variational function is suitable to treat the quadratic angular terms of the potential.

When only the first three channels ($\alpha = 1 \div 3$) are included, the binding energy is underestimated by about $0.6 \div 0.9$ MeV so that the structure of the system is described in a rather rough way. When the successive four channels, corresponding to $\ell_\alpha = L_\alpha = 2$, are also included, the binding energy improves by about 0.5 MeV; the four channels with odd partial waves, $\ell_\alpha = L_\alpha = 1$, allow for components with antisymmetrical radial dependence, and increase the binding by 0.1 MeV, approximately.

The problem of improving the variational wave function so determined can be afforded in different ways. In ref. 14 it has been found that, in the case of a central potential, the difference between the variational Euler wave function and the exact one can be expanded in terms of a small number of correlated harmonic-oscillator (H.O.) basis functions. When the same approach is used for realistic interactions, the function $F_\alpha(x_i, x_j, x_k)$ to be inserted in eq. (2.3) is taken of the form

$$x_i^{\ell_\alpha}\, y_i^{L_\alpha} F_\alpha(x_i, x_j, x_k) = f_\alpha(x_i)g_\alpha(x_j)g_\alpha(x_k) \times$$

$$\times \left\{ a_0 x_i^{\ell_\alpha}\, y_i^{L_\alpha} + \sum_{n_a n_b} a_{n_a n_b} R_{n_a \ell_\alpha}(\beta_\alpha x_i) R_{n_b L_\alpha}(\beta_\alpha y_i) \right\}, \tag{2.7}$$

where the functions f_α, g_α are those obtained by the Euler-type procedure and $R_{n\ell}$ are harmonic oscillator radial functions. The non-linear parameters β_α can be chosen at the best for the various channels, however, in the cases we have exploited there is a small

Model	Method	B (MeV)	$<T>$ (MeV)	$P_{S'}$	P_D	P_P	R (fm)
RSCV8	E	7.50	51.76	1.23	9.59	0.07	1.76
	E + CHO	7.57	52.23	1.27	9.72	0.08	1.75
	CHH	7.58	52.10	1.35	9.60	0.08	1.76
	Faddeev[5]	7.59	52.2				1.76
	GFMC[5]	7.54(2)	54.2(2)				
AV14	E	7.57	45.23	0.99	8.86	0.070	1.78
	E + CHO	7.64	45.75	1.06	8.97	0.075	1.76
	CHH	7.66	45.60	1.11	8.92	0.070	1.76
	CRC $- 12^{13}$	7.678		1.13	8.96	0.075	
	CRC $- 26^{13}$	7.684	45.68	1.13	8.97	0.076	
	Faddeev[4]	7.67		1.12	8.96	0.08	

Table II. *Results obtained for the three–nucleon ground–state in correspondence to the two model potentials specified in column 1 and including in the wave function the twelve channels listed in table I, by the Euler (E), Euler + CHO (E + CHO) and CHH method. The corresponding values of the Faddeev [4,5], GFMC [5] and CRC [13] techniques have been reported; the row denoted CRC − 12 (CRC − 26) lists the values obtained by including 12 (26) channels in the wave function. B, $\langle T \rangle$ and R are the total binding energy, the average kinetic energy and the mass radius; P_s', P_D and P_P are the percentage of the mixed–symmetry S–states, of the D–states and of the P–states, respectively.*

dependence on the precise β_α values and the common value $\beta = 1.0$ fm^{-1} has been used. It has to be noticed that the correlation factor for the harmonic-oscillator functions in eq. (2.7) has been taken equal to the Euler function calculated for the corresponding channel: such a choice should provide a correct wave function for small interparticle separations.

For a given channel α, the HO functions form a complete set and the linear parameters a_0 and $a_{n_a n_b}$ can be determined by solving a generalized eigenvalue problem. The number of correlated harmonic–oscillator (CHO) functions corresponding to a given number N_Q of oscillator quanta

$$N_Q = 2n_a + \ell_a + 2n_b + L_\alpha ,\qquad(2.8)$$

is increased to get the convergence in the calculated binding energy. In practice, the choice $N_Q = 10$ for the first eigth channels and $N_Q = 8$ for the remaining four channels listed in table I is adequate since larger N_Q values do not produce any appreciable improvement.

The results obtained for the various quantities, when all the twelve channels are taken into account, are reported in the rows denoted by E+CHO of table II. For both the potentials considered, the calculated binding energy differs from the Faddeev value by about 0.02 MeV which proves the good accuracy of the method. The contribution from other channels different from those listed in table I has been estimated [13] to be approximately 0.006 MeV. Therefore the missing binding energy in the Euler + CHO approach is spreaded out over a large variety of CHO states with high oscillator quantum numbers and a resummation of them appears to be problematic.

Motivated by these results, we have explored another class of basis functions, namely the Hyperspherical Harmonics (HH) functions which have been used by many authors to investigate a variety of few-body systems [11,15,21-23]. One of the advantages of the HH expansion is that the "best" dependence on the hyperradius $\rho = (x_i^2 + y_i^2)^{1/2}$ is determined by solving a set of coupled differential equations; then, we can expect that the HH expansion be able to converge faster than the HO expansion.

In the CHH expansion the i-th radial amplitude is written in the form

$$x_i^{l_\alpha} y_i^{L_\alpha} F_\alpha(x_i, x_j, x_k) = f_\alpha(x_i) g_\alpha(x_j) g_\alpha(x_k) \sum_{n=1}^{M_\alpha} U_n^{(\alpha)}(\rho)^{(2)} P_n^{l_\alpha, L_\alpha}(\phi_i) \rho^{l_\alpha + L_\alpha} , \quad (2.9)$$

with $x_i = \rho \cos \phi_i$ and

$$^{(2)}P_n^{l_\alpha, L_\alpha}(\phi) = N_n^{l_\alpha, L_\alpha}(\sin \phi)^{L_\alpha} (\cos \phi)^{l_\alpha} P_n^{L_\alpha + 1/2, l_\alpha + 1/2}(\cos 2\phi) , \quad (2.10)$$

where $P_n^{(\alpha, \beta)}(x)$ are Jacobi polynomials and $N_n^{l_\alpha, L_\alpha}$ are normalization constants. The functions f_α, g_α are obtained by the Euler procedure previously outlined but, as discussed in the next section, they can be derived also by a simpler procedure. The equations satisfied by the functions $U_n^{(\alpha)}(\rho)$ entering into eq. (2.9) are determined by requiring the stationarity of the functional $\langle \Psi \mid H - E \mid \Psi \rangle$. For each channels α, M_α specifies the number of HH functions included in eq. (2.9). Since the HH basis forms a complete set, by increasing M_α the proper radial structure is generated. In practice, only four equations for the first eight channels and three equations for the last four are sufficient to obtain convergence for the binding energy value.

The results obtained for the three-nucleon system with the CHH expansion are presented in the rows labelled by CHH of table II. All the values have been obtained by including the twelve channels listed in table I. The close agreement between our results and those obtained by the Faddeev [4,5], GFMC [5] and Coupled Rearrangement Channel [12] (CRC) techniques confirms the good accuracy obtainable with the CHH method.

In order to improve further the method, the flexibility of the trial radial functions can be increased by including a larger number of channels; however, we can be satisfied of the capacity of the CHH approach and apply it to the study of four nucleon systems, where the results available for realistic potentials have a rather poor accuracy; this point will be discussed in the next section. We only recall that the HH formalism can be derived very generally for a generic N-body system and thus the CHH method is almost directly appliable to systems with a number of particles greater than three: the main difficulty encountered is related to the numerical evaluation of the multidimensional integrals.

3. WAVE FUNCTION FOR THE FOUR-BODY SYSTEM.

In this section, the extension of the CHH method to study the four-nucleon problem is presented. In the previous section, it has been shown that the CHH expansion seems to be more flexible and to converge faster than the CHO one, at least for the three-body case. For these reasons, we have decided to study the four–particle system only with the correlated HH expansion.

For a system of four particles the hyperradius ρ and the hyperangular variables ϕ_3, ϕ_2 are defined as

$$\rho = \sqrt{x_i^2 + y_i^2 + z_i^2} \, , \quad \cos\phi_3 = x_i/\rho \, , \quad \cos\phi_2 = y_i/(\rho\sin\phi_3) \, , \qquad (3.1)$$

where x_i, y_i and z_i are the Jacobi vectors. The wave function is then written in terms of the variables ρ and Ω, where Ω stands for the set of variables ϕ_3, ϕ_2, $\hat{x}_i$, $\hat{y}_i$ and $\hat{z}_i$. The HH functions are defined as

$$Y_{[l_1,l_2,l_3,K',K]}(\Omega) = Y_{l_1}(\hat{z}_i)Y_{l_2}(\hat{y}_i)Y_{l_3}(\hat{x}_i)^{(2)}P_{K'}^{l_2,l_1}(\phi_2)^{(3)}P_K^{l_3,K'}(\phi_3) \, , \qquad (3.2)$$

where the functions P are defined by eq. (2.10).

The wave function of a nuclear system must be completely antisymmetric. For a system of four particles, the trial wave function adopted here has a rather simple structure and it is a product of an antisymmetric function $\chi_a(S,T)$, depending on the spin and isospin coordinates, and a function depending on the spatial coordinates and involving state–dependent correlations too. The explicit form of the wave function is

$$\begin{aligned}
\Psi_4 = \prod_{j>i=1}^{4} f(r_{ij}) \Bigg\{ &\sum_{K,K'} U_{K,K'}^s(\rho)S\Big[Y_{[0,0,0,K',K]}(\Omega)\Big] \\
&+ \sum_{K,K'} U_{K,K'}^m(\rho)S\Big[Y_{[0,0,0,K',K]}(\Omega)(1 + \vec{\sigma}_i\cdot\vec{\sigma}_j)\Big] \qquad (3.3) \\
&+ \sum_{K,K'} U_{K,K'}^d(\rho)S\Big[Y_{[0,0,0,K',K]}(\Omega)S_{ij}\Big] \Bigg\}\chi_a(S,T) \, ,
\end{aligned}$$

where S is a symmetrizing operator with respect to the particle indices, S_{ij} is the tensor operator and $f(r)$ is a suitably chosen correlation factor. When the operators $\vec{\sigma}_i\cdot\vec{\sigma}_j$ and S_{ij} act on $\chi_a(S,T)$, a mixed symmetry S–wave component and a D–wave component are generated. The inclusion of isospin–dependent correlations would not produce new components due to the relation $\vec{\tau}_i\cdot\vec{\tau}_j\chi_a = (-2 - \vec{\sigma}_i\cdot\vec{\sigma}_j)\chi_a$.

When the wave function is expanded in channels in analogy to the three–body case, the dependence of the wave function given by eq. (3.3) on the spin operators corresponds to a three channel approximation. For $A = 3$, the three–channel approximation does not allow for a detailed description of the structure of the system, therefore we can expect that the wave function specified by eq. (3.3) will be useful mainly as a starting and reasonable approach to the problem.

At present the work is in progress and only the results for central potentials are presented here. We have considered two versions of the Malfliet–Tjon nucleon–nucleon interaction, namely the spin–independent MT V and the spin–dependent MT I-III version [20]. In both cases, the interaction is supposed to be active in all the partial waves; the parameters of the potentials are listed in table III.

For the spin–independent case the calculations have been performed by retaining only the first term on the right hand side of the wave function specified in eq. (3.3).

Potential	Channel	V_1 (MeV)	μ_1 (fm^{-1})	V_2 (MeV)	μ_2 (fm^{-1})
MTV		1458.047	3.11	−570.089	1.55
MT I − III	s	1438.720	3.11	−513.968	1.55
	t	1438.720	3.11	−626.885	1.55

Table III. *Potential parameters used in the calculations; for both potentials the radial dependence is $[V_1 \exp(-\mu_1 r) + V_2 \exp(-\mu_2 r)]/r$. s (t) denotes the spin–singlet (spin–triplet) component of the potential. $\hbar^2/m = 41.47$ MeV·fm^2.*

Potential	Method	B (MeV)	$\langle T \rangle$ (MeV)	R (fm)	$P_{S'}$
(a) MTV	CHH	31.355	69.63	1.409	
	CRC	31.357			
	ATMS	31.36			
	GFMC	31.3(2)			
(b) MT I − III	CHH	31.99	70.52	1.41	2.12

Table IV. *Results obtained for the four–nucleon system interacting via (a) the spin–independent MT-V potential and (b) the spin–dependent MT I-III potential. CHH denotes the results obtained in the present work; for the case (a) the values obtained by the CRC [25], ATMS [26] and GFMC [27] have been also reported. B, $\langle T \rangle$, R and $P_{S'}$ are the total binding energy, the average kinetic energy, the mass radius and the percentage of the mixed–symmetry S–states, respectively.*

The correlation function $f(r)$ has been determined by a simple procedure rather than by solving an Euler–type equation as in the case discussed in section 2. $f(r)$ satisfies the zero–energy Schroedinger equation

$$\left[-\frac{\hbar^2}{m} \nabla^2 + V(r) + \lambda(r) \right] f(r) = 0 , \qquad (3.4)$$

where $V(r)$ is the pair potential and $\lambda(r)$ is an additional term of the form $\lambda(r) = \Lambda \exp(-\gamma r)$; the depth Λ is chosen so as to satisfy the boundary condition $rf(r) \rightarrow 1$ for large r and the parameter γ can be used as a variational one, but the results have a little dependence on it. Finally, the unknown function $U^*(\rho)$ are determined by solving a set of coupled equations. The results are presented in table IV(a) where also the values obtained by other methods are reported for the sake of comparison. We can observe that the CHH results are in good agreement with the ones obtained by the ATMS [25], GFMC [26] and CRC [27] methods.

For the spin–dependent MT I-III potential, the wave function contains the first two terms in the right hand side of eq. (3.3). The function $f(r)$ satisfies the eq. (3.4) but in this case the potential $V(r)$ is the average of the singlet and triplet components. The values obtained for the binding energy and the mixed–symmetry S'–wave percentage are reported in table IV(b).

4. CONCLUSIONS.

We have investigated the bound–state properties of the triton by a rather sophisticated variational technique. Two realistic model potentials have been used, namely the so–called Reid V8 and Argonne V14 soft–core interactions.

In the first step of the approach, the spatial part of each channel included in the wave function is taken as a product of unidimensional functions which are determined by solving a set of Euler integro–differential equations. For both the potentials considered, the triton binding energy is underestimated by about 0.1 MeV.

To determine the full details of the radial functions a greater flexibility is required. To this end we have expanded the missing wave function on a set of correlated basis function. Two different bases have been chosen, namely the Harmonic Oscillator and the Hyperspherical Harmonic functions. By inclusion of a limited number of these correlated states, a significant improvement of the binding energy has been obtained. Moreover, an overall very good agreement has been found between our twelve channel CHH calculation and the corresponding one of the CRC method.

In conclusion, the method of expanding the wave function on a set of correlated basis functions is succesfull for the three–nucleon problem, and it appears quite promising also for studying larger nuclear systems, in particular the four–nucleon one.

REFERENCES.

1) G. L. Payne, J. L. Friar, B. F. Gibson and I. R. Afnan, Phys. Rev. **C22** (1980) 823;

2) C. Hajduk and P. U. Sauer, Nucl. Phys. **A369** (1981) 321;

3) S. Ishikawa, T. Sasakawa, T. Sawada and T. Ueda, Phys. Rev. Lett. **53** (1984) 1877; T. Sasakawa and S. Ishikawa, Few-Body Systems **1** (1986) 3;

4) C. R. Chen, G. L. Payne, J. L. Friar and B. F. Gibson, Phys. Rev. **C31** (1985) 266; **C33** (1986) 1740;

5) J. Carlson, Phys. Rev. **C36** (1987) 2026; **C38** (1988) 1879;

6) A. D. Jackson, A. Lande and P. U. Sauer, Phys. Lett. **35B** (1971) 365; S. N. Yang and A. D. Jackson, ibid. **36B** (1971) 1;

7) Y. Akaishi, M. Sakai, J. Hiura and H. Tanaka, Prog. Theor. Phys. Suppl. **56** (1974) 6;

8) M. A. Hennel and L. M. Delves, Nucl. Phys. **A246** (1975) 490 and references therein;

9) P. Nunberg, D. Prosperi and E. Pace, Nucl. Phys. **A285** (1977) 58;

10) J. Carlson, V. R. Pandharipande and R. B. Wiringa, Nucl. Phys. **A401** (1983) 59; R. Schiavilla, V. R. Pandharipande and R. B. Wiringa, ibid. **A449** (1986) 219;

11) J. L. Ballot and M. Fabre de la Ripelle, Ann. Phys. **127** (1980) 62;

12) M. Kamimura, Phys. Rev. **A38** (1988) 629;

13) H. Kameyana, M. Kamimura and Y. Fukushima, Phys. Rev. **C40** (1989) 974;

14) A. Kievsky, S. Rosati and M. Viviani, Nucl. Phys. **A501** (1989) 503;

15) A. Kievsky, S. Rosati and M. Viviani, Few Body Syst. **9** (1990) 1; A. Kievsky, S. Rosati and M. Viviani, submitted to Phys.Rev.C, preprint 1991;

16) A. Kievsky, S. Rosati and M. Viviani, Few Body Systems, in press;

17) R. V. Reid, Ann. Phys. (N.Y.) **50** (1968) 411;

18) R. B. Wiringa, R. A. Smith and T.L. Ainsworth, Phys. Rev. **C29** (1984) 1207;

19) S. Fantoni, L. Panattoni and S. Rosati, Il Nuovo Cim. **69A** (1970) 88;

20) R. A. Malfliet and J. A. Tjon, Nucl. Phys. **A217** (1969) 161;

21) J. L. Ballot and J. Navarro; J. Phys. **B8** (1975) 172; C. D. Lin, Phys. Rev. bf A12 (1975) 493;

22) Y. A. Simonov, Sov. J. Nucl. Phys. **3**, (1966) 461; A. M. Badalyan and Y. A. Simonov, Sov. J. Nucl. Phys. **3** (1966) 755;

23) G. Erens, J. L. Visschers, and R. van Wageningen, Ann. of Phys. **67**, (1971) 461; M. Beiner and M. Fabre de la Ripelle, Nuovo Cim. Lett. **1** (1971) 584; V. F. Demin and Y. E. Pokrovsky, Phys. Lett. **B47**, (1973) 394; Y. A. Simonov, in **"The Nuclear Many–Body Problem"** (F. Calogero and C. Ciofi degli Atti Eds.), Bologna 1973;

24) J. L. Ballot, Few Body Syst. Suppl. **1** ed. C. Ciofi degli Atti, O. Benhar, E. Pace and G. Salmè (Springer Verlag, Wien 1987), p. 140;

25) M. Kamimura and H. Kameyama, Nucl. Phys. **A508** (1990) 17c;

26) Y. Akaishi, Few Body Syst. Suppl. **1** ed. C. Ciofi degli Atti, O. Benhar, E. Pace and G. Salmè (Springer Verlag, Wien 1987), p. 120;

27) J. G. Zabolitzky, K. E. Schmidt and M. H. Kalos, Phys. Rev. **C25** (1982) 1111.

Few-Body Systems, Suppl. 6, 573—580 (1992)

Few-
Body
Systems
© by Springer-Verlag 1992

DISCRETE ANALOGS OF HYPERSPHERICAL HARMONICS AND THEIR USE FOR THE QUANTUM MECHANICAL THREE BODY PROBLEM

Vincenzo Aquilanti and Simonetta Cavalli

Dipartimento di Chimica dell' Università

06100 Perugia, Italy

Abstract. An algorithm for the solution of Schrödinger equation in a discrete basis is illustrated with reference to the three body problem. It exploits the explicit construction of discrete analogs of spherical harmonics and leads to sparse matrix representations of the kinetic energy operator and a diagonal representation of the interaction potential.

Introduction

In the application of quantum mechanics to few body problems, the solutions of Schrödinger equation are often sought by expanding the unknown eigenfunctions on a suitable basis, so to obtain an algebraic problem. Convergence of the procedure is the main practical limitation.

The technique here described is founded on the exploit of discrete analogs of orthonormal bases usually defined on continuous variables. These bases are orthonormal polynomial sets, and our representation requires the discrete counterparts of orthogonal polynomials, i.e. polynomials orthogonal on lattice points. We specialize the following presentation to the fundamental case of hyperspherical harmonics, thus generalizing the case of spherical harmonics, discussed previously [1]. From the *mathematical* wiewpoint, the harmonics belong to the class of Jacobi polynomials, which are functions of the hypergeometric, $_2F_1$, family. Their discrete analogs belong to higher hypergeometric families: in our applications, we will use Hahn polynomials, $_3F_2$ hypergeometric functions of unit argument. The latter discrete polynomials can be identified, in particular cases, with the Clebsch-Gordan coefficients or vector coupling coefficients in the quantum theory of angular momentum. This fact provided the first motivation [2] of our work and leads to a *physical* picture of the technique, whereby the discretization is interpreted as a quantization of an artificial hyperangular momentum: therefore in the previous paper [1] we introduced as a natural definition of this procedure the name of *hyperquantization* algorithm.

Discrete analogs of hyperspherical harmonics

We use in the following notations and results from Ref. [3], where properties and use of hyperspherical harmonics for the few body quantum mechanical problem were presented. We introduce a discrete representation in which the Jacobi polynomials $P_n^{\alpha,\beta}(\cos\vartheta)$ [4] and their special cases (Gegenbauer and Legendre polynomials and Wigner functions) are replaced by the Hahn polynomials $Q_n(k,\alpha,\beta,N)$ [5], [6] which are functions of a discrete variable k and are orthonormal on a set of $N+1$ points. The Hahn polynomials are terminating generalized hypergeometric functions with unit argument $_3F_2(1)$ [6]:

$$Q_n(k,\alpha,\beta,N) = {}_3F_2(-n,-k,n+\alpha+\beta+1;\alpha+1,-N;1) \tag{1}$$

For α and β fixed, the set of N Hahn polynomials is a finite system of orthogonal polynomials which are the discrete analog of an infinite set of Jacobi polynomials. They enjoy many properties [6] such as the orthonormality with respect to the discrete variable k and to the degree of the polynomial n.

By the comparison of the definition of Hahn polynomials (eq. (1)) and the definition of the generalized 3j symbol [7], it can be shown that they coincide up to the normalization and weight

$$\begin{pmatrix} \frac{N}{2}+\frac{\alpha}{2} & \frac{N}{2}+\frac{\beta}{2} & n+\frac{\alpha+\beta}{2} \\ \frac{N}{2}-\frac{\alpha}{2}-k & -\frac{N}{2}-\frac{\beta}{2}+k & \frac{\alpha+\beta}{2} \end{pmatrix} = \frac{\exp\{i\pi(2n+\beta-k)\}}{(2n+\alpha+\beta+1)^{\frac{1}{2}}}\overline{Q}_n(k) \tag{2}$$

$\overline{Q}_n(k)$ indicates the Hahn polynomial multiplied by the proper normalization factor and weight function. When α and β are integer numbers, the parameters which define the $_3F_2(1)$ are integers as well (c.f. eq. (1)) and the generalized 3j symbol in eq. (2) degenerate into a usual 3j symbol or Clebsch-Gordan coefficient.

The hyperspherical expansion

The Schrödinger equation for a system of three particles in hyperspherical coordinates is:

$$\left\{-\frac{\hbar^2}{2\mu}\left[\rho^{-5}\frac{\partial}{\partial\rho}\rho^5\frac{\partial}{\partial\rho}+\frac{\Lambda^2}{\rho^2}\right]+V(\rho,\Omega_5)\right\}\Psi(\rho,\Omega_5) = E\Psi(\rho,\Omega_5) \tag{3}$$

where ρ is the hyperradius of a hexadimensional sphere S^5 parametrized by five angles denoted collectively by Ω_5, μ is the three body reduced mass, $V(\rho,\Omega_5)$ and E are the potential and total energy respectively, and $\Lambda^2(\Omega_5)$ is the Casimir operator for the group O(6). Alternative angular parametrizations [8] are available for the Jacobi vectors **r** and

R. Given three particles A, B, C the vector **r** joins A with B and the vector **R** joins C with the center of mass of the molecule AB. We choose the hyperspherical parametrization in which Ω_5 indicates the five angles $\chi, \vartheta_1, \vartheta_2, \phi_1, \phi_2$, where $\chi = \arctan \frac{R}{r}$ and the couples of angles ϑ_1, ϕ_1 and ϑ_2, ϕ_2 are the orientations in the space fixed frame of the vectors **r** and **R** respectively. The Casimir's operator Λ^2 takes the form:

$$\Lambda^2 = -\sin^{-1} 2\chi \frac{\partial}{\partial \chi} \sin 2\chi \frac{\partial}{\partial \chi} + \frac{j^2}{\cos^2 \chi} + \frac{\ell^2}{\sin^2 \chi} \qquad (4)$$

being j and ℓ the standard symbols to indicate the rotational and orbital angular momentum numbers respectively, their projection on the space fixed frame quantization axis are m_j and m_ℓ. The hyperspherical harmonics $Y_{\lambda j \ell}(\Omega_5)$ are the eigenfunctions of the Casimir operator

$$\Lambda^2 Y_{\lambda j \ell}(\Omega_5) = -\lambda(\lambda + 4) Y_{\lambda j \ell}(\Omega_5) \qquad (5)$$

where $\lambda = 2n - j - \ell$ is the grand angular momentum quantum number and $n = 0, 1, \ldots$. The quantum numbers j and ℓ and their projections take the following allowed value: $j, \ell = 0, 1, \ldots, \lambda$, $-j \le m_j \le j$ and $-\ell \le m_\ell \le \ell$. The hyperspherical harmonics $Y_{\lambda j \ell}$ are given by the following expression:

$$Y_{\lambda j \ell}(\Omega_5) = \mathcal{N}_{\lambda j \ell}^{3;3} \cos^{j+\frac{1}{2}} \chi \sin^{\ell+\frac{1}{2}} \chi P_{\frac{\lambda - j - \ell}{2}}^{\ell+\frac{1}{2}, j+\frac{1}{2}}(\cos 2\chi) Y_{j m_j}(\vartheta_1, \phi_1) Y_{\ell m_\ell}(\vartheta_2, \phi_2) \qquad (6)$$

$\mathcal{N}_{\lambda j \ell}^{3;3}$ is the normalization factor [3].

The eigenfunctions, $Y_{\lambda j \ell}^{JM}$, of the total angular momentum J and of its projection M on the quantization axis are a combination of the hyperspherical harmonics:

$$Y_{\lambda j \ell}^{JM}(\Omega_5) = H_{\lambda j \ell}^{3;3}(\chi) \mathcal{Y}_{j \ell}^{JM}(\vartheta_1, \vartheta_2, \phi_1, \phi_2) \qquad (7)$$

where $\mathcal{Y}_{j \ell}^{JM}$ is a bipolar harmonic

$$\mathcal{Y}_{j \ell}^{JM}(\vartheta_1, \vartheta_2, \phi_1, \phi_2) = \sum_{m_j, m_\ell} \langle j m_j \, \ell m_\ell \mid JM \rangle Y_{j m_j}(\vartheta_1, \phi_1) Y_{\ell m_\ell}(\vartheta_2, \phi_2) \qquad (8)$$

and $H_{\lambda j \ell}^{3;3}(\chi)$ indicate:

$$H_{\lambda j \ell}^{3;3}(\chi) = \mathcal{N}_{\lambda j \ell}^{3;3} \cos^{j+\frac{1}{2}} \chi \sin^{\ell+\frac{1}{2}} \chi P_{\frac{\lambda - j - \ell}{2}}^{\ell+\frac{1}{2}, j+\frac{1}{2}}(\cos 2\chi) \qquad (9)$$

From now on we therefore work at fixed angular momentum J. In the notation, it will be omitted for simplicity. In order to solve the Schrödinger equation for the motion of three particles in the space in the adiabatic representation, the wavefunction $\Psi(\rho, \Omega_5)$ is to be expanded in series of the adiabatic eigenfunctions $\Phi_i(\rho, \Omega_5)$:

$$\Psi(\rho, \Omega_5) = \rho^{-\frac{5}{2}} \sum_i \Phi_i(\rho, \Omega_5) F_i(\rho) \qquad (10)$$

The function $\Phi_i(\rho, \Omega_5)$, at a fixed value of the hyperradius ρ, satisfies the following equation:

$$\left\{ -\frac{\hbar^2}{2\mu\rho^2} \Lambda^2 + V(\rho, \Omega_5) \right\} \Phi_i(\rho, \Omega_5) = \epsilon_i(\rho) \Phi_i(\rho, \Omega_5) \qquad (11)$$

$\epsilon_i(\rho)$ being the energy of the adiabatic state i at a given value of ρ. The adiabatic eigenfunctions are expanded in series of hyperspherical harmonics:

$$\Phi_i(\rho, \Omega_5) = \sum_{\lambda j \ell} t_{i\lambda j\ell}(\rho) Y_{\lambda j\ell}^{JM}(\Omega_5) \tag{12}$$

Inserting eq.(12) into eq.(11) and integrating over the angular variables the following system of coupled equations is obtained:

$$\sum_{\lambda' j' \ell'} \left\{ \frac{\lambda'(\lambda'+4) + 15/4}{2\mu\rho^2} \delta_{\lambda\lambda'} \delta_{\ell\ell'} \delta_{jj'} + V_{\lambda j\ell,\lambda' j' \ell'}(\rho) \right\} t_{i\lambda' \ell' j'}(\rho) = \epsilon_i(\rho) t_{i\lambda j\ell}(\rho) \tag{13}$$

where

$$V_{\lambda j\ell,\lambda' j' \ell'}(\rho) = \int d\Omega_5 Y_{\lambda j\ell}^{JM*}(\Omega_5) V(\rho, \chi, \vartheta) Y_{\lambda' j' \ell'}^{JM}(\Omega_5) \tag{14}$$

is an element of the interaction matrix V. In this representation the kinetic energy is diagonal and the coupling is in the potential energy. It is not necessary to perfom the integration over all the four angles $\vartheta_1, \vartheta_2, \phi_1, \phi_2$ because the potential energy depends only on the internal angle ϑ which is the angle between the two vectors $\mathbf{R}$ and $\mathbf{r}$. If we consider the basis set $\mathcal{Y}_{j\Omega}^{JM}$ labeled by the numbers j and Ω and related by a transformation matrix G^J [9], [10] to the basis $\mathcal{Y}_{j\ell}^{JM}$ we transform eq.(14) into the following relation:

$$V_{\lambda j\ell,\lambda' j' \ell'}(\rho) = (1/4) \int_{-1}^{1} d\cos 2\chi \, H_{\lambda j\ell}^{3,3}(\chi) V_{j\ell,j' \ell'}(\rho, \chi) H_{\lambda' j' \ell'}^{3,3}(\chi) \tag{15}$$

where

$$V_{j\ell,j' \ell'}(\rho, \chi) = \sum_{\Omega,\Omega'} G_{j\ell\Omega}^{J} \left(\int_{-1}^{1} d\cos\vartheta \, Y_{j\Omega}^{*}(\vartheta, 0) V(\rho, \chi, \vartheta) Y_{j'\Omega'}(\vartheta, 0) \right) G_{j\ell\cdot\Omega'} \delta_{\Omega,\Omega'} \tag{16}$$

and $Y_{j\Omega}(\vartheta, 0)$ is a standard spherical harmonic.

Problems arising in the applications of the above formalisms are the calculations of these multidimensional integrals and the necessity to compute, store and manipulate very large matrices. This is due to the fact that the quantum number λ is very "bad", in the sense that the hyperspherical expansion is in all cases of interest very slowly convergent. Therefore we are now interested in a representation for which the interaction matrix is diagonal, its elements being the values of the potential at fixed values of the angular variables χ, ϑ. Such a representation is obtained by the discretization of χ and ϑ. The discrete hyperspherical harmonics are built up by multiplying the discrete analog of the classical orthogonal polynomials. In the following we show how to diagonalize the matrix of the potential V by using the discrete orthogonal polynomials of the previous section.

For a given value of the total angular momentum J, we fix a positive integer I, where $(I+1)$ specifies the number of points which have been selected for the variable 2χ, each point being labeled by τ:

$$\cos 2\chi_\tau = \frac{-2\tau}{I+1} \tag{17}$$

where $-I/2 \leq \tau \leq I/2$. The infinite set of the orthonormal functions $H_{\lambda j\ell}^{3,3}(\chi)$ is replaced by the finite set of $I+1$ orthonormal Hahn polynomials $\overline{Q}_{\frac{\lambda-j-\ell}{2}}(\frac{\ell}{2} + \tau, \ell + 1/2, j + 1/2, I)$ or generalized 3j symbols. In order to simplify the formalism, in the following we refer to the order of the Hahn polynomials by replacing $\frac{\lambda-j-\ell}{2}$ with n, which is the degree of the discrete polynomial and varies between 0 and I. For each value of n we choose $K+1$ couples of values of j and ℓ. The rotational quantum number j takes the values $0, 1, \ldots, j_{max}$ and the orbital quantum number ℓ takes values which are compatible with the values of n and J previously fixed. The elements of the matrix V can be written as follows:

$$V_{nj\ell,n'j'\ell'} = \sum_\tau G^I_{n\tau} V_{\tau,j\ell,j'\ell'} G^I_{n'\tau'} \tag{18}$$

where from eq.(16) we have:

$$V_{\tau,j\ell,j'\ell'} = V_{j\ell,j'\ell'}(\rho, \chi_\tau) \tag{19}$$

$V_{\tau,j\ell,j'\ell'}$ are the elements of a matrix $\mathbf{V}^K$ and $G^I_{n\tau}$, are the elements of the orthonormal matrix $\mathbf{G}^I$:

$$G^I_{n\tau} = \exp\{i\pi(\frac{I}{2} + \tau - \lambda - \ell + \frac{1}{2})\}$$

$$(\lambda + 2)^{\frac{1}{2}} \begin{pmatrix} \frac{\ell}{2} + \frac{\ell}{2} + \frac{1}{4} & \frac{\ell}{2} + \frac{i}{2} + \frac{1}{4} & \frac{\lambda}{2} + \frac{1}{2} \\ -\tau - \frac{\ell}{2} - \frac{1}{4} & \tau - \frac{i}{2} - \frac{1}{4} & \frac{i+\ell}{2} + \frac{1}{2} \end{pmatrix} \tag{20}$$

The $\mathbf{G}^I$ matrix has dimension $(I+1)(K+1) \times (I+1)(K+1)$ and is made of $(I+1)(I+1)$ blocks each one of dimension $(K+1)(K+1)$. Each block has different from zero only the elements along the main diagonal. Inside any bloch τ and n are fixed and the quantum numbers j and ℓ assume all the possible $K+1$ values. Blocks adjacent horizontally differ for the value of τ and blocks adjacent vertically differ for the value of n. The $\mathbf{V}^K$ matrix, of dimension $(I+1)(K+1) \times (I+1)(K+1)$, is block diagonal; only the $(I+1)$ blocks, of dimension $(K+1)(K+1)$ along the main diagonal contain elements different from zero, inside any block the number τ is constant. In the discrete representation each block of $\mathbf{V}^K$, indicated by v^K, can be written as the product of five matrices of dimension $(K+1)(K+1)$ and its elements are defined by:

$$v^K_{j\ell j'\ell'} = \sum_{\Omega\nu} g^J_{j\ell\Omega} g^{j_{max}}_{j\Omega\nu} v^{KI}_\nu g^{j_{max}}_{j'\Omega\nu} g^J_{j'\ell'\Omega} \tag{21}$$

where

$$g^J_{j\ell\Omega} = (-)^{j+\Omega} \sqrt{\frac{2}{1+\delta_{0n}}} \langle j\Omega \; J - \Omega \mid \ell 0 \rangle$$

and

$$g^{j_{max}}_{j\Omega\nu} = (-)^{\frac{i_{max}-\Omega}{2} - \nu - j} \langle \frac{j_{max} - \Omega}{2} \nu, \frac{j_{max} + \Omega}{2} - \nu \mid j0 \rangle$$

The range of ν, $\frac{-j_{max}+\Omega}{2} \le \nu \le \frac{i_{max}-\Omega}{2}$, depends on the value of Ω. For each value of Ω, $j_{max} - \Omega + 1$ values of the angle ϑ, labeled by ν, are selected:

$$\cos \vartheta_\nu = \frac{-2\nu}{j_{max} + 1} \tag{22}$$

The matrix $\mathbf{g}^J$ allows to turn the basis labeled by j and ℓ into the basis labeled by j and Ω, its elements are Clebsch-Gordan coefficients and have been defined in [9], [10]; ℓ and Ω are the row and column labels respectively. The elements of the matrix $\mathbf{g}^{j_{max}}$ have been defined in [11], they are Clebsch Gordan coefficients as well; j and ν are the column and row labels respectively. This matrix transforms the basis $\mid j\Omega\rangle$ into the discrete basis $\mid \Omega\nu\rangle$. The matrices $\boldsymbol{g}^J$ and $\boldsymbol{g}^{jmax}$, of dimension $(K+1)(K+1)$, are block diagonal with respect to j and Ω respectively. The blocks of $\boldsymbol{g}^J$ are $(2j+1)(2j+1)$ dimensional and those of $\boldsymbol{g}^{jmax}$ are $(j_{max} - \Omega + 1)(j_{max} - \Omega + 1)$ dimensional. The matrices $\boldsymbol{g}^J$ and $\boldsymbol{g}^{jmax}$ are τ independent and therefore are the same for the all $I+1$ blocks $\boldsymbol{v}^K$. The elements of the diagonal matrix $\boldsymbol{v}^{KI}$ are the values of the potential energy at the points selected by ν for a given value of τ.

The set of coupled equations in eq.(13), in a *finite basis representation*, can be written in the following way:

$$\{\sum_{n'j'\ell'} \frac{4n'(n'+\ell'+j'+2) + (\ell'+j')(\ell'+j'+4) + 15/4}{2\mu\rho^2} \delta_{nn'}\delta_{\ell\ell'}\delta_{jj'}$$

$$+ \sum_{\tau\Omega\nu} G^I_{n\tau} G^J_{j\ell\Omega} G^{jmax}_{j\Omega\nu} V^{IK}_{\nu\tau} G^{jmax}_{j'\Omega\nu} G^J_{j'\ell'\Omega} G^I_{n'\tau}\} t^{IK}_{in'\ell'j'}(\rho) = \epsilon_i(\rho) t^{IK}_{inj\ell}(\rho) \tag{23}$$

The $\boldsymbol{G}^J$ and $\boldsymbol{G}^{jmax}$ matrices are block diagonal and are obtained assemblying the $(I+1)$ $\boldsymbol{g}^J$ and $\boldsymbol{g}^{jmax}$ matrices along the main diagonal. The matrix $\boldsymbol{V}^{IK}$ is diagonal, its elements are the values of the potential $V(\rho, \chi_\tau, \vartheta_\nu)$. The above expression takes the following expression in a matricial form:

$$\left(\boldsymbol{K} + \tilde{\boldsymbol{G}}^I \tilde{\boldsymbol{G}}^J \tilde{\boldsymbol{G}}^{jmax} \boldsymbol{V}^{IK} \boldsymbol{G}^{jmax} \boldsymbol{G}^J \boldsymbol{G}^I\right) \boldsymbol{T}^{IK} = \boldsymbol{T}^{IK}\boldsymbol{\epsilon} \tag{24}$$

$\boldsymbol{K}$ is the kinetic energy matrix, $\boldsymbol{\epsilon}$ is the matrix of the adiabatic eigenvalues and $\boldsymbol{T}^{IK}$ is the matrix of the eigenvectors in the finite representation.

In order to obtain the Schrödinger equation in the *discrete representation* we perform the following transformation:

$$\boldsymbol{T}^{IK} = \tilde{\boldsymbol{G}}^I \tilde{\boldsymbol{G}}^J \tilde{\boldsymbol{G}}^{jmax} \boldsymbol{Z}^{IK} \tag{25}$$

where $\boldsymbol{Z}^{IK}$ is the eigenvector matrix in the discrete representation. Inserting this equation in eq. (24) and multiplying the left-hand side of eq.(24) for $\boldsymbol{G}^{jmax} \boldsymbol{G}^J \boldsymbol{G}^I$ we obtain the desired discrete representation in which the potential energy is diagonal and the coupling is tranferred to the kinetic energy:

$$\left(\boldsymbol{G}^{jmax} \boldsymbol{G}^J \boldsymbol{G}^I \boldsymbol{K} \tilde{\boldsymbol{G}}^I \tilde{\boldsymbol{G}}^J \tilde{\boldsymbol{G}}^{jmax} + \boldsymbol{V}^{IK}\right) \boldsymbol{Z}^{IK} = \boldsymbol{Z}^{IK}\boldsymbol{\epsilon} \tag{26}$$

The elements of the kinetic energy matrix $\boldsymbol{K}^{IK}$ in the discrete representation are:

$$K^{IK}_{\Omega\tau\nu\Omega'\tau'\nu'} = \sum_{nj\ell} G^{jmax}_{j\nu\Omega} G^J_{j\ell\Omega} G^I_{\tau nj\ell} K_{nj\ell,nj\ell} \tilde{G}^I_{\tau'nj\ell} \tilde{G}^J_{j\ell\Omega'} \tilde{G}^{jmax}_{j\nu'\Omega'} \tag{27}$$

In the above expression the sum over n can be performed analiticaly by making use of the fact that the matrix elements $G^I_{\tau nj\ell}$ satisfy a difference equation [6]. This sum couples elements with $\tau' = \tau, \tau \pm 1$ and therefore with respect to the index τ the kinetic energy matrix is tridiagonal. Its elements can be eveluated by making use of the difference equation and of the orthogonality properties of the Hahn polynomials. The sums over j and ℓ are to be made numerically, but once for all systems. The kinetic energy matrix $\boldsymbol{K}^{IK}$ of the discrete representation results to be block tridiagonal.

The case of zero total angular momentum

When the total angular momentum J is zero, then the rotational j and the orbital ℓ quantum numbers are equal. For this important special case the bipolar harmonic (see eq.(8)) reduces to a Legendre polynomial:

$$\mathcal{Y}_\ell = (-)^\ell \frac{(2\ell+1)^{\frac{1}{2}}}{4\pi} P_\ell(\cos\vartheta) = (-)^\ell \frac{1}{2\sqrt{\pi}} Y_{\ell 0}(\vartheta,\phi) \tag{28}$$

and the function $H_{\lambda j\ell}^{3,3}(\chi)$ becomes a a Gegenbauer polynomial. In this case the eigenfunction of the total angular momentum reduces to a harmonic of R^4 [3]:

$$Y_{\frac{\lambda}{2}\ell 0}(2\chi,\vartheta,\phi) = (-)^\ell \frac{1}{\sqrt{\pi}} \mathcal{N}_{\frac{\lambda}{2}\ell}^3 \sin^{\ell+\frac{1}{2}} 2\chi\, C_{\frac{\lambda}{2}\ell}^{\ell+\frac{1}{2}}(\cos 2\chi) Y_{\ell 0}(\vartheta,\phi) \tag{29}$$

The above function, without the phase factor, corresponds to a tree with a free branch on the left side and a branch with three leaves on the right side. Such a tree represents the Fock parametrization of R^4. As shown in a previous paper[12], another parametrizaion, called symmetric, is also available. The latter is illustrated by a tree with two leaves on each branch. The link between the two alternative parametrizations has been established [12], they are connected by a SU(2) Clebsch-Gordan coefficient. The functions corresponding to this tree are Wigner functions:

$$Y_{\frac{\lambda}{2}\frac{\ell}{2}0}(2\Theta,2\Phi,\phi) = \left(\frac{\lambda+2}{16\pi^3}\right)^{\frac{1}{2}} d_{\frac{\ell}{4}\frac{\ell}{4}}^{\frac{\lambda}{4}}(4\Theta) \exp(i\sigma\Phi) \tag{30}$$

The discrete harmonics for the Fock representation are:

$$
\begin{aligned}
Y_{\frac{\lambda}{2}\ell 0}^{I,N}(2\chi,\vartheta,\phi) =\ & [(I+1)(N+1)]^{\frac{1}{2}} \frac{(-)^{\frac{N}{2}-\nu}}{\sqrt{\pi}} \exp\{i\pi(\frac{I}{2}+\tau-\lambda+l-\frac{1}{2}) \\
&\times (\lambda+2)^{\frac{1}{2}}
\begin{pmatrix}
\frac{I}{2}+\frac{\ell}{2}+\frac{1}{4} & \frac{I}{2}+\frac{\ell}{2}+\frac{1}{4} & \frac{\lambda}{2}+\frac{1}{2} \\
-\tau-\frac{\ell}{2}-\frac{1}{4} & \tau-\frac{\ell}{2}-\frac{1}{4} & \ell+\frac{1}{2}
\end{pmatrix}
\begin{pmatrix}
\frac{N}{2} & \frac{N}{2} & \ell \\
\nu & -\nu & 0
\end{pmatrix}
\end{aligned} \tag{31}
$$

where $\cos 2\chi_\tau = \frac{-2\tau}{I+1}$ and $\cos 2\vartheta_\nu = \frac{-2\nu}{N+1}$. The discrete harmonics of the symmetric representation are:

$$
\begin{aligned}
Y_{\frac{\lambda}{2}\frac{\ell}{2}0}^{I,N}(2\Theta,2\Phi,\phi) =\ & [(I+1)N]^{\frac{1}{2}}(-)^{\frac{I}{2}+\tau-\frac{\lambda}{2}} \left(\frac{\lambda+2}{\pi}\right)^{\frac{1}{2}} \\
&\times
\begin{pmatrix}
\frac{I}{2} & \frac{I}{2}+\frac{\alpha}{4} & \frac{\lambda}{4} \\
-\tau & \tau-\frac{\alpha}{4} & \frac{\alpha}{4}
\end{pmatrix}
\left\{
\begin{matrix}
\cos\frac{\sigma 2\pi\nu}{2N} \\
\sin\frac{\sigma 2\pi\nu}{2N}
\end{matrix}
\right.
\end{aligned} \tag{32}
$$

where $\cos 4\Theta_\tau = \frac{-2\tau}{I+1}$ and $\Phi_\nu = \frac{2\pi\nu}{N}$. The kinetic energy matrix in this case is simpler then before.

580

Conclusions

The previous theory is formulated with in mind as specific applications those quantum mechanical three body problems that arise in the description of elementary chemical reactions and large amplitude vibrations of triatomic molecules. Modifications to treat cases where particle spin effects are important, and extensions from three to few body problems appear feasible.

The main advantages of the procedure, from the viewpoint of applications to *exact* (and therefore unavoidably large scale) computations, are the sparseness of the kinetic energy matrix representation, its full generality in terms of simple sums, and the diagonal representation of the potential energy (i.e. no integrals).

Finally, the technique is also interesting because through the discrete analogs of hyperspherical harmonics one introduces new quantum numbers as their labels. These quantum numbers can be given a physical interpretation and therefore used as alternative basis representations for the scattering matrix. An example is the ν quantum number (see above), which has been shown [11] to be related to the directional properties of scattering events, and therefore referred to as the *steric* quantum number. The *approximate* conservation of these quantum numbers under special circumstances is being investigated, also as a route to simplifying the numerical efforts.

References

1. Aquilanti, V., Cavalli, S., Grossi, G.: Theor. Chim. Acta **79**, 283 (1991)

2. Aquilanti, V., Grossi, G.: Lett. Nuovo Cimento **42**, 157 (1985)

3. Aquilanti,V., Cavalli, S., Grossi,G.: J. Chem. Phys. **85**, 1362 (1986)

4. Abramovitz, M., Stegun, I.A.: Handbook of Mathematical Functions. New York: Dover 1965

5. Gasper,G. in: Askey, R.A. (ed.), Theory and application of special functions. p.375, New York: Accademic Press 1975

6. Karlin, S., McGregor, J.L.: Scripta Math. **26**, 33 (1961)

7. Raynal J.: J.Math.Phys. **19**, 467 (1978)

8. Aquilanti, V., Cavalli, S.: J. Chem. Phys. **85**, 1355 (1986)

9. Walker, R.B., Light, J.C.: Chem. Phys. **84**, 7 (1975). Launay, J.M.: J. Phys. B **9**, 1823 (1976)

10. Aquilanti, V., Beneventi, L., Grossi, G., Vecchiocattivi, F.: J. Chem. Phys. **89**, 751 (1988)

11. Aquilanti, V., Cavalli, S., Grossi, G., Anderson, R.W.: J. Phys. Chem., in press

12. Aquilanti, V., Grossi, G., Laganà, A.: J. Chem. Phys. **76**, 1587 (1982)

Few-Body Systems, Suppl. 6, 581—586 (1992)

Few-
Body
Systems
© by Springer-Verlag 1992

MANY-BODY COULOMB SYSTEMS ABOVE THE THRESHOLD
FOR TOTAL BREAK-UP

Hubert Klar
Fakultät für Physik
Hermann-Herder-Strasse 3
7800 Freiburg / Germany

INTRODUCTION

The simplest nontrivial many-body Coulomb problem consists of
three charged particles. Situations leading to three charged
particles in their continua are (i) double photoionisation of
an atom, and (ii) single ionisation of an atom, for instance
by an electron. Alternative cross sections may be observed
experimentally. In non-coincidence experiments one measures
total, singly and doubly differential cross sections. In these
cases one obseves only one of the escaping fragments, often an
electron. Coincidence experiments, on the other hand, observe
two of the fragments. This may be done with or without
analysis of the energy distribution between these fragments.
The most sensitive probe for a collision model needs
comparison with a triply differential cross section, a
coincidence measurement including the measurement of the
energy distribution. Only in this case the kinematic of the
reaction is fully determined. Theoretical work needs then
neither averages nor integrations before comparison with the
experiment can be made. Such triply differential cross
sections (TDCS) have been measuered since 1969 for electron
impact ionisation of atoms, socalled (e,2e) measurements, see
/1/. Triply differential crosss sections for double photoionisation have been measured only very recently for the first
time /2/.

From the theoretical view point the problem of treating three
or more charged continuum particles is fare from being solved.
The main difficulty emerges from the long range Coulomb
interaction which forbids free particles at large particle
separations. Already the two-body problem shows this
difficulty with its immediate and well known consequences:
Diverging logarithmic phases, diverging density of states at
threshold, break-down of perturbation theory. Similar and even
more difficulties we expect for many bodies where analytic
solutions are not known.

THEORY

The first theoretical (e,2e) work was based on first order perturbation theory /3/. This Bethe theory, in our terminology now a first Born approximation, should work at high incident energies and small momentum transfer. As long as total cross sections were considered it was believed that the Bethe theory is applicable already at energies of a few hundred eV above threshold. TDCS at the same energy calculated within the Bethe theory are however only in qualitative agreement with the experiments. This became particularly obvious for atomic hydrogen as target /4/ where the Bethe matrix element can be evaluated analytically. During the last two decades much theoretical work has been done to improve results. Several techniques of scattering theory have been applied, see /5/ for pre-1989 papers, but all these methods (second Born approximation, R-matrix, pseudostate expansions, distorted wave approximations) suffer from the lack of Coulomb boundary conditions at large particle separations — a key ingredient to describe many charged continuum particles.

The exact boundary conditions for any number of charged "free" particles are known since a long time, see /6/. For the case of two protons and one elctron a globally defined wave function which satisfies this Redmond asymptotic was first written down but not employed numerically in /7/. For an arbitrary three-body Coulomb system this wave function consists of a plane wave factor to describe the free particles accompanied by three Coulomb distortion factors, one for each two-particle subsystem. Each of these distortion factors given by a confluent hypergeometric function takes into account the long range Coulomb interaction in the corresponding two-particle subspace. This wave function has been shown to factorise in suitably defined six-dimensional parabolic coordinates /8/.

RESULTS

During the last three years this wave function has been employed within the frame of a T-matrix formulation to calculate TDCS for electron impact ionisation of atomic hydrogen, see /9,10/. Fig. 1 shows a typical example. Here the incident electron energy is E =250 eV, the ejected electron energy is E =5 eV, the scattering angle of the fast outgoing electron is 3°. The solid curve is the result of Ref. 9 using a wave function satisfying exact boundary conditions, the dotted curve is the Born approximation, and the dashed curve shows the same TDCS for positron impact. The points are experimental data /1/. We remark that the Born approximation as a first order theory predicts the same TDCS both for electron and positron impact. At decreasing incident energies the agreement between theory and experiment is less perfect. This is to be expected because at lower energy the correlated motion within the whole system 'target + projectile' becomes increasingly important also at smaller interparticle separations where the wave function of Ref. /8-10/ is less accurate. Departures between theory and experiment are e. g. seen at the incident energy E =54.4 eV, see /10/. One may expect that the results become increasingly

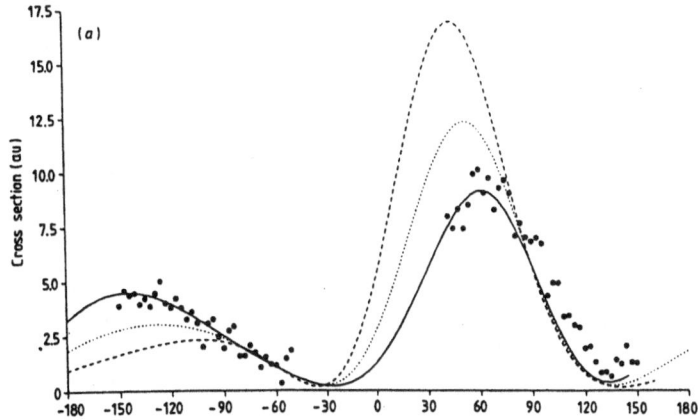

Fig.1: TDCS for electron impact ionisation of H(1s), see text.

worse for decreasing energy. This however is not generally the case. Let us look for the situation close to threshold. It is eassily seen /10/ that the Wannier threshold law /11/, see also below, is not reproduced by /9,10/. We believe that the reason for this lack has to do with the fact that we do not take into account the coupling of the low continuum states to the high Rydberg states. Angular correlations at very low energies, however, do not seem to depend sensitively on the Wannier phenomenon. Ref. /10/ shows examples for this at incident energies of only 15.6 eV and 17.4 eV.

We have also investigated the high energy behaviour of TDCS. At small values of the momentum transfer, the regime of the Born approximation, we find one escaping electron much faster than the other. The Coulomb distortion factors for the two interactions fast electron – ion and fast electron – slow electron may then be replaced by its high energy asymptotics, see /12/ for details and results. This high energy approximation still satisfies exact boundary conditions, and may be considered as an improved Born approximation. Calculations at a few keV incident energy indicate convergence to the Born approximation at increasing energies, but this convergence is extremely slow.

THRESHOLD LAWS

The ionisation of an atom by charged particle impact close to threshold constitutes an extremely complicated and essentially unsolved theoretical problem. We mention here only that no fully convincing solution to that problem is available. In particular no quantum theoretical description leading to a result in agreement with experiments exists. E. g. the above mentioned theory /8-10/ in contradiction to the observation predicts an exponential threshold law, see /10/ for details whereas all experimental data show a power law with fractional exponents, see below.

The only theory with a result consistent with modern and accurate experiments was developed long ago by Wannier /11/.

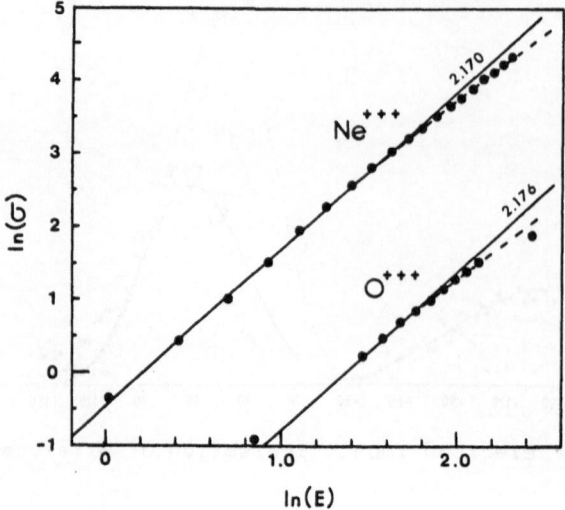

Fig.2: Triple photoionisation of Ne and O close to threshold, from /18/, see also text.

This theory is based on classical mechanics and contains as a key ingredient the hypothesis of an ergodic motion of the two escaping electrons. This assumed ergodicity within the negative complex (atom + electron) leads with help of additional assumptions to a uniform energy distribution among the electron, i. e. each energy sharing is equally likely. Then Wannier shows in his pioneering work the slope of the total ionisation cross section as a function of the excess energy E to be given by a power law

$$\sigma \propto E^{n}$$

where the exponent depends only on masses and charges of the escaping fragments. The following table summarises experimentally observed exponents and compares these data with theoretical predictions based on Wannier's theory and extensions of it:

reaction	n (exp)	theory	reference
$e^{-} + He \rightarrow He^{+} + 2e^{-}$	1.131 + .019	1.127	13
$h\nu + H^{-} \rightarrow H^{+} + 2e^{-}$	1.15 + .04	1.127	14
$e^{-} + Ne \rightarrow Ne^{+} + 2e^{-}$	1.13 + .02	1.127	15
$e^{-} + Ar \rightarrow Ar^{+} + 2e^{-}$	1.10 + .04	1.127	16
$h\nu + He \rightarrow He^{2+} + 2e^{-}$	1.05	1.056	17
$h\nu + Ne \rightarrow Ne^{3+} + 3e^{-}$	2.170	2.162	18, 19
$h\nu + O \rightarrow O^{3+} + 3e$	2.176	2.162	18, 19

It is clearly seen that all measurements are consistent with the Wannier theory. In particular the triple escape /18/ is well described by this theory /19/, see also Fig. 2. A convincing derivation of these fractional power laws from quantum theory is one of the outstanding problems of modern atomic theory.

ACKNOWLEDGEMENT

Support by Deutsche Forschungsgemeinschaft (SFB 276) is gratefully acknowledged.

REFERENCES

/1/ Ehrhardt, H., Jung, K., Knoth, G., Schlemmer, P., Z. Phys. D1, 3 (1986)

/2/ Mazeau, J., Selles, P., Weymel, D., Huetz, A., J. Phys. B in press (1991)

/3/ Bethe, H. A., Ann. Phys.(Leipzig) 5, 325 (1930)

/4/ Ehrhardt, H., Knoth, G., Schlemmer, P., Jung, K., Phys. Lett. 110A, 92 (1985)

/5/ Byron, F. W., Joachain, C. J., Phys. Rep. 179, 211 (1989)

/6/ Rosenberg, L., Phys. Rev. D8, 1833 (1973)

/7/ Garibotti, C. R., Miraglia, J. E., Phys. Rev. A21, 572 (1980)

/8/ Klar, H., Z. Phys. D16, 231 (1990)

/9/ Brauner, M., Briggs, J. S., Klar, H., J. Phys. B 22, 2265 (1989)

/10/ Brauner, M., Briggs, J. S., Klar, H., Broad, J. T., Rösel, T., Jung, K., Ehrhardt, H., J. Phys. B 24, 657 (1991)

/11/ Wannier, G., Phys. Rev. 90, 817 (1953)

/12/ Berakdar, J., Klar, H., Brauner, M., Briggs, J. S., Z. Phys. D 16, 91 (1990)

/13/ Cvejanovic, S., Read, F., J. Phys. B 7, 1841 (1974)

/14/ Donahue, J. B., Gram, P. A. M., Hynes, H. V., Hamm, R. W., Frost, C. A., Bryant, H. C., Butterfiled, K. B., Clarke, D. A., Schmitt, W. W., Phys. Rev. Lett 48, 1538 (1982)

/15/ Hink, W., Kees, L., Schmitt, H. P., Wolf, A., Innershell and X-ray Physics of Atoms and Solids, ed. Fabian, D. J., Kleinpoppen, H., Watson, L. M., Plenum New york, 327-330 (1981)

/16/ Hippler, R., Klar, H., Saeed, K., McGregor, I., Duncan,
     A. J., Kleinpoppen, H., J. Phys. B 16, L617 (1983)

/17/ Kossmann, H., Schmidt, V., Andersen, T., Phys. Rev.
     Lett. 60, 266 (1988)

/18/ Samson, J. A. R., Angel, G. C., Phys. Rev. Lett. 61, 1584
     (1988)

/19/ Klar, H., Schlecht. W., J. Phys. B 9, 1699 (1976)

Few-Body Systems, Suppl. 6, 587—594 (1992)

NON RELATIVISTIC FIELD THEORY IN FEW-BODY ATOMIC PHYSICS.

E. Ficocelli Varracchio

Department of Chemistry

University of Bari

70126 Bari - ITALY

A Field-Theoretic formulation of e^+-atom scattering is considered, leading to a closed equation for the optical potential of the formalism. An RPA⊕Ladder approximation scheme is suggested and this is applied, numerically, to the e^+-He system.

1. INTRODUCTION.

The conventional approach to the study of bound state and scattering problems, for few body systems, in atomic (and molecular) physics, generally proceeds through a preliminary determination of the (approximate) quantum mechanical wave function, for the full system, and the subsequent extraction, from this, of the observables of interest. On a closer examination, such an approach is at variance with the picture generally adopted for an analysis of experimental data, in the atomic domain. This usually considers, in fact, mechanisms based on the excitation of a "limited" number of particles of the full system, out of some reference (or "vacuum") state. Such a picture would then seem to point out that much of the information contained in the total wave function is apparently "redundant", from the point of view of performing a comparison with experiment itself.

A closer contact, with the previous point of view, may be obtained, on the other hand, by adopting, at the very beginning, the formalism of second quantization, that more readily leads to an analysis of processes, in terms of "amplitudes", representing the creation (or destruction) of a limited number of particles out of (or into) a vacuum state. Such formulations concentrate directly on the specific quantum mechanical amplitudes of interest, thereby bringing into closer contact theory with experiment. Besides being more appealing on such grounds, the formalism of second quantization, and the related Field Theoretic (FT) techniques (immediately following from this), offer some definite advantages, when explicit numerical applications have to be performed. As an example, the full machinery of Feynman diagrams, and related techniques of summation of infinite diagrammatic series, readily become available, as powerful alternatives to the more conventional approaches of first quantization. In such a language, atomic theory then appears more directly in line with similar formulations of high energy and nuclear physics.

During the last few years we have been considering the application of (non relativistic) FT techniques both to scattering problems [1] and to the determination of bound states [2], in few-body atomic physics. In this paper we wish to present some extensions of the formalism, for collisions of positrons (e^+) on atomic targets. In particular, we shall concentrate on elastic collisions, in the low energy regime (i.e. up to a few times the ionization potential of the target). In such an energy interval, the collisional dynamics is essentially determined by target polarization effects and by couplings to the (virtual or real) channel of Positronium (Ps) formation. In the present formulation, we shall then point out that the Random Phase Approximation (RPA) and Ladder schemes are, among all possible Feynman diagrams, those most directly responsible for a correct description of the dynamics and we shall outline schemes of summation of the corresponding infinite diagrammatic series.

The plan of the paper is, then, as follows. In Section 2 the basic equations of the FT approach will be summarized. The RPA $\oplus$ Ladder approximation scheme of the theory will be outlined in Section 3, where an explicit numerical application to the e^+-He system, will be considered. This will be followed by a concluding summary, in Section 4.

2. FIELD THEORY OF e^+-ATOM SCATTERING.

In this Section we shall summarize the basic FT equations, presiding over a full description of the elastic dynamics, for e^+-atom systems. In the development of the theory, we shall essentially refer to the literature [3] for details on the FT formalism, while we shall concentrate, more directly, on those aspects of the theory, of relevance for the atomic situation of interest.

We wish to consider processes of the type

$$e^+(k_i) + A(i) \ \text{---}> \ e^+(k_i') + A(i) \tag{1}$$

where "i" labels the initial electronic state (usually the ground state) of the A atomic species and $|k_i| = |k_i'|$ (for a recent review of the experimental situation, see Ref. [4]). It is easy to show that, in the FT formalism, the theoretical information on the cross section, for the process (1), is contained into the one-positron propagator, defined according to

$$S(2,2') = (1/i) < \Psi_i | T[\phi(2)\phi'(2')] | \Psi_i > \tag{2}$$

where T is the time ordering operator, $|\Psi_i>$ represents the target wave function, involved in the process (1), ϕ ($\phi^\dagger$) is a destruction (creation) field operator for the e^+ species, whose space-spin-time coordinates are compactly denoted by integers (e.g. $2 \equiv (r_2,\sigma_2,t_2) \equiv (r_2,t_2)$). It is not difficult to establish, for the S amplitude, an integral equation of the type

$$S(2,2') = S_0(2,2') + \int d4d4' \ S_0(2,4)\Lambda(4,4')S(4',2') \tag{3}$$

In (3), S_0 represents the free particle propagator (scattering in the absence of any interaction), while the Λ "optical potential" contains the sum of all infinite diagrammatic series, contributing to the elastic dynamics, in (1).

In order to make the interaction between the positron and target electrons (e^-) more explicit, it is now convenient to express Λ in the form

$$\Lambda(2,2') = -i \ \delta(2-2') \int d1 \ V_p(1,2)G(1,1^+) \ + \tag{4}$$

$$i \int d1d3d3'd4 \ V_p(1,2)S(2,4)G(1,3)\Gamma_p(43,3'2')G(3',1)$$

where, in particular, $V_p(1,2) = -\delta(t_1-t_2)/|r_1-r_2|$ represents the bare (Coulomb) e^+-e^- attractive interaction potential, while G is the one-electron propagator, defined similarly to (2), in terms of electron destruction (creation) field operators. An equation for such a G propagator could easily be derived, by using similar techniques, but, for our present purposes, we shall simply assume that G is known to the Hartree-Fock (HF) level of accuracy (e.g. $G \approx G^{HF}$ will finally be enforced in the numerical application of the formalism). In the language of atomic physics, the first term, on the right hand side (rhs) of (4), pictures, then, an elastic scattering process off a target "frozen" in the $|\Psi_i>$ initial state, while the rightmost term, in (4), containing the Γ_p "vertex interaction", introduces the effects of "dynamical polarization" of the target, along with couplings to the Ps formation channel, mentioned above.

The Γ_p vertex is the quantity that we would like to approximate, in the end, according to the RPA $\oplus$ Ladder scheme of the present theory. In order to do so, it is, anyway, first necessary to obtain an exact equation for such an amplitude. It turns out that, in the FT formalism, such a closed expression can be obtained, at the expense of having to introduce "functional differentiations" into the theory. At any rate, this is not to be considered as a drawback of the formalism, in that, as will be evident later on, the equations of the full theory will completely determine, at the same time, useful approximation schemes for the functional differentiations themselves. Without going into the details of the derivation, we shall then simply state the final results of the present FT approach. In particular, it can be shown that the Γ_p vertex is explicitly given by

$$\Gamma_p(21,1'2') = \Xi_p^{(1)}(21,1'2') - \qquad (5)$$

$$\int d3d5d3'd5' \ \Xi_p^{(1)}(23,3'2')G(3',5)\Gamma_e(15,5'1')G(5',3) \ +$$

$$\int d4d6d4'd6' \ \Xi_p^{(2)}(24,4'2')S(4',6)\Gamma_p(61,1'6')S(6',4)$$

with the $\Xi_p^{(1)}$ and $\Xi_p^{(2)}$ amplitudes defined, in turn, in terms of the functional differentiation equations

$$\Xi_p^{(1)}(23,3'2') = \delta\Lambda(2,2')/\delta G(3',3) \qquad (6a)$$

$$\Xi_p^{(2)}(24,4'2') = \delta\Lambda(2,2')/\delta S(4',4) \qquad (6b)$$

It will be noticed, in particular, that in the second term, on the rhs of (5), a new amplitude, Γ_e, appears under the integral sign. The notation used for such a quantity is suggestive of the fact that Γ_e is actually the vertex interaction for a system containing only electrons, where it plays the same role that Γ_p has, for the present system, containing both e^+ and e^- species. The corresponding term, in (5), then brings into the elastic dynamics all the relevant information about the target structure (e.g. excited states of the target). Finally, the $\Xi_p^{(1,2)}$ quantities are defined in terms of the functional expressions (6), that will be discussed, in more detail, in the next Section.

Equations (3)-(6) completely define the full, non perturbative structure, of the present theory, so that, in the next Section, we shall concentrate on obtaining a convenient approximation scheme, to such equations.

3. THE RPA ⊕ LADDER APPROXIMATION.

As pointed out in (3), we essentially need an approximation to the Λ optical potential, which is defined, in (4), in terms of Γ_p. Now the full equation for Γ_p contains a term (third on the rhs of (5)), that actually explicitly makes it into an integral equation. This term depends also on the $\Xi_p^{(2)}$ amplitude that, as shown by (6b), is defined in terms of the functional variation of Λ with respect to the S propagator. A more careful consideration of such a functional differentiation shows that, even in lowest order of perturbation theory, $\Xi_p^{(2)}$ will be of the second order in V_p and, therefore, probably negligible in a first approach to the formalism. By neglecting the contribution of the term containing $\Xi_p^{(2)}$ (third term on the rhs of (5)), the equation for Γ_p then more simply becomes (in a convenient "matrix type" notation)

$$\Gamma_p = \Xi_p^{(1)} - \Xi_p^{(1)} \, G \, \Gamma_e \, G \tag{7}$$

which will be completely determined once approximations to $\Xi_p^{(1)}$, G and Γ_e will be specified.

We turn our attention, next, to the $\Xi_p^{(1)}$ amplitude, defined in (6a). By considering the functional dependance of Λ on G (see (4)), it is immediate to verify that $\Xi_p^{(1)}$ may be expressed according to

$$\Xi_p^{(1)} = -iV_p + iV_p \, S \, G \, \Gamma_p + iV_p \, S \, \Gamma_p \, G + O(\delta\Gamma_p/\delta G) \tag{8}$$

with the rightmost term denoting quantities of the order of $\delta\Gamma_p/\delta G$. Eq. (8) then shows that, to lowest order, $\Xi_p^{(1)}$ simply reduces to the bare Coulomb attraction potential, V_p. The second and third terms, on the rhs of (8), picture instead, two-body collisions for e^+-e^- pairs, corresponding to excitation processes that, in the FT language, can be described as being of the particle-particle and particle-hole type, respectively [1]. In the atomic situation of interest it is expected that particle-particle processes will be the leading ones, so that we may finally approximate $\Xi_p^{(1)}$ according to

$$\Xi_p^{(1)} = -iV_p + iV_p \, S \, G \, \Gamma_p \tag{9}$$

that amounts to bringing, into the collisional dynamics, the contribution of the Ladder diagrammatic series. Eqs. (7) and (9) represent, then, the final set of coupled equations to be solved, in the present scheme. In order to do so we have to settle, next, on some approximation scheme for Γ_e, to be used on the rhs of (7). A convenient approach, for this purpose, amounts to keeping the RPA diagrams [5], in the description of the atomic target (e.g. $\Gamma_e \sim \Gamma_e^{RPA}$ on the rhs of (7)). If we enforce, next, in both (7) and (9), $G \sim G^{HF}$ and $S \sim S_0$, we obtain

$$\Gamma_p^{RPA\oplus L} = \Xi_p^{(1)RPA\oplus L} - \Xi_p^{(1)RPA\oplus L} \, G^{HF} \, \Gamma_e^{RPA} \, G^{HF} \tag{10a}$$

$$\Xi_p^{RPA\oplus L} = -iV_p + iV_p \, S_0 \, G^{HF} \, \Gamma_p^{RPA\oplus L} \tag{10b}$$

as the final equations introducing the RPA $\oplus$ Ladder approximation, in the scattering process (1). Once eqs. (10) have been solved, the $\Gamma_p^{RPA\oplus L}$ vertex can be used into eq. (4), thereby defining the corresponding $\Lambda^{RPA\oplus L}$ approximation to the optical potential.

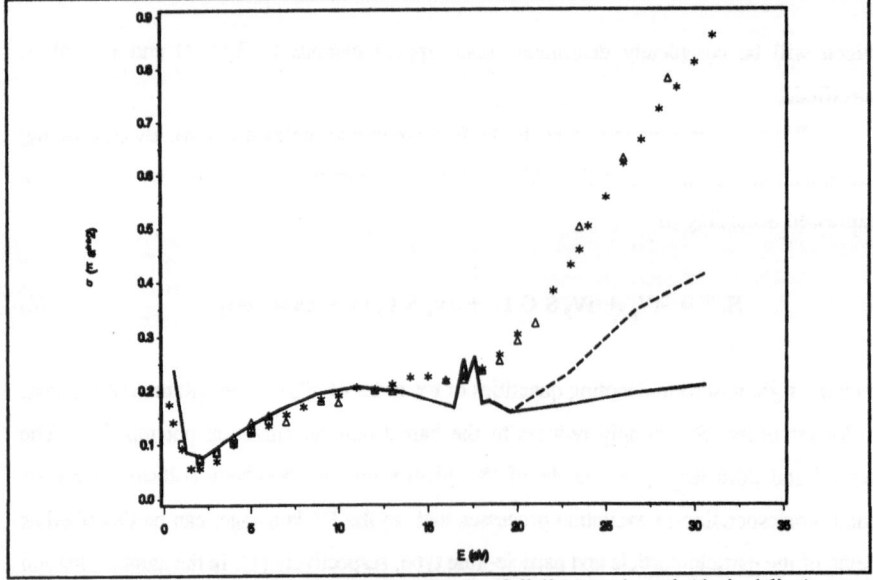

Figure 1 RPA⊕Ladder approximation to elastic (full line) and total (dashed line) cross sections, for the e^+-He system (see text for details).

We have solved, numerically, the coupled set of eqs. (10) (details of the computations will be given elsewhere), obtaining the RPA⊕Ladder cross section values presented in Fig. 1. In particular, the full and dashed lines, in this Figure, correspond to the theoretical "elastic" and "total" cross sections, for the e^+-He system. It should be pointed out that the "total" theoretical cross section curve has been calculated from the optical theorem, by using the imaginary component of the Λ optical potential, in (4), as obtained from the present solution of eqs. (10). The evaluation of such a quantity is necessary, at this stage, since the experimental results, denoted by the different symbols in Fig. 1, refer, in turn, to "total" cross section values [6-8].

4. SUMMARY AND OUTLINE OF FUTURE DEVELOPMENTS.

The theoretical values presented in Fig. 1, clearly show the relevance of the contribution of the Ps formation channel, that, in the case of He, becomes energetically available around 17.7 eV. The structure in the theoretical cross sections, centered around 17.5 eV, is essentially due to a contribution of the Ladder diagrams, in the present theory. In fact, if the rightmost term of (9) is neglected (corresponding to an enforcement of the simpler RPA approximation), the structure in the theoretical cross sections completely disappears, showing that RPA theory, alone, is not able to lead to a coupling with the Ps formation channel.

On the other hand, theoretical values of "total" cross sections (dashed curve in Fig. 1), lie below the experimental results, thus showing a deficiency in the present scheme. We believe that such a lack of agreement is partly due to numerical approximations, enforced in our present solution of eqs. (10) (these will be discussed in more detail elsewhere), and partly to our neglect, in this stage, of the third term on the rhs of (8), representing particle-hole contributions to the Ladder series.

We can then finally say that the present calculations seem to indicate the general validity of the approach, outlined in this paper, but that, in order to obtain a quantitative agreement with experiment, the theory will have to be augmented, along the lines discussed above. Such generalizations of the formalism, and its extension to a consideration of different types of processes, are currently under study.

594

REFERENCES.

1. Ficocelli Varracchio, E.: in Atomic Physics with Positrons, Humberston, J.W., Armour
E.A.G. (eds.): NATO ASI Series B: Physics Vol. 169, New York-London: Plenum 1987.
2. Ficocelli Varracchio, E.: Few-Body Systems 8, 65 (1990).
3. Kirzhnits, D.A.: Field Theoretical Methods in Many-Body Systems. Oxford-New-York:
Pergamon 1967.
4. Charlton, M., Laricchia, G.: J. Phys. B: At. Mol. Opt. Phys. 23, 1045 (1990).
5. Ficocelli Varracchio, E.: J. Phys. B: At. Mol. Opt. Phys. 23, L109 (1990).
6. Stein, T.S., Kauppila, W.E., Pol, V., Smart, J.H., Jension, G.: Phys. Rev. A17, 1600
(1978).
7. Canter, K.F., Coleman, P.G., Griffith, T.C., Heyland, G.R.: J. Phys. B: At. Mol.
Phys. 6, L201 (1973).
8. Sinapius, G., Raith, W., Wilson, W.G.: J. Phys. B: At. Mol. Phys. 13, 4079 (1980).

Few-Body Systems, Suppl. 6, 595—604 (1992)

GREEN'S FUNCTIONS FOR MOLECULES

L.S. Cederbaum

Theoretische Chemie, Physikalisch–Chemisches Insitut

Universität Heidelberg

Im Neuenheimer Feld 253, D–6900 Heidelberg, FRG

The properties of a quantum system and the physical processes which can be described by Green's functions are briefly outlined. A general approximation scheme to evaluate Green's functions and propagators is discussed. Illustrative applications to molecules are presented and related to experiment.

1.Introduction

The method of Green's functions has been widely used to study quantum properties of several– and many–particle systems. The general theory (see, for instance, /1/ and /2/) has been applied, for example, to nuclei /3,4/, to solids /5,6/ and to atoms and molecules /7,8/. In each field different approximations to the exact Green's functions have to be introduced taking into account the specific interactions characteristic of the system under investigation. The long–range interaction in solids and the short–range interaction in nuclear matter allow to derive meaningful approximations by considering only partial summations of particular types of Feynman diagrams. These comprise the RPA (random phase approximation) and the ladder diagrams, respectively /1,9/. For atoms, molecules and possibly nuclei, the finite range of the system generally prevents a single series of diagrams from dominating the others /8,10/. To achieve reliable results it is, therefore, necessary to handle all Feynman diagrams up to a given order of perturbation theory on the same footing and to estimate the higher orders in a systematic way. This is explicitly attempted in the algebraic diagrammatic construction (ADC) scheme /11,12/. In the n—th order scheme all diagrams up to n—th order are included as well as all series of diagrams arising from them. The ladder diagrams of second and third order, for example, thus generate the whole series of ladder diagrams. The ADC has been applied to atoms and, in particular, to numerous molecules /13/. We would like to mention that other approximations to Green's functions are available which have also been applied to atoms and molecules. An incomplete list of references includes /14,15,16,10/.

In the present paper we briefly review the theory of Green's functions and relate these functions to observable properties of the system under investigation. A few illustrative applications to molecules are discussed.

2.Green's functions

There exists a hierarchy of Green's functions (GF). These functions are defined as the N—electron ground—state expectation value of a time—ordered product of annihilation and creation operators. The simplest and probably most relevant member of the hierarchy is the one—particle GF

$$G_{k\ell}(t,t') = -i\langle \Psi_0^N | T\{a_k(t)a_\ell^\dagger(t')\} | \Psi_0^N \rangle \qquad (2.1)$$

where Ψ_0^N is the exact ground state, a_k and $a_\ell^\dagger$ denote annihilation and creation operators, respectively, and T is the Wick time—ordering operator /1,2/. The next member of the hierarchy is the two—particle GF which contains two annihilation and two creation operators and depends on four time variables. The higher members are defined analogously.

For many applications one only needs a particular component of the full GF. For example, the particle—particle (p—p) propagator /17/, II, is obtained from the two—particle GF by equating the times of both creation operators and also of both annihilation operators.

Numerous properties of physical relevance can be deduced from the GF's. This becomes more evident if the GF's are Fourier transformed to energy space. Noticing that the one—particle GF and the p—p propagator are time—translational invariant, the transformation is readily performed:

$$G_{k\ell}(\omega) = \sum_n \frac{x_k^{(n)} x_\ell^{(n)*}}{\omega + (E_n^{N-1} - E_0^N) - i\eta} + \sum_m \frac{x_k^{(m)} x_\ell^{(m)*}}{\omega + (E_0^N - E_m^{N+1}) + i\eta} \qquad (2.2)$$

$$\Pi_{k\ell,k'\ell'}(\omega) = \sum_n \frac{x_{k\ell}^{(n)} x_{k'\ell'}^{(n)*}}{\omega + (E_n^{N-2} - E_0^N) - i\eta} + \sum_m \frac{x_{k\ell}^{(m)} x_{k'\ell'}^{(m)*}}{\omega + (E_0^N - E_m^{N+2}) + i\eta} \qquad (2.3)$$

Here E_0^N and E_n^M denote the exact energies of the N–electron ground state and of the n–th M–electron state. η is a positive infinitesimal. The first and second parts of G describe the elimination of an electron from and the attachment of an electron to the system, respectively. The first part contains all cationic states and exhibits explicitly as poles the ionization potentials (IP) of the system. Analogously, the electron affinities (EA) appear as poles of the second part which is associated with anionic states. The structure of Π is very similar to that of G. This propagator describes the simultaneous annihilation of two electrons from the system as well as the simultaneous attachment of two electrons to the system. In particular, the pole positions of the first part of Π represent double ionization potentials (DIP).

Not only the pole positions are related to experiment. The residue amplitudes $x_k^{(n)}$ and $x_{k\ell}^{(m)}$ contain relevant information on the single and double ionization and attachment (or scattering) processes. For instance, the partial–channel photoionization cross section for production of molecular ions in the n–th state can be determined in the sudden approximation /18,19/. Furthermore, the ground state expectation value of an arbitrary one–particle operator as well as the ground state energy itself can be computed using G /1,2/. Analogously, the knowledge of Π suffices to calculate the expectation value of two–particle operators.

Bell and Squires /20/ were the first to show that the optical potential for elastic scattering may be identified with the self–energy of the one–particle GF. The self–energy $\Sigma(\omega)$ connects the GF with the free GF via the Dyson equation /1,2/. In matrix notation this equation reads

$$\underline{G}(\omega) = \underline{G}^0(\omega) + \underline{G}^0(\omega)\underline{\Sigma}(\omega)\underline{G}(\omega) \qquad (2.4)$$

The free GF $\underline{G}^0(\omega)$ is defined in analogy to the GF, but with an unperturbed Hamiltonian instead of the full Hamiltonian. The self–energy can be viewed as an effective energy–dependent one–particle potential caused by correlation effects. If we neglect these correlation effects, the optical potential reduces to the well–known static–exchange potential /21/ evaluated with respect to the Hartree–Fock (HF) potential. The self–energy consists of a static part $\Sigma(\infty)$ not depending on ω and a dynamic part depending on ω:

$$\Sigma(\omega) = \Sigma(\infty) + M(\omega) \tag{2.5}$$

Interestingly, the static part can be evaluated once the dynamic self–energy is known /8,13/. The static part has a simple interpretation. In spatial representation it takes on the following appearance /22/

$$\Sigma(r,r',\infty) = W + \delta(r{-}r') \int d\bar{r} \, \frac{\rho(\bar{r},\bar{r})}{|r{-}\bar{r}|} - \frac{\rho(r,r')}{|r{-}r'|} \tag{2.6}$$

where ρ is the exact one–particle density and W is the interaction of the projectile with the nuclei. The static part can thus be viewed as the static–exchange interaction of the incoming electron with the <u>correlated</u> target. An analysis of the physical origin of the dynamic part and its connection to the polarisation potential has been given by Meyer /23/.

3. Evaluation of Green's functions

It is well–known that GF's can be expanded in terms of Feynman diagrams /1,2/. The algebraic diagrammatic construction (ADC) is a scheme to carry out in a systematic manner infinite partial summations of Feynman diagrams. The n–th order scheme, ADC(n), is complete to n–th order, i.e., it contains <u>all</u> the Feynman diagrams up to n–th order and infinitely many diagrams of higher orders. The choice of the latter diagrams is automatically dictated by the former ones and the analytic structure of the exact GF or dynamic self–energy part. The scheme can be applied to any GF or to a single component of it. In the present work we apply it directly to the dynamic self–energy as an example. For more details and for explicit working equations up to ADC(4) we refer to /11/. The ADC(3) working equations for the p–p propagator are given in /24/ and the polarisation propagator is discussed in /12/.

The dynamic self–energy $M(\omega)$ can be written as the sum of two parts $M^{(I)}$ and $M^{(II)}$ analytical in the upper and lower half of the complex ω–plane, respectively:

$$M(\omega) = M^{(I)}(\omega) + M^{(II)}(\omega) \tag{3.1}$$

Each of these parts has a spectral representation. Consequently, we may write

$$\underline{M}^{(I)} = \underline{m}^{(I)\dagger}(\omega\underline{1} - \underline{\Omega}^{(I)} + i\eta)^{-1}\,\underline{m}^{(I)} \tag{3.2}$$

where $\underline{\Omega}^{(I)}$ denotes the diagonal matrix having the pole positions of $\underline{M}^{(I)}$ as elements. An analogous result holds for $\underline{M}^{(II)}$.

The self–energy is given in eq. (3.2) in the diagonal representation where its poles are explicitly displayed. A general non–diagonal representation is obtained by inserting unity $\underline{Y}\,\underline{Y}^{\dagger} = \underline{1}$, where $\underline{Y}$ is a unitary matrix, twice into (3.2). The result takes on the following appearance

$$\underline{M}^{(I)} = \underline{U}^{(I)\dagger}(\omega\underline{1} - \underline{K}^{(I)} - \underline{C}^{(I)})^{-1}\underline{U}^{(I)} \tag{3.3}$$

where

$$\underline{U}^{(I)} = \underline{Y}^{\dagger}\underline{m}^{(I)} \qquad \underline{K}^{(I)} + \underline{C}^{(I)} = \underline{Y}^{\dagger}\underline{\Omega}^{(I)}\underline{Y} \tag{3.4}$$

From now on we shall drop the superscripts (I) and (II) whenever unnecessary. The ADC makes use of the perturbation expansion of the effective coupling matrix $\underline{U}$ and the effective interaction matrix $\underline{C}$ in terms of the electron–electron repulsion:

$$\underline{U} = \underline{U}^{(1)} + \underline{U}^{(2)} + \underline{U}^{(3)} + \ldots \tag{3.5a}$$

$$\underline{C} = \underline{C}^{(1)} + \underline{C}^{(2)} + \underline{C}^{(3)} + \ldots \tag{3.5b}$$

It is noted that the expansion of the effective coupling and effective interaction matrices begins in first order of perturbation theory. $\underline{K}$ is defined as the zeroth order of the r.h.s. of (3.4). The perturbation expansion of the self–energy itself immediately follows from (3.5) and

$$(\omega\underline{1} - \underline{K} - \underline{C})^{-1} = (\omega\underline{1} - \underline{K})^{-1}[\underline{1} + \underline{C}\,(\omega\underline{1} - \underline{K})^{-1} + \ldots] \tag{3.6}$$

The result reads

$$\underline{M}(\omega) = \underline{U}^{(1)\dagger}\,(\omega\underline{1} - \underline{K})^{-1}\,\underline{U}^{(1)} + \underline{U}^{(1)\dagger}\,(\omega\underline{1} - \underline{K})^{-1}\,\underline{C}^{(1)}\,(\omega\underline{1} - \underline{K})^{-1}\,\underline{U}^{(1)}$$
$$\tag{3.7}$$
$$+ \quad \underline{U}^{(2)\dagger}\,(\omega\underline{1} - \underline{K})^{-1}\,\underline{U}^{(1)} + \underline{U}^{(1)\dagger}\,(\omega\underline{1} - \underline{K})^{-1}\,\underline{U}^{(2)} + \ldots$$

where all terms up to third order are displayed explicitly. In the ADC the expansions of $\underline{U}$ and $\underline{C}$ are determined by comparing (3.7) with the expansion of the self–energy via Feynman diagrams. Equating the Feynman diagrams of second order and the first term on the r.h.s. of (3.7) allows the determination of $\underline{U}^{(1)}$. The Feynman diagrams of third order can be divided into two sets according to their dependence on $(\omega \underline{1} - \underline{K})^{-1}$. Since $\underline{U}^{(1)}$ is already known, one set of diagrams determines $\underline{U}^{(2)}$ and the other allows the evaluation of $\underline{C}^{(1)}$ (see second term on the r.h.s. of (3.7)). Higher orders are computed analogously. The Feynman diagrams up to n–th order are used to evaluate $\underline{U}$ and $\underline{C}$. The resulting quantities are then inserted into (3.3) to obtain the dynamic self–energy part in ADC(n).

It should be noted that the resulting self–energy and Green's function are in ADC(n) correct up to n–th order and contain <u>infinitely</u> many terms of higher orders. On the other hand, the perturbation expansion (3.7) used to determine $\underline{U}$ and $\underline{C}$ only includes terms up to n–th order, and, more importantly, exhibits an incorrect dependence on ω not shared by the exact and ADC self–energy parts which are both subject to a spectral representation. Finally, we mention that the size of the configuration space needed in ADC(n) is restricted. The configuration space required by ADC(n) is substantially smaller than that which is required by Rayleigh–Schrödinger perturbation theory for obtaining the same energies accurate to n–th order. For a detailed discussion of this relevant aspect of ADC, see /11/.

4. Illustrative applications

By far most of the applications of GF's are connected with the one–particle GF. A comprehensive list of the applications of this GF to atoms and molecules up to 1984 can be found in /13/. In the following we briefly discuss a recent application of the one–particle GF to the ionization of CN dimers and one application of the much less studied p–p propagator.

Van der Does and Bickelhaupt /25/ claimed to have reported the synthesis of CNNC, a centrosymmetric isomer of the well–known cyanogen NCCN. A photoelectron spectrum (PES) of the synthesized molecule has been recorded /26/. High–resolution infrared and microwave spectra /27/ as well as our calculations on the PES (presented in /28/) showed beyond doubt that the species measured experimentally is not CNNC, but rather the non–symmetric CN dimer CNCN.

Our GF results obtained using the ADC(3) are depicted in Fig. 1 together with a schematic drawing of the experimental PES /26/. Shown are sticks at the

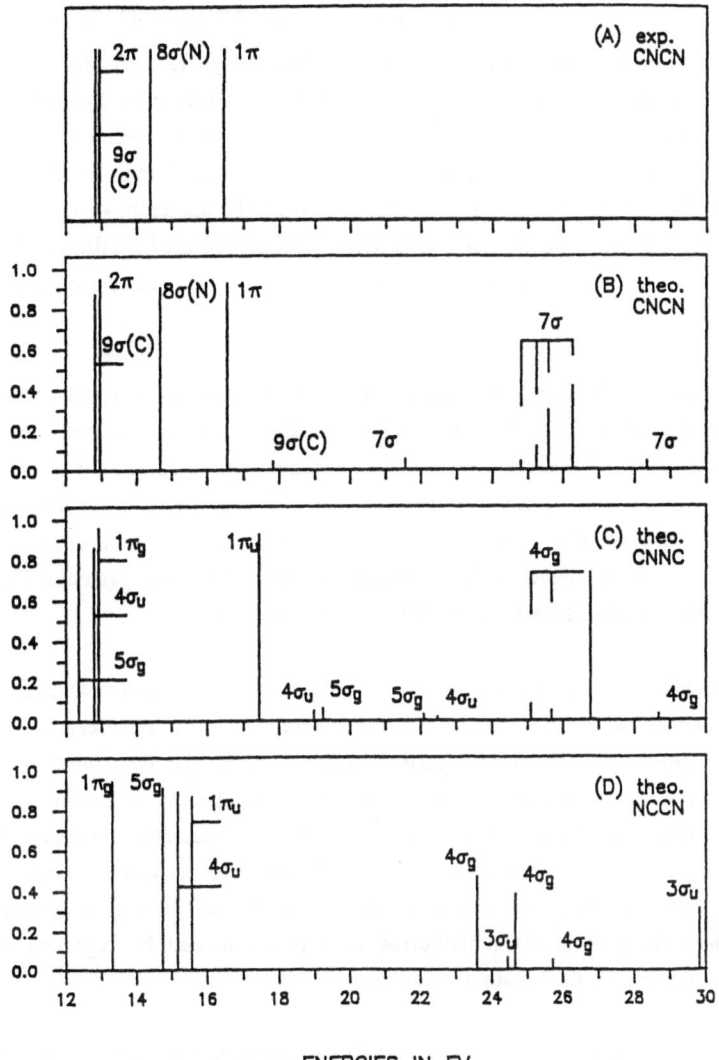

Figure 1:   Schematic drawing of the experimental photoelectron spectrum of
CNCN(A) and calculated spectra of CNCN(B), CNNC(C) and
NCCN(D). The calculations are discussed in detail in /28/.

positions of the IP's. The height of a stick indicates the corresponding pole strength $|x_k^{(n)}|^2$. The calculation on CNNC predicts that the three peaks associated with the ejection of electrons from the $5\sigma_g$, $4\sigma_u$ and $1\pi_g$ orbitals lie very close to each other. These peaks are well separated in energy from the one corresponding to the $1\pi_u$ orbital. The computation on CNCN predicts, on the other hand, a peak lying energetically just in the middle of the peakless gap in the spectrum of CNNC and shows that the experimental spectrum is of CNCN. The computed results on CNCN agree very well with the measured ones. The details of the claculations are published elsewhere /28/.

In addition to the main lines in the spectrum associated with the outer valence orbitals the calculations predict interesting satellite structures as well as the appearance of multiplet structure in the inner valence region. In the inner valence region a main line may cease to exist. Its intensity can be distributed over many ionic states giving rise to the *breakdown of the orbital picture of ionization*. This phenomenon has been found to be common to most molecular systems. For a thorough discussion of this phenomenon we refer to /19/.

Auger spectroscopy provides valuable information on the electronic structure of molecules and of atoms in different chemical enviroments /29/. To interpret Auger spectra one needs the DIP's and rates associated with the contributing final dicationic states. For atoms and atom—like molecules the density of doubly ionized states of experimental relevance is relatively small and it is reasonable to expect that the observed Auger peaks can be interpreted in terms of individual states. The situation is different for polyatomic molecules. Even on the level of independent particles the density of states contributing to a spectrum can be high and may considerably increase by final state interactions.

As an example we briefly discuss here the Auger spectrum of benzene. Many dicationic states and the Auger spectrum of this molecule have recently been computed via the p—p propagator on the ADC(2) level /30/. No other theoretical results are available on benzene. The calculated and experimental spectra are shown in Fig. 2. It is relevant to note that as many as 226 computed dicationic states have been found to contribute to the lower energy part of the spectrum. The high energy wing has not been evaluated since many more states are required and the numerical effort grows rapidly. It has become clear that the Auger spectra of larger molecules are very complex spectra. Although a few relatively well resolved peaks are observed, each of these may contain several dicationic states. In benzene the calculations predict that the first peak in the spectrum already contains three states. Seven states are found to contribute to the second and sixteen to the third peak.

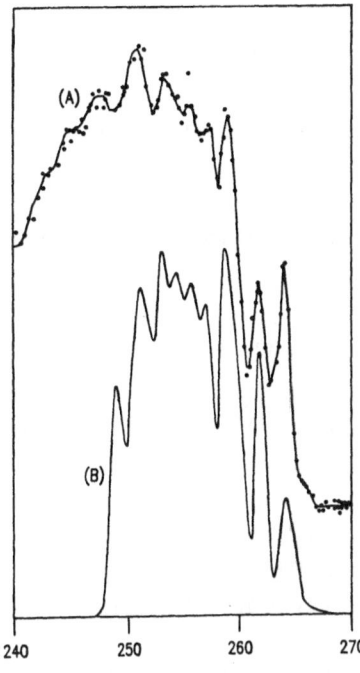

Figure 2: Experimental (A) and computed (B) Auger spectrum of benzene. For details see /30/.

Due to the very high density of dicationic states, correlation effects show up already at very low energies, causing a substantial spreading of Auger intensity over hundreds of double ionization transitions (for a detailed discussion of these effects see /31/). It is, therefore, essential to have a theoretical model which incorporates electron correlation and makes feasible the calculation of such large numbers of dicationic states. GF methods provide such a theoretical framework.

Acknowledgements

The author would like to express his sincere thanks to H.–D. Meyer, J. Schirmer, A. Sgamellotti, F. Tarantelli and W. von Niessen for numerous fruitful collaborations and discussions.

References

/1/ A.L. Fetter and J.D. Walecka, Quantum Theory of Many–Particle Systems, McGraw–Hill, New York (1971)

/2/ D.J. Thouless, The Quantum Mechanics Of Many–Body systems, Academic Press, N.Y. (1972)

/3/ A.B. Migdal, Theory of Finite Fermi Systems and Applications to Atomic Nuclei, Interscience Publishers, N.Y. (1967)

/4/ P. Ring and P. Schuck, The Nuclear Many–Body Problem, Springer (1980)

604

/5/   G.D. Mahan, <u>Many—Paricle Physics</u>, Plenum Press, N.Y. (1981)

/6/   E.N. Economou, <u>Green's Functions in Quantum Physics</u>, Springer, N.Y. (1979)

/7/   J. Linderberg and Y. Öhrn, <u>Propagtors in Quantum Chemistry</u>, Academic Press, N.Y. (1973)

/8/   L.S. Cederbaum and W. Domcke, Adv. Chem. Phys. <u>36</u>, 205 (1977)

/9/   D. Pines, <u>The Many—Body Problem</u>, Benjamin/Cummings Publishing Company, Reading, Massachusetts (1962)

/10/   L.S. Cederbaum, Theoret. Chim. Acta <u>31</u>, 239 (1973)

/11/   J. Schirmer, L.S. Cederbaum and O. Walter, Phys. Rev. A <u>28</u>, 1237 (1983)

/12/   J. Schirmer, Phys. Rev. A <u>26</u>, 2395 (1982)

/13/   W. von Niessen, J. Schirmer and L.S. Cederbaum, Computer Physics Reports <u>1</u>, 57 (1984)

/14/   G. Csanak, H.S. Taylor and R. Yaris, Adv. Atom. Mol. Phys. <u>7</u>, 287 (1971)

/15/   M.F. Hermann, K.F. Freed and D.L. Yeager, Adv. Chem. Phys. <u>48</u>, 1 (1981)

/16/   Y. Öhrn and G. Born, Adv. Quant. Chem. <u>13</u>, 1 (1981)

/17/   N. Fukuda, F. Iwamoto and K. Sawada, Phys. Rev. A <u>135</u>, 932 (1964)

/18/   L.S. Cederbaum, Mol. Phys. <u>28</u>, 479 (1974)

/19/   L.S. Cederbaum, W. Domcke, J. Schirmer and W. von Niessen, Adv. Chem. Phys. <u>65</u>, 115 (1986)

/20/   J.S. Bell and E.J. Squires, Phys. Rev. Lett. <u>3</u>, 96 (1959)

/21/   N.F. Lane, Rev. Mod. Phys. <u>52</u>, 29 (1980)

/22/   H.G. Weikert and L.S. Cederbaum, Few—Body Systems <u>2</u>, 33 (1987)

/23/   H.–D. Meyer, J. Phys. B <u>21</u>, 3777 (1988)

/24/   A. Tarantelli and L.S. Cederbaum, Phys. Rev. A <u>39</u>, 1656 (1989)

/25/   T. van der Does and F. Bickelhaupt, Angew. Chem. <u>100</u>, 998 (1988)

/26/   O. Grabant, C.A. de Lange, R. Mooymen, T. van der Does and F. Bickelhaupt, Chem. Phys. Lett. <u>155</u>, 221 (1989)

/27/   F. Stroh and M. Winnewisser, Chem. Phys. Lett. <u>155</u>, 21 (1989)

/28/   L.S. Cederbaum, F. Tarantelli, H.G. Weikert, M.K. Scheller and H. Köppel, Angew. Chem. <u>101</u>, 770 (1989); J. Electr. Spectr. <u>51</u>, 75 (1990)

/29/   M. Thompson, M.D. Baker, A. Christie and J.F. Tyson, <u>Auger Electron Spectroscopy</u>, Wiley, N.Y. (1985)

/30/   F. Tarantelli, A. Sgamellotti, L.S. Cederbaum and J. Schirmer, J. Chem. Phys. <u>86</u>, 2201 (1987)

/31/   E. Ohrendorf, F. Tarantelli and L.S. Cederbaum, J. Chem. Phys. <u>92</u>, 2984 (1990)

Few-Body Systems, Suppl. 6, 605—614 (1992)

BOUND STATES AND MOLECULAR STRUCTURE OF SYSTEMS WITH HYPERONS

Yoshinori AKAISHI

Department of Physics, Hokkaido University, Sapporo 060, Japan

Abstract

Microscopic calculations are done for Σ-hypernuclear few-body systems by a method named ATMS. Among two- to five-body systems, only the $^4_\Sigma He(0^+)$ and $^4_\Sigma H(0^+)$ hypernuclei are expected to be bound: The binding energy and the width of the former are calculated to be $3.7 \sim 4.6$ MeV and $4.5 \sim 7.9$ MeV, respectively. The observation of $^4_\Sigma He$ at KEK is in good agreement with the above prediction. The nucleus-Σ potential has a strong Lane term and a repulsive bump at short distance. The Lane term makes the system bound and the bump suppresses the $\Sigma N \rightarrow \Lambda N$ conversion. X-ray measurement of level shifts in the $^4 He-\Sigma^-$, $^3 He-\Sigma^-$ and $^3 H-\Sigma^-$ atoms can provide another information on the Lane term.

In ^{208}Pb, there may exist a peculiar state, Coulomb-assisted (atom-nucleus) hybrid state, where Σ^- is trapped in the surface region by the strong interaction with the aid of the inner centrifugal repulsion and the outer Coulomb attraction.

An analysis is given for new data of $\Xi^- \cdot {}^{12}C$ atomic or nuclear systems from the emulsion-counter experiment at KEK. The double-Λ hypernucleus formation rate is calculated for a stopped Ξ^- on ^4He. A high branching ratio of 37% is obtained for the $^4_{\Lambda\Lambda}H$ formation from a $\Xi^- \cdot {}^4He$ atom. The detection of about 2.3 MeV neutron is proposed to search for lightest double-Λ hypernucleus $^4_{\Lambda\Lambda}H$.

1. Sigma-Hypernuclei

1.1. INTRODUCTION

In 1980, the observation of narrow peaks due to the Σ production was reported in the strangeness exchange $^9Be(K^-, \pi^-)$ reaction at CERN [1]. The widths are about 8 MeV, which are unexpectedly narrow compared to anticipating theoretical predictions of 20 ~ 30 MeV. Since then, data of such narrow peaks have been accumulated for $(K^-, \pi^{\pm})$ reactions [2]: All peaks lie in continuum region above the Σ-emission threshold. In some better-statistics experiments, peak-like structures disappeared in the pion spectra. In spite of many theoretical efforts [2] it still remains a puzzle why such narrow peak structures could exist in the Σ-continuum region. In order to confirm the existence of Σ-hyhernuclei, more definitive data must be supplied.

Harada et al.[3] showed theoretically the possible existence of $^4_\Sigma H$ and $^4_\Sigma He$ below the Σ-emission threshold on the basis of four-body calculations with realistic ΣN potentials (SAP-1,2) which simulate the Nijmegen model-D potential [4]. The $^4_\Sigma He$ hypernucleus is in a bound state with $J^\pi = 0^+$ and T ~ 1/2 (99%). Its binding energy and width are calculated to be 3.7 ~ 4.6 MeV and 4.5 ~ 7.9 MeV, respectively. We use terminology "bound" in the sense that the Σ channel is closed, though the conversion Λ channel is open.

In 1989, Hayano et al.[5] presented an evidence of the bound Σ-hyper-nucleus. They observed $^4_\Sigma He$, below the $t+\Sigma^+$ threshold by $3.2^{+0.4}_{-1.4}$ MeV with a width of $4.6^{+2.1}_{-1.8}$ MeV, in the ^4He(stopped K^-,π^-) reaction experiment at KEK. The energy and the width are in good agreement with the above theoretical prediction.

"Is there a bound $^4_\Sigma He$?"[6] The in-flight K^- experiment at the magic incident momentum (450 MeV/c) could be the presently best way to definitely answer the above question. Recently the $^4He(K^-, \pi^-)$ in-flight experiment has been done with 600 MeV/c at BNL by Hayano, Hungerford and others to get a final answer to the existence of $^4_\Sigma He$.

1.2. FOUR-BODY CALCULATION

In order to make microscopic four-body calculations tractable, we construct a complex ΣN potential by eliminating the ΛN channel. Its radial form taken to be of two-range Gaussian form,

$$V_{TS} = v_C \exp\{-(r/a_C)^2\} + \{v_A + i\ w_A\} \exp\{-(r/a_A)^2\},$$

where the imaginary part describes the $\Sigma N \to \Lambda N$ conversion process. The potential parameters [3] are determined so as to reproduce S-matricies of of the Nijmegen model-D potential [4] at low energies. It is referred to as SAP-1 (Sigma Absorptive Potential #1). SAP-2 is a weaker conversion case which may not be rejected from present experimental two-body data. The ΣN potential is strongly isospin-spin dependent. It is much attractive in the (isospin,spin)=(1/2,1) and (3/2,0) states. No imaginary part appears in the isospin=3/2 states because Λ has no isospin degree of freedom.

The microscopic calculation of the ΣNNN four-body system is done as follows by ATMS (Amalgamation of Two-body correlations into Multiple Scattering process) [7]. The four-body wave function is taken to be

$$| \Psi > = \prod_{i>j=1}^{3} f(r_{ij}) \prod_{k=1}^{3} g(r_{k\Sigma}) \ | T, S > ,$$

where f and g are NN and ΣN radial functions. When once a type of the wave function is set up, the constituents $f(r)$ and $g(r)$ are determined by the stationary condition of the variation

$$< \hat{\delta\Psi} | \hat{H} - \lambda | \Psi > = 0,$$

where δ denotes the variation with respect to f or g. Since the ΣN potential has the imaginary part, the eigenvalue λ becomes complex,

$$\lambda = E - i \frac{\Gamma}{2} ,$$

where Γ is the $\Sigma N \to \Lambda N$ conversion width of the ΣNNN system.

The results for $_{\Sigma}^{4}He$ are followings: Its energy and the width are

$$(B_{\Sigma^+}, \Gamma) = (4.6 \text{ MeV}, 7.9 \text{ MeV}) \quad \text{for SAP-1},$$
$$= (3.7 \text{ MeV}, 4.5 \text{ MeV}) \quad \text{for SAP-2}.$$

The ΣN distribution extends outside compared to the NN distribution. The rms distance of ΣN is 2.9 fm for SAP-1 and 3.1 fm for SAP-2, which is longer by about 0.5 fm than that of NN, 2.5 fm.

1.3. CHARACTERISTICS OF NUCLEUS-Σ POTENTIAL

The nucleus-Σ potential is written as

$$\hat{U}_{nucl-\Sigma} = U^0 + U^\tau \vec{T}_c \vec{t}_\Sigma .$$

The first term is repulsive at short distance. The second is the Lane term and is sufficiently strong. The Lane term plays an essential role to make the nucleus-Σ system bound. In fact, if the coupling potential coming from the Lane potential is switched off, neither h + Σ^0 state nor t + Σ^+ state can be bound. Figure 1 shows the nucleus(3N)-Σ potentials in the T=3/2 state and the T=1/2 state. The potential for T = 3/2 is strongly repulsive and has no bound state. The bound $^4_\Sigma$He state appears in the T = 1/2 state. Since the conversion width is proportional to the overlap between the imaginary potential and the Σ distribution, the width is reduced to about half by the repulsion at short distance.

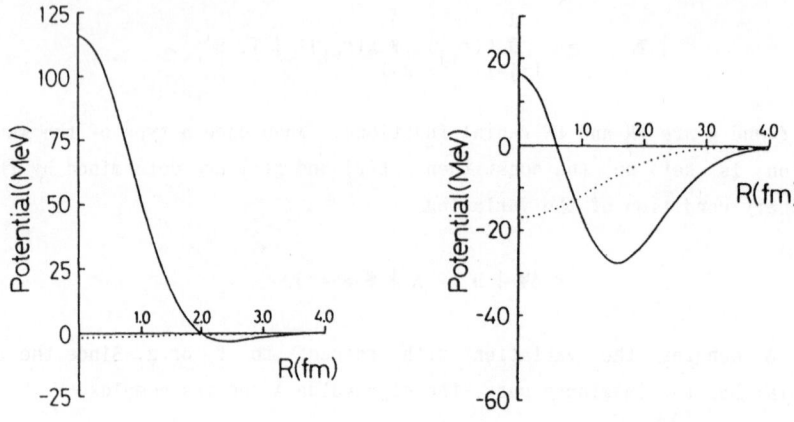

Fig. 1.
The nucleus-Σ potentials for T=3/2 and T=1/2 cases. The solid and the dotted lines are the real and the imaginary parts, respectively.

Similar properties of nucleus-Σ potentials can be seen in other light Σ-hypernuclear systems. A strong central repulsion appears in the α-Σ potential: The α particle is of isospin 0, and the α-Σ potential has no Lane term which could bring about strong attraction. This explains why there exists no α-Σ bound state. For $^8_\Sigma$Li ($^8_\Sigma$Be), its clustering is investigated by Okabe et al.[8] and a rotating α + $^4_\Sigma$H (α + $^4_\Sigma$He) di-molecular structure is discussed by Langanke et al.[9]

A level shift of the ^{4}He·Σ^- atomic 2p state was observed from X-ray measurement at CERN [10]. Level-shift data for ^{3}He·Σ^- and ^{4}He·Σ^- atoms can give another important information on the nucleus-Σ strong interaction. By using SAP-1 potential we obtain following results for the complex level-shifts (energy shift, width) by strong interaction [11]:

$$(-33 \text{ eV, } 97 \text{ eV}) \quad \text{for } ^3\text{He} \cdot \Sigma^- \text{ 2p state,}$$
$$(-13 \text{ eV, } 8 \text{ eV}) \quad \text{for } ^4\text{He} \cdot \Sigma^- \text{ 2p state.}$$

The difference between the ^{3}He case and the ^{4}He case is due to the Lane term. The Lane term makes the energy shift of ^{3}He about three times as large as that of ^{4}He. It also connect the ^{3}He·Σ^- 2p state to the ^{3}H + Σ^0 open channl, and thus, the width of 97 eV consists of 73 eV Σ^0-escape width and 24 eV conversion width. The ^{3}He·Σ^- 3d→2p X-ray (12.503 keV) observation can give information on the Lane term of the nucleus-Σ strong interaction.

1.4. Σ-HYPERNUCLEAR STATES IN HEAVIER NUCLEI

If the narrow-width phenomenon is not limited to some particular light nuclei but is extended to heavier nuclei, the Σ-nuclear study would arouse much more interest.

The nucleus-Σ potential for ^{208}Pb derived from the realistic ΣN interaction has a repulsive bump near the nuclear surface [12]. The ΣN interaction consists of repulsive and attractive parts which are not so different from one another. The appearance of the repulsive bump is due to a delicate balance of the repulsion and attraction of the ΣN interaction and is a common feature for light to heavy Σ-hypernuclei.

The concept of a Coulomb-assisted hybrid state was proposed by

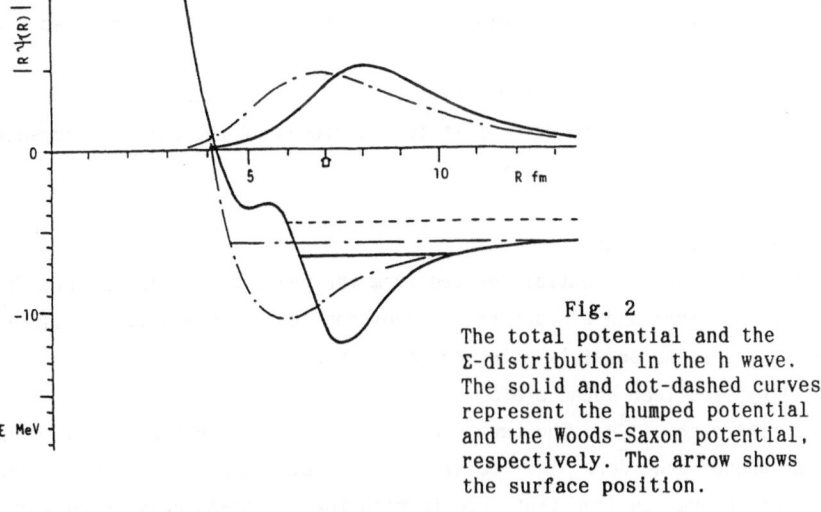

Fig. 2
The total potential and the Σ-distribution in the h wave. The solid and dot-dashed curves represent the humped potential and the Woods-Saxon potential, respectively. The arrow shows the surface position.

Yamazaki et al.[13] Figure 2 shows a h-wave hypernuclear state, where Σ^- is trapped in the nuclear surface region by the peculiar-shaped strong interaction with the aid of the inner centrifugal repulsion and the outer Coulomb attraction. It is just a typical example of the Coulomb-assisted hybrid state. This hybrid state has a narrow width assured by the strong centrifugal repulsion in the h wave.

The in-flight ^{208}Pb(K^-,π^+) reaction is investigated at incident K^- momentum 300 MeV/c [14]. Under this recoilless condition a substitutional state $(h_{11/2})_p^{-1}(h_{11/2})_\Sigma$ is selectively excited. The h wave peak is sharp and prominent and is well separated from the spectrum of the quasi-free Σ^- production. Thus, the Coulomb-assisted hybrid state with a high orbital angular momentum is found to be a favorable candidate for future experimental observation of heavy Σ hypernuclei, though to get high-statistics we need high-intensity kaon-beam facilities.

In the π^- spectrum of the ^{9}Be(K^-,π^-) reaction at CERN [1] two peaks were reported in the Σ-continuum region. Ikeda and Yamada [15] proposed the concept of a coupled isotriplet in order to explain the lower peak. The explanation is based on the three Σ-hypernuclear states ^{8}Li $-\Sigma^+$, ^{8}Be$^*-\Sigma^0$ and ^{8}B $-\Sigma^-$, of which core nuclei form the T=1 isotriplet. A strong Lane potential mixes the three states and forms a good-isospin state below the lowest threshold of the three. Thus the escape width is suppressed.

A possible structure of the second peak was discussed by Khin Swe Myint et al.[16] The Coulomb interaction prepares an atomic bound state in the ^{8}B $+\Sigma^-$ channel region. In the internal region the Lane term of the strong interaction plays a dominant role and recovers the isospin symmetry broken in the atomic state. Then the internal state is connected to the ^{8}Be$^*+\Sigma^0$ open channel and gets the escape width. This coupled resonance state is called an atom-nucleus hybrid state.

Unfortunately the statistics of the ^{9}Be(K^-,π^-) experiment at CERN is very poor and we cannot confirm the existence of the peaks at present. Future reexperiments are requested.

1.5. CONCLUDING REMARKS

The nucleus-Σ potential derived from the realistic ΣN interaction has a strong Lane term and a short-range repulsion in its real part. We suggest three types of the possible existence of Σ-hypernuclei.

1) Hypernuclear bound state:
It is the case of $^4_\Sigma$He. The Lane term of the nucleus-Σ potential makes the system bound and does the total isospin almost pure. The short-range repulsion plays an important role in reducing the $\Sigma N \rightarrow \Lambda N$ conversion width.

The present theoretical analysis shows that the $^4_\Sigma$He hypernucleus observed at KEK is a bound state with $J^\pi = 0^+$ and $T \sim 1/2$ (99%), whose narrow width is due to the central repulsion of the nucleus-Σ potential.

2) Coulomb-assisted hybrid bound state:

There is a possibility that Σ-hypernuclear states with narrow widths exist even in heavy nuclei like ^{208}Pb under the cooperation of the strong interaction, the Coulomb interaction and the centrifugal potential. The Coulomb and the centrifugal potentials close up the surface region of the nucleus, where the peculiar-shaped strong interaction works. The state may be useful for the investigation of Σ-hypernuclear ℓs splitting which is proportional to the large angular momentum ℓ.

3) Coupled-isomultiplet resonance state:

Atom-nucleus hybrid resonance state:

They might be the case of $^9_\Sigma$Be. The Lane term plays essential roles in forming Σ-resonance states in continuum. In the atom-nucleus hybrid case the Coulomb interaction gives a foundation of the existence and the strong interaction builds up a peculiar character of the state with escape width.

Lastly, we can conclude that the observation of $^4_\Sigma$He at KEK and BNL is a mile stone of the study of rich field of sigma-hypernuclear systems.

2. Strangeness=-2 Hypernuclei

2.1. INTRODUCTION

S=-2 hypernuclei are of particular interest: What disturbance would be raised when strange particles doubly enter into a nucleus without obeying the Pauli exclusion principle? What nature would be revealed about the $\Lambda\Lambda$ interaction?

Recently, an event of double-Λ hypernucleus and two events of Ξ-hyper atom or nucleus have been reported through analyses of the emulsion-counter hybrid experiment at KEK by Imai et al.[17]. At BNL a counter experiment is proposed to detect the double-Λ hypernucleus formation from stopped Ξ^- on ^{6}Li [18].

Our knowledge of double-Λ hypernuclei is still very limited: There exist only three emulsion data of $^{10}_{\Lambda\Lambda}$Be [19] and $^6_{\Lambda\Lambda}$He [20] observed in early 1960's and of $^{13}_{\Lambda\Lambda}$B [17,21] at KEK. Thus our aim is to produce avandant double-Λ hypernuclei enough to investigate the nature of the $\Lambda\Lambda$ inter-action.

Our idea is following. The reaction Q value 28.33 MeV of $\Xi^- + p \rightarrow \Lambda + \Lambda$ is accidentally very close to the binding energy 28.30 MeV of the alpha particle. If the alpha particle is broken up after the absorption of a stopped Ξ^-, the Q value is almost exhausted and two produced Λ's cannot get enough energy to escape. Thus, there is a large chance to form a double-Λ hypernucleus. The hypernucleus is $^4_{\Lambda\Lambda}$H, of which possible existence was theoretically predicted [22].

2.2. FORMATION RATE OF DOUBLE-Λ HYPERNUCLEUS

We calculate the branching ratio of the double-Λ hypernucleus formation from a $\Xi^- \cdot {}^4$He atom. All the possible processes are followings:

$$
\begin{aligned}
\Xi^- + {}^4\text{He} &\rightarrow \Lambda + \Lambda + p + n + n + 0.03 \text{ MeV}, &&\text{negligible} \\
&\rightarrow \Lambda + \Lambda + d + n &+ 2.25 \text{ MeV}, &&\text{negligible} \\
&\rightarrow \Lambda + \Lambda + t &+ 8.51 \text{ MeV}, &&18\ \% \\
&\rightarrow \Lambda + {}^3_{\Lambda}\text{H} + n &+ 2.38 \text{ MeV}, &&2\ \% \\
&\rightarrow \Lambda + {}^4_{\Lambda}\text{H} &+ 10.55 \text{ MeV}, &&43\ \% \\
&\rightarrow {}^4_{\Lambda\Lambda}\text{H} + n &+ 2.83 \text{ MeV}, &&37\ \%.
\end{aligned}
$$

The last figures are the branching ratio of the respective processes obtained by Kumagai et al.[23], where the binding energy of $^4_{\Lambda\Lambda}$H is taken to be 2.80 MeV [22]. The small Q values favor two-body breakup processes, and the formation rate of the double-Λ hypernucleus becomes high.

In the case of ^{6}Li the Q values are large compared with those of the ^{4}He case. We show only two as examples:

$$
\begin{aligned}
\Xi^- + {}^6\text{Li} &\rightarrow \Lambda + \Lambda + \alpha + n &+ 24.63 \text{ MeV}, &&\sim 100\ \% \\
&\rightarrow {}^6_{\Lambda\Lambda}\text{He} + n &+ 35.53 \text{ MeV}, &&\text{small}^{*)}.
\end{aligned}
$$

The ^{6}Li nucleus has a $\alpha + d$ cluster structure. A Ξ^- hyperon at some atomic orbit meets first a p-shell proton. The deuteron breakup loses only 2.2 MeV energy, while the $\Lambda\Lambda$ sticking to α gains 10.9 MeV energy. Thus the Q value is large for the double-Λ nucleus formation. For large Q's the four-body breakup becomes dominant. *)Zhu et al's result is not small but 3% [24].

A stopped Ξ^- on ^{4}He target is able to form the double hypernucleus $^4_{\Lambda\Lambda}$H with high probability: Its branching ratio amounts to 37 %. The formation of $^4_{\Lambda\Lambda}$H emits a neutron with a definite energy of 2.3 MeV. Thus, by the neutron detection the double Λ hypernucleus can be identified [18]. It is noted that the breakup of the alpha particle or alpha cluster profits the formation of double Λ hypernuclei. In this respect, ^{9}Be also seems to be a preferable target for the double-Λ hypernucleus production.

The observation of $^4_{\Lambda\Lambda}$H, if succeed, has much importance. Theoretically, $^4_{\Lambda\Lambda}$H is predicted to be the lightest double-Λ hypernucleus. The observation of its existence and its weak decay could provide a severe restriction on the existence of the H dibaryon [25].

2.3. Ξ-HYPERATOM OR Ξ-HYPERNUCLEUS

Aoki et al.[26] reported the production of two single-Λ hyper-fragments by Ξ^- capture in the emulsion experiment at KEK [17]. A Ξ^- hyperon was captured at rest on ^{12}C and two hyperfragments were emitted;

$$\Xi^- + {}^{12}C \longrightarrow {}^4_\Lambda H + {}^9_\Lambda Be.$$

The two fragments decayed as

$$^4_\Lambda H \longrightarrow \pi^- + {}^4He \quad \text{and} \quad {}^9_\Lambda Be \longrightarrow {}^4He + {}^2H + p + 2n, \text{ etc.}$$

From the data the binding energy of Ξ^- is estimated to be $- B_\Xi = -0.53\pm0.33$ MeV, which deviates from the atomic levels, -1.13 MeV (1s) and -0.28 MeV (2s,2p), in the pure Coulomb interaction.

We introduce strong interaction between Ξ^- and ^{12}C by employing Dover and Gal's Woods-Saxon type potential with $V_0 = -28$ MeV, $R_0 = 1.1 \times A^{1/3}$fm and a = 0.65 fm [27]. Finite-size effect of the Coulomb potential is taken into account. The results are followings:

$$1s: -14.2 \text{ MeV (2.1 fm)}, \quad 2s: -0.63 \text{ MeV (11 fm)},$$
$$2p: -3.73 \text{ MeV (3.4 fm)}, \quad 3p: -0.25 \text{ MeV (24 fm)},$$

where the figures in the parentheses are the r.m.s. distance between Ξ^- and ^{12}C. The above data suggests that Ξ^- was captured from the 2s atomic state shifted by the strong interaction.

Recently, Chung et al.[28] has found another similar event in the emulsion at KEK. It is considered that Ξ^- stopped again on ^{12}C and the same fragments $^4_\Lambda H$ and $^9_\Lambda Be$ appeared. The binding energy is preliminarily estimated to be 3.74 MeV. This energy corresponds to the 2p state obtained above which is not an atomic but a nuclear state as judged from the r.m.s. distance. Although there remains a possibility of $^9_\Lambda Be^*$(3.08 MeV [29]) + $^4_\Lambda H$ fragmentation with $B_\Xi = 0.66$ MeV, this second event could be a first reliable observation of the Ξ-hypernucleus because of the identification of the Ξ^- track in the emulsion.

Thus, the S=-2 hypernuclear physics is coming to an exciting stage.

References

1. Bertini,R., et al.: Phys. Lett. **B90**, 375 (1980)

2. Dover,C.B., Millener,D.J., Gal,A.: Phys. Rep. **184**, 1 (1989)

3. Harada,T., Shinmura,S., Akaishi,Y., Tanaka,H.:
   Nucl. Phys. **A507**, 715 (1990)

4. Nagels,M.M., Rijken,T.A., de Swart,J.J.:
   Phys. Rev. **D12**, 744 (1975); **D15**, 2547 (1977); **D20**, 1633 (1979)

5. Hayano,R.S., et al.: Phys. Lett. **B231**, 355 (1989)

6. Dalitz,R.H., Davis,D.H., Deloff,A.: Phys. Lett. **B236**, 76 (1990)

7. Akaishi,Y.: Lect. Notes in Phys. **273**, 324 (1986)

8. Okabe,S., Harada,T., Akaishi,Y.: Nucl. Phys. **A514**, 613 (1990)

9. Grund,B., Langanke,K., Warmann,T.:

10. Baird,S., et al.: Nucl. Phys. **A392**, 297 (1983)

11. Mukai,S., et al.: private communication

12. Khin Swe Myint, Tadokoro,S., Akaishi,Y.:
    Prog. Theor. Phys. **82**, 112 (1989)

13. Yamazaki,T., Hayano,R.S., Morimatsu,O., Yazaki,K.:
    Phys. Lett. **B207**, 393 (1988)

14. Tadokoro,S., Akaishi,Y.: Phys. Rev. **C12**, 2591 (1990)

15. Ikeda,K., Yamada,T.: Int. J. Mod. Phys. **A3**, 2339 (1988)

16. Khin Swe Myint, Akaishi,Y.: Prog. Theor. Phys. **79**, 454 (1988)

17. Imai,K.: Nucl. Phys. **A527**, 181c (1991)
    Aoki,S., et al.: Prog. Theor. Phys. **85**, 1287 (1991)

18. May,M: Nuovo Cimento **102A**, 401 (1989)

19. Danysz,M.,et al.: Phys. Rev. Lett. **11**, 29 (1963)

20. Prowse,D.J.: Phys. Rev. Lett. **17**, 782 (1966)

21. Dover,C.B., Millener,D.J., Gal,A., Davis,D.H.: preprint (1991)

22. Nakaichi-Maeda,S., Akaishi,Y.: Prog. Theor. Phys. **84**, 1025 (1990)

23. Kumagai,I., Khin Swe Myint, Akaishi,Y.: Soryusiron-Kenkyu **82**, 42 (1990)

24. Zhu,D., et al.: preprint BNL-46135

25. Kerbikov,B.O.: Sov. J. Nucl. Phys. **39**, 516 (1984)

26. Aoki,S., et al.: preprint KEK 91-104

27. Dover,C.B., Gal,A.: Ann. of Phys. **146**, 309 (1983)

28. Chung,S.H.: private communication

29. May,M., et al.: Phys. Rev. Lett. **51**, 2085 (1983)

LIST OF PARTICIPANTS

Adam, R.
Department of Physics
Univ. of South Africa
P.O. Box 392
Pretoria, 0001, South Africa

Fax: 012-429-3221/3434
E-mail: 076MAART.WITSVMA@f4.n7104.z5.fidonet.org

Akaishi, Y.
Department of Physics
Hokkaido University
Kita-ku Kita-10 Nishi-8
Sapporo 060, Japan

Fax: 011-746-5444
E-mail: E13061G@JPNCCKU

Alt, E.O.
Institut für Physik
Universitaet Mainz
Staudingerweg 7
D 6500 Mainz - FRG

Fax: (6131) 392991
E-mail: ALT@VIPMZF.Physik.Uni-Mainz.DE

Aquilanti, V.
Dipartimento di Chimica
Università di Perugia
06100 Perugia
Italy

Fax: 075-5855606
E-mail: PGKIN@IPGUNIV

Arenhoevel, H.
Institut für Kernphysik
Universitaet Mainz
D-W-6500 Mainz
Germany

Fax: 49 6131 392964
E-mail: arenhoevel@vkpmza.kph.uni-mainz.de

Arvieux, J.
Laboratoire Nationale Saturne
F-91191
Gif-sur-Yvette Cedex
France

Fax: 33169082970
E-mail: ARVIEUX@FRCPN11

Bakker, B.L.G.
Department of Physics and
Astrophysics- Vrije Universiteit
De Boelelaan 1081, NL-1081 HV
Amsterdam - The Netherlands

Fax: 020-461459
E-mail: blgbkkr@nat.vu.nl

Barone, V.C.
Dipartimento di Fisica
Università di Perugia
Via A.Pascoli
06100 Perugia - Italy

Fax: 075-44666
E-mail: BARONE@PERUGIA.INFN.IT

Bawin, M.
University of Liége
Institut de Physique B.5
Sart Tilman
B-4000 Liege 1 - Belgium

Fax: 32-41-56.23.55
E-mail: U216204@BLIULG11

Belyaev, V.B.
JINR
Lab. of Theor. Phys.
141980 Dubna
URSS

Fax: 7095-200-2283
E-mail: BELYAEV@LJAP.JINR.DUBNA.SU

Bernstein, A.M.
Massachusetts Institute of Technology
Bldg. 26-419
Cambridge, Mass. 02139
USA

Fax: 617-258-6923
E-mail: Bernstein@MITLNS

Bijtebier, J.
Vrije Universiteit Brussel
Fakulteit der Wetenschappen
Theoretische Fysica
Pleinlaan 2 B 1050 - Brussel, Belgium

Fax: +32-2-6412276
E-mail: z02601@bbrbfu01

Boffi, S.
Università di Pavia
Dipartimento di Fisica Nucleare e Teorica
Via Bassi, 6
I-27100 Pavia - Italy

Fax: (0382) 423241
E-mail: BOFFI@PAVIA.INFN.IT

Bogdanova, L.N.
Institute for Theoretical and
Experimental Physics
ITEP, Moscow
117259 USSR

Fax: 7-095-123-65-84
E-mail: pliv@pliv.kiae.su

Boschitz, E.
Universität Karlsruhe
Inst. für exp. Kernphysik
Postfach 3640
7500 Karlsruhe 1 - Germany

Fax: 7247-823548/825070
Telex: 17724716kfk

Bruno, M.
INFN - Sezione di Bologna
Via Imerio 46
40126 Bologna
Italy

Fax: 051/247244
E-mail: BRUNO@BOLOGNA.INFN.IT

Cameron, J.M.
Indiana University Cyclotron Facility
2401 Milo Sampson Ln.
Bloomington, IN 47401
USA

Fax: 812 855 6645
E-mail: CAMERON@IUCF

Carbonell, J.
Institute des Sciences Nucleaires
53 Av. des Martyrs
38026 Grenoble Cedex
France

Fax: 33 76 284004
E-mail : Carbonel@FRCPN11

Cederbaum, L.S.
Theoretische Chemie
Physikalisch-Chemisches Institüt
Im Neuenheimer Feld 253
D-6900 Heidelberg, Germany

Fax: 49-6221-563199
E-mail: H35@DHDURZ1

Ceuleneer, R.
Université de Mons-Hainaut
Faculté des Sciences - Ave. Maistriau 19
B-7000 Mons
Belgium

Fax: 65 37 30 54
E-mail: SCEULE @ BMSUEM11

Ciofi degli Atti, C.
Dipartimento di Fisica
Università degli Studi
Via A.Pascoli
06100 Perugia - Italy

Fax: 075-44666
E-mail: CDA@PERUGIA.INFN.IT

Cisbani, E.
INFN - Sezione Sanità
Istituto Superiore di Sanità
V.le Regina Elena, 299
00161 Roma - Italy

Fax: 39-6-4462872
E-mail: CISBANI@SANITA.INFN.IT

Close, F.
Rutherford - Appelton Laboratory
Chilton
Didcot OX110QX
United Kingdom

Fax: 44-235-446733
E-mail: FEC@UKACRL

Coon, S.A.
Physics Department
New Mexico University
Box 30001, Dept. 3D
Las Cruces, New Mexico 88003
USA

Fax: 505-646-1934
E-mail: SCOON@NMSU.EDU

Damjanovic, M.
Faculty of Farmacy
Dept. of Physics and Mathematics
P.O.Box 146, Subotica 8
Beograd, Yugoslavia

Fax: (0)11 657 837
E-mail: EDAMJANO@YUBGEF51

Danilin, B.V.
Kurchatov Institute of Atomic Energy
Kurchatov sq. 1,
Moscow 123182
USSR

Fax:
E-mail: ZHUKOV@AOGLOB.KIAE.SU

Decker, M.
Physikalisches Institut
Universitaet Bonn
Endenicher Allee 11-13
D-5300 Bonn - Germany

Fax: 49228737869
E-mail: UNP018@DBNRHRZ1

De Sanctis, E.
INFN - Lab. Nazionali di Frascati
Casella Postale 13
I-00044 Frascati (Roma)
Italy

Fax:
E-mail:DESANCTIS@FRASCATI.INFN.IT

De Sanctis, M.
INFN - Sezione di Roma
Università "La Sapienza"
P.le A.Moro, 2
00185 Roma - Italy

Fax: 06-4957697
E-mail: DESANCTIS@ROMA.INFN.IT

de Swart, J.J.
Institute for Theoretical Physics
University of Nijmegen
Toernooiveld I
6525 ED Nijmegen - The Netherlands

Fax:
E-mail: U634999@HNYKUN11

de Witt Huberts, P.K.A.
NIKHEF-K
P.O.Box 41882
1009 DB Amsterdam
The Netherlands

Fax: 31-20-592 2165
E-mail: marijke @nikhefk.nikhef.nl

Didelez, J.P.
Institute de Physique Nucleaire
Bat 100
91406 Orsay Cedex
France

Fax: 33 1 69416470
E-mail: DIDELEZ@FRIPN51

Dolbilkin, B.
Institute for Nuclear Research
Moscow   Fax:
USSR   E-mail: dolbikin@inucres.mosk.su

Doleschall, P.
Central Research Institute for Physics
H-1525 Budapest   Fax: 36-1 1696 567
P.O. Box 49   E-mail:
Hungary

Drago, A.
Dipartimento di Fisica
Università di Ferrara
Via Paradiso, 12   Fax: 0532 - 762057
44100 Ferrara - Italy   E-mail: DRAGO@FE.INFN.IT

Düren, M.
Max Planck Institut für Kernphysik
Postfach 103 980
D-6900 Heidelberg   Fax: -6221-516540
Germany   E-mail: DUE@CRUXNHD3

Efros, V.D.
I.V.Kurchatov Institute
of Atomic Energy
Moscow 123182   Fax: 095-196132
USSR   E-mail: EFROS@jbivn.kiae.su

Ellerkman, G.H.
Physikalisches Institut
Universitaet Bonn
Endenicher Allee 11-13 - AVZ I   Fax: 49228737869
D-5300 Bonn - Germany   E-mail: UNP051@DBNRHRZ1

Fabre de la Ripelle, M.
Institute de Physique Nucleaire
91406 Orsay Cedex   Fax:
France   E-mail: FABRE@FRCPN11

Fabrocini, A.
Dipartimento di Fisica
Università degli Studi
P.zza Torricelli, 2   Fax:39-50-48277
56100 Pisa - Italy   E-mail: FABROCINI@PISA.INFN.IT

Faessler, A.
University of Tübingen
Institute of Theor. Physics
Auf der Morgenstelle 14
7400 Tübingen, Germany

Fax: 07071-296400
E-mail: faessler@mailserv.zdv.uni-tuebingen.de

Fearing, H.W.
TRIUMF
4004 Wesbrook Mall
Vancouver, B.C.
Canada V6T 2A3

Fax: 604 222 1074
E-mail: Fearing@triumfcl

Ficocelli Varracchio, E.
Department of Chemistry
University of Bari
70126 Bari
Italy

Fax: 080-242129
E-mail: FICO@IBACSATA

Fiedeldey, H.J.
Department of Physics
University of South Africa
P.O.Box 392 Pretoria, 001
South Africa

Fax: 27 12 4293434
E-mail: 076MAART.WITSVMA@f4.n7104.z5.fidonet.org

Fonseca, A.C.
Centro Fisica Nuclear
Univ. of Lisbon
Av. Gama Pinto 2
1699 Lisboa, Portugal

Fax: 765622
E-mail: fonseca@ptifm

Frascaria, R.
Institute de Physique Nucléaire
BP1 91406 Orsay Cedex
France

Fax: 69416470
E-mail: FRASCARI@FRIPN51

Friar, J.L.
Los Alamos National Laboratory
MS B283
Los Alamos NM 87545
USA

Fax: (505)-665-4055
E-mail: FRIAR@LAMPF

Frullani, S.
Laboratorio di Fisica
Istituto Superiore di Sanità
V.le Regina Elena, 299
00161 Roma - Italy

Fax: 39-6-4462872
E-mail: FRULLANI@SANITA.INFN.IT

Garcon, M.
DAPNIA/SNN, Bat. 472
CEN/Saclay
91191 Gif-sur-Yvette
France
Fax: 33-1-69088643
E-mail: MICHEL@FRSAC11

Garibaldi, F.
Laboratorio di Fisica
Istituto Superiore di Sanità
V.le Regina Elena, 299
00161 Roma - Italy
Fax: 39-6-4462872
E-mail: GARIBALDI@SANITA.INFN.IT

Ghio, F.
Laboratorio di Fisica
Istituto Superiore di Sanità
V.le Regina Elena, 299
00161 Roma - Italy
Fax: 39-6-4462872
E-mail: GHIO@SANITA.INFN.IT

Giebeler, J.
Physikalisches Institut
Universitaet Bonn
Endenicher Allee 11-13 - AVZ I
D-5300 Bonn - Germany
Fax: 49228737869
E-mail: UNP047@DBNRHRZ1

Golak, J.
Jagellonian University
Institute of Physics
ul. Reymonta 4
30-059 Krakow - Poland
Fax: 4812337086
E-mail: UFGOLAK@PLKRCY11

Hajdas, W.
Eidgenoessische Technische Hochschule
Hoenngerberg - Inst. fuer
Mittelenergiephysik
HPK IMP G15 - CH-8093 Zürich
Switzerland
Fax: CH-01-371- 2665
E-mail: HAJDA@CAGEIR5A
HAJDA@CVAX.PSI.CH for EARN

Henley, E.M.
University of Washington
Physics Department, FM-15
Seattle, WA 98195
USA
Fax: (206) 685-0635
E-mail: Henley@UWAPHAST

Hersbach, J.P.T.
Institute Voor Theoretische Fysica
Princetonplein 5
P.O.Box 80.006
3508 TA Utrecht - The Netherlands
Fax: (0)30-531601
E-mail: HERSBACH@HUTRUU51

Hersman, F.W.
University of New Hampshire
Department of Physics
Durham, NH 03824 (USA)

Fax: 603-862-2998
E-mail: hersman@curie.unh.edu

Iachello, F.
Yale University
Center for Theor. Physics, Sloane Lab.
New Haven, Connecticut 06511
USA

Fax: 1(203)432-5741
E-mail:

Ishikawa, S.
Institut für.Theoretische Physik II
Ruhr-Universität Bochum
Universitätsst. 150
P.O.Box 102148
D-4630 Bochum 1 - Germany

Fax: 234-700-2001
E-mail: M4A2S2C@JPNTOHOK

Jans, E.
NIKHEF-K
Postbus 4395
1009 AJ Amsterdam
Holland

Fax: (0)20-5922165
E-mail: eddy@nikhefk.nikhef.nl

Jodice, M.
INFN - Sezione Sanità
Istituto Superiore di Sanità
V.le Regina Elena, 299
00161 Roma - Italy

Fax: 39-6-4462872
E-mail: MAURO@SANITA.INFN.IT

Johansson, A.
The Svedberg Laboratory
Box 533 - 75121
Uppsala
Sweden

Fax: 018-183833
E-mail: JOHANSSON@TSL.UU.SE

Kalashnikova, Y.S.
Institute of Theoretical
and Experimental Physics (ITEP)
B.Cheremushkinskaya 25
USSR 117259 Moscow

Fax: +7-095-1236584
Telex: 411059 cerii su

Kamuntavicius, G.P.
Vytautas Magnus University
Daukanto 28
Kaunas, 233000
Lithuania

Fax: +7-0127-203-858
Telex: 269857 VYTUN SU

Karl, G.
Department of Physics
University of Guelph
Guelph, Ontario
N1G 2W1 - Canada
Fax: 519-836-9967
E-mail: PHYGKARL@VM.UoGuelph.CA

Kermode, M.W.
University of Liverpool
Department of Applied Mathematics
and Theor. Physics
P.O.Box 47, Liverpool L69 3BX, UK
Fax: 051 708 6502
E-mail: SX09@Liverpool.AC.UK

Kersting, R.
Physikalisches Institut
Universitaet Bonn
Endenicher Allee 11-13 - AVZ I
D-5300 Bonn - Germany
Fax: 49228737869
E-mail: UNP05B@DBNRHRZ1.

Kievsky, A.
INFN - Sezione di Pisa
Dipartimento di Fisica
Piazza Torricelli, 2
56100 Pisa - Italy
Fax: 39-50-48277
E-mail: KIEVSKY@IPIINFN

Kierchbach, M.
Institute for Nuclear Physics
Technical University
D-6100 Darmstadt, Germany

Klar, H.
University Freiburg
Fakultat für Physik
Hermann-Herder-Str. 3
D-7800 Freiburg - Germany
Fax: 0049-761-203-4527
E-mail: HUKL at DFRRUF1

Kobbe, A.
Physikalisches Institut
Universität Bonn
Endenicher Allee 11-13 - AVZ I
D-5300 Bonn - Germany
Fax: 49228737869
E-mail: UNP073@DBNRHRZ1.

Kok, L.P.
Inst. f. Theor. Phys.
P.O.Box 800
9700 AV Groningen
The Netherlands
Fax: +3150634947
E-mail: lpkok@hgrrug5 (bitnet)

Kondratyuk, L.
Institute for Theoretical
and Experimental Physics
B.Cheremushvinskaya 25
Moscow 117259 USSR

Fax: (007)(095)1236584
Telex: 411059 cerii su

Krivec, R.
Department of Theoretical Physics
Jamova 39, P.O. Box 100
YU-61111 Ljubljana
Yugoslavia

Fax: +38 61 219 385, +38 61 273 677
E-mail: KRIVEC@IJS.AC.MAIL.YU
        KRIVEC%CATHY@YUBGEF51

Krug, J.
Institut für Experimentalphysik I
der Ruhr-Universität Bochum
Universitätsstraße 150
P.O.B. 102148
D-4630 Bochum 1 Germany

Fax: 0049-234-700-3552
E-mail: P160305@DBORUB01

Kvitsinsky, A.A.
University of Leningrad and
Inst. des Sciences Nucléaires
53 Av. des Martyrs,
38026 Grenoble Cedex - France

Fax: 33-76284004
E-mail: KVITSIN@FRCPN11

Lee, T.-S.H.
Physics Division
Argonne National Laboratory
Argonne, ILL.60439
USA

Fax: 708-972-3903
E-mail: Lee@ANLPHY

Leeb, H.
Institut für Kernphysik
Technische Universität Wien
Wiedner Hauptsraße 8-10/142
A-1040 Wien, Austria

Fax: (0222) 564203
E-mail: E142005@AWITUW01

Le Goff, J.M.
CEA/DPhN/SEPN
CEA CEN Saclay
91191 Gif sur Yvette
Cedex - France

Fax: 33-1-69087584
E-mail: LEGOFF@FRSAC11

Leidemann, W.
Dipartimento di Fisica
Università di Trento
I-38050 Povo (Trento)
Italy

Fax: 0461/881696
E-mail: LEIDEMANN@ITNVAX.INFN.IT

Lovitch, L.
Dipartimento di Fisica
Università degli Studi
Via Paradiso 12
44100 Ferrara - Italy
Fax: 0532-762057
E-mail: lovitch@fe.infn.it

Manayenkov, S.
Leningrad Nuclear Physics Institute
188350 Gatchina, Leningrad
Leningrad District
USSR
Fax: 7-81271-37196
E-mail:Teren@LNPI.SPB.SU

Markushin, V.
I.V.Kurchatov Institute
of Atomic Energy
123182 Moscow
USSR
Fax: 007 (095) 1969889
E-mail: pliv@pliv.kiae.su.

Mathiot, J.-F.
Division de Physique Theorique
Institut de Physique Nucleaire
F-91406 Orsay Cedex
France
Fax: (33) 1 69 28 58 97
E-mail: MATHIOT@FRCPN11

McCarthy, I.E.
Flinders University of South Australia
School of Physical Sciences
GP0 Box 2100,
Adelaide 5001, Australia
Fax: 61 8 201 2905
E-mail: Ian@esm.cc.flinders.edu.au

Melezhik, V.S.
JINR
Head Post Office
P.O.Box 79
Moscow
USSR
Fax:
E-mail: MELEZH@LCTA.JINR.DUBNA.SU

Merkuriev, S.
Leningrad State University
Leningrad
119178 U.S.S.R.
Fax:
E-mail: merc@hq.lgu.spb.su

Milana, J.
Physics Department
College of William & Mary
Williamsburg, VA 23185
USA
Fax: 804-221-3540
E-mail: milana@cebafvax

Morgenstern, J.
CEN Saclay
DPhN/SEPN - BP2
F 91191 Gif-sur-Yvette Cedex
France

Fax: 1-69 08 75 53
E-mail: MORGEN at FRSAC11

Morita, H.
Sapporo Gakuin University
Bunkyo-dai 11 Ebetsu
Hokkaido 069
Japan

Fax: (011)-386-8113
E-mail: E13061G@JPNCCKU

Mosconi, B.
Dipartimento di Fisica
Università di Firenze
Largo E.Fermi 2
50125 Firenze - Italy

Fax: 055-229330
E-mail: MOSCONI@FI.INFN.IT

Mougey, J.Y.
CEBAF
12000 Jefferson Avenue
Newport News, VA 23606
USA

Fax: 1-804-2497363
E-mail: MOUGEY@CEBAFVAX

Narodetskii, I. M.
Institute for Theoretical
and Experimental Physics
B.Cheremushvinskaya 25
Moscow 117259 USSR

Fax: (007)(095)1236584
Telex: 411059 cerii su

Nikolaeva, R.
Joint Institute for Nuclear Research
Lab. of Theoretical Physics
Head Post Office, P.O.Box 79 Dubna
Moscow, USSR

Fax:
E-mail: RUNIK@THEOR.JINRC.DUBNA.SU

Obersteiner, P.
Institute for Theoretical Physics
University of Graz
Universität platz 5
A-8020 Graz, Austria

Fax: (316) 384091
E-mail: OBERSTEINER@EDVZ.UNI-GRAZ.ADA.AT

Orlandini, G.
Dipartimento di Fisica
Università di Trento
I-38050 Povo (Trento)
Italy

Fax: 0461/881696
E-mail: ORLANDINI@ITNVAX.INFN.IT

628

Pace, E.
Dipartimento di Fisica
Università di Roma "Tor Vergata"
Via E.Carnevale   Fax:
I-00173 Roma - Italy   E-mail: FISTEOR@IRMISS

Pauschenwein, J.
Institute for Theoretical Physics
University of Graz
Universitätplatz 5   Fax: (316) 384091
A-8020 Graz, Austria   E-mail: PAUSCHENWEIN@EDVZ.UNI-GRAZ.ADA.AT

Peña, M.T.
Centro de Fisica Nuclear
da Universidade de Lisboa
Av. Gama Pinto 2   Fax: 351-1-765622
1699 Lisboa Codex- Portugal   E-mail: TERESA@PTIFM

Petratos, G.G.
Stanford Linear Accelerator Center
Bin 20, P.O.Box 4349
Stanford, CA 94309   Fax: (415)3233626
USA   E-mail: GGP@SLACVM

Plessas, W.
Inst. for Theor. Phys.
Univ. Graz   Fax: 316 384091
Universitätsplatz 5   E-mail: B6241DAC@AWIUNI11
A-8010 Graz   PLESSAS@EDVZ.UNI-GRAZ.ADA.AT

Pokrovsky, Yu.
I.V.Kurchatov Institute
of Atomic Energy
Moscow 123182   Fax: 095-1961632
USSR   E-mail: pokr@jbivn.kiae.su

Ponomarev, L.
I.V. Kurchatov Institute
of Atomic Energy
Moscow 123182   Fax: 095-1961632
USSR   E-mail: pliv@pliv.kiae.su

Prosperi, D.
Università di Roma I
Dipartimento di Fisica
Piazza Aldo Moro, 2   Fax: 06-4957697
00185 Roma - Italy   E-mail: PROSPERI@ROMA1.INFN.IT

Reichelt, T.
Physikälisches Institut
Universität Bonn
53 Bonn Nussalle 12
Germany

Fax:
E-mail: REICHELT@DBNPIB5

Revai, J.
Central Research Institute for Physics
H-1525 Budapest
P.O. Box 49
Hungary

Fax: 36-1 1696 567
E-mail: H745REV@ELLA.HU

Richter, J.
Physikalisches Institut
Endenicher Allee 11-13
D-5300 Bonn 1
Germany

Fax: 228-737869
E-mail: UNP064@DBNRHRZ1

Rosati, S.
Dipartimento di Fisica
Università di Pisa
P.zza Torricelli 2
56100 Pisa - Italy

Fax:
E-mail: THEO@ICNUCEVM

Rossi, P.
INFN - Sezione Sanità
V.le Regina Elena 299
00161 Roma
Italy

Fax: 39-6-4462872
E-mail: rossi@lnf.infn.it

Round, G.D.
University of Surrey
Dept. of Physics
Guilford, Surrey
GU2 5XH England

Fax: ++44 483 304212
E-mail: AR3@UKACRL

Salmé, G.
INFN - Sezione Sanità
V.le Regina Elena, 299
00161 Roma
Italy

Fax: 39-6-4462872
E-mail: GSLM@SANITA.INFN.IT

Sandhas, W.
Physikalisches Institut
Endenicher Allee 11-13
D-5300 Bonn 1
Germany

Fax: 228-737869
E-mail: UNP064@DBNRHRZ1

630

Sasakawa, T.
University of Library and
Information Service
Kasuga 1-2, Tsukuba-shi, Ibaraki-ken
305 Japan

Fax: 81-298-52-4326
E-mail: SASAKAWA@ULIS.AC.JP

Schellingerhout, N.
Institute for Theoretical Physics
University of Gröningen
P.O.Box 800
9700 AV Gröningen, The Netherlands

Fax: 31 50 634947
E-mail: SCHELLIN@TH.RUG.NL

Scoccola, N.N.
Niels Bohr Institute
Blegdamsvej 17
DK-2100 Copenhagen
Denmark

Fax: +45-31-42-10-16
E-mail: SCOCCOLA@NBIVAX.NBI.DK

Scopetta, S.
INFN - Sezione di Perugia
Via Elce di Sotto
06100 Perugia
Italy

Fax: 39-75-44666
E-mail:

Servadio, S.
Department of Physics
University of Pisa
Piazza Torricelli 2 - Pisa
Italy

Fax: 39-6-48277
E-mail:

Seth, K.K.
Northwestern University
Physics Department
Evanston, IL 60208
USA

Fax: 708-491-9982
E-mail: SETH@FNAL or FNAL::SETH

Sgamellotti, A.
Dipartimento di Chimica
Università di Perugia
Via Elce di Sotto, 8
06100 Perugia - Italy

Fax: 075-5855516
E-mail: CHIMT3@IPGUNIV

Shebeko, A.V.
Institute of Physics & Technology
The Ukrainian Academy of Sciences
Kharkov 310108
USSR

Fax: (057)235-17-38
Telex: 115175 DEKAN SU

Sick, I.
Institut für Physik
Ulingelbergstr. 82
CH4056 Basel
Switzerland

Fax: ++41-61-2673784
E-mail: SICK%URZ.UNIBAS.CH@CERNVAX

Simula, S.
INFN - Sezione Sanità
V.le Regina Elena, 299
00161 Roma
Italy

Fax: 39-6-4462872
E-mail: SIMULA@IRMISS

Skopik, D.M.
Saskatchewan Accelerator Laboratory
University of Saskatchewan
Saskatoon, Saskatchewan,
Canada S7N0W0

Fax: 306-966-6058
E-mail: SKOPIK @Skatter.usask.ca

Sofianos, S.A.
University of South Africa
Dept. of Physics
University of South Africa
P.O. Box 392 Pretoria, 0001 South Africa

Fax: 27 12 4293434
E-mail: 076MAART.WITSVMA@f4.n7104.z5.fidonet.org

Speth, J.
Forschungszentrum Jülich GmbH
Institut für Kernphysik
Postfach 1913
D-5170 Jülich 1 - Germany

Fax: 02461-613930
E-mail: KPH119@DJUKFA11

Stadler, A.
Institut für Theoretische Physik
Universität Hannover
Appelstrasse 2, D-3000 Hannover 1
Germany

Fax: (511) 762-4877
E-mail: BCASTAD@DHVRRZN1

Takibayev, N.Zh.
Inst. of Nucl. Physics of Kazakh
SSR Academy of Sciences
Alma-Ata 480082
USSR

Fax: 7-327-2-636634 (690660)
    7-327-2-631207 (690660)
Telex: 251258 NAUKA US

Tarducci, R.
INFN - Sezione di Perugia
Dipartimento di Fisica
06100 Perugia
Italy

Fax: 075-44666
E-mail: TARDUCCI@PGINFN.

Thomas, A.W.
Dept. of Physics, Univ. of Adelaide
P.O.Box 498
S.A. 5001 Adelaide   Fax: 61-8-224 -0464
Australia   E-mail: athomas@physics.adelaide.edu.au

Tomusiak, E.
University of Saskatchewan
Dept. of Physics
Saskatoon, Saskatchewan,   Fax: 306 966 6400 / 306966 6058
Canada S7N0W0   E-mail: Tomusiak@skatter.usask.ca

van Dantzig, R.
NIKHEF-K
P.O.Box 41882
1009 DB Amsterdam   Fax: 31-20-592 2165
The Netherlands   E-mail: rvd@nikhefk.nikhef.nl

Van de Vyver, R.
Nuclear Physics Laboratory - RUG
Proeftuinstraat 86
B-9000 Gent   Fax: 32-91-646699
Belgium   E-mail: ROBERT@inwphys.rug.ac.be

van der Hart, H.W.
Zeeman Laboratory
University of Amsterdam
Plantage Muidergracht 4   Fax: 31 20 5255802
1018 TV Amsterdam - The Netherlands   E-mail: HvdHart@SARA.NL

Viviani, M.
INFN - Sezione di Pisa
Dipartimento di Fisica
P.zza Torricelli, 2   Fax: 050/589047
56100 Pisa - Italy   E-mail: THEO@ICNUCEVM

Walcher, T.
Institut für Kernphysik
der Universitat Mainz
Postfach 3980 J.J. Becherweg 45   Fax: 49 61 31 39 2964
D-6500 Mainz - Germany   E-mail: Walcher@DMZNAT51

Williamson, C.F.
Massachusetts Institute of Technology
Bldg. 26-431
Cambridge, MA 02139   Fax: 617-245-0901
USA   E-mail: CFW@MITBATES

Witala, H.
Institute of Physics
Jagellonian University
Reymonta 4
PL-30059 - Cracow (Poland)

Fax: 004812-337086
E-mail: UFWITALA@PLKRCY11

Zhukov, M.V.
Kurchatov Institute
of Atomic Energy
Kurchatov sq.1, Moscow 123182
USSR

Fax:
E-mail: ZHUKO@AOGLOB.KIAE.SU

Author Index

K.-M. Schmitt, H. Arenhövel

Complete Atlas of Polarization Observables in Deuteron Photodisintegration Below Pion-Threshold

(Few-Body Systems/Supplementum 4)

1991. VIII, 298 pages.
Cloth DM 170,-, öS 1190,-
Reduced price for subscribers to the journal
"Few-Body Systems": Cloth DM 153,-, öS 1071,-
ISBN 3-211-82320-4

Prices are subject to change without notice

For the first time, a complete calculation of all 288 polarization observables of deuteron photodisintegration for polarized photons and an oriented deuteron target is presented for energies below π-production threshold. The observables are calculated within a nonrelativistic framework but with inclusion of lowest-order relativistic effects. Explicit meson exchange currents and isobar configurations as manifestation of subnuclear degrees of freedom are included in the calculation. The sensitivity of the various polarization observables with respect to subnuclear degrees of freedom, to electric and magnetic multipole contributions and to a variety of realistic potential models are systematically investigated.

Thus this atlas provides the most detailed and systematic survey on polarization observables of this important process. It allows to analyse the different dynamical properties of the np-system as contained in the various observables and, therefore, will be useful for both theoretical studies and for the planning and evaluation of experiments as well. It serves in addition as an important supplement to the recent general review on deuteron photodisintegration by A. Arenhövel and M. Sanzone (Few-Body Systems, Suppl. 3).

Springer-Verlag Wien New York

H. Arenhövel, M. Sanzone

Photodisintegration
of the Deuteron

A Review of Theory and Experiment

(Few-Body Systems / Supplementum 3)

1991. 94 figures. VII, 183 pages.
Cloth DM 125,–, öS 875,–
Reduced price for subscribers to the journal
"Few-Body Systems": Cloth DM 113,–, öS 788,–
ISBN 3-211-82276-3

Prices are subject to change without notice

The two-body system has played a major role in nuclear physics and still holds the stage. Over the last fifty years a steady flow of experimental and theoretical papers on deuteron photodisintegration and its inverse reaction demonstrates the continuing interest in this fundamental process.

This book contains an almost exhaustive review of the theoretical approaches employed to evaluate two-body photodisintegration of the deuteron at low and intermediate energies.

Moreover, a critical survey of all the experiments concerning this process is given and the most accurate data are selected in order to compare the experimental results with the theoretical ones. Many information can be taken from this review, in particular that the current theory is able to describe the experimental data pretty well in the energy region below the pion production threshold, while more precise experimental data and improved theoretical calculations are needed in and above the delta-resonance region. Moreover, the role of meson exchange currents and isobar degrees of freedom is clearly seen in this process, the importance of different approximations in the calculations is clarified and the necessity of studying polarization observables is stressed.

Springer-Verlag Wien New York